Food, Pharmaceutical & Bioengineering Division 2020

AA008609

Held at the 2020 AIChE Annual Meeting

Online
16 – 20 November 2020

ISBN: 978-1-7138-2309-4

Printed from e-media with permission by:

Curran Associates, Inc.
57 Morehouse Lane
Red Hook, NY 12571

CURRAN ASSOCIATES INC.
proceedings.com

Some format issues inherent in the e-media version may also appear in this print version.

Printed with permission by Curran Associates, Inc. (2021)

For permission requests, please contact AIChE
at the address below.

AIChE
120 Wall Street, FL 23
New York, NY 10005-4020

Phone: (800) 242-4363
Fax: (203) 775-5177

www.aiche.org

Additional copies of this publication are available from:

Curran Associates, Inc.
57 Morehouse Lane
Red Hook, NY 12571 USA
Phone: 845-758-0400
Fax: 845-758-2633
Email: curran@proceedings.com
Web: www.proceedings.com

TABLE OF CONTENTS

607397 Transformer Networks for Fast and Improved Protein Structure Prediction

Abstract:

We put forward a novel neural-network based protein structure prediction pipeline which maps the protein sequence to structure problem as a language translation problem (i.e. used by Google translator). It uses a transformer network at the heart of its core learning module. Transformer networks have shown to be capable of 'paying attention' to the entire amino acid sequence (rather than fragments) while discerning biophysical and biochemical rules that decide its structure. We have trained our network over every single protein structure deposited in public databases. We have used it for successfully recapitulating structures of 'in-house' de novo proteins, which were not a part of the training set, with sub 4-angstrom RMDSs. Our future plan is to use a top-down implementation of this tool to identify all possible sequences that will not hamper the structural integrity of a given protein. This method can then be used to re-design enzymes with enhanced cofactor/ substrate specificity, designing high-affinity high-avidity nanobodies to target disease causing antigen proteins, and also for redesigning porins for use in drug-delivery devices.

609441 Homogeneous Megamolecules for Targeting Heterogeneity in Breast Cancer

Abstract:

Tumors are composed of a collection of genetically and phenotypically distinct cancer cells. For example, expression of cancer cell surface markers, such as HER2, are heterogeneously expressed, making targeted therapies ineffective at completely eradicating all cancer cells. To improve therapeutic activity in heterogeneous tumors, we have developed nanobody-enzyme megamolecules to deliver chemotherapeutics to the tumor microenvironment. Megamolecules are proteins with site-specific crosslinking that enable homogeneous protein-protein conjugates with precise covalent bonding. We synthesized nanobody-enzyme megamolecules and showed specific binding of the anti-HER2 nanobody domain to HER2+, but not HER2-, cancer cells. Nanobody binding delivered the enzyme yeast cytosine deaminase, which elicited a cytotoxic effect by activating the non-toxic prodrug 5-fluorocytosine to the cytotoxic 5-fluorouracil. We varied the number of domains of nanobody and enzyme in the conjugate independently and measured their effect on cancer cell viability in both adherent and spheroid culture. Increasing the number of enzyme domains from one to two increased drug activation by allowing for intramolecular dimerization of the enzyme. The increased enzyme activity resulted in a 4- and 40- fold decrease in cytotoxicity in adherent and spheroid culture, respectively. Increasing the number of nanobody domains from one to two increased cancer cell affinity by 4-fold, but did not increase cytotoxicity significantly due to lowered enzyme activity. The most potent nanobody-enzyme megamolecule was cytotoxic to HER2- cells when greater than 30% HER2+ cells where present in an adherent co-culture, eliciting a bystander killing effect. This study determines design rules for constructing nanobody-enzyme conjugates for cancer therapy.

610115 Antibody-Recruiting Protein-Catalyzed Capture Agents As a Rapid Therapeutic Countermeasure Against the Novel Coronavirus Sars-Cov-2

Abstract:

The novel coronavirus (SARS-CoV-2) emerged in late 2019 in Wuhan, China, and quickly spread worldwide, infecting millions and causing a high death toll alongside devastating economic damage. A gap in our ability to address the SARS-CoV-2 outbreak is the lack of an effective therapy. Here, we apply a versatile all-synthetic platform, called the antibody-recruiting protein-catalyzed capture agents (AR-PCCs), to rapidly generate highly targeted anti-viral compounds against SARS-CoV-2. AR-PCCs are composed of two molecular motifs: an epitope-specific peptide macrocycle (the PCC) that binds the pathogen, and a hapten (the AR moiety) that recruits endogenous antibodies. We hypothesized that AR-PCCs that adsorb to functional regions of SARS-CoV-2 would recruit antibodies that disable viral function and promote immune clearance. Bioinformatics, structural, and molecular dynamics analyses prioritized several target epitopes on the SARS-CoV-2 spike protein, which mediates cellular entry of the SARS-CoV-2 virion. Subsequent PCC screens against these epitopes yielded cohorts of consensus peptide ligands with high sequence homology, which were then vetted for high-affinity binding to full-length spike in high-throughput Luminex and ELISA platforms. Lead PCC binders were subjected to medicinal-chemistry type modifications to provide molecular-level insights into binding avidity, as well as to enhance *in vivo* performance metrics. The most promising PCCs were conjugated with an antibody-recruiting hapten, 2,4-dinitrophenyl, and interrogated by viral neutralization assays using live virus. *In vivo* testing of AR-PCCs in SARS-CoV-2 animal models is on-going. The work here demonstrates that AR-PCCs can be developed on an accelerated timescale of a few months, making the technology capable to address rapid pandemic outbreaks such as SARS-CoV-2.

601624 Computationally Engineering SARS Antibodies for COVID-19 Therapeutics

Abstract:

The emergence of Corona Virus Disease 2019 (COVID-19) has resulted in a pandemic with nearly 1,000,000 global cases as of April 1, 2020. There is a dire need for possible therapeutic agents or vaccines. The causative agent of this outbreak is the Betacoronavirus SARS-CoV-2, which has high genomic and structural homology with the SARS-CoV virus that caused the Severe Acute Respiratory Syndrome (SARS) pandemic in 2003-2004. Given the similarity between the two viruses, previous research on SARS immune responses can be leveraged to develop therapeutic treatments for COVID-19.

Antibodies play an important role in the protection against and recovery from viral infections. Neutralizing antibodies have been looked into as potential strategy for treating patients with SARS and other viral infections, because of their ability to neutralize the biological effects of the viral particles. SARS-CoV-2 uses the spike (S) protein to interact with the ACE2 cell receptor to enter host cells, similar to SARS-CoV. Thus, SARS-CoV specific antibodies are a promising potential therapeutic approach against the current disease. However, three antibodies (M396, S230, and 80R) which bind the S protein of SARS-CoV and neutralize the virus failed to show measurable binding to SARS-CoV-2 in early experiments.

We have employed computational methods to understand why those three antibodies have lost binding to SARS-CoV-2 and predict mutations to recover it. Our results show that the loss of binding is due to SARS-CoV-2 mutations that disrupt antibody "hotspot" interactions. Interestingly, almost all of the lost interactions in the three antibodies were salt bridges with the remainder being disruptions of hydrophobic patches. Based on this finding, a comprehensive interface mutational analysis was conducted to identify antibody mutations that can introduce novel hotspots with SARS-CoV-2. For all three antibodies, salt bridge and hydrophobic patch mutations were identified that are predicted to recover binding to SARS-CoV-2. Initial experiments were being conducted at the time this abstract was prepared. This presentation will describe the computational findings for why binding was disrupted and how it can be recovered, with experimental results hopefully included pending progress in the coming weeks and months.

597370 Biosynthesis of Alkyne, Alkene, and Halogenated Amino Acids for Expanded Genetic Codes

Abstract:

Living systems have been able to construct an enormous range of functions from a relatively limited set of functional groups, especially compared to the functional group and chemical reaction diversity that is available to synthetic chemists. As such, several bio-orthogonal reactions have been developed; taking advantage of the absence of certain functional groups inside the cell that can then be used as a specific chemical handle for a reacting partner when incorporated into a metabolite or macromolecule of interest. One of the most useful of these bio-orthogonal functional groups has been the terminal alkyne, which can be used within the complex cellular milieu to react selectively with another bio-orthogonal group, the azide, to ligate the two reacting partners in a biocompatible copper-catalyzed azide alkyne cycloaddition (CuAAC or "Cu-Click") reaction. Since its discovery, the CuAAC has found broad utility in a wide range of applications such as attachment of fluorescent probes, pull down and discovery of small molecule and protein binding partners, ligation to bioactive payloads (e.g. antibody-drug conjugates), or modifying macromolecular solubility and stability (e.g. PEGylation or tethering to solid supports). More recently, "Click" reactions for selective ligation of terminal alkenes with tetrazines and tetrazoles have also been developed. However, the application of these reaction to living systems has thus far been constrained by the requirement that both substrates must be prepared by chemical synthesis and supplied exogenously.

In this talk, I present the discovery and characterization of the natural, biosynthetic pathway for a terminal alkyne amino acid, β-ethynylserine (βes), from *Streptomyces cattleya*. Comparative genomics and comparative metabolomics were used to identify the genomic cluster and pathway intermediates, while analytical techniques were used to characterize the structure of individual chemical species. The enzymes uncovered in this pathway allow for the production of various halo, terminal alkene, and terminal alkyne L-amino acids from basic carbon sources, that can be genetically encoded and leveraged for novel applications. For example, we envision using elements of this pathway as proteomic tags that can be spatially and temporally targeted in living, complex hosts. Additionally, the βes pathway opens up the possibility of engineering industrial hosts that can seamlessly integrate halo, alkene, and alkyne amino acids into biomolecules, such as enzymes and natural product therapeutics, *at scale*. Taken together, this work presents an effort to bridge together advances in chemical biology (bioorthogonal reactions and genetic code expansion) with synthetic biology approaches of host engineering.

Marchand, J. A., Neugebauer, M. E., Ing, M. C., Lin, C.-I., Pelton, J. & Chang, M. C. Y. Discovery of a pathway for terminal-alkyne amino acid biosynthesis. *Nature.* (2019) doi:10.1038/s41586-019-1020-y.

Neugebauer, M. E., Sumida, K. H., Pelton, J. G., Mcmurry, J. L., **Marchand, J. A.** & Chang, M. C. Y. (2019). A family of radical halogenases for the engineering of amino-acid-based products Monica. *Nat. Chem. Biol.* (2019). doi: 0.1038/s41589-019-0355-x

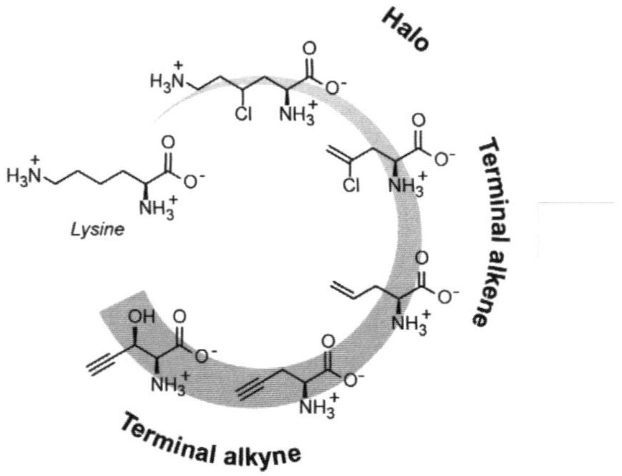

613362 Self-Assembled Protein Hexamers for Intracellular Delivery of Antibodies (INVITED SPEAKER)

Abstract:

Proteins can provide therapeutic functions simply not possible with small molecule drugs, but their large size and folded structure present critical challenges in terms of delivery, stability and activity. We take advantage of protein size, structure and the ability to interact with other proteins, in order to create therapeutic protein materials via self-assembly routes not available for small molecules. The ability to control assembly of therapeutic proteins is essential to manipulating the final physical properties of the material, ensuring retention of protein activity, and directing the interactions between materials and cells. In this presentation, I will describe the development, characterization and performance of Hex nanocarriers, which are self-assembled complexes capable of intracellular protein delivery. The Hex carriers seek to open up the intracellular space for antibodies and antibody variants as an approach to aim at "undruggable" intracellular protein targets. We applied a rational protein design strategy, combining a self-assembling hexamer barrel with six to twelve antibody binding domains that can bind the constant region of any antibody, making it a platform technology. Additionally, we have designed several sequence variations with different properties and features. We have performed extensive molecular characterization to understand the loading, dynamics and stability of our assemblies. We have demonstrated intracellular delivery of both non-specific fluorescent antibodies and specific functional antibodies and have developed insight on the uptake, trafficking and fate of internalized antibodies. The current status of application of these carriers towards therapeutic disease targets will also be shared.

610348 Pre-Clinical Manufacturing and Process Intensification of a Subunit Vaccine for COVID-19 (Faculty Candidate)

Abstract:

Rapid manufacturing and distribution of vaccines at global scale is a key tenant of response to the COVID-19 pandemic, which spread globally on the order of weeks. Elimination of the disease will require several billion doses of a vaccine, based on the basic reproductive number of at least 2-3 estimated by current models. A majority of these doses need to be deployed in low- and middle-income countries, which are especially vulnerable to the disease, and could serve as epicenters for resurgence. Subunit vaccines are enticing candidates because they are potentially compatible with existing platforms for the manufacturing of similarly sized proteins at global scale, like the production of billions of yearly doses of Hepatitis B vaccine and insulin in the yeast *Komagataella phaffii* (*Pichia pastoris*). The first six vaccines for COVID-19 to enter human trials, however, did not include a subunit vaccine. What is needed are platforms for the rapid generation of subunit vaccine candidates for pre-clinical testing that are subsequently amenable to scaled up manufacturing.

We expressed the SARS-CoV-2 receptor binding domain (RBD) in *K. phaffii*. Speed of the yeast host, combined with InSCyT, our benchtop scale, continuous, end-to-end manufacturing platform, enabled purification and characterization of RBD in quantities and qualities sufficient for pre-clinical studies in less than one month. During pre-clinical studies, we used modern diagnostic tools like RNAseq and rapid genome engineering tools like CRISPR to improve specific and volumetric productivity of our host and process by >5x. Our intensified process demonstrates rapid development of a manufacturing process compatible with existing centralized and distributed manufacturing platforms, and within the low cost necessary for global distribution.

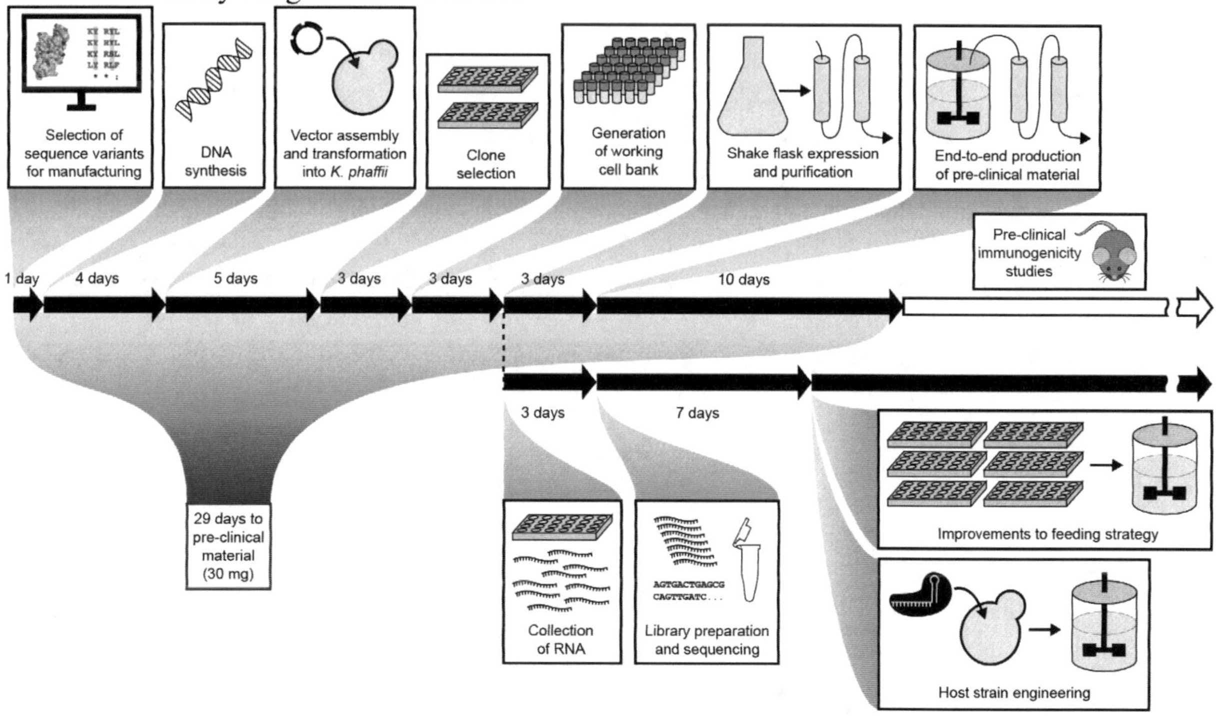

Enhanced Mixing and Oxygen Transfer in a Continuous Bioreactor with Optimized Spiroid

Rithvija Avvari[1], Paul W. Todd[2], and Thomas R. Hanley[1]
[1]Department of Chemical Engineering, Auburn University, AL 36849
[2]Magnaquant, 11760 Commonwealth Drive, Louisville, KY 40299

Abstract

Oxygen transfer is a key issue when culturing cells in a bioreactor. Conventional bioreactors are operated in batch mode and can employ impellers for increased mixing and mass transfer which results in shear stress. The high shear stress from mechanical impellers typically results in greater cell death, especially for mammalian cells. Researchers in bioprocess and pharmaceutical industries are focusing on developing methods to increase mixing and maintain shear stress at optimal levels for cells to survive and proliferate in the reactor. A reactor environment that best provides low shear and high oxygen transfer along with controlled growth conditions will be suitable for cell cultures at a larger scale.

A novel rotating continuous bioreactor with an internal spiroid can provide enhanced mixing and oxygen transfer for improved cell viability. The spiroid is embedded in the cylindrical wall of the bioreactor to enhance oxygen transfer to the liquid phase which fills two-thirds of the bioreactor. The bioreactor is a horizontal cylinder rotated on a roller bed. The rotation rate of the reactor can be adjusted to control the flow of gas and liquid in the spiroid. When the partially filled reactor is rotating, the spiroid picks up slugs of gas and liquid near the reactor exit and delivers them to the reactor entrance. Thus, the spiroid increases the gas-liquid contact areas within the reactor to increase oxygen mass transfer. The bioreactor is produced using rapid prototyping and can be operated in either batch or continuous modes with inlet flows via rotary unions available to provide medium and oxygen and outlet flows for waste and in-line analysis. Computational fluid dynamic simulations were utilized to model flow and oxygen transfer in the spiroid at various operating conditions. The rotational rate of the reactor and the diameter and length of the spiroid were key parameters explored for optimization of the spiroid. Models were generated using CFD studies for calculating the exact value of volumetric oxygen transfer in the spiroid. Cell culture experiments using *Saccharomyces cerevisiae* were conducted under various conditions to optimize the spiroid and reactor productivity. The amount of oxygen transfer with and without the spiroid was calculated for these cells and compared to published values. Based on these observations, changes to reactor design for better efficiency and performance were also made.

Keywords: spiroid, cell culture, computational fluid dynamics, continuous bioreactor

Introduction

Continuous bioprocessing is emerging in upstream processing and is preferred over traditional batch processing. One important area of investigation is the design and operation of bioreactors. A bioreactor provides a suitable environment (adequate nutrients and oxygen) for cell cultures to proliferate under controlled conditions. Increasing productivity while reducing shear is one common goal for these reactors. Most bioreactors provide mixing with the use of impellers resulting in high shear that can cause cell death. A reactor environment that best provides low shear and high oxygen transfer along with controlled growth conditions will be suitable for cell cultures at a larger scale.

The current research focused on developing a rotating continuous bioreactor with an internal spiroid embedded on the outer wall that enhances oxygen transfer to the liquid phase by increasing gas-liquid interfacial area. When the partially filled reactor is rotating, the spiroid picks up slugs of gas and liquid near the reactor exit and delivers them to the entrance of the reactor. The rotational rate of the reactor can be adjusted to control the flow of gas and liquid in the spiroid. The bioreactor is produced using rapid prototyping and can be operated in either batch or continuous mode. Also, the effect of spiroid in the reactor was analyzed at several rotational rates and medium flow rates.

Material & Methods

The continuous bioreactor used for this study is a scaled-up version of an earlier bench-scale bioreactor (Fang et al., 2018) and consists of an outer cylindrical shell with a spiroid. The spiroid is embedded onto the inner wall of this cylinder to increase gas-liquid contact areas which helps in enhancing oxygen transfer within the reactor. The bioreactor can be rotated horizontally with the help of a roller bed as shown in Figure 1. The two inlet and two outlet ports present at either end of the reactor help in providing medium and oxygen continuously to the reactor and facilitate in-line and off-line sampling. This bioreactor is prototyped using a 3D printer.

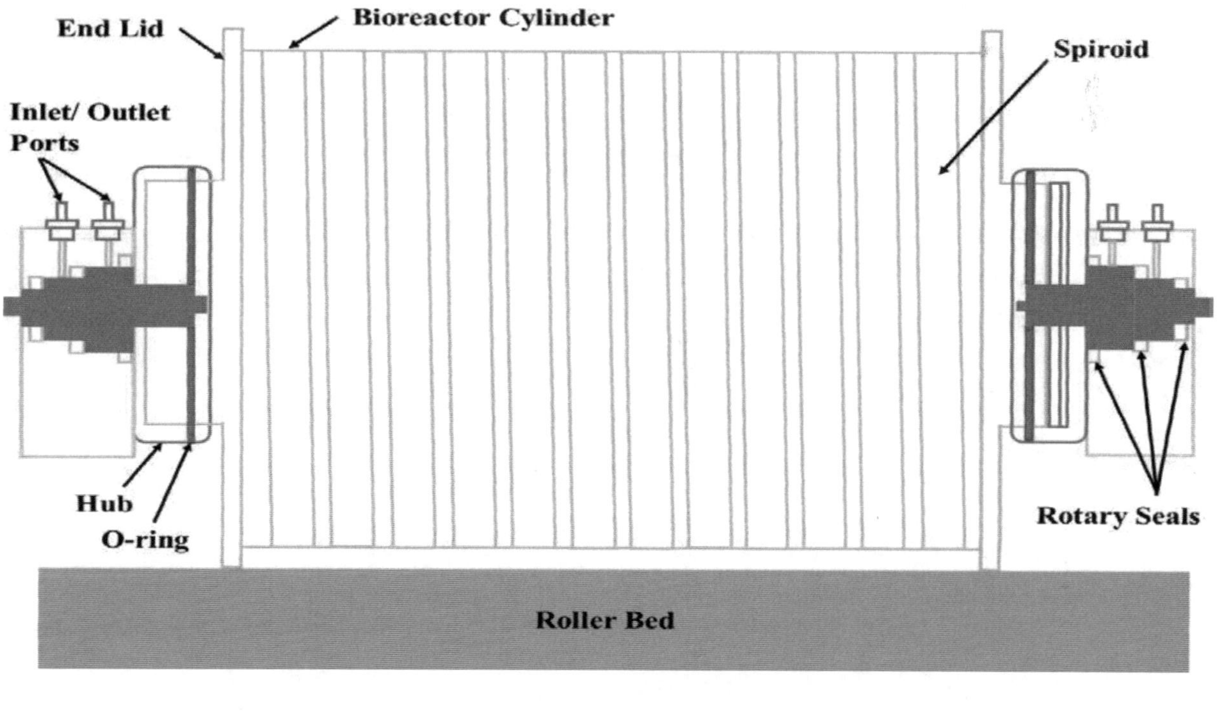

Figure 1 - Bioreactor Schematic (Fang et al., 2018)

The dimensions of the scaled-up bioreactor and original bioreactor are summarized in Table 1.

Table 1 - Bioreactor dimensions

Parameter	**Original Bioreactor**	**Scaled-up Bioreactor**
Chamber length (m)	0.100	0.193
Chamber Inner diameter (m)	0.0444	0.0889
Spiroid Length (m)	1.09	3.12
Spiroid Inner Diameter (m)	0.00476	0.00953
Spiroid turns	8	12
Volume (10^{-4} m^3)	1.11	9.86

Computational Fluid Dynamics

The optimization of spiroid is a major factor which will enhance the oxygen transfer in the bioreactor. The length and diameter of the spiroid must be optimum to provide the highest amount of oxygen transfer in the reactor. ANSYS workbench 19.2 was used in this research to carry out the CFD simulations. The basic steps for performing any simulation in ANSYS workbench are as follows:

1. Create the geometry (Design Modeler was used in this research)
2. Mesh the geometry using ANSYS meshing tool.
3. Create 'Named selections' for the domain of study.
4. Setup the model needed to illustrate the flow phenomenon
5. Run the simulations
6. Analyze and predict the results using CFD-post processing.

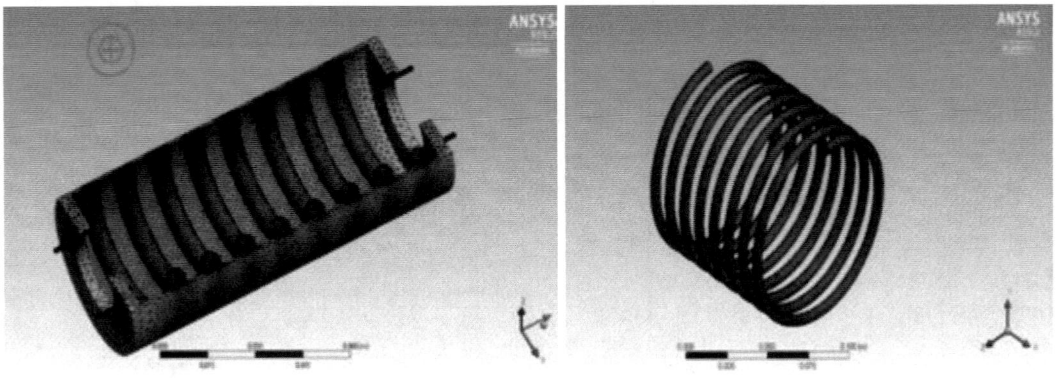

Figure 2 – Bioreactor cross-section and independent spiroid showing FEA mesh pattern.

(ANSYS 19.2)

Cell culture – Strain and Medium

The culture used in this research was *Saccharomyces cerevisiae* (yeast) obtained from ward's science (VWR life sciences, USA). The growth medium used for yeast was YPD medium, which contained 20g peptone, 10g yeast extract, 20 glucose per L of distilled water (VWR life sciences, USA). The pH of the medium was 6.5 and it was autoclaved for 15 minutes or sterile filtered before use. The final concentration of glucose in the medium during experiments was $7gL^{-1}$.

Operating Conditions

Saccharomyces cerevisiae was precultured in a shaker incubator at 200 rpm and 30°C for 12 hours. The optical density of the culture was then adjusted to 0.1 before transferring into the bioreactor. The reactor was operated in batch mode first for four hours, and then continuous mode with different rotation rates (6rpm, 8rpm) and fresh medium flowrates (1.5ml/min and 3ml/min). The process flow diagram for continuous mode experiments is shown in figure 3. Oxygen was provided to the reactor periodically along with venting to maintain constant levels. After batch mode of 4 hours, fresh medium was pumped into the reactor and the same amount was removed to maintain constant volume. The feed was maintained sterile.

Analytical Methods – Optical Density measurements

The growth of yeast in the reactor was monitored by measuring OD_{600} of the samples using Flex station 3 Micro plate reader (Molecular Devices) at different time intervals during the experiment. Glucose consumption in the reactor was measured by using a glucose monitor (GM100, Bioreactor sciences, Inc., Lawrenceville, GA, USA).

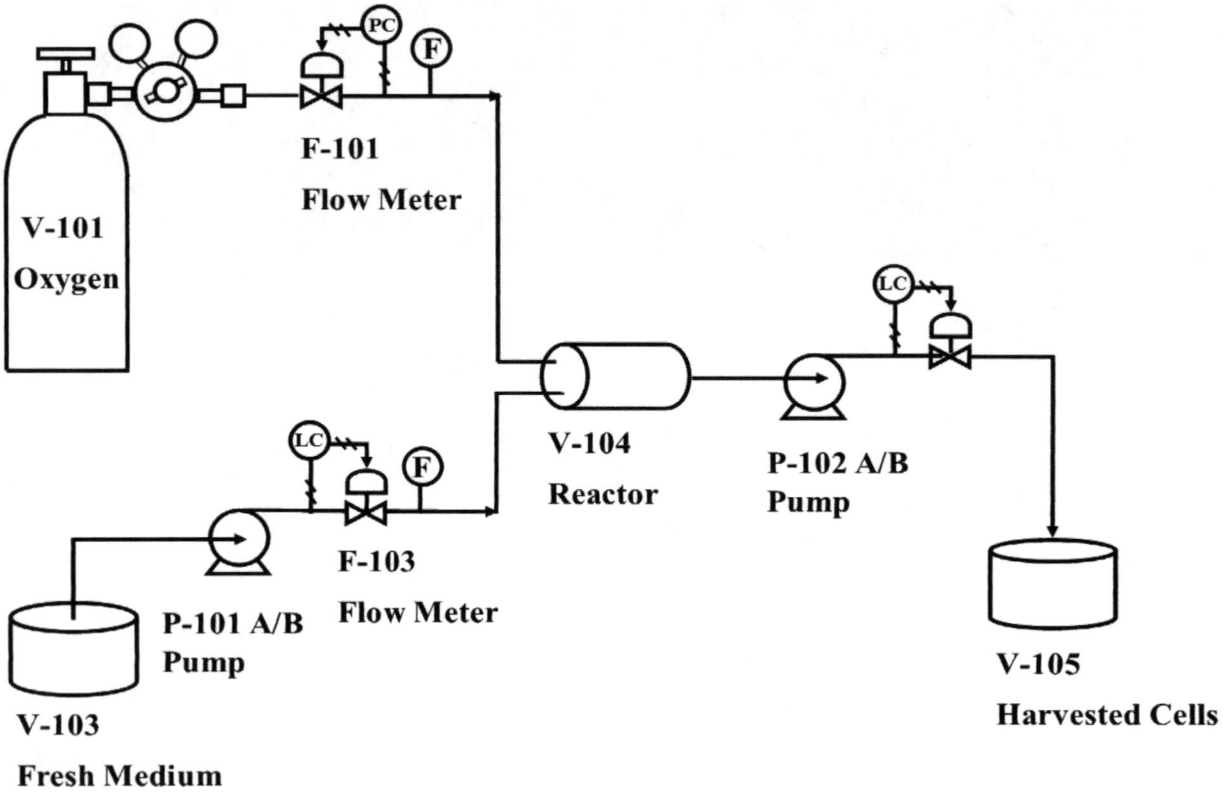

Figure 3 – Process flow diagram: Continuous cell culture experiments adapted from (Fang et al)

Results and Discussion

CFD simulations

In this research, a pressure-based model was used. The simulations were carried out at steady-state and gravity was disabled. The VOF method is used in this research to simulate flow through the spiroid. The interface between the phases was applied using the sharp/dispersed option with implicit formulation. Water was set as primary phase (blue in Figure 4) and air as secondary phase (red in Figure 4). Phase interactions like surface tension were enabled at standard conditions. The boundary conditions were appropriately set for each phase and the pressure in the headspace of the spiroid was assumed to be 1.0 atmosphere.

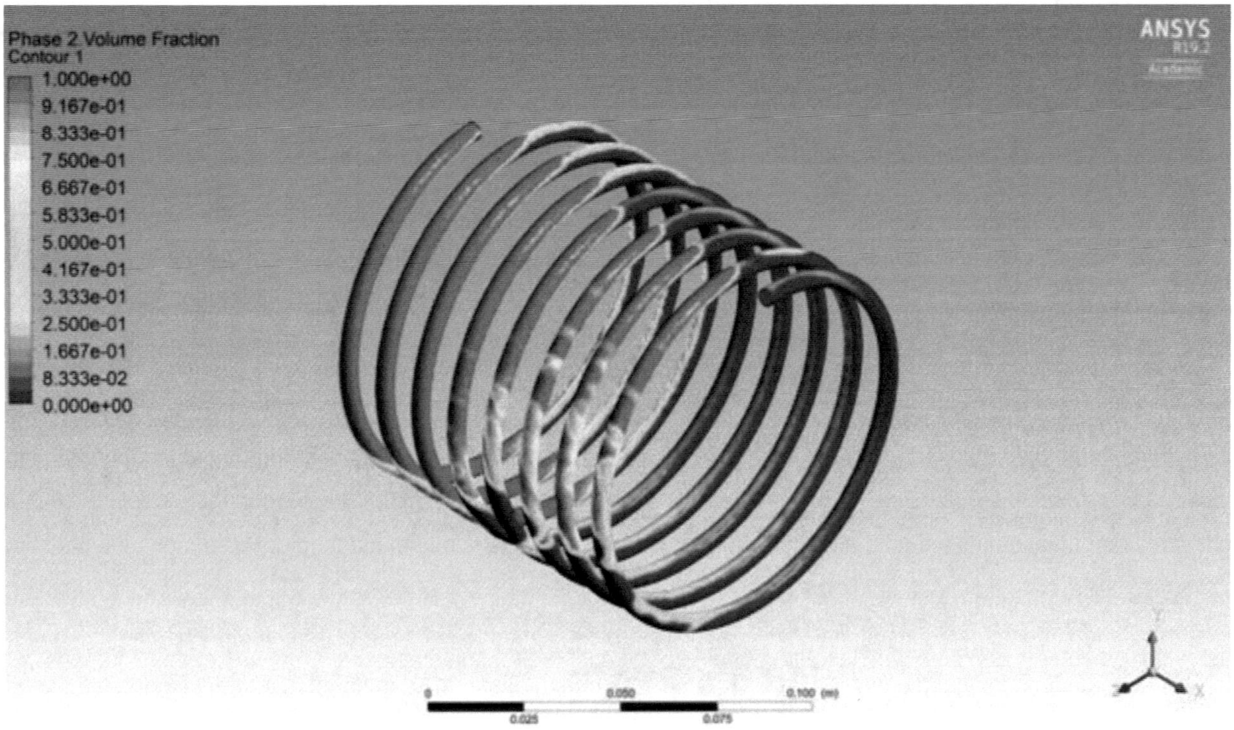

Figure 4 - Multiphase contours of water and air using the VOF method

8 RPM with reactor two-thirds full

The criterion for convergence was checked by using the residual profiles. In case the solution does not converge even after long iterations, convergence conclusions can be made if the residuals continue to remain almost the same or decrease at a very small rate. Mass conservation for each was also checked to validate the convergence. Figure 4 shows the formation of thin liquid film (areas of yellow to green) from which it can be interpreted that the interfacial area in the volumetric mass transfer coefficients consists of film area and gas-liquid contact area.

Cell Culture Results

The growth of *Saccharomyces cerevisiae* in the bioreactor was relatively low at lower rotation rates which could be due to low oxygen delivery or poor mixing. For batch mode, the reactor was operated for a period of 12 hours to analyze the growth of yeast at 8RPM with and without spiroid. Samples were collected every 2 hours. Optical density measurements along with glucose consumption was monitored. Figure 5 shows the effect of spiroid on cell dry weight measurements at 8RPM. The cell growth in the bioreactor with spiroid was 21% more than the reactor without spiroid (0.58 g/l over 0.37 g/l) over a period of 12 hours. The overall operating time for the reactor decreased by 27% using the spiroid in batch mode.

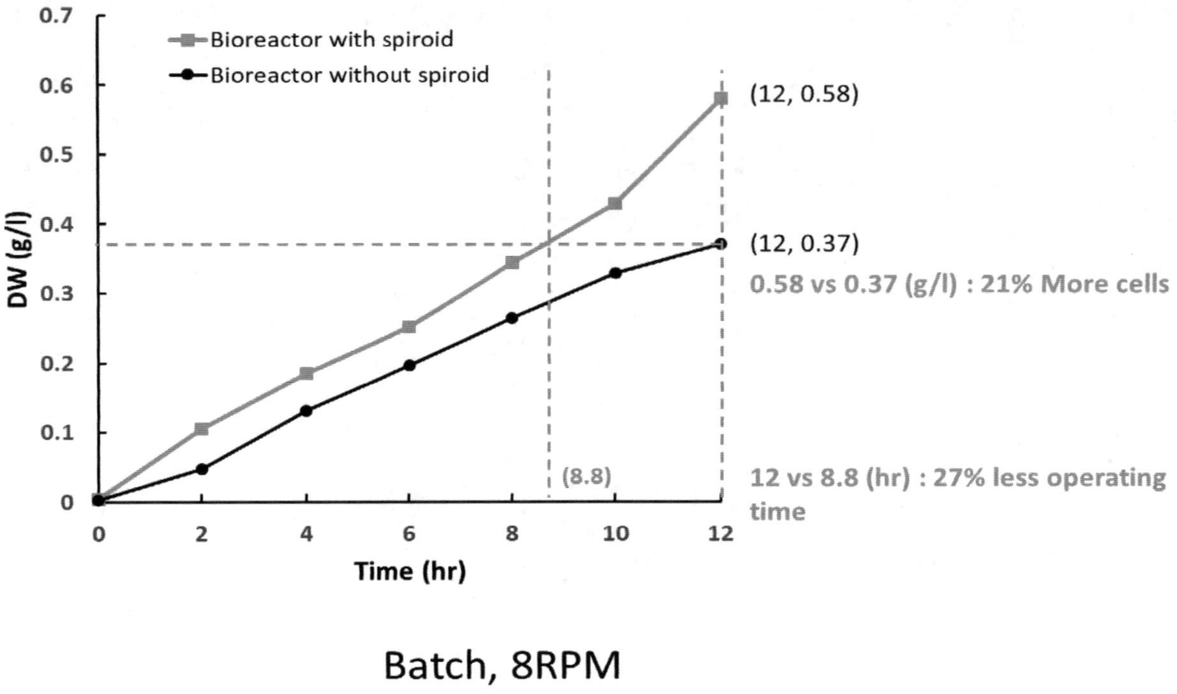

Figure 5 – Cell dry weight – Batch mode, 8RPM (with and without spiroid)

For continuous mode, experiments were conducted at two different rotation rates (6RPM, 8RPM) and two medium flowrates (1.5ml/min and 3ml/min) to demonstrate different steady state concentrations. Duplicate experiments were conducted at each condition. Cells grew in exponential phase in batch mode for 4 hours and eventually reached steady state concentration with continuous flow in both growth curve and glucose consumption profiles. Figure 6 represents steady state concentrations controlled by different flow rates. The growth curves were consistent in batch mode and the steady state concentration decreased with increased flowrates. This could be due the incoming fresh media. Increase in cells at lower flowrates showed a decrease in substrate level concentration as seen in glucose profile.

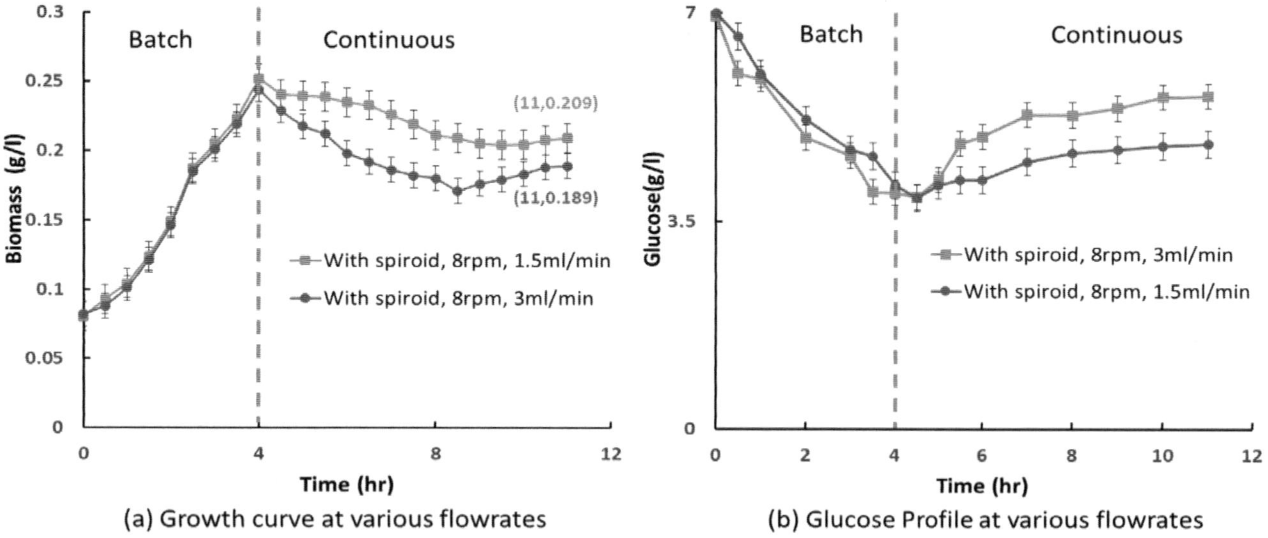

Figure 6 – Growth curve & glucose profile at different flowrates using spiroid

Figure 7 compares the growth of yeast at two different rotational rates. Cell growth at 6 RPM was slower in both batch and continuous modes when compared to 8RPM. Enhanced oxygen delivery at higher rotational rates was observed, as in the CFD simulations. Higher steady- state concentration was observed at high rotational rate indicating the increased oxygen delivery capacity.

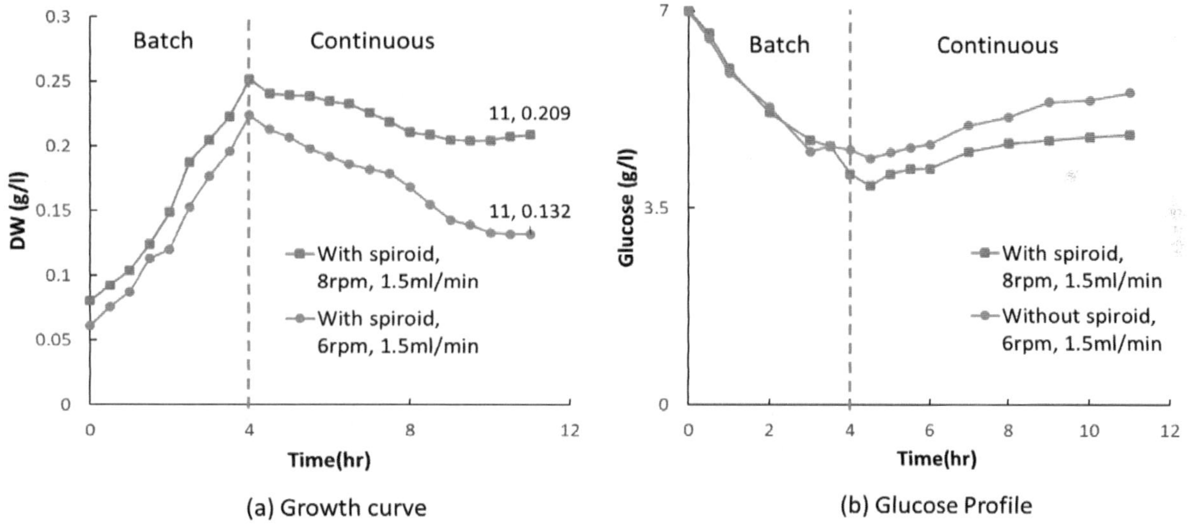

Figure 7 – Growth curve & glucose profile at different rotational rates using spiroid

From Figure 8, the effect of using spiroid could be clearly seen. The increase in cell dry weight in batch mode and continuous mode could be attributed to the use of spiroid as the rest of the parameters were kept constant. Yeast grew faster in batch mode with spiroid when compared to without spiroid. Increase in steady state cell concentrations were noticed with the use of spiroid. The concentration change was minor when the system reached steady state in both the conditions. This increase is clearly due to the use of spiroid in the bioreactor.

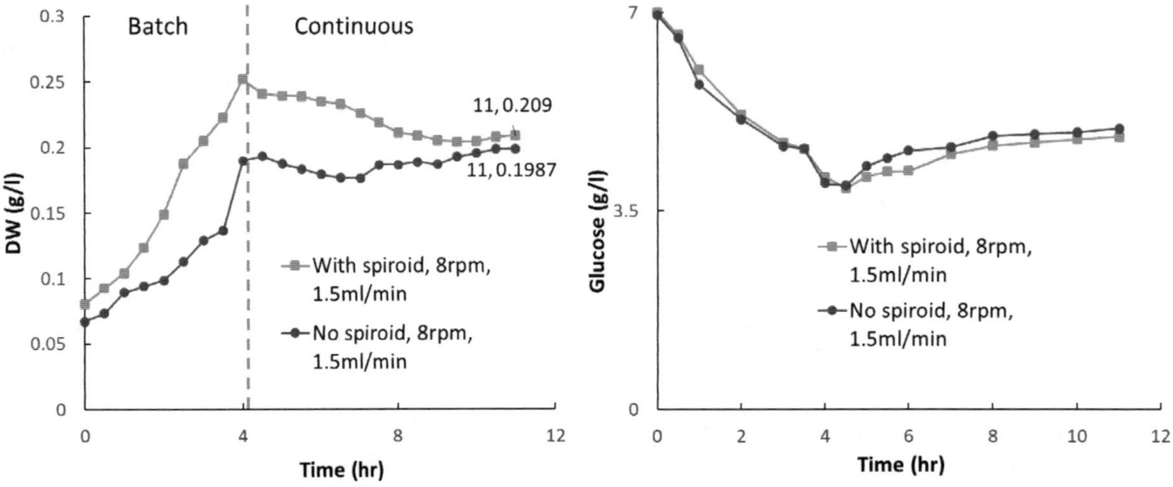

(a) Growth curve

(b) Glucose Profile

Figure 8 – Growth curve & glucose profile at 8RPM with & without spiroid

Conclusions

The use of spiroid in the bioreactor increased oxygen transfer and mixing. This increase in oxygen transfer led to an increase in cell population. The bioreactor could be operated both in batch and continuous modes for longer durations without leakages, contrary to traditional batch bioreactors. Various steady state concentrations for different medium flowrates were obtained. Increase in cell concentration and decrease in operating time was observed for the bioreactor with spiroid.

References

Augenstein DC, Sinskey AJ, Wang DIC. 1971. Effect of Shear on the Death of Two Strains of Mammalian Tissue Cells. *Biotechnol. Bioeng. 13:409–418.*

Avvari, R., Todd, P., & Hanley, T. (2019, November 11). Spiroid Design Optimization in a Rotating Continuous Bioreactor for Enhanced Oxygen Mass Transfer. *2019 AIChE Annual Meeting Proceedings*, Orlando, FL, November 2019.

Berson, R. E., and Friederichs, G. (2008). A self-feeding roller bottle for continuous cell culture.*Biotechnology Progress*, *24*(1), 154–157.

Childs, P. R. N. (2011). Rotating Cylinders, Annuli, and Spheres. In *Rotating Flow*.

Dhanasekharan, K. M., Sanyal, J., Jain, A., and Haidari, A. (2005). A generalized approach to odel oxygen transfer in bioreactors using population balances and computational fluid dynamics. *Chemical Engineering Science*, *60*(1), 213–218.

Eagle, H. (1973). The effect of environmental pH on the growth of normal and malignant cells. *Journal of Cellular Physiology*, 82(1), 1–8.

Elias, C. B., Desai, R. B., Patole, M. S., Joshi, J. B., and Mashelkar, R. A. (1995). Turbulent shear stress—Effect on mammalian cell culture and measurement using laser Doppler anemometer. *Chemical Engineering Science*, 50(15), 2431–2440.

Fill, B., Gartner, M., Johnson, G., Horner, M., and Ma, J. (2008). Computational fluid flow and mass transfer of a functionally integrated pediatric pump-oxygenator configuration. *ASAIO Journal*, 54(2), 214–219.

Furukawa, K., and Ohsuye, K. (1998). Effect of culture temperature on a recombinant CHO cell line producing a C-terminal α-amidating enzyme. *Cytotechnology*, 26(2), 153–164.

Gabardo, S., Rech, R., Rosa, C. A., and Ayub, M. A. Ô. Z. (2014). Dynamics of ethanol production from whey and whey permeate byimmobilized strains of Kluyveromyces marxianus in batch andcontinuous bioreactors. *Renewable Energy*, 69, 89–96.

Galaction, A. I., Cascaval, D., Oniscu, C., and Turnea, M. (2004). Enhancement of oxygen mass transfer in stirred bioreactors using oxygen-vectors. 1. Simulated fermentation broths. *Bioprocess and Biosystems Engineering*, 26(4), 231–238.

Garcia-Ochoa, F., and Gomez, E. (2009). Bioreactor scale-up and oxygen transfer rate in microbial processes: *An overview. Biotechnology Advances*, 27(2), 153–176.

Gardel, L. S., Correia-Gomes, C., Serra, L. A., Gomes, M. E., and Reis, R. L. (2013). A novel bidirectional continuous perfusion bioreactor for the culture of large-sized bone tissue-engineered constructs. *Journal of Biomedical Materials Research* - Part B Applied Biomaterials, 101(8), 1377–1386.

Gareau, T., Lara, G. G., Shepherd, R. D., Krawetz, R., Rancourt, D. E., Rinker, K. D., and Kallos, M. S. (2014). Shear stress in fl uences the pluripotency of murine embryonic stem cells in stirred suspension bioreactors. (June 2012), 268–278.

Garza-García, L. D., García-López, E., Camacho-León, S., Del Refugio Rocha-Pizaña, M., López-Pacheco, F., López-Meza, J., … Alvarez, M. M. (2014). Continuous flow micro-bioreactors for the production of biopharmaceuticals: The effect of geometry, surface texture, and flow rate. *Lab on a Chip*, 14(7), 1320–1329.

Gòdia, F., Albiol, J., Montesinos, J. L., Pérez, J., Creus, N., Cabello, F., … Lasseur, C. (2002). MELISSA: A loop of interconnected bioreactors to develop life support in Space. *Journal of Biotechnology*, 99(3), 319–330.

Hutmacher, D. W., and Singh, H. (2008). Computational fluid dynamics for improved bioreactor design and 3D culture. *Trends in Biotechnology*, *26*(4), 166–172.

Liow, K. Y. S., Tan, B. T., Thouas, G. A., and Thompson, M. C. (2009). CFD modeling of the steady-state momentum and oxygen transport in a bioreactor that is driven by an aerial rotating disk. *Modern Physics Letters B*, *23*(2), 121–127.

Niño, L., Peñuela, M., and Gelves, G. R. (2018). Gas-Liquid Hydrodynamics Simulation using CFD in a Helical Ribbon Impeller Applied for Non-Newtonian Fluids. *International Journal of Applied Engineering Research*, *13*(11), 9353–9359. Retrieved from http://www.ripublication.com

Sergio S. de Jesus, Edgar Leonardo Martinez, Aulus R.R. Binelli, Aline Santana, R. M. F. (2013). CFD Simulation of Hydrodynamic Behaviors and gas liquid mass transfer in a stirred airlift bioreactor. *International Journal of Chemical Materials Science and Engineering*, *7*(12), 165–169.

Salari, R., & Salari, R. (2017). Investigation of the Best Saccharomyces cerevisiae Growth Condition. *Electronic Physician*, 9(1), 3592-3597. doi:10.19082/3592

S. Fang, P. W. Todd, T. R. Hanley. 2017. Enhanced oxygen delivery to a multiphase continuous bioreactor. *Chem. Eng. Sci*. 170, 597–605.

600854 Continuous Protein Purification: Utilizing Multi-Stage Liquid-Liquid Extraction

Abstract:

Aqueous Two-Phase Systems (ATPS) are a liquid-liquid extraction (LLE) technique which can be used as a cheap and continuous purification technique for therapeutic proteins. Using both cheaper and continuous techniques for the manufacture of therapeutic proteins has been encouraged by regulatory bodies, including the FDA, as it is expected to increase the quality and consistency of products as well as reducing the manufacturing costs. However, for the process to be taken up by industry ATPS must compete with the current batch purification procedures, which can be done by extending ATPS into a multi-stage operation. To this end, to both reduce the reliance on individual expertise and trial and error, there is a need to develop models to aid in system design.

In this study we present single stage equilibrium data which was determined experimentally for a model protein, haemoglobin, in a case study ATPS of 13.0 % w/w polyethylene-glycol (PEG) and 11.2 % w/w potassium-phosphate buffer at pH 8.0. The equilibrium data is then used in the development of a model to describe the behaviour of haemoglobin in a multi-stage counter-current ATPS with both: liquid-liquid distribution and liquid-interface-liquid distribution. The liquid-liquid distribution model is based upon work from Rosa et al. (2009a, b), Lui et al. (2018) and Chandler et al. (2019). The liquid-interface-liquid model has the added consideration of accounting for the material which precipitates at the interface of the ATPS. These models are then both compared against an experimental case study of a three stage counter-current ATPS which was spiked with haemoglobin.

It is found that using multi-stage ATPS the yield of haemoglobin in the model system is increased from 61.8 in a single stage to 85.3% in three stages. Single stage equilibrium data shows that material immediately partitions into all three regions, top phase, bottom phase and interface, as opposed to only two regions until the system reached saturation. As a result, the liquid-interface-liquid model describes the case study system better than the liquid-liquid distribution model, reducing error from 40% to 11%.

603949 Using a High Throughput Cell Culture Workflow for Process Development and Process Characterization for Therapeutic Proteins

Abstract:

Application of high throughput technologies for biologics allows new molecules to transition from bench-scale to at-scale manufacturing and to the clinic at a faster rate. We have established a high throughput workflow for upstream cell culture bioreactor experiments, sample preparation, sample submission, tracking and analytical testing to increase efficiency across all process development and process characterization projects.

This workflow begins with the use of ambr15™ and ambr250™ (Sartorius) bioreactor systems. Traditionally, bench-scale bioreactor systems (2-L to 5-L) were used for process development and process characterization, but these bioreactors have a longer turnaround time and require more resources for preparation and study execution. Use of disposable ambr™ bioreactor systems enable scientists to perform 24 to 48 bioreactor runs in parallel, with significantly lower resource need. Sample preparation and sample handling with high throughput bioreactor systems can be challenging. We have implemented an approach for sample preparation using a sample numbering system, TubeWriter, plate centrifuges, Opentron liquid handler and sample tracking using barcodes. This workflow and procedures have been harmonized across all groups in the department.

We will present our approach for establishing this workflow, resource and throughput comparison between traditional bioreactor system and ambr™ systems, efficiency gain in sample preparation and sample handling using the high throughput technologies. We consider high throughput as a means for continuous improvement to meet our goal of Gene to GMP in 9 months.

606205 Process Development, Characterization, and Understanding in an Integrated Continuous Monoclonal Antibody Manufacturing Testbed

Abstract:

Process and product understanding is at the root of manufacturers' efforts to produce safe and effective medicines. The 2009 ICH and FDA Guidance for Industry Q8(R2) Pharmaceutical Development describes a quality-by-design (QbD) approach that "emphasizes product and process understanding and process control, based on sound science and quality risk management" (1). During the product development stage, the critical quality attributes (CQAs) required to ensure product quality are defined, and the relationship of these CQAs to the critical process parameters (CPPs) that impact them are determined. Manufacturers can use a risk-based approach to define a design space in which the combination of process inputs and process parameters are demonstrated to provide assurance of product quality (2,3). Development of a control strategy that maintains the process within the bounds of the design space will ensure consistent product quality.

Researchers in our group at MIT are building novel first-principles- and data analytics-based mathematical modeling tools for the manufacturing of biopharmaceuticals. To experimentally validate these modeling tools and to fully understand the impact of model choice on product quality, a fully instrumented and integrated continuous testbed for the manufacturing of monoclonal antibodies (mAbs) was constructed. The testbed consists of 4 parallel upstream systems including 4 perfusion devices, with one reactor assembly integrated with a fully continuous downstream system including Protein A chromatography, in-house designed viral inactivation, and ion exchange chromatography. The testbed is equipped with instrumentation to fully characterize the process, including in-reactor probes for Raman spectroscopy (Kaiser RamanRXN2), viable cell density (Aber Futura), and optical density (Optek). To provide further at-line process and product characterization, each upstream assembly is equipped with two MAST Sample Pilots for automated sampling of both the reactor contents and the perfusate. The MAST system delivers samples to a Nova FLEX2 cell culture analyzer for key metabolite quantification as well as verification of in-reactor sensors including pH. For at-line characterization of CQAs, the cell-free perfusate samples are collected in a Gilson GX-271 liquid handler, purified using at-line purification using Protein A chromatography, and delivered automatically via MAST to either an Agilent 1260 Bio-Inert HPLC for assessment of aggregation and titer or to an Agilent 6545XT LC/QTOF for characterization of glycosylation profiles using mass spectrometry.

This presentation describes the instrumentation and discusses data collection and process integration within the testbed. The experimental data generated by this highly instrumented integrated continuous biomanufacturing testbed enable the evaluation, development, and validation of modeling methods and control strategies, and ultimately contribute to improved process understanding for better informed risk-based decisions during manufacturing campaigns.

References:

1. Pharmaceutical development Q8(R2). International Conference on Harmonisation of Technical Requirements for Registration of Pharmaceuticals for Human Use, ICH Harmonised Tripartite Guideline (2009).
2. Anurag S. Rathore and Helen Winkle. Quality by design for biopharmaceuticals. *Biotechnology* 27(1):26-34 (2009).
3. Mo Jiang, Kristen Severson, J. Christopher Love, Helena Madden, Patrick Swan, Li Zang, and Richard D. Braatz. Opportunities and challenges of real-time release testing for biopharmaceutical manufacturing. *Biotechnology and Bioengineering*, 114(11):2445-2456 (2017).

606346 AMBR®250 Scale-DOWN MODEL Limitations and MASS Transfer Characterization

Abstract:

Cell culture process development and process characterization studies leverage bench-scale bioreactor systems as scale-down models for large scale manufacturing. The ambr®250 micro-scale bioreactor platform is a bench-scale system that offers customizable automation and individualized bioreactor control to support high-throughput workflows. While ambr®250 bioreactors provide an excellent option to facilitate expansive studies, there are some scale-dependent limitations that must be addressed to develop an appropriate scale-down model. Power input and mass transfer are inherent scale-dependent principles which are directly impacted by the mixing and gassing strategies implemented at small scale, and small-scale operational parameters must be defined with the key study outcomes in mind.

During the development of an ambr®250 scale-down model for process characterization studies, we observed limitations that caused sharp, transient drops in the dissolved oxygen (DO) level. This phenomenon was found to correlate with the particular set of bioreactor parameters, which led us to perform additional bioreactor characterization studies for mass transfer of dissolved gases. Our results expand upon previous works by incorporating the mass transfer contribution from the transient headspace gas composition, assess multiple bioreactor fill volumes, and leverage representative cell culture media. In our ambr®250 kLa calculations we found that both the transient headspace gas composition and the bioreactor working volume contribute significantly to overall mass transfer. Then the magnitude and duration of impacts on ambr®250 DO control resulting from antifoam additions, feed additions, and bioreactor uncapping at various time points during a production run were compared to theoretical modeling based on our kLa calculations. We conclude by providing alternative approaches for scaled-down process parameters to achieve consistent DO control in ambr®250 bioreactors.

612928 Non-Disruptive Approaches to Estimate Volumetric Mass Transfer Coefficient (kLa) in Real-Time to Obtain Oxygen Uptake Rates (OUR) for Feed Control (Invited Speaker)

Abstract:

Robust and efficient methods to determine the volumetric mass transfer coefficient (kLa) in real-time, that uses only common industrial sensors, are desired. kLa can change in a complex, nonlinear manner during cultures, as it is dependent many parameters, including stir speed, impeller shape and dimension, bioreactor shape and dimensions, media chemical characteristics, cell secretions, and temperature. Estimating real-time kLa without filtering or latency effects allows for the calculation of oxygen uptake rates (OUR) that can be used to control glucose feeding. Currently, the industry standard for feed control is set offline, where the protocol may be adjusted by an operator in response unexpected online or offline measurements, but rarely are online automation used to control feeding to optimal rates. For *Escherichia coli*, balancing the carbon flux of glycolysis and the tricarboxylic acid (TCA) cycle is critical to minimize acetate accumulation (a waste product) and maximize final cell densities. In this work, an online, state estimator was developed that uses standard industrial sensors to predict the oxygen transfer rate (OTR) in bioreactors, where OTR is directly related to OUR. OUR is an indicator of the metabolic state of the culture and the carbon flux balance. Recursive least squares was used to match a model of OTR to the estimated inlet oxygen concentration, measured dissolved oxygen, and measured off-gas concentrations. The kLa estimator was used to control glucose feeding for *E. coli* cultures without operator interaction, to culture temperature and induction strength changes. For these *E. coli* cultures, high-cell density conditions were reached (100 g dry cell weight per L), out competing exponential feeding. Also, initial work has been conducted with Chinese hamster ovary (CHO) cell cultures. For CHO cell cultures, process corrections, such as antiform additions and base additions for pH adjustment, caused disturbances in the off-gas measurements. Work is ongoing to adapt the non-disruptive kLa estimator to CHO cell cultures to match the success of the *E. coli* cultures.

611644 Calcium Phosphate Nanocarriers Produced By Flame Spray Pyrolysis for High-Loading Delivery of Biological Drugs

Abstract:

Nanoparticle-based drug delivery systems have been considered as an auspicious alternative for efficient delivery of biological drugs (such as proteins and peptides) [1], which is deemed challenging due to large size and enzymatic degradation susceptibility of biologics. However, the clinical translation of such therapeutic nanoformulations is still in its infancy due to (i) complex synthesis methods of nanocarriers, and (ii) poor loading efficiency of biologics in comparison to clinically relevant therapeutic doses [2].

Here, we aim to address these challenges by engineering calcium phosphate (CaP) nanocarriers with tailored properties by flame spray pyrolysis (FSP) [3,4]. FSP is a scalable process with high production rates and also allows for fine tuning of nanoparticles properties. CaP nanoparticles with different specific surface areas (SSA) are obtained by varying precursor concentration in the flame. Our main goal is the synthesis of nanoparticles with high loading capacity of biological drugs (proteins and peptides). Bovine serum albumin (BSA) and bradykinin are used as model protein and peptide, respectively, and are physisorbed on the as-prepared nanoparticles. Their concentration along with the incubation time are varied in order to investigate their influence on loading efficiency. High loading values, reaching 350 mg/g and 600 mg/g of CaP for BSA and bradykinin, respectively, are obtained.

Furthermore, as a proof of concept, we apply the developed protocol and physisorb LL-37 antimicrobial peptide on CaP nanoparticles. Maximum loading achieved here (~800 mg/g) outperforms all LL-37 loading values found in the literature both for inorganic and organic nanocarriers [5]. Typical fractal-like structure of flame made agglomerated/aggregated nanoparticles promotes high drug loading efficiency. Moreover, it is shown that CaP nanoparticles protect LL-37 from enzymatic degradation by Protease K *in vitro* and that the peptide retains its antimicrobial activity against *Escherichia coli* and *Streptococcus pneumoniae* after being physisorbed on nanoparticles. The antimicrobial activity of LL-37 loaded nanoparticles, as assessed by monitoring bacterial growth, outperforms that of the free peptide for *E. coli*, whereas similar behavior is observed for free LL-37 and LL-37 loaded CaP nanoparticles for *S. pneumoniae*.

In a nutshell, the potential of FSP for fabrication of nanocarriers with high drug loading capacity of biologicals is shown in this study, which will improve therapeutic efficiency by delivering high concentration of biological drugs at the diseased sites.

References

1. Soares S., Sousa J., Pais A., Vitorino C., Nanomedicine: Principles, Properties, and Regulatory Issues. *Chem.* 2018;6:360.
2. Hua S., de Matos M.B.C., Metselaar J.M., Storm G., Current Trends and Challenges in the Clinical Translation of Nanoparticulate Nanomedicines: Pathways for Translational Development and Commercialization. *Pharmacol.* 2018;9:790.
3. Mädler L., Kammler H.K., Mueller R., Pratsinis S.E., Controlled synthesis of nanostructured particles by flame spray pyrolysis. *Aerosol Sci.* 2002;33:369-389.
4. Loher S., Stark W.J., Maciejewski M., Baiker A., Pratsinis S.E., Reichardt D., Maspero F., Krumeich F., Günther D., Fluoro-apatite and Calcium Phosphate Nanoparticles by Flame Synthesis. *Mater.* 2005;17:36-42.
5. Tsikourkitoudi V., Karlsson J., Merkl P., Loh E., Henriques-Normark B., Sotiriou G.A., Flame-made calcium phosphate nanoparticles with high drug loading for delivery of biologics. *Molecules* 2020;25:

609237 Programming Biomaterial Self-Assembly to Advance Molecular Robotics and Gene Delivery

Abstract:

Synthesizing biomaterials using nature's building blocks enables engineers to construct highly programmable devices with unprecedented control over chemical and physical properties, providing a basis for next generation smart materials with applications in biology, manufacturing, and medicine. Here, I will highlight my advances using engineered biomaterials to build 1) controllable nanoscale kinematic joints and mechanisms and 2) charge-driven polymer assemblies for nucleic acid delivery. First, we leveraged the unique structural properties of DNA to create mechanical joints and multi-joint mechanisms with defined 2D and 3D motion. We then use these to demonstrate multiple actuation methods for conformational control of dynamic nanostructures. Second, we took advantage of the strong negative charge of therapeutic nucleic acids to sequester them inside a nanoparticle core using cationic polypeptides, while a neutral polymer shell protects them from nucleases and immune response. We probed the molecular details of each component to establish structure-property relationships and to understand their effect on the stability of the nanoparticles, improving efficient design and construction capabilities for therapeutic delivery devices. I plan to use this groundwork to advance the construction of complex engineering devices in the future, incorporating aspects from both assembly methods in a largely unexplored fashion to translate molecular interactions to larger-scale biohybrid materials.

608038 Understanding the Complex Rheology of Supramolecular Hydrogels for Designing Injectable Drug Delivery Materials

Abstract:

Supramolecular hydrogels have emerged as an exciting class of biomaterials for the development of injectable therapeutic strategies and extrudable materials for 3D printing. They are comprised of physically associated and dynamic crosslinking motifs that imbue them with complex rheological behaviors (e.g. yield stress, shear-thinning, thixotropy). Most notably, supramolecular hydrogels may demonstrate solid-like behaviors at rest and liquid-like behaviors in response to shear, allowing for both injection and printing processes. Emerging advances in the formulations of supramolecular hydrogels for a broad range of applications has resulted in a pressing need to better elucidate key property–function relationships between rheological properties and extrudability. However, defining extrudability for a broad range of design constraints (e.g. needle vs. catheter) *a priori* remains an important challenge in the field.

Herein, I present a model for the extrusion of supramolecular hydrogels through high gauge needles to predict and validate corresponding injection pressures. A generalizable engineering design strategy is shown for defining target flow properties. The complex rheological behavior of supramolecular hydrogels at the shear rates relevant for injection will be presented, highlighting the need for comprehensive rheological characterization up to shear rates in the range of $10^3 – 10^6$ s^{-1}. I will discuss resulting strategies for tuning the solid-like properties of supramolecular hydrogels without compromising their injectability or printability. I show that increasing the concentration and altering the composition of the supramolecular crosslinker has a large impact on the hydrogel's properties at rest while having minimal impact on the flow properties at shear rates operative during printing and injection. Through the discovery of design rules for supramolecular hydrogels and development of property-function relationships, I hope to create engineering design strategies to accelerate the materials discovery and design process for the next generation of biomaterials for injectable drug delivery and 3D printing.

603941 A Microfluidic Model to Assess Subcutaneous Transport and Pharmacokinetics *in Vitro*

Abstract:

Subcutaneous injections are one of the most common routes of medication administration and they offer higher patient adherence and the benefit of self-administration compared to intravenous injection. Estimating the bioavailability via the subcutaneous route is thus an essential component of drug development. However, the *in vivo* estimation of such pharmacokinetics in a pre-clinical setting is far from reliable due to injection site-to-site variations and poor correlation between human and animal data. An ideal *in vitro* model can address these issues using a properly controlled microenvironment that is physiologically relevant. Nevertheless, such models are not currently available and those that are available are void of critical cellular components. In this talk, we introduce a microfluidics-based approach to mimicking the subcutaneous physiology and eventually evaluating pharmacokinetics.

Our microfluidic device incorporates critical elements of the subcutaneous space including multiple relevant cell types and vascular flow. Transport properties measured in this device can be translated to pharmacokinetics obtained from *in vivo* studies. We validated that the measured diffusivities in our device are on the same order of magnitude as those measured *in vivo*. We were able to capture the effects of molecular weight, shape and charge on the rate of subcutaneous diffusion. By changing cell types, we also compared data between human and mice subcutaneous environment. Our device demonstrates the potential of cellular microfluidic devices as a controlled and predictable means for studying drug transport in the subcutaneous space.

594226 Treating Cystic Fibrosis Lung Infections with Bacteria-Inspired Nanoscale Drug Delivery Systems

Abstract:

A cardinal feature of Cystic Fibrosis (CF) lung disease is bacterial colonization with multidrug resistant pathogens such as Pseudomonas aeruginosa (PA). Combinations of synergistic antimicrobials can help overcome resistant infections. However, controlling dosing and delivery of multiple drugs can prove challenging. Additionally, CF lung mucus barriers create significant obstacles to effective antibiotic delivery. To address these critical needs, we are developing antimicrobial-loaded Bacteria-Inspired Nanoscale Drug Delivery Systems (BINDDS), inspired by bacterial properties of mucus penetration and bacterial adherence. By permeating throughout the mucus, adhering to bacteria, and delivering combinations of synergistic antimicrobials as well as mucolytic agents, BINDDS will facilitate antimicrobial activity against CF pathogens at lower doses than antibiotic alone. BINDDS are functionalized using layer-by-layer (LbL) electrostatic assembly. LbL assembly was used to load tobramycin (Tob) and antimicrobial silver nanoparticles (AgNP) onto BINDDS with high loading efficiencies. AgTob-BINDDSs demonstrated synergistic antimicrobial activity against PA bacterial strains. AgTob-BINDDS are being delivered in CF mouse models of PA lung infection, and bacterial load and mouse survival will be analyzed to evaluate therapeutic efficacy. BINDDS are further functionalized with mucolytic agents such as DNAse (Pulmozyme), and we are studying how the shape, size, and charge of BINDDS influence their permeation into mucus barriers and subsequent mucolytic activity in order to improve therapeutic efficacy. Taken together, these studies are providing immense insight into methods for overcoming delivery barriers to treat CF lung infections, and will initiate the development of a translatable material to improve CF patient outcomes and quality of care.

a

b

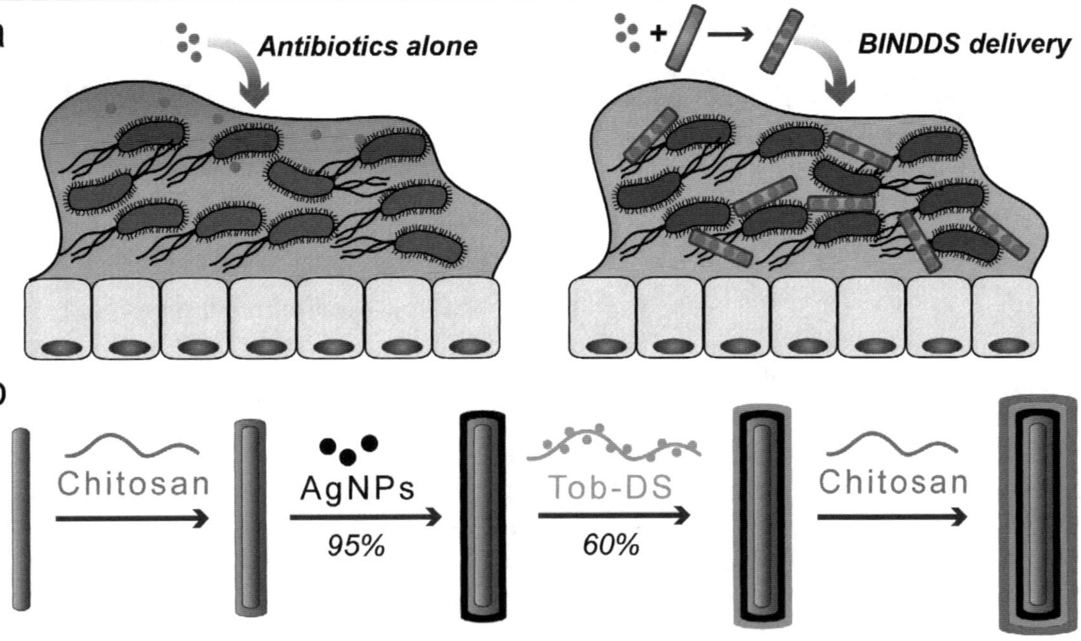

c

Treatment	Tob MIC	AgNP MIC	AgNP LD50
Tob	*4.0*	--	--
AgNP	--	*14*	*39.4*
Tob+AgNP	*0.5*	*0.88*	*34.4*
AgTob-BINDDS	*2.0*	*3.5*	*>75*

Bacteria-inspired nanoscale drug delivery systems (BINDDS) for Cystic Fibrosis drug delivery. (a) Schematic of traditional antibiotic delivery for CF infection treatment, which suffers from poor mucus penetration, fast clearance times, and limited activity. BINDDS allow for high mucus permeation, sustained drug release, and have tunable properties to enhance activity. (b) BINDDS are functionalized via layer-by-layer (LbL) electrostatic assembly with silver nanoparticles (AgNPs) and tobramycin (Tob) for synergistic antimicrobial activity. (c) Summary of minimum inhibitory concentrations (MIC) and 50% lethal dose (LD50) values for AgTob-BINDDS against *Pseudomonas aeruginosa PA14* and A549 lung cells, respecitvely. BINDDS display synergistic antimicrobial activity and mitigate cytotoxic effects of AgNPs, thus providing a wide therapeutic winidow for improved CF drug delivery. MIC and LD50 values reported in μg/ml.

594481 High Quality Protein Structure Prediction Using Equivariant Convoluted Networks with Applications in Drug Design and Next Generation Biomaterials

Abstract:

Over the past six decades, researchers have been able to discern and report three dimensional geometries of proteins in labs using experimental techniques like cryo-electron microscopy, nuclear magnetic resonance and X-ray crystallography. However, each method depends on a lot of trial and error, which lead to both time and monetary overhead (often thousands of dollars per protein structure). This is why biologists are turning to AI-based methods as a proxy to this cumbersome process for proteins. A sophisticated protocol will enable accelerated research in drug design and even determining novel biocatalytic scaffolds, and other biomaterials. To this end, there exists three state of the art tools (as of March 2020) – AlphaFold (Google, DeepMind), and trRosetta (David Baker Lab), and RGN (from our lab by Mohammed AlQuraishi) which are able to use convoluted and/or recursive geometric neural networks to predict most probable dihedrals from a sequence, which is when used to build a protein structure.

However, none of these methods are able to resolve (a) loop structures, and (b) cis/ trans orientations of amino acids (because it is difficult two learn from the bi-modal distribution of w (*omega*), which is 0° for cis and 180° for trans amino acids). We hereby propose a novel method which will learn rotations and torsions, in addition to inter-residue distances and dihedrals to predict a distogram which not only encodes information between residue *i* and *i+1,* but also information about all possible NC2 information using an equivariant neural network (see Figure 1). Subsequently, the best fit Ca-trace would be obtained that meet the distance, dihedral, rotational, and torsional constraints. Finally, in-built functions of PyRosetta would be used to build a PDB structure of the protein with appropriate rotamer-repacking to obtain a lowest energy structure.

As potential application of this tool, I envision applications ranging from therapeutic drug design, to next generation materials for CAR-T cell therapy or for precise bioseparations and drug delivery using stable ensembles of block copolymers and designed channel proteins.

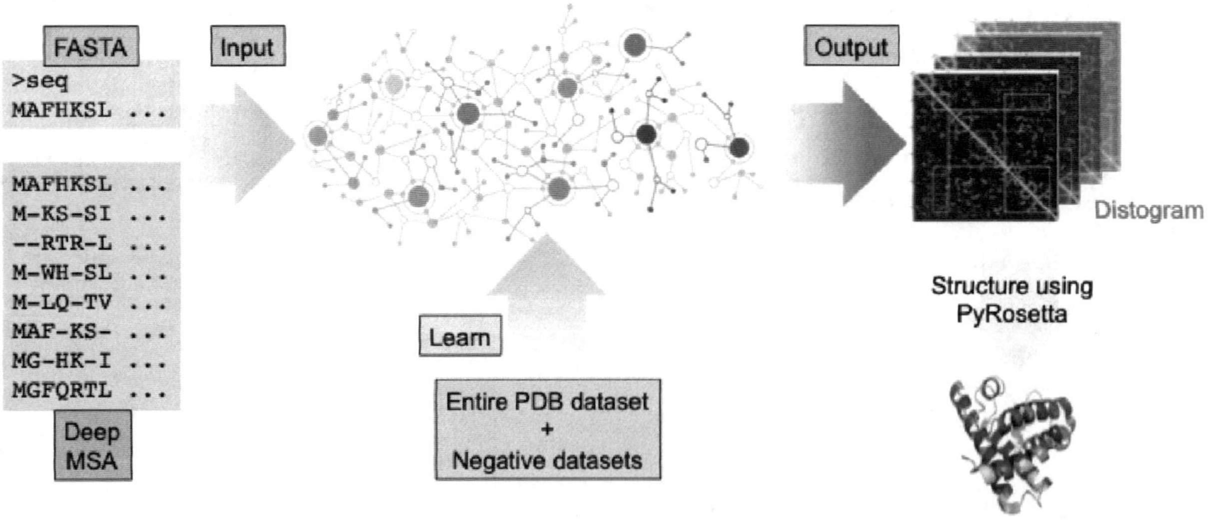

595933 Accelerating the Discovery of Polymeric Vehicles for Gene Editing through Combinatorial Synthesis and Statistical Learning

Abstract:

Complexity is a hallmark of interactions between synthetic materials and the biological environment, but frequently poses barriers to the hypothesis-guided discovery of new materials. Developing polymers for gene delivery is one such multi-dimensional design challenge that requires an intricate balance between intermolecular forces implicated in the packaging, cellular uptake and release of genetic payloads from polymer-nucleic acid complexes. Lacking a complete mechanistic picture of the effects of polymer composition and architecture on its efficiency and toxicity, polymer chemists rely on iterative "one-polymer-at-a-time" approaches that are often plagued by trial and error. Though gene editing has made rapid progress, the paucity of chemically defined, safe and efficient delivery vehicles has hindered its clinical translation. To accelerate polymer discovery for vital clinical applications such as gene editing, we need to replace traditional iterative processes with alternative experimental strategies that leverage advances in parallel synthesis and data analytics. In this talk, I will describe how combinatorial polymer libraries, high-throughput experimentation and statistical models have been combined to yield a powerful predictive platform for the discovery and optimization of polymeric vehicles for CRISPR/Cas9 delivery. A chemically diverse library of well-defined statistical copolymers was synthesized through RAFT polymerization and physicochemical properties such as polymer composition, molecular weight, zeta potential, pKa, polyplex diameter evaluated. Polymer candidates were screened after complexation with clinically relevant genetic payloads such as plasmid DNA and ribonucleoprotein complexes (RNP) comprising single guide RNA (sgRNA) and Cas9 protein. The hit polymer identified through this study outperformed commercial transfection reagents by achieving nearly 60% editing efficiency via non-homologous end-joining. We suggest that this polymer could not have been accessible through human intuition and hypothesis-driven approaches. Further, for each payload, we identified the structural drivers for nucleic acid internalization, cell viability and transfection efficiency and mined structure-function correlations and predictive relationships that can aid experimentalists engaged in polymer design and synthesis for diverse gene therapy applications.

607998 Development of a Sustainable Bioprocess Platform Technology for Biomass Improvement of Biotherapeutic Strain *Lactobacillus Reuteri*

Abstract:

The recent market research report has predicted the global probiotic market would be worth USD 32.b billion by 2021, with the Europe and Asia accounting for nearly 42 and 30% of the total revenues, respectively. The Food and Agriculture Organization (FAO) and International Scientific Association of Probiotics and Prebiotics had defined 'probiotics' as live microorganisms which will confer a health benefits on the host when administered in adequate amounts. The emergence of probiotic therapeutics had driven major challenges for the probiotic industries. First, selection of a most promising probiotic strain as the industries are looking for multi-potential probiotic bacterial strains. Among numerous species of *Lactobacillus* are widely used, heterofermentative *Lactobacillus reuteri* could be most suitable candidate as probiotic industrial workhorse as this bacterium identified to have multiple beneficial effects on host health including prevention and/or amelioration of diverse disorders. In addition, their 'heterolactic' fermentation and co-metabolism over wide range of carbon sources has induced synthesis of various organic acids and other valuable bio-active compounds. Yet, scaling up microbial fermentation process is being another challenge for probiotic industries. The selected strains need to be improved to achieve a cost-effective scale up process with several fold improvement in titers. Thus, in this present work, bioprocess strategy was applied to achieve higher cell biomass production of this probiotic bacterium in the semi-industrial scale platform. The strategy involved medium optimization, batch cultivation, optimization of physical parameters and feeding of limiting nutrients. Medium optimization was performed using combinations conventional and statistical methods involving RSM platform. Batch cultivation was performed in 6L stirred tank bioreactor to optimize physical parameters including types of aeration, agitation and controlled or uncontrolled pH cultivations. Further, a fed-batch operation was employed to achieve high cell density with continuous feeding of limiting of nutrients. The key findings were: (i) maximal biomass of 3.56 g L^{-1} was produced in the optimized medium compared to 1.67 g L^{-1} biomass in the unoptimized MRS medium; (ii) highest biomass obtained under microaerophilic condition; (iii) highest biomass of 6.78 g L^{-1} was obtained with further optimization in fed batch cultivation. These results support the industrial use as an alternative to produce probiotics low-cost ingredients to obtain high biomass concentrations in bioreactors.

601545 Using *Yarrowia Lipolytica* As a New Biomanufacturing Platform for Producing High-Value Products from Lipid Substrates

Abstract:

The oleaginous yeast *Yarrowia lipolytica* is able to grow on medium with any lipids as the sole carbon source, which provides us opportunities for biomanufacturing of high-value products from cost-effective agriculture feedstocks such as plant oils and animal fats. The United States produces more than 20 million tons of plant oils and animal fats every year, a value that is about twice as much as total sugars produced. While sugars are used as the primary carbon source in current biotechnology industry to make value-added fuels, chemicals, and pharmaceuticals, oils and fats are usually considered as common agriculture commodities for food and feed applications and generate low or limited economic values. In our research group at UMass Lowell, we are trying to establish a new yeast biomanufacturing platform that converts oils and fats into a series of high-value products such as omega-3 fatty acids, carotenoids, wax esters, and many other lipid-derived or –assisted products. One of the major challenges for using oils/fats for large-scale biomanufacturing is the mixing and mass transfer limitations caused by the insolubility of lipids in aqueous medium. Here we will present how the multi-phase CFD simulation was used for analysis of mixing and mass transfer behaviors and design of bioreactors. The simulation results were further validated by real fermentation experiments under various bioreactor design and operating conditions. To facilitate the extracellular fatty acid uptake and intracellular fatty acid conversion, the *Y. lipolytica* cells were engineered to maintain certain size and morphology to accelerate the bioprocesses. Several product examples including citric acid, wax esters, and omega-3 EPA produced from a vegetable oil by the wild-type and engineered *Y. lipolytica* in our lab will also be presented.

Strategic Preparations of DPP-IV Inhibitory Peptides from Val-Pro-Xaa and Ile-Pro-Xaa Peptide Mixtures

Changge Guan[1,2] · Shun Iwatani[2] · Xin-hui Xing[1] · Naoyuki Yamamoto[2]

Accepted: 29 September 2020

Abstract
Various peptides with inhibitory activity against human dipeptidyl peptidase-IV (hDPP-IV), which degrades glucagon-like peptide 1 (GLP-1) and decreases insulin release, were isolated from synthetic tripeptide mixtures, having sequences Val-Pro-Xaa (VPX) or Ile-Pro-Xaa (IPX). All peptides isolated by reversed-phase high performance liquid chromatography (HPLC) analysis were measured for hDPP-IV inhibitory activity. VPV and VPI prepared from the VPX mixture and IPI isolated from the IPX mixture, showed the highest hDPP-IV inhibitory activity. The dissociation constants for hDPP-IV and IC_{50} values of the peptides were analyzed to understand the reason for the potent inhibitory activity of these tripeptides. The IC_{50} and *Ki* values of VPV, VPI, and IPI were found to be 20.2, 22.2, and 46.7 μM and 10.8, 11.3 and 21.4 M^{-1}, respectively. Further, degradation of the peptides by the proteolytic action of hDPP-IV was analyzed to understand the stability of these peptides. Peptides with lower *Ki* and IC_{50} values showed relatively slower degradation when incubated with hDPP-IV. These results suggested that a peptide might have higher hDPP-IV inhibitory activity because of its higher affinity and stronger resistance to hDPP-IV, which is caused by the introduction of a hydrophobic amino acid at the C-terminal end of its VP or IP sequence.

Keywords Val-Pro-Xaa (VPX) · Ile-Pro-Xaa (IPX) · hDPP-IV inhibitory peptide · Synthetic tripeptides

Introduction

Many kinds of food derived bioactive peptides were previously isolated and reviewed (Chakrabarti et al. 2014; Maqsoudlou et al. 2019; Pessione and Cirrincione 2016; Marcone et al. 2017). Many of these peptides were found to have certain effects on human health, in many clinical trials (Hill and Newburg 2015; Akazawa et al. 2008; Dobenecker et al. 2018). Proteins obtained from major food materials were considered safe based on a history of consumption for a long period of time without any adverse effects, and advantageous due to their low cost and stable supply. The commonest food derived proteins were obtained from milk, soy, wheat, eggs, rice, oat, meat, and fish (Rutherfurd-Markwick 2012; Fields et al. 2009; Moller et al. 2008).

Currently, various peptides with DPP-IV inhibitory activity are developed from food derived protein hydrolysates to reduce appetite and diet uptakes (Silveira et al. 2013; Nongonierma et al. 2019, 2017). The induction of incretins, GLP-1 and glucose-dependent insulinotropic peptide (GIP), which bind to receptors on pancreatic cells, shows promising effects of promoting secretion of insulin from the pancreas (Orskov 1992; Lambeir et al. 2003). However, GLP-1 can easily be hydrolyzed by intestinal DPP-IV into its inactive form and partially by vascular endothelial DPP-IV in the blood. DPP-IV is a membrane bound ectopeptidase anchored to the surface of epithelial and endothelial cells in the intestine, the bone marrow, and the kidney (Schmidt et al. 1985). Hence, inhibition of DPP-IV increases GLP-1 levels in the circulation, which, in turn, increases insulin release, when needed. In recent years, some DPP-IV inhibitory drugs were found to be efficacious in diabetic subjects in clinical trials (Hermansen et al. 2007; Utzschneider et al. 2018). However, these drugs were not available for pre-stage diabetic and obese subjects.

✉ Naoyuki Yamamoto
n-yamamoto@bio.titech.ac.jp

1 MOE Key Laboratory for Industrial Biocatalysis, Department of Chemical Engineering, Institute of Biochemical Engineering, Tsinghua University, Beijing 100084, China

2 School of Life Science and Technology, Tokyo Institute of Technology, Yokohama, Kanagawa 226-8501, Japan

Published online: 07 October 2020

The most potent DPP-IV inhibitory peptides reported were Ile-Pro-Ile (Diprotin A: $IC_{50} = 3.5$ μM) and Val-Pro-Leu (Diprotin B: $IC_{50} = 15.8$ μM) (Umezawa et al. 1984). Various hydrolysates of food proteins with enzymatic action were used to isolate DPP-IV inhibitory peptides, which were, then, purified by HPLC and identified by mass spectrum analysis (Ji et al. 2017; Harnedy et al. 2015). However, the large number of hydrolyzed and partially hydrolyzed peptides made the purification and identification process complicated, thus, making the screening of active peptides, a time-consuming process. As shown in Fig. 1, previous research suggested that the preferred method of identification of DPP-IV inhibitory peptides with sequences, Trp-X, X-Ala, BCAA-Pro, and Thr-X, is based on the similarities between these inhibitory peptides and other synthetic dipeptides (X: the 20 different kinds of amino acids) (Lan et al. 2015). Moreover, DPP-IV has active sites (Ser630, Asp708, His740) similar to those of X-prolyl dipeptidyl aminopeptidase (XPDAP; Ser348, Asp468, His498), which can cleave the N-terminal of the X-Pro sequence (Rigolet et al. 2005). Interestingly, tripeptides that possessed a Proline at the second position showed potent DPP-IV inhibitory activity (Nongonierma et al. 2018). This suggested that peptides with sequences VPX and IPX may have properties of potent DPP-IV inhibitory activity. To prepare all VPX and IPX peptides, amino acid mixture was added at the C-terminal ends of VP and IP to reduce the time and cost for the synthesis. In this regard, we tried to isolate the most potent DPP-IV inhibitory peptides with higher stability and stronger affinity to DPP-IV, from two mixtures, one of twenty kinds of VPX peptides and another of twenty kinds of IPX peptides (Fig. 1).

Materials and Methods

Reagents and Peptides

Human dipeptidyl peptidase-IV (hDPP-IV, > 200 U/ml) was obtained from the ATGen company (South Korea); Glycyl-proline-p-nitroanilide-p-tosylate (Gly-Pro-pNA) was obtained from the Cayman Chemical Company (Michigan, The United States of America); Peptides Ile-Pro (IP), Val-Pro (VP), Val-Pro-Val (VPV), Val-Pro-Ile (VPI), Ile-Pro-Tyr (IPY), Ile-Pro-Lys (IPK), and Ile-Pro-Gln (IPQ) with purity > 95%; Ile-Pro-Leu (IPL) with purity = 94.9%, and Ile-Pro-Ile (IPI) (Diprotin A) with purity = 99%, and VPX and IPX mixed peptide mixtures were synthesized by GenScript (Tokyo, Japan); VPL (Diprotin B) was bought from BioVision (San Francisco, The United States of America); trifluoroacetic acid (TFA) and acetonitrile were obtained from the Nacalai Tesque (Kyoto, Japan). *o*-phthalaldehyde was obtained from Setareh biotech, LLC (OR, USA).

HPLC Analysis

The VPX and IPX peptide mixtures were dissolved in water containing 0.1% (v/v) TFA (1 mg/ml VPX and IPX mixtures). The peptides in the VPX and IPX mixtures were then fractionated using HPLC- SHIMADZU LC-20AT (Kyoto, Japan). Mobile phase A was ultrapure water containing 0.1% (v/v) TFA, produced by the Milli-Q Integral Water Purification System, and mobile phase B was acetonitrile containing 0.1% (v/v) TFA. For peptide purification, 1 ml of peptide solution was injected into an HPLC column

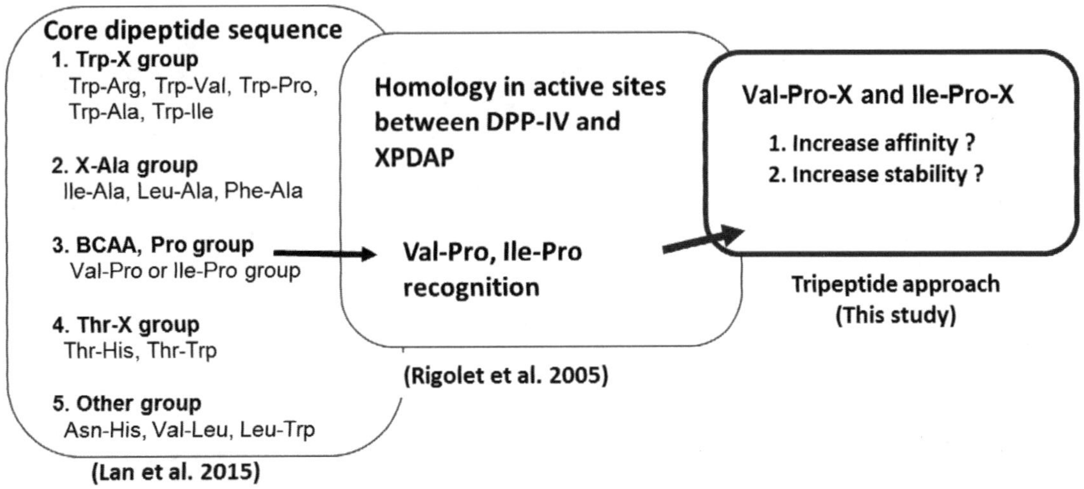

Fig. 1 Strategic screening of DPP-IV inhibitory tripeptides during the synthetic method. Important sequences of some types of dipeptides from the study (Lan et al. 2015), Val-Pro and Ile-Pro from the study (Rigolet et al. 2005), and predicted idea for Val-Pro-X and Ile-Pro-X in the present study

(5C18-AR-II, 4.16 ID×250 mm, COSMOSIL Packed Column, Nacalai Tesque, Kyoto, Japan). The peptides were then eluted according to the following gradient schedule: 0–30 min: 0–20% B; 30–37 min: 20–100% b; 37–39 min: 100% A; 39–40 min: 100–0%; the flow rate was 1 ml/min. The peptides were monitored by measuring the absorbance at 214 nm. Each peptide was eluted for 10–40 min and then, collected by a single peak. The samples were freeze-dried and then dissolved in 1 ml of phosphate-buffered saline (PBS) for further analysis.

DPP-IV Inhibitory Assay

DPP-IV inhibitory activity was measured according to a slightly modified version of the method described previously (Nongonierma et al. 2016). hDPP-IV was used for the triplicate measurements. Peptide solution (20 µl), substrate solution- Gly-Pro-pNA (2 mM; 20 µl), and Tris–HCl buffer (pH 8.0; 20 µl) were pre-mixed in a 96-well microplate. A reaction was initiated by adding 40 µl of hDPP-IV (final concentration 0.025 units/ml) in 100 µl of the above mixture. The samples were then incubated at 37 °C for 60 min. Absorbance at 405 nm was measured using a microplate reader, Varioskan Lux (Thermo Scientific, Tokyo, Japan). The negative control mixture included Tris–HCl buffer (pH 8.0), Gly-Pro-pNA, hDPP-IV, and PBS. One unit (U) of hDPP-IV inhibition was defined as the activity of the sample, showing half the absorbance, compared to the absorbance of the control. The hDPP-IV inhibitory unit of the sample was calculated according to the following Eq. (1):

$$InhibitionUnit(U) = \frac{AbsControl - AbsSample}{AbsControl} \times 2 \tag{1}$$

where *Abs Control*: the absorbance of the control sample with PBS added; *AbsSample*: the absorbance of the peptide sample. Specific activity of each peptide was evaluated by the measurement of DPP-IV activity and peptide concentration by OPA method (Roth 1971).

Identification of the Peptide

The peptides were desalted and concentrated using a Zip Tip C18 microchip (Merck Millipore, Germany). For time-of-flight mass spectrometry (TOF MS) analysis, 2-cyano-3-(4-hydroxyphenyl) acrylic acid was used as a matrix for sample preparation. Mass spectra were analyzed by an UltrafleXtreme TOF/TOF MS (Bruker Daltonics GmbH, Bremen, Germany), operating in positive reflection ion mode between m/z 0 and 3000 Da. The parameters were set as follows- ion source 1: 25 kV; ion source 2: 21.5 kV; lens: 10 kV; reflector 1: 25.5 kV; reflector 2: 14.19 kV; pulsed ion extraction: 10 ns; and detection gain: 9.16.

Stability Study of the Peptides

Each peptide was isolated from the HPLC sample after injecting 1 mg of both VPX and IPX solutions. The peptide was then, dried and dissolved in 1 ml of PBS. Fifty microliters of hDPP-IV (0.0625 U/ml) was added to 200 µl of the peptide, and the mixture was incubated at 37 °C for 1 and 2 h. The peak area of the peptide was quantified by HPLC analysis for incubation at 0, 1, and 2 h. For the control, 50 µl of PBS was added to 200 µl of the peptide, instead.

Determination of the Mode of Inhibition

The mode of DPP-IV inhibition was assessed by using of Lineweaver and Burk double reciprocal representation, previously described by Nongonierma et al. (2016). In this analysis, the concentration of Gly-Pro-pNA ranged from 0.2 to 0.8 mM (final concentration). The negative control samples did not contain any inhibitory peptides. Synthesized DPP-IV inhibitory peptides (VPI, VPV, VPL, IPI, IPL, IPQ, IPY, IPK, and IPN) were evaluated at concentrations 5, 5, 10, 20, 20, 20, 20, 20, and 20 µM corresponding to their IC50 value divided by 4 to prepare the plot as described in previous report (Nongonierma et al. 2016). After 60 min of incubation at 37 °C, the absorbance at 405 nm was recorded using a microplate reader.

Statistics Analysis

The hDPP-IV inhibition experiments were conducted triplicated data. The statistics analysis was then, used to test the significance.

Results

Selection of DPP-IV Inhibitory Peptides from VPX Mixture

To select the potent DPP-IV inhibitory tripeptides with sequence VPX, a mixture of twenty kinds of VPX peptides was synthesized by introduction of amino acid mixture at the C-terminal end of VP. Each peptide peak in the tripeptide mixture was eluted with acetonitrile containing 0.1% (v/v) TFA during HPLC and the DPP-IV inhibitory activity of each peptide was measured using a linear gradient. Sixteen major peaks were fractionated (from VPX-1 to VPX-16) (Fig. 2a). Some peaks showed overlapping of peptides in a single peak (Fig. 2a). Before the identification of peptide in each fraction, DPP-IV inhibitory activity of the fraction was tentatively measured and compared by inhibitory units (one unit fraction contains 50% inhibitory peptide). The average of the DPP-IV inhibitory activity

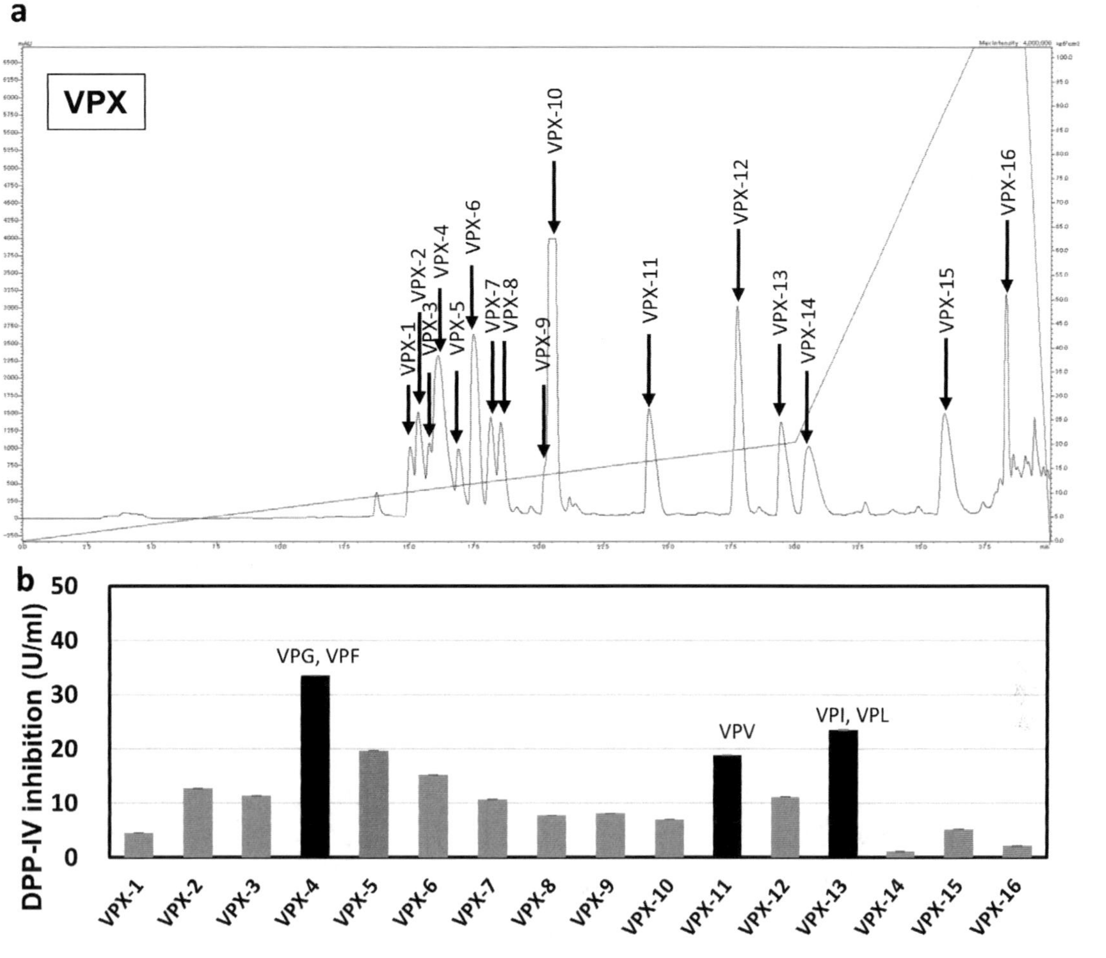

Fig. 2 Chromatograms of Val-Pro-X tripeptide mixture by HPLC analysis (**a**) and DPP-IV inhibitory activity of each collected peptide (**b**) (n=3)

values of the sixteen peptides was 11.06 U/ml. The activity of VPX-4 was the highest (33.5 U/ml) (Fig. 2b). Fractions VPX-5, VPX-11, and VPX-13, also showed DPP-IV inhibitory activity over 14 U/ml, and were selected, along with VPX-4, for further study.

TOF-Mass analysis was conducted for each of the selected peptides to predict its peptide sequence. The sequences for peptides in VPX-4, VPX-11, and VPX-13 were predicted as Val-Pro-Gly (VPG) + Val-Pro-Phe (VPF), Val-Pro-Val (VPV), and Val-Pro-Ile (VPI) + Val-Pro-Leu (VPL), respectively (Fig. 2b). The sequence of peptide in VPX-5 fraction could not be determined due to low yield. To obtain potent DPP-IV inhibitory peptide, peptide with high specific activity (activity/mg peptide) was screened. VPX-4 showed the highest activity, but contained highest amount of peptide (Fig. 2a and Table 1). This suggests lower specific activities of VPG and VPF. For further study, all selected peptides with sequence VPV in VPX-11, and VPI and VPL in VPX-13, were synthesized and their expected retention times were confirmed by HPLC analysis (data not shown).

Selection of DPP-IV Inhibitory Peptides from IPX Mixture

To select the potent DPP-IV inhibitory tripeptides with sequence IPX, a mixture of twenty kinds of IPX peptides was synthesized by introduction of amino acid mixture at the C-terminal end of IP. Seventeen major peptide peaks were fractionated by reversed-phase HPLC and the DPP-IV inhibitory activity of each peptide was measured (Fig. 3a). The small number of fractionated peaks suggested overlapping of some of the peptides. The average of the DPP-IV inhibitory activity values of the seventeen peptides was 12.15 U/ml. Peptides in IPX-3 and IPX-14 showed potent activities, 42.75 and 48.8 U/ml, respectively (Fig. 3b). However, the specific activities (activity/mg peptide) of IPX-3 and IPX-10 were thought to be not high because of higher amount of

Table 1 DPP-IV inhibitory activity of peptide in each fraction by HPLC

Fraction	Expected peptide	Peptide concentration* (mM)	hIC50 (mM)
VPX-4	Val-Pro-Gly, Val-Pro-Phe	3.79	0.77
VPX-11	Val-Pro-Val	3.61	0.70
VPX-13	Val-Pro-Ile, Val-Pro-Leu	2.97	0.38
IPX-1	Ile-Pro-Asn, Ile-Pro-Lys, Ile-Pro-Gln	2.56	0.84
IPX-3		6.35	1.24
IPX-10	Ile-Pro-Val	3.31	1.20
IPX-12	Ile-Pro-Tyr	1.68	0.82
IPX-14	Ile-Pro-Ile or Ile-Pro-Leu	0.53	0.14
IPX-15	Ile-Pro-Ile or Ile-Pro-Leu	2.87	1.14

*Peptide concentration was measured by OPA method (Roth 1971)

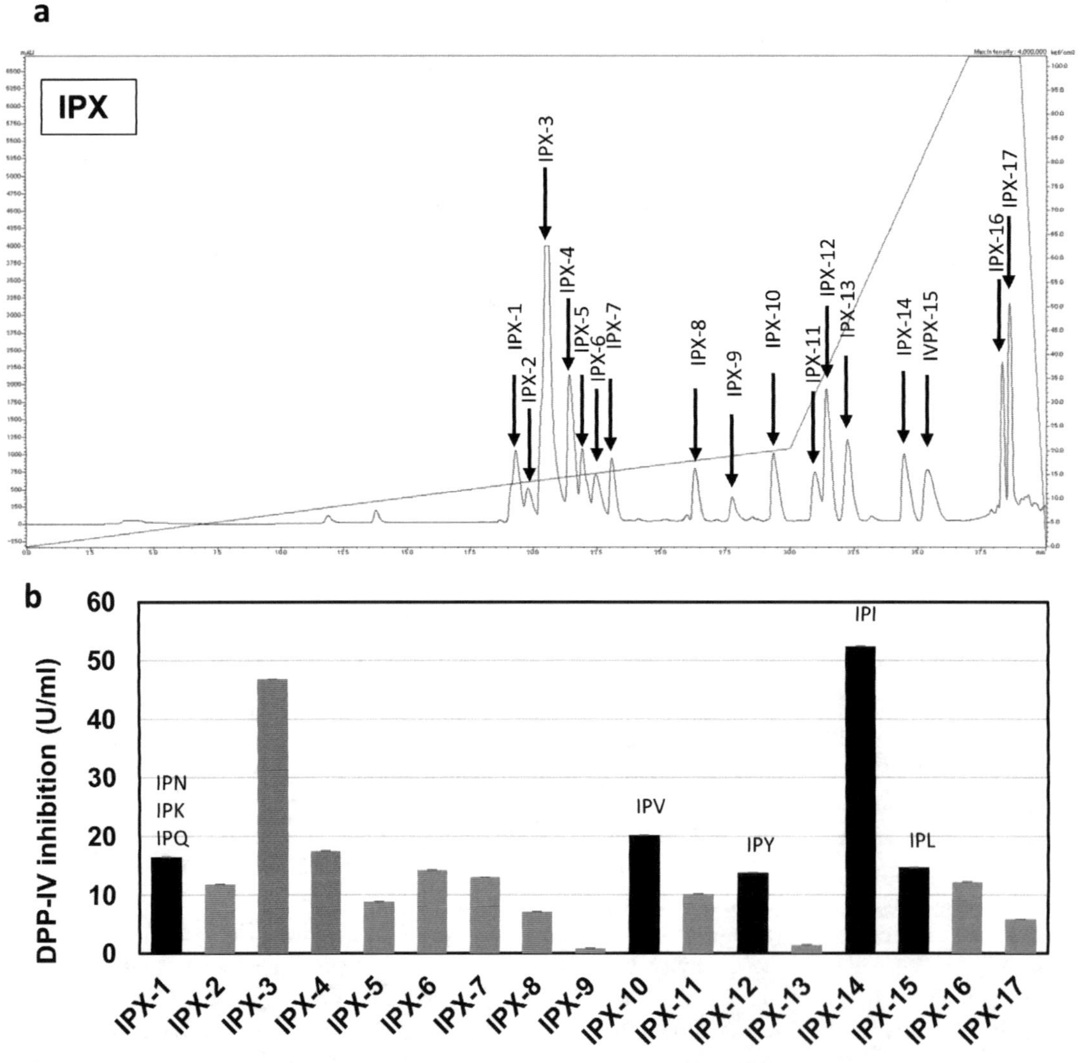

Fig. 3 Chromatograms of Ile-Pro-X tripeptide mixture by HPLC analysis (**a**) and DPP-IV inhibitory activity of each collected peptide (**b**) (n = 3)

peptide in the fractions (Table 1). The sequence of peptide in IPX-4 fraction could not be determined due to low yield. So, peptides in fractions IPX-1, IPX-12, IPX-14, and IPX-15 showed higher specific DPP-IV inhibitory activities (Fig. 3b) were selected for further study. All DPP-IV inhibitory activities for the selected peptides were over 14 U/ml.

TOF Mass analysis was used to predict the sequences of peptides in the peaks IPX-1, IPX-10, IPX-12, IPX-14, and IPX-15. The sequences were predicted as Ile-Pro-Asn + Ile-Pro-Lys + Ile-Pro-Gln, Ile-Pro-Val, Ile-Pro-Tyr, Ile-Pro-Ile or Ile-Pro-Leu, and Ile-Pro-Ile or Ile-Pro-Leu, respectively. All predicted peptides in the peaks IPXs were then, synthesized and their expected retention times were confirmed by HPLC analysis (data not shown). IPI and IPL with same molecular weight were assigned to IPX-14 and IPX-15, respectively (Fig. 3b).

DPP-IV Inhibitory Activity

In this study, peptide with higher specific DPP-IV inhibitory activity with lower amount of peptide was screened. VPX-4 in Fig. 2 and IPX-3 and IPX-10 in Fig. 3 showed higher DPP-IV inhibitory activities, but were considered to be lower specific activities because of higher amount of peptide was suggested in the fraction (Table 1). So, three kinds of DPP-IV inhibitory peptides selected from the VPX-11 and VPX-13 and six kinds of DPP-IV inhibitory peptides selected from the IPX-1, IPX-12, IPX-14, and IPX-15 were synthesized and their characteristic features were compared to understand the potency of their activity. First, all peptides were evaluated at minimum concentration of the peptide to demonstrate 50% inhibition of hDPP-IV (hIC_{50}). Tripeptides VPX and IPX showed relatively higher DPP-IV inhibitory activity than VP and IP, respectively (Table 2). The VPV sequence predicted in VPX-11 also showed potent DPP-IV inhibitory activity (hIC_{50} = 22.2 μM). Sequences IPI in IPX-14 and VPL in VPX-13, previously reported as Diprotin A and B (Umezawa et al. 1984), showed high inhibitory activity (hIC_{50} = 46.7 and 114.1 μM, respectively) (Table 2). However, these values were higher than those reported with porcine DPP-IV (pDPP-IV) (pIC_{50} = 3.5 and 15.8 μM, respectively) (Umezawa et al. 1984). VPI sequence in VPX-13 showed the highest DPP-IV inhibitory activity among the tested peptides (hIC_{50} = 20.2 μM).

Various Kinetics Analysis

Lineweaver and Burk double reciprocal plot was prepared to understand the mode of inhibition of the selected peptides against hDPP-IV. The nine tested peptides showed competitive hDPP-IV inhibitory activity (Fig. 4). The *Ki* values of each peptide, reflecting the affinity of the peptide to the enzyme, were calculated (Table 2). The *Ki* value was relatively lower for the peptide with higher DPP-IV inhibitory activity (lower hIC_{50} value). VPI with the highest inhibitory activity (hIC_{50} = 20.2 μM), showed the lowest *Ki* value (10.8 M^{-1}). This result demonstrated that the introduction of Ile or Val into the C-terminal end of VP (Nongonierma 2016) and that of Ile into the C-terminal end of IP (Nongonierma 2016) could increase the affinity of the peptide to hDPP-IV and thus, increase its DPP-IV inhibitory activity.

Table 2 Characterization of the DPP-IV inhibitory tripeptides with synthetic peptides

Fr No in HPLC	Peptide	hIC50 (μM)	*K*i Value (M^{-1})	Reported pIC50 (μM)	Stability* (min)	References
–	Val-Pro	377.3	146.63	380.3		Nongonierma et al. (2016)
VPX-11	Val-Pro-Val	22.2	11.3	6.6		Nongonierma et al. (2019)
VPX-13	Val-Pro-Ile	20.2	10.8	–	500	
VPX-13	Val-Pro-Leu (Diprotin B)	114.1	48.4	15.8		Umezawa et al. (1984)
VPX-14	Not predicted	–	–	–	61	
–	Ile-Pro	322.6	134.8	149.6		Nongonierma et al. (2016)
IPX-1	Ile-Pro-Asn	135.9	52.8	57.5		Uenishi et al. (2013)
IPX-1	Ile-Pro-Lys	195.2	64.3	–		
IPX-1	Ile-Pro-Gln	94.3	40.7	–		
IPX-12	Ile-Pro-Tyr	113.5	46.2	–		
IPX-13	Not predicted	–	–	–	48	
IPX-14	Ile-Pro-Ile (Diprotin A)	46.7	21.4	3.5	79	Umezawa et al. (1984)
IPX-15	Ile-Pro-Leu	95.5	43.2	–		

*Stability was represented by Half-life (min)

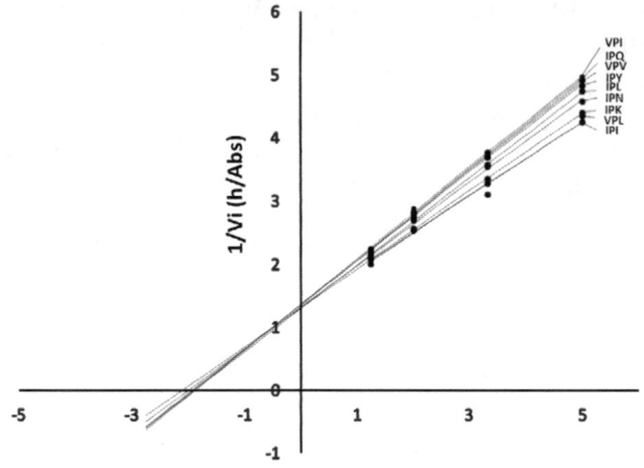

Fig. 4 Lineweaver and Burk double reciprocal plots for DDP-IV inhibition by nine kinds of tripeptides- VPI, IPQ, VPV, IPY, IPL, IPN, IPK, VPL, and IPI

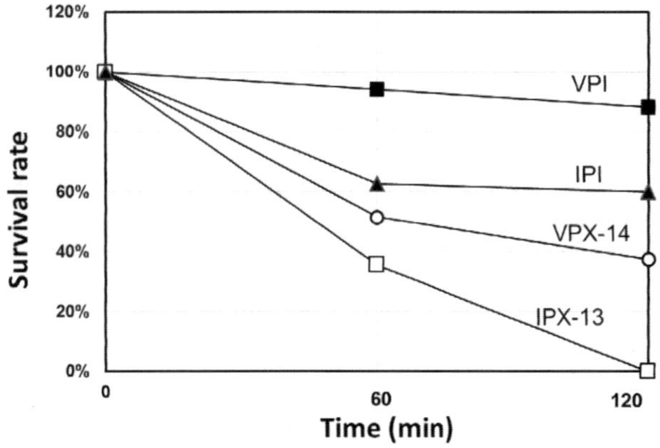

Fig. 5 Degradation of VPI and IPI and peptides purified in the fractions of VPX-14 and IPX-13 by HPLC, on incubation with hDPP-IV at 37 °C for 1 and 2 h

Stability of the Peptide

The resistance of the tripeptide to hDPP-IV, which can cleave off the N-terminal dipeptide from the tripeptide, was analyzed. On incubating each of the four peptides with hDPP-IV for 60 and 120 min, a higher resistance to hDPP-IV was observed for peptides with higher hDPP-IV inhibitory activity, such as VPI and IPI, whereas quick degradation of peptides was observed for peptides with lower hDPP-IV inhibitory activity, such as those in IPX-13 and VPX-14 (Fig. 5). These results showed that the rate of degradation of a peptide by hDPP-IV is related to the hDPP-IV inhibitory activity of the peptide. In other words, potent hDPP-IV inhibitory peptides have higher affinity and resistance to hDPP-IV (Table 2).

Discussion

In this study, strategic approaches were applied to obtain potent DPP-IV inhibitory peptides from synthetic tripeptides. The study was based on finding homologous sequences in active sites of DPP-IV and XPDAP (Lan et al. 2015). The XP sequence in peptide was considered as the core sequence for XPDAP and DPP-IV recognition, but for stable binding with the enzyme, the sequence needed to be longer. Thus, tripeptides that could add a third amino acid to the XP peptide were prepared. By adding of amino acid mixture at the C-terminal ends of VP and IP, the cost for the synthesis was reduced to about 10%. As shown by the results, tripeptides with sequences VPI and VPV, with potent DPP-IV inhibitory activity, were successfully screened from two mixtures of twenty kinds of VPX and IPX peptides each.

This study demonstrated a promising approach to screen potent DPP-IV inhibitory peptides efficiently and to understand the role of the third amino acid in VPX and IPX sequences. For instance, hDPP-IV showed higher affinity to tripeptides due to the introduction of a particular third amino acid at the C-terminal end of the VP or IP sequence (decrease of *Ki* value) (Table 2). This result also indicated slower degradation of VPV and VPI than other VPX peptides, by proteolytic action of hDPP-IV. This observation helped us to understand the potency of the inhibitory activity of the peptides. By introducing Ile at the C-terminus of the IP sequence, the resultant peptide showed higher affinity to DPP-IV and greater inhibitory activity against DPP-IV. Hydrophobic amino acids (Ala, Ile, Leu, Pro, Trp, and Val) at the third position of both VPX and IPX tripeptides made these peptides the most effective for DPP-IV inhibitory activity (Table 2). Quantitative structure activity relationship (QSAR) outcomes showed the importance of hydrophobic amino acids in bioactive peptides in potent inhibition of metabolic enzymes (Nongonierma et al. 2016). To understand the relationship between structural features and different inhibitory effects of VPX and IPX, extensive analysis of the impact of the third amino acid on the affinity and resistance of the peptide to DPP-IV must be conducted.

Six tripeptides with high hDPP-IV inhibitory activity that were prepared in this study have been previously reported. Five peptides, VPI, LPK, IPQ, IPY, and IPL, were newly detected these DPP-IV inhibitory activities in this study. Among them, VPI had the highest hDPP-IV inhibitory activity, which was also higher than that of Diprotin A (IPI) that has been previously reported as the most potent DPP-IV inhibitory peptide (24). IC_{50} values of VPV and IPI measured using pDPP-IV (porcine DPP-IV) were reported as 6.6 μM and 3.5 μM (Nongonierma et al. 2019; Umezawa et al. 1984). In this study, the IC_{50} value of VPV measured using hDPP-IV (human DPP-IV) was lower than that of IPI, indicating that VPV has a higher hDPP-IV inhibitory activity than IPI (Table 2). hDPP-IV and pDPP-IV sequences showed similarity in their primary amino acid sequences up to 88% but differed in amino acid sequences around the active sites. pDPP-IV also showed one possible disulfide bond between its cysteine residues. To understand the different inhibitory effects of peptides on porcine and human DPP-IV, the inhibitory effects of sixty-two different kinds of peptides on both enzymes were previously compared (Lacroix et al. 2015). The two enzymes were inhibited by short peptides similarly, but pDPP-IV was inhibited with greater potency than hDPP-IV (Lacroix et al. 2015). This suggested that the peptides that strongly inhibited pDPP-IV might be less effective on hDPP-IV, unlike the peptides screened in this study.

The various short DPP-IV inhibitory peptides used in this study showed increased DPP-IV inhibitory activity by prolonging the length of the dipeptide. The potency (IC_{50}) of two hundred, twenty-two kinds of DPP-IV inhibitory peptides derived from food proteins were reviewed based on their grand average of hydropathicity values (GRAVY) and molecular weights (Liu et al. 2019). Over 88.4% of the reported peptides had molecular weights less than 1000 Da, and more than 50% had molecular weights less than 500 Da (tetrapeptide). Among tested peptides, 53% of peptides were hydrophilic with GRAVY lower than 0 and the remaining 47% were hydrophobic. In another report, the In Silico approach was used to study the potency of IPI analogues (Nongonierma et al. 2018). Thus, peptides with 3 to 4 amino acids showed relatively higher DPP-IV inhibitory activity.

Certain potent hDPP-IV inhibitory peptides, including VPI and VPV, were successfully screened in this study. DPP-IV inhibitory peptides released in Atlantic salmon skin gelatin hydrolysate, showed anti-diabetic effect and hence, was found to be effective at a dosage of 300 mg per day for 5 weeks. During the test period, the hydrolysate was effective in lowering blood glucose levels, inhibiting plasma DPP-IV activity, and increasing the plasma GLP-1 and insulin levels (Hsieh et al. 2015). The impact of these peptides on GLP-1 production, food uptake, body weight control, and insulin secretion must be evaluated in different dosages in an animal study. Moreover, these potent food-derived peptides need to be prepared for human use. To do this, some food proteins can be hydrolyzed by food grade proteases. These peptides can then be used in clinical trials.

In conclusion, potent DPP-IV inhibitory peptides, VPV, VPI, and IPI, were efficiently selected from synthetic peptide mixtures with VPX and IPX sequences, based on previously reported information on similarity of sequences of DPP-IV and XPDAP. Increased affinity and resistance of peptides VPV, VPI, and IPI, to DPP-IV were found to be important for potent inhibition.

Acknowledgements Changge Guan gratefully acknowledges the financial support provided by the China Scholarship Council (No. 201806210223), who supported his study in Tokyo Institute of Technology.

Compliance with Ethical Standards

Conflict of interest All authors declare that no conflict of interest.

Ethical Approval This article does not contain any animal and clinical studies performed by the authors.

References

Akazawa N, Hamasaki A, Tanahashi K et al (2008) Lactotripeptide ingestion increases cerebral blood flow velocity in middle-aged and older adults. Nutr Res 53:61–66

Chakrabarti S, Jahandideh F, Wu J (2014) Food-derived bioactive peptides on inflammation and oxidative stress. Biomed Res Int 2014:608979

Dobenecker B, Reese S, Jahn W et al (2018) Specific bioactive collagen peptides (PETAGILE®) as supplement for horses with osteoarthritis: a two-centred study. J Anim Physiol Anim Nutr (Berl) 102:16–23

Fields K, Falla TJ, Rodan K et al (2009) Bioactive peptides: signaling the future. J Cosmet Dermatol 8:8–13

Harnedy PA, O'Keeffe MB, FitzGerald RJ (2015) Purification and identification of dipeptidyl peptidase (DPP) IV inhibitory peptides from the macroalga *Palmaria palmata*. Food Chem 172:400–406

Hermansen K, Kipnes M, Luo E et al (2007) Efficacy and safety of the dipeptidyl peptidase-4 inhibitor, sitagliptin, in patients with type 2 diabetes mellitus inadequately controlled on glimepiride alone or on glimepiride and metformin. Diabetes Obes Metab 9:733–745

Hill DR, Newburg DS (2015) Clinical applications of bioactive milk components. Newburg Nutr Rev 73:463–476

Hsieh CH, Wang TY, Hung CC et al (2015) Improvement of glycemic control in streptozotocin-induced diabetic rats by Atlantic salmon skin gelatin hydrolysate as the dipeptidyl-peptidase IV inhibitor. Food Funct 6:1887–1892

Ji W, Zhang C, Ji H (2017) Purification, identification and molecular mechanism of two dipeptidyl peptidase IV (DPP-IV) inhibitory peptides from Antarctic krill (*Euphausia superba*) protein hydrolysate. J Chromatogr B 1064:56–61

Lacroix IME, Li-Chan ECY (2015) Comparison of the susceptibility of porcine and human dipeptidyl-peptidase IV to inhibition by protein-derived peptides. Peptides 69:19–25

Lambeir AM, Durinx C, Scharpe S et al (2003) Dipeptidyl-peptidase IV from bench to bedside: an update on structural properties, functions, and clinical aspects of the enzyme DPP IV. Crit Rev Clin Lab Sci 40:209–294

Lan VT, Ito K, Ohno M et al (2015) Analyzing a dipeptide library to identify human dipeptidyl peptidase IV inhibitor. Food Chem 175:66–73

Liu R, Cheng J, Wu H (2019) Discovery of food-derived dipeptidyl peptidase IV inhibitory peptides: a review. Int J Mol Sci 22:E463

Marcone S, Belton O, Fitzgerald DJ (2017) Milk-derived bioactive peptides and their health promoting effects: a potential role in atherosclerosis. Br J Clin Pharmacol 83:152–162

Moller NP, Scholz-Ahrens KE, Roos N et al (2008) Bioactive peptides and proteins from foods: indication for health effects. Eur J Nutr 47:171–182

Maqsoudlou A, Sadeghi Mahoonak A, Mora L et al (2019) Controlled enzymatic hydrolysis of pollen protein as promising tool for production of potential bioactive peptides. J Food Biochem 43:e12819

Nongonierma AB, Cadamuro C, Le Gouic A et al (2019) Dipeptidyl peptidase IV (DPP-IV) inhibitory properties of a camel whey protein enriched hydrolysate preparation. J Food Chem 279:70–79

Nongonierma AB, Mazzocchi C, Paolella S et al (2017) FitzGerald R.: release of dipeptidyl peptidase IV (DPP-IV) inhibitory peptides from milk protein isolate (MPI) during enzymatic hydrolysis. J Food Res Int 94:79–89

Nongonierma AB, Dellafiora L, Paolella S et al (2018) *In Silico* approaches applied to the study of peptide analogs of Ile-Pro-Ile in relation to their dipeptidyl peptidase IV inhibitory properties. Front Endocrinol 9:329

Nongonierma AB, Richard J, FitzGerald DJ (2016) Structure activity relationship modelling of milk protein-derived peptides with dipeptidyl peptidase IV (DPP-IV) inhibitory activity. Peptides 79:1–7

Orskov C (1992) Glucagon-like peptide-1, a new hormone of the enteroinsular axis. Diabetologia 35:701–711

Pessione E, Cirrincione S (2016) Bioactive molecules released in food by lactic acid bacteria: ncrypted peptides and biogenic amines. Front Microbiol 7:876

Rigolet P, Xi XG, Rety S et al (2005) (2005) The structural comparison of the bacterial PepX and human DPP-IV reveals sites for the design of inhibitors of PepX activity. FEBS J 272:2050–2059

Roth M (1971) Fluorescence reaction for amino acids. Anal Chem 43:880–882

Rutherfurd-Markwick KJ (2012) Food proteins as a source of bioactive peptides with diverse functions. Br J Nutr 108:S149–S157

Schmidt WE, Siegel EG, Creutzfeldt W (1985) Glucagon-like peptide-1 but not glucagon-like peptide-2 stimulates insulin release from isolated rat pancreatic islets. Diabetologia 28:704–707

Silveira ST, Martínez-Maqueda D, Recio I et al (2013) Dipeptidyl peptidase-IV inhibitory peptides generated by tryptic hydrolysis of a whey protein concentrate rich in β-lactoglobulin. Food Chem 141:1072–1077

Uenishi K, Kabuki T, Teshima T (2013) Dipeptidyl peptidase IV inhibitors. Japanese Patent. P2014–1223A

Umezawa H, Aoyagi T, Ogawa K et al (1984) Diprotins A and B, inhibitors of dipeptidyl aminopeptidase IV, produced by bacteria. J Antibiot 37:422–425

Utzschneider KM, Tong J, Montgomery B et al (2018) The dipeptidyl peptidase-4 inhibitor vildagliptin improves beta-cell function and insulin sensitivity in subjects with impaired fasting glucose. Diabetes Care 31:108–113

596115 Modelling and Dynamic Optimization of Lager Beer Fermentation for Optimal Flavor Profile and Operation

Abstract:

Beer production is an intricate chemical process. From the same critical ingredients (a starch source, yeast, hops and water), a highly complex mixture of chemical species is obtained. The fluctuating combinations of these compounds, also called off- and on-flavors, are responsible for the unique taste of each beer. Thus, the ability to improve any stage of production will have a significant effect on profitability and the ultimate success or failure of a brewery.

Therefore, it is important to carefully describe the different stages of the beer production process, in order to optimize process conditions and beer flavour. To do so, in this study, a comprehensive fermentation model and optimization problem were formulated. This model includes a complete yeast growth model, production of ethanol and flavor related desirable and undesirable chemical species, including temperature dependency of such processes as well.

Beer flavor reflects the combination of a large number of chemical species, mostly resulting from the yeast metabolism. Therefore, in this work, the goal is to determine how an up-to-date industrial beer production should optimally operate. To achieve this, a two-step procedure is applied: (i) a comprehensive fermentation model is developed, including a yeast growth model, ethanol and by-products production (fusel alcohols, esters, VDKs, acetaldehyde and sulfidic compounds); and, (ii) a dynamic optimization problem is formulated and implemented with the objective of maximizing the conversion of substrate to ethanol, while dealing with the final allowable concentrations of the by-products as strict constraints.

Preliminary results show that each by-product flavor threshold affects process performance and beer flavor in a distinctive manner, which is also supported by published literature. Results indicate that the maximum acceptable concentrations of off- and on- flavors, like diacetyl and ethyl acetate, have considerable impact on the flavor profile, as well as on batch duration.

A comprehensive fermentation model was developed based upon kinetic models. Furthermore, in order to optimize the flavor profile and process conditions, a dynamic optimization was formulated and solved. Thus, we believe, that this work contributes to a comprehensive understanding of the impact of different chemical species (off- and on-flavors) on the optimal beer flavor profile, as well as in the optimization of process conditions.

608819 How Temperature during Fermentation Affects Quality and Flavor in Cocoa Liquor

Abstract:

The cocoa tree belongs to the genus *Theobroma cacao* L., it has four categories Criollo, Forastero, Trinitario and Clones (these are generated by human selection, to increase desired features. This tree produces a well-known product, cocoa beans, which are used by different industries such as food, cosmetic and pharmaceuticals because of its physical and chemical properties. To use cocoa beans by the industries, they need to pass across different transformation processes, the first one is the harvest from the tree, followed by the cut of the pods, fermentation of the beans and the final step is drying the beans.

Fermentation is a spontaneous process that occurs in wooden boxes, it usually takes between five to seven days; during this period the temperature varies between 25°C to 50 °C, this range allows the growth of different microorganisms (e.g. yeast, lactic acid bacteria and acetic acid bacteria), they metabolize the sugars and other compounds in the beans to produce volatile compounds which are linked to the aroma and flavor in the cocoa liquor. The importance of several strains of yeast and their presence during the first 48 hours of cocoa fermentation to develop a quality fermentation had been reported in the literature. The drying process of the beans takes place slowly and progressive under sunlight, the humidity of the bean needs to be lower than 7%, to avoid the presence of fungi or insects.

Colombian Cocoa has been recognized by the International Cocoa Organization (ICCO) as fine or flavor (95%), this means that cocoa has aroma and flavor of fruits, flowers, nuts, and malt. However, due to the dissimilar ecological, genetic, microclimatic conditions and fermentation process (e.g. time, mass, temperature) in the country; the cocoa produced has a high variability of the chemical compounds that determine aroma and flavor. On the other hand, one of the most cultivated varieties in Colombia is the clone CCN51, because of its high productivity and resistance disease. However, this clone lacks pronounced specific flavors that make it common cocoa. Though, this clone is interesting for this project because the changes in its flavor could be more noticeable.

The main goal of this research is to identify the influence of different temperature profiles during fermentation of cocoa clone CCN51 in the quality of dry beans, sensory profile, volatile compound profile of cocoa liquor and the presense of yeast in the first 48 hours of fermentation.

Fermentation quality is reported by the fermentation index, after a cut test to the dry bean to check the physical changes, this is an indirect tool to predict the potential flavor of the beans. In the cut test, the beans can exhibit a fine, medium and deficient fermentation, that is represented by a color range from dark brown, light brown and violet, respectively. Also, fungi or insect bites are checked in the open bean. Index values around 70 to 100 percent indicate a good fermentation process.

The sensory profile consists of an evaluation of the cocoa liquor (obtained from the fermented and dry beans) by professional cocoa tasters. There is a scale from 0 to 10 for basic, specific and acquired attributes; basic attributes are acid, bitter, sweet and astringent, specific correspond to cocoa, flower, fruit, and nuts, and acquired to mold, chemical, green, smoked and wood. A sensory profile is used to determine the acceptance in the market. The sensory profile of clone CCN51 has been determined as common due to the high presence of acidity, astringent and low notes of specific flavors.

Volatile compounds contribute and are responsible for the aroma and sensorial profile of the cocoa liquor, this analysis is carried out using headspace solid-phase microextraction-gas chromatography-mass spectrometry (HS-SPME GC-MS). This technique is used to identify and quantify the accumulation of volatile compounds in cocoa liquor.

The cocoa plantation used for the fermentation, lab scale, was hybrid CCN51, it was harvested from San Vicente de Chucurí, Santander, Colombia. They were in storage from 4 to 5 days since the harvest time and manually open, then the pulp was placed in contact with the pods for 3 to 4 hours allowing the inoculation with native microorganism and transferred into acrylic boxes permitting the drainage, the fermenting mass was turned into a different acrylic box at 48, 72, 96, 120 hours until it completed 6 days of fermentation. The boxes were incubated at three different temperature profiles. The first temperature profile was the same reached on fermentation boxes at San Vicente de Chucurí; the other two temperature profiles were found in literature about controlled laboratory cocoa fermentation. After the fermentation process, the beans were dried at sun exposure until they reached less than 7% humidity. The fermented and dried beans were sent to the Federación Nacional de Cacaoteros (FEDECACAO), to evaluate the fermentation index by the cut test of beans and the sensorial analysis by an expert panel. Finally, the volatile compound in the cocoa liquor was extracted and quantified by solid-phase micro extraction-gas chromatography-mass spectrometry (SPME-GC-MS).

The three temperature profiles showed different fermentation indexes and flavor in the sensory analysis, for both basic and specific flavors. Comparing these results with previous fermentations reported in the literature for cocoa CCN51 from the same region, it is notable the influence of temperature to improve the flavor of cocoa.The presence and maximum amount of yeast were different in all the temperature profiles, it is important to remember that the yeasts are responsible for metabolized volatile compounds associated with flavor in the cocoa liquor. On the other hand, the cocoa liquor for each temperature profile exhibits similar volatile compounds but in different amounts for each one.

These results let us know the importance of fermentation temperature to obtain quality dry beans that can be transformed into cocoa liquor that has a sensory profile with characteristic organoleptic desired in the food industry and can be compared with the presence and quality of volatile compounds developed through the fermentation process.

608004 Production of Heavy Oil Liamocin in Fermentation By *Aureobasidium Pullulans*: Kinetics, Process Optimization and Economic Analysis

Abstract:

Aureobasidium pullulans is a ubiquitous fungus that is well known for producing exopolysaccharide pullulan, polymalic acid (PMA), and many useful industrial enzymes such as xylanase and lipase. Some strains of *A. pullulans* can also produce extracellular heavier-than-water oils known as liamocins. Liamocins are esters composed of a single mannitol headgroup O-acetylated with 3 or 4 3,5-dihydroxydecanoic ester groups, and can be used as biosurfactant, antimicrobial agent against some *Streptococcus* spp., anti-cancer agent, and biofuels. The objectives of this research were to study the liamocin fermentation kinetics and scale up the process in bioreactors for industrial application. We have screened several liamocin-producing strains. In shake-flasks, the best strain *A. pullulans* NRRL 62031 produced liamocin from glucose at a final titer of 29.5 g/L and yield of 0.30 g/g in a medium containing 12% (w/v) glucose and the C/N (w/w) ratio of 175 in 11 days. Fermentation was scaled up in a 5-L stirred-tank bioreactor, achieving a 0.20 g/g yield in 216 hours. Our preliminary process simulation and cost analysis showed that liamocin can be produced at a cost of $3.2/kg, which is significantly lower than the market price for biosurfactants. The effects of dissolved oxygen, pH, and glucose feeding strategies on liamocin fermentation in bioreactors and production cost will be reported in this paper.

613034 "Invited Talk" Biotech in the Flavor Industry

Abstract:

In modern age, biotech has incredible values in the flavor industry. The market demand for natural flavoring has been skyrocketing since people pay more and more attention to what they eat. Fermentation as one of the a few technologies approved by FDA to make natural flavors has gained its spot light in meeting consumers' craving for health and well-being. In addition, equipped with rapid development in metabolic engineering tools, a wide arrange of fine chemicals which was only available from specific plants before, now can also be achieved by enhanced fermentation with considerably lower costs, and much shorter production time. Yet, biotech also has inevitable limits to have broader applications in flavors. First, plant based flavor molecules often come with complex synthetic pathways, which are difficult to be balanced in microbes, not mention with decent productivity. To make things even more challenging, global food regulations on non-GMO and natural have become increasingly restricted. In this review, we'll look into the current status of the exploitation of biotech in the flavor industry, also provide foresight on strategies to overcome technical hurdles while maintain its advantages in food labeling.

610308 Synergy of an HDAC Inhibitor and Chemotherapy Delivered Via Lipid Nanoemulsions for TNBC

Abstract:

Triple negative breast cancers (TNBCs) represent a heterogenous disease with high patient mortality relative to other breast cancers. Doxorubicin (Dox), a DNA intercalating drug, is widely used as a chemotherapy for TNBC. However, it often accompanies severe side effects that limits the effective therapeutic dose. We hypothesized that the combination of Dox with a histone deacetylase (HDAC) inhibitor, panobinostat (Pan), may improve therapeutic efficacy by attacking rapid cell proliferation and abnormal histone deacetylation. Herein, pH-responsive, lipid targeting nanoemulsions (pLNEs) were designed to direct both Dox and Pan to TNBC cells that overexpress lysophosphatidic acid receptor 1 ($LPAR_1$) via the ratio of lysophosphatidic acid to lysophosphatidylcholine (LPA/LPC) and trigger Dox release via complexation with docosahexanoic acid (DHA) at low pH. We demonstrated that PAN and Dox exhibited a synergistic effect on the cell viability of TNBC cells at a specific ratio. The pLNE with a 1:1 ratio LPC and LPA had significantly higher tumor uptake efficiency than pLNEs with either LPC or LPA alone. pLNEs were used to encapsulate Pan and Dox. We performed in vivo studies that demonstrated a significant reduction in tumor burden relative to controls (including a Doxil mimic, Dox-LP) (Fig.1A). Additionally, survival curves demonstrated that Dox/Pan pLNEs had significantly greater survival than controls (Fig. 1B). Molecular analysis revealed that TNBC growth inhibition was mediated by re-expression of thioredoxin-interacting protein (TXNIP) and suppression of PI3K/Akt. axis. Dox/Pan-pLNEs resulted in a significant reduction in tumor burden and enhanced survival using a synergistic combination of HDAC inhibitor and the chemotherapeutic Dox. The reduced Dox concentration may reduce side effects. This lipid targeting, Dox/Pan encapsulating pLNE represents a platform for treating TNBC with greater therapeutic efficacy.

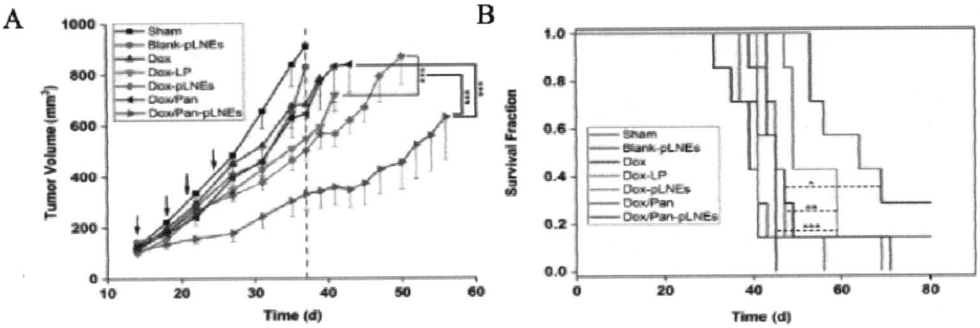

Figure 1. (A) TNBC orthotopic tumor growth (MDA-MB-231) treated with Dox/Pan-pLNEs and controls. (B) Kaplan-Meier survival analysis of TNBC orthotopic tumor model study treated with Dox/Pan-pLNEs and controls.

611937 Mesenchymal Stem Cells Anchored with Thymidine Phosphorylase for Doxifluridine-Mediated Cancer Therapy

Abstract:

Introduction: Many tumors express thymidine phosphorylase (TYMP) with various levels, however due to tumor heterogeneity the amount of TYMP is usually not enough to convert prodrug doxifluridine to toxic drug fluorouracil (5-FU). Since human mesenchymal stem cells (hMSCs) has unique features of tumor-tropism and low immunogenicity, the purpose of this study is to use mesenchymal stem cells as carriers to deliver TYMP to cancer cells and then trigger their death by administrating doxifluridine.

Materials and Methods: First, TYMP gene sequence and core streptavidin (coreSA) were constructed into pET-30a plasmid. After bacterial transformation and colony screening, TYMP-coreSA fusion protein was expressed by IPTG induction and purified by immobilized metal affinity chromatography. The concentration and enzyme activity of purified TYMP-coreSA fusion protein were determined separately by Bradford assay and prodrug conversion assay. The characterized TYMP-coreSA was further anchored on the cell membrane of biotinylated hMSCs via biotin-streptavidin binding. hMSCs anchored with TYMP-coreSA were then co-cultured with adenocarcinoma A549 cells and treated with 200 uM prodrug doxifluridine over the course of five days. Two ratios of TYMP-coreSA-tethered hMSCs to A549 cells in the co-cultured wells were tested for anticancer efficacy.

Results and Discussion: The expression and purification of TYMP-coreSA fusion protein was visualized with a clear band in the target 75 kDa region by SDS-PAGE and Western blot analysis. TYMP-coreSA level and prodrug conversion were confirmed by enzyme activity assay. For the co-culture of TYMP-coreSA-decorated hMSCs and A549 cells, our results showed that 2:1 ratio led to A549 cell death on day 2, in comparison with cell death on day 3 for 1:1 ratio. Moreover, the eradication of A549 cancer cells was shown after 5 days for 2:1 ratio of TYMP-coreSA-tethered MSCs to A549 cells for prodrug therapy.

Conclusions: Harnessing hMSCs as cell carriers for the delivery of TYMP enzyme to cancer cells could lead to significant cell death post-treatment of prodrug doxifluridine.

607114 Calcium Phosphate-Polymeric Nanoparticle System for Co-delivery ofmicroRNA-21 Inhibitor and Doxorubicin

Abstract:

The development of nano-scaled carriers as a drug delivery platform has made tremendous progress in the fight against cancer. Despite the progress, drug resistance remains one of the major challenges of nanotechnology-based cancer therapeutics. Due to gradually developed resistance of traditional anti-cancer drugs, new drugs that can target specific carcinogenic pathways are being sought based on molecular biology, such as small molecules, antibodies, RNAs, and DNAs. Among these, microRNA inhibitors (miRi) hold great promise in cancer therapeutics by down-regulating the proteins promoting multi-drug resistance. miRNA is a short non-coding RNA that can effectively inhibit protein translation or degrade specific mRNA transcripts. One unique feature of miRNA is that one miRNA can regulate multiple gene expressions. A key miRNA can act as a master switch for multiple pathways. An approach of a NP design carrying miRNA-21 inhibitor (miR-21i) and a chemotherapeutic agent is expected to effectively treat the current challenges of drug resistance. However, the synthesis of a co-delivery system encapsulating miRNA inhibitor and a chemotherapeutic agent is quite challenging, due to the opposite nature of the hydrophilicity of RNAs and the hydrophobicity of anti-cancer drugs.

In this study, we developed a calcium phosphate (CaP)-coated nanoparticle (NP) formulation to co-deliver miR-21i along with Dox. This NP design can be used to deliver the two agents with different physiochemical properties. The NP formulation was optimized for particle size, polydispersity, Dox loading, and miR-21i loading. The NP formulation was confirmed to downregulate miR-21 levels and upregulate tumor suppressor gene levels. The cytotoxic efficacy of the combined miR-21i and Dox-containing NPs was found to be higher than that of Dox. Therefore, the CaP-coated hybrid lipid-polymeric NPs hold potential for the delivery of miR-21i and Dox..

605186 Antibody Drug Nanoparticle Induces Synergistic Treatment Efficacies in Breast Cancer Cells

Abstract:

Chemotherapeutic drugs suffer from non-specific binding, undesired toxicity and poor blood circulation which contribute to poor therapeutic efficacy. The goal of this work is to engineer antibody-drug nanoparticles (ADNs) comprising of pure anti-cancer drug nanorods (NRs) in the core with a therapeutic monoclonal antibody on the surface of the nanoparticles for specific targeting and synergistic treatments of cancer cells. The ADNs were designed by first synthesizing ~95 nm diameter × ~500 nm long paclitaxel (PTX) NRs using nano precipitation method (Figure 1a, 1b) . The surface of PTXNRs were functionalized at 2¢ OH nucleophillic site using carbonyldiimidazole and conjugated to Trastuzumab (TTZ) through the lysine group interaction forming PTXNR-TTZ conjugates. The percentage conjugation efficiency was >95% with a drug to antibody mass ratio of 10. *In vitro* cytotoxicity analysis showed PTXNR-TTZ inhibited >80% of human epidermal growth factor receptor 2 (HER2) positive BT-474 breast cancer cells more efficiently than individual PTX and TTZ treatments alone for 72 h. A combination index analysis indicated synergistic combination of PTXNR-TTZ compared with the doses of each drug alone. The molecular mechansims of PTXNR-TTZ were investigated using cell cycle and Western blot analyses. Cell cycle analysis showed PTXNR-TTZ arrested >80% of BT-474 breast cancer cells in G2/M phase (Figure 1c), while >70% of untreated cells were found in G0/G1 phase indicating that G2/M arrest influenced apoptosis. A similar percentage of G2/M arrested cells were found to induce caspase-dependent apoptosis in PTXNR-TTZ treated BT-474 cells as revealed using Western blot analysis. PTXNR-TTZ treated BT-474 cells showed 1.3, 1.4 and 1.6-fold higher expressions of cleaved caspase-9, cytochrome C and cleaved caspase-3, respectively than untreated cells, indicating up-regulation of caspase dependent apoptotic pathways (Figure 1d).

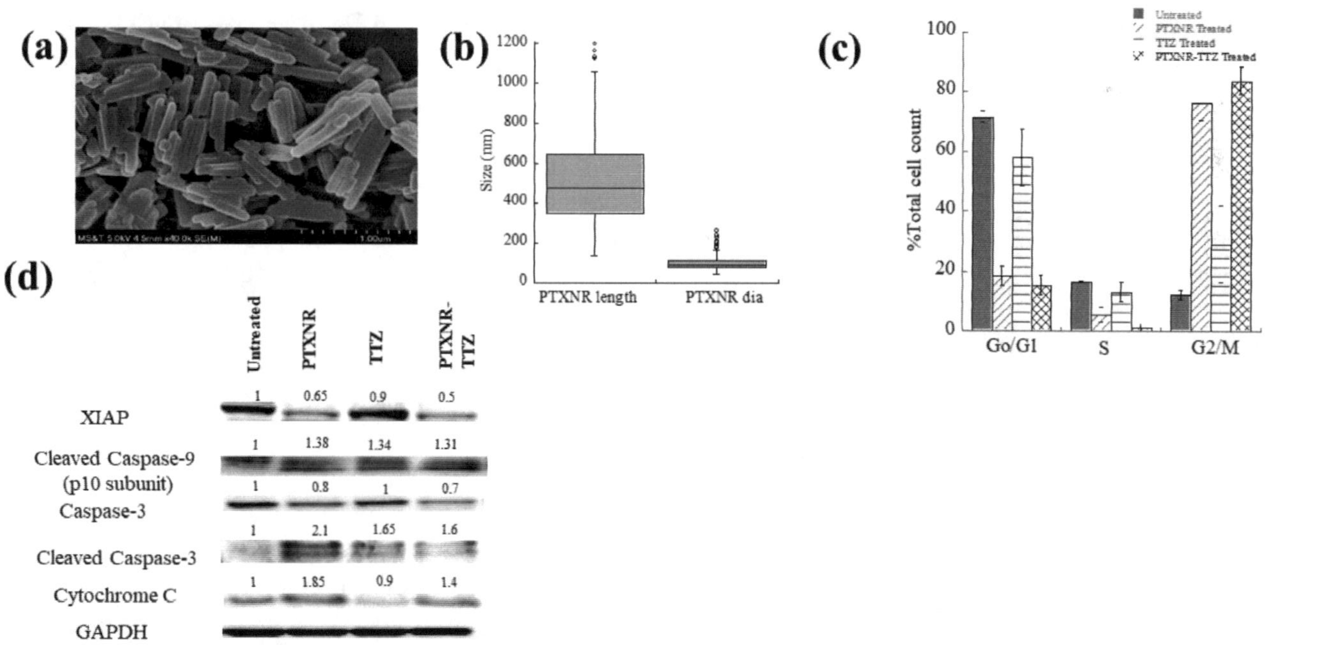

Figure 1. Characterization of PTXNR nanoparticles and G2-M cell cycle arrest by combination drug PTXNR-TTZ. (a) scanning electron microscopy (SEM) image of PTXNR nanoparticles (b) percentile plot of the PTXNR particle length and diameter (c) cell cycle analysis after 72 hr in BT-474 breast cancer cells (d) western blot analysis in BT-474 breast cancer cells.

611413 ICAM-1 Nanobody Density on Liposomes Affects Selectivity for Triple Negative Breast Cancer and Inflamed Endothelium

Abstract:

Introduction: Triple negative breast cancer (TNBC) is a subtype of breast cancer characterized by the absence of ER, PR, and HER2 receptors and a high rate of metastasis. There are currently no commercially available targeted therapies for the treatment of TNBC. To address this deficiency, we synthesized liposomes conjugated with nanobodies specific to ICAM1, a receptor that is overexpressed on both tumor-associated vasculature and TNBC. We hypothesized that liposome drug delivery vehicles (DDVs) with higher anti-ICAM1 nanobody densities will bind a greater number of ICAM1 receptors to achieve stronger and more specific binding to target cells and increase vascular permeability through downstream ICAM1 signaling. We have employed a covalent conjugation chemistry that enables the DDV to retain binding specificity and biological activity in the presence of mouse plasma. Our data will show that ICAM1 nanobody density may be optimized to greatly improve TNBC targeting and penetration of tumor vasculature, resulting in improved tumor accumulation and growth inhibition.

Materials and Methods: Initial studies were performed with liposomes that were synthesized using lipids displaying nitriloacetic acid (NTA) and DiO, enabling conjugation of anti-ICAM1 nanobodies containing a his-tag and fluorescent detection of the liposomes respectively. Covalently-conjugated liposomes were synthesized using a published chemistry[1]. The liposomes displayed nanobody densities ranging from 3,000 SdAb/μm^2 to 44,000 SdAb/μm^2. Binding experiments were performed at 4°C for 30 minutes to prevent internalization. Cell binding was analyzed by measuring DiO fluorescence of cells by flow cytometry. Endothelial cell (EC) monolayer penetration was assessed by performing transwell

experiments with TNF-α-stimulated and unstimulated confluent EC monolayers. To evaluate the downstream effect of ligand density on VE-cadherin dissociation from tight junctions, TNF-α stimulated HUVEC monolayers were treated with liposomes for 30 minutes at 37°C, followed by immunofluorescent (IF) staining for VE-cadherin and subsequent imaging on a confocal microscope.

Results and Discussion: Liposomes were synthesized with ligand surface densities ranging from 3,000 SdAb/μm^2 to 44,000 SdAb/μm^2. Cell binding studies with ICAM1-targeting NTA-liposomes indicated that the maximum relative binding to TNF-α-stimulated ECs and TNBC cells over corresponding control cell lines occurred at 33,000 SdAb/μm^2, where liposomes displayed approximately 4 and 3 fold higher binding respectively. Interestingly, the 33,000 SdAb/μm^2 formulation achieved the greatest penetration of both stimulated and unstimulated EC monolayers. IF staining of VE-cadherin on liposome-treated monolayers revealed that ligand densities of 22,000 SdAb/μm^2 and 33,000 SdAb/μm^2 resulted in 73% and 53% dissociation of VE-cadherin from tight junctions in TNF-α-stimulated EC monolayers, respectively.

Conclusions: We show that binding specificity for both TNBC and cytokine-stimulated ECs over non-diseased counterparts, EC permeability, and ICAM1 downstream signaling can be enhanced through optimization of anti-ICAM1 nanobody density on the surface of a liposome DDV, where the optimal surface density for all phenomena occurs at 22,000 – 33,000 SdAb/μm^2. This data can be used in the design of DDVs that permeabilize tumor-associated endothelium and bind cancer cells. Future research will be dedicated to elucidating the molecular mechanisms responsible for these observations and validating their influence on tumor targeting and therapeutic efficacy in vivo.

References:

1 Cong, Y. *et al.* Site-specific PEGylation at histidine tags. **23**, 248-263, (2012).

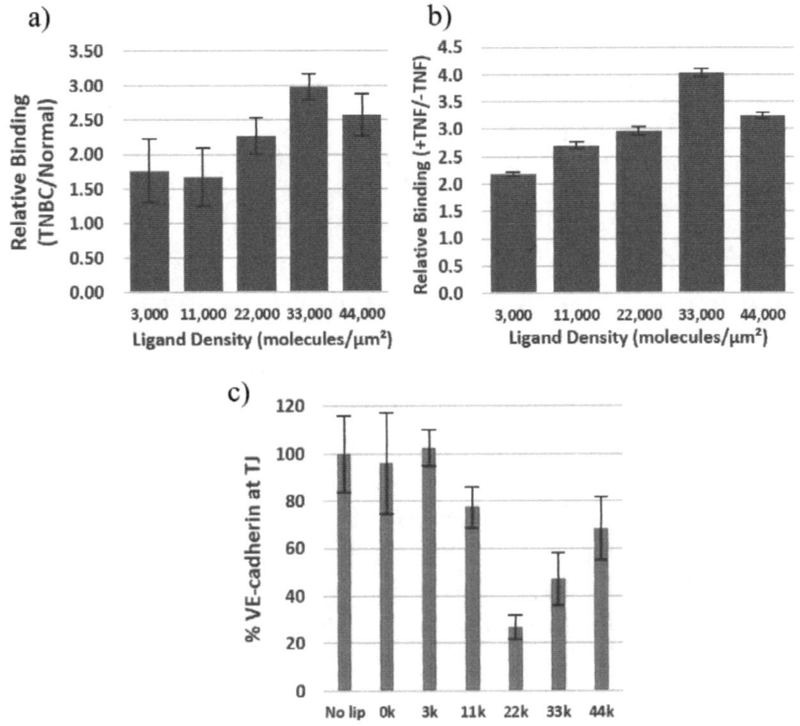

Figure 1. Relative binding of liposomes to (a) ICAM1-overexpressing TNBC cells (MDA-MB-436) over non-neoplastic breast cells with low ICAM1 expression and (b) TNF-α-stimulated ECs over unstimulated ECs. (c) % VE-cadherin remaining at tight-junctions after liposome treatment.

614312 Peptide Engineering for Targeted, Intracellular Delivery of siRNA and Proteins (INVITED SPEAKER)

Abstract:

Protein nanostructures with tunable chemistries would be extremely useful in a wide variety of applications ranging from biocatalysis to biopharma. Biopharmaceutical products are especially well-posed for impact: drugs such as monoclonal antibodies, peptides, and enzymes are rapidly transforming the global pharmaceutical industry, and biologics represented 35% of all new FDA approvals between 2010 and 2016. Moreover, the pipeline has a dearth of intracellular protein drug candidates, even though intracellular proteins comprise 61% of the proteome and have mulitple predicted therapeutic applications. In fact, essentially all protein drugs to date have been restricted to vascular targets or targets on the cell surface, referred to as the 'accessible target space.'

We are designing peptide functionalization approaches in protein nanostructures that will enable improved control over intracellular targeting and delivery of siRNA and protein cargoes via improved control over the display of peptide ligands. Our specific goals are to demonstrate the use of versatile and modular protein engineering approaches (*e.g.*, 'Tag/Catcher' chemistries; unnatural amino acid [UAA]/'click' chemistries) to create protein nanostructures that can be targeted and delivered with high efficiency, cell specificity, and no activity loss. Herein, we will describe approaches to enhance targeted protein delivery into inflammatory breast cancer (IBC) cells by tuning the number and density of epidermal growth factor receptor (EGFR) ligands, and by altering the size of the nanostructures used for delivery.

608538 Curating a Comprehensive Set of Enzymatic Reaction Rules for Efficient Novel Biosynthetic Pathway Design (Industry Candidate)

Abstract:

Enzyme promiscuity, where enzymes may catalyze a range of side reactions in addition to its main reaction, is a widely recognized yet still largely unexplored phenomenon in biological systems. This could open up vast possibilities for bioproduction of valuable chemicals, where novel biosynthetic pathways leading to molecules of interest can be constructed based on enzymes with desired promiscuous reactions. In order to design these biosynthetic pathways, we curated a set of generalized enzymatic reaction rules capable of describing all common enzymatic transformations, which can be applied in the Biochemical Network Integrated Computational Explorer (BNICE) platform to predict the entire space of possible metabolic reactions. We have created a minimum number of these rules verified to 1) comprehensively cover known reactions across metabolic databases, 2) describe reactions with the maximum level of promiscuity, and 3) represent unique enzymatic transformations. Based on these rules, a cheminformatics workflow is developed to efficiently predict and prune novel biosynthetic pathways towards molecules of interest, and systematically identify the most promising ones for experimental validation. By leveraging the entire knowledge of possible metabolic reactions through enzymatic reaction rules, we are able to accelerate novel pathway design and enable bioproduction towards a wide range of new molecules.

602501 Modeling Growth Kinetics and Metabolism of *Clostridium Acetobutylicum/Clostridium Ljungdahlii* Co-Culture with Cell Fusion (Industry Candidate)

Abstract:

Clostridia organisms have been of interest for decades due to their unique ability to ferment a wide range of carbon sources to useful bioproducts. The metabolic repertoire of these anaerobes has been further expanded in co-cultures due to the diversity of substrates they can consume and unexpected syntrophic behaviors that are still being discovered. One such example lies in the syntrophic co-culture of *Clostridium acetobutyllicum* (*C. ac*) and *C. ljungdahlii* (*C. lj*). In addition to the discovery of an upregulation of *C. lj sadh* and *23bdh* gene expression in the presence of *C. ac* (allowing *C. lj* to convert acetone and acetoin produced by *C. ac* to isopropanol and 2,3-butanediol, respectively), *C. ac* and *C. lj* cells were recently shown to fuse membranes and exchange proteomes. This work aims to characterize the resultant change in growth kinetics due to the observed fusion/protein exchange event using a kinetic model which characterizes both the pure and mixed-proteome *C. ac* and *C. lj* growth rates and the cell fusion/proteome exchange rate. The parameterized kinetic model is used to inform the construction of a community genome-scale metabolic model of pure and mixed-proteome *C. ac* and *C. lj* cells using the dynamic multi-species metabolic modeling (DMMM) framework, and characterize the dynamic shift in co-culture metabolism and redox state related to the observed fusion event required to support the experimentally measured isopropanol and 2,3-butanediol production. Single organism strain design tools (i.e. optKnock, optForce) are being adapted to support the inclusion of multi-organism models in order to understand how the *C. ac* and *C. lj* genomes can be manipulated to maximize the production of fermentation products of interest (i.e isopropanol, 2,3-butanediol) under the newly discovered *C. ac/C.lj* co-culture conditions.

608260 Optimization-Based Analysis of Comparative Metabolomic Data Identifies Regulatory Mechanisms in *Staphylococcus Aureus* mutants (Faculty Candidate)

Abstract:

Recent trends have recorded a significant rise in community-associated methicillin-resistant *Staphylococcus aureus* (CA-MRSA) infections. Additionally, most clinical isolates can develop into recalcitrant multiple-antibiotic-resistant and host immune-tolerant biofilm communities, necessitating novel therapeutic strategies for combating *S. aureus* infections. Therefore, there has been a steady interest and focus towards understanding how staphylococcal metabolism relates to antibiotic resistance and pathogenesis. While previous efforts have been made to understand the regulatory networks for the antibiotic mechanism of action to *S. aureus*, genetic perturbations can also play a major role in understanding the effectiveness of antibiotic killing and inhibition functions. Some key questions in this regard are still unanswered, including i) what are the core carbon and nitrogen metabolic pathways that are crucial for staphylococcal growth and survival; ii) how specific mutations alter the metabolic landscape, growth rate and survival of *S. aureus*; iii) would a systems-level metabolic model be able to accurately predict growth and metabolite changes that occur within defined mutants.

To answer these questions, we performed a metabolomic analysis of 8 mutants in the central metabolism of *Staphylococcus aureus*. These mutants include *pyc* (pyruvate carboxylase), *citZ* (citrate synthase), *sucA* (2-oxoglutarate dehydrogenase), *ackA* (acetate kinase), *gudB* (glutamate dehydrogenase), *ndhA* (NADH dehydrogenase), *menD*(menaquinone biosynthesis protein), and *atpA* (ATPase subunit). These mutants were selected for their potential in identifying carbon and nitrogen redirection pathways as they affect important physiological processes including glycolysis, TCA cycle activity, gluconeogenesis, Electron Transport Chain (ETC), cellular redox potential, overflow metabolism and nitrogen metabolism. Quantitative metabolite levels of the *S. aureus* mutants grown in chemically defined media (CDM) were compared to the wild-type *S. aureus* strain to identify the statistically significant fold changes. three mutants (*menD, atpA, ackA*) in exponential phase and seven mutants (*pyc, citZ, sucA, gudB, ndhA, atpA, ackA*) in the post-exponential phase, along with the wild-type following growth in CDM. The metabolite levels were also estimated in the wild-type strain in aerobic, anaerobic and micro-aerobic conditions. In total, 53 intermediate metabolites in the central carbon metabolism were measured by LC-MS/MS analysis.

We integrated the metabolomic data with our recently published genome-scale metabolic model of *S. aureus*[1] that combines genome annotation data, reaction stoichiometry and thermodynamic information, and regulation information from biochemical databases and previous strain-specific models [2-8], which were validated though existing and new experimental results on gene essentiality, auxotrophy, and metabolite excretion. From the metabolite pools, we performed a maximum and minimum flux-sum analysis[9] for each of the reactions in the model upon each of the mutations. From the deviation of flux-sum ranges of reaction in the mutants compared to wild-type, we identified a set of reactions that are modulated by each of the mutations. We then employed a multi-level optimization framework to identify a minimal number or reactions which must be regulated in the each of these mutations. By studying the metabolic consequences of these different mutations, we achieved greater resolution of the known pathways utilized by *S. aureus*, as well as pinpointed to the regulatory influences caused by these mutations. This will allow us to identify the core metabolic pathways and enzymes that are used by *S. aureus*, but also fringe pathways that are activated following a block in the core metabolic pathways. The results from our analysis highlight the adaptive capabilities of *S. aureus* and identify the metabolic bottlenecks in wild-type and mutant growth. This will lead to the identification of additional growth-inhibiting single- and double-knockouts that can potentially provide a solution to the prevailing antibiotic resistance of this organism.

References:

1. Islam MM, Thomas VC, Van Beek M, et al. An integrated computational and experimental study to investigate Staphylococcus aureus metabolism. *NPJ Syst Biol Appl.* Jan 30 2020;6(1):3.
2. Becker SA, Palsson BO. Genome-scale reconstruction of the metabolic network in Staphylococcus aureus N315: an initial draft to the two-dimensional annotation. *BMC microbiology.* Mar 07 2005;5:8.
3. Bosi E, Monk JM, Aziz RK, Fondi M, Nizet V, Palsson BO. Comparative genome-scale modelling of Staphylococcus aureus strains identifies strain-specific metabolic capabilities linked to pathogenicity. *Proc Natl Acad Sci U S A.* Jun 28 2016;113(26):E3801-3809.
4. Henry CS, DeJongh M, Best AA, Frybarger PM, Linsay B, Stevens RL. High-throughput generation, optimization and analysis of genome-scale metabolic models. *Nature biotechnology.* Sep 2010;28(9):977-982.
5. Schellenberger J, Park JO, Conrad TM, Palsson BO. BiGG: a Biochemical Genetic and Genomic knowledgebase of large scale metabolic reconstructions. *BMC bioinformatics.* Apr 29 2010;11:213.
6. Kanehisa M, Goto S. KEGG: Kyoto Encyclopedia of Genes and Genomes. *Nucleic acids research.* Jan 1 2000;28(1):27-30.
7. Pence HE, Williams A. ChemSpider: An Online Chemical Information Resource. *J Chem Educ.* Nov 2010;87(11):1123-1124.
8. Ren Q, Chen K, Paulsen IT. TransportDB: a comprehensive database resource for cytoplasmic membrane transport systems and outer membrane channels. *Nucleic Acids Res. //* 2007;35:D274-D279.
9. Chung BK, Lee DY. Flux-sum analysis: a metabolite-centric approach for understanding the metabolic network. *Bmc Syst Biol.* 2009;3:117.

610708 Elucidation of Trophic Interactions in an Unusual Single Cell Nitrogen-Fixing Symbiosis Using Metabolic Modeling (Industry Candidate)

Abstract:

Marine nitrogen-fixing microorganisms are an important source of fixed nitrogen in oceanic ecosystems. The colonial cyanobacterium *Trichodesmium* and diatom symbionts were thought to be the primary contributors to oceanic N_2 fixation until the discovery of the unusual uncultivated symbiotic cyanobacterium UCYN-A (*Candidatus Atelocyanobacterium thalassa*). UCYN-A has atypical metabolic characteristics lacking the oxygen-evolving photosystem II, the tricarboxylic acid cycle, the carbon-fixation enzyme RuBisCo, and *de novo* biosynthetic pathways for a number of amino acids and nucleotides. Therefore, it is obligately symbiotic with its single-celled haptophyte algal host. UCYN-A receives fixed carbon from its host and returns fixed nitrogen, though the exact identities of the exchanged metabolites and extent of metabolic flows are unknown. In order to investigate how this syntrophy is coordinated, we reconstructed bottom-up genome-scale metabolic models of UCYN-A and its algal partner to explore possible trophic scenarios associated with nitrogen fixation and biomass synthesis. Since both partners are uncultivated and only the genome sequence of UCYN-A is available, we used the phylogenetically related *Chrysochromulina tobin* as a proxy for the host. Through the use of flux balance analysis, we determined the minimal set of metabolites and biochemical functions that must be shared between the two organisms to ensure growth and activity. We quantitatively investigated the metabolic characteristics that facilitate daytime N_2 fixation in UCYN-A and possible oxygen-scavenging mechanisms needed to create an anaerobic environment to allow nitrogenase to function. This is the first application of a flux balance analysis framework to examine the tight metabolic coupling between single-celled microbes in marine symbiotic communities. Experimental investigations into this system have so far remained elusive due to difficulties in culturing UCYN-A and/or its host. This has rendered computational studies on metabolic models an exciting route for deciphering possible trophic scenarios.

599472 Predicting Nash Equilibria and Limiting Cross-Feedings for Microbial Metabolic Interactions

Abstract:

Microbial metabolic interactions impact ecosystems, human health and biotechnology profoundly. Understanding metabolic interactions is a fundamental task in microbiome sciences. Despite continual advances in experimental approaches for detecting interactions within microbial communities, we still need governing principles and predictive models to predict and explain microbial community metabolism in the context of ecology and evolution.

Inspired by the evolutionary game theory, we formulated a bi-level optimization framework termed NECom for the prediction of Nash equilibria of microbial community metabolic models. NECom is free of a long hidden 'forced altruism' formulation in previous static algorithms while allowing for 'sensing and responding' between microbial members that are missing in dynamic methods. We showed enhanced predictive accuracy of NECom over existing classes of algorithms. Using NECom, We successfully predicted several classical games in the context of metabolic interactions that were falsely or incompletely predicted by existing methods, including prisoner's dilemma, snowdrift game and mutualism. The results provided insights into why mutualism is favorable despite seemingly costly cross feeding metabolites, and demonstrated the potential to predict heterogeneous phenotypes among the same species. We then applied NECom to a reported algae-yeast co-culture system that shares typical cross-feeding features of lichen, a model system of mutualism. More than 1200 growth conditions were simulated including conditions corresponding to 3221 experimental data points. Without fitting any ad-hoc parameters, an overall 63.5% and 81.7% reduction in root-mean-square error in predicted growth rates for the two species respectively was achieved when compared with flux balance analysis. Through NECom-enabled shadow price analysis, we identified limiting cross feeding metabolites and explained a predicted frequency-dependent growth pattern, offering insights into population-stabilizing microbial interactions.

Predicting Nash equilibria and limiting crossfeedings in microbial communities

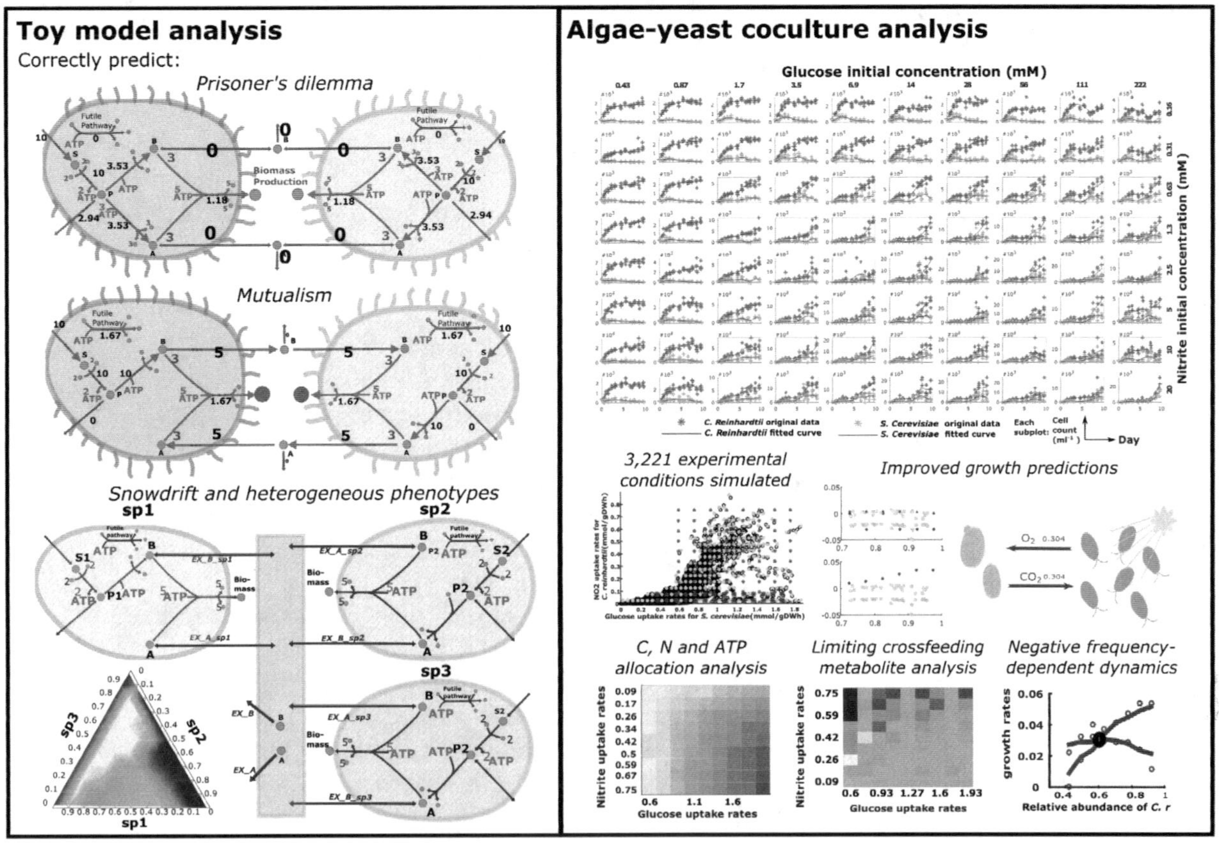

608842 Understanding the Effects of Cooperation and Competition on Microbial Growth Pattern through Mathematical Modeling and Experiments (Faculty/Industry Candidate)

Abstract:

Most biological functions are performed by complex multicellular structures that act in perfect unison in a coordinated manner. These structures are formed from a few cells in the beginning, reproducing and responding to their extracellular environments. Hence, it is essential to understand the role of the factors affecting the formation of these structures. Relatively simpler microbial communities in well controlled experimental setups are suitable for studying this phenomenon of structure or pattern formation (Chen et al., 2014). Moreover, it has been seen that secretion of 'public goods' in systems such as the one used in this study leads to multicellularity itself (Koschwanez et al., 2013). In this work, we experimentally study how opposing strategies such as cooperation and competition lead to interesting microbial growth patterns and develop a mathematical model based on these observations that can help us understand the underlying dynamics and potentially, in future, help in making design choices that will lead to desirable pattern formations tailored to specific biotechnological applications. Budding yeast (also known as baker's or brewer's yeast), Saccharomyces cerevisiae hydrolyzes disaccharides such as sucrose into easier-to-metabolize monosaccharides such as glucose and fructose in the scarcity of extracellular glucose or fructose. This conversion is catalyzed by the enzyme invertase in the periplasmic space (area between the cell wall and plasma membrane) of the 'producer' cells. The greater part of the hydrolyzed monosaccharides diffuses away before being consumed by the cell that produces it, and 'cheater' cells (cells without the gene that produces the hydrolyzing enzyme invertase) in the same medium, can survive on this glucose and fructose. We study the growth of two such strains in a co-culture. An interesting growth dynamics arises in this reaction-diffusion system, from the production, diffusion and consumption of the monosaccharides. In this case, these strains have also been genetically modified to apply an additional diffusion gradient to the system that leads to competition between the two cell types. The presence of these two opposing interactions results in interesting growth patterns. This system is studied by manipulating aspects of the system that controls the availability of the limiting nutrients to the cells, such as, the initial population density, the initial ratio of producer and cheater cells and the initial concentrations of the limiting nutrients. A dynamic mathematical model is developed to simulate the multiple reaction-diffusion phenomena in the system. The concentration profiles of the limiting nutrients and the two cell types over the entire culture plate are tracked over time using a system of PDEs (partial differential equations). The model predicted growth patterns are compared with the experimental ones to calibrate the model. Then, the model is further used to predict the final pattern based on the initial conditions. It is also utilized to provide important insight into the growth dynamics of such microbial communities such as what kind of initial conditions lead to what regimes of growth and how multiple factors such as initial population density, initial ratio of the two types of cells and initial concentrations of limiting nutrients favor growth of one type of cell over the other and lead to different growth patterns.

References

Chen L, Noorbakhsh J, Adams RM, Samaniego-Evans J, Agollah G, Nevozhay D, et al. (2014) Two-Dimensionality of Yeast Colony Expansion Accompanied by Pattern Formation. PLoS Comput Biol 10(12): e1003979. https://doi.org/10.1371/journal.pcbi.1003979

Koschwanez, J. H., Foster, K. R. & Murray, A. W. (2013). Improved use of a public good selects for the evolution of undifferentiated multicellularity. eLife 2013; 2:e00367 DOI: 10.7554/eLife.00367

613018 Development of Machine Learning Methods for Prediction of Microbial Metabolisms and Biosynthesis Performance (Invited Speaker)

Abstract:

Mechanistic based metabolic models have been widely used to predict cell physiologies under simplified bioprocess conditions. For highly complex biological systems, machine learning models may be able to decode the relationship among genetic, enzymatic, and environmental factors that lead to specific metabolic states. To fulfill this hypothesis, we have attempted to develop and employ data driven models to offer predictions on microbial physiologies. In the first case study, we have shown that machine learning can reasonably predict fluxomes as a function of bacterial species, substrate types, growth rate, oxygen conditions, and cultivation methods. Specifically, ^{13}C-metabolic flux analysis (^{13}C-MFA) is a powerful tool to experimentally measure *in vivo* enzyme reaction rates in microorganisms. We leveraged published data from approximately 100 ^{13}C-MFA papers on heterotrophic bacterial metabolisms and manually extracted features. Then we tested three machine learning methods (Support Vector Machines, k-Nearest Neighbors, and Decision Trees) to capture the sophisticated relationship between environmental/genetic factors and metabolic flux configurations. We also implemented quadratic programming to adjust flux profiles predicted from machine learning to satisfy stoichiometric (mass balance) constraints. The resulting model could forecast bacterial fluxome under denoted conditions and facilitate the metabolic analysis procedures.

In the second case study, we integrated data-driven methods with genome scale metabolic model for assessment of microbial bio-production (yield, titer and rate). Using *Escherichia coli* or *Yarrowia lipolytica* as examples, we organized and curated data sets comprising recent biomanufacturing papers. We then augmented the features (e.g., product and substrate types, bioreactor conditions, and genetic background) extracted from literature with additional features derived from genome-scale model simulations. To alleviate the challenges of sparse data sets, data augmentation and ensemble learning were employed. The hybrid framework demonstrated a reasonably high cross-validation accuracy for prediction of cell factory performance metrics under presumed bioprocess and pathway conditions. These predictions could be: 1. used to assess and rank these influential factors on bio-production; 2. integrated with technoeconomic analysis for prior estimation of cell factories outcomes (i.e., serve as the useful risk assessment tool); or 3. employed in conjunction with genome scale modeling to improve computational design tools.

603353 Organotypic Brain Slices As a Platform for Measuring Response to Biological Stimuli

Abstract:

Introduction: Chronic diseases are the leading cause of death and disability, affecting 6 in 10 adults in the United States. Developing novel therapeutics is necessary to combat chronic disease and reduce this burden, but the process to take a drug from discovery to marketplace is slow and costly. Failure to translate outcomes in animal models to success in human clinical trials is a major contributor to the time and cost associated with bringing drugs to market and is in part due to the inability of animal models to recapitulate the physiology of the disease in question. 2D *in vitro* cell culture models are simple and inexpensive, but lack contact between neighboring cells and native tissue architecture. While *in vivo* models are the most physiologically relevant, there are multiple disadvantages to this approach, including hefty animal care/housing expenses and inability to gain mechanistic insights, particularly when using larger, non-mouse models. Therefore, there is motivation to develop platforms that accurately recapitulate *in vivo* physiology, but are higher throughput and tailorable. Organotypic slice models have advantages over primary cell culture, as multicellular interactions are retained and bias in cell type is eliminated. Additionally, tissue slices retain 3D architecture and the slices mature *in vitro* in a similar manner to native conditions, allowing slices to be used long term in culture without sacrificing physiological relevance. A single animal can yield numerous healthy organotypic slices, which reduces the required number of animals required of *in vivo* animal testing. Relative to *in vivo* models, the extracellular environment can be better controlled, which allows for high throughput screening of various therapeutics or to study the biological response to any stimulus, including toxins and injury/disease models. Here, we demonstrate the use of organotypic brain slice cultures to probe cellular response to oxygen-glucose deprivation (OGD), an injury model for hypoxia and ischemia, and subsequent treatment of that injury with a therapeutic enzyme (superoxide dismutase).

Methods: Postnatal day (P) 7 Sprague-Dawley rats were sacrificed and the brain was promptly extracted. Hemispheres were separated and cut coronally into 300-micron slices. Slices were separated in ice cold dissection media (HBSS, 5 g/L D-glucose) and plated onto 0.4 um semi-porous membranes (3 slices per membrane) in a 6 well plate with fresh media (50% HBSS, 44% MEM, 5% horse serum, 2 mM GlutaMAX, 1% penicillin streptomycin) and cultured for 1 week. Media was replaced daily and supernatant was immediately stored at -80 °C. Media samples were analyzed by lactate dehydrogenase (LDH) assay to assess total cellular death as a function of culture time. Slices were stained with propidium iodide (PI), washed thoroughly, and fixed in 10% phosphate buffered formalin on days 1, 2, 3, 5, and 7. These slices were permeabilized with Triton X-100 (1%), counter stained with DAPI (1:10,000) and imaged by confocal microscopy to assess cellular death in the cortex, thalamus, and hippocampus. To analyze cell-cell interaction, bolistic transfection of neurons was used. Gene gun particles were prepared by adding 10 mg gold (1-micron size), 50 ul of 0.05M spermidine, 20 ug mCherry plasmid, and 50 uL of 1M $CaCl_2$, incubated, and the pellet washed with 100% ethanol. Bullets were formed using Tefzel tubing and then dried with nitrogen, cut with the Helios® Tubing Cutter and stored at 4 °C with a desiccation pellet until use. Slices are bombarded using a Helios® Gene Gun. A helium pulse at 200 psig propels mCherry-coated, 1-micron gold particles at the tissue to transfect individual cells on the tissue. Following 2 days of incubation, slices are stained with Cd11b+ (1:500) for microglia. OGD was performed on 3 DIV (days *in vitro*) slices by: 1) removing brain slices from glucose-treated media; 2) depriving slices of oxygen by purging a sealed chamber containing the well plates with nitrogen, and 3) incubating at 37 °C for 0.5h to 3h.

Following injury, non-transfected slices were cultured in normal culture media doped with superoxide dismutase (0.1 mg/slice) for 24 h. Culture media was aspirated and stored at -80 C for use with LDH assay. Slices were stained with PI and imaged to assess cellular death in the cortex, thalamus, and hippocampus. The response protein level was measured with ELISA. Following injury, mCherry-transfected slices were transferred into a perfusion chamber/confocal microscopy system and culture media is perfused at 20 ul/h using a 2-way syringe pump. Images containing neurons and microglia were acquired every 15 min for 6 h and the interactions between neurons and microglia were analyzed.

Results: We found that the optimal time to begin the OGD injury model for P7 rats was at 3 DIV, based on cell death data from LDH assay and PI staining. In the first few days of culture, slices undergo an inflammatory response due to the stress of slicing. As a result, cell death is high initially and the conditions do not make for an adequate experimental window to measure the response of a stimulus until initial death has subsided. By varying the total OGD time, we demonstrated the ability to precisely control a stimulus, measuring a dose-dependent response to the injury. Slices subjected to OGD without any treatment showed significantly more cellular death than to the healthy control group, as expected based on previous OGD studies with 2D cell culture and tissues. Additionally, OGD slices treated with superoxide dismutase (SOD) showed significantly less cellular death than the untreated OGD slices. This was also expected, as SOD is a therapeutic enzyme that works against reactive oxygen species (ROS). Although ROS are present in healthy tissue, OGD causes their presence to increase above the basal level, which leads to oxidative stress and downstream cellular damage. We demonstrated that organotypic brain slices can be cultured with high viability for several hours in a controlled perfusion environment compatible with multi-channel confocal imaging. The combination of time lapse imaging with biolistic transfection on organotypic brain slices to probe multicellular interactions has not previously been reported in the literature.

Conclusions: Organotypic brain slices can be paired with novel time-lapse imaging techniques to probe multicellular interactions with physiologically relevant cytoarchitecture and precise control of the extracellular environment. While we demonstrate the use of OGD with the organotypic brain slice platform, there are a myriad of other possible stimuli, including other injury and disease models, toxins, and therapeutics. Furthermore, the severity of the stimulus is controllable; thus, one can correlate the response of a disease or toxin to multiple levels of severity. With this approach, we can gather essential information about the underlying mechanism, using that data to drive our research direction toward a therapeutic that will be safe and effective. Finally, we show that organotypic brain slices can be used to screen therapeutics, and in this study, evaluated SOD. Because many brain slices are garnished from a single animal, this platform is compatible with the screening of multiple drugs at the same time without the need for excessive sacrifice and animal housing costs. We anticipate that whole hemisphere brain slices will be used to further the development and translation of therapeutics for combatting neurological disease.

604194 Metabolomics Approach to Improving CHO Cell Productivity

Abstract:

Mammalian cell culture using Chinese hamster ovary (CHO) cells is a key process in the industrial manufacturing of therapeutic biopharmaceutical proteins such as monoclonal antibodies (mAb). With process optimization, CHO cells can produce recombinant proteins on the scale of 10 g/L of culture, including post-translational modifications that are compatible for use in humans. This optimization may include various next-generation techniques, such as very high cell density culture in perfusion bioreactors and online process analytics. Another approach to improving titer (and hence volumetric productivity) is to enhance cell-specific productivity (qP). Much of the titer increases seen in the past few years have been due to increased viable cell density (VCD) and sustained viability. In comparison, qP has been more difficult to understand or control, especially as many cell lines are generated using random integration of the plasmid carrying the mAb of interest. Productivity is not only impacted by the process but also varies among parental cell lines and individual clones. This study used a metabolomics approach to investigate the metabolic factors influencing clone productivity, with the aim of identifying metabolic indicators that can be used to improve qP or expedite clone selection. Greater productivity in production bioreactors can lead to lower costs, smaller footprints, and more facility flexibility.

A library of 12 clones from the same host cell line, six producing mAb A and six producing mAb B, was selected based on preliminary data to represent a wide variety of growth and qP profiles. Fed-batch experiments were performed in 5-liter bioreactors using the same production process for all 12 clones, resulting in a range of peak viable cell densities (VCDs), titers and qPs as expected (*Fig. 1)*.

Supernatant samples collected on day 7 (early stationary phase) were analyzed using four different combinations of liquid chromatography (LC) and untargeted mass spectrometry (MS) methods. The MS experiments were performed on a time-of-flight (TOF) LC-MS system (TripleTOF 5600+, AB SCIEX, Framingham, MA). Over 4,000 LC-MS features were detected, including duplicates between methods (i.e., same metabolite detected two or more times by different LC and MS combinations) and adducts. Using a previously developed automated annotation tool (Biologically Consistent Annotation, BioCAn) tool, 99 unique metabolites were putatively identified. Pearson and Spearman correlation coefficients were calculated between each of these putatively identified metabolites and peak VCDs (representing growth) and qPs. Early stationary phase measurements for 80 of these metabolites were significantly correlated ($p<0.05$) with qP, growth, or both. In general, metabolites significantly correlating with qP had positive associations, while those significantly correlating with growth had negative associations (*Fig. 2A*). Only 3 metabolites had negative correlations with qP and 1 metabolite had a positive correlation with growth, and none of these were significant. Pathway enrichment analysis using Metaboanalyst (*Fig. 2B*) showed that metabolites having significant correlations with qP were enriched in alanine, aspartate, and glutamate metabolism. Other enriched pathways included arginine biosynthesis, glyoxylate and dicarboxylate metabolism, and the TCA cycle.

To test whether any of these significantly correlated metabolites could be used as biomarkers of productivity, another fed-batch study was performed with a new library of 12 clones from the same host cell line as above, all producing a third antibody (mAb C). Again, a wide range of growth and productivity profiles was observed. Samples from day 7 of bioreactor culture were collected for a targeted LC-MS experiment (multiple reaction monitoring, MRM). The target analytes comprised a subset of the significantly correlated metabolites identified from the untargeted experiment. In this analysis, compound A was found again to have a significant, positive correlation with qP, confirming an association from the untargeted experiment (*Fig. 3A, Fig. 3C*). Additionally, compound B, which was positively correlated with qP in the untargeted experiment but did not reach the $p<0.05$ level of significance, was found in this second set of clones to have a significant, positive correlation (*Fig. 3B, Fig. 3D*).

Next, a study was performed in fed-batch 250-mL bioreactor cultures to see whether metabolites with significant correlations with qP were directly responsible for improved qP or whether they were byproducts of pathways that are active in a high productivity metabolic state. Select compounds from the enriched pathways were added back as boluses in varying concentrations to cultures of multiple cell lines during mid-exponential phase (day 3). One of these compounds, compound C, showed dose-dependent improvements to qP in five out of the six cell lines tested (*Fig. 4A*). While the highest dose (6 mM) slightly reduced VCD, overall there was an increase in volumetric titer of 15-107% for four of the cell lines (*Fig. 4B*).

These studies demonstrate value in using untargeted metabolomics to better understand qP. In this work, we were able to identify potential biomarkers that can be used to shorten cell line development timelines and a metabolite that increased qP and titer when added to bioreactor cultures. Because these studies were conducted across multiple clones, these metabolites and the pathways they belong to may be indicative of a metabolic phenotype that supports a high level of protein production.

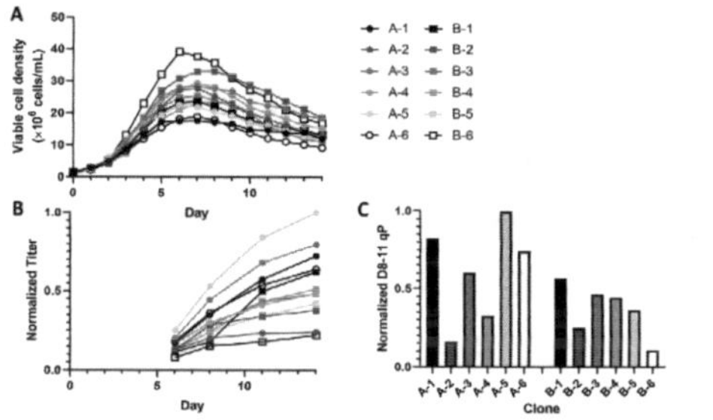

Figure 1. Twelve cell lines producing one of two monoclonal antibodies were cultured in fed-batch bioreactors under identical process conditions. Viable cell density (VCD) (A) was recorded daily. Titer (B) was measured in samples collected on days 6, 8, 11, and 14. Cell-specific productivity (C) was calculated from integrated VCD and titer.

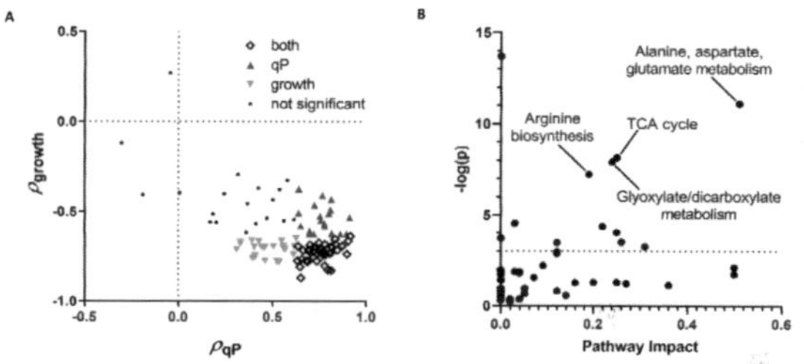

Figure 2. (A) Correlation analysis on BioCAn-annotated features showed significant correlations for 80 metabolites with growth, qP, or both. (B) The set of metabolites correlated with qP, including those also correlated with growth, were revealed through pathway analysis to be enriched in four main pathways.

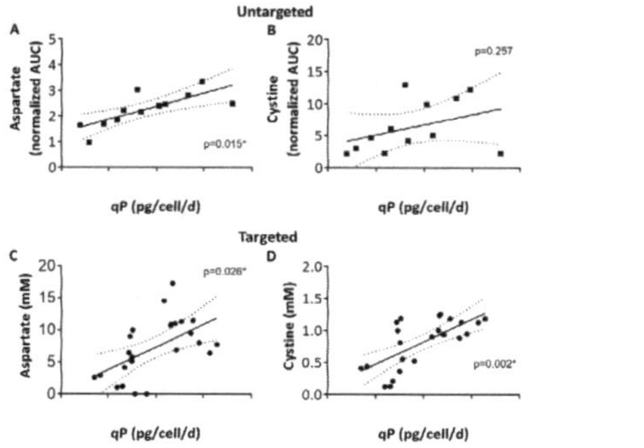

Figure 3. Metabolites found to be correlated with qP in the untargeted experiment (A, B) with the first set of 12 clones were used in a targeted experiment (C, D) with a different set of 12 clones. Compounds A and B were found to be significantly positively correlated with qP in the targeted experiment.

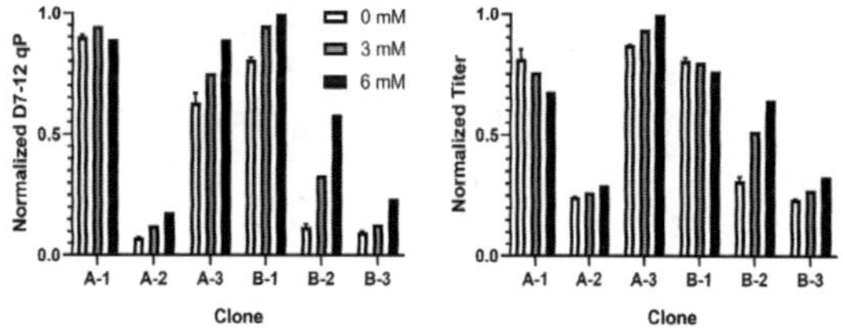

Figure 4. (A) Five out of six cell lines, each expressing one of two monoclonal antibodies, showed a dose-dependent increase in specific productivity upon addition of compound C in bioreactors. (B) Although some cell lines experienced a slight decrease in peak VCD, four of the cell lines showed an overall increase in final volumetric titer.

602300 Intercellular Communication between Breast Cancer Cells and Adipose-Derived Stem Cells Using a Novel 3D-Printed Plate Insert Interfacing with an Agarose Hydrogel Layer

Abstract:

Intercellular communication is a physiological phenomenon crucial to development, disease progression, and wound healing in both multicellular and unicellular organisms. Cells secrete signaling molecules (e.g. growth factors, cytokines, hormones) into the extracellular space which are detected by neighboring cells to influence proliferation, migration, or apoptosis. Intercellular communication has been primarily investigated using the Transwell assay by co-culturing two different cell types in two separate chambers. This approach has several limitations including: (i) an inability to directly visualize the cells, (ii) the invasion of the cells in the top chamber into the polycarbonate membrane, and (iii) very low cellular yields for post-experimental analysis for PCR and western blot analysis. While microfluidic devices have garnered significant attention in recent years to overcome these limitations, they are still limited by total cellular yield which limits their use for common downstream analysis. To overcome these drawbacks, we developed a novel co-culture approach that utilizes a 3D-printed plate insert for a 10 cm petri dish that incorporates an agarose hydrogel layer to serve as a semi-permeable membrane to create two distinct culture chambers. The agarose hydrogel is coated with poly-D-lysine to facilitate cellular attachment, spreading, and growth of adherent cell lines in the top layer which physically separates the two culture regions, yet chemically connects them by allowing the mass transfer of biomolecules between the top and bottom layer. The approach design is user-friendly, inexpensive, and allows for dynamic and tunable co-culture conditions while preserving similar growth trends as compared to commonly used tissue culture plastic (TCP). Mass transfer experiment and COMSOL simulations confirm the rapid diffusion of biomolecules and oxygen across the hydrogel approaching a steady-state within ~12-18 hours. The plate insert approach allowed for the direct visualization of cell lines in both the top and bottom chambers facilitating the study of independent cell growth and morphology. A proof-of-concept study using the plate insert was performed to investigate intercellular communication between breast cancer cells and adipose-derived stem cells (ASCs). Recent studies have shown that ASCs play a critical role in promoting metastasis and drug resistance in breast cancer patients. The plate insert was used to study the interactions between two different breast cancer cell lines, MDA-MB-231 (triple-negative)and MCF7 (ER+), co-cultured with ASCs. Using this novel co-culture approach, enhanced cellular growth and overexpression of EMT-associated genes were identified in the MDA-MB-231 cell line when they were co-cultured with ASCs. Moreover, increased drug resistance in the MCF7 cell line in the presence of ASCs was observed through increased cell proliferation. The co-culture approach allowed for the independent harvesting of both cell lines in numbers sufficient for downstream PCR and Western Blot analysis to identify changes in the genotypic and phenotypic signature of cancer cells while this approach does not result in the infiltration of cells from the top layer to the bottom, preventing the cross-mixing of cells.

598619 Characterization of Anti-Fusogenic Monoclonal Antibodies Against Human Coronaviruses

Abstract:

Coronaviruses (CoVs) cause a variety of respiratory and/or enteric tract infections and are noted for their zoonotic potential. The pandemic caused by the severe acute respiratory syndrome CoV-2 (SARS-CoV-2) as well as outbreaks caused by SARS-CoV and MERS-CoV have highlighted the dangers that these zoonotic CoVs pose to human health. CoVs are especially noted for their ability to transmit across species due to variety of strategies they possess in infecting cells. As an example, the CoV S protein, which facilitates receptor binding and membrane fusion, can be cleaved by distinct proteases to expose a fusion peptide (FP) that directly mediates fusion at the plasma membrane surface or at an endosomal membrane. Recently solved Cryo-EM structures have shown that the pre-fusion CoV FP region is exposed and can elicit broadly neutralizing antibodies as it is a well conserved domain across the CoV family. Thus, we designed a novel S antigen incorporating the SARS-CoV pre-fusion FP region as a strategy to generate neutralizing antibodies. This antigen was injected into mice to elicit an immune response and collect antibodies. Functional characterization of these antibodies has determined that they only block SARS-CoV pseudoparticle infection when present during the viral-host cell membrane fusion step. Furthermore, since the CoV FP is well conserved across the CoV family, these antibodies were shown to be cross-reactive against MERS-CoV pseudoparticle infection. We believe this antigen design may present novel strategies towards addressing SARS-CoV-2 and future CoV outbreaks.

608607 Light-Inducible Regulation of Autophagy in Chinese Hamster Ovary Cells for Production of Therapeutic Proteins

Abstract:

Autophagy is a potential tool to significantly increase the production of therapeutic proteins. As a major pathway in the cellular energy balance, autophagy is an evolutionarily conserved process by which proteins, organelles, and lipid matter are recycled into amino acids and free fatty acids to synthesize new cellular components. Several groups have demonstrated that modulating autophagy significantly influences therapeutic protein production[1,2]. However, precise regulation of autophagy for higher titers remains challenging. Current methods of modulating autophagy lack temporal control or specificity to the autophagy pathway, and the way in which autophagy controls protein synthesis is not well-understood. Thus, developing a method for sensitive control over the process is critical for elucidating the relation between autophagy and protein production and effectively regulating autophagy for higher titers. We are constructing light-inducible systems for transcriptional and direct regulation of autophagy in Chinese Hamster Ovary (CHO) cells, which are the major mammalian cell line for industrial therapeutic protein production.

The transcriptionally regulated system enables light-inducible gene expression by using Clustered Regularly Interspaced Short Palindromic Repeats-dCas9 (CRISPR-dCas9), a modular system that can be used to regulate any gene of interest[3]. We have demonstrated the tunability of the system using green fluorescent protein (GFP), achieving up to five-fold higher GFP expression for activated cells (Figure 1). We are currently using the system to express constitutively active mutants of AMP-activated protein kinase (AMPK) and mechanistic target of rapamycin 1 (mTORC1), which are well known inducers and inhibitors of autophagy, respectively.

We have also designed a light-inducible system which specifically regulates subcellular mTORC1 localization. Localization of mTORC1 to the lysosome has been shown to be sufficient for its activation[4], after which it inhibits autophagy. Thus, photoregulation of mTORC1 localization to the lysosome can be used to control autophagy. Compared to transcriptional regulation, direct regulation of protein activity has the advantage of eliminating lag-times of the system due to the time taken to express the protein when turning the system on and the half-life of the protein when turning the system off.

Using the two light-inducible systems, we are investigating the effects of AMPK and mTORC1 activity on autophagy rate and protein synthesis. The specificity of the systems allow precise control over autophagy, which will facilitate maximizing titers and advance knowledge on how autophagy regulates protein production.

References

1. Nasseri, S. S. *et al.* Increased CHO cell fed-batch monoclonal antibody production using the autophagy inhibitor 3-MA or gradually increasing osmolality. *Biochem. Eng. J.* (2014) doi:10.1016/j.bej.2014.06.027.
2. Lee, J. S. & Lee, G. M. Rapamycin treatment inhibits CHO cell death in a serum-free suspension culture by autophagy induction. *Biotechnol. Bioeng.* **109**, 3093–3102 (2012).
3. Polstein, L. R. & Gersbach, C. A. A light-inducible CRISPR-Cas9 system for control of endogenous gene activation. *Nat. Chem. Biol.* **11**, 198–200 (2015).

4. Sancak, Y. *et al.* Ragulator-Rag complex targets mTORC1 to the lysosomal surface and is necessary for its activation by amino acids. *Cell* **141**, 290–303 (2010).

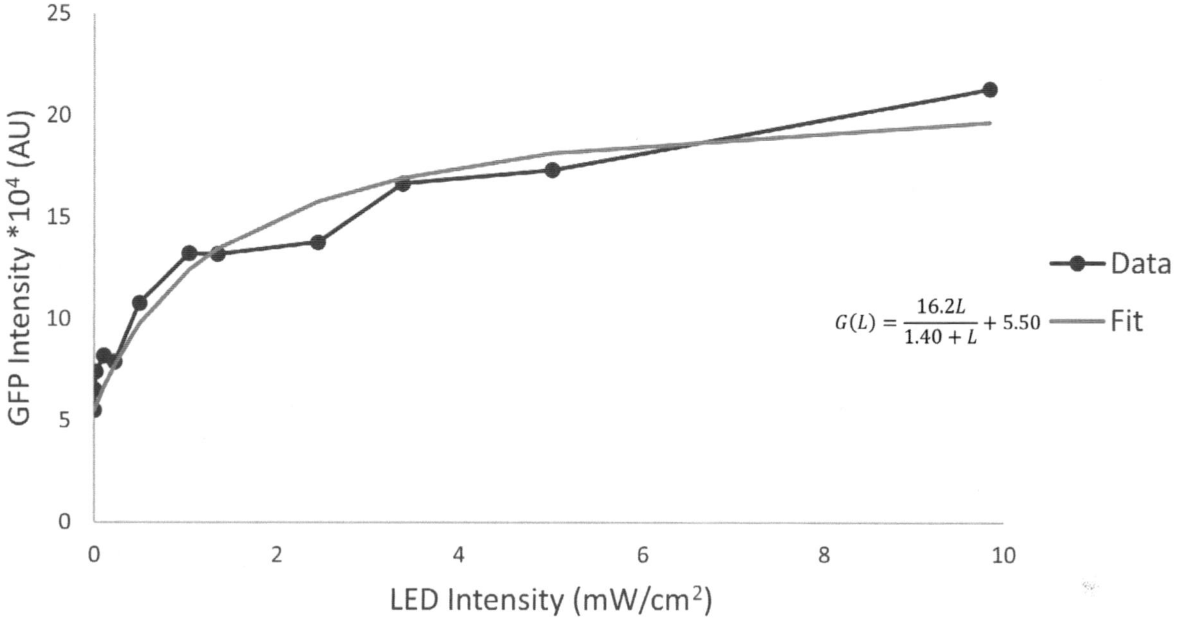

Figure 1. Steady-state GFP fluorescence intensity as a function of LED intensity.

610429 Accelerated Virus-Less Generation of Stable Insect Sf9 Cell Lines for High-Yield Production of Influenza Vaccines

Abstract:

Current manufacturing of vaccines in insect Sf9 cell line relies on the Baculovirus Expression Vector System, which is limited by virus-induced cell lysis. Thus, there is an interest in the field towards virus-less gene expression to simplify the costly product purification process. Current virus-less methods are based on random integration of the gene of interest into the genome, which generates a heterogeneous cell population with a wide distribution in both protein expression and growth rate. This population is unstable over time hampering the reproducibility of vaccines production. Therefore, a high-yield stable cell line is typically isolated by limiting dilution, a low-throughput process that can take up to 8 weeks. To accelerate the generation of stable cell lines, a potential solution is to use site-specific genome editing tools such as CRISPR to rapidly create more homogenous, high-yield populations. However, the lack of genomic information of the Sf9 cell line has hindered the design of key CRISPR components and limited its utilization.

Taking advantage of the recently published Sf9 genome, we not only knocked out the fused lobes gene to enable expression of human-like glycoprotein, but also achieved targeted insertion of a short DNA sequence. As in other higher eukaryotes, however, a substantial proportion of CRISPR-mediated double stranded breaks resulted in non-templated repair by nonhomologous end joining (NHEJ). To increase the insertion efficiency, we aimed to knock out this competing repair pathway. We successfully employed bioinformatics tools and identified hypothetical genes responsible for NHEJ. More importantly, we confirmed their biological function, as their knockout via CRISPR resulted in stable Sf9 cell lines with improved efficiency in site-specific gene integration through homologous directed repair (HDR). These engineered Sf9 cell lines have enabled rapid generation of high-yield stable cell lines without the need for limiting dilution. This system facilitates high-throughput production of recombinant HA to develop improved vaccines for each year's influenza season and offers an attractive platform to advance the development and manufacturing of other protein therapeutics.

614784 Single-Cell Multi-Omics Landscape of CD19 CAR-T Cell Infusion Products Predicts Long-Term Responses in Pediatric B Cell Acute Lymphoblastic Leukemia (Invited Speaker)

Abstract:

Abstract: Chimeric antigen receptor (CAR)-engineered T cell therapy has demonstrated high response rate in the treatment of B cell acute lymphoblastic leukemia (B-ALL), but ~10% of patients still do not respond and more than 50% of patients relapse within a year. How to predict the clinical response prior to treatment remains a challenge for CAR-T cell therapy. We investigated single-cell transcriptome and surface epitome sequencing of CD19 CAR-T cell infusion products at the basal state, upon antigen-specific stimulation, upon CD3/CD28 TCR stimulation, and under a control stimulation with mesothelin antigen presenting cells. We sequenced 75 samples including replicates and total 101,326 single cell transcriptomes with 18 proteins from 12 pediatric ALL patients treated with CD19-BB-z CAR-T therapy at UPenn. Computational data analysis revealed a highly heterogenous activation landscape of infusion products and identified multiple molecular signatures to not only predict non-responders from responders but also long-term durable responders vs recurrent patients (relapse). These signatures are mainly associated distinct T cell subset cytokine profiles such as Th1, Th2, and inhibitory factors as well as T cell memory and metabolic activities. How to predict long term responses is a major challenge in the field of CAR-T therapy. Our study provides an informative platform to discover predictive biomarkers and identify the first cellular signature that can predict relapse in patients with B-ALL.

607468 Controlled Release of NKT Cell Agonist and Non-Replicating Pathogen for Single-Dose Vaccination

Abstract:

Introduction. Many vaccines require multiple doses scheduled months or years apart in order to impart successful immunization [1]. This is not compatible with the immediate vaccination timelines of potential biological warfare or worldwide pandemics. Plus, is not feasible for people living in poverty-stricken areas. To address the need for single-dose vaccines, this work examines two avenues: (1) use of an adjuvant, alpha-Galactosylceramide (alpha-GalCer), which can stimulate activity of natural killer T (NKT) cells. NKT cells are an unusual lymphocyte subset that can be therapeutically activated with superagonists to potently boost vaccine-specific CD8 cytotoxic T lymphocytes (CTLs) with long-lasting memory and the production of antibodies; and (2) use of polymeric microparticles for controlled release of adjuvant and antigen for primer and booster doses.

Methods. In vivo studies in mice were performed where mice were injected with PK8 (influenza) and alpha-GalCer separately and combined to determine the effectiveness of alpha-GalCer as an adjuvant. Intraperitoneal and intramuscular injection methods were used and compared. Blood samples were collected every 15 days following vaccination and analyzed for virus specific antibodies by ELISA. PLGA microparticles were synthesized via a double emulsion method [2]. Microparticles were synthesized with encapsulated fluorescein (as an adjuvant model) and gold nanoparticles (as a vaccine model). Microparticles were then analyzed for size using microscopy and ImageJ sizing analysis, loading via indirect method, and release. To analyze release, samples were suspended in PBS and aliquots were assayed using a fluorometer or spectrometer at specific time points. Parameters of the double emulsion were varied and analyzed for their effect on size, loading, and release. Then, microparticles with encapsulated H1N1 vaccine and alpha-GalCer were synthesized and analyzed. H1N1 release was measured via ELISA and alpha-GalCer release was measured using LC-MS/MS as in reference [3]. Microparticles were also imaged using scanning electron microscopy at different timepoints during release to access particle breakdown.

Results. Alpha-GalCer was shown to enhance IgG1 levels when co-administered with PR8, indicating that alpha-GalCer stimulates Th2 immune responses (see Figure 1).

Average loading and sizes can be seen in Table 1. For fluorescein loaded microparticles, 65:35 PLGA resulted in higher loading than 50:50 PLGA. Lower Pluronic F68 in the W2 phase resulted in higher loading (0.25%). 65:35 PLGA resulted in larger microparticles whereas amount of Pluronic in the W2 phase did not significantly affect particle sizes.

The 50:50 PLGA released all encapsulated fluorescein after approximately 130 days. Amount of pluronic F68 used in the formulation did not significantly affect the release profile, which makes sense since it also did not have a significant effect on microparticle size. The 65:35 PLGA has only been releasing for 83 days and has not yet released all of its encapsulated payload.

For microparticles loaded with gold nanoparticles, the amount of gold nanoparticles loaded in the W1 phase was shown to not affect loading significantly. 0.5% pluronic F68 in the W2 phase resulted in the highest loading. Release for gold nanoparticles was not shown in the 60 days it was measured, indicating that the particles will not release any gold nanoparticles until they are fully degraded.

Alpha-GalCer release was shown to have an initial "burst" release followed by minimal diffusional release for 80 days (see Figure 2).

Vaccine release was shown to also have an initial "burst" release followed by minimal diffusional release for 35 days as shown in Figure 3. The plots were not significantly different based on the concentration of vaccine in the W1 phase (0.25-1 mg/ml). Release is still being measured and quantified.

SEM images shown in Figure 4 before release and after 65 days showed that pores begin to form in the microparticles.

Conclusions and future work. Alpha-GalCer was shown to stimulate IgG1 immune responses *in vivo* when co-administered with PR8. Microparticles were shown to load and release all molecules when using 0.25% pluronic F68 as a stabilizer in the double emulsion. Microparticles were approximately 1-5 μm in size. Release profiles following an initial release of therapeutic followed by minimal passive release. All of the fluorescein was released within 130 days. Release of other molecules, vaccine and alpha-GalCer is still ongoing. SEM images showed that pores begin to form in the microparticles after 65 days of release. After initial release profiles are acquired, the microparticle synthesis will be optimized using different PLGA ratios or drug loading to achieve desired release profiles of burst release, followed by another burst release after 1 month. Optimized microparticle formulations will be tested *in vivo*.

References:

[1] S. Lu, "Heterologous prime-boost vaccination," *Curr. Opin. Immunol.*, vol. 21, no. 3, pp. 346–351, 2009, doi: 10.1016/j.coi.2009.05.016.

[2] N. Y. Martinez, P. F. Andrade, N. Durán, and S. Cavalitto, "Development of double emulsion nanoparticles for the encapsulation of bovine serum albumin," *Colloids Surfaces B Biointerfaces*, vol. 158, no. 1, pp. 190–196, 2017, doi: 10.1016/j.colsurfb.2017.06.033.

[3] K. Kaneko, A. McDowell, Y. Ishii, and S. Hook, "Selective quantitation of the incorporation of the immunomodulator α-galactosylceramide in liposomes using LC-MS/MS," *Int. J. Mass Spectrom.*, vol. 392, pp. 96–101, 2015, doi: 10.1016/j.ijms.2015.09.016.

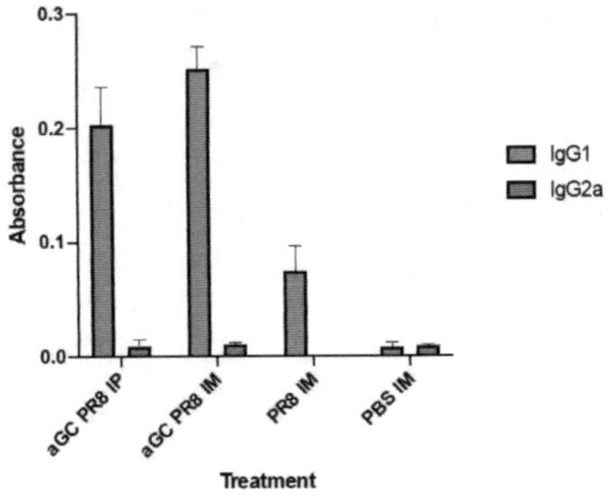

Figure 1. IgG levels induced in mice with administration of (1) alpha-GalCer and PR8 by intraperitoneal route; (2) alpha-GalCer and PR8 by intramuscular route; (3) PR8 only by intramuscular route; (4) PBS control

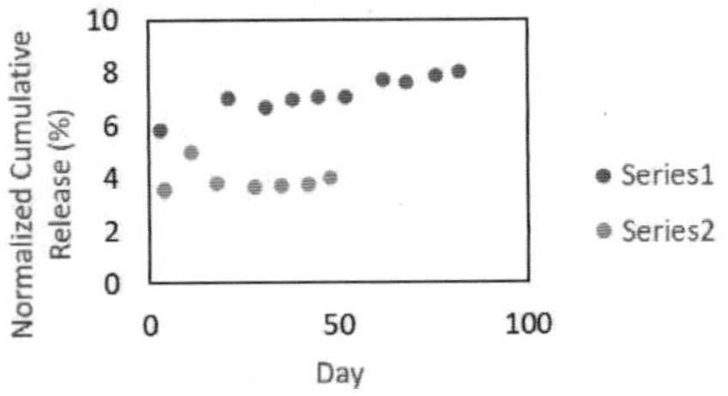

Figure 2. alpha-GalCer release from PLGA microparticles. Series 1 and 2 represent 2 independent samples.

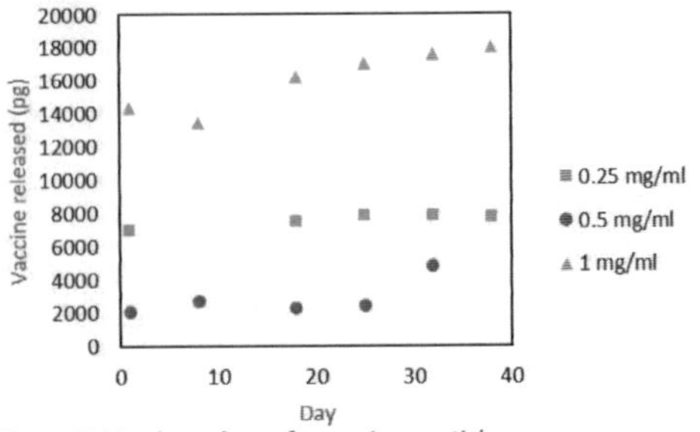

Figure 3. Vaccine release from microparticles.

Table 1. Average loading and size data for PLGA microparticles with fluorescein.

	Average Loading (%)	Standard Deviation Loading (%)	Average Size (µm)	Standard Deviation Size (µm)
50:50 PLGA 0.25% F68	25.8	1.8	5.5	4.7
50:50 PLGA 0.5% F68	26.1	4.6	4.9	3.7
50:50 PLGA 1% F68	19.6	2.4	5.1	4.6
65:35 PLGA 0.25% F68	91.5	5.9	7.0	7.8

Figure 4. PLGA microparticles before release (A) and after 65 days of release (B).

609537 A Single-Shot Vaccination Using Stimuli-Responsive Nanoparticles for the Co-Delivery of Multi-Functional Payloads

Abstract:

In addition to the prolonged immune activation using polymeric antigen carriers, multi-faceted stimulation is crucial to the development of efficient and safe vaccines that reduce or circumvent the need for multiple doses. In this study, we developed a single-shot vaccine platform using stimuli-responsive, biodegradable nanoparticles made of acid-cleavable polyketals for the encapsulation of a broad range of antigens such as GFP, OVA, CBU1910 from *Coxiella burnetii*, and HA's from influenza viruses. Surface-initiated photopolymerization synthesized the vaccine formulation that trapped multi-functional payloads of antigenic proteins and immunostimulatory adjuvants in the polymer network via attractive electrostatic interactions or specific affinity between His-tag (proteins) and Nickel-NTA (polyketal polymers). This vaccine design enabled the antigen release into the cytoplasm from the mildly acidic phago-lysosomes in antigen presenting cells (APCs), resulting in efficient and directed antigen presentation to T cells and the production of immunostimulatory cytokines (e.g., interferon-gamma [IFN-γ] and interleukin-4 [IL-4]), upon a single-shot administration. The novel vaccines developed in this study activated both humoral and cellular immunities and are capable of inducing desired immune response to bacterial and viral infections and can potentially be extended to other infectious diseases. This presentation will cover vaccine design, biomaterials synthesis, vaccine carrier preparation and characterization, and in vitro and in vivo validation on immune activation, and highlights the roles of interdisciplinary approaches and tool sets from engineering, chemistry, and biology in addressing a key challenge in human health.

609917 Inhibition of Glycolysis in the Presence of Antigen Generates Antigen Specific Treg Responses in Rheumatoid Arthritis

Abstract:

Introduction: T cells regulate immune homeostasis by generating specific inflammatory or anti-inflammatory immune responses. However, immune dysregulation and over activation of inflammatory T cells can induce autoimmune diseases such as Rheumatoid Arthritis (RA). Antigen-specific enrichment of immunosuppressive regulatory T-cells (Tregs) and reduction of antigen-specific T-effector cells (Teff) can prevent the progression of RA. Notably, dendritic cells (DCs) are effective modulators of antigen-specific T-cell responses. Interestingly, activated DCs and T-effector cells, such as T-helper 1 (Th1) and T-helper 17 (Th17), utilize glycolysis for their energy needs while Tregs rely on fatty acid oxidation for their energy needs. Current RA therapies do not target the root cause of RA or address the antigen specific immune responses against synovial tissue. We hypothesize that blocking DC glycolysis and simultaneous antigen expression may lead to the generation of tolerogenic DCs (tDCs) and subsequent antigen specific Treg induction. Here we show a method where local delivery of an antigen specific therapy can generate antigen-specific Tregs and reduce antigen-specific Th17, Th1 and autoantibodies in a collagen-induced arthritis (CIA) mouse model of RA.

Materials and Methods: Metabolite-based polymeric particles were generated by oil in water emulsions. Scanning Electron Microscopy and Dynamic Light Scatter determined the size of microparticles (MPs). Nuclear Magnetic Resonance and release kinetics assessed the rate of polymer degradation. In vitro studies of immunofluorescence, enzyme-linked immunosorbent assay and flow cytometry determined MP modulation of adaptive immune responses. A DBA/1j CIA mouse model was injected with 1mg of MPs for prevention and treatment studies. These mice were then sacrificed on day 30, and the blood, spleen and lymph nodes (LNs) were harvested to analyze the presence of antigen specific autoantibodies, T cells, and antigen specific T cells.

Results and Discussion: The MPs were determined to be about 1µm and were shown to degrade in a sustained manner. Flow cytometry confirmed that the intracellular delivery of PFK15 and bc2 via paKG particles induced tDCs. In vivo studies demonstrated that the PFK15+bc2 treatment group had lower thickening of the ankle, less antigen specific autoantibodies and an increased population of proliferating Tregs. Additionally, the PFK15+bc2 prevention group had an increased population of antigen specific Tregs and number of proliferating Tregs in the popliteal LNs. Notably, the no treatment group and PFK15+bc2 treatment groups had elevated levels of antigen specific Th1 and Th17 pro-inflammatory cells compared to the prevention group, which had undetectable levels.

Conclusions: The immune non-activating metabolite-based MPs modulate DC function and subsequently alter antigen specific T cell immune responses in mice. This data suggests that the local delivery of PFK15+bc2 via paKG MPs is a viable strategy to prevent RA progression and development.

602696 Epigenetic Modulation of Inflammatory T Cells in Autoimmune Disease

Abstract:

Introduction: A shared hallmark of gut autoimmunity is the marked decrease in the concentration of short chain fatty acids (SCFAs), which are metabolic products of gut microbiota and important immunomodulators that regulate gut health. The four and five carbon SCFAs, butyrate and pentanoate respectively, are two of the primary immune modulating agents in the gut, and potently inhibit the activity of histone deacetylases (HDAC) and thereby regulate gene transcription of key transcription factors and cytokines by modulating chromatin structure in a concentration dependent manner. HDAC inhibition by SCFAs suppress the production of pro- and enhance anti-inflammatory cytokine production by T cells in the gut. The characteristic inflammation in gut autoimmune disorders is caused, in part, by the upregulation of pro-inflammatory T cells, often accompanied by a deficiency in the number and function of immunosuppressive regulatory T cells (T_{regs}). An increase in the concentration of pro-inflammatory cytokines such as IL-6, interferon-gamma (IFN-γ) and tumor necrosis factor-alpha (TNF-α) leads to differentiation of naïve CD4+ T cells into pro-inflammatory IL-17A producing T helper (Th) 17, and IFN-γ-producing Th1 cells. We hypothesized that promoting HDAC inhibition with SCFA supplementation might restore the regulation of cytokine production by T cells, downregulate the pro-inflammatory mediators and restore the balance in T cell subsets.

Materials and Methods: To test the hypothesis, we isolated C57Bl/6J mouse T cells from the spleen and cultured them in media containing butyrate and pentanoate. T-cell media, consisting of RPMI cell media prepared with 10% Fetal Bovine Serum was used to culture the cells. In order to simulate the inflammatory milieu, the wells of T cells were activated with anti-CD3 and, anti-CD28 antibodies in media with Th17-inducing cytokines. Cells were counted at predetermined time intervals and stimulated at predetermined time intervals with phorbol myristate acetate and ionomycin to induce cytokine expression. Cytokine transport was blocked with brefeldin A and pro and anti-inflammatory cytokine expression was analyzed using flow cytometry. Tregs and Th17 cells were quantified based on FoxP3 and ROR-γt expression respectively.

Results and Discussion: Butyrate reduced IL-17A expressing CD4+ mouse T cells to 10%, compared to 17% in Th17-inducing control conditions. In Th17-inducing conditions alone, 25% of cells expressed pro-inflammatory TNF-α. Addition of Butyrate to the cell culture medium suppressed this expression to 15%. We measured the concentration-dependence of T cell expansion on pentanoate in non-inflammatory conditions. With 1 mM pentanoate, we measured 5-fold expansion over 4 days. 2 mM pentanoate induced 11-fold expansion of T-cells over 4 days, compared to 2.5-fold expansion in control conditions. 3 mM induced 6-fold expansion over 4 days, indicating that there is an optimal concentration for maximizing T cell proliferation. In non-inflammatory conditions, addition of 1 mM butyrate resulted in 40% expression of IL-10, an immunosuppressive cytokine that is associated with T_{reg} function, compared to ~1% in the control group. Pentanoate induced 20% IL-10 expression. However, upregulation of IL-10 was abrogated in Th17 inducing conditions, with IL-10 expression falling to 2%. To restore the production of anti-inflammatory IL-10, we tested combinations of epigenetic modulators by introducing all-trans retinoic acid (ATRA), a clinically used DNA methylating agent. Strikingly, ATRA stabilized the T_{reg} phenotype, increased the proportion of T_{regs} 3.5-fold compared to the control group in Th17 inducing conditions, and reduced IL-17A expression to 10% compared to 28% in control conditions. ATRA and butyrate in combination reduced the expression of IL-17A to 1%.

Conclusions: Our findings suggest that SCFAs potently modulate the activity of T cells and may enhance the functionality of T_{regs}. The metabolic modulation by pentanoate is responsible for both increased expansion and IL-10 expression, an effect which could be useful in combination with butyrate which can enhance T_{reg} differentiation. However, the abrogation of SCFA mediated IL-10 production in simulated inflammatory conditions suggests that restoring immune homeostasis in established disease may require a multifactor approach. The combination of ATRA and butyrate significantly promotes the differentiation of naive T cells to T_{regs} and inhibits IL-17 production in Th17 inducing conditions and may be a means to restore immune balance in gut-based autoimmune disorders.

609116 Erythrocyte Anchored Systemic Immunotherapy: Metastasis Driven Cancer Immunotherapy for Local and Systemic Tumor Suppression

Abstract:

Immunomodulation in tumor microenvironments is a promising strategy to employ body's own immune system to treat tumors in patients with advanced cancer stages like metastasis. However, achieving site-specific local immunomodulation post systemic administration has been a challenge. Here we describe a new strategy, Erythrocyte Anchored Systemic Immunotherapy (EASI), that enables local immuno-restoration in hard-to-reach cancer metastatic sites. Briefly, by engineering their
interactions with erythrocytes, ImmunoBait (chemokine nanoparticles) anchor onto erythrocytes, hitch
a ride and are precisely deposited in the vicinity of target metastatic sites. ImmunoBait gradually releases chemokine, modulates the local microenviroment by restoring the chemokine gradient, subsequently leading to a significant enhancement in the infiltration of effector immune cells. Our in vivo results based on a spontaneous breast cancer lung metastasis model showed that EASY remarkably inhibited the progression of lung metastasis and significantly extended the animal survival.
Moreover, the immune response induced by EASY significantly suppressed the growth of re-challenged
distant tumors. Our findings indicate that EASY is a potent strategy for immuno-modulation to combat metastatic cancers.

613340 Biomimetic Nanoparticles for Multi-Antigenic Vaccine Delivery (INVITED SPEAKER)

Abstract:

The recent success of immunotherapies has highlighted the power of leveraging the immune system in the fight against cancer. In order for most immune-based therapies to succeed, T cell subsets with the correct tumor-targeting specificities must be mobilized. When such specificities are lacking, providing the immune system with tumor antigen material for processing and presentation is a common strategy for stimulating antigen-specific T cell populations. While straightforward in principle, experience has shown that manipulation of the antigen presentation process can be incredibly complex. Herein, I report on a biomimetic nanoparticulate anticancer vaccine formulation capable of activating multi-antigenic immunity. The design leverages the unique advantages of recent nanoparticle technology, delivering both syngeneic cancer material along with a potent immunological adjuvant in a format that promotes effective antigen presentation. The final formulation is capable of generating strong antitumor responses in vivo and can work together with other immunotherapies such as checkpoint blockades to help control tumor growth.

604884 Microbiome-Based Mixed-Acid Fermentation of Lignocellulosic Biomass and Wastewater Activated Sludge Under Mesophilic and Thermophilic Conditions

Abstract:

Lignocellulosic biomass has been considered as an abundant renewable source for the production of biofuels and biochemicals. Robust microbial communities in activated sludges could convert the carbohydrates from lignocellulosic biomass to medium chain carboxylic acids, which can be subsequently utilized as precursors to produce more valuable bioproducts. In this study, wastewater activated sludges were inoculated to pure sugar solutions and sulfite prehydrolysate for anaerobic fermentation, respectively. The fermentation has been conducted under both mesophilic (35 °C) and thermophilic (55 °C) conditions. The result indicated that lactic acid was the major product for pure sugar fermentation at 55 °C, while more medium chain carboxylic acids (such as acetic acid, propionic acid and butyric acid) were generated at 35 °C. Interestingly, it has been found that the addition of peptone could significantly promote the carboxylic acid (especially acetic acid and propionic acid) production rate and affect the acid compositional profiles. For the sulfite prehydrolysate fermentation, similar stimulation effect was observed on acid production at 35 °C, however, the sugars consumption was significantly limited at 55 °C. The reason is that the carbonyl inhibitors in the prehydrolysate showed a strong inhibitory effect at a higher temperature. Additionally, we plan to use the sulfite pretreated aspen as the substrate to investigate anaerobic digestion of woody biomass in our future studies. It is hypothesized that the supplementary of cellulase could accelerate substrate digestion, as well as acid fermentation. To better understand the microbiome-based fermentation process, whole genome sequencing (WGS) analysis will be performed to identify the predominant microbial communities associated with the carboxylic acid productions.

604659 On the Quantitative Evaluation of the Impact of the Concentration and Distribution of Fat on the Microbial Dynamics within Viscoelastic Triphasic Food Model Systems

Abstract:

Bacteria play a key role in food production for many different reasons. Some food related bacteria are beneficially used as starter cultures in the dairy industry, other spoilage bacteria cause organoleptic changes to the food product and other pathogenic bacteria are harmful and can lead to severe disease or death. To achieve microbiologically safe food products, it is important to understand the bacterial dynamics in single culture, but also in co-culture, as in a food system multiple bacteria can co-exist and bacterial communities behave vary differently to single cultures (Moons, Michiels and Aertsen, 2009; Fox *et al.*, 2014; Zilelidou and Skandamis, 2018). Additionally, the bacterial behaviour is influenced by the environment. Food products are generally very complicated biochemically, structurally and biophysically. Therefore, to obtain a fundamental understanding of intra-species interactions as affected by the food environment, it is important to design simpler *in vitro*models. Most quantitative studies on food related microbial kinetics (growth or inactivation) in *in vitro*models are conducted in liquid broths and very few in simple mono-phase homogenous gel systems (with gelling agents such as agar, gelatine etc.) (Fröhling*et al.*, 2012; Baka*et al.*, 2017;Fei*et al.*, 2020). Those studies have showed that the bacterial behaviour is different in liquid systems compared to solid systems, as the availability of nutritional components and oxygen changes and so does the bacterial communication (Wimpenny*et al.*, 1995; Smet*et al.*, 2017;Heir*et al.*, 2018). Most food products are solid or solid-like and it is therefore important to account for more complex composition and structural properties of food models.

In this study, building on our recently developed biphasic protein/polysaccharide food model (Costello*et al.*, 2018, 2019), we have developed triphasic food models enriched with fat, as studies in real food have revealed that the fat concentration has a significant impact on the bacterial behaviour (Brocklehurst and Wilson, 2000; Skandamis and Nychas, 2012).

Our novel triphasic systems consist of Xanthan gum, Whey protein and vegetable oil. Various oil concentrations in the range 10-60% were considered (range similar to real food products). A microscopic comparison showed that these novel triphasic systems are structurally similar to real foods; like soft cheeses or meat patés. Furthermore, rheological analysis of the food models showed that the oil contentis a major factor affecting the rheological and structural behaviour, as well as the physicochemical properties and oil droplet size.

Quantitative analysis of the microbial dynamics in the tri-phase systems with oilconcentration of 0%, 20% and 60% was conducted. More specifically, the growth of the foodborne gram-positive pathogen *Listeria monocytogenes*, the gram-negative pathogen *Escherichia coli*, and the spoilage bacteria *Pseudomonas aeroginosa* was monitored at 37 °C, while the growth of the starter organism *Lactococcus lactis* was monitored at 30 °C. Furthermore, advanced confocal laser scanning microscopy(CLSM) was used for the spatial determination of the food model components and for the analysis of the distribution of bacterial colonies on the food models.

Overall, the oil concentration in the triphasic model did not affect the macroscale/macroscopic microbial growth kinetics (as evaluated by the Baranyi Roberts growth model) for all bacteria under study. However, at the microscopic scale, generally, the bacterial colonies reduced in size with increasing oil concentration. The higher oil concentration resulted in space limitations and diffusional limitations of nutrients. Furthermore, in terms of growth location most bacterial colonies/aggregates were located close to the oil droplets. Due to these microscopic differences, different levels of cell-cell and colony-colony interactions take place for different structural configurations of our models. This can allow bacteria to send signalling molecules, such as the so-called quorum sensing molecules which can lead to different levels of stress adaptation. Additionally, the size of the colonies can have a direct impact on the efficiency of mild preservation methods.

In conclusion, our results indicate the importance of accounting for food biochemical composition and microstructural complexities when monitoring food related bacteria. A multi-level complex analysis on both macro and microscale enables a better prediction of bacterial interactions: whilst the macroscopic microbial kinetics can be unaffected by small environmental changes, on a microscale level small alterations of the environment can result in substantial changes in the bacterial interactions, which can further impact the efficiency and the design of food decontamination treatments.

ACKNOWLEDGEMENTS:

This research was supported by the Doctoral College and the Department of Chemical and Process Engineering of the University of Surrey, the EPSRC and the Royal Society. E.V. is grateful for a Royal Academy of Engineering Industrial Fellowship.

REFERENCES:

BARANYI, J. et al. (1994). 'A dynamic approach to predicting bacterial growth' International Journal of Food Microbiology, **23**, 277–294. doi: https://doi.org/10.1016/0168-1605(94) 90157-0.

Baka, M. *et al.*(2017) 'Dynamics of Listeria monocytogenes at suboptimal temperatures in/on fish-protein based model systems: Effect of (micro)structure and microbial distribution', *Food Control.* Elsevier Ltd, 73, pp. 43–50. doi: 10.1016/j.foodcont.2016.06.031.

Brocklehurst, T. F. and Wilson, P. D. G. (2000) 'The role of lipids in controlling microbial growth', *Grasas y Aceites*, 51(1–2), pp. 66–73. doi: 10.3989/gya.2000.v51.i1-2.408.

Costello, K. M. *et al.*(2018) 'Modelling the microbial dynamics and antimicrobial resistance development of Listeria in viscoelastic food model systems of various structural complexities', *International Journal of Food Microbiology*. Elsevier, 286(February), pp. 15–30. doi: 10.1016/j.ijfoodmicro.2018.07.011.

Costello, K. M. *et al.*(2019) 'A multi-scale analysis of the effect of complex viscoelastic models on Listeria dynamics and adaptation in co-culture systems', *AIChE Journal*, (August), pp. 1–15. doi: 10.1002/aic.16761.

Fei, C. *et al.*(2020) 'Nonuniform growth and surface friction determine bacterial biofilm morphology on soft substrates', pp. 1–11. doi: 10.1073/pnas.1919607117.

Fox, E. M.*et al.*(2014) 'Phylogenetic profiles of in-house microflora in drains at a food production facility: Comparison and biocontrol implications of listeria-positive and -negative bacterial populations', *Applied and Environmental Microbiology*, 80(11), pp. 3369–3374. doi: 10.1128/AEM.00468-14.

Fröhling, A. *et al.*(2012) 'Atmospheric pressure plasma treatment of Listeria innocua and Escherichia coli at polysaccharide surfaces: Inactivation kinetics and flow cytometric characterization', *Innovative Food Science and Emerging Technologies*. Elsevier Ltd, 13(JANUARY), pp. 142–150. doi: 10.1016/j.ifset.2011.11.002.

Heir, E. *et al.*(2018) 'Listeria monocytogenes strains show large variations in competitive growth in mixed culture biofilms and suspensions with bacteria from food processing environments', *International Journal of Food Microbiology*. Elsevier, 275(March), pp. 46–55. doi: 10.1016/j.ijfoodmicro.2018.03.026.

Moons, P., Michiels, C. W. and Aertsen, A. (2009) 'Bacterial interactions in biofilms', *Critical Reviews in Microbiology*, 35(3), pp. 157–168. doi: 10.1080/10408410902809431.

Skandamis, P. N. and Nychas, G. J. E. (2012) 'Quorum sensing in the context of food microbiology', *Applied and Environmental Microbiology*, 78(16), pp. 5473–5482. doi: 10.1128/AEM.00468-12.

Smet, C.*et al.*(2017) 'Impact of food model (micro)structure on the microbial inactivation efficacy of cold atmospheric plasma', *International Journal of Food Microbiology*. Elsevier B.V., 240, pp. 47–56. doi: 10.1016/j.ijfoodmicro.2016.07.024.

Wimpenny, J. W. T.*et al.*(1995) 'Submerged bacterial colonies within food and model systems: their growth, distribution and interactions', *International Journal of Food Microbiology*, 28(2), pp. 299–315. doi: 10.1016/0168-1605(95)00065-8.

Zilelidou, E. A. and Skandamis, P. N. (2018) 'Growth, detection and virulence of Listeria monocytogenes in the presence of other microorganisms: microbial interactions from species to strain level', *International Journal of Food Microbiology*. Elsevier, 277(October 2017), pp. 10–25. doi: 10.1016/j.ijfoodmicro.2018.04.011.

608875 Predictive Modelling of Minimal Food Processing Techniques

Abstract:

In recent years, the focus of food microbiology has been shifting from simply ensuring food safety and extending its shelf life to finding minimal processing techniques to preserve key food characteristics such as nutritional content, taste, texture and natural colour (Shah et al., 2019; Troy et al., 2016). Among novel minimal treatments that are being studied for the deactivation of harmful pathogens and food spoilage organisms are natural antimicrobials (e.g., nisin), ultrasound and cold atmospheric plasma (CAP) (Costello et al., 2019, 2018). The efficacy of each of these treatments is dependent on many factors such as food type, concentration (in the case of antimicrobials) or intensity of treatment as well as storage conditions (i.e. temperature and/or protective atmosphere). Moreover, treatments can be combined to achieve the desired level of decontamination (Berdejo et al., 2019; Ross et al., 2003). These aspects pose substantial challenges due to the range of potential food products, bacterial pathogens, treatments and conditions that must be explored to find optimal processing strategies while ensuring minimal detriment to food characteristics.

Most previous studies focused on the inactivation of pathogens by natural antimicrobials such as nisin, applied individually or in combination with ultrasound or CAP, have been conducted on real foods or in laboratory broths (for example, Koda et al., 2009; Stratakos and Grant, 2018). However, studies in real food products are useful only for the system studied, and liquid broth models do not account for structural effects of solid-like foods (Smet et al., 2018).

Our previous contributions addressed this issue through the development of a range of monophasic and biphasic food model systems using Xanthan gum (XG) and/or Whey protein isolate (WPI) and/or fat that mimic a variety of real food products and are stable at a range of temperatures. The advantage of these model food systems consists in high reproducibility of their structural and rheological properties that can also be readily tuned to mimic those of a wide variety of real food products.

Systematic studies of the combined effects of artificially added nisin and food model structure under different temperatures on the growth kinetics and spatial organization of *Listeria* revealed the strong influence of microscopic structural features (i.e., the distribution of the protein-polysaccharide phases) on colony size and distribution, apart from the more readily observable population-scale growth rate of the bacteria (Costello et al., 2019, 2018).

Despite the growing amount of evidence on the efficacy of different treatment strategies the available data and growth models for specific cases reported thus far are lacking a predictive capability across different foods, treatments and storage conditions.

In this work we develop a predictive phenomenological model describing the growth of *Listeria* in a unified formulation across a wide range of food products under a variety of conditions including treatments with nisin, ultrasound and CAP as well as temperatures. The main differentiating feature of this approach is that the effects of structural composition and corresponding rheological properties are incorporated into the model. In addition, the modelling approach can be readily extended to other pathogenic bacteria using a relatively small set of quantitative experimental observations under specific conditions.

The predictive modelling framework has the potential to accelerate the development of novel microbial decontamination treatments for food products, better assess food condition based on storage history as well as optimise storage and distribution protocols to ensure food safety and minimise costs.

References

Berdejo, D., Pagán, E., García-Gonzalo, D., Pagán, R., 2019. Exploiting the synergism among physical and chemical processes for improving food safety. *Curr. Opin. Food Sci.* 30, 14–20.

Costello, K., Gutierrez-Merino, J., Bussemaker, M., Ramaioli, M., Baka, M., Van Impe, J.F., Velliou, E.G., 2018. Modelling the microbial dynamics and antimicrobial resistance development of *Listeria* in viscoelastic food model systems of various structural complexities. *Int. J. Food Microbiol.* 286, 15–30.

Costello, K., Gutierrez-Merino, J., Bussemaker, M., Smet, C., Van Impe, J.F., Velliou, E.G., 2019. A multi-scale analysis of the effect of complex viscoelastic models on *Listeria* dynamics and adaptation in co-culture systems. *AIChE J.* e16761.

Koda, S., Miyamoto, M., Toma, M., Matsuoka, T., Maebayashi, M., 2009. Inactivation of *Escherichia coli* and *Streptococcus mutans* by ultrasound at 500 kHz. *Ultrason. Sonochem.* 16, 655–659.

Ross, A.I.V., Griffiths, M.W., Mittal, G.S., Deeth, H.C., 2003. Combining nonthermal technologies to control foodborne microorganisms. *Int. J. Food Microbiol.* 89, 125–138.

Shah, U., Ranieri, P., Zhou, Y., Schauer, C.L., Miller, V., Fridman, G., Sekhon, J.K., 2019. Effects of cold plasma treatments on spot-inoculated *Escherichia coli* O157:H7 and quality of baby kale (*Brassica oleracea*) leaves. *Innov. Food Sci. Emerg. Technol.* 57.

Smet, C., Baka, M., Dickenson, A., Walsh, J.L., Valdramidis, V.P., Van Impe, J.F., 2018. Antimicrobial efficacy of cold atmospheric plasma for different intrinsic and extrinsic parameters. *Plasma Process. Polym.* 15, 1–12.

Stratakos, A.C., Grant, I.R., 2018. Evaluation of the efficacy of multiple physical, biological and natural antimicrobial interventions for control of pathogenic *Escherichia coli* on beef. *Food Microbiol.* 76, 209–218.

Troy, D.J., Shikha, K., Kerry, J.P., Tiwari, B.K., 2016. Sustainable and consumer-friendly emerging technologies for application within the meat industry : An overview. *Meat Sci.* 120, 2–9.

604141 Gut Bacterial Product of Flavonoid Metabolism Exhibits Enhanced Aryl Hydrocarbon Receptor Agonist Activity in Caco2 Human Intestinal Epithelial Cells

Abstract:

Flavonoids, with over 6,000 unique structures occurring in nature [4] are the largest class of polyphenolic phytochemicals. Abundant in plants considered health-promoting, flavanoids' health benefits are associated with their antioxidant and anti-inflammatory properties [1], the potency of which vary widely even among flavonoids of the same subclass. Further, there is no consensus regarding the molecular mechanisms underlying flavonoid bioactivity.

In the intestine, one potential molecular target is the aryl hydrocarbon receptor (AhR), a nuclear receptor capable of binding both endogenous metabolites and xenobiotic chemicals to regulate inflammatory pathways [3]. Recent studies from our lab and others have shown that flavonoids can exhibit agonist or antagonist activities to induce or inhibit AhR-responsive genes [5]. Some flavonoids, however, show minimal activity, which is consistent with findings that the immunomodulatory effects of flavonoids are structure dependent.

As dietary compounds that are only partially absorbed, flavonoids are available for metabolism by gut microbiota, leading to diverse products. In addition to hydrolyzing glycosidic bonds to release flavonoids from their glycoside forms, gut bacteria can perform a variety of reactions, e.g., ring fission, that cannot be formed through host metabolism alone. In this regard, exploiting the bioactivities of flavonoids requires a thorough understanding of their interactions with gut bacterial enzymes, resultant structural modifications, and the effects of these modifications on activation of host cellular receptors such as the AhR. The present study aims to computationally predict the microbial metabolites of dietary flavonoids, identify the enzymes that catalyze these transformations, and investigate their immunomodulatory effects as AhR agonists.

Based on the genomes of operational taxonomic units (OTUs) detected by 16S rRNA sequencing in anaerobic batch cultures of murine cecum contents, we assembled a model of metabolic reactions catalyzed by bacterial enzymes in these cultures. We curated the OTU table by eliminating species that are unlikely to be found in the murine intestine. The remaining species were searched against KEGG, UniProt and GenBank to retrieve their annotated genome sequences and match gene orthologs with corresponding enzyme commission (EC) numbers. Using a previously developed algorithm [6], the reactant-product pair data associated with the model's EC numbers were used to learn a set of biotransformation operators for the murine cecum culture. Excluding reactions with molecular oxygen, more than 12,000 unique operators were derived. These operators were applied to 19 dietary flavonoids representative of the major subclasses that have been shown to exhibit anti-inflammatory activities. Our prediction results showed that unique microbial flavonoid metabolites were mostly a result of C-glycoside hydrolysis, de/methylation, de/hydroxylation and ring fission. The predictions regarding microbial metabolites overlapped with recently published *in vitro* experimental results. However, the enzymatic mechanisms have not been investigated in great detail. Our correlation analysis on predicted metabolites and corresponding microbial enzymes indicates that transferases, rather than oxidoreductases, likely catalyze the de/hydroxylation reactions in the presence of appropriate co-factors. Whereas the transferases were distributed widely across the bacterial families in our model, ring fission enzymes for specific flavonoid subclasses were found in only a few strains. One particularly interesting finding was the ring fission of naringenin to naringenin chalcone, which, according to reaction definitions in KEGG, requires an enzyme known to occur only in plants.

To validate the predictions, varying doses of the flavanone naringenin were added to the murine cecum contents culture and samples collected at different times were analyzed for the predicted metabolic products. Untargeted LC-MS experiments detected a dose- and time-dependent increase in one of the predicted metabolites, naringenin chalcone (Figure 1). Gene expression assays for induction of CYP1A1, CYP1B1, and UGT1A1 in human epithelial (Caco-2) cells showed that the chalcone product is an AhR ligand. Moreover, naringenin chalcone exhibited higher induction of UGT1A1 gene expression. Further structure activity relationship studies showed that other chalcone isomers, including 2,2'-dihydroxychalcone, 2,2',4'-trihydroxychalcone and 2,2',5'-trihydroxychalcone, also show significant AhR activity in Caco-2 cells (Figure 2).

In plants, flavanone ring fission is catalyzed by chalcone isomerase (CHI). However, none of the species in the murine cecum microbiota model have a gene for CHI (or a homolog). An alternative enzyme is chalcone synthase (CHS), which reacts 4-coumaroyl-CoA with malonyl-CoA to form naringenin chalcone. This enzyme has been shown to also bind naringenin as a substrate at an active site that is highly conserved across species [2]. To explore CHS catalysis as an alternative reaction for chalcone formation in the cecum culture, we created a phylogenic tree with all bacterial strains in the cecum culture model that have genes homologous to CHS from *Medicago sativa* (a legume expressing CHS with known naringenin binding activity) with a cutoff value of 80%. We found two species, *Paenibacillus lactis* and *Bacillus subtilis*, encoding CHS homologs with a high degree of similarity to *M. sativa* CHS. Enzyme docking simulations showed that CHS from *M. sativa* and *P. lactis* have similar binding scores (-7 and -4, respectively) when naringenin was used as the ligand, suggesting that bacterial CHS could catalyze ring fission of flavanones to form chalcone products that appear to be more potent AhR agonists than the parent compounds.

Taken together, these findings suggest that ring fission of flavanones can be catalyzed by microbial CHS and the resulting molecules can exhibit enhanced AhR agonist activity. In ongoing work, we are investigating if monocultures of *Paenibacillus lactis* and *Bacillus subtilis* can catalyze the ring fission of naringenin under anaerobic conditions. Prospectively, the prediction and validation tools developed in this work could facilitate the identification of bioactive metabolites of dietary flavonoids and their dependence on the enzymatic functions of the gut microbiota.

Figure 1. A schematic representation of the experimental design (a) Microbial metabolite predictions of selected flavonoid compounds. (b) *In vitro* validation of predictions using untargeted metabolomics (c) Phylogenic tree of the CHS sequences of *Medicago sativa* and modeled gut microbiome community members (cutoff >80%) (d) Enzyme docking simulation of CHS from *Paenibacillus lactis* with naringenin as the ligand

Figure 2. Induction of UGT1A1 expression by naringenin and naringenin chalcone in Caco-2 cells. Results are expressed as relative to UGT1A1 expression as a response to DMSO treatment. TCDD: 2,3,7,8-tetrachlorodibenzo-*p*-dioxin. DMSO: Dimethyl sulfoxide

1. **1. Chirag M Lakhani1 2, Arjun K Manrai1, 3, Jian Yang4, 5, Peter M Visscher#4, 5,*, and Chirag J Patel#1, 1Department, B.T.T.**, "乳鼠心肌提取 HHS Public Access," *Physiology & behavior*, **176** (3), pp. 139–148 (2019).
2. **2. Ferrer, J.L., J.M. Jez, M.E. Bowman, R.A. Dixon, and J.P. Noel**, "Structure of chalcone synthase and the molecular basis of plant polyketide biosynthesis," *Nature Structural Biology*, **6** (8), pp. 775–784 (1999).
3. **3. Neavin, D.R., D. Liu, B. Ray, and R.M. Weinshilboum**, "The role of the aryl hydrocarbon receptor (AHR) in immune and inflammatory diseases," *International Journal of Molecular*

Sciences, **19** (12), (2018).
4. **4. Panche, A.N., A.D. Diwan, and S.R. Chandra**, "Flavonoids: An overview," *Journal of Nutritional Science*, **5**, (2016).
5. **5. Park, H., U. Jin, A.A. Orr, *et al.***, "HHS Public Access," **32** (11), pp. 2353–2364 (2020).
6. **6. Yousofshahi, M., S. Manteiga, C. Wu, K. Lee, and S. Hassoun**, "PROXIMAL: A method for prediction of xenobiotic metabolism," *BMC Systems Biology*, **9** (1), pp. 1–17 (2015).

Figure 1

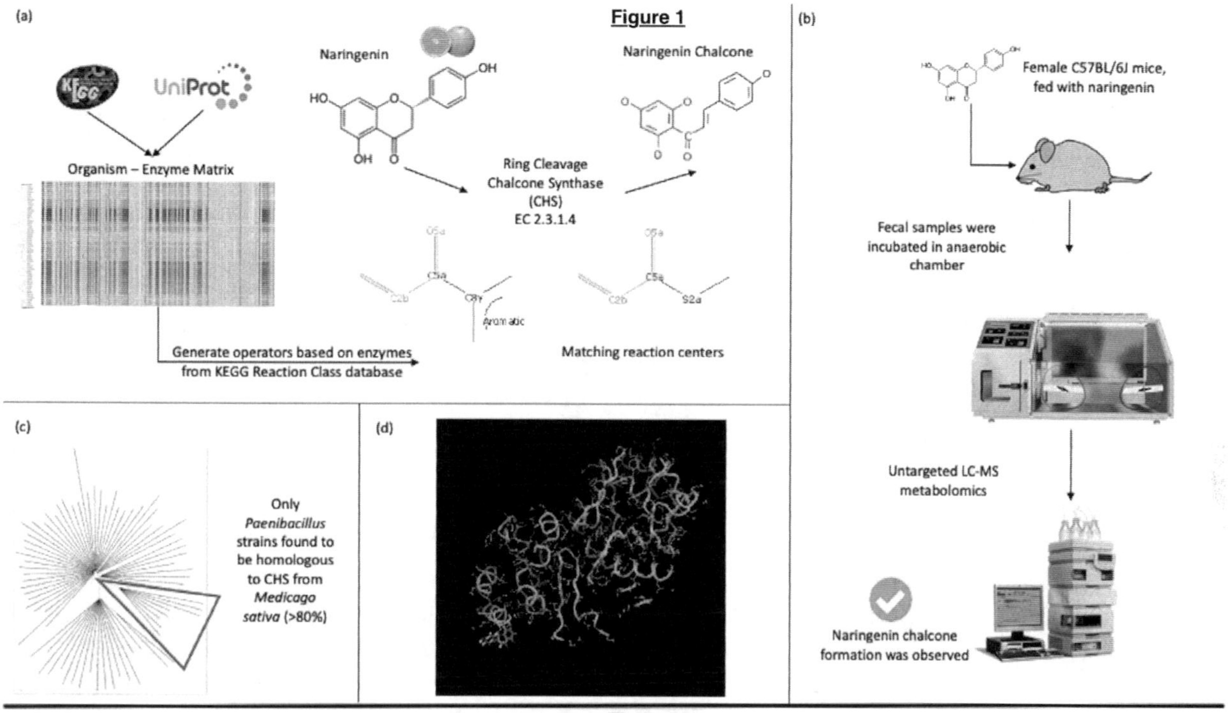

Figure 2

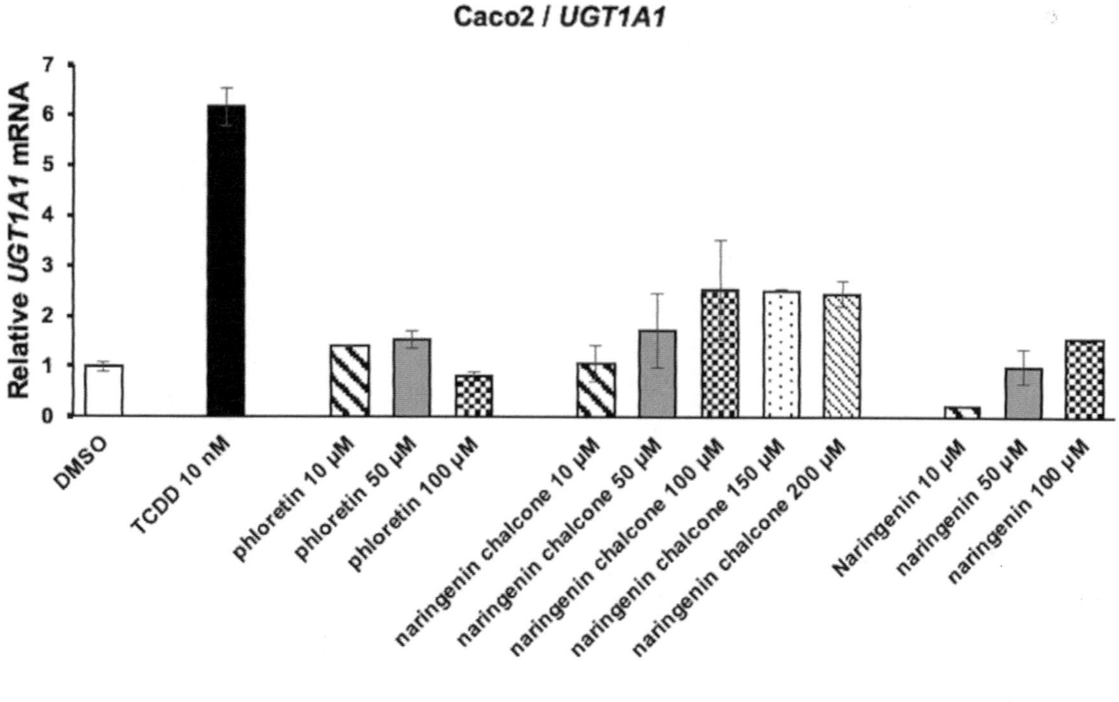

607806 Predicting Microbial Community Assembly and Function in Synthetic Human Gut Microbiomes

Abstract:

Microbial communities are complex systems that exist in nearly every observed environment on earth and carry out chemical transformations with critical impacts to society, including oceanic CO_2 fixation impacting climate change, production of molecules in the plant rhizosphere impacting crop growth, and degradation of dietary substrates in the human gut impacting host health. The microbes in these communities have evolved to optimize their own growth in such environments, often times with undesired outcomes from the human perspective. Thus, microbiome engineering efforts have begun to investigate how we can design and control microbially communities for desired outcomes. Most studies to date have focused on understanding the ecological principles that determine microbial community composition (i.e. how much of each species), neglecting prediction of the functional outputs of these communities (e.g. chemical transformations).

Here, we have applied a model-guided iterative experimental approach to understand the key factors impacting production of butyrate by the human gut microbiome, an important microbial community function that protects the human host from a wide range of diseases, including diabetes, colitis, and colorectal cancer. Our empirically informed model includes a dynamic growth component to quantify how interspecies interactions impact the growth of each species and a statistical model to quantify how interspecies interactions impact the relative butyrate yield for each species. We began by minimally parameterizing our model using experimental observations of the growth and butyrate production of 25 human gut bacterial isolates in monoculture and two-species communities and used this model to design new experiments with increasing community complexity, improving our model at each iteration. We finally used the model to design a set of complex communities (10-25 species) with a wide range of butyrate production (0-50 mM) and used the model parameters to improve our understanding of the key interactions that impact butyrate production in the human gut microbiome. This approach is easily generalizable to any measurable microbial community function and could be used to develop solutions to many critical problems related to human health, energy, and environmental sustainability.

601671 Demonstrated Neuroprotective and Antioxidant Effects of Sugarcane Derivatives: *In Vitro* Rotenone-Induced Oxidative Stress on Neuroblastoma Cells (SH-SY5Y)

Abstract:

Plant-derived extracts from traditionally medicinal plants have proven to be an important source of bioactive compounds, many of which have been the basis for the development of novel pharmaceuticals. In Latin America, sugarcane (*Saccharum officinarum*) has been extensively used ancestrally to combat disease. In particular, non-centrifugal sugarcane (NCS), one of its primary products in the region, is traditionally used for the treatment of skin infections and malnutrition [1]. Some studies have determined the importance of the manufacturing process in the nutraceutical and antioxidant activities of non-centrifugal sugar [2] [3] [4]. The nutritional components enclose bioactive substances such as minerals, vitamins, and amino acids that are involved in beneficial contributions to human health [5]. The most interesting and potentially therapeutic activities of sugarcane and its derivatives (juice, syrup and non-centrifugal sugar) are their neuroprotection and antioxidant effects, which can be mainly attributed to the presence of a large amount of phenolic and flavonoid compounds in sugar cane juice [4] [6] [7]. The aim of this study is to confirm that these effects are observable in an *in vitro* neurodegenerative model where oxidative stress is induced by rotenone on neuroblastoma (SH-SY5Y). The inhibitory effect of rotenone is due to the blockage of the Complex I of the mitochondrial electron transport chain. In turn, this can ultimately cause degeneration of dopaminergic neurons by an exacerbated production of ROS [8] [9] [10]. The samples of sugarcane-derived products analyzed in the present study were obtained under controlled thermal processing conditions by Corporación Colombiana de Investigación Agropecuaria (Agrosavia), and were compared to commercially available NSC products fabricated under traditional methods. Antioxidant activity present in all five sugarcane derivatives was confirmed by DPPH and ABTS radical scavenging activity assays. In addition, neuroprotective effects were determined on SH-SY5Y cells treated with 150μM rotenone and cultured in a CO2 incubator for 24 hours. The cell response was evaluated by looking at changes in the mitochondrial membrane potential after exposure to a range of concentrations of juice, syrup, non-centrifugal sugar, pulverized sugarcane stalk and commercially available non-centrifugal sugar prepared in DMEM and incubated for an additional 24 hours. Changes in mitochondrial membrane potential were determined by JC-1 quantification using confocal microscopy and by analyzing a total of 25,000 events in a FACS CANTO flow cytometer. According to the results shown in Table 1, all the samples exhibit radical free scavenger activity. Syrup and non-centrifugal sugar, exhibited a significant DPPH radical scavenging activity, which can be related to their polyphenolic content. The observed trend was in accordance with the total phenolic content quantified by Folin–Ciocalteu method, where the highest concentration of the phenolic compounds corresponded to syrup and NCS. It is hypothesized that some of the phenolic content is lost during the lyophilization of the assayed samples. Furthermore, confocal and flow cytometry results allowed for the quantification of mitochondrial dysfunction and provided insights as to if the neuroprotective effect of sugarcane derivatives was mediated by biological events on the mitochondria. Treatment with sugarcane-derived samples led to a noticeable preventive effect against rotenone-induced mitochondrial malfunction in all the concentrations assessed as shown in Figure 1A. Confocal microscopy results shown in Figure 1B focused on the changes in red to green fluorescence (R/G) ratios, which were 3.759±0.977 for the control, 1.310±0.29 for traditional NCS, 2,219±0.784

for Agrosavia NCS, 2.979 ±0.758 for syrup, 2.267±0.537 for juice, and 3.455±0.442 for sugarcane. In rotenone-treated cells, red fluorescence decreased while green fluorescence increased, which strongly suggests loss of mitochondrial membrane potential. However, the R/G ratio of sugarcane derivatives, particularly of juice, demonstrated that mitochondrial membrane potential remained very close to that of the control (untreated samples).

In the present work, we found that sugarcane derivatives can protect neuroblastoma (SH-SY5Y) cells against rotenone-induced oxidative stress. Antioxidant compounds present in all samples appear capable of modulating ROS levels and thereby maintaining mitochondrial membrane potential, which suggests their prospective use in the treatment of neurodegenerative diseases. In conclusion, natural antioxidants are promising candidates as chemopreventive agents for neurodegenerative disease treatment.

References:

1. Karthikeyan J, Samipillai SS. "Sugarcane in therapeutics," *J Herb Med Toxicol* 4:9–14 (2010).
2. Meerod K, Weerawatanakorn M, Pansak W. "Impact of Sugarcane Juice Clarification on Physicochemical Properties, Some Nutraceuticals and Antioxidant Activities of Non-centrifugal Sugar," *Sugar Tech* 21:471–80 (2019). https://doi.org/10.1007/s12355-018-0646-7.
3. Vera-Gutiérrez T, García-Muñoz MC, Otálvaro-Alvarez AM, Mendieta-Menjura O. "Effect of processing technology and sugarcane varieties on the quality properties of unrefined non-centrifugal sugar," *Heliyon* 5 (2019). https://doi.org/10.1016/j.heliyon.2019.e02667.
4. Jaffé WR. "Nutritional and functional components of non centrifugal cane sugar: A compilation of the data from the analytical literature," *J Food Compos Anal* 43:194–202 (2015). https://doi.org/10.1016/j.jfca.2015.06.007.
5. Jaffé WR. "Health Effects of Non-Centrifugal Sugar (NCS): A Review," *Sugar Tech*14:87–94 (2012). https://doi.org/10.1007/s12355-012-0145-1.
6. Weerawatanakorn M, Asikin Y, Takahashi M, Tamaki H, Wada K, Ho CT, et al. "Physico-chemical properties, wax composition, aroma profiles, and antioxidant activity of granulated non-centrifugal sugars from sugarcane cultivars of Thailand," *J Food Sci Technol* 53:4084–92 (2016). https://doi.org/10.1007/s13197-016-2415-5.
7. Payet B, Sing ASC, Smadja J. "Assessment of antioxidant activity of cane brown sugars by ABTS and DPPH radical scavenging assays: Determination of their polyphenolic and volatile constituents," *J Agric Food Chem* 53:10074–9 (2005). https://doi.org/10.1021/jf0517703.
8. Xicoy H, Wieringa B, Martens GJM. "The SH-SY5Y cell line in Parkinson's disease research: a systematic review," *Mol Neurodegener* 12:1–11 (2017). https://doi.org/10.1186/s13024-017-0149-0.
9. Cabezas R, El-Bachá RS, González J, Barreto GE. "Mitochondrial functions in astrocytes: Neuroprotective implications from oxidative damage by rotenone," *Neurosci Res* 74:80–90 (2012). https://doi.org/10.1016/j.neures.2012.07.008.
10. Xiong N, Long X, Xiong J, Jia M, Chen C, Huang J, et al. "Mitochondrial complex I inhibitor rotenone-induced toxicity and its potential mechanisms in Parkinson's disease models," *Crit Rev Toxicol* 42:613–32 (2012). https://doi.org/10.3109/10408444.2012.680431.

Table 1. Values obtained for antioxidant evaluation of the studied samples. Ascorbic acid and Trolox are used as reference for DPPH and ABTS assays, respectively.

Samples	Total phenolic content [a]	DPPH[b]	ABTS[c]
SC	2583.8	4.81± 0.69	ND
Juice	743.9	4.33 ± 0.35	2.49± 0.02
Syrup	2318.5	4.36 ± 0.37	1.734±0.07
Agrosavia NCS	3623.6	4.28 ± 0.33	2.39±0.074
Traditional NCS	5892.1	2.12± 0.33	1.13±0.10

[a] ppm caffeic acids [b] radical scavenging activity (IC_{50} mg/mL); [c] % ABTS scavenging activity = [1 - (A_{sample} - $A_{control}$)/$A_{control}$] × 100

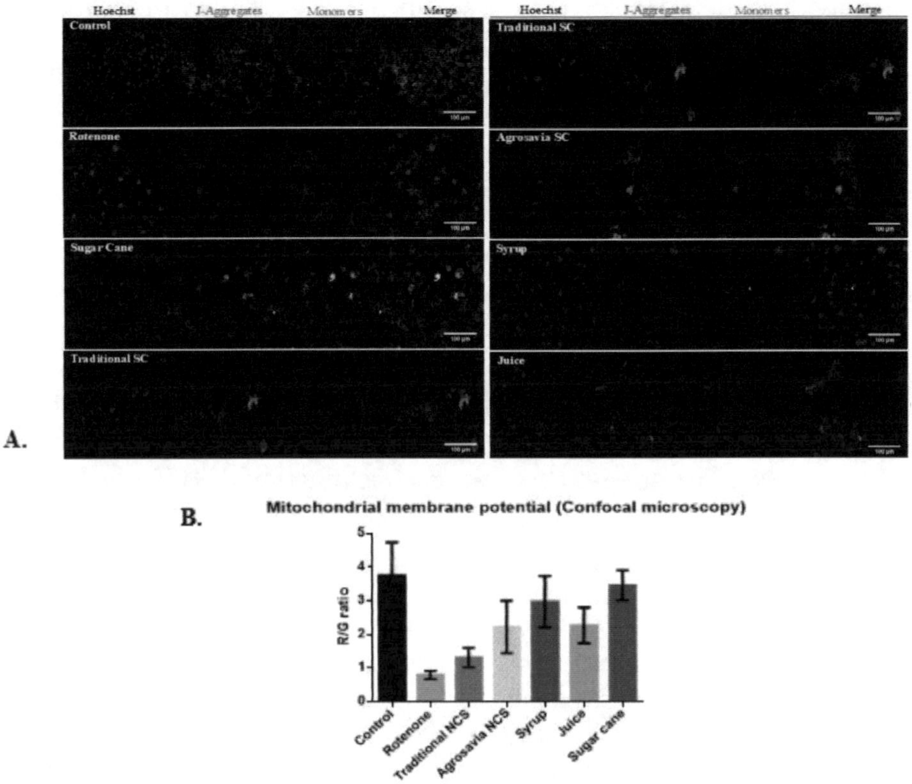

Figure 1. ***Microscopy analysis of mitochondrial membrane potential with the JC-1 dye.*** *(A) SH-SY5Y cells were treated with 200 mM rotenone for 24 hours to cause oxidative stress. Once incubation was complete, cells were washed with PBS 1x to remove any excess rotenone and further incubated with all five sugarcane-derived samples and a control with no treatment. Untreated cells, emitted red fluorescence and very little green (decrease in mitochondrial membrane potential). (B) Red to green fluorescence ratios were determined for each sample group. Cells treated with sugarcane derivatives evidenced mitochondrial depolarization and were not completely protected; however, the red fluorescent response is greater than the green, thereby indicating that sugarcane derivatives are capable of reducing oxidative stress in vitro. (R/G ratios plotted are as follows: 3.759±0.977 for control, 1.310±0.29 for traditional NCS, 2,219±0.784 for Agrosavia NCS, 2.979 ±0.758 for syrup, 2.267±0.537 for juice and 3.455±0.442 for sugar cane)*

609424 "Invited Talk" Exhes Study Reveals the Impact of Prenatal Exposure to Metals, PFOS, PFOA, Organophosphates, and Organochlorines on Early Child Development.

Abstract:

Unravelling the exposome may be a cornerstone towards precision prevention in public health. The aim of this study was to established links between metabolic pathway perturbations and clinically observed phenotypes of neonates. The introduction of exposome studies to research started to make clear that the timing of exposure to environmental factors could be even more critical than the dose. Given the complexity of the individual lifetime exposome assessment, a broad array of technologies must be employed. The presented approach, developed in the framework of HEALS EU project, was applied to 50 mother-child pairs from Spain. The samples were collected in the framework of the HEALS pilot European Exposure and Health Examination Survey (EXHES).

Urine samples were thawed in controlled conditions. A quantity of 600 μL urine samples was centrifuged (10000 rpm for 10 min) to remove all precipitates. 500 μL of supernatant was placed on autosampler vials and diluted with 1000 μL of LC-MS water. Furthermore, a quantity of 50 μL was used for the preparation of the pooled quality control (QC) sample following the same procedure as the samples. The same procedure was followed for the preparation of the QC pooled samples taking a quantity of 50 μL of each sample. Serum samples were thawed in stable conditions at 4°C. Immediately the samples were vortexed, and 1:3 methanol was added to perform protein precipitation. After centrifugation, at 15000rpm for 15 min, 300μl of the supernatant was transferred to dryness through a sample concentrator using N_2. The dry extracts were reconstituted with 100% water LC-MS grade with internal standards and vortex for 5 min. A final centrifuge of the samples at 15000rpm for 10 min was performed before transferring them to autosampler vials for analysis. A pooled QC sample was prepared following the procedure mentioned above.

An Agilent 1290 Infinity HPLC LC System coupled to an Agilent 6540 HRMS-QTOF/ LCMS system used for urinary and serum untargeted metabolomics analysis. Mobile Phase A was LC-MS grade water with 0.1% formic acid and mobile phase B was acetonitrile with 0.1% formic acid. The flow rate was set at 400 μL/min and 250 μL/min in case of urine and serum samples analysis, respectively. Chromatographic separations were achieved using an Acquity UPLC HSS T3 column (100 x 2.1 mm, 1.8 μm, Waters, Milford, MA, USA) maintained at a constant temperature of 40 °C. For the urine samples, the following gradient was used for both positive and negative mode: 1% B at 0 min, 1% B at 1 min, 15% B at 3 min, 50% B at 6 min, 95% B at 9 min, 95% B at 10 min, 1% B at 10.1 min and 1% B at 15 min. The following gradient was used for serum samples: 0% B at 0 min, 0% B at 5 min, 100% B at 20 min, 100% B at 30 min, 0% B at 33 min, 0% B at 35 min. The injection volume was 5 μL. The data are acquired between 50 and 1000 m/z at a scan rate of 1.5 spectra/s in centroid mode at a resolution of 40,000 FWHM. For urine samples analysis the following conditions were used: gas temperature 330 °C, drying gas 7 L/min, nebulizer 50 psig, fragmentor 200 V, skimmer 65 V, and capillary voltage 3500V or -3500V in positive or negative mode, respectively. The source conditions for serum samples analysis were as follows: gas temperature 300 °C, drying gas 7 L/min, nebulizer 50 psig, fragmentor 250 V, skimmer 65 V, and capillary voltage 3500V or -3500V in positive or negative mode, respectively.

Spectral processing was performed using the Bioconductor R - based packages XCMS and CAMERA. Briefly, following the import of the raw data in .mzML format using the ImportRawMSData function, an initial data inspection was performed using the base peak chromatograms of the samples, boxplots representing the distribution of total ion currents per file, and heatmaps. These plots were, also, USED to inform the setting of parameters downstream (e.g. the noise level, peak width, s/n). Chromatographic peak detection was performed based on the centWave algorithm. Internal standards were added to the QC samples to evaluate the expected mass error (ppm) and mzdiff, the most critical parameters. The Obiwarp method was used for alignment, and the retention time adjustment map was used as a diagnostic plot. The featureDefinitions function was used for the correspondence. We used the fillChromPeaks method to fill in intensity data for missing values from the original files due to false negatives. The PerformPeakAnnotation function used for isotope and adduct annotation using the CAMERA package. Also, An R package, called xMSannotator was used for Network-Based annotation. More specifically, the funtion multilevelannotation was used for metabolites annotation. The function uses a multi-level scoring algorithm to annotate features by comparing the MS/MS from the analyzed samples and the MS/MS from the online databases (HMDB, KEGG, and LipidMaps) and in-house build database. The querying was strict to metabolites that can be found in urine or blood To increase the confidence level of annotation. The adduct list used for database matching was set at c("all") for all possible positive or negative adducts. Mass tolerance in ppm for database matching was set based on the optimized calculated mass errors. Finally, in the case of HMDB only the "Detected" metabolites were considered. Then, a batch effect correction procedure was performed. The 80% rule was applied to the QC samples to obtain consistent variables. The instrument and overall process variability were then determined by calculating the median RSD for all endogenous metabolites for the cohorts under study. Enrichment and pathway analyses were performed using GeneSpring, which mapped significant biomarkers to known biochemical pathways based on the information contained in public databases (MetaCyc, Wikipathways, and KEGG). The Exposome-Wide Association Study (EWAS) approach was adopted to comprehensively and systematically explore and associate multiple exposure factors and modifiers discovering and replicating robust correlations with metabolites levels and dysregulated pathways. The 'X-Wide Association Analyses (XWAS)' package in R was used for this. Volcano plots and correlation globes were used for data visualization.

According to the results of the chemical analyses on the biosamples collected, the mean serum level of PFOS was higher than that of PFOA in analyzed samples. In urine samples of the 1st, 2^{d} and 3^{d} trimester, 2-diethylamino-6-methylpyrimidin-4-ol was the most abundant organophosphate. Among organochlorines, 4,4'-DDE and PCB101 showed respectively the highest and lowest mean serum value in the 1st-trimester samples. Also, in the 2^{d} and 3^{d} trimester samples, 4,4'-DDE was the most abundant organochlorine compound. The most abundant metals in urine samples of the 1st, 2^{d} and 3^{d} trimester were Zn, As and Se. In these samples, V, Cr and Mn showed the lowest levels. In serum samples, Zn, Cu, and Se had the highest levels while V, Co and Cr had the lowest levels.

Metabolite identification revealed that the total number of unique annotated metabolites in urine and serum samples analysis using LC-HRMS was 751, and 7830, respectively. The biomarkers of interest detected in the Spain samples through serum untargeted metabolomics analysis were the citric acid, which is participating in the TCA cycle, the amino acids L-Citrulline and L-Arginine, which are involving in the Citrulline-Nitric Oxide cycle Pathway, the L-Glutamine found in S-methyl-5thio-Alpha-Dribose 1 -Phosphate Degradation Pathway. Finally, S-Adenosylhomocysteine, and 5-Hydroxyindoleacetic acid, are both metabolites of the Folate metabolism Pathway. The urinary biomarkers of interest detected in the samples from Spain were the Citric acid, Oxoglutaric acid, trans-Aconitate, and Isocitrate, which are involving in the TCA cycle, the amino acids L-Tyrosine and 5-Hydroxy-L-tryptophan, the hippuric acid participating in the Phenylalanine metabolism, the

N6,N6,N6-trimethyl-L-lysine, which is one of the Serotonine's degradation Pathway metabolites. Moreover, hexanoglycine was detected, as a metabolite of Fatty acid Pathway, the N-Acetylglutamic acid involving in the S-methyl-5thio_Alpha-Dribose 1 -Phosphate Degradation Pathway, and the 5-Hydroxyindoleacetic acid, which is involved in the Serotonine Degradation Pathway. These metabolites are linked either with health outcomes, such as neurodevelopmental disorders, overweight, obesity and diabetes, or with exposure to endocrine disruptors, according to previously published data.

The detected metabolites on serum samples were mapped on 246 pathways, while urinary metabolites on 163.

According to EWAS analysis, birth weight is positively affected by S-Adenosylhomocysteine levels during the first trimester of pregnancy and negatively associated with the levels of Citrulline, and DEAMPY, at delivery. In addition, higher exposure levels to Hexachlorocyclohexane (HCH), 2,2',4,5,5'-Pentachlorobiphenyl (PCB101) and 2.4'-DDT, can lead to height increasement. The same outcome is associated with citric acid levels. Head circumference is positively associated with exposure to 4.4'-DDT at the first trimester. Overall, functionally coupling advanced bioinformatics algorithms applied to omics data with exposome-derived information on exposures and health indicators can support the high-dimension-biology-based association of environmental exposures and adverse health outcomes in early life.

Overall, functionally coupling advanced bioinformatics algorithms applied to omics data with exposome-derived information on exposures and health indicators can support the high-dimension-biology-based association of environmental exposures and adverse health outcomes in early life.

600369 Development of an Electroactive Platform for Detection of Virus Fusion to Host Membranes

Abstract:

Every year, new virus outbreaks are reported and have potential to cause pandemics with devastating socio-economic impacts, with the coronavirus disease 19 (COVID-19) serving as a prime example. Recent viral outbreaks in this decade have alluded to the importance of quick and accurate viral sensing strategies so that the infected individual can begin their treatment and limit contact with uninfected individuals quicker, mitigating the impact of an outbreak. PCR, one of the most common viral detection methods, is a specific and sensitive detection technique, but requires trained personnel and sophisticated machinery, which can slow response times. An ideal detection method should integrate the specificity and sensitivity with the speed necessary to diagnosis patients rapidly.

Enveloped viruses (such as influenza virus, coronavirus) enter host cells by binding to a receptor and then fusing its viral membrane to the host cell membrane to insert its genome. We propose a novel sensing platform that leverages the virus's propensity to bind specifically and fuse to a host cell membrane. We can couple a biomimetic membrane on top of an electrically conductive polymer (PEDOT:PSS) to probe the electrical responses of the system once viral fusion has occurred. Successful viral fusion to the membrane in a matter of minutes may cause a detectable change in electrical response, serving as the readout of a viral infection. Using influenza virus as a proof-of-principle, we first show that we can form the biomimetic membrane incorporating the influenza viral receptor on PEDOT:PSS. We then visualize influenza single virion fusion on PEDOT:PSS and demonstrate that the kinetics of this fusion event are comparable to influenza fusion on a non-conductive polymer. Lastly, using electrical impedance spectroscopy, we provide evidence that influenza viral fusion causes a detectable change in electrical response, demonstrating that it is possible to exploit viral membrane fusion as a diagnostic strategy.

599943 Engineering DNA-Based Materials for the Analysis of Live Single Cells

Abstract:

The ability to probe the composition of live cells is important for learning about chemical processes inside of cells as well as for creating diagnostic platforms based on molecular profiling. However, current strategies for detecting analytes in single living cells are severely limited both in terms of probe design and sensing strategy. While fluorescent proteins and nucleic acid-based aptamers have revolutionized live-cell imaging, these sensors must be genetically encoded into the cells and therefore have limited applications in a clinical setting. Many techniques require cell lysis or a bulk population of cells to yield a detectable quantity of the target analyte. Additionally, a major impediment to the development of intracellular probes is the inability of most probes to traverse the cell membrane. Consequently, these probes require the use of transfection reagents or physical methods that perturb the cell membrane, which in turn can interfere with the measurements by changing the analyte's expression profiles. Furthermore, existing strategies for signal transduction to indicate the presence of a given analyte, such as fluorophore-quencher-based systems and FRET-based methodologies, are limited and susceptible to false-positive signal. Therefore, the development of a versatile platform for intracellular analyte detection in live cells is an outstanding challenge.

To address this challenge, I have first developed DNA-based aptamers that utilize a novel sensing strategy. Specifically, these aptamers are chemically modified at a single base with a viscosity-sensitive dye. Target binding to the aptamer changes the conformation of the aptamer and forces the dye to intercalate (or FIT) between the DNA base pairs, turning its fluorescence on. This "FIT" strategy avoids false-positive signals common to most fluorophore-quencher-based systems, quantitatively outperforms FRET-based techniques, gives up to 20-fold fluorescence enhancement in buffer, and allows for the detection of analytes in complex biological milieu such as human serum.

To apply this technique to the analysis of intracellular analytes, I have then developed a new platform based on protein spherical nucleic acids (ProSNAs). ProSNAs are composed of a protein core surrounded by a DNA shell. The dense functionalization and radial arrangement of the DNA around the protein core allows efficient cellular uptake of the ProSNA without the use of transfection reagents or physical stimuli. By using a FIT-aptamer as the DNA sequence, this design allows for the detection of intracellular analytes. As an example, I demonstrate that using an i-motif sequence that undergoes pH-dependent conformational changes allows for the detection of intracellular pH. As more than 500 aptamers are known for various targets, our nano-construct is a powerful platform for the detection of intracellular analytes. Moreover, by using a functional protein as the core, it is possible to detect analytes for which aptamers do not exist and further expand the range of analytes that can be detected. Using glucose oxidase as the core, I show that glucose can be detected (for which is there is no known aptamer with a biologically relevant binding affinity) in 8 different cell lines. Taken together, these constructs represent a plug-and-play platform for detecting a wide variety of intracellular analytes with great potential in the field of biodiagnostics.

References

†**Equal Contribution**

Samanta, D.†; **Ebrahimi, S. B.**†; Mirkin, C. A., "*Nucleic-Acid Structures as Intracellular Probes for Live Cells,*" *Advanced Materials* **2020**, *32*, 1901743.

Ebrahimi, S. B.†; Samanta, D.†; Cheng, H. F.; Nathan, L. I.; Mirkin, C. A., "*Forced Intercalation (FIT)-Aptamers,*" *Journal of the American Chemical Society* **2019**, *141*, 13744.

Samanta, D.; Iscen, A.; Laramy, C. R.; **Ebrahimi, S. B**.; Bujold, K. E.; Schatz, G. C.; Mirkin, C. A., "*Multivalent Cation-Induced Actuation of DNA-Mediated Colloidal Superlattices,*" *Journal of the American Chemical Society* **2019**, *141*, 19973.

608845 Molecular Engineering of a Colorimetric Nanogel Sensor for Clinical Radiotherapy and Trauma Monitoring

Abstract:

Radiotherapy using ionizing radiation (e.g. X rays) remains a mainstay of treatment modalities for different types of cancers. Overexposure to radiation in the clinic or as part of an adverse event can induce tissue toxicity and damage, whereas underdosing can lead to poor efficacies of tumor ablation. Effective radiation sensors are critical in ensuring optimum delivery to patients undergoing radiotherapy and for patient stratification in cases of radiation exposure. Currently existing sensors possess several disadvantages in rapid radiation detection because of their fragility, sensitivity to light and heat, dose-rate dependence, long processing times, lack of conformation to tissue morphologies and / or high costs. Commonly used Nanodots™ are expensive, require separate read-out devices and do not conform to tissue morphologies. We developed a novel colorimetric nanogel sensor, based on the formation of maroon-colored gold nanoparticles from their colorless precursor salt formulations, upon exposure to different levels and types of ionizing radiation. By measuring the intensity of color developed with a simple bench-top absorbance spectrophotometer we were able to detect doses typically administered in fractionated cancer radiotherapy (i.e. ranging from 0-5 Gy doses) and beyond. Topographical distribution of delivered radiation doses was qualitatively and quantitatively determined following addition of quenching agents, and the efficacy of the nanogel sensor was demonstrated for different sources of ionizing radiation including protons, electrons, photons and radioactive isotopes. The translational potential of the sensor was demonstrated using anthropomorphic phantoms and in live canine patients undergoing clinical radiotherapy treatments. Our results demonstrate a new generation of nanogel sensors for colorimetric detection of radiation for improving patient safety in clinical radiotherapy and trauma care.

609804 Unique Electrochemical Detection of Sepsis Using Triplex Biomarker Detection Panel with IL-6, IL-8 and IL-10 in Blood Plasma

Abstract:

Sepsis is a life-threatening disease that is caused by the body's extreme reaction to an infection which could lead to tissue damage, multiple organ failure or even death. It is one of the leading causes of death worldwide, estimated to affect more than 30 million people every year according to the world health organization. Despite scientific developments in health care technologies in the past three decades, a substantial challenge continues to persist for early rapid diagnosis of sepsis. Thorough and timely diagnosis plays a substantial role in increasing the survival chances for patients with adequate treatment, thus reducing mortality. However, current gold standard blood culture techniques require large patient sample volumes and provides delayed results (6 hours to up to 5 days). This may lead to clinical ambiguity in treatment measures, propelling the body's immune response to spiral uncontrollably into septic shock or severe sepsis. This fails to deliver the required speed and accuracy to enable accurate diagnosis. To overcome existing challenges, we tailored a specific biosensor to screen for sepsis as a rapid point-of-care (POC) technology enabling multiplexed detection capabilities. The novelty presented in this research combines the effectiveness of specific multiplexed biomarkers for detection of sepsis with electrochemical impedance spectroscopy (EIS) technique to enable clinically efficient decision-making process. This work leverages the detection of pro and anti-inflammatory cytokine biomarkers such as interleukin 6 (IL-6), interleukin-8 (IL-8) and interleukin-10 (IL-10) in blood plasma. Researchers suggests good correlation between timing of pro and inflammatory cytokine release to degree of sepsis, associated with deleterious effects. The developed label-free, point-of-care biosensor is surface engineered with specific capture probes that utilizes EIS to measure the capacitive impedance change reflecting binding interactions between the capture probe and target biomarker enabling simultaneous triplex detection. Specificity of the POC biosensor was validated using cross-reactive studies, which displayed insignificant interference from non-specific biomarkers. The biosensor also displays stable and repeatable performance. Key advantage of the triplex biosensor is the use of undiluted blood plasma volume of less than 40 μL, providing result within 15 minutes. Additionally, the triplex biosensor technology was validated with 20 patient samples that were clinically tested positive for sepsis. This unique sensor technology would allow medical practitioners to facilitate targeted interventions for septic patients as a rapid prognostic approach, preventing complications arriving from sepsis.

598873 Harnessing the Protein Corona Towards Carbon Nanotube-Based Sensor Design

Abstract:

Unpredictable protein adsorption on nanoparticles remains a considerable challenge towards effectively applying nanotechnologies, such as nanosensors, in biological environments. However, if well understood and controlled, proteins adsorbed on nanoparticles can be exploited to develop protein-based nanosensors.

Herein, we employ our knowledge of spontaneous "protein corona" formation on single-walled carbon nanotubes (SWCNTs) in biofluids towards rational nanosensor design based on protein-SWCNT complexation.[1] SWCNTs are implemented as the sensor platform due to their desirable optical and physical characteristics, including photostable fluorescence in the near-infrared tissue transparency window[2] and high aspect ratio that confers delivery capabilities.[3] We first study the protein corona composition formed on SWCNT-based probes within relevant biofluids such as blood plasma and cerebrospinal fluid.[4] Then, we take advantage of the intrinsic binding affinity of specific proteins on the carbon nanotube to selectively adsorb proteins that, in turn, bind to their natural ligand partners. Thus, the adsorbed protein acts as the recognition moiety that interacts with the environment and detects the analyte of interest, tethered to the SWCNT that serves as the signal transducer. This nanosensor platform combines the sensitive and selective binding capabilities of proteins for their ligand partners with the near-infrared fluorescence readout of SWCNTs that is modulated upon protein-ligand binding. In this manner, we demonstrate that although formation of the nanoparticle protein corona can impede the intended nanobiotechnology application, it also presents an opportunity to create improved protein-nanoparticle architectures by manipulating selective protein adsorption on the nanoparticle surface.

References

1. Pinals, R. L., Yang, D., Lui, A., Cao, W. & Landry, M. P. Corona Exchange Dynamics on Carbon Nanotubes by Multiplexed Fluorescence Monitoring. *J. Am. Chem. Soc.* (2019) doi:10.1021/jacs.9b09617.
2. Boghossian, A. A. *et al.* Near-Infrared Fluorescent Sensors based on Single-Walled Carbon Nanotubes for Life Sciences Applications. *ChemSusChem* **4**, 848–863 (2011).
3. Demirer, G. S. *et al.* High aspect ratio nanomaterials enable delivery of functional genetic material without DNA integration in mature plants. *Nat. Nanotechnol.* **14**, 456–464 (2019).
4. Pinals, R. L. *et al.* Protein Corona Composition and Dynamics on Carbon Nanotubes in Blood Plasma and Cerebrospinal Fluid. *bioRxiv* 2020.01.13.905356 (2020) doi:10.1101/2020.01.13.905356.

607484 Point-of-Care Cancer Biomarker Detection System Integrating Surface Acoustic Wave Streaming and Metal-Enhanced Fluorescence

Abstract:

Point-of-care cancer biomarker detection system integrating surface acoustic wave streaming and metal-enhanced fluorescence

Yuqi Huang, Shuangming Li and Venkat R. Bhethanabotla

yuqi@usf.edu

shuangming@mail.usf.edu

bhethana@usf.edu

Department of Chemical and Biomedical Engineering, University of South Florida, Tampa, Florida 33620-5350, USA

Abstract:

Immunofluorescence is one of the primary techniques for microbiological assays. It can help to visualize, identify, and even quantify the interaction between antibody and antigen, some even used for DNA and protein detection.[1,2] However, there is a clinical need of increasing its limit of detection, specificity and portability. In this work, a portable optical fluorescence system was designed to be combined with a direct digital synthesizer for surface acoustic wave (SAW) generation and metal-enhanced fluorescence (MEF) to lower the detection limit in immunofluorescence assays.[3,4] MEF was achieved by depositing silver nanoparticles with covered silica film, which can plasmonically enhance fluorescence intensity. Rayleigh wave streaming was generated on a piezoelectric device for surface acoustic wave streaming purpose, which can prevent nonspecific binding on the biological interface.[5] The miniaturized prototype, integrating acoustic wave removal of interfering proteins combined with optical detection, was developed to realize immune detection on a portable platform with limit of detection at nanogram per milliliter level in detecting carcinoembryonic antigen (CEA) from human plasma.

Recent work shows that the MEF substrate fabricated with the above method can lower the detection limit of CEA from PBS to 100pg/mL level. However, testing the detection limit for CEA from human plasma has raised to 500ng/mL. Non-specific proteins, dusts and electrolytes in human plasma are primary suspects to raise background noise which increases the limit of detection. Further tests show that by adding SAW streaming to the sample tested on MEF substrate, the detection limit brings back down to 50pg/mL. A reasonable inference to this detection limit drop is the removal of non-specific protein by streaming surface acoustic wave to the sample surface, where the non-specific absorbents are removed by shear force from the wave, and the background noise is reduced to lower the limit of detection for CEA from human plasma.

1. K. Aslan, I. Gryczynski, J. Malicka, E. Matveeva, J. R. Lakowicz, and C. D. Geddes, "Metal-enhanced fluorescence: an emerging tool in biotechnology," Current Opinion in Biotechnology, vol. 16, pp. 55-62, **2005**.
2. R. A. Goldsby, T. J. Kindt, B. A. Osborne, and J. Kuby, "Immunology," ed: New York: WH Freeman, **2003**.
3. S. Li and V. Bhethanabotla, "Design of a Portable Orthogonal Surface Acoustic Wave Sensor System for Simultaneous Sensing, Removal of Nonspecifically bound Proteins and Mixing", Sensors, vol. 19(18), **2019**.
4. Liu, J.; Li, S.; Bhethanabotla, V. R., Integrating Metal-Enhanced Fluorescence and Surface Acoustic Waves for Sensitive and Rapid Quantification of Cancer Biomarkers from Real Matrices. *ACS Sensors*, *3* (1), 222-229, **2018**.
5. S. Cular, D. W. Branch, V. R. Bhethanabotla, G. D. Meyer, and H. G. Craighead, "Removal of nonspecifically bound proteins on microarrays using surface acoustic waves," IEEE Sensors journal, vol. 8, pp. 314-320, **2008**.

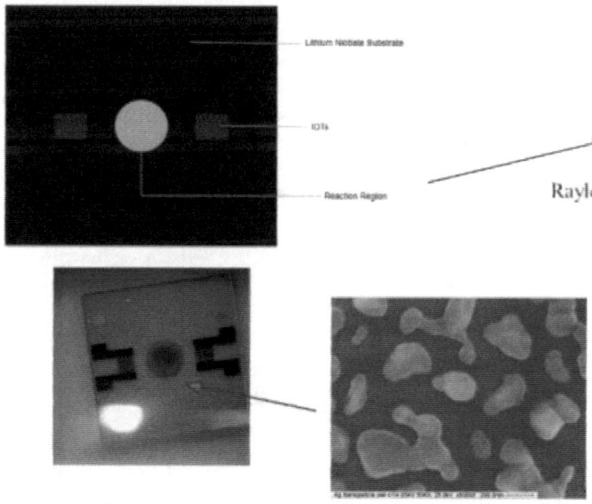

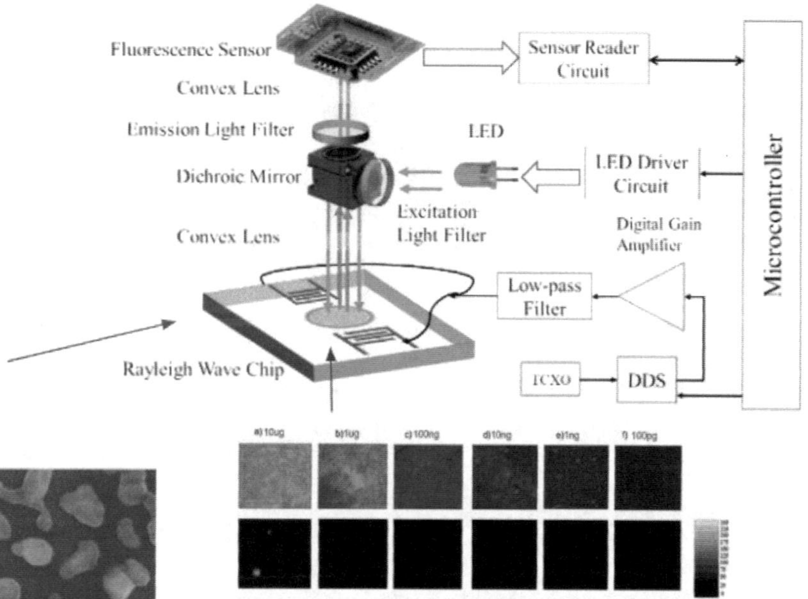

Figure 2-5. Fluorescence images for CEA diluted in PBS, top row are the fluorescence images with silver nanoparticles of 25000-45000 nanometer square. The bottom row serves as control groups without silver nanoparticles. The concentrations of CEA antigens in the images are (a) 10 μg/ mL, (b) 1 μg/mL, (c) 100 ng/mL, (d) 10 ng/mL, (e) 1 ng/mL, and (f) 100 pg/mL

609302 High-Throughput Quantification of Influenza A Virus RNA Using Novel Drop-Based qRT-PCR Analysis

Abstract:

Influenza A virus (IAV) causes seasonal epidemics and sporadic worldwide pandemics that have resulted in over 100 million deaths during the past century. IAV populations are dynamic with high mutation rates and short replication times. A better understanding of IAV diversity at the single cell level would provide greater insight into how heterogeneity of infection affects population-level dynamics. Investigating IAV and host cell dynamics is typically performed using bulk culture and may not fully capture cell-to-cell differences in virus production. Thus, there is a need for high-throughput quantification of the IAV output, or burst size, from single cells to elucidate individual host-cell dynamics at the population scale. Here, we have developed a novel drop-based quantitative reverse transcriptase polymerase chain reaction (qRT-PCR) method to measure the amount of IAV RNA contained in individual drops. Concentrations of RNA are encapsulated in tens of thousands of picoliter-sized drops. These drops are processed on a microfluidic device which splits a small portion of the drop and merges it with qRT-PCR solution. The merged drops are thermocycled using a standard qPCR machine. Drops are randomly sampled at different cycles to measure drop fluorescence. Using a novel qPCR analysis method, we quantify the amount of RNA contained in the drops. Our method is the first to measure IAV RNA using a high-throughput drop-based assay, which will allow for future work in investigating the heterogeneity of IAV infection.

597969 Microfluidic Paper-Based Analytical Devices Using Plasma Processes

Abstract:

Disability adjusted life years (DALY) are the sum of years of life lost due to premature mortality and years lived with a disability. In 2010, DALY were approximately 2.4 billion. It is estimated that ~90% of DALY are due to communicable and non-communicable diseases. This problem is even more severe in developing countries where the major stumbling block for disease control is lack of proper disease diagnostic methods. Point-of care testing (POTC) methods in which users do not require specialized equipment and sophisticated lab facilities for disease diagnostics will be ideal for developing countries. Therefore, the World Health Organization (WHO) has stipulated a set of criteria for POCT in resource-limited settings called ASSURED criteria—affordability, sensitivity, specificity, user-friendly, rapid and robust, equipment-free and deliverable to those who need it. Due to the necessity and increasing demand of personalized medicine, the global POTC market is expected to grow from US \$23.16 billion in 2016 to US \$36.96 billion in 2021 at the compound annual growth rate of 9.8% from 2016 to 2021.

Microfluidic paper-based analytical devices (μ-PAD) have shown considerable promise to meet the ASSURED criteria set by the WHO. Paper has been patterned using various fabrication methods to allow distribution of liquid samples to targeted locations where subsequent reactions with reagents deposited in the paper are performed. These reactions develop color signals that can then be read using a smartphone. Low cost, portability, and the absence of external power source requirements make these devices desirable for POTC in less developed areas. It has been more than a decade since the first μ-PAD was invented. Yet these μ-PADs are not yet commercialized because of multiple technical problems including 1) contamination and evaporation of samples because of the lack of proper μ-PAD packaging, 2) inability to carry out complex tests that require multiple reagent additions, and 3) loss of sample in flow channels.

To address the issue of contamination and evaporation, a fully enclosed paper microfluidic device has been fabricated using pentafluoroethane (PFE) plasma deposition followed by O_2 plasma etching. Structures with one and two non-interacting, fully enclosed hydrophilic channels were generated in a single paper sheet using metal masks. Furthermore, by performing an additional O_2 plasma step with a secondary mask, pinholes were created at the reaction zones for reagent loading. Finally, to demonstrate the functionality of the device, a glucose assay was performed. Quantitative analysis of glucose assays showed that the device can be used for clinically relevant range of glucose. To our knowledge, this is the first time that such structures have been fabricated without paper stacking. Multi-layer devices have enhanced functionality relative to a single channel device, because the design space for creating networks of channels within the paper substrate is greatly expanded. The fluid-filled channels in the fabricated device are isolated, thereby preventing contamination due to handling and environmental exposure. Fluid evaporation can be inhibited by sealing the device with adhesive tape without contaminating the enclosed channels hence preventing loss of sample due to evaporation; a critical factor while working with precious samples.

To address the challenge of fluid flow control in μ-PADs, we extended our work on fully enclosed channel fabrication inside paper substrates to incorporate flow control functionality into the channels. Using pentafluoroethane plasma deposition and subsequent O_2 plasma etching with metal masks, fully enclosed hydrophilic channels with varying wettability were designed and fabricated to precisely control fluid flow in μ-PADs. Flow control is one of the most critical functionalities in μ-PADs because it allows controlled implementation of multiple sequential reactions on a single device with minimal user interference. Potential benefits of flow control include improved limit of detection, ability to carry out complex diagnostic tests, reduced assay time, and decreased likelihood of human error. We studied the effect of controlled fluorocarbon removal on the surface of cellulose fibers on channel wettability and demonstrated that this behavior can be exploited to significantly vary the fluid velocity in the flow channels. Furthermore, by simply covering the flow channels with adhesive tape, we demonstrated that the effect of evaporation on the fluid flow can be significantly decreased. Finally, a device with three channels converging to a single reaction zone was fabricated and sequential delivery of fluid to the reaction zone was demonstrated. The devices made by the methods described in this paper offer significant benefits for in-field diagnostic testing where multistep protocols are required.

A device fabricated by this method is estimated to cost between 10 and 50 ¢ and the plasma processes used to fabricate these devices are compatible with roll-to-roll technology, making it suitable for large scale production. This alternative fabrication approach contributes significantly toward commercialization of low-cost μ-PADs that meet the WHO ASSURED criteria.

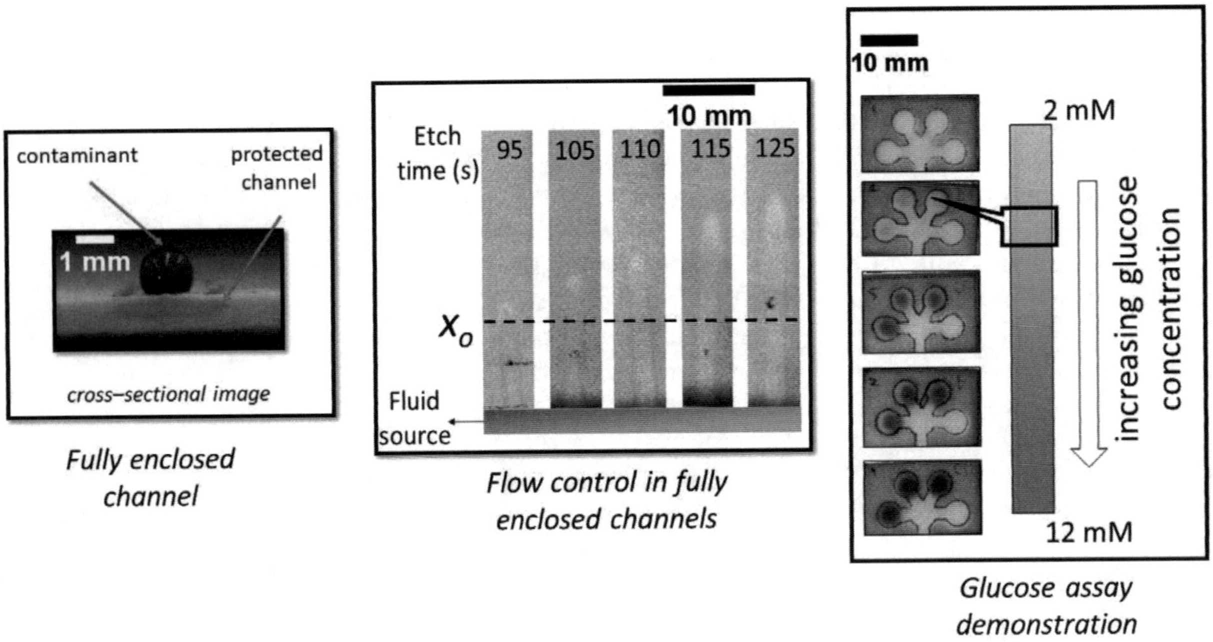

601627 From Genes to Cell Types to Behavior: Molecular Classification of Zebrafish Retinal Ganglion Cells

Abstract:

In vertebrates, visual information is detected and conveyed from the eye to the brain by multiple retinal ganglion cell (RGC) types, which can be subdivided based on anatomical, physiological and molecular characteristics. Intrinsic features as well as selective connectivity with other neuronal cell types render each RGC type selectively sensitive to a distinct visual feature, and characterizing these RGC types is essential to further understand the complex transmission of visual information. Recent studies have used high-throughput single cell RNA sequencing (scRNA-seq) to generate comprehensive molecular atlas of RGC types in several vertebrate species. It is challenging, however, to associate molecularly defined RGC types with the behaviors their activation elicits. To address this, the zebrafish is a particularly attractive model system as the larvae are transparent, allowing for detailed *in vivo* imaging of structure and imaging. Additionally, they develop rapidly and display a diverse range of visual behaviors as early as 5 days post fertilization. Here, we used scRNA-seq to classify larval and adult zebrafish RGCs at the transcriptional level, identifying 33 putative RGC types in the adult stage and 29 types in larva. These types possessed unique molecular signatures and most could be uniquely labeled based on selective expression of a single gene. This molecular atlas allowed us to further explore the expression of transcription factors, neuropeptides, and cell surface molecules across RGC types and their role in establishing unique cell identities. To investigate the extent to which RGC diversification is complete in the larval stage, we were able to assign adult identities to larval RGCs using a supervised classifier trained on larval clusters. The results indicated a high degree of transcriptional correspondence between adult and larval clusters and provide insight into the diversification of RGC types.

Using CRISPR genome engineering and intersectional genetics, reporter lines were generated that labeled RGCs based on their expression of select transcription factors. These lines allowed us to characterize the anatomical and physiological characteristics of molecularly defined RGC types. Fluorescent imaging revealed that different molecular types project to different areas of the brain. Using *in vivo* calcium imaging, we showed that RGCs respond in stereotyped patterns to visual stimuli that included bright and dark flashes, moving gratings, and prey-like stimuli. We functionally characterized a small set of RGC types defined by selective expression of the transcription factor *eomesa* to find that they exhibit response profiles that encode ambient luminance levels. Finally, we performed loss-of-function experiments to show that *eomesa*+ RGCs, in relation with their physiology, specifically regulate light-seeking navigation. Taken together, this molecular taxonomy of RGC types is comprehensive and allows us to examine RGC types from molecular composition to anatomical and physiological properties to behavior.

606191 Elucidating Gender Differences in the Immune Response in the Zebrafish Melanoma Model (Industry Candidate)

Abstract:

Approximately 600,000 cancer deaths occurred in the United States each year. Even when a patients undergoes an apparently successful treatment, the disease sometimes returns more than ten years later, a phenomenon called tumor dormancy and recurrence. In order to capture the complex dynamics of tumor growth, we propose a population balance model that uses power law size-dependent parameters representing tumor growth, reduction and metastasis to predict how the distribution of tumor sizes in a cohort of patients changes with time.

Previously, we used the zebrafish melanoma animal model to obtain data under different growth conditions. We examined gender differences in immune-competent and immune-deficient fish orthotopically-injected with a non-syngeneic, fluorescence-labelled zebrafish melanoma cell line. We have developed a fitting procedure to obtain unambiguous values for the three rate parameters by determining each from separate data. While gender differences in the determined tumor growth and metastasis parameters were not significant, we detected a strong difference between male and female immunity parameters, particularly in its power law exponent that expresses its size dependence. The upshot was that it predicted that female fish would have better outcomes than males for larger tumors. Examples of gender disparities in cancer outcomes – generally males do far worse than females – exist in other animals and humans. We followed up these initial experiments with longer term, larger inoculation-size experiments in immune-competent fish of both genders.

In this presentation, we will compare theoretical predictions with new experimental data and present a detailed analysis of male and female disparities found in the new data. We use our model-based simulations of cancer progression in fish and speculate as to the potential clinical consequences of this this type of gender difference in human melanoma. These include mortality rate, disease duration and the probability of recurrence after treatment.

607318 Analysis of Transcriptomic Data for CD4 Naïve T Cells Stimulated By Cytokines

Abstract:

Novel cancer therapeutics rely on the immune system to attack a pathogen of interest. Most pre-exhaustive therapies deplete the immune system of vital T cells, rendering the immune system un-prepared for swiftly destroying pathogens. Some of the most important T cells are naïve CD4+ T cells, which, upon coming into contact with a pathogen, will proliferate and differentiate into Helper T, Regulatory T or Follicular Helper T cells [1]. Some of the same cellular signaling pathways which govern differentiation of naïve CD4+ T cells also govern the proliferation of T cells, which is a desirable effect given the state of most immunodepleted cancer patients. Cytokines have a strong interdependent relationship with T cells and can both regulate and be produced from T cells in order to sponsor proportional immune responses, both in proliferation and differentiation [2]. Three cytokines: IL-2, IL-7 and IL-15, have been suggested in literature as having a profound impact on the proliferation of naïve T cells [3]. Here, we have engaged in a design of experiments to understand the influence of each of the three cytokines to gain insight into how to best stimulate proliferation of Naïve CD4+ T cells. We have studied the Naïve CD4+ T cell transcriptome over time using RNA Sequencing. Using EdgeR, we are able to sort through and identify genes with high magnitudes of change in the differential gene expression [4]. The output of this sorting technique is the input for further consolidation of genes of interest using principal component analysis (PCA). We identify outliers which may have unique impact to the pathways inducing proliferation in the T cells. From the PCA analysis, we use GeneMania software to find connections between the genes of interest, and the function of these genes in how they relate to proliferative pathways [5]. We also link these genes to a genome scale model for CD4+ T cells to overlap these genes and isolate the most important genes associated with key pathways [6]. We find that most of the genes linked with IL-2 stimulation are associated with cell chemotaxis and cell migration. Genes linked with IL-7 stimulation are associated heavily with electrolyte transport and chemotaxis, and genes linked with IL-15 stimulation are associated with metal ion response and regulation of growth. We can make associations using the transcriptome to highlight why the combination of cytokines we have chosen happens to be a useful path to T cell proliferation. Finally, we also correlate the growth rate time profile with the transcriptome time profile of genes sampled at the same time points to understand whether these genes can be positively correlated with the rate of proliferation of the T cells. Our findings lay the groundwork for further work in regulating CD4+ naïve T cell growth with cytokines in clinical setting.

Reference

1. *Different Subsets of T Cells, Memory, Effector Functions, and CAR-T Immunotherapy.* **Golubovskaya, V. and Wu, L.** 36, Richmond : Cancers, 2016, Vol. 8.
2. *Molecular pathways regulating CD4+ T cell differentiation, anergy and memory with implications for vaccines.* **Ahlers, J. D. and Belyakov, I. M.** 10, Bethesda : Trends in Molecular Medicine, 2010, Vol. 16.
3. *IL-2, IL-7 and IL-15: Multistage regulators of CD4+ T helper cell differentiation.* **Read, K. A., et al.** Blacksburg : Experimental Hematology, 2016, Vol. 44.
4. *edgeR: a Bioconductor package for differential expression analysis of digital gene expression data.* **Robinson, M. D., McCarthy, D. J. and Smyth, G. K.** 1, Victoria : Bioinformatics, 2010, Vol. 26.
5. *The GeneMANIA Prediction server: biological network integration for gene prioritizatio, predicting gene function.* **Warde-Farley, D., et al.** Toronto : Nucleic Acids Research, 2010, Vol. 38.
6. *Genome-Wide metabolic model to improve understanding of CD4+ T cell metabolism, immunometabolism and application in drug design.* **Han, F., et al.** 431, Kunming : Molecular Biosysems, 2016, Vol. 12.

609098 Fractional Re-Distribution Among Motility States during Ageing

Abstract:

Ageing can be defined as the accumulation of dysfunctions with the passage of time that limits the ability of organisms, tissues and cellular systems to absorb and rebound after perturbations and stresses. As integrators of molecular signals, cells offer a meso-scale view of ageing, with cellular dysfunctions occurring prior to the manifestation of age-related diseases at the clinical level. Since cells typically display dynamic phenotypes both in health and disease, we postulated that analyzing dynamic cellular movements may provide significantly more mechanistic insights relative to the standard "snapshot" data from fixed cells.

Typically, cell motility is measured in bulk, and when measured at the single-cell level, results are reported as average changes in speed and persistence. Here, we develop a novel single-cell analysis approach that defines cell motility states, and classifies cells as a function of age-related spatio-temporal motility patterns. This analysis revealed an unforeseen shift in cellular heterogeneity with age. Using primary dermal fibroblasts derived from healthy donors, we measured the motility patterns for hundreds of single cells spanning an age range of 2-92 years. Here we show that the global age-dependent decrease in bulk motility is not due to decreased motility for all cells, but rather from the fractional re-distribution among motility states.

Using the anisotropic persistent random walk together with hierarchical clustering analysis, we first defined spatial clusters that describe spatial movement patterns of cells. From this analysis we found that: (1) Age does not solely define spatial clusters, but each cluster harbors a mixture of cells from each age, (2) donors separate into polarized spatial clusters as a function of age, and (3) there is a significant decrease in the cellular heterogeneity (measures by the Shannon Entropy) for older adults.

Secondly, to define the cellular activity we converted the x-y trajectories into one-dimensional profiles to define temporal patterns of motion. This analysis identified four activity clusters, that describe the bursty vs. consistent movements of cells. We found that cells from young donors were highly enriched for activity clusters showing consistent movement with infrequent, short movement bursts. In contrast, cells from older adults were enriched for clusters showing longer, frequent bursts of movements.

Because of the complimentary information provided by the spatial and activity clusters, we co-classified each cell and computed the abundance of cells within each of the 32 motility states (8 spatial x 4 activity). Analysis of the abundances per age group showed not only significant enrichments and depletions for various motility states, but also a significant decrease in the heterogeneity of cells derived from older adults. This work reveals that ageing information is encoded within the intrinsic cellular variations in motility.

Together, our study presents a general framework to analyze single-cell motility based on spatial- and activity-patterns of cells. This analysis provides novel insights into the ageing process, with potential applications for precision health via the development of robust biomarkers for stratification in health and disease.

609314 Proteomics-Informed Signal Transduction Modeling of Valve Interstitial Cell Activation

Abstract:

Calcific aortic valve disease (CAVD) is the most common valvular heart disease in the Western world, resulting in fibrosis and calcification of the aortic valve that can only be treated surgically. The resident valve interstitial cells are tissue-specific fibroblasts under normal conditions but become activated to myofibroblasts early in CAVD development. These myofibroblasts secrete both new extracellular matrix components and proteases that degrade the extracellular matrix, thereby remodeling the structure of the aortic valve. Targeting this myofibroblast transition with pharmaceutical therapies is expected to reverse or halt CAVD progression; however, drug targets for CAVD remain elusive.

One primary driver of the myofibroblast transition in CAVD is transforming growth factor beta (TGFβ). TGFβ can be released from the extracellular matrix in response to matrix degradation and mechanical deformation and is also newly synthesized by myofibroblasts in response to a variety of external stimuli. The relevant signaling network is comprised of both canonical and non-canonical components and many uncertainties remain. Therefore, we have begun to model the signal transduction network stemming from TGFβ to better understand the effects of this key stimulus on the myofibroblast transition in valve interstitial cells.

Models of signal transduction networks are classically informed by a priori deciding which proteins and post-translational modifications are important and then designing experiments to collect these selected measurements. This approach usually greatly limits the number of experiments that can be conducted and limits the size of the model being developed. However, newer proteomics strategies can measure tens of thousands of different peptides, including those with post-translational modifications, from thousands of proteins in a single experiment without the requirement of specifying specific proteins or modifications beforehand. Therefore, this paper uses a label-free data independent acquisition proteomics strategy to collect data from canonical and non-canonical TGFβ signaling to inform a mathematical model of the myofibroblast transition in valve interstitial cells.

610365 Genome-Wide Screening of Epigenetic Modifiers during Cell Reprogramming

Abstract:

A slight mis-interaction of an enhancer-gene pair could give rise to fatal problems, if the gene is expressed in wrong position at a wrong time. To prevent that, chromatins are often divided into "closed" heterochromatin and "open" euchromatin regions, such that cell-type inappropriate genes remain silenced even in the presence of activators. However, heterochromatin causes problems during cell reprogramming. For example, if some of the key neuronal genes are located in the heterochromatin of fibroblast cells, the neuronal genes won't be activated even in the presence of all the activators, resulting in poor reprogramming efficiency. Using the combination of genome-wide "seq" techniques like ChIP-seq and RNA-seq and quantitative imaging, we sought to elucidate the mechanism of heterochromatin maintenance and characterize the kinetics of gene expression upon depletion of heterochromatin regulators during direct cell reprogramming from fibroblast cells into hepatic cells. We identified ~100 proteins that are closely associated with the maintenance of heterochromatin domain. We found that RNAi knockout of each protein leads to de-repression of key hepatic genes in the presence of hiHep transcription factors that are known to activate hepatic genes. To analyze whether such de-repression of genes is a direct consequence of heterochromatin de-compaction, we used ChIP-seq and immunostaining to visualize H3K9me3 and H3K27me3 marked domain in the genome. We found that H3K9me3 and H3K27me3 marked heterochromatin regulate distinct set of genes. With our imaging and genomics assays, we hope to provide better cell reprogramming strategies to achieve higher efficiency.

612968 A Chinese Hamster Reference Genome for the Community, and Applications to Understanding and Predicting the Quality of Biopharmaceuticals (Invited Speaker)

Abstract:

The Chinese hamster genome provides a critical resource for the biotechnology community and is used in numerous ways. Recent improvements to the reference genome are enabling a wide range of studies that benefit the manufacturing of biopharmaceuticals. These impacts include ways to model and predict glycosylation as well as opportunities to impact host cell protein profiles.

609846 Synthesis and Characterization of Degradable Nanoparticles for Controlled Delivery of Growth Factors for Bone Regeneration

Abstract:

Fractures that fail to heal naturally due to severe trauma, otherwise known as nonunion bone fractures, pose a significant burden on a patient's lifestyle. Bone tissue engineering systems have recently been explored in an effort to overcome the limitations of current gold standard treatment options for nonunion fractures. Delivery of growth factors in such systems has been of particular interest due to their ability to enhance cell recruitment and promote osteogenic differentiation and angiogenesis. However, current growth factor delivery methods come with their own challenges and have yet to see commercial success. Growth factors are subject to rapid enzymatic degradation and diffusion from the defect site, resulting in the need to deliver supra-physiological doses. Encapsulation of growth factors within particulate carrier systems has been explored as a method to improve protein stability and to provide greater control over release kinetics. Here, we have developed polymeric, hydrolytically degradable nanoparticles to provide controlled release of growth factors for bone tissue engineering applications.

Customizable poly(lactic acid)-*b*-poly(ethylene glycol) dimethacrylate crosslinking agents were synthesized through a ring opening polymerization reaction, and were subsequently incorporated into poly(methacrylate-co-methacrylic acid) nanoparticles. To achieve a range of degradation profiles, the PLA and PEG chain length in the crosslinker was varied. The crosslinked, co-polymeric nanoparticles were synthesized using UV-initiated free radical emulsion polymerization. Fourier-transform infrared spectroscopy was used to confirm the composition of the resulting nanoparticles and dynamic light scattering and electrophoretic light scattering were used to assess the particle size and zeta potential, respectively. Nuclear magnetic resonance spectroscopy was used to study the influence of polymer composition on degradation kinetics. The particle loading capacity and release profiles were analyzed using chymotrypsin as a model for bone morphogenetic protein-2, a widely studied growth factor for bone regeneration. Ultimately, our work shows that variation of nanoparticle degradation rates can control the release rate of growth factors and provide sustained delivery within a therapeutic timeframe for bone regeneration. In addition, our system shows promise for the potential to sequentially deliver multiple growth factors with physiologically relevant profiles.

The work was supported in part by a grant from the National Institutes of Health (R01 EB022025).

607965 Effect of Hydrogel Formulation on Vitamin E Uptake and *in Vitro* Drug Release Kinetics for Improved Drug Release from Contact Lenses

Abstract:

Introduction. The most prevalent delivery method for common ophthalmic medications is eye drops [1,2]. However, eye drops have a low bioavailability of approximately 5% and a low average residence time in the eye of two minutes [1,3,4]. The rest of the drug is absorbed into the conjunctivitis and nose, allowing most of the drug to enter the blood stream and reach systemic circulation, raising concerns about side effects and toxicity [1,5]. Eye drops pose an additional problem for individuals who wear contact lenses as they must first remove their contact lenses, apply the eye drops, wait for the drug to be absorbed, and then reinsert their lenses. Ultimately, poor compliance to frequent dosing regimens and a low bioavailability make eye drops an inefficient solution for delivering ophthalmic drugs, and there is a need for improved drug delivery to the eye.

Thus, the use of contact lenses to deliver drugs has been explored. Contact lenses made from silicone or hydroxyethyl methacrylate (HEMA) hydrogel can be can be loaded with drugs either through dissolving the drug into the water phase of the lens [6,7] or through binding the therapeutic to the polymer matrix [8,9]. Contact lenses do not require surgery or frequent application, and it is expected that they will improve patient compliance as it is estimated that 45 million people in the US wear contact lenses[10]. A remarkable advantage of drug release from contact lenses compared with eye drops is the drug released from the contact lens has a longer residence time in the post lens tear film (POLTF) than eye drop residence time in the tear film, leading to higher flux into the cornea [11]. Plus, contact lenses will reduce the drug inflow into the nasolacrimal sac, which will reduce uptake into the bloodstream [12]. Yet, drug release only occurs for a few short hours from commercial lenses loaded by drug soaking [13–20]. To address the issues of low drug loading and short release times, our group has extensively researched how the addition of vitamin E barriers to contact lenses can extend the release of drugs to days or weeks [7,21,22].

The current study sought to determine the effect of lens composition on vitamin E uptake and drug release kinetics in order to improve the future use of contact lenses for extended drug delivery.

Methods. Molds for lenses were 3D printed. The silicone hydrogels were prepared by polymerizing a high ion permeability hydrophilic monomer along with a high oxygen permeability silicone monomer. Dimethylacrylamide (DMA) was used as the hydrophilic monomer, 3-methacryloxypropyltris(trimethylsiloxy)silane (TRIS) as the hydrophobic monomer and bis-alpha,omega-(methacryloxypropyl) polydimethylsiloxane as the macromer to ensure solubilization of these two components. Additionally, 1-vinyl-2 pyrrolidone (NVP) was added to increase water content and ethylene glycol dimethacrylate (EGDMA) was added for controlled crosslinking. Silicone hydrogels of several different compositions were prepared by free radical bulk polymerization of the monomers using UV photoinitiation with 2,4,6-trimethylbenzoyl-diphenyl-phosphineoxide (TPO). The various silicone formulations measured are shown in Table 1.

HEMA lenses were prepared by adding HEMA, water, and EGDMA in different ratios. Plus, vitamin-

E barriers were added to the different formulations. This was done by soaking the lens in a vitamin-E ethanol solution for 24 hours, followed by soaking in water for 3 hours and then dipping quickly in ethanol to remove surface bound vitamin-E. The water content of the lenses was measured by weighing their dry weight and hydrated weights. Vitamin-E loading was also measured by weighing dry lenses before and after vitamin-E loading protocol. Cyclosporine, a model hydrophobic drug, was loaded into the lenses by soaking the lenses in a cyclosporine-PBS solution until equilibrium. Then the lenses were transported to PBS and the release of cyclosporine was measured over time using UV-Vis spectrometry.

Results. The vitamin E loaded and water uptake of the various lens formulations can be seen in Table 2. Form 2 has most DMA (hydrophilic monomer) and highest water content of silicone lens formulations. Form 1 has the most TRIS (hydrophobic monomer) and highest vitamin E loading of silicone lenses. Comparing Form 6 to 8, 8 has increased NVP. There was not a significant difference in vitamin E or water content (two tailed t-test alpha 0.05). Form 3 had the lowest DMA also lowest water content.

Release plots are shown in Figure 1. For release, forms 1-8 all were still releasing up to 50 days. Forms 2 and 4 which both had low amounts of TRIS released the most cumulative cyclosporine at 70 μg. By adding the vitamin-E barriers, release was significantly slowed down and only ~20 μg was released from the formulations in 40 days compared to ~50-70 μg without vitamin-E. HEMA lenses released all loaded cyclosporine (35 μg) in 12 days. However, release was extended from HEMA lenses when vitamin-E barriers were added, and were still releasing at 30 days. Release plots are being analyzed for partition and diffusivity coefficients.

Conclusion. By manipulating the formulations for the hydrogels, vitamin E uptake and cyclosporine release can be significantly changed. More cyclosporine loading can be achieved along with slower release profiles. Future work will focus on calculating diffusivity coefficients of cyclosporine from the lens to quantitatively evaluate the lens property effects on release. In addition, other lens properties such as oxygen and ion transport will also be evaluated. Moreover, a model hydrophilic drug (e.g. timolol) will also be analyzed for release.

References

1. Schultz, C. Safety and Efficacy of Cyclosporine in the Treatment of Chronic Dry Eye. *Ophthalmol. Eye Dis.* **6,** 37–42 (2014).
2. Negri, L., Ferreras, A. & Iester, M. Timolol 0.1% in Glaucomatous Patients: Efficacy, Tolerance, and Quality of Life. *J. Ophthalmol.* **2019,** 4146124 (2019).
3. Hsu, K. H., Gause, S. & Chauhan, A. Review of ophthalmic drug delivery by contact lenses. *J. Drug Deliv. Sci. Technol.* **24,** 123–135 (2014).
4. Peng, C. C., Kim, J. & Chauhan, A. Extended delivery of hydrophilic drugs from silicone-hydrogel contact lenses containing Vitamin E diffusion barriers. *Biomaterials* **31,** 4032–4047 (2010).
5. Farkouh, A., Frigo, P. & Czejka, M. Systemic side effects of eye drops: A pharmacokinetic perspective. *Clin. Ophthalmol.* **10,** 2433—2441 (2016).
6. Kim, J., Conway, A. & Chauhan, A. Extended delivery of ophthalmic drugs by silicone hydrogel contact lenses. *Biomaterials* **29,** 2259–2269 (2008).
7. Peng, C. C., Burke, M. T. & Chauhan, A. Transport of topical anesthetics in vitamin E loaded silicone hydrogel contact lenses. *Langmuir* **28,** 1478–1487 (2011).
8. Dutta, D., Cole, N., Kumar, N. & Willcox, M. D. P. Broad spectrum antimicrobial activity of melimine covalently bound to contact lenses. *Investig. Ophthalmol. Vis. Sci.* **54,** 175–182 (2013).

9. Willcox, M. D. P., Hume, E. B. H., Aliwarga, Y., Kumar, N. & Cole, N. A novel cationic-peptide coating for the prevention of microbial colonization on contact lenses. *J. Appl. Microbiol.* **105,** 1817–1825 (2008).
10. Cope, J. R. *et al.* Risk behaviors for contact lens–related eye infections among adults and adolescents — United States, 2016. *Morb. Mortal. Wkly. Rep.* **66,** 4032–4047 (2017).
11. Gulsen, D. & Chauhan, A. Ophthalmic drug delivery through contact lenses. *Investig. Ophthalmol. Vis. Sci.* **45,** 2342–2347 (2004).
12. Xinming, L. *et al.* Polymeric hydrogels for novel contact lens-based ophthalmic drug delivery systems: A review. *Contact Lens Anterior Eye* **31,** 57–64 (2008).
13. Sedlácek, J. Possibility of the application of ophthalmic drugs with the use of gel contact lenses. *Cesk. Oftalmol.* **21,** 509–12 (1965).
14. Hillman, J. S. Management of acute glaucoma with Pilocarpine-soaked hydrophilic lens. *Br. J. Ophthalmol.* **58,** 674–679 (1974).
15. Ruben, M. & Watkins, R. Pilocarpine dispensation for the soft hydrophilic contact lens. *Br. J. Ophthalmol.* **59,** 455–8 (1975).
16. Schultz, C. L., Nunez, I. M., Silor, D. L. & Neil, M. L. Contact lens containing a leachable absorbed material. (1995).
17. Dixon, P. *et al.* Controlled delivery of pirfenidone through vitamin E-loaded contact lens ameliorates corneal inflammation. *Drug Deliv. Transl. Res.* **8,** 1114–1126 (2018).
18. Fathalla, D., Soliman, G. M. & Fouad, E. A. DLatanoprost Liposomes for Glaucoma Treatment Development and in vitro/in vivo Evaluation of Liposomal Gels for the Sustained Ocular Delivery of Latanoprost. *J. Clin. Exp. Ophthalmol.* **6,** 390 (2015).
19. Ciolino, J. B. *et al.* A drug-eluting contact lens. *Invest. Ophthalmol. Vis. Sci.* **50,** 3346–52 (2009).
20. Ciolino, J. B. *et al.* A Prototype Antifungal Contact Lens. *Invest. Ophthalmol. Vis. Sci.* **52,** 6286–6291 (2011).
21. Peng, C. C. & Chauhan, A. Extended cyclosporine delivery by silicone-hydrogel contact lenses. *J. Control. Release* **154,** 267–274 (2011).
22. Hsu, K. H., Fentzke, R. C. & Chauhan, A. Feasibility of corneal drug delivery of cysteamine using vitamin e modified silicone hydrogel contact lenses. *Eur. J. Pharm. Biopharm.* **85,** 531–540 (2013).

Table 1. Formulations of silicone lenses synthesized in this study.

Formulation	Macromer	Tris	DMA	NVP	EGDMA
1	0.8	0.8	0.8	120	100
2	0.8	0.4	1.2	120	100
3	1.2	0.6	0.6	120	100
4	1.2	0.4	0.8	120	100
5	1.225	0.4	0.8	120	75
6	1.25	0.4	0.8	120	50
7	1.275	0.4	0.8	120	25
8	1.13	0.4	0.8	240	50

Table 2. Average vitamin E loadings and water contents.

FORM	average g vitamin/g lens	standard dev g vit E/g lens	water content percent	std dev percentage
1	0.16	0.04	24.63	2.88
2	0.05	0.02	38.72	1.98
3	0.02	0.01	17.69	4.74
4	0.06	0.03	29.06	3.85
5	0.03	0.02	26.95	1.40
6	0.04	0.02	28.24	2.94
7	0.05	0.00	31.68	1.93
8	0.05	0.03	29.95	3.03
HEMA	0.13	0.11	46.76	4.66

Mean ± standard deviation for n≥3

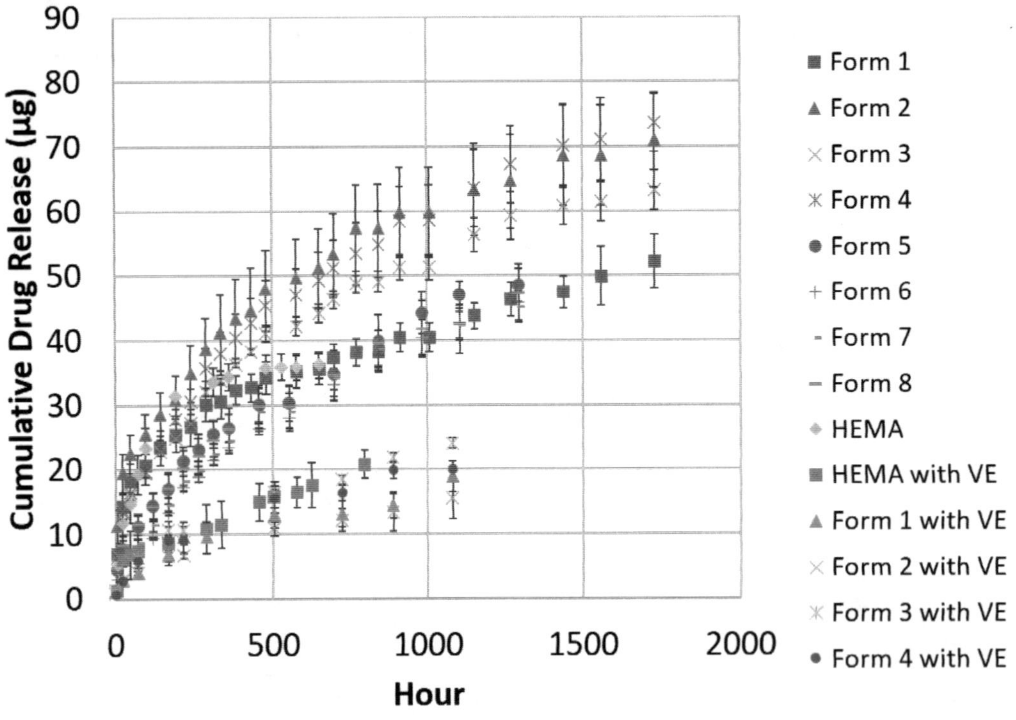

Figure 1. Cyclosporine release plots for various hydrogel formulations.

603938 Evaluation of Ammonium-Based Ionic Liquids As Novel Chemical Permeation Enhancers for Transdermal Drug Delivery

Abstract:

Room temperature ionic liquids (ILs) are gaining growing interests in the area of transdermal drug delivery research thanks to several experimental evidence that they can enhance the permeation of macromolecules across the skin barrier[1,2,3,4]. By definition, ILs are salts governed by ionic interactions as opposed to conventional chemical permeation enhancers which are molecular liquids. ILs are ideal candidates for transdermal permeation enhancement due to their low volatility, high recyclability and what's more important, high tunability. Despite their vast parameter space available to synthesize and explore, our current knowledge of ILs is limited to a few specific ones[4]. In this talk, we aim to investigate the fundamental difference between ILs and conventional chemical enhancers in terms of skin permeation and provide a general guideline to future studies in this area.

Fourier-transform infrared (FTIR) spectroscopy has been shown as a powerful tool to examine solvent-skin interactions which significantly influence the permeability of drugs in transdermal applications. By comparing the FTIR spectra of 40 conventional chemical permeation enhancers and 40 novel ILs, we demonstrate superior performance of ILs in terms of high permeation enhancement and low irritation potential. This general observation can be explained by the fundamental difference between ILs and their molecular counterparts: polar interactions contribute to a significant portion of the cohesive energy density in ILs due to their ionic nature as compared to covalent materials. This property is preferable when a chemical enhancer acts to extract skin lipids and therefore increases drug permeability. While the search for molecular liquids with improved permeation performance has reached a bottleneck, our investigation of ILs as a novel class of permeation enhancers opens up new opportunities for transdermal drug delivery applications.

1. Qi, Q. M., & Mitragotri, S. (2019). Mechanistic study of transdermal delivery of macromolecules assisted by ionic liquids. *Journal of Controlled Release, 311-312*, 162–169.
2. Tanner, Eden E. L., et al. "Design Principles of Ionic Liquids for Transdermal Drug Delivery." *Advanced Materials*, vol. 31, no. 27, 2019, p. 1901103.
3. Tanner, E. E., Ibsen, K. N., & Mitragotri, S. (2018). Transdermal insulin delivery using choline-based ionic liquids (CAGE). *Journal of Controlled Release*, *286*, 137-144.
4. Ibsen, K. N., Ma, H., Banerjee, A., Tanner, E. E., Nangia, S., & Mitragotri, S. (2018). Mechanism of Antibacterial Activity of Choline-Based Ionic Liquids (CAGE). *ACS Biomaterials Science & Engineering*, *4*(7), 2370-2379.

610598 Sprayable, Thermoreversible Hydrogels for Improved Treatment of Burn Wound Infections

Abstract:

Due to the high rate of infection in hospitalized burn wounds and the painful treatment of applying antimicrobial creams and ointments, the goal of this project is to develop a sprayable, antimicrobial hydrogel to use as a treatment for burn wound infections. Sprayable hydrogels are expected to be less painful to apply than current topical, antimicrobial products, will sustain drug release over longer periods of time, and will keep the wound bed under optimal conditions (moist, absorbing wound exudate, etc.). Pluronic F127, a commercially available thermoreversible hydrogel, and hyaluronic acid, an active healing agent and hydrogel, were combined in various percentages to test which formulations have desirable spray patterns, gelation temperatures, drug diffusion, and gel dissolution. Methylene blue was incorporated into the gel for visualization and to act as a model drug in diffusion studies. Ciprofloxacin, an antimicrobial agent, was added after preliminary studies using methylene blue. To determine spray characteristics of the formulations, cold hydrogels were loaded into Aptar Pharma sprayers and sprayed four, five, and six inches away from a target. The targets were photographed and analyzed to determine the spray spread, and the sprays were characterized via high-speed imaging to examine the plume differences from different nozzles. The gelation temperature and viscosity of each gel composition was determined using a rheometer, with a desirable gelation temperature between 21 and 32°C to avoid gelation while spraying but allowing gelation on skin. Diffusion and dissolution experiments were conducted in vitro to determine how quickly the gel compositions release drug and dissolve in saline solution. Our results show that changes in the spray pattern are largely controlled by the choice of nozzle and the gel viscosity. These results will be important for application purposes because we want to cover large, irregularly shaped wounds quickly and completely. Drug and other gel additives were shown to alter the gelation temperature, and several gel formulations with drug additives have been made that transition to a gel within the desired range for application. Preliminary diffusion and dissolution experiments showed that increasing the gel percentage prolonged drug release and dissolution in saline solution. These results will influence the amount of drug that will need to be loaded into the gel to maintain effective dosing and the number of dressing changes required for the system. Future studies will include antimicrobial testing against three common burn wound pathogens and transport studies across skin.

607615 Priming of Live Therapeutic Bacteria for Improved Survival in Gastric Fluid

Abstract:

There is increasing interest in the use of bacteria as live cell therapies, as they can produce and secrete biologics, sense and therapeutically respond to their environments, and modulate the composition of the human microbiome. However, the acidic conditions of the stomach significantly reduce the viability of live therapeutic bacteria (LTB) during gastric transit, presenting a delivery challenge for LTBs. It is known that pathogenic bacteria activate specific transcriptional programs during transit through the gastrointestinal tract (GIT) that improve their viability in the stomach. Therefore, we propose that LTBs can be primed to activate similar transcriptional programs *in vitro* to improve their resistance to gastric pH. To test this, we primed the candidate LTB *E. coli* Nissle 1917 (EcN) to acid conditions during overnight culture in acidic media (pH = 5.0). We show that overnight priming significantly improves the survival of EcN during a challenge in simulated gastric fluid (SGF, pH = 2.8) ($p < 0.01$). We found that the acid-resistance of EcN is dependent on its growth phase, with improvements in survival during the stationary phase relative to the exponential phase ($p < 0.01$). By screening EcN colonies that were viable after an SGF challenge ('survivors') and growing them in neutral media (pH = 7.0), we show that acid-resistance is transient, as the survivors have no improvement in viability compared to an SGF-naïve control during the *in vitro* challenge. This suggests that acidic priming activates transcriptional programs that improve survival, rather than triggering the formation of stable genetic mutants. The reversibility of EcN acid-resistance makes it an attractive delivery strategy, as it enables improvement in LTB survival during short-term gastric transit without permanently altering the EcN genome or phenotype. Collectively, this approach allows for the study of LTB adaptations to the GIT in a controlled, *in vitro* environment. By priming LTBs to intestinal challenges prior to oral administration, this delivery approach could provide a low-cost and rapid method for improving gastric survival and colonization of LTBs, enhancing their therapeutic potential.

607765 Exploring *in Vitro* and *in Vivo* Contributions to Lignocellulose Degradation By Multi-Domain Enzymes from Extremely Thermophilic *Caldicellulosiruptor* Species (Faculty/Industry Candidate)

Abstract:

Caldicellulosiruptor species are anaerobic bacteria that are the most thermophilic (T_{opt} = 70-78°C) microorganisms currently known to degrade lignocellulosic biomass. Large (~170-230 kDa) multi-domain cellulases and hemicellulases, found in a single genomic locus termed the glucan degradation locus (GDL), are of interest for dissecting the biomass degradation ability of these species. Using recently developed genetic techniques in *C. bescii*, a replicating vector was developed to over-express enzymes in *C. bescii* for purification and characterization, while maintaining their native glycosylation. All six GDL enzymes from *C. bescii* (CelA-CelF), and four additional novel GDL enzymes from *C. danielii* and *C. morganii,* were expressed and characterized using this system. A simplex centroid mixture experimental design was used to examine enzyme cocktails containing the six *C. bescii* GDL enzymes *in vitro,* and a model relating enzyme composition to glucan conversion was constructed. In parallel, genetic tools were used to construct knockout strains of *C. bescii* with deletion of various combinations of the 6 GDL genes to assess their role *in vivo*. *C. bescii* strain RKCB123 *ΔcelA-celE* was not able to degrade microcrystalline cellulose and was extremely impaired in its ability to degrade lignocellulosic substrates switchgrass and poplar. Finally, enzyme mixtures optimized by the *in vitro* model were examined alone and when added to cultures containing knockout strain RKCB123 (*ΔcelA-celE*). Optimized enzyme cocktails alone performed only 20% the microcrystalline cellulose degradation of wild type *C. bescii*, while with strain RKCB123 they restored degradation to wild type levels. This suggests mass action effects caused by sugar fermentation performed by *C. bescii* cells helps to drive biomass degradation by these enzymes. The ability of *C. bescii* to efficiently degrade lignocellulosic biomass using the enzymes of the GDL, along with our ability to manipulate this species genetically, make *C. bescii* of interest as a platform organism for industrial biotechnology applications. Recent work, combining a *C. bescii* strain engineered to produce ethanol with transgenic poplar lines showed nearly 90% carbohydrate conversion and fermentation, serving as a proof-of-concept for the conversion of renewable biomass feedstocks by *C. bescii* to bio-based fuels and chemicals.

References

[1] Conway JM, Crosby JR, McKinley BS, Seals NL, Adams MWW, Kelly RM. 2018. Parsing *in vivo* and *in vitro* contributions to microcrystalline cellulose hydrolysis by multi-domain glycoside hydrolases in the *Caldicellulosiruptor bescii* secretome. *Biotechnol Bioeng*. 115:2426-2440.

[2] Conway JM, McKinley BS, Seals NL, Hernandez D, Khatibi PA, Poudel S, Giannone RJ, Hettich RL, Williams-Rhaesa AM, Lipscomb GL, Adams MWW, Kelly RM. 2017. Functional analysis of the Glucan Degradation Locus (GDL) in *Caldicellulosiruptor bescii* reveals essential roles of component glycoside hydrolases in plant biomass deconstruction. *Appl Environ Microbio*l. 83:e01828-01817.

[3] Conway JM, Crosby JR, Hren AP, Southerland RT, Lee LL, Lunin VV, Alahuhta P, Himmel ME, Bomble YJ, Adams MWW, Kelly RM. 2018. Novel multidomain, multifunctional glycoside hydrolases from highly lignocellulolytic Caldicellulosiruptor species. *AIChE J*. 64:4218-4228.

[4] Straub CT, Khatibi PA, Wang JP, Conway JM, Williams-Rhaesa AM, Peszlen IM, Chiang VL, Adams MWW, Kelly RM. 2019. Quantitative fermentation of unpretreated transgenic poplar by Caldicellulosiruptor bescii. *Nat Commun* 10:3548.

608796 Modulation of Membrane Lipid Parameters Enabled Prediction of Cell Growth in Combined Inhibitory Conditions

Abstract:

Robust biocatalysts support bio-production at industrially relevant titers. Engineering of the microbial cell membrane engineering to increase microbial robustness has shown promising outcomes. Characterization of the effects of representative inhibitors on the cell membrane can contribute to rational design strategies. In this study, we engineered *E. coli* strains for a range of membrane lipid metrics and properties when the cells were cultured in several distinct inhibitory conditions. We used experimental data from single culture conditions to identify correlations between membrane composition metrics, membrane physical properties, and growth of the strain. This information for distinct inhibitors was used to estimate trends for culture conditions with multiple inhibitors. We experimentally validated the response to inhibitors to various levels by combining 2% v/v ethanol, 2.5 mM furfural, and 42 ºC in minimal media. Membrane hydrophobicity in the combined condition was found to be a meaningful feature that enabled the prediction of cell growth in the combined condition relative to estimations based on the growth rates from individual conditions. This work demonstrates that the modulation of membrane features, specifically the lipid composition, can be utilized to engineer microbial robustness for inhibitory conditions.

609945 Understanding and Eliminating the Detrimental Effect of Thiamine Deficiency on the Oleaginous Yeast *Yarrowia Lipolytica* (Industry Candidate)

Abstract:

Thiamine is a vitamin that functions as a cofactor for key enzymes in carbon and energy metabolism for all living cells. While most plants, fungi and bacteria can synthesize thiamine *de novo*, the oleaginous yeast, *Yarrowia lipolytica*, cannot. In this study, we used proteomics together with physiological characterization to understand key metabolic processes influenced and regulated by thiamine availability and identified the genetic basis of thiamine auxotrophy in *Y. lipolytica*. Specifically, we found thiamine depletion results in decreased protein abundance of the lipid biosynthesis pathways and energy metabolism (i.e., ATP synthase), attributing to the negligible growth and poor sugar assimilation observed in our study. Using comparative genomics, we identified the missing 4-amino-5-hydroxymethyl-2-methylpyrimidine phosphate synthase (THI13) in the *de novo* thiamine biosynthesis of *Y. lipolytica,* and discovered an exceptional promoter, P3, that exhibits strong activation or tight repression by low and high thiamine concentrations, respectively. Capitalizing on the strength of our thiamine-regulated promoter (P3) to express the missing gene from *Saccharomyces cerevisiae* (scTHI13), we engineered the thiamine-prototrophic *Y. lipolytica.* By comparing this engineered strain to the wildtype, we unveiled the tight relationship linking thiamine availability to lipid biosynthesis and demonstrated enhanced lipid production with thiamine supplementation in the engineered thiamine-prototrophic *Y. lipolytica.*

610671 Spatiotemporal Control of Protein Localization and Intracellular Metabolic Flux with an RNA Based, High Affinity, Dynamic Scaffold (Industry Candidate)

Abstract:

In biotechnology, platform organisms such as E. coli and S. cerevisiae are used produce small chemicals of interest through non-native enzymatic pathways. Non-native pathway productivity is traditionally optimized through metabolic pathway manipulation and/or directed evolution of the target enzymes, which require multiple iterations and do not always yield the desired results. Alternatively, nonnative enzymes can be clustered together in the cell to maximize metabolic flux through the nonnative pathway. If the clustering is not dynamic, metabolite imbalances can occur which are detrimental to maximizing product titer. To address this issue, a dynamic high affinity scaffold to colocalize enzymes was developed. The scaffold utilizes small RNA sequences and orthogonal Cas6 proteins taken from the CRISPR/Cas Type I systems. The Cas6 proteins bind with high affinity and sequence specificity to short RNA hairpin sequences. Extending the 5' region of the short RNA sequences allows assembly of the scaffold via base pair hybridization. Scaffold disassembly can occur via addition of a short 6-18 nt toehold sequence on one of the hybridized strands. Upon expression of a third RNA strand, the trigger, the scaffold disassociates via toehold mediated strand displacement (TMSD). Scaffold assembly and disassembly via trigger strand expression has been demonstrated using the split luciferase reporter system. Scaffold assembly has been demonstrated to occur only when all correct scaffold components are expressed. Scaffold disassembly has been demonstrated using synthetic triggers and native sRNA sequences (RyhB). Furthermore, a turn ON system which relies on the expression of the trigger to assemble the scaffold has also been developed using the split luciferase system. Finally, the toolkit has been applied to increase the product yield of model metabolic pathways. The expanded capabilities of this scaffold toolkit increase the scope of its application and allows users more freedom to design a scaffold that directly fits their needs.

608046 The Self-Assembling Property of the Major Shell Proteins of the 1,2-Propanediol Bacterial Microcompartment Is Necessary for Compartment Formation (Faculty Candidate)

Abstract:

From small viruses to large eukaryotic cells, biological structures contain proteins that self-assemble into intricate constructs to perform diverse biological functions. The assemblies can be composed of a single protein, as in virus-like-particles, or a more complex mixture of multiple proteins, as in bacterial microcompartments (MCPs). MCPs compartmentalize enzymes and cofactors of a metabolic pathway within a protein shell. This compartmentalization offers solutions to the common bottlenecks of cellular systems used for biomolecule productions, such as reduced flux through the pathway, unwanted side-reactions, and low yield due to cell toxicity from toxic intermediates. The 1,2-propanediol utilization (Pdu) MCP is one of the most studied MCPs. Yet, how the eight shell proteins —PduA, PduB, PduB', PduJ, PduK, PduN, PduU, and PduT – assemble to form the MCP shell and the critical properties necessary for the assembly is not yet known. This knowledge gap prevents researchers from engineering the native system and harnessing this powerful strategy of subcellular organization. In this work, we fill some of that gap by investigating the role of the self-assembling property of the Pdu shell proteins on the MCP formation. We demonstrate that hexameric shell proteins PduA and PduJ self-assemble into long rods and induce a cell division defect when overexpressed. This property is crucial for MCP formation, and we show that the presence of at least one of the two shell proteins is necessary for the MCPs to properly assemble and cannot be substituted by other shell proteins with no self-assembly property. In the process of our study, we also discovered a new control for the Pdu MCP system, where all enzymes are expressed but no protein shell assemble to encapsulate them. This control will facilitate further characterization of the Pdu MCP shell properties, assembly, and function. The improved understanding will enable engineering and wider application of MCPs, such as in metabolic engineering, that will fully harness the benefits offered by subcellular compartmentalization.

613132 Tools to Accelerate Development of Non-Conventional Microbes (Invited Speaker)

Abstract:

Non-model microbes offer many attractive properties for metabolic engineering including efficient plant biomass deconstruction, increased product tolerance, broad simultaneous substrate utilization and much more. However, many remain undeveloped due to a lack of tools to manipulate and engineer these species. In particular, non-model species offer unique challenges such as genome compositional biases, and a lack of well-developed genetic tools that stymie strain development even with the availability of powerful CRISPR-based tools. To address this need, my lab develops novel synthetic biology tools to engineer non-conventional organisms. In my talk, I will introduce our work with environmentally-isolated anaerobic fungi to illustrate specific challenges faced in strain development. I will then describe our work with prokaryotic Argonautes, programmable DNA endonucleases, that have the potential to rival CRISPR-based gene editing tools in species with extreme biases in genome composition. We have developed *in vivo* screens that allow for the rapid characterization of putative Argonautes and used these tools to validate previously unrecognized structural insights into the function of controversial NgAgo. Finally, I will provide a brief overview of an emerging platform in my lab to rapidly prototype environmentally-sensitive gene expression in a species-agnostic manner. Together, these tools will accelerate the development of emerging non-model species and facilitate their use for biomanufacturing.

605158 Mechanistic Understanding of the Biological Responses to Polymeric Nanoparticles

Abstract:

Polymeric nanoparticles have demonstrated tremendous potential for theranostic and nanomedicine applications due their central role in the delivery of a wide range of therapeutic and imaging agents [1-4]. Nonetheless, increasing studies have reported the failures of these nanomaterials in realizing their designed objectives because of a single or combination of factors, notably non-specific biomolecule adsorption, inefficient cellular internalization, triggered immune response, and fast body excretion, which have collectively challenged the merit of nanomedicine for disease theranostics [5-7]. Although numerous efforts have attempted to regulate the biological behaviors of polymeric nanoparticles by controlling their physicochemical properties, a large number of the demonstrated studies have focused solely on either the biological effects induced through the regulation of a single nanoparticle parameter or the separate molecular, cellular, and *in vivo* behaviors of nanoparticles. Crucially, the interdependency of multiple physicochemical properties of polymeric nanoparticles in determining their biological effects from molecular to organism levels is still largely unclear, which warrants a thorough investigation.

In this work, we prepared a series of polymeric nanoparticles consisting of various biocompatible polymeric nanocarriers (i.e., polystyrene (PS), poly(lactic-*co*-glycolic acid) (PLGA), and polyethylene glycol (PEG)) encapsulating organic fluorophores [8]. Subsequent to the synthesis of these nanoparticles, we evaluated their physicochemical properties, such as surface morphology, size distribution, zeta potential, lipophilicity, optical absorbance, and fluorescence. Next, through a range of experimental characterization and molecular simulation, we systematically probed the interdependent effect of nanoparticle lipophilicity, zeta potential, and size on the formation of protein corona around the nanoparticles as well as the *in vitro* cellular uptake and *in vivo* biodistribution of nanoparticles in a zebrafish model.

We observed that nanoparticle lipophilicity influenced the recruitment of non-specific biomolecules, where the adsorption of biomolecules could be reduced by decreasing nanoparticle lipophilicity and vice versa. For a given lipophilicity, the nanoparticle size had a more significant effect in determining biomolecule adsorption than the nanoparticle surface charge. However, irrespective of size, the surface charge of nanoparticles was noted to affect their endothelium and macrophage uptake as well as circulation lifetime. More clearly, the negatively charged nanoparticles could be internalized preferentially by endothelial cells without the need for targeting ligands. Additionally, these nanoparticles were minimally internalized by macrophages and showed a much prolonged *in vivo* circulation lifetime. Based on our experimental data, we proposed a two-step framework to rationally design a single polymeric nanoparticle system with highly regulated *in vitro* and *in vivo* biological behaviors.

In summary, the optimization of the interplay between nanoparticle lipophilicity and surface charge is necessary to control the overall biological behaviors of polymeric nanoparticles. Our study has offered a strong basis for the engineering of enhanced polymeric nanoparticle systems to regulate multiple biological events simultaneously for *in vivo* theranostic and nanomedicine applications.

References

[1] Elsabahy et al., *Chemical Reviews* 115 (2015), 10967-11011.

[2] Ulbrich et al., *Chemical Reviews* 116 (2016), 5338-5431.

[3] Kenry et al., *Advanced Materials* 30 (2018), 1802394.

[4] Kenry et al., *Accounts of Chemical Research* 52 (2019), 3051-3063.

[5] Wilhelm et al., *Nature Review Materials* 1 (2016), 16014.

[6] Rosenblum et al., *Nature Communications* 9 (2018), 1410.

[7] *Nature Nanotechnology* 14 (2019), 1083.

[8] Kenry et al., *ACS Nano* 14 (2020), 4509-4522.

605108 Alterations in the Membrane Lipid Composition of Live Cells Affects the Ability of Cells to Internalize Nanomaterials

Abstract:

Over the past decades, engineered nanoparticles have been extensively used to deliver drugs or imaging agents to mammalian cells. Nanoparticles can be internalized into the cells through passive diffusion or a number of endocytic pathways. Previous studies on nanoparticle endocytosis have been primarily focused on elucidating the endocytic pathways, and the membrane receptors involved in each pathway. However, the role of the lipid microenvironment in regulating nanoparticle endocytosis has remained unexplored. In this study, live cell-lipid exchange was used to replace the lipids of the outer leaflet in A549, alveolar epithelial cells, with lipids of choice and examine their role in the uptake of silica nanoparticles (50 nm).

Methyl-alpha-cyclodextrin (MαCD) was used to alter the lipid composition of outer leaflet of cell plasma membrane. For the exchange studies, MαCD was pre-loaded with sphingomyelin (18:0, 18:1, and 24:0), 1,2-dioleoyl-sn-glycero-3-phosphocholine (36:2 PC, DOPC), and 2-((2,3-bis (oleoyloxy)propyl)dimethylammonio) ethyl hydrogen phosphate (36:2 inverse PC, DOCP). In each case, the exchange of cellular lipids with these synthetic lipids was confirmed by chromatography and mass spectrometry and cell viability after exchange was examined using the MTT assay. Control and lipid-exchanged cells were then exposed to fluorescent silica nanoparticle in serum-free medium or supplemented with fetal bovine serum (FBS) and incubated for one hour at 37 °C. The intensity of fluorescent nanoparticles, associated with or internalized into cell plasma membrane, was measured using flow cytometry. Lysosome and early endosome marker were used to study the intracellular fate of nanoparticles using confocal microscopy.

All exogenous lipids maintained cell viability; however, lipid-exchanged cells showed significant differences in their ability to endocytose nanoparticles in serum-free medium. DOCP and SM 18:0-exchanged cells showed significantly reduced nanoparticle uptake, SM 18:1-exchanged cells showed increased uptake, while DOPC-loaded cells showed a comparable level of uptake to control cells. Confocal microscopy confirmed the differences in observed in flow cytometry experiments, while showing differences in intracellular localization of nanoparticles in control and lipid-exchanged cells. No difference in nanoparticle uptake was observed in FBS-supplemented medium, likely due to the fact that serum proteins resulted in significant aggregation of nanoparticles, rendering them too large for endocytic uptake. These findings clearly indicate an important role for membrane lipid composition in nanoparticle uptake and are important in understanding the differences in uptake in cells with different membrane lipid compositions.

606677 Morphology-Based Transport of Gold Nanoparticles in Mature Plant Leaves

Abstract:

Plant genetic engineering has the potential to mitigate the effects of climate change and rising populations via the creation of high-yield crop varieties and climate-resilient plants. To this end, we require a plant genetic engineering toolset that is (i) plant-species independent, (ii) able to bypass physical barriers present in mature plant tissues, and (iii) capable of controlled localization and cargo release. A principal challenge in this field remains the efficient delivery of biological cargo across a plant's cell wall and cell membrane. Existing tools for plant genetic engineering suffer limitations in species range, cargo diversity, tissue damage, efficiency, and transgene integration resulting in regulatory oversight. Recently, nanoparticles (NPs) have emerged as promising materials for use as cargo carriers into plant cells. Owing to their highly tunable chemical and physical properties, NPs can be synthesized and surface functionalized to achieve targeted localization and cargo release.

Current techniques either use pathogens or biolistic force to deliver biomolecules to plants. There is interest in unassisted (non-biolistic, force-independent) delivery of genetic cargoes into plant cells, in which nanocarriers may need to remain below the cell wall size exclusion limit. However, there lacks a set of heuristics for NP design for unassisted DNA, RNA, and protein delivery in plants.

Herein, we utilize a library of NA-coated gold nanoparticles as model materials to elucidate the effect of NP morphology on transport in plant tissues and on cargo protection against nuclease degradation. Our study suggests that nanoparticle morphology impacts both the uptake pathway, timescale, and extent of nanoparticle internalization into plant cells. The workflow presented in this work can be utilized to probe the effect of other nanoparticle-based design parameters such as size and surface chemistry on biomolecule delivery efficacy, which can enable the creation of NP design heuristics for force-independent internalization. This work serves as an initial platform that can be built on to understand how design parameters affect delivery in plants for more targeted applications of bionanotechnologies in plant genetic engineering and more broadly in environmental and agricultural science.

596314 Silver Nanoparticle Surface Chemistry Affects Protein Interactions, Cell Viability, and Gene Expression in Human Liver Cells

Abstract:

Silver nanoparticles (AgNPs) are increasingly being used in consumer and biomedical applications for their antimicrobial and plasmonic properties. Yet, detailed characterizations of physiological responses to AgNPs, as well as the influence of particle properties (such as surface chemistry) and interactions with other biomolecules on these responses are needed before their wide application. Furthermore, increasing concerns of frequent and cumulative exposures point to a need for cumulative risk assessment studies of AgNP exposure. Our work uses AgNPs with a range of surface coatings to evaluate the role of surface charge in protein interactions, cellular cytotoxicity, and gene expression. Conformational changes in the model protein human serum albumin (HSA) upon AgNP interaction, as monitored via circular dichroism spectroscopy, were most pronounced for cationic particles. In contrast, HSA-AgNP binding constants were highest for negatively charged AgNPs, though the differences in affinities were not significant across different surface coatings. Consistent with the trends in HSA-AgNP binding affinities, HSA-mediated changes in AgNP hydrodynamic radius and oxidative dissolution were highest for anionic particles. Cytotoxicity experiments carried out using hepatic cells revealed that none of the AgNP preparations studied showed significant toxicity at low nanoparticle concentrations. At higher concentrations, however, cationic AgNPs showed significant impacts on cell viability, both with and without HSA coating. Consistent with the cytotoxicity studies, results of transcriptomics analysis demonstrated that AgNPs altered gene transcription of HepG2 cells in a time-dependent manner, with the most dramatic effects seen for cationic AgNPs. Furthermore, universal to all surface-coatings, AgNPs treated cells responded by inactivating proliferation and enabling cell cycle checkpoints. Taken together, these results demonstrated the significance of AgNP surface charge in both biophysical HSA-AgNP interactions and cytotoxicity and gene expression, with cationic AgNP causing greatest disruption to HSA structure and cell responses.

609405 Inert Nanoparticles for Enhancing the Survival of Primary Macrophages.

Abstract:

Cell-based therapies have recently emerged as a new platform for treating a spectrum of chronic degenerative diseases, with serious advances made especially in hematopoietic stem cell and reactive lymphocyte therapies. Macrophages are leukocytes responsible for phagocytosing foreign objects, bacteria, and apoptotic cells, maintaining homeostasis, and bridging the innate and adaptive immunity. Macrophage-based cell therapy has been identified as a promising therapeutic approach to treat chronic immune dysfunctions, owing to the functionality and phenotype plasticity of macrophages. However, a major challenge in the advancement of macrophage-based cell therapies is the difficulty to maintain adequate cell viability *ex vivo* following macrophage isolation from tissue. Current attempts of macrophage *in situ* activation or autologous cell therapy following *ex vivo* stimulation fail to elicit effective therapeutic responses, due to low cell survival rates upon reintroduction. Therefore, there is a need to develop new approaches to regulate primary macrophage survival in both *in vivo* and *ex vivo* environments.

We attempted to extend the viability of *ex vivo* macrophage cultures through treatment of inert poly(ethylene glycol) (PEG) diacrylate (PEGDA)-based nanoparticles. We observe that *ex vivo* bone marrow-derived macrophages (BMMs), which typically undergo apoptosis within 2 weeks under appropriate culture conditions, can persist and maintain functionality for up to 6 months after being treated with a *single* dose of PEGDA nanoparticles. We find that nanoparticle uptake significantly delays primary macrophage cell death through the downregulation of caspase-dependent apoptosis pathways. Furthermore, we observe no major changes to macrophage phenotype following nanoparticle uptake. However, phagocytosis of PEGDA nanoparticles leads to increased expression of major histocompatibility complex (MHC) class II uniquely in the absence of other co-stimulatory molecules. The enhanced survival phenomenon is also applicable to *ex vivo* cultures of terminally differentiated macrophages isolated from tissue, including peritoneal and alveolar primary cells. Overall, this work demonstrates for the first time the ability of inert nanoparticles to suppress apoptotic signaling and prolong viability in primary macrophages extracted from different tissues. This work could eliminate a major obstacle standing in the way of developing macrophage-based cell therapies, as well as provide mechanistic insight into the implications of nanoparticle phagocytosis on cell longevity and function.

612225 Peptides As Surface Coatings of Nanoparticles That Penetrate Human Cystic Fibrosis Sputum and Uniformly Distribute In Vivo Following Pulmonary Delivery

Abstract:

In mucosal-associated diseases, drug and gene delivery systems have to traverse multiple biological barriers to achieve therapeutic efficacy. In particular, in select pulmonary diseases such as cystic fibrosis (CF), therapeutic delivery remains a significant challenge due to concentrated viscoelastic mucus, which prevents effective penetration and transmucosal transport of drugs and particles. To address this problem, we used combinatorial peptide-presenting phage libraries with next-generation DNA sequencing to identify hydrophilic, net-neutral charged peptides as coatings that enable penetration of human CF mucus *ex vivo* with ~600-fold better penetration than control. In addition, these peptide coatings improve uptake into lung epithelial cells compared to uncoated or standard poly(ethylene glycol) (PEG)-functionalized nanoparticles and exhibit enhanced uniform distribution and retention in the mouse lung airways. These peptide coatings address multiple delivery barriers and effectively serve as excellent alternatives to standard PEG surface chemistries to achieve mucus penetration and address some of the challenges encountered using these chemistries. This biomolecule-based strategy can address multiple delivery barriers and hold promise to advance efficacy of therapeutics for diseases like CF.

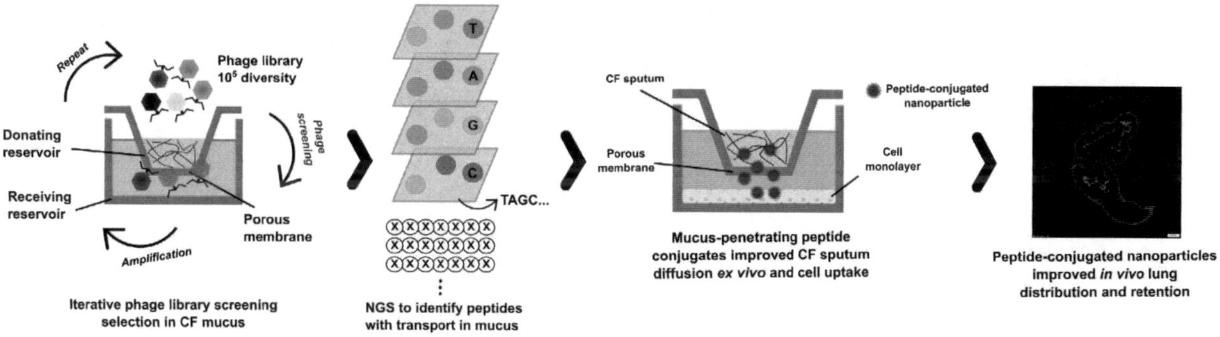

601987 Producing Predictive Chemical Process Models from a State of Incomplete Mechanistic Knowledge and Small Data Sets

Abstract:

ABSTRACT: In this work, we explore the possibility of developing a quantitative and predictive model for chemical processes, without having to first gain a complete understanding of the underlying mechanisms. We find that this can be done efficiently combining: (1) expectations for the functional expression of the process—extracted from a general, *qualitative* understanding on how the chemical process ought to behave—and, (2) small data sets providing examples of the process outputs resulting from varying process inputs. The strategy discussed is labeled knowledge-constrained machine learning. The main example we take is on reaction kinetic models—and, here, the expectations are written in terms of constraints on the model predictions over time, and the data is a series of time-point measurements for the reaction outcome under different conditions. But, the intent of the presentation is to discuss a general methodology that may be used to develop predictive models across the range of unit operations found in the processes used to produce synthetic drug substance API's.

594384 Identifying the Processing Space of Continuous Granulation: A Case Study of Extended Release Tablets

Abstract:

While continuous manufacturing (CM) has been employed for directly compressible tablets, the continuous wet granulation has also been used as an alternative in continuous tableting lines. To date, no matrix-based extended release tablets manufactured by a CM technology has been approved. Compared to batch granulation, continuous twin screw granulation exerts higher shear to produce denser granules with broader size distribution within very short time (seconds). Changing the screw configuration, process, and formulation parameters can alter the physical properties of the granules, e.g., shape, flow, compressibility, density, size distribution, etc. In twin screw granulation, complex interactions among these parameters manage the granule properties. This study aimed at identifying the critical material attributes and process parameters and their interactions to be controlled for switching a batch process to a CM granulation process.

Using batch high shear granulation, we evaluated the formulation and process parameters that can produce metoprolol succinate extended release tablets. These formulation and process parameters were successfully transferred to continuous twin screw granulation to produce the same dissolution properties of the finished product. The effects of the screw configuration parameters on the characteristics of the granules were further investigated. The interaction between the processing, e.g., liquid to solid ratio, screw speed, barrel fill volume, etc., and the significant screw parameters was studies to unveil the processing space and to identify the failure modes.

Similar granule quality attributes were obtained for most of the formulations; although, the conditioned bulk density and compressibility of the granules determined the dissolution properties of the tablets. The number of kneading and sizing elements had the most significant impact on granule characteristics. The effects of the staggering angle and distance between the kneading zone of the screws were negligible.

This presentation will then highlight the processing space of screw granulation to produce extended release tablets. Some of the failure modes will be discussed along with the feasibility of implementing PAT tools to monitor the granulation process and granules properties.

DISCLAIMER

This presentation reflects the views of the presenter and should not be construed to represent FDA's views or policies.

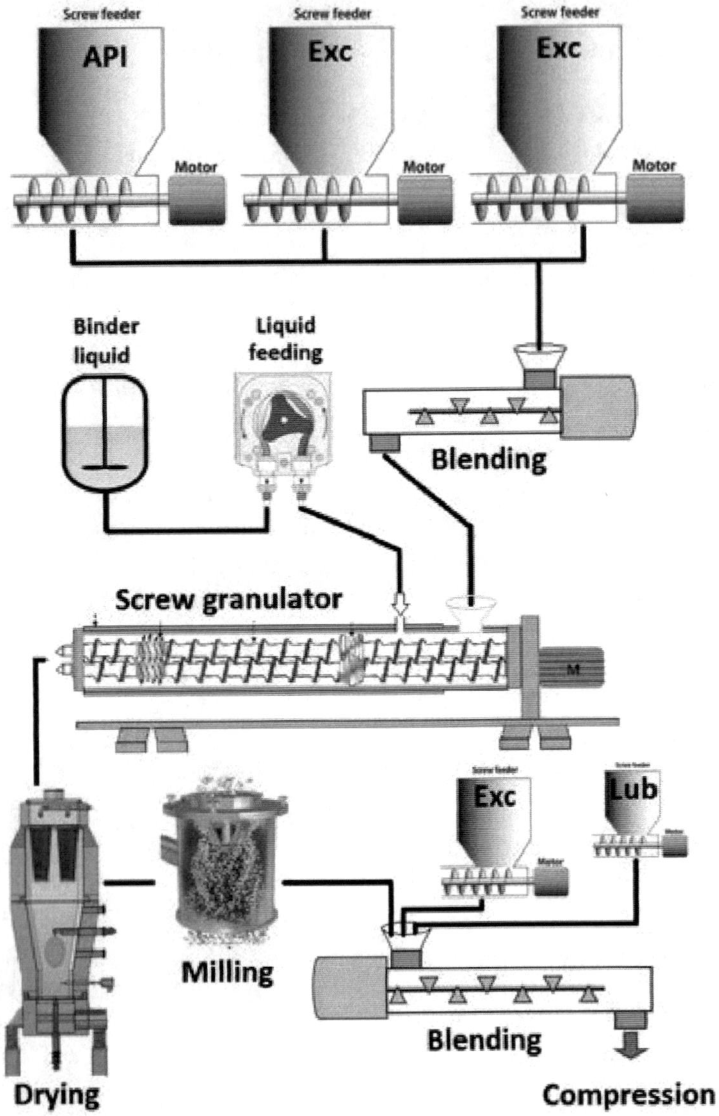
Screw feeder
API
Motor
Screw feeder
Exc
Motor
Screw feeder
Exc
Motor
Binder
liquid
Liquid
feeding
Blending
Screw granulator
M
Exc
Lub
Milling
Blending
Drying
Compression

608665 Dynamic Optimization Strategies Towards Addressing the Challenges in Protein Therapeutics

Abstract:

Protein therapeutics is an important field of the pharmaceutical industry. Current challenges in this field include meeting the increasing market demands, handling the entrance of biosimilar products and complying with quality by design (QbD) principles [1-3]. To tackle these challenges effectively, efficient use of existing tools, methods and solution approaches together with the development of novel ones is necessitated. This work provides an overview of our recent efforts to address some of the aforementioned challenges using model-based dynamic optimization strategies.

Model-based approaches are powerful tools towards facilitating process development and advancing process performance. They offer a promising alternative to costly and time-consuming experimentation [1], and can have a profound effect on process sustainability and profitability. Recent advances in cell culture modeling (with cell cultures representing a typical production platform for many of the products of this class) have enabled significant insight to the system's dynamics and the underlying production mechanisms, and have paved the way towards developing dynamic optimization strategies for process intensification (cf., [4-6]).

Existing numerical methods and optimization software for solving the resulting dynamic optimization problems rely almost exclusively on local techniques. Due to the noncovexity of the described complex biological systems, the vast majority of these problems exhibits the presence of multiple local minima. Nevertheless, deriving suboptimal local solutions can have direct quality and cost impacts. Despite the significant progress in deterministic global optimization [7], it is not yet tractable for the considered systems. Hence, alternative methods to overcome the hazards of suboptimal local optimality are required. Stochastic and heuristic approaches can be advantageous to this end. In our previous work [8], we have shown how incorporating biological process knowledge into the optimization problem formulation can facilitate the derivation of superior local solutions. In a similar direction, we illustrate how proper decomposition of the optimization problem can outweigh certain advantages [9].

Beyond increasing product yield, monitoring and controlling product quality is essential. Regulatory authorities including FDA are increasingly recommending consideration of QbD to develop processes that consistently meet quality specifications [3, 10]. This requires further attempts in modeling, as well as reconsiderations in the formulation of the dynamic optimization problem to effectively account for direct quality attributes. Yet, current methods in this direction are still in an early stage. The recent work in [11] demonstrates a successful example of model-based optimization to maximize galactosylation (a glycosylation related metric, with glycosylation presenting a critical quality attribute for many therapeutic proteins [12, 13]) content for an antibody case study. In our recent study [14], we have combined model simulations and design of experiments to investigate the effects of nutrients and nucleotide sugars feeding in protein glycosylation. Building on this basis, we have developed an optimization framework to maximize process performance while meeting quality specifications, and thus provided a successful example for utilizing dynamic optimization to apply the QbD paradigm in protein therapeutics [15].

Acknowledgements:
This work has received funding from the European Union's Horizon 2020 research and innovation programme under the Marie Skłodowska-Curie grant agreement no.675251.

References:

[1] Farzan, P., Mistry, B., & Ierapetritou, M. G. (2017). Review of the important challenges and opportunities related to modeling of mammalian cell bioreactors. AIChE Journal, 63(2), 398-408.
[2] Shukla, A. A., Wolfe, L. S., Mostafa, S. S., & Norman, C. (2017). Evolving trends in mAb production processes. Bioengineering & Translational Medicine, 2(1), 58-69.
[3] Collins, P. C. (2018). Chemical engineering and the culmination of quality by design in pharmaceuticals. AIChE Journal, 64(5), 1502-1510.
[4] Dhir, S., Morrow Jr, K. J., Rhinehart, R. R., & Wiesner, T. (2000). Dynamic optimization of hybridoma growth in a fed-batch bioreactor. Biotechnology and Bioengineering, 67(2), 197-205.
[5] Kiparissides, A., Pistikopoulos, E. N., & Mantalaris, A. (2015). On the model-based optimization of secreting mammalian cell (GS-NS0) cultures. Biotechnology and Bioengineering, 112(3), 536-548.
[6] Kappatou, C. D., Mhamdi, A., Campano, A. Q., Mantalaris, A., & Mitsos, A. (2018). Model-based dynamic optimization of monoclonal antibodies production in semibatch operation—Use of reformulation techniques. Industrial & Engineering Chemistry Research, 57(30), 9915-9924.
[7] Chachuat, B., Singer, A. B., & Barton, P. I. (2006). Global methods for dynamic optimization and mixed-integer dynamic optimization. Industrial & Engineering Chemistry Research, 45(25):8373–8392.
[8] Kappatou, C. D., Mhamdi, A., Campano, A. Q., Mantalaris, A., & Mitsos, A. (2017). Dynamic optimization of the production of monoclonal antibodies in semi-batch operation. In Computer Aided Chemical Engineering (Vol. 40, pp. 2161-2166). Elsevier.
[9] Kappatou, C. D., Altunok, O., Mhamdi, A., Mantalaris, A., & Mitsos, A. (2019). Sequential and Simultaneous Optimization Strategies for Increased Production of Monoclonal Antibodies. In Computer Aided Chemical Engineering (Vol. 46, pp. 1021-1026). Elsevier.
[10] Pais, D. A., Carrondo, M. J., Alves, P. M., & Teixeira, A. P. (2014). Towards real-time monitoring of therapeutic protein quality in mammalian cell processes. Current Opinion in Biotechnology, 30, 161-167.
[11] Kotidis, P., Jedrzejewski, P., Sou, S. N., Sellick, C., Polizzi, K., del Val, I. J., & Kontoravdi, C. (2019). Model-based optimization of antibody galactosylation in CHO cell culture. Biotechnology and Bioengineering, 116(7), 1612-1626.
[12] Hossler, P., Khattak, S. F., & Li, Z. J. (2009). Optimal and consistent protein glycosylation in mammalian cell culture. Glycobiology, 19(9), 936-949.
[13] Blondeel, E. J., & Aucoin, M. G. (2018). Supplementing glycosylation: A review of applying nucleotide-sugar precursors to growth medium to affect therapeutic recombinant protein glycoform distributions. Biotechnology Advances, 36(5), 1505-1523.
[14] Ehsani, A., Kappatou, C. D., Mhamdi, A., Mitsos, A., Schuppert, A., & Niedenfuehr, S. (2019). Towards model-based optimization for quality by design in biotherapeutics production. In Computer Aided Chemical Engineering (Vol. 46, pp. 25-30). Elsevier.
[15] Kappatou, C. D., Ehsani, A., Niedenführ, S., Mhamdi, A., Schuppert, A., & Mitsos, A. Quality-targeting dynamic optimization of monoclonal antibody production. (Submitted)

608965 Integrated Supply Chain Network Design and Inventory Management for Autologous Cell Therapy.

Abstract:

Personalized medicine is gaining a lot of attention in the pharmaceutical industry as it can treat diseases that earlier had no cure. In the past few years, we witnessed exponential growth in the development of cell and gene therapy products. Cancer Research Institute reported that 753 cancer cell therapies are currently under development all over the world, of which 375 are in clinical trials [1]. There has been an increase in the growth and demand for autologous therapy due to the recent approval from the FDA [2]. Therefore, autologous therapy is undergoing the transformation from laboratory to commercial scale.

The autologous therapy starts with a collection of cells from the patient, which are cryopreserved and sent to the manufacturing site. The cells are then formulated and enriched. The formulated cells are again cryopreserved and transferred back to the patient, where they are injected back. Unlike traditional pharmaceutical drugs that are produced in bulk, these drugs are patient-centric and have to be manufactured for each patient separately. In addition, the process also deals with living cells that go around different nodes in the supply chain. Therefore, the cells have to be cryopreserved in storage device and transported across the supply chain.

An integrated approach is considered for supply chain network design and inventory management of the raw materials. We consider a network with a set of hospitals where the patients arrive. The facilities to be set up are distribution centers and manufacturing centers. The manufacturing center receives the samples from the patient and engineers the cells to create the drug. The distribution centers have the inventory for the storage containers and are installed for the intermediate halt for the sample transportation between the hospital and manufacturing center. To account for the transit times for the material transfer between facilities, we define the parameter accounting for both transportation time and waiting time for different transportation modes as well as the processing times at each of the node. We also simultaneously optimize inventory policy. We use periodic-review policy for the raw materials inventory at the distribution center [3]. Under this policy, if the inventory is below the defined level, the order is placed to restore the inventory to level . Since the storage devices are reusable, the policy is modified to account for the returning devices after the delivery of the drug.

Given the demand distribution, the potential locations of distribution and manufacturing centers, costs and time for processing, inventory, and transportation, we determine the location of different centers, inventory levels at the distribution centers, response time, and cost of delivery for each sample, with objective function of minimizing the cost and response time. The problem is formulated as a single objective optimization where response time is multiplied by a factor and added to the cost function. A case study of China is considered, with 20 hospitals situated in different provinces and a total of 130 patients over a period of 1 month. Also, the location of 20 potential locations of the distribution centers and 20 manufacturing centers are given. We found out that the majority of the cost is incurred in setting up manufacturing and distribution centers. The cost accounting for the response time depends on the multiplicative factor we choose. In addition, we also explored the importance of batching the samples at the distribution centers and transporting to the manufacturing center subjected to the response time constraint. Decisions related to inventory of raw material are critical to have an efficient supply chain. Also, sensitivity analysis is conducted to study the effect of different parameters on the optimal network design and overall cost.

References:

1. Tang, J., et al., *The global landscape of cancer cell therapy.* Nat Rev Drug Discov, 2018. **17**(7): p. 465-466.
2. Bach, P.B., S.A. Giralt, and L.B. Saltz, *FDA approval of tisagenlecleucel: promise and complexities of a $475 000 cancer drug.* Jama, 2017. **318**(19): p. 1861-1862.
3. Brunaud, B., et al., *Inventory policies and safety stock optimization for supply chain planning.* AIChE Journal, 2019. **65**(1): p. 99-112.

600490 Two-Stage Control Using Model-Based Reinforcement Learning and Predictive Control for Fed-Batch Bioreactor

Abstract:

As the demand for biomass and pharmaceutical materials is increasing in the market, various genetically modified cells are tested in a batch bioreactor in order to increase production rate [1]. At the same time, a robust and efficient control strategy is required to successfully produce the product with uncertainty of cellular behaviors [2]. However, limitation of the information on the system dynamics makes it challenging problems to optimize the operating conditions of a bioreactor and to implement an automatic controller. In general, the biosystem is less reproducible and shows higher nonlinearity compared to other engineering systems. In particular, most of the parameters in the first-principle model are lumped type and these values would be varied not only in a single batch operation with changes of the states but in a batch-to-batch fashion. Therefore, implementing an automatic control to track the optimal control trajectories obtained off-line is not a robust approach to produce fine products for a large number of batch reactors.

In this work, we propose a two-stage control strategy to optimally control a fed-batch type bioreactor that produces penicillin. The proposed strategy utilizes model-based Reinforcement Learning (RL) off-line to optimize the operating trajectories of state and input. In on-line, the Moving Horizon Estimation (MHE) is applied to estimate the parameter values, and an adaptive Model Predictive Control (MPC) is utilized to simultaneously modify and track the trajectories obtained in off-line. First, the first-principle model of the system dynamics is represented in Stochastic Differential Equations (SDEs) form which could theoretically handle the white noise in a continuous system. In addition, the variance of the process noise could shift a mean value of drifting terms of the system so that the changes in parameter values could be presented by adjusting the variance of the process noise [3]. These SDEs are numerically integrated and served as a virtual plant. The next step is obtaining the operating trajectories of the reactor off-line. The optimal control method could be classified in three classes, direct, indirect and RL (dynamic programming-based method). The direct method is the one mostly applied in practice since the problem formulation and the procedure of obtaining the solution is intuitive. On the other hand, RL has several advantages over the direct method. RL provides a closed-loop solution of the optimal control problem so that the control policy could be obtained instead of just a control trajectory [4]. This is a more robust approach since RL evaluates the control cost for the case where the states have deviated from the optimal trajectories. In addition, value based RL provides not only the optimal trajectories but also the value of state which is an expected total cost of the control policy starting from the state [4]. Although this value is just a by-product to find the optimal control policy, it is useful information when the trajectories are modified on-line.

The trajectories suggested by RL could serve as a baseline of optimal control. However, due to the change in parameter values and disturbances, tracking those trajectories might not be feasible which requires a modification of trajectories. For this purpose, MHE and adaptive MPC could be utilized to estimate the parameters and modify the trajectories on-line. However, even though changes in parameter values are successfully estimated by MHE, the modification of trajectories by MPC for the entire horizon could not be feasible by the lack of computational time. This forces MPC to solve the optimal control problem with a limited length of prediction horizon where the terminal state of the bioreactor is not considered. This becomes a problem because most of the important cost is imposed only at the terminal state of the bioreactor. Therefore, providing a proper terminal cost of MPC becomes a crucial point to properly modify the trajectories. At this point, the value calculated from the RL could be used as the terminal cost of MPC. Although this value would not be exact by the changes of parameter values, the simulation results suggest that this approximation is valid for MPC to stably calculate the new trajectories.

The key to this combination of MPC and RL is selecting the proper length of the prediction horizon. Under this scheme, the length of the prediction horizon is not only determined by the computational limits, but it is related to the reliability of the parameter values. The long prediction horizon implies that MPC is more rely on the newly estimated parameter values than the one used for RL. The presence of the parameter changes is one support on using the longer prediction horizon. However, the reliability of the estimated parameter values is related to both the quality and quantity of measurement data. Therefore, the choice of prediction horizon length should be carefully determined by evaluating the reliability of the parameter values. We propose one method for evaluating such reliability and propose a method of adjusting the prediction horizon length.

[1] C. Larroche, Sanromán Maria Ángeles, G. Du, and A. Pandey, Current developments in biotechnology and bioengineering. Amsterdam: Elsevier, 2017.

[2] Shuler and Kargi, Bioprocess engineering. United States: Academic Internet Publishers, 2007.

[3] H.-H. Kuo, Introduction to stochastic integration. New York: Springer, 2006.

[4] R. S. Sutton, Reinforcement learning: an introduction. Cambridge, Mass: MIT Press, 2018.

608716 Sustainable API Development: Developing a Strategy for Measuring and Reducing Environmental Impact in the Pharmaceutical Industry

Abstract:

In January 2020, AstraZeneca announced its ambition to eliminate greenhouse gas emissions from its sites and fleet by 2025 and to become carbon negative across the entire value chain by 2030. The development and manufacture of active pharmaceutical ingredients (API) contributes significantly to AstraZeneca's carbon footprint and process engineers have a responsibility to play a leading role in achieving the zero-carbon target.

This presentation will share the approach taken to implement a strategy for measuring and reducing environmental impact during the development and manufacture of APIs. This approach comprises of influencing ways of working during API process development, implementing robust measurement of process environmental impact and identifying opportunities for innovation through science and technology to minimise AZ's carbon footprint.

This approach describes developing a procedure to appropriately consider sustainability alongside other key drivers for process design such as product quality, manufacturability and regulatory factors. Similarly, creating a culture to deliver continuous improvement against established metrics for sustainability in every aspect of API development and manufacturing.

This case also describes initiation of novel technology investigations to significantly reduce the carbon footprint of AstraZeneca's API manufacturing processes. This will be a leading factor in achieving a 50% reduction in carbon footprint for API development and the zero-carbon ambition of the company.

606173 An Improved Method for Predicting Autoignition Temperatures Based on First Principles

Abstract:

Methods for predicting autoignition temperatures (AIT) have been historically inaccurate and are rarely based on the underlying physical phenomena leading to observed AIT. Previous attempts at predicting AIT have been developed based on Quantitative Structure-Property Relationships (QSPR) and additive group contribution methods. This work presents an improved method for predicting AIT based on the method originally developed by the late Dr. William H. Seaton. The method of Seaton is described in detail including its use of first principles to predict AIT. Improvements on the method of Seaton and underlying principles are presented and discussed.

611821 Water Washing of Startup Crude Oil

Abstract:

During startup of cold sub-sea wells, methanol is injected at the wellhead to prevent the formation of hydrates in the pipeline/riser system. Methanol injection continues until the fluid arrival temperature exceeds hydrate-formation temperature at the operating pressure, and the typical duration is a few days. Methanol contamination is problematic, and hence the methanol specification of the crude oil is usually less than 20 ppmw. Water or brine wash is often used to reduce the methanol content of the startup crude oil. This paper studies the design of these systems using process simulation. The enabling need is for a reliable and accurate thermodynamic model. The model must describe the three-phase system containing the aqueous and oil liquid phases and the vapor phase, and the key needed feature is description of the distribution coefficient of methanol between the oil and aqueous phases. A model has been developed based upon the ElectrolyteNRTL model in Aspen Plus, and the presentation describes the key data sources and the parameter development. The model enables design of an optimum process flowsheet, and a two-stage countercurrent has been determined to be the optimum option.

607019 Universal Correlation for the Binary Interaction Parameter of Hydrogen with Petroleum Fractions in Refinery Hydroprocessing Applications

Abstract:

Hydrotreating and hydrocracking applications in oil refineries use large quantities of hydrogen. Hydrotreating removes hetero-atoms of sulfur, nitrogen and oxygen and can saturate aromatic ring structures, producing more desirable transportation fuels, such as ultra-low sulfur diesel. Hydrocracking is a more severe reaction, which breaks carbon-carbon bonds and improves the ratio of hydrogen to carbon in the fluid, again, making more suitable transportation fuels. Hydrogen is expensive and is either made on-site, perhaps through steam reforming of methane, or is bought from outside as a utility.

To design new hydroprocessing facilities and to improve the performance of existing plant, it is essential to be able to predict the solubility of hydrogen in typical hydrocarbon fluids. The hydrogen has to be separated from the product streams in high and low pressure separators and is then recycled to the reactor feed, where it is combined with hydrogen make-up.

Although Detailed Hydrocarbon Analysis (DHA) enables us to define the composition of reactor feedstocks in terms of a larger slate of defined chemical components, it is still common practice to represent the heavier portion of the feed in terms of undefined components, or petro fractions, characterized by narrow cut boiling points and densities from a petroleum assay.

Cubic equations of state (EoS), such as Soave-Redlich-Kwong and Peng-Robinson are the workhorse thermodynamic methods for calculating vapor liquid equilibrium (VLE) in refinery processes. From the study of the VLE of binary systems of hydrogen as a solute in a defined hydrocarbon solvent, it is evident that a significant correction to the basic Van der Waals mixing rule in the EoS is required if we are to predict accurately the composition of hydrogen in the liquid phase. Moreover, we can show that this correction, through a binary interaction parameter (k_{ij}) needs to be temperature-dependent. While we can correlate k_{ij} for hydrogen binaries with defined components through regression of experimentally measured VLE data, the problem is how we can predict the interaction between hydrogen and any petro fraction? We observe that increasingly large values of k_{ij} are required as the molecular weight of the hydrocarbon increases and any petro fractions are likely to be in the higher molecular weight range. Ignoring the interaction parameter between hydrogen and a petro fraction will result in significant inaccuracy in any prediction of solubility.

In this study, we will show how the wealth of binary VLE data for hydrogen with defined hydrocarbon solvents can be reduced through regression and conversion to Henry's Law constants that correlate with the reduced temperature and critical pressure of the solvent [1,2]. We propose a new universal correlation for these Henry's Law constants that will allow the subsequent prediction for k_{ij} between hydrogen and a petro fraction, based only on knowledge of the critical temperature, critical pressure and acentric factor of the petro fraction that can be calculated from well-established estimation methods. The validity of this approach will be shown by comparing the predicted solubility of hydrogen in a well-analyzed diesel stream with a base case using methods currently described in the Technical Data Book of the American Petroleum Institute [3].

[1] Trinh TKH, de Hemptinne JC, Lugo R, Ferrando N, Passarello JP, "Hydrogen Solubility in Hydrocarbon and Oxygenated Organic Compounds." J. Chem. Eng. Data 2016,61,119-34

[2] Fernández-Prini R, Alvarez JL, Harvey AH, "Henry's Constants and Vapor–Liquid Distribution Constants for Gaseous Solutes in and at High Temperatures "J. Phys. Chem. Ref. Data, 3012, 32, 903-912

[3] API Tech Data Book 10, 2019

600821 Distillation Curve and Vapor Pressure Predictions of Alcohol-Gasoline Mixtures Using an Advanced Equation of State

Abstract:

Due to federal environmental regulations, such as the Energy Policy Act of 2005 in the United States, ethanol is frequently blended into gasoline in an effort to reduce tailpipe emissions. Moreover, there is a shift towards increasing the use of more sustainable energy resources such as biofuels, which includes ethanol, as transportation fuels. A fundamental understanding of how blending alcohol into a gasoline stream affects the properties of said stream is of critical importance. Specifically, the change in the distillation curve and vapor pressure is essential from a manufacturing and engine performance perspective. In the plant setting, there exists empirical correlations to predict these changes. However, these are typically very limited in scope as a correlation is only applicable to a specific plant gasoline mixed with a specific alcohol in a certain blend ratio.

Process simulators of petroleum process rely on equation of state (EoS) predictions of the phase behavior of petroleum pseudo-components defined by normal boiling point T_b, specific gravity *SG,* and *MW*. For gasoline blending these pseudo-components are typically generated from D86 curves. Hence, any method to be widely used in the prediction of alcohol / gasoline blending in refining processes should be accurate for petroleum pseudo-components.

In this work, we present a fundamental methodology for alcohol / gasoline blend property prediction using the polar PC-SAFT EoS[1] combined with the recently developed petroleum parameterization methodology, TAPPS (formerly referred to as EMPETRO)[2-6]. TAPPS is a highly accurate methodology used to determine hydrocarbon PC-SAFT parameters for petroleum species defined solely by T_b, *SG,* and *MW.* In addition to highly accurate predictions for pure component vapor pressures and densities, TAPPS accounts for polar features of petroleum such as the fact that aromatic species receive hydrogen bonds. This makes it well suited to describe alcohol / gasoline blending.

The approach presented here employs the base gasoline D86, average gasoline SG, and dry vapor pressure equivalent (DVPE) to generate gasoline pseudo-components. Using these pseudo-components, it is demonstrated the PC-SAFT / TAPPS approach allows for the quantitative prediction of the D86 and DVPE of alcohol / gasoline mixtures. We validate the approach by reproducing both literature[7,8] and ExxonMobil refinery data[9] for various blends. This work demonstrates the strength of PC-SAFT / TAPPS in predicting the physical properties of non-traditional blends of petroleum streams and lays the foundation for the development of a predictive blending tool that will be valuable when considering emerging biofuel markets.

References

[1] J. Gross; G. Sadowski, *Ind. Eng. Chem. Res.*, **40** (2001) *1244.*

[2] B. D. Marshall; C. P. Bokis, *Fluid Ph. Equilibria*, **489** (2019) *83.*

[3] B. D. Marshall; C. P. Bokis, *Fluid Ph. Equilibria*, **478** (2018) *34.*

[4] B. D. Marshall, *Fluid Ph. Equilibria*, **493** (2019) *153*.

[5] B. D. Marshall, *Fluid Ph. Equilibria*, **497** (2019) *79*.

[6] B. D. Marshall, *Fluid Ph. Equilibria*, **507** (2020) *112420*.

[7] V. F. Andersen; J. E. Andersen; T. J. Wallington; S. A. Mueller; O. J. Nielsen, *Energy Fuels,* **24** (2010) *2683*.

[8] V. F. Andersen; J. E. Andersen; T. J. Wallington; S. A. Mueller; O. J. Nielsen, *Energy Fuels,* **24** (2010) *3647*.

[9] J. R. Vella; B. D. Marshall, *Ind. Eng. Chem. Res.* (2020) *accepted*.

606540 Development of Mean Ionic Activity Coefficients for Modification of the Krichevsky – Ilinskeya Equation for Methylimidazolim – Type Ionic Liquids

Abstract:

Removal of H_2S from oil and gas streams has been a priority, oftentimes a necessity, for upstream crude oil/gas facilities for decades. However, the advances in production, namely horizontal drilling and hydraulic fracking, have produced oils and natural gases that contain higher concentrations of H_2S than previously seen. Crude oils containing > 0.5 % sulfur are considered sour and must undergo separation and/or conversion processes to lower the H_2S content prior to entering the midstream pipelines and railcars. H_2S scavengers can be categorized into two types: non-regenerative and regenerative. Non-regenerative scavengers, such as triazines, use an irreversible reaction to convert H_2S forming sulfur-containing hydrocarbons. With regenerative scavengers, such as alkanolamines, absorb H_2S using reversible absorption with stripping recovery processes.

Ionic Liquids (ILs) are a new family of absorbent materials that are being sought to replace amine-type reversible absorbents due to lower production of volatile organic carbons (VOCs) and reduced costs of the regenerative stripping phase. The ILs that have been chosen for this work were 1-butyl-3-methylimidazolium hexafluorophosphate, 1-butyl-3-methylimidazolium chloride, and 1-butyl-3-methylimida-zolium tetrafluoroborate. In this work, the mean ionic activity coefficient and the adjustable parameters for the Pitzer Ion Interaction Model have been determined. An infinite dilution procedure has been employed where a concentrated mixture of H_2S in IL was stripped using helium gas. The decrease in H_2S concentration (measured through gas chromatography) over time was used to determine the pertinent thermodynamic parameters.

In addition, the Henry's constant and partial molar volume of H_2S in IL were determined. The Krichevsky – Kasarnovsky equation has been shown to be applicable over a wide range of temperatures, pressures, and solute/solvent systems. However, in some cases (i.e. where T is near T_c) the molar volume ( ) is not constant, and there are strong deviations between the measured solubility and the calculated Henry's constants. From preliminary estimates, the Krichevsky – Ilinskeya equation (a modification of the Krichevsky – Kasarnovsky equation) does not model the phenomenon well. Ultimately, this work seeks to establish a new thermodynamic model that replaces the 2-suffix Margules component of the Krichevsky – Ilinskeya equation with the mean ionic activity coefficient found in Pitzer Ion Interaction model. The experimental results and model development will be presented.

607700 From Skin to Nervous System – Epidermal Neural Crest Stem Cells, a Novel Autologous Candidate for Therapeutic Applications

Abstract:

Neural crest (NC) cells play a central role in forming the peripheral nervous system, the craniofacial skeleton, and the pigmentation of the skin during development due to their broad multilineage differentiation potential into neurons (Neu), Schwann cells (SC), melanocytes (MC), and smooth muscle cells (SMC). Recently, we identified an easily accessible source of pluripotent NC stem cells from human inter-follicular keratinocyte (KC) cultures (KC-NC) isolated from glabrous neonatal foreskin. Using small molecules FGF2 and IGF1 as well as inhibition of Tgf-β1 we are able to maintain SOX10+/FOXD3+ KC-NC in vitro (Fig. 1a). Genome wide transcriptomic analysis, as well as single cell RNA-seq provided evidence for the upregulation of NC specific genes and differentiation trajectories (Fig. 1b). The clonal multipotentency of the diverse KC-NC population was verified with in vitro clonal assays, further differentiated into NC specific lineages (Neu, SC, MC, SMC, Fig. 1c). Each differentiation was successfully subjected to functional assays involving electrophysiological tests for neurons, L-Dopa assays for MC and gel compaction assays for SMCs. In ovo transplantation further verified the multipotency of KC-NC with GFP+ KC-NC migrating into a variety of areas in the transplanted embryos, including axonal alignment with Tuj1+ nerve axons. Transitioning onto the more clinically relevant model of KC-NC from adult skin, we showed that, apart from slightly longer doubling times, adult KC-NC showed no other signs of cellular senescence, maintained their multipotency in vitro and migrated along stereotypical pathways and contributed to multiple NC derivatives in ovo (Fig. 1d), making them a promising candidate for autologous cell transplantation therapeutic strategies. In order to test the in vivo functionality of our KC-NC we adopted a CNS myelin deficient mouse model (shi/shi) and observed migration of KC-NC out of the lesion site in the corpus callosum and proliferation after the first weeks of transplantation (Fig. 1e). Taken together our work suggests that Epidermal Neural Crest Stem cells are a promising source of autologous transplantation to be tested for a variety of nervous system degenerative disorders.

607738 Regulating Endodermal Cell Lineage through Genetic Manipulation of Developmental Regulatory Gene Circuits

Abstract:

Human Pluripotent stem cells (hPSC's) are a very useful tool for regenerative medicine allowing for the replacement of damaged tissues by implantation of functional tissue or cells. Of specific interest is the ability to properly differentiate liver, pancreas, and lung cells which all derive from endoderm using hPSC's. One issue that arises is a result from the heterogeneity in the gene expression of hPSC-derived endoderm which can severely compromise the downstream fate of differentiation. Improper differentiation protocols of embryonic stem cells have also been shown to produce unwanted tissue during transplants including skeletal muscle, bone cell subpopulations, gut epithelial-like structures and columnar cells (Kubo et al. Development and Disease 2004). Hepatocyte Nuclear Factor 3B (Foxa2) is known as a Pioneer Transcription Factor (TF) due to its ability to bind albumin enhancers and open normally inaccessible heterochromatin, effectively priming such for a cascade of regulatory elements to further differentiation through necessary gene activation. Foxa2 has been shown to be required for developmental processes such as transitioning to air breathing at birth (Huajing Wan, Jeffery A. Whitsett 2004.), differentiation of pancreatic α-cells (Cathrine Lee, Klause H. Kaestner 2005), and control of differentiation of goblet and enteroendocrine cells in mice (Diana Z, Klause Kaestner 2009). In combination with Foxa2, Hepatocyte Nuclear Factor 3A (Foxa1) has been shown to equally vital. To better understand how Foxa2 and Foxa1 regulates developmental master regulatory gene circuits (DRGC) for induction, maintenance, and differentiation of endoderm towards liver, we provide a model using human hepatocellular carcinoma (HepG2) cells as well as Human Stem Cells (BXS). HepG2 being a well-studied cell line is also known to express alpha-fetoprotein (AFP) and albumin (Alb). RNA interference is used to transiently knockdown both Foxa1/2 with siRNA and a permeant knockdown cell line was created with short hairpin RNA (shRNA). Using qPCR, it was observed that the transient approach resulted in downregulation of AFP, Alb, and Hhex, the double knockdown of foxa1/2 also resulted in upregulation of Sox7, CXCR4, NANOG, and Pax6. shRNA knockdown in HepG2 resulted in Foxa1/2 $^{-/-}$ stable cell lines with permanent Foxa1/Foxa2 knockdown, confirmed by qRT-PCR and Western Blot analysis. These cell lines demonstrated downregulation of Alb, Hhex, HNF4A, HNF1B, PDX1, and SOX9. These data suggest that the Foxa1/2 regulation of albumin enhancer and HNF4A present in liver development may also be modeled in HepG2 cells, suggesting the DRGC's may be similar when compared to hepatic endoderm cells as well as suggesting possible de-differentiation back to a definitive endoderm like state. Futher analysis of lineage and metabolic changes are explored though RNA-seq. Preliminary data of double knockdown of Foxa1/2 in PSC derived gut tube progenitors reduces the development of liver like morphology. Future work will include analysis of combined effects of Foxa1/2 knockdown and Sox17 overexpression as Sox17 has been shown to be vital for endoderm induction as this TF regulates and maintains endoderm while suppressing differentiation into liver (Pfister et al., Int. J. Dev. Biol. 2010). Future work will also include use of dox inducible Foxa1/2 $^{-/-}$ iPSC cell lines to study the effects of time dependent knockdown on differentiation of hPSC-derived endoderm.

609584 Role of FcRn-Mediated Recycling in IgG Transport across an *in Vitro* Blood-Brain Barrier

Abstract:

Immunoglobulin G (IgG)-based therapies for numerous neurodegenerative indications are being evaluated in clinical trials as first-in-class disease-modifying agents. Although promising, poor brain delivery of IgG (~0.01% of the administered dose) – attributed to the restrictive phenotype of brain endothelial cells (BECs) that form the blood-brain barrier – necessitates dosing regimens that raise concern regarding treatment cost and accessibility. Successful strategies to improve IgG brain delivery, such as antibody engineering, must first identify the processes influencing IgG transport across BECs. To overcome the experimental complications of in vivo models/methods that have contributed to conflicting hypotheses, mainly concerning the neonatal Fc receptor (FcRn) that regulates IgG recycling and transport in peripheral endothelial cells, IgG transport and trafficking was systematically evaluated using an in vitro model of brain endothelium derived from human pluripotent stem cells. Complementary super-resolution and live-cell fluorescence microscopy were used to correlate intracellular distributions and transport rates for IgGs exhibiting disparate FcRn engagement. Relationships between processing and transport were further explored using modified IgG and other macromolecules, which revealed the impact of endocytoic pathways on transport rates. Collectively, our findings shed light on the factors influencing the rate of brain entry for conventional therapeutic IgG and also provide possible routes to engineer improved variants with enhanced brain delivery.

603192 Human Pluripotent Stem Cell Expansion and Endoderm Differentiation in an Automated Stirred-Suspension Bioreactor

Abstract:

Human pluripotent stem cells (hPSCs) are a potentially limitless source of cellular therapeutics and cells for drug screening. Consequently, a robust, scalable, and xeno-free stem cell expansion and differentiation process is necessary to meet clinical demand. We demonstrate the expansion of hPSCs in a controlled stirred-suspension bioreactor as aggregates and their subsequent specification to definitive endoderm (DE) as an example of differentiation towards a germ layer, which can then be differentiated into medically relevant types such as pancreatic cells and hepatocytes. This process allows the production of hPSCs in a homogenous and regulated environment meeting the 10^8-10^9 cells/patient demand of most stem cell-based treatment protocols.

H9 human embryonic stem cells (hESCs) and IMR90-4 induced PSCs (hiPSCs) were seeded in a 250 mL instrumented bioreactor followed by the formation of aggregates. Mitigation of shear stress, which was accomplished through the gradual increase in the agitation rate after seeding, was found to be essential for maintenance of appropriate aggregate size, viability and pluripotency. H9 and IMR90 cells were propagated for five and six days, respectively, in xeno-free medium. Culture for longer periods between passages led to decreased pluripotency. The pH in the bioreactor was maintained at 7.4±0.2 and dissolved oxygen was kept constant with controlled headspace aeration. The H9 and IMR90 cells grew 24- and 10-fold, respectively with corresponding viability of greater than 87-90%. Continuous cell proliferation resulted in increases in the mean aggregate size from 29.44±8.91 μm and 35.06±8.39 μm (day 0) to 192.57±72.60 μm and 179.07±57.56 μm (days 5-6), i.e. size that does not hamper oxygen transfer. Cultured aggregates retained their pluripotent state with 70.21±1.48% of H9 cells and 79.31±2.74% of IMR90 cells being $NANOG^+$/$POU5F1^+$ on day 5 and 6 as determined by flow cytometry. After the expansion stage, the culture was switched to xeno-free priming medium for a day. Cells were further differentiated for three (IMR90) or four (H9) days. By day 3, there were 87.36±6.10% FOXA2/$SOX17^+$ IMR90-derived cells, whereas 69.36±2.80% FOXA2/$SOX17^+$ cells were noted by day 4 in H9 hESC cultures. H9 aggregate differentiation had to be optimized since IMR90 aggregates differentiate to definitive endoderm more readily while H9 cell require an extra day of differentiation and inhibition of the GSK-3β and still do not achieve the same efficiency of differentiation. Agitation ensured proper mixing without adverse effects on aggregates, which maintained normal morphology through the expansion and differentiation stages. In parallel, data was collected for cells seeded at the same density and cultured in 6-well suspension plates with orbital shaking. While there were differences in the rate of hPSC growth in the two systems, differentiation in the bioreactor yielded a higher percent of DE cells compared to that in well suspension culture.

Ongoing work focuses on further coaxing hPSC-derived DE cells toward PDX1/$NKX6.1^+$ cells in the bioreactor. We have successfully differentiated a hESC cell line, $INS^{GFP/W}$ MEL1, in spinner flasks (as bioreactor surrogates) obtaining 33.34±2.79% GFP^+ (INS^+), 43.42±8.58 $NKX6.1^+$, and 69.64±3.48 $PDX1^+$ cells. Given the scalability of stirred-suspension cultivation systems, the successful expansion and directed differentiation of hPSCs in a fully instrumented stirred-suspension bioreactor paves the way for production of hPSC-derived progeny in commercially relevant scales while minimizing batch-to-batch variation. Moreover, the xeno-free maintenance and specification of hPSCs eliminates the risk of cross contamination with animal-derived components. Our findings will contribute to biomanufacturing technologies for hPSC-based therapies.

Acknowledgements: This work is supported by the grants CBET-1743367 (EST) and CBET-1743407 (MH) from the National Science Foundation (NSF).

597601 Metabolic Characteristic and Immunomodulation of Adipose-Derived Stem Cells during *in Vitro* Culture Expansion

Abstract:

Human mesenchymal stem cells (hMSCs) are known for their potential in regenerative medicine due to their ability to differentiate into multiple cell lines, secrete trophic factors, and regulate immune responses in damaged tissue. Preclinical studies have shown promising results for endogenous regeneration after hMSC transplantation in various diseases. To keep up with the cell demand for clinical trials, hMSCs, usually isolated from bone marrow, must be expanded *in vitro* to reach large quantity, which eventually leads to cellular senescence and loss of therapeutic functions. An increasing wealth of literature indicates that metabolic plasticity in bone marrow-derived hMSCs is associated with cellular senescence during *in vitro* culture. The redox cycle NAD+/NADH and Sirt-1 axis play an important role in regulating aging-related functional decline. To address the challenging isolation of hMSCs from bone marrow, other sources such as adipose tissue-derived stem cells (hASCs) have been considered as a promising alternative for cell therapy due to easy access and large tissue enrichment. However, differences have been observed for hASCs in context of metabolic characteristics and response to *in vitro* culture stress. The relationship between metabolic profiles and stem cell functions, especially the immunomodulatory ability of hASCs, is largely unknown compared to the thoroughly investigated bone marrow derived hMSCs. Therefore, in this study, we thoroughly evaluated changes in metabolism and redox cycles in hASCs during extensive culture expansion. From passage 5 (P5) to passage 12 (P12), hASCs did not respond to culture stress as significantly as bone marrow derived hMSCs. The results indicated the limited cellular senescence with well-preserved stem cell characteristics. In addition, we found that the NAD+/NADH redox cycle as well as metabolic homeostasis are well maintained in hASCs, indicating that hASC senescence may be regulated through an alternative mechanism rather than NAD+/-Sirt aging pathways. For immunomodulation, however, hASCs at high passage still exhibited a decline of potentials in response to an inflammatory environment. Moreover, we employed both proteomics and genomics analyses to identify the key metabolic pathways that may regulate immunomodulatory potentials in hASCs during culture expansion. Together, this study advances our understanding of metabolism and senescence in hMSCs from different tissue sources and provides guidance for *in vitro* biomanufacturing of hASCs for potential clinical applications.

597624 Chemically-Defined Generation of Aorta-Gonad-Mesonephros (AGM)-like Hematopoietic Stem and Progenitor Cells from Human Pluripotent Stem Cells

Abstract:

Multipotent self-renewing hematopoietic stem and progenitor cells (HSPCs) regenerate the adult blood system after transplantation, which is a curative therapy for numerous diseases including immune-deficiencies, leukemia, and other cancers. Although substantial effort has been applied to generating *de novo* HSPCs from human pluripotent stem cells, chemically-defined, serum-free, or feeder cell-free conditions for HSPC derivation have not been reported. Here we describe the development of a defined, albumin and feeder-free culture, with synergistic modulation of Activin and Wnt signaling during the hemato-vascular mesoderm formation, supports the generation of multipotent HSPCs that go through the intermediate aorta-like vessels, similar to hematopoiesis in the aorta-gonad-mesonephros (AGM). Importantly, the resulting cells displayed lymphoid, myeloid and erythroid potential *in vitro*, and homed to fish caudal hematopoietic tissue (CHT) *in vivo* after transplantation, mimicking aspects of human AGM hematopoiesis. Our findings provide a significant advance in defining critical components for the induction of homogenous definitive hematopoiesis *in vitro*. By reducing the complexity of hematopoietic induction, our approach will offer enhanced understanding of human embryonic hematopoiesis and expedite the development of hPSC-derived hematopoietic and immune cell products for therapeutic purpose.

613359 Molecular Elucidation and Engineering of Stem Cell Fate Decisions (INVITED SPEAKER)

Abstract:

As the field progressively learns more about the mechanisms that control stem cell behavior – in particular their two hallmark properties of self-renewal and differentiation – deeper insights are gained into their potential roles and applications in organismal development, adult homeostasis, and regenerative medicine. The niches in which stem cells reside present them with a spectrum of signals to control their behavior, and it has become increasingly evident that not only are local biochemical signals important, but biophysical features of the environment that modulate the presentation of this biochemical information can impact cell behavior. For example, spatial and temporal variation in the presentation of cues is important information that modulates fate decisions and tissue structure. Furthermore, the tissue matrix has variable bulk mechanical properties and surface topographical features depending on how its assembled, both of which modulate cell behavior.

We have created several technology platforms to investigate the biophysical regulation of stem cell behavior, in particular to understand and control the differentiation of adult neural stem cells and human pluripotent stem cells into neurons. First, we are developing and harnessing optogenetics systems to investigate how cellular signaling dynamics impact fate decisions. Second, we have been developing bioactive, synthetic material systems to investigate the effects of cell-matrix and cell-cell interactions on cellular function. Finally, we work towards translating the basic information that emerges from both of these efforts into safe, scaleable, fully defined, robust culture and implantation systems for stem cell based regenerative medicine efforts to treat human disease.

609303 An Integrated Process Automation System in API Manufacturing: Key Characteristics and Case Study

Abstract:

Safety is always first priority and a critical concern in the chemical industry. Jiangsu Provincial Emergency Management Administration (JSEMA) in China released <Basic Requirements for Diagnosis and Improvement of Instinct Safety> (Su Emer. NO.53) in June 2019, which specifies the requirements for the main aspects of chemical production. The concept of "integrated process automation" in Su Emer. No. 53 poses a great challenge to API manufacturing. It emphasizes mandatory safety and intensified automatic implementation. This paper illustrates the key characteristics in the integrated process automation of API manufacturing and gives some case studies.

Such automation system includes sophisticated solution and cutting-edge tools. According to the recipe and process of the product, the consultant optimizes the control strategic with expertise to enhance the safety and meet the regulation requirements. The process control modules or batch control program are deployed in DCS system for the purpose of multifunctional and flexible manufacturing. In case study it gives example of the mixing control strategic in the reactors which is critical and should consider all concerned factors. Usually the geared motor electricity is ignored to monitor, the absence may interfere the cooling system activation in time, its real time monitoring and interlock with cooling media are necessary; another example is the utility facilities configuration and interlock with automatic systems which should be tuned duly to eliminate the hidden trouble. The integrity of safety instrumented functions (SIFs) should be determined through risk assessment tools and rating tools, such as HAZOP (Hazard and Operability Analysis) and LOPA (Layer of Protection Analysis). The related case study reports sharing control components between SIS and DCS will increase the possibility of failure, which should be validated. The allocation of center control room and controller assignment should be concerned with HSE guideline. It also discusses the roadmap of information technology and MES (Manufacture Execution System) for API manufacturing. Our implementation shows that the integrated process automation system can help to reach instinct safety, which increase the productivity and improve the quality of products with regulatory compliance.

605455 A Quantitative Assay of Sodium Triacetoxyborohydride

Abstract:

Sodium triacetoxyborohydride (STAB) is a common reducing agent for reductive amination reactions. Due to sensitivities to water and air, the potency of this reagent tends to degrade over time. Traditional analytical assays for STAB fail to accurately assess the content of the active borohydride species, and often report assay purities of 97% or greater, when the true potency is lower. An incorrect or unknown potency of the STAB being used may lead to significant challenges when scaling up reductive aminations. An excess of STAB may be required and reactions may not achieve full conversion with the amount of STAB charged, requiring additional charges. The use of an excess of STAB comes with greater raw material costs and the thermal and H_2 evolution hazards associated with quenching the excess reagent.

This presentation outlines a simple assay, based on an aldehyde reduction, that was developed to determine the active borohydride content of STAB. The HPLC assay yield of a salicylaldehyde reduction has been shown to accurately determine this potency and has been validated against an H_2 evolution method, as well as yields obtained from a reductive amination.

607708 Delivery Strategies for Live Therapeutic Bacteria

Abstract:

The microbiome is a collection of bacteria, viruses and fungi that perform essential functions for human health. Dysbiosis, or the dysregulation of the microbiome, can contribute to chronic inflammation, metabolic disorders, neurological diseases, and cancer. Clinically, one of the only methods for altering the composition of the microbiome is with broad-spectrum antibiotics, which contribute to dysbiosis and create selective pressure for antibiotic resistant pathogens. As such, live therapeutic bacteria (LTB) have been identified as an attractive alternative to antibiotics, as they can modulate the composition of the microbiome without eliminating commensal species. However, LTBs are commonly delivered orally and experience significant challenges during gastrointestinal (GI) transit, including survival in the harsh conditions of the stomach, adherence to the intestinal lumen, and navigation of interpatient variability in the microbiome due to factors such as diet. Few engineered drug delivery platforms have been rationally developed to address these challenges. In this work, we describe two complementary platforms that separately adapt LTBs to improve viability during gastric transit and modify LTB surfaces to aid in adhesion along the GI tract. In the first system, we sought to understand the influence of factors such as diet on LTB adaptation and colonization. We delivered an LTB to female BALB/c mice on two distinct diets and found that diet significantly influenced both the appearance of a small colony variant of the LTB, as well as its distribution along the GI tract. As adaptation appeared to significantly affect LTB colonization, we developed an approach to control the adaptation of LTBs to specific environmental insults during GI transit. We found that growth in an *in vitro* adaptation media can prime LTBs for the acidic conditions of the stomach, significantly improving their viability during a challenge in simulated gastric fluid. Following transit through the stomach, LTBs must adhere to the intestinal lumen to persist and colonize in the GI tract. Therefore, for our second system, we developed a platform to chemically conjugate targeting ligands to the surface of LTBs to improve their adhesion to the GI mucosa. This platform significantly improves both the rate of colonization, as well as the concentration of LTBs in the GI tract of female BALB/c mice. Collectively, this work represents a simple approach to adapt LTBs to the intestinal environment during *in vitro* growth, while enabling subsequent modification of the LTB surface for improved intestinal adherence. Together, these delivery techniques for LTBs may help overcome challenges associated with GI transit, colonization and adaptation.

609030 In Vitro Biomimetic Tumor Model with Cancer-Specific Gene Regulators and Green Fluorescence Reporter for High Throughput Drug Screening

Abstract:

Although much scientific effort has been expended for cancer treatment, this disease is still one of the leading causes of death around the world. The limitations and shortcomings of current therapies highlight the urgent need for more effective therapeutic strategies. So far, most of the chemotherapy drugs act negatively on the cell division without selection, leading to serious adverse effect on normal cells. The current mainstream of anti-cancer drug development is to find the drugs that have specific targets (DNA, mRNA or proteins). Those targets are overexpressed in cancerous cells and are critical for carcinogenesis. In this study, we engineered breast and lung cancer cells to express the enhanced green fluorescent protein (EGFP) under cancer-specific gene promoters for three known carcinogenesis genes - survivin, Bcl-2 and hTERT, which enable online high-throughput screening of drugs that may up- or down-regulate these cancer genes as indicative of efficacy in inhibiting breast and lung cancers. Additionally, the mechanism of drugs and the crosstalk between cancer genes can be elucidated at the early stage of drug discovery. Another issue in the pharmaceutical industry is the high drug failure rate owing to poor predictability of early drug assays in a two-dimensional cell culture system. In such traditional assay, multiple cell-cell and cell-matrix interactions are deficient, leading to incomprehensive cellar responses to drug agents. To improve the drug assay predictability, our assay has focused on the construction of three-dimensional biomimetic cancer cell models with the recapitulation of multiple elements of tumor microenvironment. While developing the new drug screening assay, we also combined the rapid cell detection approach to achieve the high-throughput capability. Our optimized drug screening process can help to discover more anti-cancer therapies.

609939 Preparation and Use of Fine Grade Engineered Excipients for Direct Compression of Binary Blends of Cohesive Drug Powders

Abstract:

Novel, surface modified excipients are prepared using a fluid energy mill (FEM) to simultaneously mill and dry coat microcrystalline cellulose (MCC) with nano-silica to produce different sizes of fine, well flowing MCC excipient [1]. The purpose of these new class of excipients is to enable direct compression (DC). DC route is one of the most efficient and economic methods for tablet manufacturing since only blending and compression operations are required, greatly simplifies the process-train. However, the popularity of direct compression is limited by its stringent requirements, such as good flowability, high packing density, and excellent compaction for excipient-active pharmaceutical ingredient (API) blends. When the blend has high drug loading with a poorly flowing API, the flowability and tabletability of blends is generally decreased sharply. In a recent work, it has been shown that surprisingly, fine excipients of 20 mm size that have improved flow and bulk density after silica dry coating, can greatly improve direct compression performance [2]. Unfortunately, it could not work for very fine cohesive APIs such as 10 mm acetaminophen (APAP) at higher drug loading. In this work, binary blends of several fine sized dry coated MCC-based excipients are considered with three different fine, cohesive APIs at high drug loadings. Bulk density, FFC, and tablet compactability for all these blends are measured and compared to determine which excipient sizes provide the best overall performance. The results are analyzed for the relative sizes of the excipients and APIs to better understand the interplay between the components of the blends and assessment of the extent of silica transfer from coated excipient powders to API powders and its effect on blend bulk density and flowability. It is expected that using such novel surface engineered fine excipients in blends with fine APIs may also help in reduced segregation tendency during overall processing. Hence, these results are expected to provide industry relevant guidelines for the use of finer surface modified excipients in DC tableting.

596229 Applying Machine Learning to Predict Therapeutic Antibody Viscosity

Abstract:

A general machine learning feature selection protocol to predict antibody stabilities is proposed. These features only require information from sequences and molecular structures of antibodies. In addition, the physically relevant features are easily interpretable and help to design experiments to understand the underlying mechanisms of antibody stabilities. Two case studies are demonstrating using this protocol. First, we applied this framework to predict aggregation behaviors of 21 marketed monoclonal antibodies at high concentration (150 mg/mL), yielding an accuracy of 97% on validation tests with only three features. Models trained using structural features, rather than sequence features, correlated better with aggregation. Second, the concentration dependent viscosity behavior of 27 FDA approved antibody drugs was measured in concentrated solutions at high concentration (150 mg/mL). Combining molecular modeling and machine learning feature selection, we found that the net charge in the antibodies and the amino acid composition in the variable region are key features which govern the viscosity behavior. We presented two predictive models for aggregation and viscosity based on machine learning trained models, facilitating early stage design.

602107 Development of Multi-Adjuvant Systems for Vaccines Against Infectious Diseases

Abstract:

The ability to elicit potent humoral and cellular immune responses is a key element required for the development of efficacious prophylactic and therapeutic vaccine formulations. For vaccines against intracellular pathogens, the induction of a balanced, long-lived, and potent $CD4^+$ and $CD8^+$ T cell activation and proliferation is critical, combined with robust antibody responses. One strategy is to activate innate immunity and proinflammatory responses through Toll-like receptor-independent cytosolic pathways. Importantly, engagement of these mechanisms has demonstrated its ability to provide protection against bacterial challenge, and several small molecules have been evaluated in preclinical and clinical studies as vaccine adjuvants against infectious diseases.

Small molecules can bind to cytoplasmic receptors triggering signaling cascades that activate innate immunity. Despite their potential as adjuvants, small molecules have some limitations that need to be overcome. For example, there may be low exposure in target tissues due to their size and charge, and are prone to enzymatic biodegradation; with some off-target toxicities. In order to address these challenges, the use of drug delivery systems is being explored.

In this work, three different polymeric nanoparticle platforms were evaluated for small molecule delivery and for their capability to act synergistically as co-adjuvants and elicit robust immune responses. In these studies, nanoparticles based on natural and synthetic polymers were fabricated using water/oil/water and oil/water emulsions. Particle characterization was performed using dynamic light scattering, scanning electron microscopy, and zeta potential measurements that confirmed the size, morphology, and charge of the resulting nanoparticles. Small molecule compounds were loaded into each of the platforms, and their loading and release efficiency were determined using liquid chromatography. Finally, these systems were evaluated using *in vivo* murine models following parenteral administration to assess their ability to induce cytokine secretion, antibody and cell-mediated immunity. The results obtained demonstrate the potential of polymeric particulate systems as dual delivery and co-adjuvant platforms for the development of small molecule-adjuvanted prophylactic and therapeutic vaccines.

603577 Development of a Harvest Small Scale Model to Simulate Continuous Centrifugation

Abstract:

Disk stack centrifugation is one of the main clarification steps used in large scale mammalian cell culture processes. During process development, pilot scale disc stack centrifuge is normally used to develop and characterize the harvest process for new cell culture processes due to the lack of smaller scale model for disc stack centrifugation. In order to increase speed of development, a small scale model (SSM) of a disk stack centrifuge is of high demand to enable harvest development in an earlier stage of process development. A small scale model can also improve throughput and reduce cost. In this study, we have developed a SSM for disc stack centrifugation using a capillary shear device combined with benchtop centrifugation. Capillary tube size, length, and flow rate were optimized to generate energy dissipation rates comparable to large scale disk stack centrifuges. Additionally, shear force from multiple centrifuges at different scales were standardized and compared. Mathematical correlations were established for energy dissipation, viable cell density, cell viability, and lactate dehydrogenase (LDH) activity. Our SSM offers an approach to understand how the separation process, such as centrifuge designs, sizes, and configurations affect filtration performance and product quality. This work can speed up harvest development during process scale-up in biopharmaceutical mammalian cell culture processes.

599615 Optimizing and Reducing Cleaning Production Loss in Pharma, Agrochemical and Specialty Chemical Companies Using a New Expert System.

Abstract:

Pharmaceutical, Biopharmaceutical, Agrochemical and Specialty chemical companies make many different products in their plants. This is true both for the active ingredient stages as well as the secondary of formulation processing. Cleaning and set up between products is critical and reduce productivity time up to 50% in some cases. The presentation will show the critical areas important in the modelling development in this expert system and how it will aid companies in scheduling their productions to reduce cleaning time and to reduce issues with cleaning.

599656 Revolutionize of Pharmaceutical Product Changeovers By Enhanced Cleaning with Microbubbles

Abstract:

Pharmaceutical, Biopharmaceutical, Agrochemical and Specialty chemical companies make many different products in their plants. This is true both for the active ingredient stages as well as the secondary of formulation processing. Cleaning and set up between products is critical and reduce productivity time up to 50% in some cases. Microbubbles have been shown to enhance many unit operations e.g. distillation. The benefits of microbubbles in cleaning will be presented along with the impact this will enable in regulatory/ registration processes.

611034 Mechano-Evolution Results in Novel Cellular Phenotypes

Abstract:

The fact that cellular properties and functions are governed by mechanical properties of the substrate has been well established in literature. Cells attain distinct set of phenotypes (like cell spreading area, focal adhesion area, nuclear height, YAP localization) and functions (like growth rate, migration speed, traction) on substrates of different rigidity and it is termed as 'phenotypic plasticity' by evolutionary biologists. Our hypothesis is that cause of this phenotypic plasticity could be natural selection by the mechanical properties of the substrate, i.e., the substrate selects for the phenotype that has higher fitness than the alternate phenotype. To test this hypothesis, we performed sustained culture of a genetically variable cell population on soft (E= 1kPa) and stiff (E= 308 kPa) polyacrylamide gels. We chose NIH 3T3 fibroblasts which have been cultured on tissue culture plastic (stiff substrate) for decades and are expected to be well-adapted on stiff substrate. As per our hypothesis, when we expose the genetically variable fibroblast populations to a novel soft substrate for a long time, the soft substrate should select for the phenotype that is optimal for survival of the population on that stiffness. This would open the possibility of engineering cells with novel properties by allowing them to evolve on biomaterials of different stiffness.

611181 Stiffness Induces Metabolic Dysfunction in Primary Hepatocytes

Abstract:

Liver is a complex organ that performs myriad functions as a center of metabolism, having key roles in synthesis, detoxification, and digestion. Liver stiffness has been established as a useful clinical metric to stage liver disease and in recent years as a prognostic tool predictive of outcomes. However, it remains unclear how tissue stiffness itself contributes to hepatic dysfunction. To study the effect of stiffness on liver function, we employed our BEASTS (Bio-Engineered Adhesive Siloxane Substrate with Tunable Stiffness) platform to recreate the range of stiffness observed in healthy to highly fibrotic environments in cell culture. A significant challenge in *in vivo* models, our approach allows for control of stiffness corresponding to various stages of fibrosis. Primary hepatocytes were cultured on BEASTS mimicking liver stiffness from healthy to highly fibrotic tissue. Key markers of hepatic function, urea and albumin production, decreased when hepatocytes were cultured on fibrotic stiffness. Enhanced glycolytic capacity and mitochondrial respiration were exhibited by hepatocytes cultured on fibrotic stiffness. Greater intracellular lipid accumulation was observed in hepatocytes cultured on fibrotic stiffness, along with dysregulated lipid metabolic genes. Hepatocytes cultured on fibrotic stiffness displayed a lower abundance and shift in the redox balance of intracellular glutathione, suggesting impairment of the oxidative stress response and a plausible mechanism for elevated oxidative stress in the fibrotic environment. Together our data suggest a role of increased liver stiffness in driving hepatocyte metabolic dysfunction and contributing to loss of organ function in the setting of liver disease. A more nuanced approach to therapeutic development for liver fibrosis can be aided by data from studying how tissue stiffness influences hepatocyte biology.

602344 Quantifying the Effects of Elevated Hydrostatic Pressure on the Barrier Function of Mammary Epithelial Cells

Abstract:

The onset of pregnancy in mammals begets dramatic changes in the mammary gland to prepare for breastmilk production. Tight junctions, intercellular junctional structures that regulate paracellular transport, are crucial for regulating pregnancy-induced morphogenesis in the mammary gland. Strong tight junctions with a high "barrier function" prevent milk from leaking out of the lactating gland into the surrounding interstitium, and weak tight junctions with a reduced barrier function coincide with involution (shrinkage) of the mammary gland during weaning. Although it is known that hydrostatic pressure exerted by accumulated milk helps to weaken tight junctions during involution, the exact mechanism is unclear. To identify how pressure regulates tight junctions, we custom-built a novel cell culture system to expose mammary epithelial cells to different pressures, up to 1.5 kPa, on their apical (luminal) sides. After exposure to pressure for 6 and 24 h, we observed significant decreases in transepithelial electrical resistance (TEER), a measure of barrier function, for pressures greater than 1 kPa (**Figure**). Additionally, we found that exposure to pressure decreases the expression of E-Cadherin at intercellular boundaries. Lastly, we found that exposure to pressure has no apparent effect on either cell proliferation or death. Ultimately, these findings may have implications for better understanding mechanisms underlying complications related to breastfeeding, such as mastitis.

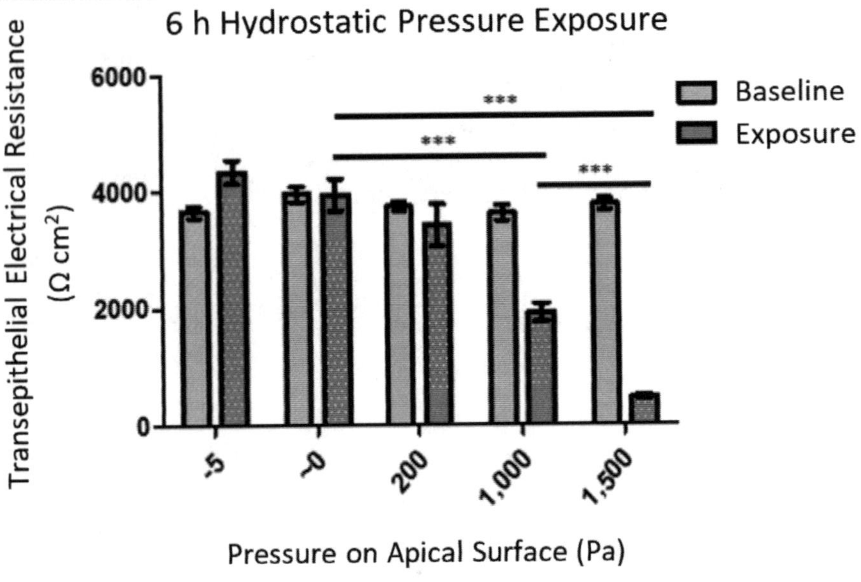

Figure: Transepithelial electrical resistance (TEER) plotted as a function of hydrostatic pressure during a 6 h exposure. The measurements for the "Baseline" group took place right before pressure exposure began. The measurements for the "Exposure" group took place right after exposure ended. Bars denote SEM. *** = $p < 0.001$, n = 4.

602363 Modulation of Adipocyte Size and Fat Pad Weight Via Resveratrol Releasing Scaffolds Implanted into Epididymal Adipose Tissue

Abstract:

Chronic overnutrition leads to lipid accumulation within the adipose tissue which has significant consequences that underlie metabolic diseases. We hypothesized that enhancing lipid catabolism within the adipose tissue is a promising strategy to combat this lipid overload. Treatment with resveratrol, a naturally occurring small molecule, activates fatty acid utilization in adipose tissue; however, a major limitation of this drug is its poor bioavailability. To overcome this challenge, we implanted resveratrol releasing poly(lactide-co-glycolide) (PLG) scaffolds into epididymal fat with the goal of enhancing local lipid catabolism. In this study, lean mice were implanted with resveratrol releasing scaffolds or PLG scaffolds with no drug payload. Resveratrol scaffolds decreased adipocyte size compared to PLG scaffolds, which was also associated with an elevation of AMP kinase activation. Furthermore, based on the results of immunohistochemistry and cell culture experiments, we propose that resveratrol scaffolds decrease adipocyte size due to an increase in lipid utilization of immune cells located at the tissue-material interface. In a follow-up study, resveratrol scaffolds decreased body weight gain, adipose tissue expansion, and adipocyte hypertrophy compared to PLG scaffolds when implanted in mice 4 weeks prior to a high fat diet challenge. Importantly, this scaffold-based strategy required a single administration compared to the daily dosing schedule that is usually required for oral administration. Collectively, this work demonstrates that the use of tissue engineering scaffolds to locally deliver small molecules to the adipose tissue may be a promising strategy to overcome poor drug bioavailability as well as to target key metabolic cells in adipose tissue as a treatment for metabolic disease.

608030 Bioprocessing to Fabricate Stem Cell-Derived β Cells for Diabetes Cell Replacement Therapy

Abstract:

INTRODUCTION: A major goal of cellular engineering is to produce mature, functional cells for therapeutic applications. Human pluripotent stem cells (hPSCs) are incredibly useful for this goal, as they have the potential to become any cell type found in the body if given the correct signals from the surrounding environment. Generation of insulin-producing β cells from hPSCs is one such application of this technology for the treatment of patients with diabetes [1]. We have developed a multi-faceted strategy involving modulating the physical microenvironment through the cytoskeleton and three-dimensional cell arrangement in addition to targeted activation and inhibition of specific cellular signal transduction networks to produce stem cell-derived β (SC-β) cells capable of rapidly reversing diabetes in mice.

RESULTS: We found that conditions that promote a disorganized actin cytoskeleton to be a strong inducer of pancreatic endocrine, including three-dimensional cell arrangement, cellular attachment onto soft surfaces, or treatment with the compound latrunculin A that depolymerizes actin. Specifically, these conditions induce the expression of the transcription factor NEUROG3, which is the master regulator of endocrine cell fate during development. We also found that combination of these conditions to modulate the actin cytoskeleton with Alk5 inhibition, NOTCH inhibition, and thyroid hormone receptor activation to be a strong inducer of SC-β cell differentiation [2]. These SC-β cells expressed many markers found in real β cells from donors, such as insulin, PDX1, and NKX6-1. They were able to function, undergoing glucose-stimulated insulin secretion, including dynamically responding to high glucose levels with both first and second phase insulin secretion kinetics. We have scaled this process to make over 100 million cells per batch with over 90% of cells characterized as pancreatic endocrine. Single-cell RNA sequencing (scRNA-seq) revealed not only SC-β cells but also other endocrine cell types found in the pancreas, including α-like and δ-like cells expressing glucagon and somatostatin, respectively. Upon transplantation into mice with severe pre-existing diabetes (>500 mg/dL blood glucose), SC-β cells rapidly restored normoglycemia in 2 weeks and could maintain blood glucose control for 1 year. The in vivo environment induced further maturation in the gene expression signature of transplanted SC-β cells to more closely resemble β cells from donors, as assessed with scRNA-seq.

CONCLUSIONS: Our approach combines modulation of the insoluble microenvironment or direct cytoskeletal modulation with activation and inhibition of developmental signaling pathways to produce SC-β cells that are functional and capable of reversing pre-existing diabetes in mice [3]. These results demonstrate the potential for SC-β cells to be used in diabetes cell replacement therapy [4].

REFERENCES:

[1] Pagliuca FW, Millman JR, Gürtler M, Segel M, Van Dervort A, Ryu JH, Peterson QP, Greiner D, Melton DA. Generation of functional human pancreatic β cells in vitro. *Cell.* 2014 Oct 9;159(2):428-39. doi: 10.1016/j.cell.2014.09.040.

[2] Velazco-Cruz L, Song J, Maxwell KG, Goedegebuure MM, Augsornworawat P, Hogrebe NJ, Millman JR. Acquisition of Dynamic Function in Human Stem Cell-Derived β Cells. *Stem Cell Reports*. 2019 Feb 12;12(2):351-365. doi:10.1016/j.stemcr.2018.12.012.

[3] Hogrebe NJ, Augsornworawat P, Maxwell KG, Velazco-Cruz L, Millman JR. Targeting the cytoskeleton to direct pancreatic differentiation of human pluripotent stem cells. *Nature Biotechnology*. 2020 Apr;38(4):460-470. doi:10.1038/s41587-020-0430-6.

[4] Maxwell KG, Augsornworawat P, Velazco-Cruz L, Kim MH, Asada R, Hogrebe NJ, Morikawa S, Urano F, Millman JR. Gene-edited human stem cell-derived β cells from a patient with monogenic diabetes reverse preexisting diabetes in mice. *Science Translational Medicine*. 2020 Apr 22;12(540):eaax9106. doi: 10.1126

613354 Hang on Tight: Modulating the Tissue Microenvironment with Micro and Nanoscale Cues (INVITED SPEAKER)

Abstract:

Cells interact intimately with complex microdomains in their extracellular matrix (ECM) and maintain a delicate balance of mechanical forces. Tissue injury results in acute degradation of the ECM and disruption of cell-ECM contacts, manifesting in loss of cytoskeletal tension, leading to pathological cell transformation and the onset of disease. Micro and nanoscale constructs have been developed to provide cells with microdomains to form focal adhesion binding sites, which enable restoration of cytoskeletal tension. These synthetic anchors can activate signaling cascades to modulate downstream gene-level transcription, making this a unique materials-based approach for altering ell behavior. By incorporating micro and nanoscale features into biomaterials, one can influence biological processes such as matrix production, immune activation, and barrier function. The understanding of how small scale topographies can influence the tissue microenvironment allows us to design platforms for applications in therapeutic delivery and tissue regeneration. In this talk, I will discuss our recent work on the on the fabrication, functionalization and implementation of these features for therapeutic and regenerative medicine applications, as well as the mechanisms underlying these interactions.

599202 Assessing the Metabolic Capabilities of the Yeast *Issatchenkia Orientalis* SD108 and Its Application to Biochemical Production (Faculty Candidate)

Abstract:

Many platform chemicals can be produced from renewable biomass by microorganisms, with organic acids making up a large number of these chemicals. Intolerance to the resulting low pH growth conditions, however, remains a challenge for the industrial production of organic acids by microorganisms. The yeast *Issatchenkia orientalis* SD108 is a promising host for industrial production because it is tolerant of low pH conditions. Here, we explore the carbon substrate utilization capacity of *I. orientalis* SD108 by quantitatively measuring growth rates. To systematically assess the metabolic capabilities of this non-model organism, we develop a genome-scale metabolic model for *I. orientalis* SD108 spanning 850 genes, 1,825 reactions, and 1,700 metabolites. In order to improve the quantitative predictions of the model, organism-specific macromolecular composition and ATP maintenance requirements were estimated from experimental data generated as part of this study by making use of chemostat data performed under carbon limited growth conditions. We examine the metabolic network's topology, including essential genes, and draw comparisons with *Saccharomyces cerevisiae.* We also employ the model to survey the growth impact of targeted gene deletions and compare to *in vivo*results we obtained using a CRISPR/cas system. Next, we propose and investigate production pathways for succinic acid through the use of the OptKnock framework to discern knockouts which link production of the targeted chemical to biomass production. Finally, we discuss implications for other industrially-relevant organic acids.

609042 Analysis of a Genome-Scale Metabolic Model of *Chromochloris Zofingiensis*

Abstract:

Metabolic modelling serves as a tool to distil meaningful information from genomic data sets and generate informed strategies for engineering an organism of interest. The metabolic modelling of algae has received significantly less attention than the modelling of heterotrophic bacteria and yeast. However, autotrophic algae have tremendous potential to use as cell factories in the production of valuable chemicals from sunlight and carbon dioxide. An automated tool for the rapid annotation of photosynthetic systems (RAPS) has been developed [Metcalf et.al. 2020 *manuscript in preparation*] and used to create a genome scale metabolic model of *Chromochloris zofingiensis,* an emerging model organism that is of interest for its lipid and astaxanthin producing capabilities. Astaxanthin is a naturally occurring pigment with powerful antioxidant properties that can be sold as a high value nutraceutical compound for an estimated $7000/kg.

Our model was used to conduct a flux balance analysis for autotrophic and two different mixotrophic growth conditions, each eliciting a varied metabolic response. Detailed experimental data on biomass composition was used to generate a different biomass objective function for each set of growth conditions. Results from this work illustrate the significant differences in biomass composition that arise in this alga when grown on varying carbon sources. These differences in biomass composition result in correspondingly different flux distributions predicted in the metabolic reaction network. These results highlight the importance of measuring biomass composition experimentally when conducting computational studies of cellular metabolism.

611547 Enabling Tools for Biomanufacturing in Anaerobic Fungi from Renewable Plant Biomass

Abstract:

Microbial chemical factories are sustainable biomanufacturing platforms that complement traditional petrochemical industries by using renewable feedstocks. However, there are few economical methods to convert plant biomass into sugars that can be more readily used by existing microbial platforms without the production of toxic byproducts. An alternative strategy is the use of non-model microbes that are adapted to the direct use of crude, untreated biomass and engineer these microbes for biomanufacturing. In my lab, we focus on one such class of organisms, early-diverging anaerobic fungi (phylum Neocallimastigomycota), due to their robust degradation of untreated plant feedstocks in the herbivore digestive tract and their novel biosynthetic capabilities. To effectively leverage these abilities, however, a synthetic biology engineering toolbox is required. My lab isolates novel anaerobic fungi, characterizes them, and pursues a number of plasmid-, genomic-, and epigenomic-strategies for tool development to domesticate anaerobic fungi for biotech. We have identified several gene regulatory elements from transcriptomic and genomic resources to drive gene expression for the first time in either the nucleus or cytoplasm of anaerobic fungi. Validation studies with flavin-based fluorescent reporters and selection markers confirm sequence dependent increases in gene expression in a controllable manner. These tools enable library-based screening approaches to construct the first stable anaerobic fungal plasmid and the introduction of CRISPR-based endonucleases for genomic integration. We have also pioneered epigenetic regulation of anaerobic fungi and identified several chemical inhibitors that can perturb the epigenome of anaerobic fungi with concurrent phenotypic changes in biomass-degrading enzyme secretion. Efforts are currently underway to capitalize on systems biology methods to rapidly elucidate and expand parts available for engineering applications. In parallel, we explore how emerging systems biology knowledge of anaerobic fungi may be leveraged in existing microbial platforms. Our growing toolbox of genetic and epigenetic strategies can be readily generalized to other non-model systems, and will soon enable direct biomanufacturing in anaerobic fungi.

608042 Metabolic Regulation and Engineering in Red Yeast, a Basidiomycete of Biotechnological Interest (Industry Candidate)

Abstract:

Basidiomycetes, one of the two major divisions of fungi, have ecological roles of biotechnological interest. As endophytes, they increase the health of forests and agricultural crops. As decomposers, they facilitate turnover of biomass and are essential components of carbon, nitrogen, and phosphorous cycles. To fulfill these roles, they have developed unique biosynthetic capabilities that could be useful starting points for metabolic engineering. Basidiomycetes are still underinvestigated compared to model organisms, and their cell factory potential is underexplored. To this end, we have identified the basidiomycete *Xanthophyllomyces dendrorhous*, or red yeast, as a potential platform host because it natively produces high titers of the carotenoid astaxanthin. Yet, a dearth of genetic tools and a limited understanding of basidiomycete metabolic regulation has fundamentally limited it as a platform host. Thus, we have engaged in genomics-driven characterization and engineering of carotenoid biosynthesis in two strains of red yeast, one wild type (CBS 6938) and an evolved astaxanthin overproducer (UBV-AX4).

Our initial experiments revealed that red yeast carotenogenesis is regulated by light, but only in wild type *X. dendrorhous*. Further investigation with an LED apparatus identified UV and blue light as key activators of carotenogenesis, while UBV-AX4 was relatively unaffected. This difference in phenotypic behavior provides an ideal scenario to elucidate natural basidiomycete optogenetic circuits. To elucidate this mechanism, we sequenced the genomes of both the wild type and UBV-AX4 strains and collected transcriptomic data for strains exposed to different wavelengths of light. This data reveals differentially regulated genes for further investigation.

We next sought to isolate and characterize basidiomycete transcriptional parts for pathway engineering. However, red yeast produces a variety of compounds that are fluorescent, limiting the use of certain fluorescent reporters. Specifically, *X. dendrorhous* naturally has high fluorescence from carotenoids. We thus evaluated various reporters for optimal characterization, including NanoLuc, GFP, BFP, and three RFPs - mCherry, mKate, and mCardinal. We determined that the RFP mCardinal is the ideal reporter for this organism, as it can be measured with flow cytometry, it exhibited little background fluorescence at the RFP excitation/emission channel, and was the brightest among the RFPs tested. Using mCardinal, we characterized 10 promoters and 10 terminators. These results point to a need to determine the ideal reporter in nonconventional yeasts, especially those that produce secondary metabolites that fluoresce and absorb at wavelengths that can interfere with flow cytometry.

In sum, this work represents a genomics-driven strain engineering effort of a nonconventional basidiomycete yeast. With high quality genomes, transcriptomic elucidation of metabolic pathway regulation, and expression part characterization capabilities, we have built a foundation for future metabolic engineering of red yeast for production of industrially relevant compounds.

605288 Efficient Site-Specific Integration of Pathways in Yarrowia Lipolytica Enabled By Serine Integrases (Faculty Candidate)

Abstract:

The oleaginous yeast *Yarrowia lipolytica* is a promising platform to produce lipids, oleochemicals and acetyl-CoA derived compounds due to its high flux towards acetyl-CoA and ability to grow on a wide range of substrates. Integrated expression of pathway genes is superior to plasmid-based expression due to concern of plasmid instability. Strain engineering of *Y. lipolytica* often requires iterative integration of gene expression cassettes into genome. Due to the limited selection markers in this yeast, integration and curing of selection marker is necessary. Existing genome integration methods such as homologous recombination and CRISPR-Cas9 mediated homology directed repair are generally time-consuming and require long homologous arms. Bacteriophage serine integrases are highly specific enzymes which mediates recombination between the short attachment sites of the phage genome and its bacterial host. In this work, we first screened a pool of serine integrases to identify the best candidate in *Y. lipolytica* in terms of integration efficiency. It was found that the marker-less integration efficiency of a green fluorescent protein (GFP) cassette reached 100%. We hypothesized that serine integrases would efficiently integrate large DNA fragments including pathways of genes. Therefore, we investigated the integration of a metabolic pathway for β-carotene biosynthesis. Last, we demonstrate rapid assembly and integration of a library of metabolic pathways for β-carotene biosynthesis. Our results indicate that serine integrases are a promising tool to accelerate the genome engineering of *Y. lipolytica*.

613125 CRISPR-Enabled Functional Genomics in the Oleaginous Yeast *Yarrowia Lipolytica* (Invited Speaker)

Abstract:

Due to its ability to synthesize and accumulate intracellular lipids to high levels, *Yarrowia lipolytica* has become one of the most studied microorganisms for the biosynthesis of oleochemicals. In this presentation, we describe our efforts in designing and creating functional genomics tools to study and engineer this yeast. The key concept that we describe is the ability to characterize the double strand break cutting efficiency of CRISPR-Cas9 across the genome of *Y. lipolytica*. The cutting score (CS) metric that results from these experiments enables high quality genome-wide screening hits for desired phenotypes as well as the identification essential genes under given growth conditions. Defining a CS for over 48,000 single guide RNAs (sgRNAs) also enabled the high confidence prediction of effective sgRNAs through machine learning. Combined, our experiments and predictive capabilities provide a thorough description of the functioning of CRISPR-Cas9 across the *Y. lipolytica* genome, creates a list of effective sgRNAs for any gene in the genome, and enables the ability to predict new sgRNAs with high confidence.

607359 Enzymatic Reactive Crystallization of Cephalexin Using Immobilized Penicillin G Acylase from *E. coli*

Abstract:

Beta-lactam antibiotics have been used extensively to treat bacterial infections for nearly a century. The two largest produced beta-lactam antibiotic classes, semi-synthetic penicillins (e.g. amoxicillin) and cephalosporins (e.g. cephalexin), constitute the largest amount of sales of antibiotics for the treatment of bacterial infections. Currently the industrial production of beta-lactam antibiotics is conducted largely via batchwise chemical synthetic pathways consisting of multiple reaction steps, use of protecting groups, harsh solvents, and very low (-40°C) temperatures. Using these methods, the chemical synthesis route produces a large amount of chemical waste with an E factor of roughly 50 and is highly energy intensive due to the operation of low temperature refrigeration equipment. Additionally, due to increases in labor and raw material costs, much of the production of beta-lactam antibiotics has moved to China and India [1]. Considering recent events regarding COVID-19, the disruption international supply lines could lead to widespread shortages of highly important antibiotics in Western countries [2]. Improving the economic viability and environmental sustainability of beta-lactam antibiotic production is paramount to globalizing the production of these important drugs to decrease the chance of shortages.

Recently, we have been working to develop a continuous enzymatic beta-lactam antibiotic production process to address many of the issues of the current batchwise chemical synthesis processes to produce cephalexin and amoxicillin. These drugs were selected because they are the two highest volume manufactured and prescribed antibiotics worldwide and have been included in the world health organization (WHO) list of essential medicines as key access antibiotics [3]. Next, the move from chemical synthesis to enzymatic synthesis decreases waste production and increases yield due to high enantioselectivity of enzymes and the ability to operate in a mild temperature aqueous environment. We have expressed, purified, and engineered penicillin G acylase from *E. coli* (EcPGA), an industrially relevant enzyme used to produce precursors for semi-synthetic beta-lactam antibiotics, to synthesize cephalexin and amoxicillin. Additionally, we have immobilized EcPGA on various solid supports via covalent attachment, allowing for the recycle of the biocatalyst and decreasing the overall cost of the catalyst for production. Next, the induction of crystallization of the API within the reaction vessel protects the desired product from hydrolysis by EcPGA into a minimally soluble byproduct. The *in-situ* crystallization of the API enables the downstream separation of the solid crystals from the mother liquor into a filtration, washing, and drying protocol.

By combining these elements, we have conducted several successful seeded batch reactive crystallization experiments to produce solid cephalexin crystals. We found that after three crystallization experiments of three hours each, the catalyst suffered a 20% decrease in specific activity. This decrease may be due to crystallization within the pores of the solid catalyst, as the catalyst experienced no measurable deactivation under reaction conditions without crystallization. We have also studied the effect of different sized support particles as well as different support materials (e.g. agarose, methacrylate, polypropylene) on the diffusive properties of cephalexin and its precursors as well as on-line stability of the immobilized biocatalyst. Additionally, during reactive crystallization, after approximately four hours, the scarcely soluble byproduct, phenylglycine, generated enough supersaturation to precipitate and contaminate the solid phase. The progress of cephalexin crystallization and growth as well as phenylglycine precipitation were detected via *in-situ* microscopy and focused beam reflectance measurement (FBRM). We are currently working to implement these same elements and use these experiments to inform the design and operation of continuous MSMPR. The shift from batch to continuous production will improve the capacity and decrease the overall footprint of the production process as well as decrease the amount of off-specification product due to more robust control during steady state operation. The implementation of process analytical technology (PAT) such as Fourier transform infrared spectroscopy (FTIR) will allow for the on-line monitoring of the liquid phase components while FBRM and *in-situ* microscopy will allow for the on-line monitoring of the solid phase components. Additionally, reverse phase HPLC will be used to confirm final product purity. With the information of the components' concentrations in both phases, we will study and develop a control strategy to maintain a "state-of-control" operation of the reactive crystallization process for the production of cephalexin and amoxicillin.

References:

[1] Elander, R. P. (2003). Industrial production of β-lactam antibiotics. *Applied Microbiology and Biotechnology*, *61*(5-6), 385–392. doi: 10.1007/s00253-003-1274-y

[2] Oehler, R. L., & Gompf, S. G. (2020). Shortcomings in the US Pharmaceutical Supply Chain. *Jama*. doi: 10.1001/jama.2020.1634

[3] WHO Model Lists of Essential Medicines. (n.d.). *The SAGE Encyclopedia of Pharmacology and Society*. doi: 10.4135/9781483349985.n433

607528 Recrystallization Development for Small Molecule API (Active Pharmaceutical Ingredient) Islatravir

Abstract:

Islatravir is a nucleoside reverse transcriptase inhibitor small molecule API (active pharmaceutical ingredient) targeted for HIV PrEP (pre-exposure prophylaxis). Islatravir has multiple crystal forms and a complex phase diagram. This talk will focus on the conversion from the monohydrate to an anhydrous form. To produce anhydrous API, we have designed a slurry conversion process, during which the less stable monohydrate form dissolves and a more stable anhydrous form crystallizes simultaneously. The solution composition also comes into play, as the process requires operating within a specific window of water-activity for effective form conversion. The slurry conversion process is also designed to proceed in a controlled manner to reduce crystallization of another kinetically favored phase. In addition, Merck has recently published a novel biocatalytic synthetic scheme for the synthesis of Islatravir*. This presentation will discuss differences in anhydrous crystallization performance (morphology, particle size, and milling performance) when starting with API from a traditional chemical synthetic route vs. the novel biocatalytic route. In this talk, we discuss the factors we can manipulate to design the most robust process, ultimately producing anhydrous API with consistent properties across various batches.

*Huffman et al. "Design of an in vitro biocatalytic cascade for the manufacture of islatravir." Science, 2019, 366 (6470), 1255-1259.

609555 Seeding Strategies for the Selective Crystallisation of Model Proteins

Abstract:

The recognition of the selective nature of biopharmaceutical drugs and the advances in biotechnology have led to over 246 approved products with cumulative revenues reaching $140 billion [1]. Upstream biomanufacture of proteins has advanced, such that the time and expense bottleneck in biopharmaceutics lies in the chromatographic purification steps [3-5]. Crystallisation is a more economic and energy efficient alternative purification step, producing crystalline proteins with higher purity and stability [3, 6]. However, this technology is yet to mature to manufacturing scale and robustness.

Here, we report our optimised crystallisation, additive and seeding strategies to crystallise model proteins and demonstrate selective crystallisation of target proteins from a binary mixture.

Mesoporous silica nanoparticles (NPs) are shown to be effective heterogenous nucleants for the selective crystallisation of lysozyme from a binary mixture of lysozyme and thaumatin. From hanging drop solution conditions which can crystallise both proteins, the 'hard template' SBA-15 NPs are found to not only promote the nucleation of lysozyme, but also suppress the crystallisation of thaumatin. These conditions are also effectively scaled up to the ml level. Further to this, the effectiveness of graphene oxides and DNA polymeric additives are also examined as seeds for protein crystallisation promotion.

Here, we report on the use of several templating strategies for the crystallisation of proteins demonstrating the feasibility for selective crystallisation of proteins for downstream separations.

ACKNOWLEDGEMENT

This study acknowledges by the SCoBiC project funded by the UK's EPSRC (EP/N015916/1).

References

1. Demain, A.L., REVIEWS: The business of biotechnology. Industrial Biotechnology, 2007. 3(3): p. 269-283.
2. Yang, H. and J.H. ter Horst, Crystal nucleation of small organic molecules, in New Perspectives on Mineral Nucleation and Growth. 2017, Springer. p. 317-337.
3. Basu, S.K., et al., Protein crystals for the delivery of biopharmaceuticals. Expert opinion on biological therapy, 2004. 4(3): p. 301-317.
4. Parambil, J.V., et al., Effects of Oscillatory Flow on the Nucleation and Crystallization of Insulin. Crystal Growth & Design, 2011. 11(10): p. 4353-4359.
5. Delmas, T., M.M. Roberts, and J.Y. Heng, Nucleation and crystallization of lysozyme: Role of substrate surface chemistry and topography. Journal of Adhesion Science and Technology, 2011. 25(4-5): p. 357-366.
6. Harrison, R.G., et al., Bioseparations science and engineering. 2015: Topics in Chemical Engineering.
7. Shah, U.V., et al., Heterogeneous nucleants for crystallogenesis and bioseparation. Current Opinion in Chemical Engineering, 2015. 8: p. 69-75.
8. Shah, U.V., D.R. Williams, and J.Y.Y. Heng, Selective Crystallization of Proteins Using Engineered Nanonucleants. Crystal Growth & Design, 2012. 12(3): p. 1362-1369.

609008 Developing a Scalable Templating-Platform for Enhancing Crystallisation of Bio-Active Proteins

Abstract:

Over the last decade, bio-active peptides have emerged as the next generation of bio therapeutics with a rapidly growing demand [1]. A lack of cost-effective and reliable processes for downstream purification is a main industrial challenge for meeting the growing demand of bio-active peptides [2]. Crystallisation of peptides is a novel strategy to debottleneck downstream purification, offering a more economical and efficient technique than expensive traditional techniques [3].

Due to their low solubility and large molecular size, one of the main challenges in peptide crystallisation is the lack of controllability of the nucleation. However, very little has been done in using templates for controlling nucleation to achieve a consistent and reproducible crystallisation process. Chen et al. recently reported that highly mesoporous inorganic surfaces, referred to as "hard templates", can remarkably induce significantly faster and more controlled nucleation, even at low supersaturation [4]. In our recent work, we have successfully demonstrated a "soft templating" strategy by using soft, organic molecules, such as amino acids, to enhance nucleation. In buffer solution, amino acids form various interactions with peptides, whereby electrostatic interactions have been identified as most important. Therefore, amino acids that exhibit strong electrically charged residual groups for the relevant buffer conditions were investigated. In this presentation, the impact of types, amount and surface charge of the amino acids on the crystallisation performance of insulin at various supersaturation will be reported. With the aim of providing a strategy suitable for high-throughput peptide crystallisation, the possibility of scaling this templating strategy up, from the microlitres to the tens of millilitres, will be evaluated.

Insulin was used as model peptide and comprises of two chains, A and B, with a combined 51 amino acids and an overall molecular weight of 5.8 kDa. The impact of amino acids on crystallisation is evaluated by means of measuring induction time, crystal yield, and size distribution as a function of time. Citric buffer with a pH between 5 and 7 was used with zinc sulphate as the precipitant. The impact of amino acids on crystallisation is evaluated by monitoring the peptide concentration with UV-vis spectroscopy every 20-30 minutes. Using this, the induction time and crystal yield are obtained. Furthermore, the pH is measured at the start and the end of the crystallisation. Images obtained with a microscope are analysed to evaluate the crystal sizes and shapes. The experiments were carried out in batch crystallisers with volumes up to 15 millilitres to demonstrate the scalability.

Our recent results show that amino acids reduce the induction time of insulin crystallisation by up to 40% when compared to the induction time with no amino acids. The faster decrease in insulin concentration when amino acids are added also implies an increase in the crystal growth rate. As amino acids accelerate the crystallisation process, a yield of 90% can be achieved in a shorter period of time in crystallisers with volumes of up to 15 millilitres. Preliminary investigations show that the type and amount of amino acid added have only a minor impact on solubility, leading to a slight increase. Instead of impacting the solubility significantly, the charged residual-groups of the added amino acids increase the pH significantly. This increase in pH leads to an increase in stability of the tertiary structure of insulin by altering its physio-chemical surface properties which result in a faster nucleation. As a result of the more controlled crystallisation with amino acids, a narrower crystal size distribution can be obtained while achieving the same mean crystal size (~20 μm) when compared to the mean crystal size obtained with no amino acids. The insulin crystals obtained with amino acids show a more uniform crystal size distribution with a smaller span for crystalliser volumes reaching from the scale of microliters to the tens of millilitres. The soft templating strategy could be successfully demonstrated for crystalliser volumes of up to 15 millilitres while identifying mixing as one of the main limiting factors of scaling-up.

It has been demonstrated that amino acids as organic templates have the potential of enhancing the controllability of insulin crystallisation even at low supersaturations and to the scales of tens of millilitres. These includes a significant reduction of residence time needed in the crystalliser to obtain a desired crystal yield by maintaining a high level of consistency. Although different polar amino acids were investigated, the magnitude of advancing crystallisation differs dramatically between them as well as the ratio of amino acids concentration to insulin supersaturation.

Bibliography

1. Lau, J.L. and M.K. Dunn, *Therapeutic peptides: Historical perspectives, current development trends, and future directions.* Bioorg Med Chem, 2018. 26(10): p. 2700-2707.
2. Yang, H., et al., *Development and Workflow of a Continuous Protein Crystallization Process: A Case of Lysozyme.* Crystal Growth & Design, 2019. 19(2): p. 983-991.
3. Roque, A.C.A., et al., *Anything but Conventional Chromatography Approaches in Bioseparation.* Biotechnol J, 2020.
4. Chen, W., et al., *High Protein-Loading Silica Template for Heterogeneous Protein Crystallization.* Crystal Growth & Design, 2019. 20(2): p. 866-873.

609313 Sensor Integrated Continuous Microfluidic Mixer for Protein Screening at Constant Supersaturation

Abstract:

Sensors are great assets to quick and sensitive detection inside miniaturized devices were the amount of sample is small. Connecting an integrated microfluidic system to a control unit increases the accuracy of the experiments due to the automated adjustments of parameters. One of the examples for the potential application of a sensor integrated microfluidic system is screening crystalline materials where there is a high demand for accurate measurement of the concentration within the crystallization zone. In our previous studies, we developed a continuous microfluidic mixer for screening crystalline materials at the constant supersaturation, where the homogeneous concentration was calculated using COMSOL Multiphysics. However, it was not possible to perform an in-situ measurement for the actual concentration to experimentally verify that. Here we have coupled the microfluidic mixer design with an electrochemical sensor to enable in-situ measurement of the actual concentration inside the mixing zone. Moreover, integration of a sensor within the continuous microfluidic platform we can control the imposed supersaturation by manipulation of the flowrates. To validate the sensor integrated platform experimentally, we have done the protein crystallization of lysozyme and NaCl. Products of this crystallization process can exhibit different morphologies according to the ratio of the starting solutions. Several different experimental conditions are explored to study the morphologies obtained at assigned initial concentration as well as the growth kinetics.

606008 Particle Size Control of a Drug Substance through Process Crystallization

Abstract:

The solid state characteristics of an Active Pharmaceutical Ingredient (API) have great impact on the Drug Product (DP) performances and its formulation development approach.

One Critical Quality Attribute (CQA) that has to be controlled in order to ensure the success of a given formulation strategy in terms of manufacturability is the API Particle Size Distribution (PSD). In this case study, the API (PSD) had to be controlled to produce a blend which was homogeneous and flowable once combined with the chosen excipients and, in parallel, to reduce the risk of segregation during the DP manufacturing process.

The initial drug substance batch available for the formulation prototype production and selection showed a D(v,0.9) around 350mm. A mechanical PSD reduction via FitzMill was required in order to tune the drug substance PSD to match the PSD of the chosen excipients to ensure proper blend flowability and homogeneity attributes as well as to mitigate the risk of segregation during the encapsulation. As the alternative, the control of the PSD via process crystallization was investigated. Ultimately an anti-solvent mediated seeded crystallization process was developed. This process was able to deliver drug substance in the desired PSD range and avoided the need for post isolation milling.

610664 Unveil Epigenetic Memory Conferring Parkinson's Disease Risks Arising from Exposure to Environmental Chemicals

Abstract:

Exposures to a variety of environmental chemicals, such as herbicide and organic solvent have been shown to promote the risk of Parkinson's Disease (PD) based on population studies leading to increased public concerns of utilizing such chemicals in various applications. Atrazine, one of the most commonly used herbicides in the agricultural industry in the United States, is one of those chemicals. About 80 million pounds of atrazine were used in the U.S. every year. Atrazine residuals at a dose above its EPA limit (3 ppb) have been reported to contaminate drinking water sources. Despite many suggested disease connections with exposure to atrazine, how early-life exposure to atrazine can result in disease on-set at a later life remains poorly understood. In this work, we screened and investigated an array of epigenetic changes as potential transmission mechanism that gives rise to disease risk due to atrazine exposure that can persist and aggravate over time. To do that, we used SH-SY5Y cell line as our model neuronal system and exposed these cells to low-dose of atrazine prior to differentiation. Epigenetic assessments were carried out using immuno-staining, live-cell tracking and qPCR along with phenotypical assessments. Our results suggest that exposure to atrazine can cause immediate changes in epigenome after exposure to atrazine. The response is nonlinearly dependent on Atrazine concentrations. The changes in epigenome can persist through the neuronal differentiation and even attenuated for selected modifications. qPCR results provide further insights about underlying epigenetic enzymes involved in the regulation. Further evidence is provided in the changes of epigenetic make-up and expression of PD-related genes, i.e, α-synuclein, in exposed neurons. Our results thus collectively provide a strong evidence-based mechanism accounting for long-term PD risks due to exposure to atrazine.

608289 Screening and Identification of a High-Affinity Peptide Ligand Specific to Human CD3e

Abstract:

CD3 molecules are expressed on T lymphocytes as parts of T cell receptor complex are responsible for antigen recognition and activation of T cells. The present study aims to identify a novel peptide ligand which binds to human CD3ε in a specific manner, and to evaluate its effects on the functions of T lymphocytes. We had screened a phage-display peptide library against human CD3ε molecule using an improved subtractive biopanning method. After 5 rounds of biopanning, we had identified 13 phage clones displaying unique peptide sequences. One dominant phage clone, which was selected by occupying 90 % of tested plaques (18 out of 20), demonstrated enhanced bindings in a flow cytometry analysis after incubation with Jurkat cells. The peptide displayed by this most enriched phage clone binds to Jurkat cells in a dose-dependent manner but not to the 2PK3 B-cell lymphoma line. We employed molecular modeling and docking simulation to predict the structure of the peptide and it's binding poses on the surface of CD3ε. By conjugating the synthesized peptide to magnetic beads, we confirm that the peptide-modified beads can induce T cell receptor signaling (calcium flux) which can be used for T cell expansion. The novel peptide has a great potential in applications related to T cell immunotherapy.

610926 Single-Molecule Fluorescence Quenching Reveals Conformational Dynamics of the AAA+ Protease ClpXP

Abstract:

AAA+ proteases maintain protein quality control and help regulate cellular processes through targeted protein degradation. ClpXP from *Escherichia coli* is a model AAA+ protease that uses ATP-powered conformational motions to unfold, translocate, and degrade proteins labeled with specific tags. ClpX serves as a gatekeeper: unfolding and translocating proteins before threading the unraveled polypeptide into the lumen of ClpP for degradation. Despite numerous structural and mechanistic studies of ClpXP, the conformational states and motions in a working ClpXP motor have not yet been characterized. Here, we developed a novel, short-range single-molecule fluorescence quenching assay to monitor the functional motions of the ClpX machine bound to a colocalized ClpP molecule. The short-range sensitivity ($R_0 \sim 2.6$ nm) of our assay paired with distance calibration measurements using a DNA scaffold permit quantitative reporting of the conformational state and transitions in a single subunit of ClpX. Our results show that ClpXP cycles through dynamic and static conformational modes under varying nucleotide and substrate conditions. Additionally, we observed conformational switching in the absence of nucleotides suggesting that some structural changes are thermally driven. This method enables real-time observation of the structural dynamics of working AAA+ machines and provides context and mechanistic insight for recently solved cryogenic electron microscopy (cryoEM) structures of ClpXP[1–3]. Our technique can be adapted to study the movements and interactions of a host of molecular motors and other biomolecules where short-range conformational changes occur.

1. Fei, X. *et al.* Structures of the ATP-fueled ClpXP proteolytic machine bound to protein substrate. *Elife* **9**, 1–52 (2020).
2. Ripstein, Z. A., Vahidi, S., Houry, W. A., Rubinstein, J. L. & Kay, L. E. A processive rotary mechanism couples substrate unfolding and proteolysis in the ClpXP degradation machinery. *Elife* **9**, 1–25 (2020).
3. Gatsogiannis, C., Balogh, D., Merino, F., Sieber, S. A. & Raunser, S. Cryo-EM structure of the ClpXP protein degradation machinery. *Nat. Struct. Mol. Biol.* **26**, 946–954 (2019).

607224 Engineering Advanced Bio-Interfaces for Stem Cell-Based Tissue Regeneration

Abstract:

My research interests lie in the fields of polymeric membranes, drug/gene delivery, tissue engineering and regenerative medicine. The main goal of my research program is to address the challenges in life sciences and medicine by using engineered materials-centered approaches and elucidating the mechanisms underlying material structure-property relationships, material-cell and cell-cell interactions. This presentation will particularly concentrate on the development of functional biomaterials-based 3D platforms and biodegradable electronic interfaces for stem cell-based regenerative therapies. We will initially discuss stem cell-based strategies to enhance peripheral nerve regeneration and will continue with the application of these strategies for central nervous system problems and muscle and skin regeneration.

The multipotent adult stem cells, obtained from various tissue resources, have demonstrated the ability to differentiate into specific cell lineages.[1] Because of the accessibility and multipotency, transdifferentiation of mesenchymal stem cells (MSCs) into Schwann cell (SC)-like phenotypes (the glial cells of the peripheral nervous system providing myelination and axonal regeneration) has recently been considered as a promising approach for peripheral nerve regeneration.[1] However, this approach has not widely transitioned into clinical use due to three main challenges: (i) non-scalable and reversible transdifferentiation protocols; (ii) lack of reliability in controlling the final fate of the implanted cell population; and (iii) challenges in designing a multifunctional conduit/scaffold platform that mimics the complex extracellular matrix microenvironment.[1] We have developed different strategies to overcome these challenges and make stem cell-laden functional biomaterials-based platforms/bio-interfaces a viable solution not only for peripheral nerve injury repair but also for other tissue regeneration applications. In the first part of my presentation, I will highlight a unique biomaterials-based approach to combine different functionalities into a single conduit system that enables cellular alignment, directional growth and mimicry of native SCs' paracrine activity.[2-3] Then, I will proceed with discussing chemical stimuli-based MSCs to SCs transdifferentiation and emphasize the influence of 3D microstructural/mechanical properties of conduit platforms on MSCs behavior.[4] As an alternative, I will introduce a novel, chemical or growth factor free, electrical stimuli-based method to transdifferentiate MSCs into SCs via inkjet or 3D printed conductive graphene-based microcircuit platforms.[5-7] Following this, we will discuss my current research efforts focusing on two novel electronic interface fabrication methods that allow room temperature processing and enable the use of various biodegradable substrate materials with controlled 3D microstructural/mechanical properties.[8-9] With these methods we can fabricate high resolution and low feature size microcircuit integrated, 3D microstructured and stem cell-laden hydrogels or biodegradable electronic interfaces to provide in situ and local control on stem cell transdifferentiation and fate commitment for both *in vitro* and potentially complex *in vivo* conditions through the synergistic effect of microstructural/mechanical and electrical cues. In the second part, we will discuss my future research plans focusing on the development of biodegradable electronic platforms and understanding cellular mechanisms behind the electrical stimuli-based stem cell differentiation and tissue regeneration. We will put a particular emphasis on the use of these platforms for muscle and skin regeneration along with nerve/skin and nerve/muscle reinnervation and interaction as well as the central nervous system problems such as spinal cord injuries, neurodegenerative diseases or retinal regeneration. These platforms have demonstrated promising results in controlling the conversion of adult stem cells into specific cell lineages and the potential to pave the way for different tissue engineering applications.

References:

1. Uz, M.; Das, S. R.; Ding, S.; Sakaguchi, D. S.; Claussen, J. C.; Mallapragada, S. K., Advances in Controlling Differentiation of Adult Stem Cells for Peripheral Nerve Regeneration. *Advanced Healthcare Materials* **2018,** *7* (14), 1701046.
2. Sharma, A. D.; Wiederin, J.; Uz, M.; Ciborowski, P.; Mallapragada, S. K.; Gendelman, H. E.; Sakaguchi, D. S., Proteomic analysis of mesenchymal to Schwann cell transdifferentiation. *Journal of Proteomics* **2017,** *165*, 93-101.
3. Uz, M.; Sharma, A. D.; Adhikari, P.; Sakaguchi, D. S.; Mallapragada, S. K., Development of multifunctional films for peripheral nerve regeneration. *Acta Biomaterialia* **2017,** *56*, 141-152.
4. Uz, M.; Buyukoz, M.; Sharma, A. D.; Sakaguchi, D. S.; Altinkaya, S. A.; Mallapragada, S. K., Gelatin-based 3D conduits for transdifferentiation of mesenchymal stem cells into Schwann cell-like phenotypes. *Acta Biomaterialia* **2017,** *53*, 293-306.
5. Das, S. R.; Uz, M.; Ding, S.; Lentner, M. T.; Hondred, J. A.; Cargill, A. A.; Sakaguchi, D. S.; Mallapragada, S.; Claussen, J. C., Electrical Differentiation of Mesenchymal Stem Cells into Schwann-Cell-Like Phenotypes Using Inkjet-Printed Graphene Circuits. *Advanced Healthcare Materials* **2017,** *6* (7).
6. Uz, M.; Donta, M.; Mededovic, M.; Sakaguchi, D. S.; Mallapragada, S. K., Development of Gelatin and Graphene-Based Nerve Regeneration Conduits Using Three-Dimensional (3D) Printing Strategies for Electrical Transdifferentiation of Mesenchymal Stem Cells. *Industrial & Engineering Chemistry Research* **2019,** *58* (18), 7421-7427.
7. Uz, M.; Hondred, J. A.; Donta, M.; Jung, J.; Kozik, E.; Green, J.; Sandquist, E. J.; Sakaguchi, D. S.; Claussen, J. C.; Mallapragada, S., Determination of Electrical Stimuli Parameters To Transdifferentiate Genetically Engineered Mesenchymal Stem Cells into Neuronal or Glial Lineages. *Regenerative Engineering and Translational Medicine* **2019**.
8. Uz, M.; Jackson, K.; Donta, M. S.; Jung, J.; Lentner, M. T.; Hondred, J. A.; Claussen, J. C.; Mallapragada, S. K., Fabrication of High-resolution Graphene-based Flexible Electronics via Polymer Casting. *Scientific Reports* **2019,** *9* (1), 10595.
9. Uz, M.; Lentner, M. T.; Jackson, K.; Donta, M. S.; Jung, J.; Hondred, J.; Mach, E.; Claussen, J.; Mallapragada, S. K., Fabrication of Two-Dimensional and Three-Dimensional High-Resolution Binder-Free Graphene Circuits Using a Microfluidic Approach for Sensor Applications. *ACS Applied Materials & Interfaces* **2020,** *12* (11), 13529-13539.

611758 α2,3 Sialylated Colon and Breast Cancer Cell Extracellular Vesicles Bind to L-Selectin Under Flow Conditions

Abstract:

Extracellular vesicles (EVs) are naturally occurring vesicles released by all cells, including cancer cells. EVs are involved in normal cellular processes such as intercellular communication, but EVs released from cancer cells and their role in cancer metastasis need further understanding. In particular, it is unclear whether EVs released from cancer cells express sialofucosylated glycans that can function as L-selectin ligands, similar to the L-selectin ligands on the cancer cells from which they were derived. Therefore, we hypothesize that breast and colon cancer cells and model EVs isolated from these cancer cells express α2,3 sialylated L-selectin ligands that interact with L-selectin under flow conditions. Giant plasma membrane vesicles (GPMVs, 5-20 μm diameter vesicles) were isolated from cancer cells as a model of naturally occurring large cancer EVs (i.e., large oncosomes). GPMVs were generated from the LS174T colon cancer cell line and BT-20 and SK-Br-3 breast cancer cell lines using chemical vesiculation agents (25 mM formaldehyde and 2 mM dithiothreitol). The parallel plate flow chamber adhesion assay was used to functionally assess LS174T colon cancer cells, BT-20 and SK-BR-3 breast cancer cells, and GPMVs isolated from these three cancer cell lines that were incubated with either *V. cholera* sialidase (VC, broad substrate specificity), *C. perfringens* sialidase (CP, specificity of cleavage: α2,3 > α2,6 sialylation), or left untreated. Treatment of cancer cells and GPMVs with VC and CP sialidase significantly reduced interactions (>50%) with L-selectin substrates (L-selectin hFc coated plates) at physiologically relevant wall shear stresses compared to the untreated cancer cells and GPMVs. In addition, flow cytometry was performed on cancer cells and GPMVs treated with VC or CP sialidase or left untreated to determine the expression of candidate L-selectin ligands. All three cell lines and their isolated GPMVs tested positive for cutaneous lymphocyte antigen (HECA-452 antigen) and sialyl Lewis X (sLeX, CSLEX1 mAb), with VC and CP treatment reducing expression of HECA-452 antigen and sLeX greater than 90% as measured by mean fluorescent intensity levels. These results indicate similar L-selectin ligands are expressed on cell lines and GPMVs. Furthermore, GPMVs interacted with human peripheral blood polymorphonucleocytes and mononuclear cells in an L-selectin-dependent manner, similar to the original cell lines. Altogether, these results indicate that LS174T colon cancer cells, BT-20 and SK-BR-3 breast cancer cells, and GPMVs isolated from these cancer cells express α2,3 sialylated L-selectin ligands that interact with L-selectin under flow conditions. Understanding how EVs released from cancer cells interact with cell adhesion molecules such as L-selectin may help elucidate the role of intercellular adhesion in metastasis, leading to the development of novel diagnostics, prognostics, and therapeutics.

610036 Biofabrication of Muscle Fibers Using Surface Chaotic Flows: Chaotic 2D-Printing

Abstract:

Multiple human tissues exhibit a fibrous nature. Therefore, the fabrication of hydrogel fibers for biomedical engineering applications is a topic of current interest. Here we present a simple strategy, based on the use of chaotic flows, to fabricate continuous, long and thin filaments of gelatin methacryloyl (GelMA) for muscle tissue engineering. The fabrication of these hydrogel fibers was achieved by chaotic elongation in a finely controlled and miniaturized version of the journal bearing (JB) system. A drop of a pre-gelled solution of GelMA is injected onto a higher-density viscous fluid and a chaotic flow is applied in an iterative process. The chaotic flow exponentially deforms and elongates the drop to generate a fiber that is then photocrosslinked under UV light exposure. The progression of the elongation process can be rigorously modeled. We showed that computational fluid dynamics (CFD) simulations closely predict the motion of the inks, as well as the lengths of the filaments obtained by this chaotic advection-based fabrication process.

This technological platform is a low-cost and simple alternative to 3D-printing and other biofabrication strategies. The hydrogel filaments produced were proven to be effective scaffolds for C2C12 myoblast cells. Cells were initially seeded on the surface of these GelMA filaments and progressively developed into coherent and highly aligned muscle-like fibers within three weeks of culture. A detailed characterization of such fibers was performed through optical and scanning electron microscopy techniques, which highlighted their potential to be used as cost-effective models for muscle tissue engineering and regenerative medicine.

In this work, we introduced a simple technique, based on the use of surface chaotic 2D-flows, which enables the facile fabrication of hydrogel filaments. We demonstrated its use to fabricate muscle-like fibers, and envision the extension of the technique to other exciting tissue engineering applications.

605018 High-Throughput Generation and Study of 3D Spheroids in a Thiol-Acrylate Hydrogel Scaffold Using a Droplet Microfluidic Trapping Array

Abstract:

Culturing cancer cells in a three-dimensional (3D) environment better recapitulates *in vivo* conditions including cell-to-cell interactions and mass transfer limitations of metabolites, oxygen, and drugs. Recent drug studies have suggested that a high rate of pre-clinical and clinical failures results from mass transfer limitations associated with drug entry into solid tumors that 2D model systems cannot predict. As such, there is a need to develop high-throughput platforms to generate and study 3D tumor spheroids for drug efficacy. Existing approaches for 3D cell culture include hanging droplet arrays and spinner flasks; however, these methods suffer from low throughput and significant heterogeneity in spheroid size and uniformity. Recently, microfluidic devices have become a popular alternative to rapidly generate 3D spheroids. A popular approach utilizes microwell arrays that have been modified to prevent cellular attachment and force cellular aggregation into 3D spheroids. These devices can rapidly generate a large number of spheroids; however, most microwell arrays cannot facilitate on-chip interrogation of 3D spheroids and suffer from disaggregation due to fluid shear stress. Similarly, the spheroids generated in the above methods suffer from significant heterogeneity since they are generated from 100-500 different cells. Droplet microfluidic devices offer a promising alternative to grow 3D spheroids from a small number (~1-10) of cancer cells to reduce spheroid heterogeneity while simultaneously allowing for on-chip interrogation and evaluation. The goal of this work was to incorporate a novel thiol-acrylate (TA) hydrogel scaffold into a microfluidic droplet trapping array to generate and study 3D spheroids. The TA hydrogel has been previously demonstrated to severe as an excellent scaffold for 3D cell culture that does not require external initiation factors such as UV light, temperature, or additional reagents. Three different overhead trapping arrays were fabricated with 70, 150, or 300 μm circular traps to study the effect of droplet size and cell seeding density on spheroid formation and growth in the TA hydrogel scaffold. The overhead trapping array consisted of 1400 traps and was capable of ~99% droplet trapping and ~50% cellular encapsulation. The TA hydrogel allowed for rapid (~30-40 min) polymerization of the scaffold followed by removal of the oil phase and replacement with complete media to initiate spheroid growth. The growth and viability of model breast cancer spheroids using MCF7 at 37°C was confirmed for up to four days under static conditions. However, after four days spheroid size stabilized suggesting the need for continuous infusion of fresh growth media. This study also identified that a minimum number of encapsulated cells (~4-6) is needed to generate a spheroid and that single encapsulated cells are less likely to growth into a full-blown spheroid. The findings from this study highlight an alternative approach into generating 3D spheroids incorporating an easy-to-use, inexpensive scaffold that can be used for high-throughput drug screening.

608061 Anticancer Drug Evaluation in MCF7 Spheroids Cultured in a 3D-Printed Miniaturized Continuous Stirred Tank Reactor (mCSTR)

Abstract:

Cancer continues to be one of the deadliest diseases worldwide, with 600,000 cancer-related deaths reported in 2019 in the USA alone. Despite enormous efforts to find new therapeutic cancer targets and cancer therapies, the current platforms for evaluating drug candidate safety and efficacy have serious limitations (i.e., cost, reproducibility, availability, and ethical concerns). Furthermore, these platforms often provide unreliable results. Improved cancer research models are therefore needed to achieve better predictive results for the efficacy of new therapeutic compounds in humans.

Here, we describe a simple and robust system for anti-cancer drug screening. The system is based on the use of a 3D-printed miniaturized (3 mL volume) continuous stirred tank mini-reactor (mCSTR) that enables the extended culture (20–30 days) of breast cancer spheroids for evaluation of anti-cancer drugs. The off-center agitation system of this CSTmR enables homogeneous chaotic mixing at low speeds (100 rpm) as confirmed by computational fluid dynamics (CFD) simulation and tracer studies under batch and continuous operation.

We have cultured populations of small MCF7 cancer spheroids, fabricated using conventional methods, under continuous flow conditions at different flow rates (to attain different hydraulic residence times HRT) and with various inlet substrate concentrations. Here, the residence time (HRT) and the inlet concentration are used to modulate the growth dynamics and glucose consumption of the spheroids. For instance, we monitored the progression of size, shape, genetic expression of relevant cancer biomarkers, and glucose utilization for 30 days in continuous cultures perfused with culture media (glucose concentration of 1.02 mg μL^{-1} mg μL^{-1}) at 2.3 $\mu L\ min^{-1}$ (HRT=22h). Under these conditions, a pseudo steady state was reached, as suggested by the nearly constant glucose concentration within the tank (and the indicators of size, shape, and expression).

We also independently evaluated the effect of three different anti-cancer drugs (docetaxel, the taxol paclitaxel and the anthracycline doxorubicin) in continuously cultured spheroid populations. We continuously fed each drug (at two different doses) from day five, while monitoring the glucose concentration within the tank as an overall indicator of the progression of cell viability with time. The endpoint of each experiment was determined when glucose consumption ceased. Endpoint evaluation of spheroids exposed to different drugs and doses included morphological observations by scanning electron microscopy (SEM), viability determined by Live/Dead assays, and drug cytotoxicity as assessed by lactate dehydrogenase (LDH) activity and changes in the expression of specific genes related to drug resistance and apoptosis (BCL2, BAX, ABCB1, ABCC1 and ABCG2).

The MCF7 spheroids showed distinctive dose-dependent responses to each drug. Our results suggest that this mCSTR system is a suitable platform for the continuous perfusion and evaluation of anti-cancer drugs. Moreover, the system is highly customizable, as a wide spectrum of experiments can be implemented under continuous flow conditions simply by varying residence times, inlet concentrations, and spheroid densities. Likewise, this system could represent an option for the culture of cancer biopsies for personalized medicine applications.

611170 Nuclear Size Changes Caused By Local Motion of Cell Boundaries Unfold the Nuclear Lamina and Dilate Chromatin and Intranuclear Bodies

Abstract:

The mechanisms by which mammalian nuclear shape and size are established in cells, and become abnormal in disease states are not understood. Here, we tracked motile cells that underwent systematic changes in cell morphology as they moved from 1-D to 2-D micro-patterned adhesive domains. Motion of the cell boundaries during cell motility caused a dynamic and systematic change in nuclear volume. Short time scales (~1 h) distinguished the dilation of the nucleus from the familiar increase that occurs during the cell cycle. Nuclear volume was systematically different between cells cultured in 3-D, 2-D and 1-D environments. Dilation of the nuclear volume was accompanied by dilation of chromatin, a decrease in the number of folds in the nuclear lamina, and an increase in nucleolar volume. Treatment of 2-D cells with non-muscle myosin-II inhibitors decreased cell volume, and proportionately caused a decrease in nuclear volume. These data suggest that nuclear size changes during cell migration may potentially impact gene expression through the modulation of intranuclear structure.

600459 Engineering Vascular Cells and Macro-Fluidic Platforms for Physiological Organ and Tissue Culture

Abstract:

The vasculature plays a critical role during embryogenesis. After birth, the vasculature continues to have a strong footprint in many essential biological processes of organs and tissues. For example, the vasculature delivers nutrients and gases to maintain tissue survival, transports biological wastes to the appropriate organs for further processing, and enables trafficking of immune cells to the injured tissues. In pathological tumors, the maladaptive vasculature invades the tumor tissues to fuel the tumor growth and metastasis. In addition, the vasculature remains an essential component in tissue engineering and translational medicine. The tissue engineering field has made significant progress to create *in vitro* organ-on-chip models to better mimic several biological organs such as skin, pancreas, lung, and brain (1,2). Despite these efforts, there are setbacks in current models of organ-on-chip. For instance, to incorporate the vasculature, efforts have been made to generate microfluidic platforms with embedded mature adult endothelial cells to generate some network of blood vessels in these organ-on-chip models. However, these mature endothelial cells lose their intrinsic capability *in vitro* to generate a robust blood vessel network and lose their ability to interact and adapt to different cell types in co-culture. As a result, the blood vessels in organ-on-chip models are poorly functional. In addition, the micro-fluidic organ-on-chips are miniaturized versions of organs, where simple diffusion of nutrients and gases is sufficiently enough to feed the cells whereas organ and tissues in the body heavily rely on blood flow carrying nutrients and gases through dedicated vascular conduits (3). Thus, the small-scale organ-on-chips render the blood vessels and fluid perfusion unnecessary. They also make it challenging to address how transported nutrients, and gases by blood vessels alter cellular behavior in organs or tissues in the body.

To address the first limitation of organ-on-chip models where mature endothelial cells do not form robust vascular network *in vitro*, we first engineered adaptive vascular endothelial cells by introducing vasculogenic transcriptional factor into mature endothelial cells. The engineered endothelial cells upregulated several genes involved in angiogenesis and vasculogenesis. Consequently, in 3D hydrogel culture, these engineered endothelial cells formed a robust and functional network of vasculature. Moreover, when placed in co-culture with several other cell types such as normal colon organoids, tumor colon organoids, and pancreatic islets, they not only survive, and adapt to the culture environment of other cell types, they also avidly interact with the organoids and islets. To overcome the limitation of dimension in traditional micro-fluidic organ-on-chips, we engineer macro-fluidic devices where dimensions of the devices are extended up to 2cm in length and width. After seeding these engineered-endothelial cells into the macro-fluidic devices and incorporating fluidic pump, we successfully generated a large scale robust and functional vasculature network in macro-fluidic devices. The vascular network in macro-fluidic device is fully functional as they support perfusion of medium and even human whole blood without clotting. To further extend the study, we investigate the function of the vascularized macro-fluidic organ-on-chip by embedding human pancreatic islets inside the organ-on-chip. Upon glucose stimulation, pancreatic islets in the vascularized macro-fluidic organ-on-chip sense glucose and release insulin similar to *in vivo* physiological glucose response of human pancreatic islets (4). Our efforts continue to investigate and expand the vascularized macro-fluidic organ-on-chips into different organotypic models such as colon, and liver organ-on-chips.

[1] **Nguyen, D.H.T.,** Stapleton, S.C., Yang, M.T., Cha, S.S., Choi, C.K., Galie, P.A., Chen, C.S.

(2013). *Biomimetic model to reconstitute angiogenic sprouting morphogenesis in vitro*. **Proceedings of the National Academy of Sciences**. 110(17): 6712-6717.

[2] **Nguyen, D.H.T.,** Esak, L., Alimperti, S., Norgard. R.J., Wong, A., Lee, J.J.K., Eyckmans, J., Stanger, B.Z., Chen, C.S. (2019). *A biomimetic pancreatic cancer on-ship reveals endothelial ablation via ALK7 signaling*. **Science Advances.** 5(8): eaav6789.

[3] Galie, P.A., **Nguyen, D.H.T.**, Choi, C.K., Cohen, D.M., Janmey, P.A., Chen, C.S. (2014). "Fluid shear stress threshold regulates angiogenic sprouting". **Proceedings of the National Academy of Sciences**. 111(22): 7968-7973.

[4] Palikuqi, B., **Nguyen, D.H.T.**, Li, G., Schreiner, R., Pellegata, A.F., Liu, Y., Redmond, D., Geng, F., Lin, Y., Gómez-Salinero, J.M., Yokoyama, M., Zumbo, P., Zhang, T., Kunar, B., Witherspoon, M., Han, T., Tedeschi, A.M., Scottoni, F., Lipkin, S., Dow, L., Elemento, O., Xiang, J.Z., Shido, K., Spence, J., Zhou, J.Q., Schwartz, R.E., DeCoppi, P., Rabbany, S.Y., Rafii, S. (2020). "Adaptable human endothelial cells for hemodynamic vascularization of normal and malignant organoids". **Nature**. ***Accepted***.

601484 Spatially Controlled Cell-Free Protein Synthesis and Glycosylation on Chip

Abstract:

Among cellular processes, glycosylation of biomolecules like proteins or lipids is critical for their proper function and efficacy. Recombinant protein production is a highly conserved, templated process, while added glycans are extremely heterogenous and proper control over their expression is important in controlling biological activity and therapeutic function. Current cell-based methods to produce glycosylated biologics rely on native machinery, which inherently produces non-specific glycan profiles, making purification difficult as well as time and labor intensive. To solve these issues, cell-free technologies have been developed to produce proteins without growth constraints and allow the addition of cell-free glycosylation pathways.

I will describe our work towards a microfluidic platform analogous to the natural eukaryotic cellular assembly line, that integrates spatially separated cell-free protein synthesis, glycosylation, and enrichment of a model glycoprotein. Microfluidics allow for tight control over environmental conditions, recyclability of tethered enzymes, and importantly, separation of reactions that has yet to have been achieved in cell-free glycosylation systems. As an initial proof-of-concept, the green fluorescent protein (GFP) is used to allow easy visualization of cell-free protein production and purification. In the first module, GFP is expressed with continuous flow using *E. coli* cell-free extract. In the second module, the GFP with a terminal glycosylation tag is passed through a microfluidic chamber with glycosylation machinery. The current system uses the integral membrane oligosaccharyltransferase PglB, which efficiently transfers a heptasaccharide glycan from its lipid donor to the protein. For continuous production of glycoprotein, PglB is biotinylated and immobilized using the biotin-streptavidin interaction. In the third module, the GFP product is captured using immobilized affinity chromatography to allow for easy capture and subsequent release of the His-tag protein. This work presents a novel solution to cell-free systems that enables fundamental studies of glycosylation as well as being used for continuous cell-free glycoprotein production.

610964 Membrane Fusion Allows Ruptureless Extreme Deformation of Nuclei during Cell Migration

Abstract:

During migration of cells through narrow spaces, the nucleus undergoes extreme deformation. Such deformation is frequently accompanied by rupture of the nuclear envelope. Nuclear rupture has been implicated in pathogenesis mechanisms, however, the mechanisms by which the nucleus ruptures are not fully understood. Here we tracked fibroblasts as they migrated through vertical PDMS obstacles. Nuclei underwent extreme deformations around the posts. Interestingly, nuclei did not rupture when opposing nuclear surfaces fused with each other. Conversely, nuclei ruptured when such fusion was not possible. Quantification of nuclear perimeter revealed that dynamic membrane fusion relieved strain in the nuclear surface during deformation which likely allowed escape from rupture. These results suggest that remodeling of the nuclear envelope can allow extreme nuclear deformations without envelope rupture.

612716 Self-assembly peptide-based micromotors controlled by 3D magnetic field for deep solid tumor

Abstract:

In vivo self-assembly peptides as important candidates are expected to be used in cancer treatment. The peptide is conducive to deep penetration since its unassembled state with a small size *in vitro*; when it penetrates into the solid tumor, self-assembly behavior occurs in the tumor microenvironment, which can enhance its accumulation and internalization. However, the initiative and selectivity of the nanodrugs are insufficient because of the lack of engine to overcome the limitations of natural physical movement in the body, resulting in poor drug effects and strong side effects.

Micro/nano-motor (MNM) is the revolutionary and innovative field to replace the current drug delivery system. MNM can convert chemical or external energy in the surrounding environment into mechanical kinetic energy to produce autonomous movement. After more than ten years of development, micro/nano-motors based on various structures, such as microtubes, microwires, and spiral micromotors have been developed. Due to the tiny structure and the ability of active propulsion, micro/nano-motors present more and more advantages in the field of active drug delivery. However, using MNM method to empower active peptide-based nanodrugs have not been studied.

This work firstly suggested the peptide-based microrobots used for active drug delivery. The active engine was mounted on the peptide drugs by magnetic particles to enable its active movements in complex body fluid environments under the control of external magnetic fields to enhance its mobility. Rapid aggregation and self-assembly of drug peptides at the target site were achieved. In this work, peptide is used as the base of propulsion for micromotors for the first time, which provides a new view for active motion on the level of peptide.

612470 Phosphatidylserine Expression as a Molecular Target in *Listeria monocytogenes* and *Escherichia coli*

Abstract:

Targeted delivery of antimicrobials, e.g. via encapsulation in liposomes, is often useful for mitigating noxious side effects or enhancing anti-microbial activity against difficult to treat pathogens occupying unique biological niches. In addition, targeted delivery that significantly enhanced a compound's antimicrobial activity could make it useful against pathogens outside of its typical spectrum of activity. We have developed a novel strategy for targeting antibiotics that is highly efficacious against both Gram-positive (*L. monocytogenes*) and Gram-negative organisms (*E. coli*). This strategy takes advantage of the fact that eukaryotic and prokaryotic cells share many mechanisms of cell stress and apoptosis. One such conserved element is the expression of phosphatidylserine (PS), an anionic membrane-bound phospholipid, in response to cell stress. This expression of PS is nearly universally conserved, being demonstrated in prokaryotes and eukaryotes. Actively externalized in response to stress, PS is the natural ligand for proteins of the annexin superfamily. Interestingly, the expression of annexin superfamily members is found in both prokaryotes and eukaryotes, demonstrating the nature of important mechanisms by which cell stress is recognized, as well as signaled.

Bacterial PS expression was targeted with its ligand, annexin A5 (ANXA5). This protein was functionalized with the covalent addition of the antibiotic ampicillin and separately with the antibiotic moxifloxacin. Functionalized ANXA5 serves as a delivery vehicle, directing antibiotic to bacterial phosphatidylserine expression. Results presented here suggest that this ANXA5-ampicillin bioconjugate participates in a positive feedback loop where phosphatidylserine, the target of the delivery vehicle ANXA5, is upregulated by the chemotherapeutic payload of the bioconjugate. Importantly, the ANXA5 delivery vehicle is not toxic to bacterial cells by itself and neither is the ANXA5 bioconjugate toxic to human vascular endothelial cells. As measured by the IC50 (half maximal inhibitory concentration), conjugation to ANXA5 resulted in increasing the antibiotic activity of ampicillin against *L. monocytogenes* and *E. coli* by more than 4 and 3 orders of magnitude, respectively, compared to free ampicillin, while activity of moxifloxacin against *L. monocytogenes* is increased by 4 orders of magnitude. Given the conservation of phosphatidylserine expression in pathologies such as oncogenesis, and other bacterial/viral/parasitic infections, we hypothesize that a therapeutic modality targeting phosphatidylserine expression may be a viable chemotherapeutic strategy in many infectious diseases.

608711 The Relevance of Simulated Lung Fluid Composition on the Drug Solubility and Predicted *in-Vivo* performance of Inhaled Drug Delivery

Abstract:

Introduction

Physiologically based pharmacokinetic modelling (PBPK) is one of the *in-silico* methods that allows establishment of *in-vitro-in-vivo* correlations (IVIVC) based on the mechanistic understanding of the physiological processes affecting drug disposition *in-vivo*. IVIVC aims to enable the correlation between the *in-vitro* behavior of the drug and its *in-vivo* performance. PBPK models aim to be able to accurately predict the drug PK in a target organ, based on different parameters such as its physicochemical characteristics, solubility, dissolution, permeability and metabolism but also on the anatomy and physiology of the target patient population[1]. However, in the case of inhaled drug delivery, the clinical relevance of the *in-vitro* methodologies currently being applied are still not well established. It is necessary, for example, to standardize a simulated lung fluid (SLF) to be used in the solubility and dissolution testing of orally inhaled products and obtain reproducible systematic sets of data able to be confidently applied to drug development. However, presently, this is not yet the case and diverse SLF media are based on the simplistic employment of phosphate buffer (PBS) in the presence of surfactants such as sodiumsodecylsulphate (SDS) or Tween 80[2]. Considering that the lung fluid is composed of surfactant lipids and proteins, PBS based media might not be able to completely capture environment in the lung[3]. Therefore, in this work by using an *in-vitro-in-silico* feedback-feedforward approach (Figure 1), we aimed to gain a systematic understanding of how different SLF components (i.e. lipids and proteins) can impact drug solubility. Two model inhaled drugs with different lipophilicities were selected to be studied, salbutamol sulphate (SS; logP=0.44) and budesonide (BUD; logP=2.52). Subsequently, the experimental values obtained for the solubility in the distinct SLFs were used as inputs in the PBPK models of the two drugs, resulting in different plasma concentration-time profiles.

Materials and Methods

The solubility of SS and BUD was investigated in PBS buffer (pH=7.40) and in SLF, by adding 5 mL of the media to an excess amount of the drug (n=3). The suspensions were conditioned at 37°C and 100 rpm, for 24h. After that, they were filtered and the drug in solution was quantified using high performance liquid chromatography (HPLC). The media simulating the lung fluid were prepared using lipids – dipalmitoylphosphatidylcholine (DPPC; Avanti Polar Lipids, USA), cholesterol (Sigma Aldrich, USA), dipalmitoylphosphatidylglycerol (DPPG; Avanti Polar Lipids, USA) and proteins – albumin (Sigma Aldrich, USA); alone or in combination.

SLF preparation

The SLF was prepared by weighing the lipids i.e. DPPC (4.60 mg/mL), DPPG (0.50 mg/mL), cholesterol (0.10 mg/mL), and dissolving them in a chloroform methanol (2:1) mixture. The solvents from the solutions were evaporated in a rotavapor (Heidolph Laborota 4000 efficient) at 50°C and 60 rpm to yield lipid bilayers. The lipid bilayers were hydrated using pure PBS buffer or PBS containing albumin. Likewise, different SLFs were obtained – SLFs containing only lipids and proteins and a combination of thereof. Albumin was tested in two concentrations, 13.0 mg/mL and 29.1 mg/mL, as present in the alveolar (AF) and bronchiolar fluid (BF), respectively[3]. This way, the systematic investigation of the impact of each component was possible.

PBPK model development and predictions

Based on the reported *in-vivo* data for the dry powder inhalers, Diskus®[4] and Turbuhaler®[5], the PBPK models for SS and BUD were respectively developed. The PCAT (Pulmonary Compartmental Absorption and Transit™) module of GP (GastroPlus; Simulations Plus, USA; version 9.7) was used to develop the PBPKs. To predict the deposition patterns of SS[6] and BUD[7], their aerodynamic size distributions and spirometry profiles were used as inputs in MPPD (Multiple-Path Particle Dosimetry, Applied Research Associates, Inc.).

Results and Discussion

Considering the necessity of having a reliable, reproducible and cost effective SLF, one of our goals was to investigate the impact of lipids and proteins on drug solubility. Solubility values of SS obtained in all the media were similar. The solubility in PBS was very high as well as in SLF containing only lipids. The presence of albumin alone or in combination with lipids (AF) in the media showed no effect on SS solubility. However, in the case of BUD, significant impact of the media components was observed and the solubility values increased in the following order: PBS < SLF with albumin AF ≈ SLF with albumin BF < SLF with lipids ≈ SLF with lipids and albumin AF. The presence of albumin alone increased the solubility almost two-fold, however no difference was observed at the two different used concentrations of the lipid (i.e. AF and BF). The use of lipids increased the solubility the most. Similar results have been reported for BUD when this was tested using the commercial natural surfactant Survanta®[8]. Consequently, it becomes evident that SS and BUD exhibit different solubility behaviors when in the presence of lipids and proteins, that is BUD is impacted and SS not. The different solubility values were used as inputs in the PBPK models of the two drugs and the impact on the PK parameters investigated. The developed pulmonary models showed a total mean absolute percentage error (MAPE) of 25% and 20% as well as a correlation factor of 0.86 and 0.89 for BUD and SS, respectively. Hence, both models were found fit to use for further predictions. As expected, negligible differences in solubilities did not impact the PK parameters of SS. This confirmed that SS disposition is not solubility limited. Nevertheless, in the case of BUD, a more pronounced impact was noted when the solubilities in different media were used. For instance, the predicted c_{max} and t_{max} had lower MAPE values when albumin was present in the system, while in case of area under the curve ($AUC_{0-\infty}$) the lowest error was seen when lipids were used. These interesting observations in the PK parameters imply that for BUD a middle value between its solubility in the alveolar and bronchiolar regions must be considered in order to more closely reflect the situation *in-vivo*.

Conclusion

The present study showed the relevance of carefully selecting the components present in the media used as SLF and highlighted how PBPK modelling can help the development of *in-vitro* methods and vice versa. We observed that for freely soluble APIs, like SS, simple media, such as PBS, are enough to investigate solubility and proceed to other stages of the biopharmaceutical evaluation. For BUD, the greater influence of the media components on solubility, and the different outcomes in terms of the predicted PK parameters, imply that in the case of poorly soluble APIs a more careful consideration must be applied to the *in-vitro* testing protocols. To support our observations, further studies should be carried out in order to screen a larger variety of molecules.

References

[1] P. Bäckman, S. Arora, W. Couet, B. Forbes, W. de Kruijf, and A. Paudel, “Advances in experimental and mechanistic computational models to understand pulmonary exposure to inhaled drugs,” *Eur. J. Pharm. Sci.*, vol. 113, no. June 2017, pp. 41–52, 2018.

[2] S. Radivojev, S. Zellnitz, A. Paudel, and E. Fröhlich, “Searching for physiologically relevant in vitro dissolution techniques for orally inhaled drugs,” *Int. J. Pharm.*, 2018.

[3] E. M. Bicer, “Compositional characterisation of human respiratory tract lining fluids for the design of disease specific simulants,” 2015.

[4] A. Moore, K. Riddell, S. Joshi, R. Chan, and R. Mehta, “Pharmacokinetics of Salbutamol Delivered from the Unit Dose Dry Powder Inhaler: Comparison with the Metered Dose Inhaler and Diskus Dry Powder Inhaler,” *J. Aerosol Med. Pulm. Drug Deliv.*, vol. 30, no. 3, pp. 164–172, 2017.

[5] S. Lähelmä *et al.*, “Equivalent lung dose and systemic exposure of budesonide/formoterol combination via easyhaler and turbuhaler,” *J. Aerosol Med. Pulm. Drug Deliv.*, vol. 28, no. 6, pp. 462–473, 2015.

[6] J. N. Seheult *et al.*, “The acoustic features of inhalation can be used to quantify aerosol delivery from a Diskus™ dry powder inhaler,” *Pharm. Res.*, vol. 31, no. 10, pp. 2735–2747, 2014.

[7] Á. Farkas *et al.*, “Numerical simulation of emitted particle characteristics and airway deposition distribution of Symbicort® Turbuhaler® dry powder fixed combination aerosol drug,” *Eur. J. Pharm. Sci.*, vol. 93, pp. 371–379, 2016.

[8] D. Palmer, S. Schürch, and J. Belik, “Effect of budesonide and salbutamol on surfactant properties,” *J. Appl. Physiol.*, vol. 89, no. 3, pp. 884–890, 2000.

Figure 1: Design of the study, with examples of solubility results and developed PBPK models

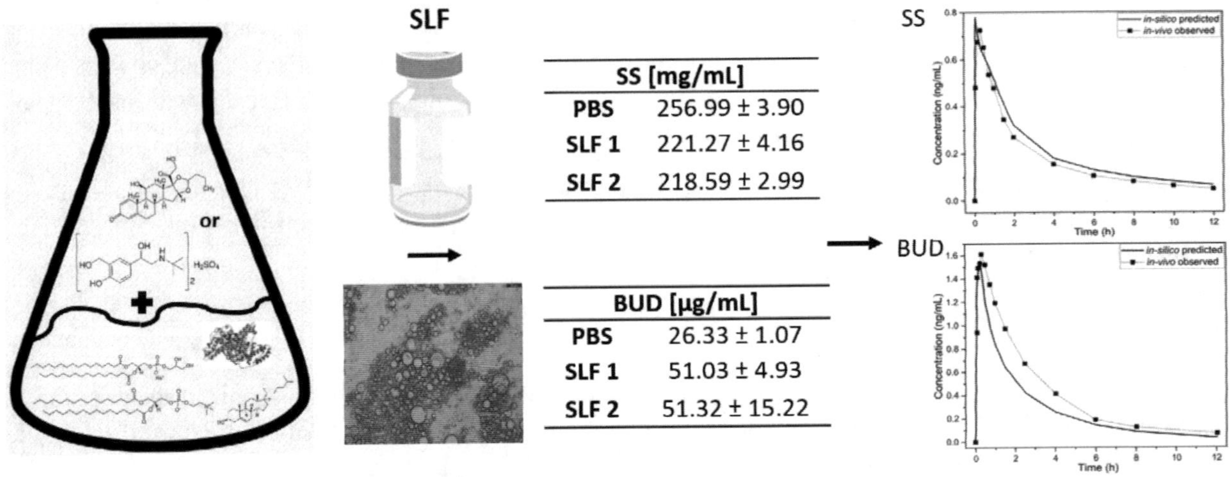

606039 Zinc Oxide Nanoparticles with an Anti-Tubercular Drug for Overcoming Mycobacterial Drug Resistance

Abstract:

Tuberculosis, caused by *Mycobacterium tuberculosis*, is one of the top ten causes of death worldwide. With the emergence of multi-drug resistant (MDR-TB) and extensively drug-resistant (XDR-TB) strains, the disease has reached serious proportions. The intrinsic drug resistance of mycobacteria, due to mycolic acid-rich cell wall and active efflux pumps, is one of the major limitations in tuberculosis treatment. These factors are responsible for low intracellular drug concentration. In addition to intrinsic drug resistance, MDR and XDR strains have also developed acquired drug resistance towards the first-line anti-TB drugs in current use. In this aspect, nanoparticles can be a useful tool for overcoming bacterial drug resistance. Metal or metal oxide nanoparticles such as zinc oxide, silver, titanium dioxide, copper are themselves, anti-microbial agents. Therefore, if combined with other anti-microbial drugs, they can demonstrate an additive or synergistic inhibitory effect. In this study, we present a nanoparticle-drug combination for enhancing the effectiveness of existing anti-tubercular drugs. To this end, zinc oxide nanoparticles (ZnO NPs) were synthesized and evaluated their antibacterial effect against mycobacteria, alone as well as in combination with the first-line anti-TB drug, rifampicin.

The ZnO NPs were synthesized using precipitation in the liquid phase. The particle size of ZnO NPs, as determined by transmission electron microscopy images, was found to be 11 nm. X-ray diffraction pattern indicated the formation of the hexagonal wurtzite phase. These ZnO NPs themselves were not inhibitory against wild-type (WT) *Mycobacterium smegmatis* up to a concentration as high as 256 μg/ml. However, a sub-inhibitory concentration of 32 μg/ml of ZnO NPs, was able to decrease the minimum inhibitory concentration (MIC) of rifampicin by four-fold against WT *M. smegmatis*. The fractional inhibitory concentration index (FICI) for the combination was found to be 0.375, which proves the synergistic interaction of the combination.

Further, to understand the mechanism of synergistic action, the effect of the NP and drug combination on various physiological parameters was investigated. ZnO NPs and rifampicin combination was observed to enhance the cell membrane permeability, which increased with time. The membrane permeability changes due to interaction with ZnO NPs with the bacterial cell membrane. The morphology of the bacteria, as observed under Cryo-SEM, displayed dramatic changes in bacterial morphology after 4 h of treatment with ZnO NPs alone as well as in combination with rifampicin. However, cells treated with ZnO NPs alone exhibited normal morphology after 12 h, suggesting eventual recovery of the cell over a longer time duration. However, in the case of the combination, it was observed to have extensive damage to the bacterial cells. These morphological translations are well correlated with membrane permeability data. Consequently, we have seen increased rifampicin uptake in a bacterial cell after treatment with a combination of ZnO NPs and rifampicin, due to underlying higher membrane permeability, which eventually results in the loss of cell viability.

Moreover, it is interesting to note that several studies have reported reactive oxygen species (ROS) production to be the primary mechanism for the anti-bacterial action of ZnO NPs. However, in our work, we have observed higher ROS production in the presence of ZnO NPs alone. In contrast, for the combination of ZnO NPs and rifampicin, ROS production was much lower compared to the corresponding concentration of ZnO NPs. Hence, we can conclude that the low ROS production did not affect the killing of the bacteria, but the combined impact of membrane damage and antimycobacterial activity of rifampicin explained the synergistic activity for the ZnO NPs and rifampicin combination.

Finally, we have also evaluated the combination of ZnO NPs and rifampicin against laboratory-generated drug-resistant strains of *M. smegmatis* as well as *M. bovis* BCG, an attenuated strain of *M. bovis and* a close relative of the virulent strain, *M. tuberculosis*. The laboratory-generated drug-resistant strains of *M. smegmatis* exhibit a MDR phenotype with two to four-fold resistance towards rifampicin. Interestingly, the combination also exhibited a synergistic interaction against all strains. Hence, with the detailed mechanistic study of the combination of ZnO NPs and rifampicin, we propose that this developed combination system can be promising for overcoming mycobacterial drug resistance. Further, we are working on mannose functionalized solid lipid nanoparticles (SLN) for delivering ZnO NPs and rifampicin to specifically target bacteria residing in macrophages.

Keywords: Drug resistance, zinc oxide nanoparticle, *Mycobacterium,* rifampicin, membrane permeability

609842 An Inexpensive, Compact, Programmable Temperature Controller and Thermocycler for Use during Device Imaging on a Microscope

Abstract:

Microfluidic devices are key enabling technologies for research and industry due to their compact size, which increases the number of integrated operations while decreasing reagent use. Common operations within these devices such as chemical and biological reactions, cell growth, or kinetic measurements often require temperature control. Commercial temperature controllers are constrained by cost, complexity, size, and especially versatility for use in a broad range of applications. Ideally, temperature control systems would be more accessible with a wide applicability. We have built and validated an inexpensive, modular, compact, and user-friendly temperature control system can function within a microscope. This system provides precise temperature acquisition and control during imaging of any arbitrary sample which can fit within a microscope slide, including microfluidic devices. The system includes two components. The first is a compact and washable Device Holder, fabricated from high temperature resistant resin to fit securely inside a microscope stage. The second is a robust Control Device, which incorporates all necessary components to program temperature settings. The system can achieve heating and cooling times between 50°C and 100°C of 32 seconds and 101 seconds, respectively. In addition to a simple user interface to program target temperature and duration, a Bluetooth enabled smartphone application has been developed for real-time data acquisition and visualization. An endpoint polymerase chain reaction (PCR) and a real time PCR performed within a microscope validated the temperature control system. This device can rapidly control the temperature on the surface of a microscope slide between ambient temperature and 100°C for an arbitrary amount of time and can perform rapid thermal cycling. Therefore, it has the potential to enable the experiments involving live cell imaging, DNA and protein analysis, in vitro drug studies, diffusion studies, crystallization, microfluidic fluid mechanics, or any other application in which temperature control and simultaneous microscopic observation is required.

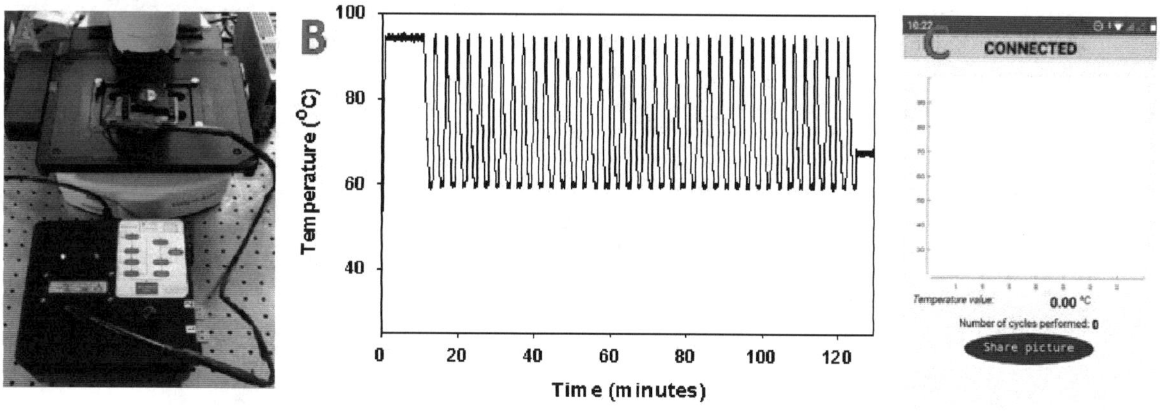

Figure 1**: Temperature control system.** A) A picture of the device prototype. B) Corresponding temperature control data acquired during a standard molecular biology reaction, polymerase chain reaction (PCR). C) Readout from the Bluetooth-based cell phone app.

612715 Genetically engineered biohybrid micromotors for high-load, highly stable, and highly active drug delivery for solid tumors

Abstract:

Cell-driven drug delivery system is a promising drug delivery strategy. This method combines living cells/bacteria with different excellent characteristics and drugs, which can effectively overcome the low bioavailability, weak targeting performance of traditional nanodrugs, and tissue penetration. Since the specific response to the target lesion, it can not only achieve efficient active targeted delivery of drugs, but also reduce toxic and side effects on normal tissues. It has been successfully used in drug presentation with promising applications in the field of disease diagnosis and treatment.

However, the use of living cells as drug carriers is still in its infancy. Considering the method of drug loading, the physical method is unstable while the chemical method is limited to the loading capacity, resulting in a low drug loading rate. In the complex fluid environment of the body, the loading stability of drugs is greatly challenged under the action of various shear forces. The propulsion of living cells for drug delivery should ensure cell activity and function. However, cell modification, integration, and drugs may affect the movement of the cells, thereby affecting its access to the lesion site and the efficiency.

This work constructed biohybrid micromotors using living cells as carriers such as spiral microalgae/magnetotactic bacteria. Free damage of cells with high capacity of drug loading was achieved through fine surface modification methods and genetic engineering. Active propulsion combined with external remote magnetic field control was used to treat solid tumors under the stimulation of low oxygen and low pH microenvironment. This work combined multi-modal control methods to achieve a conceptual verification of a new control method for cell/bacteria active drug delivery, which provides new ideas for the efficient treatment of solid tumors in deep tissue.

599508 A Systemic Approach to Understanding Compartmentalized NADPH Metabolism Under Mitochondrial Oxidative Stress

Abstract:

NADPH, a reducing power that prevents oxidative stress and promotes reductive biosynthesis, is considered highly compartmentalized in eukaryotes. Dynamic measurements of NADPH within different compartments of living cells have been challenging using available methods, limiting our understanding of the roles of NADPH in subcellular level. Here, we have systemically modulated production rates of hydrogen peroxide (H2O2) in mitochondria and assessed dynamics of both mitochondrial and cytosolic NADPH. iNap sensors revealed mitochondrial NADPH decreases during excessive generation of mitochondrial H2O2 with its ratio to $NADP^+$ decreasing 67-fold. The mitochondrial NADPH pool was observed to decrease when cytosolic H2O2 increased, whereas the cytosolic NADPH pool was maintained and did not change during increased mitochondrial H2O2. Glucose isotopic tracer analysis further demonstrated that increased activity in the pentose phosphate pathway (PPP) accompanies small decreases in the mitochondrial NADPH pool, whereas larger reductions induce both PPP activity and glucose anaplerosis. In summary, our study demonstrates that mitochondrial oxidative stress elevates a cytosolic PPP activity and glucose anaplerosis upon reduction of mitochondrial NADPH, indicating a communication between mitochondria and cytosol that supports maintenance of mitochondrial NADPH homeostasis.

611325 The Interplay of Ion Transporters and Cytoskeleton in Breast Cancer Migration and Metastasis

Abstract:

The primary cause of death among cancer patients is tumor metastasis. Cell migration is a key step in the metastatic cascade of events, as it enables tumor cells dissociating from a primary tumor to navigate through interstitial tissues and ultimately colonize distant organs. Cell motility is generally governed by cell-ECM interactions and the actomyosin cytoskeleton as well as cell volume regulation. Cell migration in confinement can persist even when F-actin is completely disrupted; a process that can be explained by the Osmotic Engine Model (OEM). According to OEM, cell locomotion is mediated by highly coordinated cycles of local isosmotic swelling at the cell leading edge and shrinkage at the trailing edge, driven by the polarization of select ion transporters and aquaporins (AQPs). We determined that the Na^{+}/H^{+} exchanger 1 (NHE1), which is polarized at the leading edge of migrating cells in the confinement, mediates regulatory volume increase, which is consistent with its role in cell protrusion. Interestingly, AQP5 is also enriched at the cell front. To date, it is unknown which ion channels and aquaporins regulate cell shrinkage at the cell trailing edge. Using a multidisciplinary approach involving state-of-the-art bioengineering, imaging tools and quantitative analysis, we have discovered that a distinct aquaporin, AQP4 polarizes at the cell rear of metastatic breast cancer cells in confinement, thereby facilitating the local volume decrease by creating an outflow of water. We have also identified a novel ion channel (called X, due to a pending patent) which is preferentially localized at the cell trailing edge of confined breast cancer cells, and works together with AQP4 to mediate cell shrinkage. Dual knockdown of NHE1 and X does not alter cell volume in confinement which is in line with their individual counteracting effect of their role in cell volume regulation. Furthermore, we demonstrate cooperative and pronounced inhibition of motility in narrow channels upon dual NHE1 and X depletion. Taken together, our data support the model by which the repeated and coordinated cycle of swelling and shrinkage at the cell poles mediated by these distinct ion transporters, regulates cell motility in confinement. This presentation will also discuss the interplay between ion transporters and cell cytoskeleton in regulating efficient cell migration in confinement. Overall, our findings extend the Osmotic Engine Model by introducing AQP4 and a novel ion transporter for confined cell migration. These findings may provide the basis for identification of novel therapeutic target to combat breast cancer metastasis.

608608 Simultaneous Quantification of Protein-DNA Contacts and Transcriptomes in Single Cells

Abstract:

The epigenome plays a critical role in regulating gene expression in mammalian cells. On the transcriptional level, packaging of DNA into chromatin can control access of transcriptional regulators to functional DNA elements such as enhancers and promoters. Higher levels of organization that contribute to the regulation of gene expression involve the spatial segmentation of the genome into compartments with transcriptionally permissive or repressive gene regulatory activities. Failure to integrate and coordinate the multi-layered regulatory control of gene expression can result in developmental defects and diseases. However, understanding how cell-to-cell heterogeneity in these epigenetic features influences gene expression variability and cellular phenotypes remains a major challenge. Therefore, to directly link heterogeneity in upstream elements such as 3D genome organization, DNA accessibility or transcriptional factor/chromatin modifying factor binding to transcriptional output, we developed a new technology to simultaneously quantify protein-DNA interactions and mRNA from the same cell (scDam&T-seq)[1,2]. scDam&T-seq enables us to directly correlate protein-DNA contacts to a particular gene expression program in a cell. By profiling lamina-associated domains (LADs) in human KBM7 cells, we reveal different dependencies between genome-nuclear lamina (NL) association and gene expression in single cells. Specifically, we find that gene expression heterogeneity is dependent on NL positioning for genomic regions that infrequently contact the lamina. Surprisingly though, for genomic regions that frequently contact the NL, we find that gene expression is not sensitive to NL positioning. In addition, we use the *E. coli* methyltransferase, Dam, as an *in vivo* marker of DNA accessibility in single cells and show that scDam&T-seq can be used to simultaneously map DNA accessibility and the transcriptome in single cells. Further, we show that scDam&T-seq can be used to map the genome-wide binding patterns of chromatin modifying factors by quantifying the genome binding landscape of a polycomb-group protein, RING1B, and the associated transcriptome. Finally, using mouse embryonic stem cells, we show that this approach can be used as a general technology to identify cell types *in silico* using the transcriptome of single cells and simultaneously determine the underlying chromatin landscape of each cell type using the corresponding protein-DNA interaction maps. Thus, scDam&T-seq is a multiomics single-cell method that enables systematic investigation of gene regulatory mechanisms on the acquisition of cell type specific transcriptional programs in heterogeneous tissues.

References:

1. Rooijers, K, [...] Dey, S.S.*, Kind, J.* *et al. Nature Biotechnology* (2019).

2. Markodimitraki, C.M., [...] Dey, S.S.*, Kind, J.* *et al. Nature Protocols* (2020).

600816 Structural Modeling Reveals Significant Differences in Fatty Acid Transporter Homologs in *Vibrio Cholerae* That May Play a Vital Role in Pathogenesis

Abstract:

Background and Significance

Vibrio cholerae is an enteric microorganism responsible for the acute intestinal infection known as cholera. Cholera is a well-known threat and has been the source of epidemics and pandemics ranging from the 1800s to this day. A reported 2.3 million cases have been reported in Yemen from 2017-2019. Furthermore, the World Health Organization (WHO) estimates that in endemic countries there are 1.3 to 4.0 million cases of cholera annually, and 21,000–143,00 deaths between 2008 and 2012. While cholera is treatable in many cases with antibiotics and rehydration, current prevention and mitigation strategies leave much to be desired.

One of the more unique and interesting characteristics of *V. cholerae* is its ability to uptake and utilize a diverse range of long-chain fatty acids (LCFAs) (Giles, et al. *Molec. Micro.* 2011). The uptake of LCFAs has traditionally been viewed as carbon source acquisition, but more recently researchers have observed the utilization of LCFAs for membrane remodeling and as signaling molecules in the virulence cascade. The long chain fatty acid transport protein FadL is a transmembrane protein that has been determined as the vector for LCFAs to be taken up by *Escherichia coli* (Van den Berg, et al. *Nature* 2004). Multiple FadL homologs exist in *V. cholerae*, and we hypothesize that this additional genetic machinery gives *V. cholerae* the ability to uptake a wider range of LCFAs than *E. coli*, as has previously been observed. We further believe that this enhanced uptake ability plays a role in *Vc*'s ability to cycle between environmental niches. Testing this hypothesis is all the more important due to the fact that similar behavior has been observed in a number of other bacterial pathogens that are hospital-acquired and/or pose a particular threat to the immunocompromised.

Methodology

Vibrio cholerae FadL homolog amino acid sequences were found using the NCBI website's Basic Local Alignment Search Tool (BLAST) and *E. coli's* FadL amino acid sequence. Three of the most prevalent homolog sequences were chosen (accession numbers: NP_230687, NP_230688, and NP_233248), and these sequences were folded into a tertiary structure conformation using the I-TASSER webserver.

The homolog tertiary structures were then visualized with Visual Molecular Dynamic (VMD) software, and comparisons of the *Vibrio cholerae* FadL homologs and the RCSB database *Escherichia coli* FadL (PDB ID: 1T16) were made using in-house scripts.

Results and Discussion

The resulting *Vibrio cholerae* FadL structures conformed to the 14-stranded β barrel ideal transmembrane protein, with discernible key structures such as the L3/L4 loops, S3 strand kink, and the N terminus hatch domain (see Figure 1). The S3 kink in particular was highly conserved between the *E. coli* crystal structure and the *V. cholerae* FadL homologs, as seen visually when superimposed. This similarity was quantified via root-mean-squared deviation relative to the *E. coli* FadL, with the

V. cholerae homologs (NP_230687, NP_230688, and NP_233248) exhibiting relatively low RMSD values of 1.69, 1.90, and 0.94 Å respectively. Further analysis showed that the S3 kink is at or close to a

neutral charge across all *V. cholerae* homologs as well as the *E. coli* crystal structure. However, there are minor dissimilarities in the location and number of polar and non-polar residues when compared to the crystal structure generated PDB.

Structural similarities are most noticeable near the base of the β barrel, with dissimilarities becoming more frequent closer to the extracellular region. This is of particular significance given that LCFAs initially bind to the extracellular L3/L4 loops and enter the channel before diffusing laterally into the outer membrane, rather than being transported into the periplasm. The observed differences in size and structure the extracellular binding loops could explain *V. cholerae*'s ability to uptake a more diverse spectrum of LCFAs than *E. coli*, and as such will be the subject of further study.

The N-terminal hatch is thought to be a ligand-gated channel, where the LCFA would bind to the high affinity binding site and cause conformational changes that open the gateway through the S2/S3 pore. Within *E. coli's* FadL, the N-terminus is located just below the high affinity binding site, allowing for binding site interactions with the gated channel. The differences in the N-terminal hatch domain are among the most glaring between the *V. cholerae* homologs and the *E. coli* FadL crystal structure. Specifically, the *Vc* N-terminal hatches contain more residues than *Ec*'s while the β barrel sizes are similar. These larger domains are observed to penetrate the S2/S3 pore, wrap around the S3 kink, or pack within the barrel. While it appears all of the *Vc* N-terminal hatch domains make contact with the high affinity binding site, these structural differences could potentially impact the diffusion of LCFAs into the membrane.

The last of the notable differences is observed in the C-terminus. In *E. coli*, the C-terminus extends away from the β barrel, whereas in the *V. cholerae* homologs this domain terminates just after the last β strand ends. *E. coli's* C-terminus is believed to play a role in the transport of LCFAs across the membrane, but its precise function remains unknown. Finally, it should also be mentioned that the LCFA diffusion channels were not observable in the *Vibrio cholerae* homologs. However, this is thought to be due to the spontaneous conformational changes in the protein structure that guide the LCFA through the channel. If there is no LCFA to be transported, the channel appears to remain closed.

Conclusions and Future Work

This work is the beginning of a comprehensive study of *Vibrio cholerae's* fatty acid transport machinery. The tertiary structures of the *V. cholerae* FadL homologs were successfully generated, and noticeable differences in regions critical for LCFA binding and diffusion were noted when compared to the *E. coli* crystal structure. These differences could shed light on V. cholerae's ability to cycle between environmental niches as well as its pathogenesis. Putative LCFA binding sites were identified visually through superimposition, and future work will refine these locations using predictive binding algorithms. Additionally, a quantitative study of the protein structures will be conducted to identify key differences (e.g., polarity, charge, conformation) compared to E. coli FadL as well as amongst homologs. This work is expected to enhance our scientific understanding of the role of LCFA uptake in bacterial pathogenesis, and could have vital implications for disease prevention and mitigation.

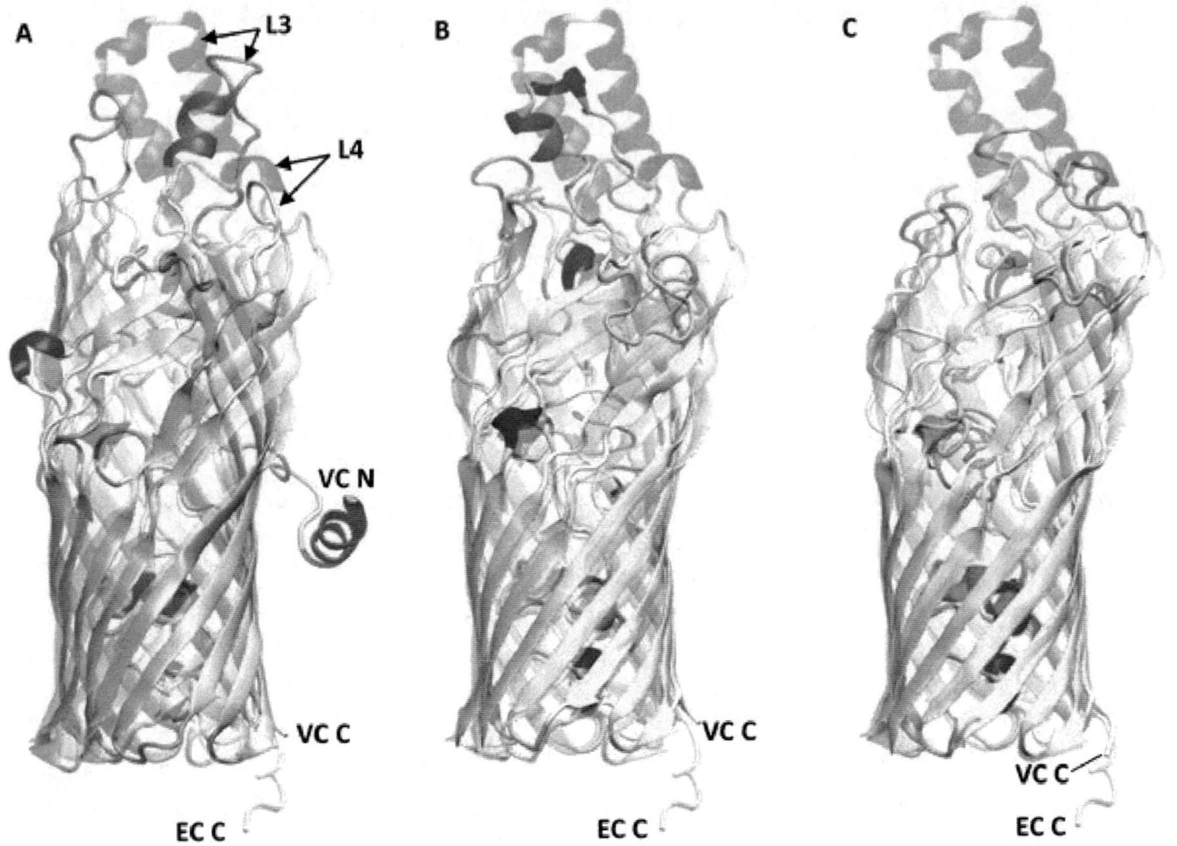

Figure 1: Superposition of *E. coli* crystal structure (transparent) onto the *V. cholerae* (opaque) homology models **(A)** NP_230687, **(B)** NP_230688, and **(C)** NP_233248. Visually discernible differences include the L3/L4 loops, the N terminus hatch domain, and the location of the C terminus (EC C versus VC C).

605357 Model-Based Assessment of Freezing Temperature Profiles for Human Induced Pluripotent Stem Cells

Abstract:

Human induced pluripotent stem (hiPS) cells are the emerging source of regenerative medicine products. Along with successful clinical studies, e.g., Parkinson's disease or spinal cord injuries, the demand is increasing. It is now urgently necessary to establish freezing processes of the cells for storage and transportation. As the technique of freezing hiPS cells, slow freezing is considered more appropriate for commercial production than vitrification because of scalability, simplicity in operation, and no direct contact with liquid nitrogen. However, the design of slow freezing processes for hiPS cells is still empirical. Towards commercial production of hiPS cells, the temperature profile of a freezer, which would have a significant impact on both cell quality and productivity, need to be established.

This work presents model-based assessment of freezing temperature profiles for hiPS cells. We developed a single-cell model by integrating models describing a temperature profile in a container, cell volume change through transmembrane transport, intracellular ice formation, and cell survival rate after thawing. This is an extension of our previously presented mechanistic model of single-cell freezing[1]. Freeze/thaw experiments of hiPS cells were performed using a programmable deep freezer in order to estimate model parameters. The objective functions were defined as the cell survival rate after thawing and the required freezing time which were the cell quality and the productivity objectives, respectively. Given the temperature profile, the developed model can produce the survival rate and the freezing time.

Using the single-cell model, we conducted the assessment of over 10,000 different temperature profiles for a programmable deep freezer using the single-cell model. Among the Pareto optimal solutions, a specific profile was identified that maximized the joint objective of quality and productivity. The profile can achieve high cell survival rate in a short freezing time, and thus could be useful in the actual process. In ongoing work, we are experimentally investigating the obtained profiles, and also assessing the impact of the choice of cryoprotective agent on the results.

Reference

1. Hayashi et al., *Comput. Chem. Eng.*, 132, 106597 (2020)

606522 The Role of Lipids on Transmembrane Protein Interactions in Viral Infections

Abstract:

We are increasingly aware of the relevance of specific lipid-protein interactions in cellular processes; not only to maintain a healthy cell, but also facilitating disease progression. In the context of viral infections, we present a molecular dynamics study that examines the interactions of the influenza A M2 ion channel in a realistic membrane model. This tetrameric viroporin is the smallest of its kind and is key for viral replication. Similar viroporins are found in other viruses such as HIV-1 (Vpu), hepatitis C (p7), and coronaviruses like the newly identified SARS-CoV-2 (E). Our simulations aim to characterize molecular interactions at the interface between viral proteins and endogenous cell membrane components. Namely, to quantify the changes in the local membrane environment due to the formation of viroporins, i.e. membrane structural and mechanical properties including lipid bilayer order and thickness, which in turn modulate channel activity. In examining this model, we also aim to identify conserved effects of viroporins on host membranes. The data from these atomistic simulations will later serve to improve a coarse-grain model of influenza A viral dynamics that accounts for membrane response to activity and aggregation of these channels.

608788 Predicting the Regulatory Role of ORM Proteins in the Sphingolipid Biosynthesis Pathway Using Ensemble Kinetic Modeling

Abstract:

Sphingolipids are an essential component of many eukaryotic cells including human and plant cell's plasma membrane and endomembranes. In plants, this class of lipids plays several functional roles including providing structural integrity to the membrane, golgi trafficking, and protein organizational domains. In addition, sphingolipids have been implicated in physiological processes such as the signaling of Programmed Cell Death (PCD) and the hypersensitive response associated with plant resistance to bacterial and fungal pathogens. The metabolic pathways associated with sphingolipid biosynthesis are tightly controlled to ensure sufficient sphingolipid availability for normal cell growth. Simultaneously, metabolic controls constrain the accumulation of sphingolipid building blocks responsible for the induction of PCD until this process is required (e.g. during the pathogen triggered hypersensitive response). Recent work has shown that the first committed step in Sphingolipid metabolism catalyzed by serine palmitoyltransferase (SPT) is the primary control point for regulation of this pathway. Furthermore, it has been found that orosomucoid-like (ORM) proteins are involved in the regulation of multiple enzymes in the pathway. Therefore, in this work, a combined computational and experimental approach was taken to mechanistically decipher the regulation of sphingolipid biosynthesis. A compartmentalized metabolic network of sphingolipid biosynthesis in *Arabidopsis thaliana* comprised of 399 metabolites and 739 reactions has been reconstructed from available literature and the KEGG database. Flux Balance Analysis (FBA) has been used to simulate the steady-state flux distribution of the network at the measured uptake rates of the starting material Sphingosine. An Ensemble Modeling (EM) framework has then been implemented to construct a kinetic model of the metabolic and regulatory network of Sphingolipid biosynthesis. The constructed model was subsequently used to predict which enzymes were regulated by ORMs as well as the metabolites regulating ORM expression. These predictions will be used to experimentally probe the role of ORMs in altering phenotypic characteristics in plants.

610553 Xanthan Gum Digestion By Human Gut Microbiota

Abstract:

The human gut microbiota is a diverse community of microbes in the lower gastrointestinal tract that is increasingly recognized as playing roles in health and disease. The historically recent inclusion of novel food ingredients could alter diet-microbiota effects on human health or be leveraged to engineer new niches in the microbiome.

Xanthan gum is a bacterial exopolysaccharide and a common food additive used for its thickening properties. Xanthan gum is a polymer of β(1,4)-glucose with alternating glucose units harboring a side chain of α(1,3)-linked mannose, β(1,2)-glucuronic acid, β(1,4)-mannose. The inner and outer mannose possess variable 6-O-acetylation and 4,6-O-pyruvylation, respectively. Although the human microbiota has been shown to degrade and consume xanthan gum, the microbes and enzymatic pathways responsible are still unknown. We surveyed a population of healthy adults and found that over a third had microbial communities that could grow on xanthan gum as a primary carbon source. Metagenomic, transcriptomic, and activity-guided fractionation and proteomic analysis of an active culture revealed a suite of enzymes that appear responsible for the full saccharification of xanthan gum. These enzymes are variably abundant across microbiota metagenomes collected from unique geographies across the world. We are currently working to fully elucidate the enzymatic pathways responsible for xanthan gum catabolism and how cross-feeding on xanthan oligosaccharides affects microbial communities in the gut. We are also investigating how these enzymes may provide utility for applications in industries that use xanthan gum, such as in drilling mud in the petroleum industry.

608132 Proteins Repurposed: Augmenting Biomaterial and Biocatalyst Function with Noncanonical Amino Acids

Abstract:

The genetic code has been deciphered, and researchers are now developing genetic methods to outfit proteins, Nature's flagship polymers, with an expanded set of noncanonical amino acids (NCAAs). Despite these advances, the unique mapping of a given amino acid sequence to a specific protein function – the "protein code" – remains mysterious. Poor understanding of the protein sequence-to-function relationship limits our ability to rationally design new proteins with predictable characteristics. What if we could use designer NCAAs to bypass this limitation, and *decide for ourselves* what a given amino acid sequence should do?

In this talk, I will present a new approach to biomaterials and biocatalyst design that uses designer NCAAs to predictably augment and transform protein function. In the realm of biomaterials, I will recount how the incorporation of a noncanonical leucine surrogate into a telechelic protein enabled precise control over polymer dynamics in model protein hydrogels, ultimately leading to the elucidation of a new and ubiquitous polymer transport mechanism in reversible macromolecular networks. In the realm of biocatalysis, I will describe our ongoing intensive efforts to design catalytically active NCAAs that effectively distill the essential reactivity of certain enzymes into single chemofunctional residues. By embedding these residues into heterochiral peptide frameworks, highly selective organocatalysts that catalyze numerous abiological transformations have been obtained. In particular, I will discuss the recent successful deployment of Pi-methyl histidine, a genetically encodable histidine analogue, as a nucleophilic catalyst for sulfamate group transfer. Taken together, these results establish a new paradigm for repurposing proteins with specific, user-defined functions.

References:

Rapp, P.B., Omar, A.K.; Shen, J.J.; Buck, M.E.; Wang, Z.G.; Tirrell, D.A. Analysis and control of chain mobility in protein hydrogels. *J. Am. Chem. Soc.*, **2017**, *139* (10), 3796–3804.

Rapp, P.B.; Omar, A.K.; Silverman, B.R.; Wang, Z.G.; Tirrell, D.A. Mechanisms of diffusion in associative polymer networks: evidence for chain hopping. *J. Am. Chem. Soc.* **2018**, *140* (43), 14185–14194.

Rapp, P.B.; Murai, K.; Ichiishi, N.; Leahy, D.K.; Miller, S.J. Catalytic sulfamoylation of alcohols with activated aryl sulfamates. *Org. Lett.* **2020**, *22* (1), 168–174.

Graphical Abstract:

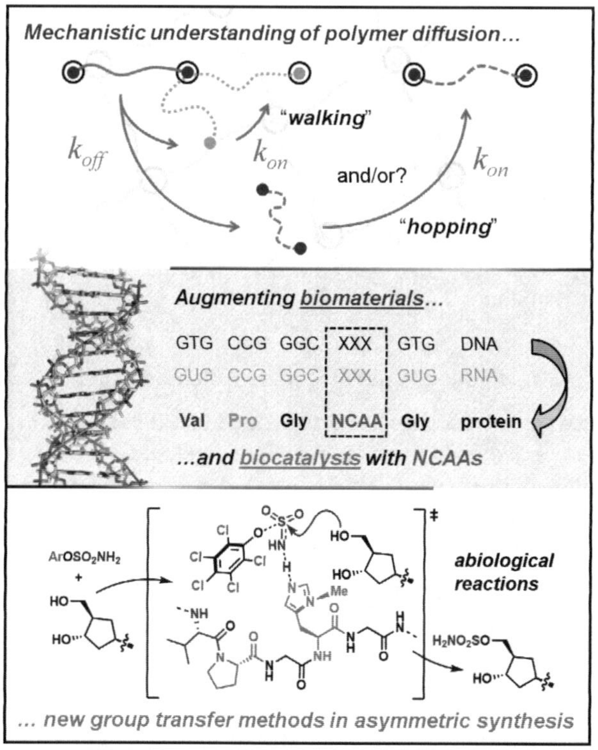

605225 *In Vivo* Multiplexed Nanodiagnostics for Assessing Biological Heterogeneity

Abstract:

Heterogeneity is a fundamental property of a biological system. The genetically homogeneous bacteria exhibit inherent phenotype heterogeneity, in which their subpopulation can always survive from viruses or antibiotic treatments. Cancer is a heterogeneous disease, in which the constituent tumor cells diversify to various molecular phenotypes. Comprehensive assessment to such heterogeneity and functional diversity of the biological components is essential for the informative disease diagnosis and the patient-specific therapeutic decision, which necessitates the use of high-dimensional and multiplexing tools. Since the expression of the biological heterogeneity is most prominent in living subjects, I aim to develop an intravital multiplexed imaging platform to simultaneously monitor as many biological components as possible.

Here I present a noninvasive and highly multiplexed *in vivo* preclinical imaging platform based on multi-color spectral palettes of surface-enhanced Raman scattering (SERS) nanoparticles[1]. I create nine spectrally distinct gold core-silica shell structured near-infrared resonant SERS (NIR-SERRS) nanoparticles, which offer both (i) high imaging sensitivity as the traditional NIR fluorescence imaging and (ii) high multiplicity originated from the narrow spectral bandwidth of SERS spectra. From the 9 NIR-SERRS nanoparticles, I select five color-encoded nanoparticles and demonstrate a noninvasive and five-plex *in vivo* imaging of the 5-color nanoparticles targeting to subcutaneous tumors in nude mice. The *in vivo* multiplexed imaging allows dynamic monitoring of the 5-color NIR-SERRS nanoparticles retention in the tumor, which demonstrates the great potential for noninvasive assessment of multiple biological targets within the tumor microenvironment.

Next, I develop biodegradable plasmonic gold supraparticles for clinical translation of the nanoparticle-based *in vivo* multiplexed imaging platform. The supraparticles are created through the assembly of gold atomic clusters. While the long-term exposure of gold nanoparticles to the liver has limited their use in humans, the gold supraparticles are biodegradable to renally excretable atomic clusters through *in vivo* enzymatic reaction. The supraparticles exhibit photoacoustic and Raman responses under the NIR excitation, which enable the noninvasive *in vivo* imaging applications.

The presented work will provide an excellent platform for the noninvasive analysis of heterogeneous biological systems such as tumors, in which multiple biomarkers should be simultaneously monitored in living subjects. It will also enable tracking multiple immune cells along with tumor biomarkers, and classifying molecular subtypes of tumors in the context of tumor-immune environments. Finally, I envision that the biodegradable design of the plasmonic supraparticles proposed here will accelerate the clinical translation of the nanoparticle-based *in vivo* multiplexed imaging platform[2,3].

1. **Yu J.H.** *et al.* Highly Multiplexed Five-Color Noninvasive Tumor Imaging of Near-Infrared Resonant Surface-Enhanced Raman Nanoparticles *In Vivo*. *submitted* (2020).
2. Park S.-m., Aalipour A., Vermesh O., **Yu J.H.** & Gambhir S.S. Towards clinically translatable in vivo nanodiagnostics. *Nat. Rev. Mater.* (2017).
3. **Yu J.H.** *et al.* High-resolution three-photon biomedical imaging using doped ZnS nanocrystals. *Nat. Mater.* (2013).

600075 Spherical Nucleic Acids As Stimuli-Responsive Synthons and Live-Cell Probes

Abstract:

Spherical nucleic acids (SNAs), nanoparticles functionalized with a dense and radially oriented layer of DNA, have emerged as a unique class of nanomaterials and opened new opportunities in materials synthesis and biodiagnostics. The presence of DNA makes these structures responsive to stimuli. In this talk, I will discuss two specific examples that show how the inherent responsiveness of the DNA sequences to a variety of different chemical species can be leveraged to (i) engineer dynamic materials assembled from SNAs and (ii) design SNA-based chemical probes for live-cell analysis.

In the first example, I will focus on how multivalent cations induce structural changes in DNA which alter interparticle distances in crystalline assemblies of SNAs. In the context of materials synthesis, the SNA platform represents a powerful synthon owing to (i) its modular structure, (ii) the specific interactions between SNAs functionalized with complementary DNA, and (iii) the ease of synthesis of sequence-defined SNAs with desired DNA lengths. SNAs can be analogized as programmable atom equivalents (PAEs) in which the nanoparticle core is the "atom" and the DNA forms the programmable "bond". Unlike atomic systems, PAEs allow the atom and bond to be varied independently. Therefore, by changing the nanoparticle size, shape, and composition and the DNA sequence, length, and density, a rich phase-space of matter can be accessed. By modulating the DNA length post-synthetically, a new class of dynamic metamaterials can be accessed. Here, I will discuss how multivalent metal cations induce the contraction and re-expansion of the DNA bonds by >85% in length, resulting in >65% changes in crystal volume, all while maintaining crystallinity. This new approach represents a powerful strategy to alter superlattice structure and stability, which can impact diverse applications through dynamic control of material properties, including optical, magnetic, and mechanical properties.

In the second example, I will describe how different DNA sequences change structure in response to specific chemical species and can, therefore, be utilized to develop SNA-based chemical probes for live-cell analysis. The detection of intracellular analytes in live single cells is a grand challenge as most current techniques require (i) the analysis of a bulk population of cells, (ii) cell lysis, or (iii) the use of transfection reagents. SNAs enable the design of highly tailorable intracellular probes because: (i) unlike linear DNA sequences, SNAs are taken up by cells without the need for transfection reagents and are resistant to nuclease degradation in the cellular environment; (ii) by judiciously choosing the DNA sequence (e.g. using aptamers that bind to specific analytes or complementary sequences that bind cellular RNA), it is possible to detect a wide range of analytes; and (iii) the modular structure allows different nanoparticle cores to be used for added functionality. I developed a new class of signaling aptamers called "FIT-aptamers". These aptamers contain a visco-sensitive dye that is forced to intercalate or FIT between base pairs upon target binding, turning the fluorescence on due to the restricted rotation of the dye. The FIT strategy reduces false-positive signals common to all fluorophore-quencher systems, provides up to 20-fold fluorescence enhancement upon target binding, and allows target detection down to nanomolar concentrations in complex milieu such as human serum. Arranging the FIT-aptamers into an SNA-architecture allows them to be used as intracellular probes. Moreover, by using a functional enzyme as the nanoparticle core, it is possible to detect analytes for which aptamers are not known. We have validated the use of these protein-based SNAs as intracellular probes by detecting intracellular pH and glucose levels in 8 different cell lines. Taken together, these advances represent a promising step toward the use of SNAs as detection tools for the molecular profiling of cells and identification of diseases based on intracellular signatures.

596226 Applying Machine Learning to Predict Therapeutic Antibody Aggregation

Abstract:

In the present study, we developed a general machine learning feature selection protocol to predict antibody stabilities. These features only require information from sequences and molecular structures of antibodies. In addition, the physically relevant features are easily interpretable and help to design experiments to understand the underlying mechanisms of antibody stabilities. Two case studies are demonstrating using this protocol. First, we applied this framework to predict aggregation behaviors of 21 approved monospecific monoclonal antibodies at high concentration (150 mg/mL), yielding an accuracy of 97% on validation tests with only three features. Models trained using structural features, rather than sequence features, correlated better with aggregation. Second, the concentration dependent viscosity behavior of 27 FDA approved antibody drugs was measured in concentrated solutions at high concentration (150 mg/mL). Combining molecular modeling and machine learning feature selection, we found that the net charge in the antibodies and the amino acid composition in the variable region are key features which govern the viscosity behavior. We presented two predictive models for aggregation and viscosity based on machine learning trained models, facilitating early stage design.

597252 Elucidating Protein Corona Composition and Dynamics on Nanoparticles in Biological Environments

Abstract:

Engineered nanoparticles are increasingly used for biological sensing, imaging, and delivery due to their distinctive optical and physical properties. Specifically, DNA functionalized single-walled carbon nanotube (DNA-SWCNT) probes operate at spatiotemporal scales necessary to capture information on chemical signaling, such as neurotransmission in the brain,[1] and can deliver nucleic acid cargo for genetic transformations, such as in living plants.[2] The critical – and often overlooked – challenge with these nanoscale tools is understanding the fundamental mechanisms of interaction between the nanoprobe and the system they are designed to query. When a nanoparticle enters a biological system, the surface becomes rapidly coated with proteins to form the "protein corona". Binding of proteins to the nanoparticle disrupts intended functionality and leads to unpredictable *in vivo* biodistribution and biocompatibility outcomes.

A comprehensive understanding of the protein corona remains a paramount barrier to successfully developing and implementing nanotechnologies within biological environments. Herein, I present multimodal characterization of (i) the protein corona composition on DNA-SWCNTs in blood plasma and cerebrospinal fluid, (ii) the kinetics of protein adsorption to DNA-SWCNTs in solution, and (iii) the colloidal morphology of proteins adsorbed to DNA-SWCNTs. I have optimized a platform to characterize protein corona composition formed on DNA-SWCNTs by proteomic mass spectrometry to determine abundant and differentially enriched vs. depleted corona proteins.[3] By varying incubation conditions of DNA-SWCNTs in biofluids, I have investigated the role of electrostatic and entropic interactions driving selective protein corona formation. To study the dynamic exchange of biomolecules on the SWCNT surface, I have developed a multiplexed fluorescence assay that enables real-time tracking of biomolecule adsorption and desorption events.[4] This corona exchange assay is generic towards the study of various biomolecules exchanging on the SWCNT surface and enables study in solution rather than on a surface-immobilized, less biologically relevant, setting.[5] Finally, I probe the morphology of DNA-SWCNT complexes in the presence of high- and low-binding model proteins (fibrinogen and albumin, respectively) using in-solution small angle x-ray scattering (SAXS).

Understanding the corona composition, timescales and driving forces of formation, and coronal geometry under relevant solution conditions informs design and synthesis of nanotechnology-based tools applied in protein-rich environments. Although corona formation can impair nanobiotechnology efficacy, it also presents an opportunity to create improved protein-nanoparticle architectures by exploiting selective protein adsorption to the nanoparticle surface. In this work, I develop techniques and analyses that directly characterize the *in-situ* protein corona microenvironment and employ this knowledge towards rational design of nanobiotechnologies.

References

1. Beyene, A. G., McFarlane, I. R., Pinals, R. L. & Landry, M. P. Stochastic Simulation of Dopamine Neuromodulation for Implementation of Fluorescent Neurochemical Probes in the Striatal Extracellular Space. *ACS Chem. Neurosci.* **8**, 2275–2289 (2017).
2. Demirer, G. S., Zhang, H., Goh, N. S., Pinals, R. L., Chang, R. & Landry, M. P. Carbon Nanocarriers Deliver siRNA to Intact Plant Cells for Efficient Gene Knockdown. *bioRxiv*

564427 (2019) doi:10.1101/564427.
3. Pinals, R. L. *et al.* Protein Corona Composition and Dynamics on Carbon Nanotubes in Blood Plasma and Cerebrospinal Fluid. *bioRxiv* 2020.01.13.905356 (2020) doi:10.1101/2020.01.13.905356.
4. Pinals, R. L., Yang, D., Lui, A., Cao, W. & Landry, M. P. Corona Exchange Dynamics on Carbon Nanotubes by Multiplexed Fluorescence Monitoring. *J. Am. Chem. Soc.* (2019) doi:10.1021/jacs.9b09617.
5. Alizadehmojarad, A. A., Zhou, X., Beyene, A. G., Chacon, K., Sung, Y., Landry, M. P. & Vuković, L. Binding affinity and conformational preferences influence kinetic stability of short oligonucleotides on carbon nanotubes. *bioRxiv* 2020.02.08.939918 (2020) doi:10.1101/2020.02.08.939918.

609819 An Autoinducible Fungus to Increase Production of Biofuels and Bioproducts

Abstract:

While bacteria have many autoinducible systems, eukaryotes, including the industrial host *Saccharomyces cerevisiae*, have few comparable systems. The most common systems use costly antibiotics (e.g. tetracycline) or special growth medias that require time-consuming preculture (e.g. galactose and sucrose). Consequentially, inducible systems are rarely utilized in commercial applications, despite the obvious benefits of decoupling the growth mode from the production mode. Decoupling is particularly valuable to the production of compounds that are toxic to the production organism, or require allocation of resources such as co-factors and ATP. Although many valuable biochemicals have toxic biosynthetic intermediates in *S. cerevisiae*, such as in the mevalonate pathway, a lack of autoinducible systems result in fermentations that rely on constitutive promoters. The goal of this work was to rewire the yeast mating signaling pathway for inducing expression of a gene (or genes) of interest. Our autoinduction system is based the engineering of three distinct modules that will set the threshold for autoinduction, prevent premature activation, and activate metabolite production. The "timing knob" module determines the cell density where activation of the target genes begins. We have developed a range of autoinducible strains integrating the *MFa1* and *STE2* genes under different promoters. Using fluorescent reporters, we observe a clear shift in the production of the GOI with increasing levels of alpha factor (ie, the inducer) at similar levels to those observed with the strongest constitutive promoters. We have cloned a variety of endogenous promoters to drive expression of alpha factor, and optimization of our expression system has the potential to improve yield, titer, process efficiency and process economics for companies producing a variety of materials, chemicals and fuels in yeast cell factories.

604727 Integrating Bisphenol a Degradation Function into *Shewanella Oneidensis*

Abstract:

Bisphenol A (BPA) is a major environmental endocrine-disrupting compound and increases concerns about its estrogenic and genotoxic effects on humans. More than one million pounds of BPA are discharged into the environment and most BPA emissions are to the aqueous environment. Also, BPA can affect the growth, reproduction, and development of aquatic organisms. Therefore, removing BPA during water treatment is critical to protect the environment as well as to mitigate human health risk. Biodegradation by microorganisms may be the most efficient method as microorganisms degrade BPA for use as a carbon and energy source. In this study, we develop *Shewanella oneidensis* MR1 as an efficient BPA degrader while enhancing electron transfer of this microorganisms by forming biofilms in bioelectrochemical systems. First, BPA biodegradation by BisdA (ferredoxin), BisdB (cytochrome P450) from *Sphingomonas bisphenolicum* AO1 was validated in *Escherichia coli* BL21 (DE3) Star strain. Our results indicated that BPA was completely degraded within 5 h, measured by LC-MS/MS. Next, we transferred those genes into *S. oneidensis* MR1 and turned their expression. BPA degradation by *S. oneidensis* strain and its biofilm formation to form bioelectrochemical system will be discussed. Establishing a BPA degradation pathway coupled with bioelectrochemical systems may be a promising strategy to efficiently remove BPA from wastewater.

609375 Biological Conversion of Methane into Polyhydroxybutyrate (PHB)

Abstract:

Methane is the second most important greenhouse gas, presenting a global warming potential 28 times higher than CO_2. It implies that methane is much more effective at trapping heat in the atmosphere, causing harmful effects to several ecosystems. On the other side, petro derivative plastics also represent a major threat to several ecosystems, especially when considered its large worldwide production and persistence in the environment due to their non-biodegradability. An alternative to solve both problems could be producing biodegradable plastics using methane as feedstock. Methanotrophs are a group of bacteria present in the environment that are capable of methane usage as carbon source, which results in natural accumulation of polyhydroxybutyrate (PHB) inside their cells as energy reserve granules. PHB is a biodegradable polymer that still presents increased production price which inhibits its entrance into the market. The main goal of this work is to find methanotrophs in the environment and evaluate the feasibility of PHB from methane. Methanotrophic bacteria were obtained after enrichment of sediment samples from mangrove sites (-23.91744, -46.21605) in 25% methane atmosphere periodically renewed for 40 days. Experiments performed to evaluate PHB production were carried out in two stages; the first one aimed at maximum cellular growth and the second one aimed at stimulating PHB production by nitrogen depletion. Nutrient deficiency in the presence of abundant carbon source is required to natural accumulation of PHB within bacterial cells. Experiments were carried out in 250 mL erlenmeyer flasks containing 50 mL of Nitrate Mineral Salts medium and bacterial inoculum of O.D. 0.1 (600 nm). The flasks were closed with rubber stoppers and incubated at 28 °C and 200 rpm for 12 days. Methane atmosphere in air at a proportion of 1:8 was used to ensure complete methane oxidation, which was replenished every 48 hours to avoid substrate limitation. At the seventh day of experiment, cells were harvested by centrifugation at 7830*g* for 10 min and transferred to nitrate free NMS to stimulate PHB production. Methane consumption was monitored by gas chromatography (GC). Lyophilized cells were subjected to propanolysis and subsequently analysed by GC for PHB quantification and characterization. Consortium MGISED was obtained from the methanotrophic enrichment. Metagenomic analysis elucidated that this was a *Methylosinus trichosporium* dominated consortium. *M. trichosporium* is a Type II methanotrophic bacterium capable of PHB accumulation. Experiments were carried out for 120 h under nitrogen depletion and the consortium achieved 20% of PHB accumulation in cell dry weight (CDW), corresponding to 0.2 g L^{-1}. Methane consumption rate throughout the experiment was 0.21 g L^{-1}, in average. This corresponds to 42% of the total carbon source available. Conversion factors of substrate in biomass and bioproduct were calculated as 0.96 g g^{-1} and 0.3 g g^{-1}, respectively. These findings will help elucidate the feasibility of the process of PHB production from methane. The data acquired from PHB production experiments using a methanotrophic consortia dominated by *M. trichosporium* provided preliminary results to study the feasibility of the process to large-scale. PHB production by the isolated strain will also be studied to achieve more accurate and profound results on this subject, but this strain has shown itself as promising for large-scale PHB production.

609563 Extracellular Carotenoid Production from Microalgae Under Increased CO_2 Concentrations

Abstract:

Carbon dioxide fixation using fast-growing photoautotrophic microorganisms, in their natural habitats or in artificial cultivation systems, provides a very promising alternative for CO_2 mitigation. Besides, it is potentially cost-effective when CO_2 conversion is associated to the production of high value bioproducts (Borowitzka, 2013). Some of the bioactive compounds from microalgae are proteins, fatty acids, vitamins, pigments and carotenoids, some of them exhibiting antioxidant activity. Among the reported antioxidants produced by microalgae, astaxanthin is the highest-value carotenoid that has achieved commercial success. Only for astaxanthin, the worldwide market in 2010 was around US$230 million, synthetic and natural, and the consumers demand for natural products make the synthetic pigments much less desirable, providing an opportunity for the production of natural pigments. Other potential carotenoids are lutein and zeaxanthin, approved as food additives for preventing macular diseases. Currently, the main sources of lutein for commercial production are marigold flowers, and there is no commercial production of lutein from microalgae (Cezare-Gomes et al., 2019).

Parachlorella kessleri was pre-selected under 30% CO_2 from water and sediment samples, isolated using appropriate techniques, identified via molecular biology and evaluated with cultivations in shake flasks under CO_2 concentrations from 5 to 30%. Besides presenting relatively fast growth and high biomass production under all tested CO_2 conditions, the microalga produced an extracellular reddish pigment with antioxidant activity. The production of antioxidants in microalgae occurs, in some cases, when they are subjected to stress. During microalgae cultivation, CO_2 enrichment is a requirement to reach high productivity. At the same time, high CO_2 levels are normally stressful to microalgae, especially for the photosynthetic apparatus, which can induce carotenoid production. Likewise, high CO_2 levels can stimulate the production of fatty acids, with applications for biofuels or as nutraceuticals or pharmaceuticals.

This work evaluated extracellular carotenoid production from the isolated *Parachlorella kessleri* cultivated under 5, 15 and 30% CO_2 in stirred tank photobioreactors. In the 10th day of cultivation, CO_2 supply was interrupted until the end of the cultivation (14th day). It is believed that this causes a stress on microalgae cells and is an imperative condition to make them produce the antioxidant. To our knowledge, this method has not been reported before. Dissolved oxygen and pH sensors allowed the evaluation of both parameters in real time during the cultivations. Growth was monitored daily by optical density measures at 750 nm and cell dry weight concentrations. Chlorophyll-a concentrations, that relates to photosynthetic activity, were calculated using the equations in (Ritchie, 2006) after extraction of the samples with methanol and absorbance measures at 652, 665 and 750 nm. In the last day of cultivation, the biomass was centrifuged, lyophilized and destined to fatty acids composition and element analyses (C,N,H) for the calculus of CO_2 fixation rates. Total lipids were extracted by the modified Bligh&Dyer method, the final chloroform phase was evaporated and the lipids reacted with methanol in acid conditions to produce the fatty acid methyl esters (FAMEs), which were extracted with hexane and detected by GC-MS. Total carotenoids were spectrophotometrically quantified in the supernatant by absorbance measures at 450 nm and using molar extinction coefficient of 2500

(Britton, 1985).

The growth curves under all CO_2 concentrations were similar, although under 30% CO_2 there was a lag phase of one day, the exponential phase was shorter (3 days comparing to 4 days in the other conditions) and after exponential phase the growth was slower than the other conditions. The pH values varied differently among the cultivations. Under 5% CO_2, pH values increased from 5.7 on the first day to 10.5 on the 5th day, decreased until the 10th day and increased again after CO_2 supply was interrupted until the end of the cultivation. The same pattern was observed on the cultivations under 15% CO_2, but the highest pH values during exponential phase were around 6.5 and reached 10.5 on the last 2 days of cultivation because of interruption of CO_2 supply. Under 30% CO_2, the pH remained on 5 to 6 during the cultivation period, which might have contributed the lower growth after exponential phase under this condition. Chloropyll-a (Chl-a) profiles exhibited the highest values of Chl-a concentrations on days 4 and 5 of cultivation (highest photosynthetic activity), decreasing until the end of the cultivation, which is in accordance to the high pH values on these days.

The calculated maximum specific velocities were 0.74 ± 0.09, 0.74 ± 0.08 and 0.77 ± 0.11 under 5, 15 and 30% CO_2, respectively, and the differences were not statistically significant ($p>0.05$). Cell dry weight concentrations on the 14th day of cultivation were 1.09 ± 0.12, 1.24 ± 0.15 and 1.09 ± 0.25 g L^{-1}, total lipids were 241 ± 11, 208 ± 5 e 183 ± 14 mg g^{-1} (d.w.) and CO_2 fixation rates were 158 ± 7, 165 ± 5 and 141 ± 3 mg $L^{-1}d^{-1}$, under 5, 15 and 30% CO_2, respectively. Minhas and coworkers (2016) reported 21.42% (d.w) of lipid content on *P. kessleri* , and de Morais and Costa (2007) reported 163 mg $L^{-1}d^{-1}$ of CO_2 fixation rate under 18% CO_2. Both values corroborate the results obtained in this work. The reddish carotenoid production was observed in all cultivations, with concentrations of approximately 0.030 ± 0.005, 0.026 ± 0.006 and 0.011 ± 0.002 μg mL^{-1} under 5, 15 and 30% CO_2, and the differences were statistically significant ($p<0.05$).

The fatty acids analysis of the biomass presented the major fatty acids: palmitic acid (C16:0), stearic acid (C18:0), oleic acid (C18:1), linoleic acid (C18:2) and α-linolenic acid (C18:3). Their relative concentrations were different depending on the CO_2 concentration. For instance, C16:0 yields were higher under 30% CO_2 (around 60%) compared to the other conditions (around 30%). In contrast, polyunsaturated fatty acids (PUFAs) were higher under 5% CO_2 (around 50%) and decreased to 28% and 22% under 15 and 30% CO_2, respectively. Another significant difference was on C18:1 production, which increased from 14% to 30%, under 5 and 15% CO_2, and decreased to 11% under 30% CO_2. The increase in oleic acid production has been previously reported as an indication of high cell density and accumulation of lipids, normally promoted because of nutrient limitations (van der Ha et al., 2012). This indicates that high CO_2 levels up to 15% can induce lipid production on this microalga.

The carotenoids produced were extracted with methanol and the methanolic extracts were characterized by UV-Vis spectrophotometry and by High Performance Liquid Chromatography (HPLC). Two fractions were separated in the HPLC, yellow and red, which presented λ_{max} of 405 and 427 nm, for the yellowish pigment, and 455, 485, 524 nm, for the reddish one. Adjacent absorbance peaks on these regions of the visible light spectrum are characteristic of carotenoids and indicate multiple chromophore groups. LC-MS/MS (Liquid Chromatography coupled to Mass Spectrometry) analysis of the reddish carotenoid fraction pointed to a carotenoid with molecular weight of 569 g

mol^{-1}, which may correspond to zeaxanthin or lutein. The MS/MS fragmentation of the molecular ion *m/z* 569 presented the normally reported mass losses for carotenoids but was not conclusive on the complete molecular elucidation.

To our knowledge, this is the first report of extracellular carotenoid production by the microalga *Parachlorella kessleri*. *Botryococcus braunii* has been reported to produce the extracellular carotenoid equinenone (Cheng et al., 2019). Extracellular production is particularly interesting considering the costs and the environmental impact of using organic solvents and harsh conditions to extract intracellular carotenoids. This could reduce the costs of the final product and increase the sustainability of the process. This work revealed that CO_2 concentrations up to 15% can be used to improve carotenoid production, also favoring lipid production, with particular interest on the PUFAs, which are high value bioproducts.

613231 Sialic-Acid Analogs for the Attenuation of A-beta Toxicity

Abstract:

Background: Amyloid-beta peptide (Aβ) is the main constituent of senile plaques and is implicated in the pathogenesis of Alzheimer's disease (AD). To that end, agents which either sequester Aβ or interfere with Aβ interaction/binding to cells have been investigated as a means to reduce the pathological effects of Aβ.

Methods: Different structural analogs of sialic acid (N-acetylneuramic acid) were used to decorate a chitosan backbone using EDC chemistry. FTIR and colorimetric assays were used to characterize the complexes. The ability of these complexes to attenuate Aβ toxicity was investigated in vitro using a model neuroblastoma cell line SH-SY5Y

Results: Oxygen substitution in ring structure is responsible for the increase in toxicity and increase in protective properties of the complexes. Also, the multi OH tail present in sialic acid is critical to attenuate toxicity. Analogs show no protective properties which reinforces the conclusion that clustering of sugars in cellular membranes play a significant role in Aβ binding.

Conclusions: Successfully produced compounds that showed varying degree of efficacy in attenuating Aβ toxicity to cells in culture. This work elucidates the impact that certain structures of sialic acid and its analogs can have on Aβ binding. It will allow for more specific and detailed improvements in the therapeutic polysaccharide structures that can be developed and modified to overcome other shortcomings of AD therapeutic development, particularly of penetrating the blood-brain barrier.

General significance: Oxygen atom plays crucial role on therapeutic effectiveness. This work can help as a general guideline for further therapeutic development.

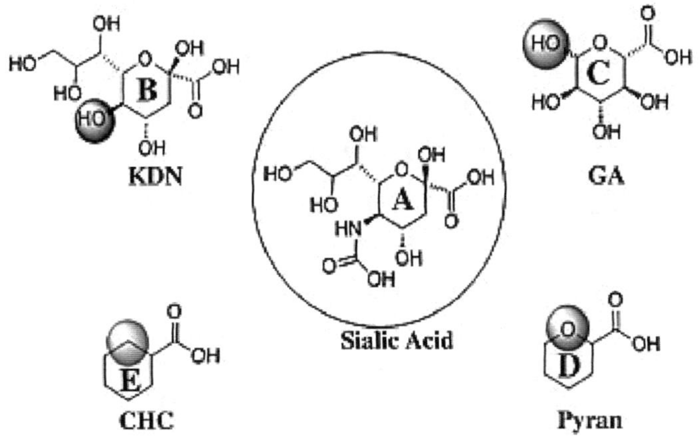

Fig. 1. Structure of analogs conjugated with chitosan. The structural difference from sialic acid is indicated by the shaded region. (A) Sialic acid (N-Acetylneuraminic acid), (B) Keto-deoxynonulosonic acid (KDN), (C) D(+)-Galacturonic acid (GA), (D) Tetrahydropyran-2-carboxylic acid (Pyran), and (E) Cyclohexanecarboxylic acid (CHC) structures are shown to elucidate these differences.

597591 Agitation in a Microcarrier-Based Spinner Flask Bioreactor Modulates Homeostasis of Human Mesenchymal Stem Cells

Abstract:

Human Mesenchymal Stem Cells (hMSCs) are well known in cell therapy due to their secretion of trophic factors, multipotent differentiation potential, and ability for self-renewal. As a result, the number of clinical trials has been steadily increasing over the last decade highlighting the need for in vitro systems capable of producing large quantities of cells to meet growing demands. However, hMSCs are highly sensitive to microenvironment conditions, including shear stress caused by dynamic bioreactor systems, and can lead to alteration of cellular homeostasis. In this study, we demonstrate a 3-fold expansion of hMSCs within a 125 mL spinner flask bioreactor system. Furthermore, we show culturing in this dynamic environment (compared to conventional 2D planar culture) results in hMSCs exhibiting higher total levels of reactive oxidative species (ROS), as well as known ROS regulator Sirtuin-3, having implications on the nicotinamide adenine dinucleotide (NAD+) metabolic pathway and metabolic homeostasis. In addition, added stress from culturing in this dynamic microenvironment increased Prostaglandin E_2 (PGE2) secretion as well as reduced the potential for Indoleamine-pyrrole 2,3-dioxygenase (IDO) secretion upon stimulus with interferon gamma (IFNγ). The results of this study have implications on engineering strategies for scalable production of hMSCs utilized in clinical trials and biomanufacturing.

598632 Retinal Photoreceptor Differentiation from Human Pluripotent Stem Cells for 3-D Bio-Printing

Abstract:

Retinal cells including light sensing photoreceptors lack regeneration potential leading to permanent vision loss. Various approaches including gene therapy, cell transplantation, and optogenetics are being studied to treat or restore the irreversible damage. Generating retinal photoreceptors using human pluripotent stem cells (iPSCs) holds great potential in treating as well developing *in vitro* models for better understanding and screening of various formulations for retinal degenerative diseases like retinitis and age-related macular degeneration. In this study, human iPSC line (iPSK3 cells) derived from skin fibroblasts cells was differentiated into retinal photoreceptors following a small molecule-based retinal induction protocol. We observed that retinal photoreceptors differentiated from the iPSCs showed marker expression for retinal progenitor cells (LHX2), ganglionic cells (PAX6), pan-photoreceptors (OTX2, CRX), and differentiated ganglionic cells (BRN3). In order to mimic the native environment during retinal development, we are optimizing suitable bio-inks in order to 3D bio-print the retinal progenitors and photoreceptors. This study provides evidence for recapitulation of the native retinal environment as well as development of photoreceptor models for drug screening purposes.

Reference

Jie Zhu, Joseph Reynolds, Thelma Garcia, Helen Cifuentes, et.al, Generation of Transplantable Retinal Photoreceptors from a Current Good Manufacturing Practice-Manufactured Human Induced Pluripotent Stem Cell Line, Stem Cells Translational Medicine, 2018;7:210-219.

613237 Compounds to Modify Plaque Formation and Toxicity of A-beta in Alzheimer's Disease

Abstract:

Alzheimer's disease (AD) is the seventh leading cause of death in the United Sates, affecting 5.5 million Americans with the number expected to double by 2050 due to the expected continued increase in average life expectancy and US population. As such, AD is expected to be a major public health problem in the upcoming decades. The current treatment (behavioral and symptom management) does not address the need to treat the underlying condition, the formation and deposition of A-beta oligomers. Since A-beta formation is a fact of existence, it becomes necessary to identify dietary and nutraceutical solutions that can be lifelong preventive measures. As such, we will look to utilize cutting-edge technologies and well-established techniques to evaluate readily available food-based options (polyphenols) to identify viable targets for such long-term preventative measures. We will utilize quartz crystal microbalance technology (QCM) and in vitro toxicity studies to determine the efficacy of such compounds. Additionally, the cross-disease utility of the techniques and devices implemented in these studies show promise for long term commercialization and effectiveness.

Therapeutic screening protocol development

We were able to implement a new evaluation procedure for determining the efficacy of a compound to disrupt A beta aggregation. We utilize quartz crystal microbalance (QCM) technology to follow the aggregation pathway of A beta in the presence and absence of target therapeutics. Changes in the aggregation pattern are stronger indicators of the potential influence of these compounds on the A beta behavior, *in vitro*. Compounds that are effective in this regime are more viable candidates as preventative measures and for treatment during the early stages of Alzheimer's disease. This combined approach of evaluating treatments for the entire Alzheimer's lifecycle is critical to what we believe is the best approach to minimizing the impact of this disease.

Data Collection

A. QCM evaluation

Initial studies with QCM proved to be more interesting and potential useful in diagnosis and compound evaluation than we had hoped. The shape of the A beta aggregation curve can be seen in Figure 1 below:

[view figure 1 below text]

Figure 1: QCM of A beta Aggregation – The curve in represents the frequency of oscillation as the protein aggregates in solution and at the surface. The yellow highlighted region represents the delay in surface aggregation. The green highlighted section represents the aggregation to the surface from solution. The difference labelled as change in frequency (Δv) represents the total change in mass due to surface aggregation

As can be seen from figure 1, there are two characteristic regimes (yellow and green highlighted sections). The surface aggregation portion (green highlight) was expected. However, the identification of the delay due to solution rearrangement is pleasant surprise. By monitoring changes in both regions with the frequency, changes in aggregation kinetics can be identified as follows:

Table 1: Expected QCM Changes from Changes in Aggregation

[see table image below text]

B. Resveratrol & EGCGBy evaluating the combination of the events for each confounder, we can determine the likelihood of a compound to have beneficial effects both *in vitro* and *in vivo*.

Resveratrol is a compound found in red wine that is known to have strong antioxidant properties and believed to be beneficial in general neuroprotection. Similarly, Epigallocatechin gallate (EGCG) is neuroprotective antioxidant found in green tea. The neuroprotective properties of these compound made them ideal candidates for evaluation. Initial results demonstrated the following:

Table 2: Qualitative Results from Initial Compound Screening

[see table image below text]

As can be seen from Table 2, the initial compounds that we have evaluated show very little natural toxicity. This is an important characteristic for any compound that we want to use as a therapeutic, for obvious reasons. As for anti-aggregation, the compounds, both of which are antioxidant polyphenols, showed no appreciable anti-aggregation characteristics in the initial studies. This finding was not surprising give then expected protective mechanism (interaction with the neuron directly). While this may seem like a lack of result, the outcome of the aggregation study was beneficial in validating the efficacy of the protocol developed. Finally, both showed some degree of neuroprotection, with resveratrol showing higher neuroprotection the ECGC. The results thus far are going to require additional testing to quantify the results with the necessary statistical accuracy, initial studies have been promising and are driving the direction of research and compound selection.

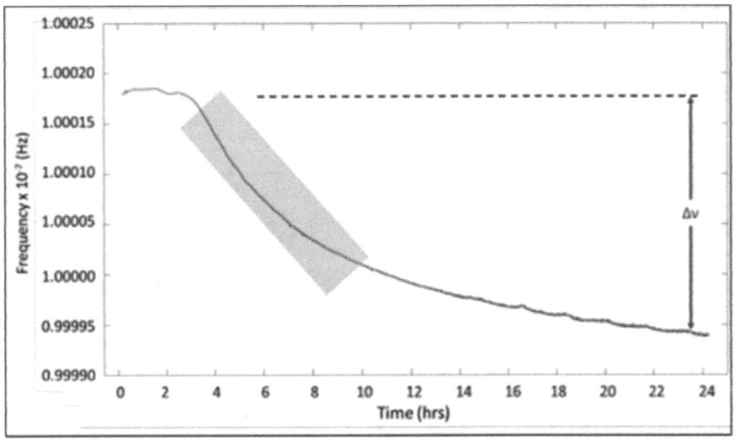

Table 1: Expected QCM Changes from Changes in Aggregation

Indicator	Increase in Toxic Species Rate	Decrease in Toxic Species Rate	Increase in Fibril Formation Rate	Decrease in Fibril Formation Rate	Total Fibril Load
Yellow Region	Shorter in length	Longer in length	No Change	No Change	No Change
Green Region	No Change	No Change	Steeper Slope	Shallower Slope	No Change
Delta Freq	No Change	No Change	No Change	Smaller	Larger

Table 2: Qualitative Results from Initial Compound Screening

Compound	Intrinsic Toxicity	Anti-aggregation	Neuroprotection
Resveratrol	Low	None	Moderate
EGCG	Low	None	Low

601717 Membrane Protein Antibody Discovery Using Yeast-Surface Display and Whole Cell Selections in Suspension

Abstract:

Membrane proteins (MPs) are important therapeutic targets as they are responsible for facilitating a wide range of biological functions critical to maintaining homeostasis. Over 60% of all clinically approved drugs target MPs, thus demonstrating the central importance of this class of proteins. Antibody discovery against complex MPs such as G-protein coupled receptors (GPCRs) is challenging due to the difficulty in recombinantly expressing glycosylated seven-pass transmembrane proteins in their native states. Here, we present and characterize a cell-cell interaction platform combining the yeast display directed evolution platform that allows for manipulation and detection of whole cells in suspension that express MP targets. We also employ a novel selection strategy that interfaces magnetic-activated cell-sorting (MACS) with our cell-cell interaction platform to enrich for specific binders to MPs within a yeast-displayed library of antibody fragments. Collectively, our new molecular engineering approaches serve as novel discovery engines for MP-targeted therapeutics.

602362 Extended 3D Cancer Cell Migration in Response to an Oscillating Chemical Gradient Breaks the Spatial Range Limitations of Conventional Chemotaxis

Abstract:

In the tumor microenvironment (TME) cancer cells are exposed to spatial and temporal pulsing of chemical gradients which influences the directed migration of cells via a chemotactic response. Traditional cellular chemotaxis *in vitro* studies utilize stable chemical gradients in two-dimensional (2D) environments; however, *in vivo* conditions result in both competing and oscillating gradients which the individual cells must sense and interpret to elicit a migratory response. Moreover, chemotaxis studies in a 2D environment do not recapitulate the spatial and temporal cues encountered by cells in three-dimensional (3D) environments. Recent studies have found that cell signaling and cell movement in 3D environments differ from those in 2D environments. Additionally, studies have shown that cells typically lack the capacity to detect external gradients under constant and high magnitude chemoattractant concentrations. Long-range chemotaxis is possible via the temporal pulsing of spatial gradients by establishing and eliminating the chemical gradient. In this study, a microfluidic approach was utilized to generate stable and tunable chemical gradients to elicit a dynamic cellular response during 3D chemotaxis in triple-negative breast cancer cells. The microfluidic device consists of three parallel channels imprinted into top polydimethylsiloxane (PDMS) device coupled with a bottom 3% (v/v) agarose slab enclosed in a Plexiglas chamber. The bottom agarose hydrogel allows for passive diffusion perpendicular to the direction of flow in the outer two channels into a center, 'flow-free' migration channel. A constant source solution with a chemoattractant was infused into one of the outermost channels to induce a linear chemical gradient across the migration channel. Time-dependent gradient oscillation was controlled by tuning the source solution flow rate between 'on' and 'off' for a defined time. The flow rate of the outer channels could be modulated to build and erase gradients allowing for temporal control over the gradient within the device. COMSOL simulation confirmed the oscillating pattern chemical gradient in the device. To study 3D chemotaxis, triple-negative breast cancer cells MDA-MB-231) were seeded into the migration channel in a collagen (2mg/mL) hydrogel to study how cancer cells respond to time-varying external chemical gradients. Preliminary results showed that the population of breast cancer cells responded to 20% FBS oscillating gradient (8 h on, 2 h off, 8 h on) by extending the migratory distance and chemotactic persistence compared to both a non-oscillatory stable chemical gradient and random migration in the absence of an external gradient. Extended cell migration in *in-vitro* chemotaxis mimics the cell behavior in an *in vivo* microenvironment that has the potential to the study of metastasis and drug screening.

604176 Tobacco Cell-Secreted Hematopoietic Growth Factors for Ex Vivo Production of Red Blood Cells

Abstract:

Ex vivo generation of clinically available red blood cells (RBCs) from hematopoietic stem cells (HSCs) represents a promising approach for overcoming the limitations associated with the use of donor's blood. Both expansion and differentiation of HSCs are highly reliant on erythropoietic growth factors (eGFs). This project aims to develop a novel plant cell-based bioproduction platform for eGFs at significantly reduced cost. In addressing the low productivity bottleneck of plant cell culture technology, we are engineering hydroxyproline (Hyp)-O-glycosylated peptides (HypGPs) that function as a molecular carrier to boost the secretion of fused proteins from culture plant cells. Each of three key eGFs essential for HSCs expansion and differentiation, erythropoietin (EPO), stem cell factor (SCF) and interleukin 3 (IL-3), were expressed in tobacco BY-2 cells as fusion with a strategically designed HypGP tag, for example $(SP)_{20}$ consisting of 20 tandem repeats of a "Ser-Pro" motif. The secreted protein yields were up to 500-fold greater than the expression of eGF controls lacking a HypGP tag. The HypGP-tagged eGF secreted from plant cells exhibited the bioactivity in stimulating the proliferation of TF1 erythroleukemic cell. Furthermore, the plant cell-derived eGFs were able to stimulate the expansion and differentiation of umbilical cord blood-derived CD34+ cell towards RBCs. This project provides a new and cost-effective bioproduction platform for eGFs, facilitating manufacturing of HSCs-derived RBCs at large scale for clinical applications.

608276 A Dynamic Elementary Flux Mode-Based Model for Antibody Producing Cell Lines and Model Based Evaluation of Fed-Batch Cultures

Abstract:

Monoclonal antibodies (MAbs) are significant reagents used extensively in diagnostic assays, therapeutic applications, and in vivo imaging. Mammalian cell cultures have become the favored production hosts for MAbs, since microbial systems are not able to carry out the complex post-translational and functional modifications of these proteins, such as glycosylation. Chinese Hamster Ovary (CHO) cells and hybridoma cells, which share similar metabolic characteristics, have been popular cell types for production of MAbs. The efficient performance of these cell cultures requires highly specialized culture media to enhance MAb yield for in vitro production in view of substantial cell death and reduced MAb productivity due to the variations in culture conditions. Although production practices have been employed for decades, cell kinetics is still under investigation to obtain quantitatively as well as qualitatively cost-effective production strategies. Creating these strategies requires understanding of cell metabolism affected by process dynamics in culture environments. Kinetic models empower us to illustrate quantitative cell growth and metabolic activity, which allows prediction of different cell phenotypes and provides better understanding of cell physiology, which is important in optimization of MAb production in animal cell cultures.

Based on elementary flux mode (EFM) analysis, a novel approach is presented for monoclonal antibody (MAb) production by GS-CHO cells. An EFM-based kinetic model is developed on the basis of a set of macro-reactions, which can predict the time-dependent concentrations of metabolites, cell growth, and MAb productivity over a range of culture conditions. The rate of each macro-reaction through the central carbon metabolism contributes to ATP production. Energy metabolism and availability of substrates determine the specific biomass and MAb synthesis rates. The model provides a framework to predict different cell phenotypes under different culture conditions with a reduced metabolic network comprised of 23 macro-reactions. The model incorporates energy metabolism with biomass and MAb formation, with the specific ATP production rate being decided by the central carbon metabolism and used for estimation of biomass and MAb synthesis rates. The reaction rate expressions are represented by Michaelis-Menten kinetics based on extracellular metabolite concentrations, which determine ATP production by glycolysis and respiratory chain. Glutamine and asparagine are considered as regulatory metabolites for GS-CHO cells. Glutamine determines asparagine utilization route and energetic state of cells, while asparagine regulates the uptake rates of aspartate and glutamate. Glutamine is considered as a regulatory metabolite for GS-CHO cell lines - its variation in the culture influences cell metabolism and growth. With glutamine-free media, asparagine is the best alternative amino acid for GS-CHO cell lines. The model was calibrated for glutamine-free and glutamine-available cases and validated for fed-batch cultures supplied with glutamine. Un-optimized fed-batch cultures have been simulated considering a constant feeding strategy and a daily feeding strategy. Higher viable cell accumulation was observed in constant fed-batch operation, due to suppression of lactate formation. The model predictions are in good agreement with the experimental data reported in literature. The proposed model can be used for design of optimal nutrient media and optimization and control of fed-batch cultures.

604647 Polymer Amphiphiles for Directed Self-Assembly Around Hepatitis B Virus Particles

Abstract:

Background: Hepatitis B virus (HBV) is a global health issue with over 250 million chronic cases world-wide, and is attributed to an estimated 800,000 deaths annually due to liver failure, or complications leading to hepatocellular carcinoma or cirrhosis[1]. Current therapies are typically life-long[1] and can lead to issues with patient compliance and cost. Chronicity of infection is due in part to the generation of non-infectious subviral particles, composed solely of the surface antigens, which are produced 1,000 to 10,000 times more than infectious particles during HBV replication[1,2]. These surface antigens contribute to the downregulation of a number of immune responses[1–3] which have been shown to recover after the antigen is removed[3]. Current HBV therapies suppress viral replication, but do not remove existing virus particles. The goal of this project is to create a system to stably encapsulate infectious and subviral HBV particles *in vivo* by the directed assembly of polymer nanoparticles. We hypothesize that this would recover proper immune function and prevent further hepatocyte infection. To create amphiphiles for directed self-assembly, antibody fragments were prepared to act as targeting ligands and conjugated to block copolymer amphiphiles.

Methods: Poly(ethylene glycol)-block-poly(lactic-co-glycolic acid) (PEG-b-PLGA) and poly(ethylene glycol)-block-poly(lactic acid) (PEG-b-PLA) were added to human blood components by solvent injection and imaged using transmission electron microscopy (TEM) to determine the effect of free polymer in the blood. Beads coated with Hepatitis B surface antigen (HBsAg) were produced to be used in place of HBV particles. Carboxylated polystyrene beads were sized by dynamic light scattering, then covalently coated with HBsAg and re-measured to confirm successful coating. Antibody fragments were prepared by digestion with pepsin and subsequently reduced with cysteamine, then purified by centrifugal filtration. Size was measured with non-reducing SDS-PAGE. The antibody fragments were conjugated to maleimide-PEG-b-PLA or maleimide-PEG-b-PLGA by thiol-maleimide reaction. An indirect enzyme-linked immunosorbent assay (ELISA) was used to measure binding activity of the antibody fragment alone and fragment-conjugated polymer to confirm binding ability of the conjugated polymer. For a non-specific binding control, un-modified PEG-b-PLA was measured by solvent injection of the polymer into a dilute solution of antigen coated beads, with size measurements taken by dynamic light scattering before solvent injection and immediately, three hours, and 24 hours after.

Results: PEG-b-PLA had no negative effect on various blood components, but PEG-b-PLGA led to disruption of plasma and platelets, indicating that it would not be suitable for injection into the blood. SDS-PAGE showed successful fragmentation of the antibody and indirect confirmation of successful antibody-polymer conjugation. The signal produced by ELISA was lower for the fragment-conjugated polymer than the unconjugated antibody fragment, but remained significantly above background levels indicating that binding still occurs. DLS results demonstrate the successful coating of antigen onto the polystyrene beads, but was determined not to be reliable for measuring non-specific encapsulation. Plans for future work include TEM imaging of the non-specific encapsulation, size-exclusion chromatography to confirm antibody-polymer conjugation, and specific encapsulation of the coated beads by solvent injection. We expect that the unconjugated polymer will show minimal encapsulation of the coated beads, instead predominantly forming empty polymersomes. In contrast, we expect that the specific encapsulation will show polymersome formation around the beads.

1 M.-F. Yuen, D.-S. Chen, G. M. Dusheiko, H. L. A. Janssen, D. T. Y. Lau, S. A. Locarnini, M. G. Peters and C.-L. Lai, *Nat. Rev. Dis. Prim.*, 2018, **4**, 1–20.

2 M. Dandri and J. Petersen, *Clin. Infect. Dis.*, 2016, **62**, S281–S288.

3 A. Bertoletti and C. Ferrari, *Gut*, 2012, **61**, 1754–1764.

604685 Cycling Aggregation and Planar Culture Extends *in Vitro* Expansion Potential of Human Mesenchymal Stem Cells

Abstract:

Mesenchymal stem cell (MSC) treatment has shown great promise in in vitro and small animal models. However, this has not been seen when applied to human clinical trials. This is due to both the availability of MSCs at the time of need and lack of viable expansion method. MSCs are very susceptible to culture adaptation-induced in vitro aging which results in loss of MSC functionality and clinical relevance. To combat this effect, this work assessed cyclical aggregation as a means of expanding MSCs while still maintaining functionality. Cyclical aggregation consists of an aggregation phase followed by dissociation and replating onto planar tissue culture plastic as a means to expand the cells. The results indicate that cyclical aggregation maintains proliferative capability, stem cell proteins, and clonogenicity, and prevents the acquisition of senescence. To determine why aggregation was responsible for this phenomenon, the integrated stress response pathway was probed with salubrial and GSK-2606414. Treatment with salubrial had no significant effect, while GSK-2606414 mitigated the effects of aggregation leading to in vitro aging. This method holds the potential to increase the clinical relevance of MSC therapeutic effects from small model systems (such as rats and mice) to humans, and may open the potential of patient-derived MSCs for treatment thereby removing the need for immunosuppression.

604846 Analyzing the Role of Chinese Hamster Ovary Extracellular Vesicles (CHO-EVs) in Extracellular Communication, Cellular State, and Protein Expression of CHO Cultures

Abstract:

Chinese hamster ovary (CHO) cells are the primary host cell line for producing protein therapeutics, resulting in the need to better understand and improve CHO cell culture conditions to increase protein production and quality. CHO cells dynamically produce and uptake extracellular vesicles (EVs) in culture to exchange small RNAs, protein material, and small DNA fragments. This exchange of protein and genetic material dynamically regulates the cellular state and protein expression of the target cell. Here, CHO cell derived EVs (CHO-EVs) harvested from various shake flask cultures are characterized in terms of size distribution, morphology, zeta potential, and RNA content. CHO-EVs are separated by differential ultracentrifugation into the heavier EV fraction and in the lighter EV fraction and the size distribution is measured by nanoparticle tracking analysis (NTA). RNA fragment analysis shows CHO-EVs are enriched in small RNAs compared to the parent cells, indicating the CHO-EVs are selectively sorting material into the CHO-EVs to regulate protein expression in target cells. Using RNA sequencing (RNA seq), the RNA profiles of the CHO-EVs were compared to the RNA profile of the CHO cells to determine which RNAs are highly enriched in CHO-EVs. To visualize the dynamic production, protein and genetic material exchange, and uptake of CHO-EVs in shake flask cultures, non-specific protein dyes (carboxyfluorescein succinimidyl ester (CFSE) and CellTracker Deep Red) were used in conjunction with correlative confocal microscopy and scanning electron microscopy (SEM). Correlating confocal images with SEM images showed the localization of exchanged CHO-EVs with the high-resolution imaging of CHO cell morphology. Additionally, flow cytometry and confocal microscopy were used to determine the proportion of CFSE and Deep Red cells in culture and the exchange of fluorescently stained CHO-EVs in culture. The characterization of CHO-EVs produced in shake flask cultures elucidates the native cell communication and material exchange mechanisms via EVs. Future steps in expanding the characterization and application of CHO-EVs include loading specific genetic cargo, such as anti-apoptotic small interfering RNAs (siRNAs), to deliver in CHO cultures to prolong CHO culture health and lifespan.

604854 Targeted Gene Therapy of Hematopoietic Stem and Progenitor Cells Using Cas9 and siRNA-Loaded Megakaryocytic Membrane Vesicles

Abstract:

There is a large spectrum of highly pervasive hematological disorders affecting red blood cells (erythrocytes), white blood cells (granulocytes), platelets (megakaryocytes) and lymphocytes. Hematopoietic stem and progenitor cells (HSPCs) can differentiate into these blood cell types, and gene editing of HSPCs can provide therapeutic benefits to patients for a variety of genetic hematological disorders, ranging from immunodeficiencies to thrombocytopenia[1-3]. For more transient gene therapy, gene expression can be modulated epigenetically using RNA interference through administration small RNAs into the target HSPCs[4,5]. Thus, by directly administering gene therapeutics into HSPCs, a significant proportion of hematological diseases can potentially be ameliorated.

Previously, it has been demonstrated that microparticles derived from megakaryocytes (MkMPs) can readily interact with and deliver cargo to HSPCs *in vitro*[6]. In our current study, we have demonstrated that these MkMPs also have the propensity to localize within murine bone marrow roughly 24-hrs after administration[7]. Furthermore, through histology, we determined that these MkMPs also interact with HSPCs and other blood cells in the bone marrow and lungs *in vivo*, thus enabling microparticles to serve as efficacious drug delivery vehicles. Thus, it is hypothesized that CRISPR Cas9 and siRNA can be delivered specifically to HSPCs through receptor-mediated endocytic pathways to HSPCs using both natural (microparticles) and semi-synthetic membrane vesicles. We have successfully loaded membrane-wrapped PLGA nanoparticles with siRNA for delivery to HSPCs *in vitro* and observed effective downregulation of an HSPC-specific gene. Likewise, CRISPR Cas9, a sequence-specific nuclease, was successfully delivered to blood cells (CHRF) via a similar membrane-wrapped polymeric nanoparticle *in vitro* and exhibited efficient gene knockout and reduced cytotoxicity in comparison to traditional plasmid-based Cas9 systems. Finally, blood-derived membrane vesicles minimize the immunogenic risk associated with other vehicles, potentially establishing cargo-loaded membrane vesicles as a safe and effective method for cell-specific *in vivo* gene editing and modulation.

REFERENCES:

1. Osborn, M. J., Belanto, J. J., Tolar, J. & Voytas, D. F. Gene editing and its application for hematological diseases. *Int J Hematol* **104**, 18-28, doi:10.1007/s12185-016-2017-z (2016).
2. Surun, D., von Melchner, H. & Schnutgen, F. CRISPR/Cas9 genome engineering in hematopoietic cells. *Drug Discov Today Technol* **28**, 33-39, doi:10.1016/j.ddtec.2018.08.001 (2018).
3. Yu, K. R., Natanson, H. & Dunbar, C. E. Gene Editing of Human Hematopoietic Stem and Progenitor Cells: Promise and Potential Hurdles. *Hum Gene Ther* **27**, 729-740, doi:10.1089/hum.2016.107 (2016).
4. An, D. S. *et al.* Stable reduction of CCR5 by RNAi through hematopoietic stem cell transplant in non-human primates. *Proc Natl Acad Sci U S A* **104**, 13110-13115, doi:10.1073/pnas.0705474104 (2007).
5. Lam, J. K., Chow, M. Y., Zhang, Y. & Leung, S. W. siRNA Versus miRNA as Therapeutics for

Gene Silencing. *Mol Ther Nucleic Acids* **4**, e252, doi:10.1038/mtna.2015.23 (2015).
6. Jiang, J., Kao, C. Y. & Papoutsakis, E. T. How do megakaryocytic microparticles target and deliver cargo to alter the fate of hematopoietic stem cells? *J Control Release* **247**, 1-18, doi:10.1016/j.jconrel.2016.12.021 (2017).
7. Escobar, C., Kao, C. Y., Das, S. & Papoutsakis, E. T. Human megakaryocytic microparticles induce de novo platelet biogenesis in a wild-type murine model. *Blood Adv* **4**, 804-814, doi:10.1182/bloodadvances.2019000753 (2020).

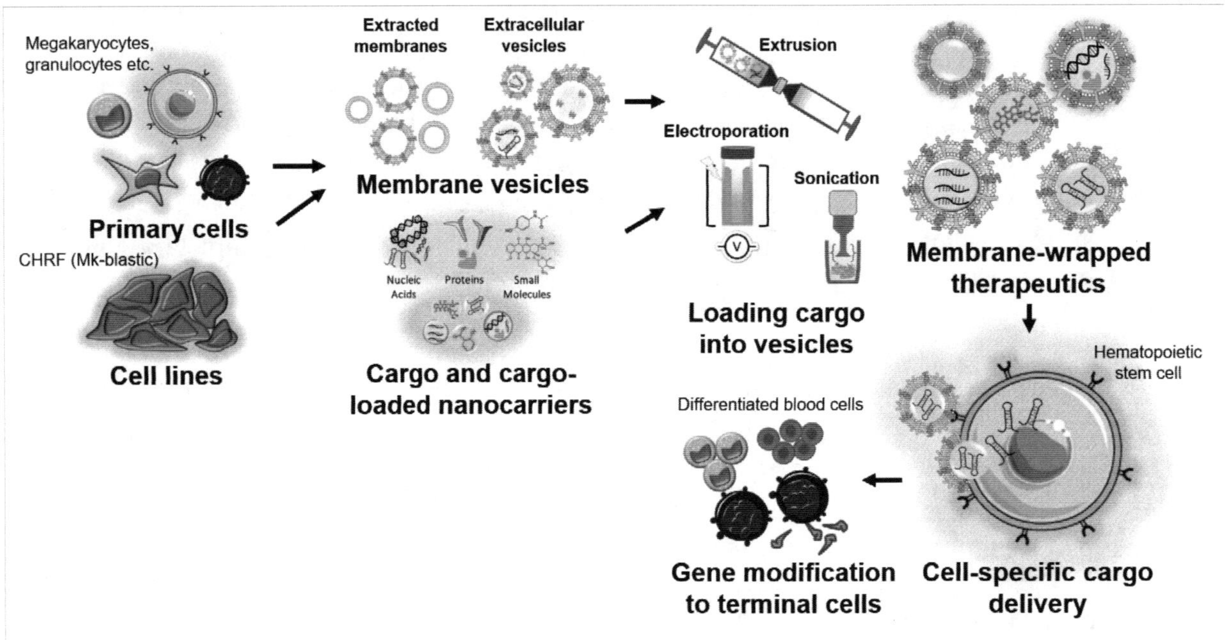

605618 How Is Hydrocephalus Treatment Dependent on Astrocyte Phenotype Expression?

Abstract:

Current understanding of the cellular response contributing to the failure of hydrocephalus cerebral spinal fluid (CSF) shunts has been limited, to date, with the evaluation of cells and tissue around the implant. Yet there is a critical knowledge gap in gene expression profiles in this tissue. When paired with cellular analysis, gene expression can lead to an understanding of the activity of the cells obstructing catheters at the time of shunt failure. This study represents the first robust investigation of the changes in gene expression levels specific to astrocyte immune response following CSF shunt implantation. From our NIH R01-funded work, we know that the tissue occluding shunts is over 80% inflammatory cells, with a more exaggerated astrocytic response when the analyzed tissue/shunt comes from a patient whose shunt was removed for obstruction. To understand how to mitigate the cell immune response to shunts, and therefore reduce mechanisms leading to shunt failure, we will perform gene expression profiling over collected CSF shunts from patients to observe whether the cells on the shunt surface are expressing two significantly different reactive astrocytes phenotype; A1 or A2. An invaluable marker to distinguish between different activation states of reactive astrocytes, is the A1 complement component C3 gene. C3 is specifically one of the most characteristic and highly upregulated genes in A1 and is not expressed by resting or A2 reactive astrocytes. To identify whether there are C3-expressing A1 astrocytes in tissues from CSF shunt regions, astrocytes that are C3-positive by fluorescence in situ hybridization (FISH) are analyzed. Then, quantitative PCR (qPCR) analysis is used to confirm upregulation of C3 in the tissue samples. We will correlate these data to ELISA characterized CSF collected from the same patients for pro-inflammatory cytokines. Therefore, we begin to shed light on astrocyte phenotype expression on shunt surfaces and root causes for implant failure. In the brain, TNFα, IL-1α and C1q combined propel resting astrocytes into a full-fledged A1 state. This study will drive investigation into whether astrocytes occluding shunts could be prevented by simply blocking secretion or action of these three cytokines to keep astrocytes out of the A1 state. FDA-approved drugs targeting TNFα and Il-1α already exist and are in use for other medical conditions.

609084 A Mechanistic Analysis of Recombinant Adeno-Associated Virus Productivity in Biomanufacturing

Abstract:

Recombinant adeno-associated viruses (rAAVs) are increasingly exploited as one of the most important vectors for *in vivo* gene therapies, including the lately US FDA-approved rAAV-based gene therapies Luxterna^{TM} (by Spark Therapeutics) and Zolgensma^{TM} (by Novartis) [1]. With the rapid growth of rAAV, current rAAV manufacturing capacity is insufficient to meet emerging demand and there is a need to improve rAAV manufacturing production to keep pace with the clinical and commercial development [2]. To date, several methods have been commonly used in rAAV manufacturing, and are based on cell cultivation and a set of genetic elements provided by either transient transfection or viral vector-based gene delivery, or integration into cell genome [3]. A key challenge with current methods is that rAAV production is prone to low yields and a large production of empty capsids. This suboptimal performance is not well understood and remains a hurdle in process productivity. To tackle these challenges, we have formulated a detailed mechanistic analysis to understand rAAV production pathways and find out bottlenecks in producing high quality products. This understanding should form a critical foundation to improve productivity and overcome the production of defective products.

In this presentation, we review the mechanisms involved in the common methods used today for rAAV production and the potential limitations and barriers in the production pathways towards high productivity. We will also review the kinetics of wild-type AAV replication, to underline some of the mechanistic differences in the rAAV production in cell cultivation that correlate with viral productivity. Based on these mechanisms, we discuss some strategies applied in literature where molecular or process designs were used to improve rAAV productivity.

Reference

1. Wang, D., P.W.L. Tai, and G. Gao, *Adeno-associated virus vector as a platform for gene therapy delivery.* Nat Rev Drug Discov, 2019. **18**(5): p. 358-378.
2. van der Loo, J.C. and J.F. Wright, *Progress and challenges in viral vector manufacturing.* Hum Mol Genet, 2016. **25**(R1): p. R42-52.
3. Penaud-Budloo, M., et al., *Pharmacology of Recombinant Adeno-associated Virus Production.* Mol Ther 2018. **8**: p. 166-180.

601662 Biocompatibility and Potential Regenerative Effects of Sugarcane Derivatives in a 2D Skin Wound Model

Abstract:

Plant extracts have been widely used for medicinal purposes throughout the world for thousands of years. Many of their applications as therapeutic treatments have been derived from tradition and often have no fundamental scientific evidence or justification based on chemical composition [1]. In many rural areas of developing countries, traditional treatments continue to be the primary source of medicine due to 53-90% of people relying on this knowledge to meet their primary healthcare needs [2] [3] [4]. In particular, non-centrifugal sugarcane is traditionally used in Colombia as a healing and regenerative ointment for superficial skin wounds. The present study aims at providing scientific evidence for this long-lasting Colombian tradition by evaluating the wound healing potential of extracts isolated from sugarcane and its derivatives, which were fabricated under meticulously controlled thermal conditions as reported by Rodríguez *et. al* [5]. Sugarcane derivatives (juice, syrup, non-centrifugal sugarcane (NCS), sugarcane stalk) produced under the previously mentioned protocol and commercially available sugarcane produced under traditional methods were lyophilized to remove up to 95% of the water content. Potentially therapeutic extracts were isolated from all five samples using supercritical fluid extraction with methanol as co-solvent [6]. All samples (sugarcane-derivatives and extracts) were evaluated at a range of concentrations (1mg/mL to 31.25 μg/mL) for their cytotoxic and regenerative effects. Cell cytotoxicity was determined by MTT assays using Vero cells (CCL-081) [7], as well as, hemolysis [8] and platelet aggregation assays using donated human blood and plasma [9]. Sample concentrations with less than 30% cytotoxicity and less than 5% hemolytic activity were used to determine the effect on cell migration in a keratinocyte 2D scratch-wound assay. For this, the total area of the scratch wound was quantified by analyzing images taken in triplicate at 3 different time points (t = 0, 12 and 24 hours) using the open-access software ImageJ [10][11]. Results showed that sugarcane juice and syrup exhibit low cytotoxicity at concentrations below 125μg/mL and 62.5μg/mL, respectively. NCS showed no evidence of major cytotoxic effects under the concentrations evaluated. Methanol-based extracts isolated from all five samples can only be used under 10% (v/v) concentrations due to the well-known methanol's cytotoxic effects *in vitro*. Moreover, none of the samples evaluated exhibited a hemolytic effect or significant platelet aggregation at the concentrations assayed. Figure 1 provides visual evidence of increased cell migration induced by the syrup sugarcane derivative when compared to the control (absence of treatment). In conclusion, the results presented in this study suggest that sugarcane and its derivatives exhibit great potential in the discovery of new therapeutic molecules for the development of novel and more effective wound healing treatments. Further studies must be performed; however, to properly identify specific therapeutic molecules, which may be responsible for the different effects hereby described.

References:

1. Reuter, J., Merfort, I., & Schempp, C. M. "Botanicals in Dermatology". American Journal of Clinical Dermatology, 1 (2010).
2. H. R. Chitme, R. Chandra, and S. Kaushik, "Studies on anti-diarrheal activity of calotropis gigantea R. Br. in experimental animals," Journal of Pharmacy & Pharmaceutical Sciences, vol. 7, pp. 70–75 (2003).
3. H.-S. Kim, "Do not put too much value on conventional medicines," Journal of Ethnopharmacology, vol. 100, no. 1-2, pp. 37–39 (2005).

4. Kowal P Chatterji S Naidoo N et al. "Data resource profile: the World Health Organization Study on global AGEing and adult health (SAGE)," International Journal of Epidemiology 41: 1639 – 49 (2012).
5. Velásquez, F., Espitia, J., Mendieta, O., Escobar, S., Rodríguez, J. "Non-centrifugal cane sugar processing: A review on recent advances and the influence of process variables on qualities attributes of final products," Journal of Food Engineering, 255, 32-40 (2019).
6. Tyskiewicz K, Konkol M, Roj E. "The application of supercritical fluid extraction in phenolic compounds isolation from natural plant materials," *Molecules* 23(10):E2625 (2018).
7. DIN EN ISO 10993-5, 2009, Biological evaluation of medical devices - Part 5: Tests for in vitro cytotoxicity.
8. Evans, B. C., Nelson, C. E., Yu, S. S., Beavers, K. R., Kim, A. J., Li, H., Duvall, C. L. "Ex-Vivo Red Blood Cell Hemolysis Assay for the Evaluation of pH-responsive Endosomolytic Agents for Cytosolic Delivery of Biomacromolecular Drugs," *Journal of Visualized Experiments : JoVE*, (73), 50166 (2013).
9. Tsujino, T., Isobe, K., Kawabata, H., Aizawa, H., Yamaguchi, S., Kitamura, Y., ... Kawase, T. "Spectrophotometric Determination of the Aggregation Activity of Platelets in Platelet-Rich Plasma for Better Quality Control," *Dentistry Journal*, 7(2), 61. (2019) doi: 10.3390/dj7020061
10. Liang, C., Park, A. & Guan, J. "In vitro scratch assay: a convenient and inexpensive method for analysis of cell migration in vitro," *Nat Protoc* 2, 329–333 (2007).
11. Ayman Grada, Marta Otero-Vinas, Francisco Prieto-Castrillo, Zaidal Obagi, Vincent Falanga, "Research Techniques Made Simple: Analysis of Collective Cell Migration Using the Wound Healing Assay", Journal of Investigative Dermatology, 137 (2017).

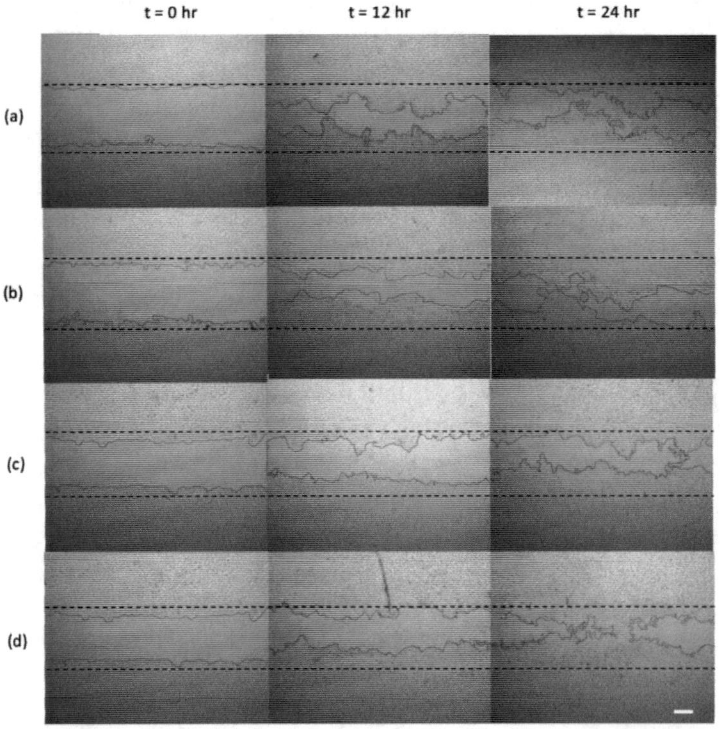

Figure 1. Scratch-wound assay in HaCat cell line. *All images were taken with 4X magnification in an inverted microscope. The columns correspond to the three time points evaluated and each row corresponds to the samples evaluated (a) juice (b) syrup (c) non-centrifugal sugarcane (NCS) and (d) control (no treatment). Images were taken immediately after inflicting the wound t=0, as well as 12 hours, 24 hours and 48 hours afterward. The images shown are representative of each of the evaluated groups (n=9). Wounds completely close after 48 hours of incubation (images not shown). Cell migration rate was determined to be the change in the distance between the two wound edges (width) per unit time (distances were highlighted in blue in the images). In order to facilitate the visualization of wound closure, the wound edges made by the initial scratch wound are represented by the black dotted lines. The white line in the right bottom corner represents the 100 μm scale bar.*

596744 Engineering Superior VEGF Antagonists to Treat Retinal Diseases

Abstract:

Neovascularization in the eye drives the pathogenesis of diseases including age-related macular degeneration and diabetic retinopathy, two of the leading causes of blindness in adults. Obstruction of the interaction between vascular endothelial growth factor (VEGF) ligands and their cognate receptors (VEGFR) shows clinical benefit in the treatment of retinal diseases. However, effective suppression of VEGF signaling and achieving sustained ocular delivery present major unmet medical challenges. The FDA-approved drug aflibercept, an Fc-fused fusion protein chimera of VEGFR that acts as a VEGF decoy, provides a good starting point for VEGF neutralization, but requires further improvements. The affinity of aflibercept for VEGF ligands is equal to that of surface-bound receptors, limiting its potency as an antagonist. Moreover, although aflibercept binds VEGF-A with high affinity, it does not interact with VEGF-C isoforms, which are often upregulated in a compensatory mechanism that leads to drug resistance. Utilizing crystallographic data of VEGF/VEGFR complexes and molecular evolution platforms, we engineered the binding domains of VEGFR to isolate high-affinity aflibercept variants that cross-react with both VEGF-A and VEGF-C isoforms. We characterized the biophysical and functional properties of these engineered variants and demonstrated their therapeutic potential in rodent models of age-related macular degeneration. Overall, our results indicate that these novel antagonists could act as superior therapeutics for the treatment of retinal diseases.

596219 Predicting Antibody Developability from Molecular Simulations and Machine Learning

Abstract:

In the present study, we developed a general machine learning feature selection protocol to predict antibody stabilities. These features only require information from sequences and molecular structures of antibodies. In addition, the physically relevant features are easily interpretable and help to design experiments to understand the underlying mechanisms of antibody stabilities. This feature selection protocol avoids fitting data with hundreds of descriptors and is readily applicable to find predictive models for various antibody stabilities. We applied this protocol to predict the aggregation behaviors of monoclonal antibodies (mAbs). We have measured the accelerated aggregation rate of 21 therapeutic mAbs at high concentration (150 mg/mL). We found that using only three features, the accuracy of predicting the aggregation rate of the 21 mAbs reaches 97%. Models trained using structural features, rather than sequence features, correlated better with aggregation. A predictive equation for screening aggregation behaviors was proposed, facilitating early stage design. This framework can be extended to train other models that predict different physical properties.

600964 Self-Assembled Cgamp-Stingδtm Signaling Complex As a Bioinspired Platform for Cgamp Delivery

Abstract:

The stimulator of interferon (IFN) genes (STING) pathway constitutes a highly important part of immune responses against various cancers and infections. Consequently, administration of STING agonists such as cyclic GMP-AMP (cGAMP) has been identified as a promising approach to target these diseases. In cancer cells, STING signaling is frequently impaired by epigenetic silencing of STING; hence, conventional delivery of only its agonist cGAMP may be insufficient to trigger STING signaling. In this work, while expression of STING lacking the transmembrane (TM) domain is known to be unresponsive to STING agonists and is dominant negative when coexpressed with the full-length STING inside cells, we observed that the recombinant TM-deficient STING protein complexed with cGAMP could effectively trigger STING signaling when delivered *in vitro* and *in vivo*, including in STING-deficient cell lines. Thus, this bio-inspired method using TM-deficient STING may present a new and universally applicable platform for cGAMP delivery.

606504 Process Analytical Technologies for Recombinant Adeno-Associated Virus-Based Gene Therapy

Abstract:

Recent advancements in viral vector development for recombinant adeno-associated virus (rAAV)—the most commonly used viral vector for in vivo gene therapies—have facilitated a surge in product development, with up to 20 approved cell and gene therapies per year expected by 2025 [1].

Supporting this surge requires tackling two current challenges in viral vector manufacturing: the increased complexity of viral vectors compared to protein therapeutics, and the required production capacity. To overcome these challenges, it is critical to establish short process development timelines, and improve the process throughput by rigorous process modeling and control. Both depend strongly on the availability of fast, accurate, and reproducible analytical technologies for process characterization. While off-line technologies are useful for collecting data that can be used to construct data-driven models and perform lot-release assays, on- and at-line technologies are required for real-time feedback control systems. However, a comprehensive evaluation of analytical technologies for viral vectors and their suitability for use in process modeling and control has not been reported yet.

Here we review the current state of process analytical technologies for rAAV, for which control strategy design is in its infancy. Given the current developments, we discuss established and upcoming new technologies targeting rAAV-specific quality attributes [2,3]. We conclude with a discussion of control strategies that would be enabled by advances in analytical technologies, and suggestions guiding future investigations into these technologies.

References

[1] Gottlieb, S. (January 15, 2019). Statement from FDA Commissioner Scott Gottlieb, M.D. and Peter Marks, M.D., Ph.D., Director of the Center for Biologics Evaluation and Research on new policies to advance development of safe and effective cell and gene therapies. U. S. Food and Drug Administration. https://www.fda.gov/news-events/press-announcements/statement-fda-commissioner-scott-gottlieb-md-and-peter-marks-md-phd-director-center-biologics

[2] Wright, J. F. (2008). Manufacturing and characterizing AAV-based vectors for use in clinical studies. Gene Therapy, 15(11), 840–848. https://doi.org/10.1038/gt.2008.65

[3] Forsberg, N., Glover, C., Hughes, J.V., Jaluria, P., King, D., Madsen, J., Moore, A., Nilsson, J., Pettit, S., Pincus, S., Sargent, B., Schwartz, C. & TreDenick, T. L. (2018). Key Considerations in Gene Therapy Manufacturing for Commercialization. The Cell Culture Dish, Inc., Fort Collins, Colorado, https://cellculturedish.com/genetherapybook/

609817 Bacterial Expression of Moringa Oleifera Seed Proteins for Virus Filtration

Abstract:

The purpose of the work presented in this poster is to recombinantly produce Moringa oleifera proteins to understand their interactions with viral capsids in the context of water filtration and aerosol filtration for PPE. 785 million people around the world lack access to safe water (1). The leading cause of acute diarrhea in children are enteric viruses, which include rotavirus and norovirus (2). Recently, we showed that functionalizing a sand filter with crude water extract from Moringa seed can increase the virus removal efficiency by 7 orders of magnitude (3). A chitin binding protein, MOCBP, is thought to be responsible for the virus removal. An antimicrobial protein, MOCP, is also thought to affect the interactions. For both MOCBP and MOCP, each with the addition of a His tag, recombinant protein production was optimized in separate BL21 *E. coli* strains and subsequently purified using a cobalt resin column. Experiments were conducted using MS2 bacteriophage, a surrogate for human enteric viruses and respiratory viruses. Each type of purified protein was tested independently and in combination to understand the individual and synergistic effect with virus capsid. Better understanding of the protein-virus interactions can help us design better columns and PPE and hopefully extend this research to other plant antimicrobial proteins.

References:

1. UNICEF, WHO (2019) *Progress on household drinking water, sanitation and hygiene 2000-2017: Special focus on inequalities* Available at: https://washdata.org/reports.
2. WHO (2008) *Weekly epidemiological record* Available at: http://www.who.int/vaccines-documents/DocsPDF02/www698.pdf.
3. Samineni L, et al. (2019) 7 Log Virus Removal in a Simple Functionalized Sand Filter. *Environ Sci Technol*:12706–12714.

610707 Machine Learning Guides Combinatorial Protein Library Design

Abstract:

Engineered protein ligands are powerful tools for therapeutic and diagnostics. The medical community would greatly benefit from increased availability of agents which can accurately detect and reliably treat aggressive diseases such as pancreatic and breast cancer. Traditionally, antibodies have played a dominant role in the development of drugs and imaging probes. However, small engineered proteins (e.g. affibody, Gp2, and monobody) have multiple advantages over antibodies including superior biodistribution properties and remarkably inexpensive production. Hence, it is critical to improve our understanding of structural and functional traits that enable the strong, selective interactions required for clinical utility. In this study, we use high-throughput selection of extremely large collections of proteins ($>10^9$ unique proteins) to generate novel binders to a diverse group of clinically relevant targets. In doing so, an important challenge that we face is deciding which protein variations to include in our initial collection of proteins. If we consider a small protein that is only 50 amino acids long, there would be over 10^{65} unique variations possible! In order to predict which variations or mutations will provide for strong binding, highly selective proteins, we make use of several resources including deep sequencing data sets obtained from our directed evolution experiments, structural data from x-ray crystallography and cryo-EM, computational docking simulations in Rosetta, Bayesian phylogenetic analysis, and mutational stability analysis. These data are then used to guide a machine learning framework to more accurately determine which protein mutations to experimentally test for optimal binding properties. Collectively, this workflow provides a more efficient approach for discovering the next generation of clinical tools for detecting and treating some of the world's most deadly diseases.

610928 Allosteric Regulation of GPCRs Driven By Engineered Protein Scaffolds

Abstract:

Improper regulation and function of G protein-coupled receptors (GPCRs) is associated with several diseases and disorders such as diabetes, obesity, and neurological abnormalities. As the largest class of receptor proteins in the human body, there are substantial opportunities for improving health by correcting the behavior of GPCRs. In this project, novel protein ligands are engineered to selectively target and regulate GPCRs. We will overcome limitations of traditional ligands (e.g. monoclonal antibodies and small molecules) by using small protein scaffolds with diverse topologies to not only bind specific proteins, but unique conformations. Conformationally selective binding proteins provide the opportunity to allosterically regulate GPCRs through precise binding interfaces to enhance or alter protein activity, as well as facilitating structural studies within this class of challenging transmembrane proteins. The high affinity binders generated in this study have the potential to a make a profound impact on the availability of new therapeutics for pervasive diseases.

The small, stable engineered proteins used in this study (e.g. affibody, Gp2 and monobody) have multiple advantages over traditional antibodies not only due to their small size, but their diverse paratope geometries and conformations enable them to recognize epitopes that are inaccessible to bulky antibodies. Moreover, small proteins have excellent biodistribution properties making them well suited for cancer and disease diagnostic agents.

This project will: 1) Identify an efficient combination of protein scaffolds having multiple distinct paratopes which can bind a diverse collection of model GPCRs; 2) Discover conformationally selective binders from within populations of strategically mutated protein scaffolds; 3) Apply the lessons learned in generating binders to the model systems to design libraries capable of binding more challenging and exciting GPCRs. The protein ligands discovered in this body of work will provide incredible opportunities for the future of high-performance therapeutics and diagnostics.

594343 Analysis of Organophosphate Residue Detection on SOLID Surface Using QCM-D As BIO-Sensing Platform

Abstract:

Detection of residual organophosphates in food is of paramount importance in current scenario. The understanding of interaction of pesticides on chemically modified surface is the backbone of biosensor fabrication. Quartz crystal microbalance with dissipation (QCM-D) is one of the most efficient techniques to analyze the interaction of organic molecules with silane modified surfaces making it desirable bio-sensing platform.

In this report, we have studied the real time adsorption of pesticide phorate and malathion on silica coated quartz crystal of QCM-D. The interaction study was first monitored on gold coated and unmodified silica coated crystal as reference. Further self-assembled monolayer was created on silica crystal using n-propyltrimethoxy silane (PTMS), octyltrimethoxy silane (OTMS), N-(3-(trimethoxysilyl)propyl) aniline and diethoxymethyl phenyl silane. The interaction study was also monitored by using atomic force microscopy (AFM). Amount of pesticide adsorbed on surface was evaluated in terms of the mass and thickness change of quartz crystal which is calculated from frequency change using Sauerbrey equation. Low value of dissipation factor (ΔD) corresponds to stronger binding of pesticide which the result is obtained for OTMS modified surface. The overall results indicate that OTMS modified surface contributes mostly to the dissipation changes (ΔD) or acoustic ratio ($\Delta D/\Delta f$). This surface shows largest amount of pesticide adsorption both for phorate and malathion which is correlated with the more hydrophobicity of the surface confirmed by contact angle measurements. The diethoxymethyl phenyl silane modified silica crystal also shows significant mass change for phorate as the phenyl group facilitates interaction with sulfur of pesticide molecule. Further verification was done by Raman spectroscopy. This interaction study may be utilized to fabricate sensors for pesticide detection.

610188 Tuning the Cell-Free Protein Synthesis System for the Development of Highly Sensitive Biosensors

Abstract:

The cell-free protein synthesis (CFPS) system has emerged as a powerful platform for advancing the ability to study and expand the potential of applied biotechnology. The unique open environment of the CFPS system has allowed for an unprecedented level of freedom to control and optimize biological systems and focus production on compound of interest rather than microbial growth, making this ***in vitro platform optimal for sensing***. Many gene circuit sensors have been designed using the cell-free system platform, but most employ a highly robust bacteriophage T7 RNA polymerase (T7 RNAP). Unfortunately, using T7 RNAP for circuits puts a limit to how many ways the user can modulate the circuit because it does not require many elements to regulate its expression, which opens opportunities for circuit leak. Using the endogenous *Escherichia coli* RNA polymerase (*Ec* RNAP) can open another level of gene regulation for circuit design due to the vast number of sigma factors and cognate promoter sequences that can be chosen from, but they must be optimized to maximize the sensitivity and output signal of the sensor. Here we employ a highly sensitive and robust sigma factor to complete the endogenous RNA polymerase holoenzyme complex. To optimize further, a strong fluorescent protein is necessary to *acquire data from low concentration analyte or not fully purified samples*. Since the 70s, Green fluorescent protein (GFP) and its variants have been used as markers of gene and protein expression, localization, and interactions, along with sensing biomarkers and assessing the quality of protein translational systems. Recently, a new fluorescent protein was discovered and *has been touted to be the brightest green/yellow fluorescent protein yet to be described.* Since cell-free sensing experiments have yet to express this new fluorescent protein, in this study, we compared it to the commonly used sfGFP for the first time, *to find the brightest and most robust fluorescent protein* expressed in the *Escherichia coli* CFPS system. The fluorescent proteins were expressed and characterized by 1) pH tolerance, 2) reaction time and temperature, 3) tolerance to matrix effects from impure samples, 4) maturation time, and 5) brightness. By combining all of these advantages, the toolbox for genetic circuitry was greatly increased. This platform as a whole can be used to create sophisticated synthetic gene circuitry networks to increase the amount of data obtained from one sample source.

611000 Engineered Heterobivalent Protein Ligands Enable Early Detection of Disease

Abstract:

Early detection of aggressive diseases leads to better patient outcomes. Currently, a major limitation for the early diagnosis of pancreatic cancer is the inability to detect exceedingly low concentrations of reliable biomarkers associated with initial onset of the disease. In early stages of the disease, physiologically relevant concentrations of circulating cancer markers may be in the femtomolar range, which are currently significantly below any reliable limit of detection. Consequently, establishing a non-invasive method for detecting subtle changes in biomarker levels with high specificity would have a profound impact on the medical community's ability to successfully treat and cure pancreatic cancer patients. To address this, we develop a heterobivalent ligand based biosensor platform. Using a combination of computational modeling and high-throughput directed evolution, our tandem protein binders can be engineered to non-competitively interact with distinct epitopes of the biomarker of interest. These binding proteins coat the surface of a signal transduction interface which then can act as a molecular actuator that closes upon binding of its target. This conformational gating mechanism then initiates a chemical reaction to trigger signal amplification. In this way, the engineered protein domains form a molecular mechanism which converts the binding of a desired biomarker into a propagating signal. This biosensor platform provides an efficient method for quantifying biomarker levels in patients. Moreover, the ease of use makes it possible to monitor trends in biomarker levels across multiple time points which is expected to be a much more reliable approach at early stages of the disease compared to current methods.

612118 Foodborne Pathogen Detection: Identification of Protein Catalyzed Capture Agents Targeting E. *coli* O157:H7

Abstract:

The World Health Organization estimates 582 million illnesses occur annually due to foodborne pathogen infections. Culture based detection methods are considered the "gold-standard" for identifying pathogen contamination, but are laborious to perform and fail to detect contamination in real-time. While antibody based sensors exhibit excellent affinity and specificity their costs, cold-chain storage requirements, and batch-to–batch variation limit their widespread use (especially in developing regions). Peptides represent ideal antibody alternatives exhibiting improved storage and shelf-life, cost-effective and reproducible chemical synthesis, and the ability to incorporated unique unnatural chemistries. This research seeks to discover novel peptide recognition elements (PCCs) targeting E. *coli* O157:H7 by the Protein Catalyzed Capture method, for applications in detecting contamination in food samples. Here, a peptide based One Bead One Compound (OBOC) library is screened against synthetic peptide epitopes (SynEps) unique to E. *coli* O157:H7. Peptide recognition elements exhibiting nM affinity are identified by an *in situ* un-catalyzed "click" reaction resulting from entropically favorable interactions between a SynEp and library bead. Edman degradation enables sequencing of the library peptide responsible for target binding. Individual PCC affinity is then assessed quantitatively by multiplex immunoassay. Magnetic beads are then modified with lead PCCs and employed to separate E. *coli* from complex samples. Ultimately, promising PCC agents will be incorporated into various sensing platforms (such as electrochemical (EIS), visual (lateral flow), or optical (SPR)) to enable wide-spread food pathogen detection.

599460 Development and Characterization of an *E. coli*-based Vitamin C Biosensor

Abstract:

Vitamin C (ascorbate) deficiency is a global public health issue most prevalent in regions that lack the appropriate resources for testing and diagnosis. The high incidence of vitamin C deficiency in resource-limited settings creates a need for an inexpensive, equipment-free, and fast-responding ascorbate detection platform. A detection tool that leverages the natural sensing machinery of bacteria has the potential to meet all of these requirements. I developed two types of *Escherichia coli*-based biosensors: whole-cell and cell-free. The performance of the two sensors was compared to assess the initial feasibility of each system for field applications.

596885 Metabolic Engineering of Microbes for the Green Production of C2-C6 Diols.

Abstract:

C2–C6 diols are very important bulk chemicals which have been widely used in a range of areas, including solvents, fuels, polymers, cosmetics, and pharmaceuticals. Bioproduction of C2–C6 diols from renewable resources is a promising route for sustainable chemical industry as demonstrated by successful commercialization of biological processes of certain diols (e.g. 1,3-propanediol an 1,4-butanediol) in the past 20 years. However, biosynthesis of many other important C2–C6 diol isomers is highly challenging due to the lack of natural synthesis pathways. Recent advances in synthetic biology have enabled the de novo design of completely new pathways to non-natural molecules from renewable feedstocks. In this study, I will present our efforts in the design of new metabolic pathways for the production of different C2–C6 diols. Especially, I will focus on new metabolic pathways and metabolic engineering strategies for the production of 1,3-propanediol, 1,3-butanediol, and 1,5-pentanediol. Our efforts show that it is very promising to develop economically competitive processes for these diols in next few years.

609066 Genome-Scale Metabolic Modeling of *Clostridium thermocellum* for Omics Integration and Modular Cell Design

Abstract:

It is widely recognized that an important step towards addressing climate change is the reduction of carbon emissions through renewable manufacturing. We can accomplish lower carbon emissions in the production of fuels and chemicals by replacing petroleum-based processes for plant biomass conversion by microbial catalysts. The anaerobic thermophile *Clostridium thermocellum* is a promising bacterium for bioconversion due to its capability to efficiently degrade untreated lignocellulosic biomass. However, the complex metabolism of *C. thermocellum* is not fully understood, hindering metabolic engineering to achieve high titers, rates, and yields of targeted molecules. In this study, we developed an updated genome-scale metabolic model of *C. thermocellum* that accounts for recent metabolic findings, has improved prediction accuracy, and is standard-conformant to ensure easy reproducibility. We illustrated two applications of the developed model. We first formulated a multi-omics integration protocol and used it to understand redox metabolism and potential bottlenecks in biofuel (e.g., ethanol) production in *C. thermocellum*. Second, we used the metabolic model to design modular cells for efficient production of alcohols and esters with broad applications as flavors, fragrances, solvents, and fuels. The proposed designs not only feature intuitive push-and-pull metabolic engineering strategies, but also novel manipulations around important central metabolic branch-points. We anticipate the developed genome-scale metabolic model will provide a useful tool for system analysis of *C. thermocellum* metabolism to fundamentally understand its physiology and guide metabolic engineering strategies to rapidly generate modular production strains for effective biosynthesis of biofuels and biochemicals from lignocellulosic biomass.

611630 Designing Microbial Co-Culture Systems for Designer Ester Biosynthesis

Abstract:

Microbial whole-cell biocatalysts perform a series of biochemical reactions to produce useful chemicals from renewable feedstocks. To optimize the target chemical production, metabolic engineering often involves overexpression of multiple enzymes of the desirable production pathway that requires multiple levels of optimization such as gene expression, enzyme solubility, metabolic burden, and metabolite toxicity. Such an optimization is not an trivial engineering problem. To tackle this, we employ the “labor division” strategy by modularly engineering a synthetic microbial consortium for efficient biosynthesis of designer bioesters as a case study. Through characterization of single and co-cultures, we not only demonstrated enhanced ester production by synthetic microbial co-culture systems but also elucidated metabolic bottlenecks encountered by a single-species fermentation.

600289 Yeast Metabolic Engineering for Fermentation of Pectin Monomers

Abstract:

Pectin-rich biomass has garnered attention as an alternative biomass source. However, some monomers derived from pectin-rich biomass, namely D-galacturonic acid, L-arabinose, and D-xylose, are not fermentable by industrial microorganisms such as *Saccharomyces cerevisiae.* The purpose of this study is to develop a *S. cerevisiae* strain capable of fermenting the pectin monomers. Expressions of eight heterologous genes and deletion of two endogenous genes, all of which were successfully completed by Cas9-based *in vivo* assembly and integration strategy, allowed the consumption of pectin monomers as sole carbon sources. To facilitate the consumption of galacturonic acid, which had the most limitations, the use of a co-substrate was tested using various sugars. As a result, we found that arabinose and xylose allowed simultaneous consumption of galacturonic acid. Based on intracellular metabolite profiling, it was concluded that the five-carbon sugars partially resolve the metabolic bottleneck of galacturonic acid. These results provides a useful strain design strategy for the development of biorefinery based on citrus peel wastes and sugarcane bagasse with high pectin content.

610787 Leveraging the Hermes Transposon to Accelerate the Development of Nonconventional Yeast-Based Microbial Cell Factories

Abstract:

We broadened the usage of DNA transposon technology by demonstrating its capacity for the rapid creation of expression libraries for long biochemical pathways, which is beyond the classical application of building genome-scale knockout libraries in yeasts. This strategy efficiently leverages the readily available fine-tuning provided by the diverse transcriptional environment surrounding each random integration locus. We benchmark the transposon-mediated integration against the non-homologous end joining-mediated strategy. The latter strategy has been demonstrated for achieving pathway random integration in other yeasts but is associated with a high false-positive rate in the absence of a high throughput screening method.

Our key innovation of a nonreplicable circular DNA platform not only increased the possibility of identifying top-producing variants to 97%. Compared to the classical DNA transposition protocol, the design of a nonreplicable circular DNA skipped the step of counter-selection for plasmid removal and thus not only reduced the time required for the step of library creation from 10 to 5 days but also efficiently removed the 'transposition escapers', which undesirably represented almost 80% of the entire population as false positives. Using one endogenous product (i.e., shikimate) and one heterologous product (i.e., (*S*)-norcoclaurine) as examples, we presented a streamlined procedure to rapidly identify high-producing variants with titers significantly higher than the reported data in the literature. We selected *Scheffersomyces stipitis*, a representative nonconventional yeast as a demo, but the strategy can be generalized to other nonconventional yeasts. This new exploration of transposon technology, therefore, adds a highly versatile tool to accelerate the development of novel species as microbial cell factories for producing value-added chemicals.

605248 Aerobic Ester Production through *Escherichia coli-Clostridium* Co-Culture

Abstract:

Butanol toxicity and oxygen sensitivity are two of the major obstacles faced in traditional acetone-butanol-ethanol (ABE) fermentation of *Clostridium*. Butanol toxicity could be resolved by converting butanol into butyl butyrate and extracting butyl butyrate simultaneously. Introducing an aerobe or facultative anaerobe into *Clostridium* culture could enable the aerobic fermentation of *Clostridium* and hence avoid the costly anaerobic pretreatment.

In this study, we developed a synthetic co-culture system consisting of an engineered *E. coli* strain and a wild-type *Clostridium* strain to allow the production of butyl butyrate at aerobic condition. In this co-culture system, facultative anaerobic *E. coli* consumes oxygen in the peripheral environment to facilitate the growth of *Clostridium*. After oxygen is depleted, butanol and butyrate are produced by *Clostridium*. Meanwhile, butyl butyrate is produced from butanol and butyrate by the engineered *E. coli* and extracted from the aqueous cell culture by the hydrophobic extractant.

To realize the conversion of butanol and butyrate into butyl butyrate, two modules were designed to be co-expressed in *E. coli*, including a butyryl-CoA-generating module and an ester-producing module. The ester-producing module was constructed by expressing an alcohol acyltransferase from strawberry (SAAT). After testing a variety of enzymes that can convert butyrate into butyryl-CoA, acetoacetyl-CoA: acetate CoA transferase (*CtfAB*) from *C. acetobutylicum* was found to be the most efficient one. Mono-culturing the *CtfAB- and SAAT-expressing E. coli with butanol and butyrate fed led to the* production of 380 mg/L butyl butyrate within 24 hours, which is the highest titer of butyl butyrate that has been produced from *E. coli*. Moreover, 110 mg/L butyl butyrate was produced within 36 hours when the *CtfAB- and SAAT-expressing E. coli* was aerobically co-cultured with *Clostridium* using glucose as the sole carbon source. To our best knowledge, this is the first time that butyl butyrate was aerobically produced by an *E. coli-Clostridium* co-culture. In addition, the butyl butyrate titer achieved here is more than twice of what was obtained from the anaerobic mono-culture of an engineered *Clostridium* strain.

As a system harnessing both the genetic tractability of *E. coli* and superior acid- and alcohol-producing ability of *Clostridium*, this co-culture system is of great potential to improve the conventional *Clostridium* fermentation and produce high value-added chemicals.

605029 Synthesis and Application of Cell-Free Synthesized Therapeutic Human Proteins

Abstract:

Proteins are highly structural polymers, performing an enormous range of roles within organisms. Unlike synthetic polymers, protein has a significant hierarchical ordering and complexity, which hinders it from being produced and used in a therapeutic purpose. Recent advancement in synthetic biology has been applied to high yielding protein production where cell extract from clarified cell lysate is used as transcription and translational machinery for protein synthesis *ex vivo*. Cell-Free Protein Synthesis (CFPS) is an advanced biosynthetic method for protein production with the ease of system manipulation for protein engineering. It is also more flexible in protein manufacturing even when the conventional cell-based protein synthesis is not efficient (e.g. antibiotics and anticancer peptides that are unfavorable to the growth of host organisms).

In this study, human skin protein Filaggrin (FLG) was selected as a model therapeutic protein. Even though its role in restoring the skin barrier integrity in the skin of patients is significant, the use of FLG as a treatment has not been studied much yet. This study aimed to increase production yield (protein mass per volume) in prokaryotic CFPS. First, the yield has increased in CFPS by optimizing (1) bacteria codon usage (codon optimization, use of strain encoding rare codon tRNA), and (2) protein synthesis condition (temperature, reaction buffer). Next, the soluble protein and total protein were quantified in both optimized CFPS and cell-based protein synthesis. The CFPS showed great potential to produce therapeutic proteins.

608251 Modifying the Operator for More Dynamic Tunable Gene Expression in *Paraburkholderia Sacchari.*

Abstract:

Recent research endeavors have turned to generating useful chemicals from biological platforms as an environmentally responsible alternative to non-sustainable sources. One such chemical is polyhydroxybutyrate (PHB), which is a biodegradable plastic. One bacterium of interest for production of PHB is *Paraburkholderia sacchari*. This non-model bacterium is a relatively high efficiency producer of PHB and can utilize both C5 and C6 sugars, making it an attractive candidate for PHB production from cheap feedstocks that may have a mixture of carbon sources. In its native form, *P. sacchari* cannot efficiently produce enough PHB for the process to be economically feasible and a metabolically engineered strain would be needed to be industrially useful. However, there are limited synthetic biology tools available to attempt metabolic engineering strategies. Here we share our development of an inducible system for gene control in *P. sacchari*. It is commonly recognized that development of promoters, specifically inducible promoters that can be turned on and off, is one of the easiest and most effective ways to control gene expression. Therefore, we sought to develop an inducible system within *P. sacchari*. The *lac* system from *E. coli* with the promoter P_{LacUV5} has been shown to achieve gene expression in *P. sacchari*, however, the off state is extremely leaky. We modified the *lac* system by changing the operator sites surrounding the promoter region to the Oid sequence shown to most strongly bind the repressor protein. This resulted in a tighter off state when used to express green fluorescent protein (GFP). Growth in the presence of varying concentrations of the inducer molecule exhibited a graded response, therefore providing a tunable and useful tool for future metabolic engineering of *P. sacchari*. In addition to the inducible system, we also tested the Anderson promoter library for constitutive (always on) expression of GFP. We saw a range of expression which provides options for constitutive control of gene expression. Current and future work within our lab is focusing on incorporating these regulatory elements into explorations of engineering *P. sacchari* for more efficient PHB production on complex feedstocks. Our findings lay a foundation for further development of a bioproduction system for PHB production; a direction of research that will provide a sustainable, environmentally responsible, and economically beneficial method for creating biodegradable plastics.

609687 Regulation of Gene Expression in E. coli Using Theophylline Sensing Riboswitches

Abstract:

Fine-tuned regulation of gene expression is highly sought after for biological engineering applications. One manner in which this ability has been achieved is through the isolation, creation, and implementation of RNA riboswitches. Riboswitches are composed of two functional domains: the aptamer and the functional domain. A functional domain alters into a different secondary conformational state as a result of the aptamer binding to its respective metabolite, which can either block or enable transcription or translation. The theophylline aptamer, one of the most studied RNA aptamers, has been attached to a switch in order to achieve gene regulation with in *Escherichia coli.* In this work, we have studied principles guiding riboswitch function in *E. coli* through use of *trans*-acting switches or *cis*-acting theophylline hammerhead ribozymes placed in the 5′ untranslated region to control expression of the *gfp* gene. These riboswitch constructs have then been analyzed for fold activation of the target gene when in the presence and absence of theophylline. Addition of spacers or anti-ribosomal binding site stems to the riboswitches greatly affected the expression of *gfp* when cells were cultured in the presence of theophylline.

601365 Acetate Switch in *Methanococcus Maripaludis* S2

Abstract:

Acetate switch is a phenomenon which involves the transition from acetogenesis to acetotrophy during the growth phase of a microorganism. This behaviour has been reported in *Escherichia coli* and several other heterotrophs. In the early growth phase, these microorganisms dissimilate (produce and secrete) acetate when the carbon flux into the cells exceeds the capacity of their central pathways or when their tricarboxylic acid cycles do not operate fully due to limited oxygen. This dissimilation allows them to generate ATP and reduce acetyl-CoA accumulation. In the late growth phase, they switch to acetate assimilation (uptake and utilisation) to adapt to the scarcity of carbon sources. To date, all reported cases have involved heterotrophs and the tricarboxylic acid cycle.

In this work, we investigated the acetate metabolism in the methanogenic autotroph *Methanococcus maripaludis* S2. This microorganism is a mesophilic, hydrogenotrophic archaeon, which uses the Wolfe cycle for its anaerobic respiration. While acetate assimilation by *M. maripaludis* is known, its ability to dissimilate acetate has not been established thus far. We performed batch culture cultivation of *M. maripaludis* in minimal medium with carbon dioxide as the sole carbon source, either ammonium or dinitrogen as the sole nitrogen sources. We observed that *M. maripaludis* dissimilated acetate in the early growth phase and assimilated it back in the late growth phase. Furthermore, the acetate concentration was generally higher in the dinitrogen-grown cultures. These results provided evidence that, (1) *M. maripaludis* can dissimilate acetate, and (2) *M. maripaludis* exhibited an acetate concentration profile similar to the acetate switch profiles in *E. coli* and other heterotrophs. This is the first evidence that autotrophs can display an acetate switch behaviour.

We further examined the genome of *M. maripaludis* S2 and determined that the protein MMP0253 may catalyse acetate formation from acetyl-CoA. This protein was found to have similar primary and secondary structures to the ADP-forming acetyl-CoA synthetase in other archaea. We cloned MMP0253 in *E. coli* strain Rosetta (DE3) using pNIC28-Bsa4 as the vector and found that the recombinant protein indeed catalysed the acetate formation from acetyl-CoA.

609226 A Rab Escort Protein in Yeast Regulates the MAPK Pathway That Controls Filamentous Growth

Abstract:

Signal transduction pathways control the response to stress and cell differentiation into specialized cell types. In yeast and other fungal species, signaling pathways control filamentous growth, which in pathogens is critical for host–cell attachment, invasion into tissues, and virulence. As commonly seen during cell differentiation, multiple pathways operate in an integrated network to regulate filamentous growth. A MAPK pathway, called the fMAPK pathway, is among the pathways that regulate filamentous growth. fMAPK pathway shares a subset of common components with two other MAPK pathways in yeast (mating and HOG pathways), including the Rho-type GTPase Cdc42, PAK Ste20, and MAPKKK Ste11. A key gap in understanding fMAPK pathway specification is connecting upstream components to downstream pathway-specific kinases, like Ste7 and Kss1. To address this gap, a genome-wide screen (of 4416 yeast genes) was performed. The screen led to identifying genes that, when overexpressed, induce a growth reporter (*FUS1-HIS3*) that responds to ERK-type MAPK pathways (Mating/fMAPK) but not p38-type MAPK pathways (HOG) in yeast. The screen analysis and target validation identified new regulators of ERK-type pathways that were not previously implicated in MAPK signaling, including *MRS6*, a Rab escort protein. Mrs6 plays an established role in protein trafficking and signaling, but has not been already connected to MAPK pathway regulation. We found out that Mrs6 overexpression stimulated phosphorylation of the MAP kinase that regulates fMAPK pathway, Kss1, but not the phosphorylation of the MAP kinases that regulate the mating or HOG pathways, indicating that it may be a specific regulator of the fMAPK pathway. Mrs6 interacted with the PAK kinase Ste20 and MAPKK Ste7 by two-hybrid analysis. The interaction between Mrs6 and Ste7 may contribute to ERK-type (*STE20->STE11->STE7*) signaling but not p38-type (*STE20->STE11->PBS2*) signaling. In mammals, loss of the *MRS6* homolog REP1/CHM leads to choroideremia, which causes progressive vision loss. Intriguingly, REP1/CHM regulates EGFR, which is a major regulator of the Grb-SOS-RAS-MEK-ERK pathway. Given that EGFR also signals through RAS-MEK-ERK, our screen may have identified a new regulator of ERK-type MAPK pathways.

610825 Genetic and Regulatory Basis of Temporal Variation in Complex Traits

Abstract:

Proliferation and metabolism are complex microbial traits controlled by large numbers of genetic loci. Exploiting the power of genome editing technology to control phenotype requires detailed models of how growth traits emerge from the genome. To this end, we have developed technologies to map the genotype-to-phenotype relationship with single-nucleotide resolution in the budding yeast *Saccharomyces cerevisiae,* using a large, inbred population of over 18,000 genotyped diploid strains. By combining this platform with high-throughput automated phenotyping, we can now measure growth rate across the mapping panel with 3-minute time resolution and capture the genetic basis of the temporal synergies and tradeoffs occurring throughout microbial culture.

In combination with allele-specific expression analyses, detailed time-resolved genetic mapping revealed the widespread contributions of protein-coding, synonymous, and both *cis*- and *trans*-acting regulatory variation to phenotype. These effects had pronounced temporal patterns, with some natural variants exhibiting antagonism across different metabolic phases of culture. *Cis*-acting regulatory variation was pervasive in metabolism and many of these variants made important contributions to phenotype. Enzymes and regulatory proteins each played key roles and opposing *cis*-regulatory effects at isozymes suggested kinetic tuning through differential expression. Understanding the molecular regulatory basis of complex traits emerging from natural genetic variation will be essential to achieve the potential of genome reading and writing in industry and medicine.

603202 Regulation of Pancreatic Regenerating Protein (Reg) Expression By Microrna-7

Abstract:

Regenerating islet-derived (Reg) proteins display pro-proliferative, anti-apoptotic and differentiation-inducing actions in normal and disease states. However, little is known about the regulation of their expression. We tested our hypothesis that Reg expression is influenced by microRNAs. *In silico* analysis identified microRNA-7 (mir7) as a candidate targeting the murine *Reg1* gene in pancreatic acinar and islet β-cells. The work opens avenues for discovering additional microRNAs influencing the expression of Reg members, thereby expanding our knowledge of the biology of Reg proteins and their role(s) in diseases such as diabetes and pancreatitis.

Rodent β-cell and acinar cell lines (ATCC, Manassas, VA) and HEK293 cells were cultured in DMEM medium with 10% fetal bovine serum (FBS) at 37 °C and 5% CO_2. Murine pancreatic islets and exocrine tissue were isolated. The levels of *reg1* and *mir7*were probed by quantitative PCR (qPCR) or immunostaining. The 3' untranslated region (3'UTR) of the *Reg1* gene as well as the mir7 sequence were cloned in expression vectors upstream of the luciferase (Luc-3'UTR-reg1) and green fluorescent protein (GFP), respectively.

Lower expression of Reg1 was observed by immunostaining and qPCR in β-cells and primary islets compared to exocrine cells. This profile was inverted for the expression of mir7. Suppression of mir7 in beta-cells, which natively exhibit high mir7 levels, resulted in an increase in reg1. Overexpression of mir7 had the opposite effect on reg1. When the Luc-3'UTR-reg1 construct was delivered with a mir7 overexpression vector to exocrine cells, the luciferase activity was lower. This finding indicated that the microRNA directly interacts with the 3'UTR of reg1. Yet, mir7 did not affect the expression of the human orthologs of murine reg1, REG1A and REG1B. Current efforts focus on the identification of other microRNA moieties that directly target human Reg genes.

This is the first account of microRNA modulation of any Reg protein member directly, and our data warrant further studies. Given the link of Reg proteins to diabetes and various cancer types, including pancreatic ductal adenocarcinoma, the identification of microRNAs regulating Reg levels has the potential to contribute to the development of pertinent therapeutics and to extend our understanding of the pathophysiological role(s) of Reg.

603564 Bioorthogonal Noncanonical Amino Acid Tagging (BONCAT) Enables Cell-Selective Proteomic Profiling in Complex Biological Systems

Abstract:

Cells within biological systems are constantly adjusting their protein synthesis in response to various environmental changes. To study the rapid cellular regulations in complex biological systems, global proteomic profiling provides important information on system-level regulations, yet physiological properties characteristic of individual cellular subpopulations could be hidden under the characterization. Instead, cell-selective proteomic profiling allows researchers to reveal the heterogeneities in biological systems with phenotypically and even genetically distinct subpopulations under different microenvironments.

The development of bioorthogonal noncanonical amino acid tagging (BONCAT) enables proteomic profiling with resolution in both space and time. BONCAT's initial role is protein labeling with temporal resolution via pulse-addition of noncanonical amino acid, which could be recognized by endogenous aminoacyl tRNA-synthetase (aaRS), into systems of interest; later on, mutant aaRSs are identified through mutant synthetase library screening, which allows for efficient incorporation of various types of noncanonical amino acids that could hardly be activated by endogenous machineries. The identification and exploitation of mutant aaRSs allow sensitive cellular selectivity during protein labeling. With unprecedented spatiotemporal resolution of BONCAT, and the advancement in high-resolution mass spectrometry and computational algorithms, BONCAT is a powerful technique for selective proteomic profiling to study physiological regulations in a wide range of complex biological systems.

We first applied the BONCAT method to cell-selective proteomic profiling in *Pseudomonas aeruginosa* biofilms. We targeted iron-starved subpopulation in biofilms and compared its proteomic profile with that of the entire system. Key gene and pathway regulations in the subpopulation are found through the analysis of the proteomic data, which suggest that iron-starved cells shift their priority towards housing keeping pathways, adapt an energy- and resources-saving mode to cope with their harsh local environmental conditions, and get prepared to disperse for better survival. Analysis of poorly studied proteins highly upregulated in the subpopulation led to the discovery of previously uncharacterized proteins that are potentially related to iron acquisition.

Next, we applied BONCAT to the study of physiological regulations in *Bacillus subtilis* K-state subpopulation. A subset of *B. subtilis* cells, typically 10% - 20% of the entire population, enter K-state in a stochastic manner. With the low level of K-state entry rate and high randomness, we challenged BONCAT to specifically capture gene and pathway regulations in K-state cells and compared the proteomic profiling with that of the entire population. Regardless of the difficulties of selective protein labeling inherent in the system, our results indicate that BONCAT has high specificity and resolution in proteomic profiling for minor subpopulations and proteins with low overall absolute abundance. We found multiple pathways and genes characteristic of K-state regulated differentially from the entire population, either significantly up- or down-regulated. Proteins that are uncharacterized or previously known for functions irrelevant of K-state are highly abundant in the subpopulation, providing new insight toward their alternative functions critical for K-state cells and future investigation directions of K-state study.

608047 Engineering the Bioavailability of Sulfur to Control Iron Redox States in *Acidithiobacillus Ferrooxidans*

Abstract:

The biomining microorganism, *Acidithiobacillus ferrooxidans,* is an acidophilic chemolithoautotroph with three major metabolic pathways to utilize iron, sulfur, and hydrogen as electron donors. Biotechnological application of this bacterium, as well as other bacteria found with it in natural consortia, is important for the global industrial biomining of approximately 5% of gold and 15% of copper[1]. The iron and sulfur metabolisms have been well-studied as the interaction of *A. ferrooxidans* with metal sulfides and other sources of metal, such as e-waste, leads to the dissolution of metal cations. However, as most studies with this organism investigate growth with a single energy source, we have exploited the ability of *A. ferrooxidans* to oxidize iron, sulfur, and pyrite simultaneously to enhance the corrosion of stainless steels. This systems-level approach has allowed us to demonstrate that the optimization of substrate oxidation via the manipulation of cultivation conditions combined with genetic engineering can be used to for the microbially influenced corrosion of complex metal containing matrices[2].

Despite the favorable bioenergetics associated with sulfur oxidation compared to ferrous iron, the oxidation of sulfur involves an extended lag phase where *A. ferrooxidans* produces the necessary extracellular polymeric substances and proteins needed to attach to the hydrophobic surface of the sulfur particle. As a result of this lag phase, when both ferrous iron and sulfur are present in solution, iron oxidation behavior precedes sulfur oxidation for iron-adapted *A. ferrooxidans* cells, leading to the long-held conclusion that *A. ferrooxidans* is an iron-preferred species. We discovered a class of biocompatible surfactants, lignosulfonates, that could be used to formulate sulfur to have more hydrophilic properties and disperse into aqueous solution.

With this novel dispersed sulfur formulation, we achieved rapid utilization of sulfur indicating that there was no substrate preference for iron in *A. ferrooxidans,* if the provided sulfur was already bioavailable[3]. The observed iron oxidation behavior in the bacteria was concentration dependent on dispersed sulfur. At higher concentrations of dispersed sulfur, the iron oxidation rates were found to be slower until at concentrations above 5 g/L, ferrous iron oxidation did not reach completion and was replaced with ferric iron reduction. We investigated this iron reduction behavior further and found that aerobic reduction of ferric iron was not dependent on pH if the metabolic state of *A. ferrooxidans* was such that the downhill iron oxidation pathway was inactive. Therefore, by manipulating the bioavailability of sulfur, the interplay between the iron and sulfur metabolisms within *A. ferrooxidans* enable the tunable control of iron redox states. This new level of control can be used to manipulate the performance of the cells for future biotechnology applications.

1 Banerjee, I., Burrell, B., Reed, C., West, A. C. & Banta, S. Metals and minerals as a biotechnology feedstock: engineering biomining microbiology for bioenergy applications. *Curr Opin Biotechnol* **45**, 144-155, doi:10.1016/j.copbio.2017.03.009 (2017).

2 Inaba, Y., West, A. C. & Banta, S. Enhanced microbial corrosion by *Acidithiobacillus ferrooxidans* through the manipulation of substrate oxidation and genetic engineering. (Submitted).

3 Inaba, Y., West, A. C. & Banta, S. Bioavailability of elemental sulfur drives substrate utilization behavior in *Acidithiobacillus ferrooxidans.* (In preparation).

610856 Persulfide and Polysulfide-Induced Posttranslational Protein Modifications in Oleaginous *Rhodococcus Jostii* RHA1

Abstract:

Persulfides and polysulfides are biosynthesized endogenously and reported to be particularly nucleophilic and reducing. The formation of posttranslational modifications (PTM) on protein thiols of these compounds may play an important role in redox signaling, regulation, mitochondrial bioenergetics, etc. Oleaginous bacteria *Rhodococcus sp.* is one of the promising candidates for lignocellulosic biomass valorization to lipid-based biofuel. It was previously reported that another level of lipogenesis regulation through redox-mediated thiol modifications. On the other hand, the catabolism of lignin and lignin-derived aromatics will induce cellular oxidative stress response as well. In this study, we aim to globally profile persulfide or polysulfide-modifications on protein thiols using mass spectrometry in *Rhodococcus jostii* RHA1. The potential landscape of sulfur-containing PTMs and their regulation may shed new insights into the regulation relevant to lipogenesis.

603886 A Mathematical Model of the Glomerular Filtration Barrier in Diabetic Kidney Disease

Abstract:

A Mathematical Model of the Glomerular Filtration Barrier in Diabetic Kidney Disease

The glomerulus has been determined to be the central tissue structure damaged within the kidney in diabetic kidney disease (DKD)(1). Due to the difficulty in acquiring experimental data regarding the progression of DKD within the glomerulus, mathematical modeling has arisen as a tool for tracking and predicting the progression of damage. There are several methods and hallmarks of damage within diabetic kidney disease, the most physically visible of which is the development of albuminuria or the leaking of the protein albumin into the urine. This development is partially due to damage caused to the glomerular filtration barrier (GFB). This barrier is composed of several layers, including the podocytes on the exterior of the glomerulus, the glomerular basement membrane, endothelial cells, and the endothelial glycocalyx. The podocyte has already been extensively studied and successfully modeled (2, 3). Here we focus on the GFB.

All components of this barrier play an integral role in blocking the passage of albumin. The endothelial fenestrations provide a hydraulic barrier, the glomerular basement membrane functions as a multi-pore mesh blocking molecules above a certain size, and the glycocalyx functions as a charge barrier against polar compounds (4). Damage to this barrier is integral in the development of albuminuria; however, the difficulty in gathering data on this barrier has resulted in a piecemeal approach to determining the progression of the diabetes associated damage.

In this work, a mathematical model is developed to track concentration, transport, and production of several damaging chemicals within diabetic kidney disease and their interactions with the GFB. Additionally, this model tracks the damage resulting from this dysregulated biochemical networks in the GFB.

References:

1. Schlondorff D. Putting the glomerulus back together: per aspera ad astra ("a rough road leads to the stars"). Kidney Int. 2014;85(5):991-8.
2. Pilvankar MR, Higgins MA, Ford Versypt AN. Mathematical model for glucose dependence of the local renin-angiotensin system in podocytes. Bull Math Biol. 2018;80(4):880-905.
3. Pilvankar MR, Yong HL, Ford Versypt AN. A glucose-dependent pharmacokinetic/pharmacodynamic model of ACE inhibition in kidney cells. Processes. 2019;7(3):131.
4. Smithies O. Why the kidney glomerulus does not clog: a gel permeation/diffusion hypothesis of renal function. Proc Natl Acad Sci USA. 2003;100(7):4108-13.

604986 Computational Modeling for Designing a Device Harboring Optogenetically Engineered Pancreatic Beta-Cells

Abstract:
Diabetes mellitus is linked to the damage or ablation of pancreatic beta-cells due to autoimmunity (type 1 diabetes) or insulin resistance (type 2 diabetes). Over 30 million people in the US are afflicted by diabetes. To this end, the use of engineered beta-cells with an increased specific rate of glucose-stimulated insulin secretion (GSIS) could aid in reconstituting blood glucose homeostasis. We implemented a drug-free approach based on optogenetics for enhancing GSIS in pancreatic β-cell through expression of a photoactivatable adenylyl cyclase (PAC) [1][2]. In this work, we employed computational modeling to design and optimize an implantable device loaded with beta-cells engineered for optogenetic modulation of their GSIS.

A reaction-diffusion model was built for a disc- or cylinder-shaped construct incorporating the LED source for photostimulated augmentation of GSIS [1][2]. The physical and optical properties of the construct were considered. The production or consumption rate of pertinent species such as glucose and oxygen were modeled with Michaelis-Menten kinetics [3]. To calculate the distribution of oxygen within the scaffold, boundary conditions were applied around the hydrogel and a no-flux condition was enforced in the area adjacent to the LED. Various numbers of cells were placed evenly in the scaffold as 100-150 µm clusters or pseudoislets (PIs). Almost 30% of PIs experienced hypoxia prompting the introduction of channels to alleviate this issues. Indeed, substantial improvement in oxygen distribution and minimal hypoxia within the hydrogel were observed when channels were positioned symmetrically around the vertical axis. The insulin flux in the current design was calculated based on the insulin secretion rate of PAC-expressing beta-cells with or without stimulation with light under different glucose concentrations.

This framework will help gain deeper insights for the cost-effective design of hydrogel constructs with appropriate geometries and cell seeding densities without compromising viability, optogenetic stimulation and function. Our approach is applicable to a range of emerging technologies capitalizing on advances in the field of optogenetics.

Acknowledgements

This work was supported by a grant (CBET-1951104) from the National Science Foundation.

References

1. Zhang, E. S. Tzanakakis, Amelioration of Diabetes in a Murine Model upon Transplantation of Pancreatic β-Cells with Optogenetic Control of Cyclic Adenosine Monophosphate, *ACS Synth. Biol.*8(10): 2248-55, 2019.
2. Zhang, E.S. Tzanakakis, Optogenetic regulation of insulin secretion in pancreatic β-cells, *Sci. Rep.*7(1): 9357, 2017.
3. Wu, et al., Oxygen transport and stem cell aggregation in stirred-suspension bioreactor cultures, *PLoS One* 9(7): e102486, 2014.

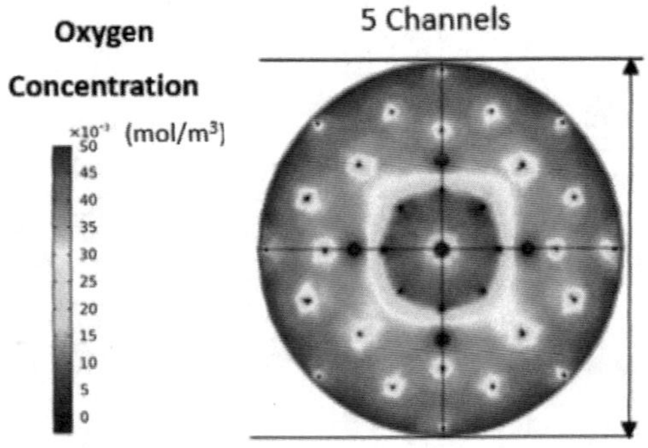

610320 Investigation of Incidents and Trends of Antimicrobial Resistance in Foodborne Pathogens in Eight Countries from Historical Sample Data

Abstract:

Antimicrobial resistance (AMR) is increasingly becoming a threat to global public health[1]. AMR-related infections limit the effectiveness of lifesaving antibiotic therapy necessary for the treatment of infectious disease and limit the success of advanced surgical procedures like organ transplants [2,3]. This results in millions of illnesses and deaths worldwide. For example, the US Centers for Disease Control and Prevention (CDC) estimated that AMR bacteria account for over 2 million illnesses and at least 23,000 deaths in the US every year [4]. Similarly, it is estimated that 426,000 infections and 33,000 deaths from antimicrobial resistant microorganisms occur in the European Union every year [5]. Furthermore, AMR places a considerable medical and financial burden on global societies and healthcare systems as a whole [1,3,6–8]. It is therefore essential to develop a better understanding of the cause and spread of AMR around the world.

In response to the need for further research into the spread of resistant foodborne pathogens and illness, the NCBI Pathogen Detection Isolates Browser (NPDIB) was created to integrate bacterial pathogen genomic sequences originating in food, patients, and environmental sources [9]. It analyzes samples from around the world and compares their genomic sequences to others already in the database. When AMR foodborne pathogens access humans, the AMR genes may be passed from the pathogens to human cells via horizontal gene transfer (HGT) through mechanisms of transformation, transduction, or conjugation[10]. This is implied by a study in which 13,514 high-confidence HGT genes were found in 308 human microbes [11]. The development of antibiotic resistance in foodborne bacteria and in HGT can significantly endanger human health. NPDIB clusters identify related sequences to uncover potential food contamination outbreaks and resistance genes. Through the database, people can search for foodborne pathogen isolates and identify pathogens with particular resistance genes. In previous literature, the database was typically only used to compare the DNA from detected pathogens to those in the database or to provide data for a historical analysis of resistance in specific pathogens [12,13]. Little research, however, has been done to examine specific AMR genes over large geographic areas.

To examine the rise and spread of antimicrobial resistance around the world, our study performs a multivariate statistical analysis of antimicrobial resistance gene data from eight different countries: the US, the UK, China, Brazil, Mexico, Canada, Australia, and South Africa. Multi-dimensional data points were projected onto a two-dimensional plane using principal component analysis and organized into a dendrogram utilizing hierarchical clustering to identify significant AMR genes and pathogens.

Given each sample's large number of dimensions, principal component analysis (PCA) was applied to each country's data to visualize the complex data matrices into two-dimensional plots, providing a clear image of the data points and their relationships. Utilizing dimensionality reduction, PCA created new individual dimensions that could retain the most information in the data matrix. These new dimensional variables, called principal components, were orthogonal vectors that represented different levels of variance in the data. In this way, the key highly occurring pathogens in each of the eight countries were identified using PCA and hierarchical clustering.

Certain outlier genes/pathogens were typically involved in high occurrences of antimicrobial resistance, and could be used to identify trends in antimicrobial resistances in the future. Statistical analysis of the data identified: (1) tet(A), aph(3")-Ib,aph(6)-Id,blaEC,blaTEM-1,qacEdelta1,sul1,sul2,and aadA1 as the nine most common AMR genes among the studied countries; (2)Salmonella enterica and E.coli and Shigella as the most common AMR foodborne pathogens; and (3) chicken as the most prevalent meat carrier of antimicrobial resistance. Our study also shows that the overall number of reported antimicrobial resistance cases in foodborne pathogens is generally rising. One potential contributing factor is the increasing antimicrobial usage in the growing livestock industry.

The large number of highly occurring AMR genes and the wide dissemination of many of these genes implies a potential rise of antimicrobial resistance. The spread of many highly occurring genes to numerous geographically distant countries with varying antimicrobial policies highlights the role of the global livestock trade in the transfer of AMR genes and foodborne pathogens. The identification of the nine globally significant highly-occurring genes, highly-occurring pathogens, and most prevalent meat carrier offer valuable insight for future research and development of new solutions for antimicrobial resistance.

References

1. Velez,R.;Sloand,E.Combating Antibiotic Resistance,mitigating future threats and ongoing initiatives. J.Clin. Nurs. 2016, 25, 1886–1889. [CrossRef] [PubMed]
2. Michael, G.B.; Freitag, C.; Wendlandt, S.; Eidam, C.; Feßler, A.T.; Lopes, G.V.; Kadlec, K.; Schwarz, S. Emerging issues in antimicrobial resistance of bacteria from food-producing animals. Futur. Microbiol. 2015, 10, 427–443. [CrossRef] [PubMed]
3. Davies, J.; Davies, D. Origins and Evolution of Antibiotic Resistance. Microbiol. Mol. Boil. Rev. 2010, 74, 417–433. [CrossRef] [PubMed]
4. Solomon, S.L.; Oliver, K.B. Antibiotic resistance threats in the United States: Stepping back from the brink. Am. Fam. Phys. 2014, 89, 938–941.
5. Friedrich, A.W. Control of hospital acquired infections and antimicrobial resistance in Europe: The way to go. Wien. Med. Wochenschr. 2019, 169 (Suppl. 1), 25–30. [CrossRef] [PubMed]
6. Roberts, R.R.; Hota, B.; Ahmad, I.; Ii, R.D.S.; Foster, S.D.; Abbasi, F.; Schabowski, S.; Kampe, L.M.; Ciavarella, G.G.; Supino, M.; et al. Hospital and Societal Costs of Antimicrobial-Resistant Infections in a Chicago Teaching Hospital: Implications for Antibiotic Stewardship. Clin. Infect. Dis. 2009, 49, 1175–1184. [CrossRef] [PubMed]
7. Lekshmi, M.; Ammini, P.; Kumar, S.; Varela, M.F. The Food Production Environment and the Development of Antimicrobial Resistance in Human Pathogens of Animal Origin. Microorganisms 2017, 5, 11. [CrossRef] [PubMed]
8. Van Boeckel, T.P.; Brower, C.; Gilbert, M.; Grenfell, B.T.; Levin, S.A.; Robinson, T.P.; Teillant, A.; Laxminarayan, R. Global trends in antimicrobial use in food animals. Proc. Natl. Acad. Sci. USA 2015, 112, 5649–5654. [CrossRef] [PubMed]
9. Delgado-Suárez, E.J.; Selem-Mójica, N.; Ortiz-Lopez, R.; Gebreyes, W.A.; Allard, M.W.; Barona-Gómez, F.; Rubio-Lozano, M.S. Whole genome sequencing reveals widespread distribution of typhoidal toxin genes and VirB/D4 plasmids in bovine-associated nontyphoidal Salmonella. Sci. Rep. 2018, 8, 9864. [CrossRef] [PubMed]
10. Davison, J. Genetic Exchange between Bacteria in the Environment. Plasmid 1999, 42, 73–91. [CrossRef] [PubMed]
11. Liu,L.;Chen,X.;Skogerbø,G.;Zhang,P.;Chen,R.;He,S.;Huang,DW.The Human Microbiome: A hotspot of microbial horizontal gene transfer. Genomics 2012, 100, 265–270. [CrossRef] [PubMed]
12. Ceric, O.; Tyson, G.H.; Goodman, L.B.; Mitchell, P.K.; Zhang, Y.; Prarat, M.; Cui, J.; Peak, L.; Scaria, J.; Antony, L.; et al. Enhancing the one health initiative by using whole genome

sequencing to monitor antimicrobial resistance of animal pathogens: Vet-LIRN collaborative project with veterinary diagnostic laboratories in United States and Canada. BMC Veter-Res. 2019, 15, 130. [CrossRef] [PubMed]
13. McMillan, E.A.; Gupta, S.K.; Williams, L.E.; Jové, T.; Hiott, L.M.; Woodley, T.A.; Barrett, J.B.; Jackson, C.R.; Wasilenko, J.L.; Simmons, M.; et al. Antimicrobial Resistance Genes, Cassettes, and Plasmids Present in Salmonella enterica Associated with United States Food Animals. Front. Microbiol. 2019, 10, 832. [CrossRef] [PubMed]

601517 Deep Learning-Aided Intelligent Flow Cytometry for Label-Free Cell Detection

Abstract:

Recent efforts are being directed to developing various label-free methods for cell detection and analysis. Besides offering simplicity and low-cost operation, a label-free method avoids any adverse effects of the staining reagents on cellular viability and functions. Nevertheless, the scope of the label-free techniques developed to date remains limited. Moreover, most of these techniques lack adequate throughput or precision. This work introduces a deep learning-leveraged label-free approach that could potentially enable cell detection and sorting at precision and throughput comparable to conventional flow cytometry. Unlike other existing label-free methods, this approach detects a cell by its autofluorescence measured using a multicolor flow cytometer. Analyses provided here indicate that autofluorescence may obscure sufficient information to discriminate different cell types. Supervised deep learning could be employed to analyze features hidden in the single-cell autofluorescence and infer cell types by mapping those features. This approach is demonstrated by classifying heterogeneous mixtures of human endothelial and breast carcinoma cells. The integration of the approach with conventional flow cytometry may enable intelligent systems for data-driven cell detection based on single-cell autofluorescence.

607999 Experimental Design Algorithm for Efficient Optimization of Culture Media

Abstract:

Design and analysis of culture media for bioprocesses often involves the use of many time-consuming experiments. However, because of the diversity of metabolites needed in many industrially relevant bioprocesses, and their nonlinear effects, design optimization is especially time-consuming and difficult to perform. This difficulty can be managed by combining a surrogate model with an optimization method to select response-optimal design parameters, often called response surface methods. Here, we use a radial basis function to learn metabolite-response interactions from benchtop scale experiments using C2C12 cells. This approximate model then feeds into a hybrid global/local stochastic optimization method to iteratively suggest optimal metabolite concentrations from a user-defined list of metabolites and objective function. This algorithm was used to design more efficient proliferation media for C2C12s in fewer experiments than traditional polynomial response surface methods, so can be used to reduce the cost of benchtop experiments in combinatorally complex experimental design problems.

596867 Whey Proteins and Whey Derivatives in the Production of Emulsions and Biodegradable Packaging Materials.

Abstract:

The environmental footprint of plastics has acquired a critical relevance in the field of food packaging. Bio-based materials are expected to mitigate the ecological impact and the dependence on fossil fuels of the food industry by enabling more sustainable packaging solutions. Biopolymers present an opportunity for the use of new, high-performance green materials to replace conventional non-biodegradable petroleum-based packaging counterparts. This has driven the search for alternative biopolymer sources and processing technologies with a special focus on the use of food manufacturing by-products. In this context, whey derivatives stand out due to the vast production of whey in the dairy industry, and the features of whey proteins as hydrocolloids, film-forming polymers and surfactant agents. Whey proteins are remarkable constituents of edible films and coatings, and stabilizers of emulsion-based packaging. Indeed, whey protein-based materials have been engineered to provide active properties to food packaging systems, such as activity against foodborne pathogens, and oxygen scavenging capacity. Moreover, whey has been used as a substrate in the microorganism-mediated synthesis of other biodegradable polymer materials with potential applications in packaging. This review discusses the main features and uses of whey proteins and whey derivatives in the development of green materials for food packaging applications, either as polymer matrices or as bioprocess substrates. Recent advances and current challenges on strategies to enhance the barrier and mechanical properties of whey protein-based films, are presented, including their chemical modification via Maillard reaction, and their combination with other polymers and nanomaterials in the form of composites.

Keywords: Whey protein, biodegradability, bio-based packaging, whey protein films, nanocomposites, Maillard reactions.

INFLUENCE OF WHEY PROTEIN EDIBLE FILM AND REFRIGERATION TEMPERATURE ON QUALITY OF ACEROLA IN NATURA DURING POSTHARVEST STORAGE

MSc. Betina Louise Angioletti, University of Blumenau, SC, Brazil*
Stefany Pergentino dos Santos, University of Blumenau, SC, Brazil
MSc. Tuany Gabriela Hoffmann, University of Blumenau, SC, Brazil
MSc. Marcel Jefferson Gonçalves, University of Blumenau, SC, Brazil
Dr. Lisiane Fernandes de Carvalho, University of Blumenau, SC, Brazil
Dr. Savio Leandro Bertoli, University of Blumenau, SC, Brazil
Dr. Carolina Krebs de Souza, University of Blumenau, SC, Brazil

**corresponding author: blangioletti@gmail.com*

Introduction

Acerola (*Malpighia emarginata* D.C.) is a small red fruit, which is a rich source of vitamin C, besides containing vitamin A, vitamin B, carotenoids and bioflavonoids. The great nutritional value of acerola has been increasing its commercial value by its great consume in the last years. However, acerola in natura is a very perishable fruit and its short shelf life difficult its distribution and commercialization. Acerola quality during postharvest depends on pre-harvest factors, mainly maturity of fruit, and postharvest factors, mainly packaging systems and storage temperature (Azeredo et al., 2012).

The best condition to acerola storage ranges from 5 to 15 °C. Storage at low temperatures decrease microbial growth, reduce respiratory rate and delay ripening (Chitarra et al., 2005). Acerola is sensitive to cold, thus storage at lower temperatures can lead to cold damage, promoting internal darkening, surface depressions and ripening failures. One of the main problems faced by acerola due to cold damages is chances in anthocyanins content, which promotes changes in color. These changes cause a visual impact in consumers and are a predominant factor in decision making for fruit consumption. (Yahia, 2011). On the other hand, storage at higher temperatures increases microbial growth, fastering senescence of acerola and promoting undesirable organoleptic changes as changes in texture and development of off-flavors.

Edible films are an emerging technology, which presents a great potential to maintain the quality of fruits. Edible films can have antimicrobial and antioxidant properties, besides giving the product a water vapor barrier. Studies suggest that films obtained from whey have antimicrobial effects, and great mechanical and physical properties, as they are transparent, have flexibility and the absence of odor and flavor, which favors consumer acceptability (Alleoni et al., 2006, Damodaran et al., 2010). In addition, the films can act as a semi-permeable barrier to moisture, oxygen and external damage, and can combat the symptoms of cold injuries.

Until the present moment, there are only few studies comparing the influence of temperature together with the application of biofilms on acerola quality. Thus, given the relevance of the factors addressed, the objective of this work was to study the influence of the application of biofilm formulated with 60% whey (no biofilm and biofilm) and storage temperature (5 and 10 °C) on the quality of acerola during storage for 12 days. Quality was evaluated through anthocyanins content, color changes (a* value) and microbial growth (mesophilic and psychrotrophic bacteria growth).

Material and Methods

Sample collecting and storage

Mature acerolas of different trees located in the city of Blumenau (Santa Catarina, Brazil), were harvested and immediately sent to Laboratory (University of Blumenau, Brazil) where they were sanitized with chlorinated water for 15 minutes. Whey was obtained after cheese production, using pasteurized milk and coagulant agent. Edible films were formulated using glycerin (5%), whey (60%) and water (35%). Acerolas were submerged in the edible film and stored in polypropylene plastic trays (14.5x13x5.3 cm) perforated on the lid. After packing, samples were submitted to refrigerated storage (5 ± 1 °C and 10 ± 1 °C) in in a BOD chamber (TECNAL, model TE-371) during 12 days.

Quality evaluation

For anthocyanin analysis, samples were prepared using 95% alcohol and 1.5 N hydrochloric acid. Total anthocyanins were determined through readings in the CIRRUS 80 ST Spectrophotometer (FEMTO, São Paulo, Brazil) at a wavelength of 535 nm. Samples Results were expressed in mg $100g^{-1}$.

Color analyses were performed using the SP60 series sphere spectrophotometer in a three-dimensional system L* (brightness), a* (red-green) and b* (yellow-blue). Results were expressed as a* (red color) (Francis, 1982, AOAC, 1997).

In microbiological analysis, mesophilic and psychrotrophic bacteria were evaluated. For quantification,10 g of acerola were diluted in 90 ml 0.1% sterile peptonized water. Dilutions were prepared in series. Mesophilic and psychrotrophic bacteria were determined using plate Count Agar (PCA, NEOGEN Culture Media) and incubation at 35 °C for 48 hours, and 7 °C for 10 days, respectively. Results were presented in CFU g^{-1} (APHA, 2012).

Statistical analysis

The Statistica 7.0 software was used to evaluate the results, through Tukey's test with 5% significance.

Results and Discussion

Anthocyanin content

Anthocyanin content in acerola at the beginning of the experiment was 3.33 mg $100g^{-1}$ of fruit. Changes during storage can be observed in Figure 1. As can be noted, at 5 °C, anthocyanin content significantly increased ($p<0.5$) until day 8 of storage, and decreased at day 12. Storage at 10 °C improved anthocyanin stability during the storage period. The decrease observed at 5 °C in the content may be explained due to cold damages, since acerola is sensitive to low temperatures. According to Alves et al. (1995), the best condition to store acerola is at 8 °C. Shin et al. (2007) studied strawberries quality and observed an increase in anthocyanin content in strawberries stored in lower temperatures (0.5 °C) and no changes in strawberries stored at 10 °C, during 3 days of storage.

At the end of storage at 5 °C, samples stored without the edible film (NEF) presented a content of 2.12 ± 0.01 mg $100g^{-1}$, while the content in samples stored with the edible film (EF) was 1.40 ± 0.01 mg $100g^{-1}$. At the of the storage at 10 °C, anthocyanins content was 3.99 ± 0.08 mg $100g^{-1}$ for NEF and

3.54 ± 0.02 mg $100g^{-1}$ for EF. The edible film was not effective in reducing anthocyanin changes at 5 °C. However, at 10 °C, the condition which had the smallest changes in anthocyanin content was acerola recovered with edible film.

Anthocyanins are very unstable pigments that can be degraded under various conditions, including temperature or during food processing and storage. Also, the higher the anthocyanin content, the better the acceptance of the product by the consumer, being important its measurement, because the yellowish colored fruits will probably be rejected by consumers.

Figure 1 – Changes in Anthocyanin content in Acerola fruit stored with and without edible film at 5 °C and 10 °C for 12 days.

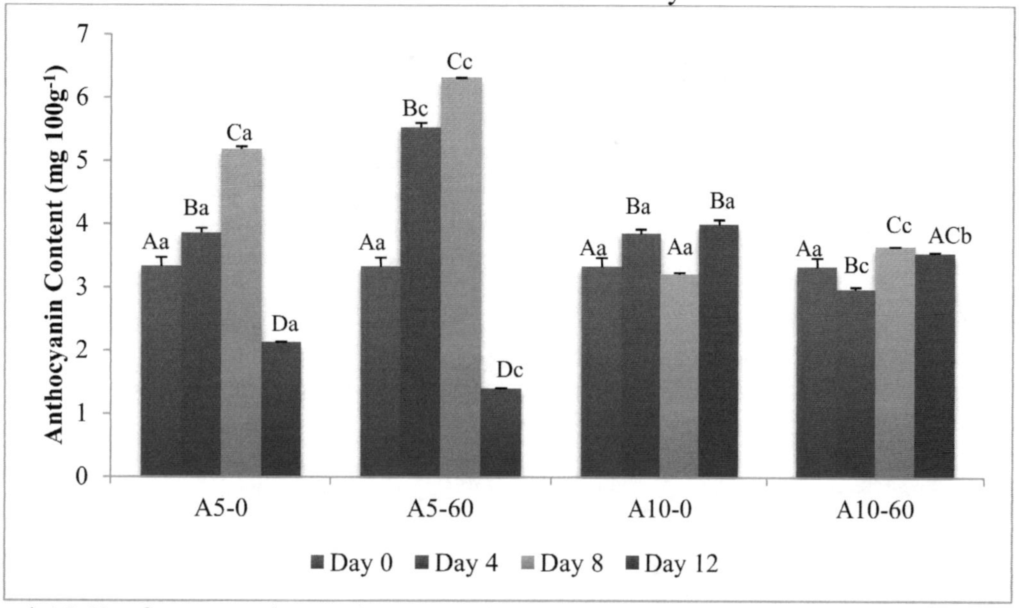

Where A5-0 and A5-60 refers to acerola stored at 5 °C and recovered without (control) and with edible film, respectively; and A10-0 and A10-5 refers to acerola stored at 10 °C and recovered without (control) and with edible film, respectively, upper case compares days of storage in the same treatment, lower case compares different edible film concentrations at the same temperature.

Red color (a value)*

Red color of acerola is seen by consumers as an indicative of quality and freshness and thus represents an important purchase decisions factor. In this work, acerola was harvest mature and presented an a* value of of 47.04 ± 3.34 at the begging of the experiment, as can be observed in Figure 2. a* values at 5 °C showed greater changes during storage than 10 °C, especially at day 12 of storage. These changes coincide with the greatest changes in anthocyanin content (day 12 of storage at 5 °C).

At the end of the storage, a* value for NEF was 32.54 ± 1.56 and for EF was 34.08 ± 4.71 stored at 5 °C, while at 10 °C, a* value for NEF was 46.68 ± 2.27 and for EF 43.00 ± 0.03. There was a significant decrease in the a* values at 5 °C. While at 10 °C no significant difference in the results was observed. At 10 °C, a* remained constant during the storage period and the use of edible film did not affect ($p>0.5$) color changes. In this work, it was observed that acerola stored at 5 °C had a greater yellowish color, while acerola stored at 10 °C had a greater reddish color (data not showed).

Figure 2 – Changes in red color (a*) of Acerola fruit stored with and without edible film at 5 °C and 10 °C for 12 days.

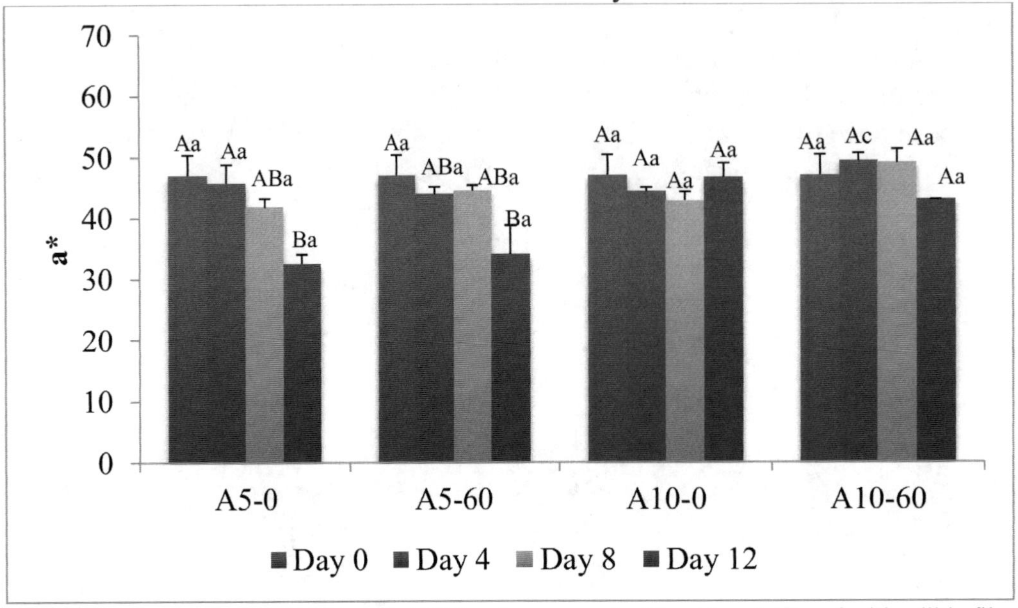

Where A5-0 and A5-60 refers to acerola fruit stored at 5 °C and recovered without (control) and with edible film, respectively; and A10-0 and A10-5 refers to acerola fruit stored at 10 °C and recovered without (control) and with edible film, respectively, upper case compares days of storage in the same treatment, lower case compares different edible film concentrations at the same temperature.

Microbial growth

Microbial growth in acerola was evaluated through Mesophilic and Psychrotrophic Bacteria growth, and results are presented in Figure 3.

Figure 3 – Mesophilic bacteria (a) and Psychrotrophic bacteria (b) growth in Acerola fruit stored with and without edible film at 5 °C and 10 °C for 12 days.

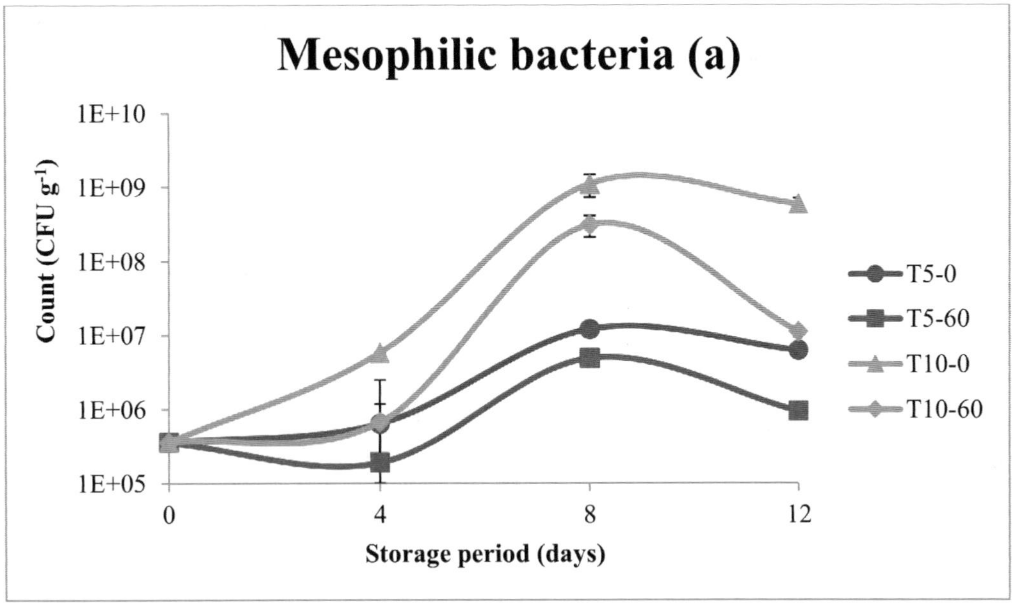

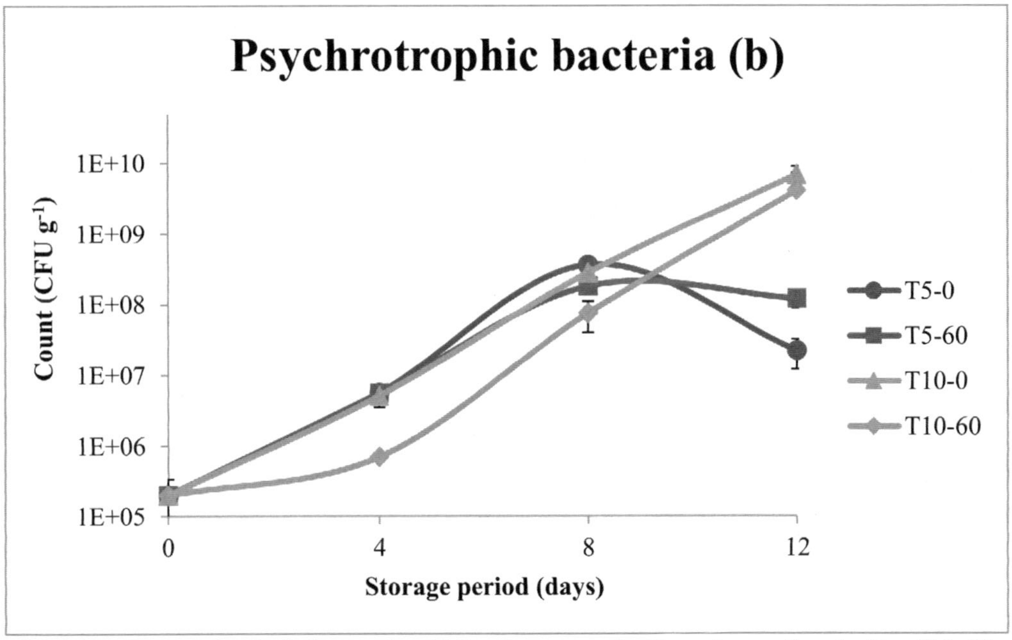

Where A5-0 and A5-60 refers to acerola fruit stored at 5 °C and recovered without (control) and with edible film, respectively; and A10-0 and A10-5 refers to acerola fruit stored at 10 °C and recovered without (control) and with edible film, respectively.

On day 0 of storage, mesophilic count was $3.6 \times 10^5 \pm 3.9 \times 10^4$ CFU g^{-1}, and psychrotrophic count was $2.0 \times 10^5 \pm 1.4 \times 10^5$ CFU g^{-1}. At the end of storage at 5 °C, mesophilic count was $6.2 \times 10^6 \pm 9.2 \times 10^5$ CFU g^{-1} for NEF and $9.4 \times 10^5 \pm 2.8 \times 10^4$ CFU g^{-1} for EF, and at the end of the storage at 10 °C, mesophilic counts were $5.9 \times 10^8 \pm 1.2 \times 10^8$ CFU g^{-1} for NEF and $1.1 \times 10^7 \pm 1.6 \times 10^6$ CFU g^{-1} for EF. As expected, mesophiles growth was lower in lower temperatures.

Comparing samples stored with and without edible film, the use of edible film extended lag phase (day 0 to day 3) and resulted in lower microbial counts at the end of storage (statistically significant). EF stored at 5 °C presented a decrease in microbial population in day 3. Schlei et al. (2020) also observed a decrease in microbial population after submitting cheese to changes in storage condition. Changing the environment by changing temperature or adding an active agent, requires a longer time of adaptation and influences bacterial growth. NEF, EF stored at 5 °C and NEF at 10 °C seemed to be in the stationary phase at the end of storage, while EF at 10 °C seemed to be in death phase of bacterial growth curse, presenting a great reduction in microbial count compared to day 8 of storage. In accordance, Damodaram et al. (2012) showed that dairy protein derivatives have antimicrobial effects.

For psychrotrophic, counts at day 12 were $2.2x10^7 \pm 9.9x10^6$ CFU g^{-1} (NEF) and $1.2x10^8 \pm 3.1x10^7$ CFU g^{-1} (EF) for storage at 5 °C and $6.9x10^9 \pm 4.2x10^8$ CFU g^{-1} (NEF) and $4.1x10^9 \pm 4.9x10^9$ CFU g^{-1} (EF) for storage at 10 °C. As can be noted, biofilm at 5 °C reduced microbial growth at day 8 °C, but increased at day 12 of storage. Moreover, use of biofilm extended lag phase in acerola stored at 10 °C, but did not influenced (statistically significant) the count at the end of storage. Moreover, microbial curve at 10 °C seemed to be in log phase until the end of storage, while at 5 °C seemed to be in stationary phase at day 12.

Conclusion

Acerola stored at 5 °C, presented a lower (statistically significant) microbial growth, for both mesophilic and psychrotrophic bacteria. However, at this temperature, lower levels of anthocyanin and a* color were noted in acerola at the end of storage, leading to a yellowish color caused by cold damage. The use of whey biofilm was effective especially in extending lag phase and reducing microbial count during storage. Moreover, the biofilm improved maintenance of anthocyanins during storage at 10 °C.

Combining higher temperatures (10 °C) with whey biofilm leaded to acerola samples with lower microbial count (compared to control samples at 10 °C) and great quality (anthocyanin and red color) after 12 days storage. This work presents an important step for the expansion of researches on acerola in natura shelf life. The authors suggest further studies in order to understand how different concentrations of whey affect acerola quality, finding the ideal concentration and optimizing the edible film.

Acknowledgements

The authors would like to thank FAPESC (Fundação de Amparo a Pesquisa de Santa Catarina) for the financial support.

References

1. Alleoni, A.C.C., Jacomino, A.P., Rosa, A.S., (2006), "Recobrimento de laranja 'Pêra' com filme de concentrado proteico de soro de leite associado a plastificantes", *Pesquisa Agropecuária Brasileira*, 41(8), pp. 1221-1226.
2. Alves, R. E., Chitarra, A.B., Chitarra M.I.F., (1995), Postharvest physiology of acerola (Malpighia emarginata DC.) fruits: Maturation changes, respiratory activity and refrigerated storage at ambient and modified atmospheres", *Acta Horticulturae*, 370, pp. 223 – 229.
3. AOAC (Association Of Official Analytical Chemists), 1997, "Official methods of analysis", *AOAC*, Gaitherburg, 16 ed.

4. APHA (American Public Health Association), (1992), "Compendium of methods for the microbiological examination of foods", APHA, Washington.
5. Azeredo, H., Miranda, K.W., Ribeiro, H., Rosa, M., Nascimento, D., (2012), "Nanoreinforced alginate–acerola puree coatings on acerola fruits", *Journal of Food Engineering,* 113(4), pp. 505-510.
6. Chitarra, M.I.F, Chitarra, A.B., (2005), "Pós-colheita de frutas e hortaliças: Fisiologia e Manuseio", *UFLA*, Lavras.
7. Damodaran, S., Parkin, K.L., Fennema, O.R., (2010), "Química de Alimentos", Porto Alegre, Artmed, 4 ed.
8. Francis, F.J., (1982), "Analysis of anthocyanins", in Markakis, P., ed., *Anthocyanins as food colors*, New York, Academic Press.
9. Schlei, K.P., Angioletti, B.L., Carvalho, L.F., Bertoli, S.L., Ratto Reiter, M.G., Souza, C.K., (2020), "Influence of temperature on microbial growth during processing of kochkase cheese made from raw and pasteurised milk", *International Dairy Journal,* 109, 104786.
10. Shin, Y., Liu, R.H., Nock, J.F., Holliday, D., Watkins, C., (2007), "Temperature and relative humidity effects on quality, total ascorbic acid, phenolics and flavonoid concentrations, and antioxidant activity of strawberry", *Postharvest Biology and Technology*, 45(3), pp. 349-357.
11. Yahia, M.E., (2011), "Acerola (Malpighia emarginata DC.)", in *Postharvest biology and technology of tropical and subtropical fruits: Açai to citrus*, Cambridge, Woodhead Publishing Limited.

ALOE VERA GEL AS NATURAL ADDITIVE TO IMPROVE OXIDATIVE STABILITY IN REFRIGERATED BEEF BURGER STORED IN AEROBIC AND VACUUM PACKAGING

MSc. Betina Louise Angioletti, University of Blumenau, SC, Brazil*
Stefany Pergentino dos Santos, University of Blumenau, SC, Brazil
MSc. Tuany Gabriela Hoffmann, University of Blumenau, SC, Brazil
MSc. Marcel Jefferson Gonçalves, University of Blumenau, SC, Brazil
Dr. Lisiane Fernandes de Carvalho, University of Blumenau, SC, Brazil
Dr. Savio Leandro Bertoli, University of Blumenau, SC, Brazil
Dr. Carolina Krebs de Souza, University of Blumenau, SC, Brazil

**corresponding author: blangioletti@gmail.com*

Introduction

Consumer demand is changing, increasing the interest in foods that are more convenient, ready-made and natural (formulated without synthetic additives). Refrigerated beef burger meets these demands, besides reducing food losses because of the greater use of meat. However, the higher manipulation and grinding process makes beef burger even more perishable than non-processed raw meat, reducing beef burger shelf life. Grinding favors the formation of unsaturated fatty acids and introduces oxygen into meat, increasing the rate of oxidative reactions.

Lipid and myoglobin (Mb) oxidation are the main oxidative reaction in meat. These processes often seem to be linked and one type of oxidation leads to formation of chemical species which may contribute with the other oxidation process. Oxidation manifests in meat through changes in color, texture and development of off-flavors, besides generating toxically compounds and promoting losses of nutrients and water (Falowo, Fayemi, & Muchenje, 2014).

In lipid oxidation, unsaturated lipids, especially phospholipids and triglycerides, act as substrate of a sequence of complex reactions whose products are mainly hydrocarbons. The reactions can be divided into three steps, namely initiation, propagation and termination, and may happen by means of different mechanisms, being autoxidation the most common. The primary consequences of lipid oxidation are organoleptic changes in aroma and texture (Falowo, Fayemi, & Muchenje, 2014; Repetto, Semprine, & Boveris, 2012).

Myoglobin (Mb) oxidation, also known as pigment oxidation, occurs due to changes in the valence state of iron ion in myoglobin. Depending on iron ligant, Mb can be converted into oxymyoglobin (OMb), deoxymyoglin (DMb) and metmyoglobin (MMb). MMb is the oxidized form of Mb and contains a ferric iron (Fe^{3+}). Typically, this pigment gives a brown color to meat, which usually leads to meat rejection by consumers. Concentration of oxygen in packaging, storage temperature, pH and storage time are important factors which influence changes in pigment (AMSA, 2012).

Oxidative stability of foods depends on concentration, composition and activity of reaction substrates, pro-oxidants and antioxidants. Shifting the balance between these factors is the key to design oxidatively stable foods. Therefore, stability of oxidative reactions can potentially be improved by the removal of oxygen (O_2), inactivation of enzymes, addition of antioxidants or protection against light (Decker, 1998). Different packaging technologies can be employed to ensure the control of these factors, changing food packaging function from a passive protective system to one which plays an important role in food quality and stability (Hoffmann et al, 2019).

In vacuum packaging, O_2 concentration is reduced, minimizing its availability to oxidative reactions and reducing microbial growth. It is one of the most employed technologies in meat packaging. However, in very low (<1.4 mmHg) and low (<7mmHg) concentrations of O_2, Mb is converted into DMb and MMb, respectively, promoting undesirable changes in meat color (AMSA, 2012). The increasing of air concentration inside the packaging during vacuum storage, due to packaging permeability, leads to undesirable changes in meat color during the firsts days of storage.

Use of active agents (in burger formulation or in packaging) is also a promising alternative to maintain food quality during storage. Antioxidants agents can act against lipid oxidation, reacting with free compounds and preventing them to form undesirable products. Moreover, it also reduces changes in meat pigment. From all active agents, natural ones are standing out because of its high efficacy and for being more environmental friendly.

Recently, Aloe vera has attracted attention in many food industry fields, being used in different formulations. This plant presents a high potential to be used as meat additive since it contains more than 70 compounds which may have antimicrobial and antioxidant properties (Shahrezaee et al., 2018). Soltanizadeh and Ghiasi-Esfahani (2015) studied the incorporation of Aloe vera (concentrations of 1, 3 and 5%) in beef burgers and found that Aloe vera acted as a hydrocolloid, minimizing undesirable changes in water content and texture, besides improving lipid stability. The authors observed that burgers formulated with 3% Aloe vera obtained the best scores in sensory analysis.

Combining vacuum packaging and Aloe vera as active agent may result in a higher quality refrigerated beef burger during storage, enhancing the effects of each technique. Given the importance, this work aimed to evaluate the influence of packaging atmosphere (aerobic and vacuum) and concentration of Aloe vera gel (2 and 4%) on the quality of beef burger during refrigerated storage at 4 °C for 7 days. Quality was evaluated by means of lipid oxidation, pigment oxidation and red color changes. Control samples were prepared for comparisons.

Material and Methods

Sample collecting and storage

Grinded beef was collected from a local market immediately after its arrival and was aseptically sent to laboratory. It was separated in portions of 40 g and Aloe vera gel (extracted from Aloe vera leaves through maceration, after removal of Aloin) was added to burger formulation in concentrations of 2% and 4% (g of Aloe vera g^{-1} of beef burger). Half of samples were stored in aerobic packaging (atmospheric pressure) and the other half in vacuum packaging, using a commercial vacuum sealer (Cetro), with vacuum capacity of 0.06 mPa. All samples were packaged in Nylon Poli packaging (3 coextruded layers of polyamide and low density polyethylene). A complete factorial experimental design 2^2 was carried out. Control samples (no additive) were also prepared and were stored in aerobic and vacuum packaging. All samples were submitted to refrigerated storage at 4 °C in a BOD chamber (Model 371, Tecnal) during 7 days.

Quality evaluation

Beef burger quality was evaluated on days 0, 3, 5 and 7 of storage, in duplicate. The parameters evaluated were lipid oxidation, pigment oxidation and red color changes (a* value). Lipid oxidation was evaluated through TBA method, according to Lemon (1985). Results were expressed in mg MDA kg^{-1} meat. Pigment oxidation was evaluated through MMb changes, measuring meat reflectance (sphere spectrophotometer, SP60 Series, X-Rite), according to AMSA (2012). Changes in red color was analyzed through changes in a* value, in CIE L*a*b color space. The combination of illuminant and observed used was a/10, since illuminant 10 favors evaluation of red color (AMSA, 2012).

Statistical analysis

The Statistica 7.0 software was used to evaluate the results, through Tukey's test with 5% significance.

Results and Discussion

Lipid oxidation

Lipid oxidation at the beginning of the experiment was 0.26 mg MDA kg^{-1} meat. Cahnges lipid oxidation in beef burger during storage can be observed in Figure 1.

Figure 1 – Lipid oxidation in beef burger formulated with Aloe vera gel (2 and 4%) and stored in aerobic (atmospheric pressure) and vacuum packaging at 4 °C for 7 days.

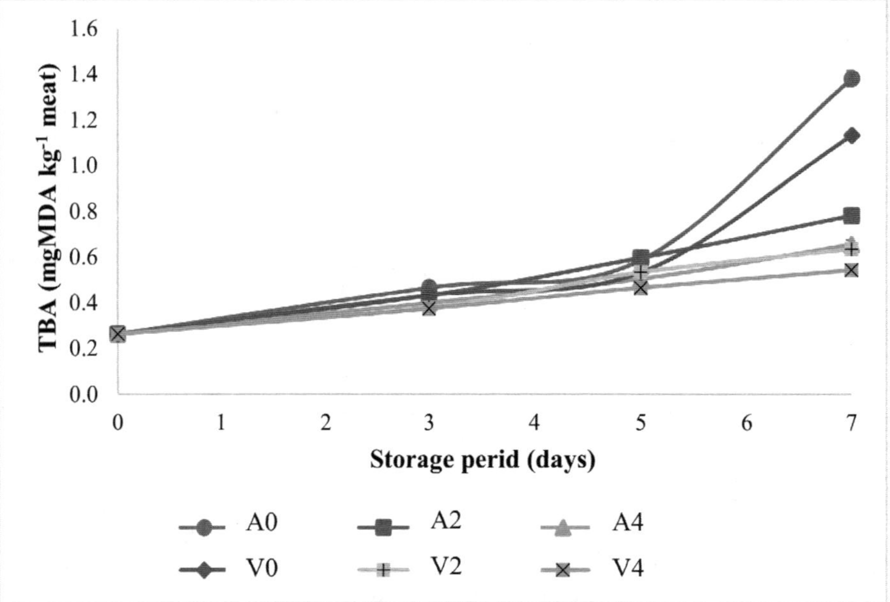

Where A0, A2 and A4 refers to samples stored in aerobic packaging (atmospheric pressure) and formulated with 0 (control), 2 and 4% of Aloe vera gel, respectively; and V0, V2 and V4 refers to samples stored in vacuum packaging and formulated with 0 (control), 2 and 4% of Aloe vera gel, respectively.

During the storage period, TBA value increased (statistical significance) for all treatments. Vacuum packaging and addition of Aloe vera gel delayed lipid oxidation, especially after day 5 of storage, and differences (statistical significance) from control samples had grown with storage period. At the end of storage, beef burgers formulated with 2% and 4% Aloe and stored in aerobic packaging (atmospheric pressure) presented a decrease in oxidation of 43.6% and 52.6%, respectively. Burgers stored in vacuum presented a decrease of 53.9% (2% Aloe vera) and 60.8% (4% Aloe vera), compared with the worst condition, which was control sample in aerobic packaging. TBA value for aerobic control sample reached 1.38 mg MDA kg^{-1} meat at day 7 of storage.

Comparing only control samples, vacuum packaging reduced lipid oxidation in 18% at the end of storage. In accordance, Khalil and Mansour (1998) demonstrated that vacuum packaging reduced lipid oxidation in meat in 20% after 7 days of storage. Vacuum packaging reduces oxidative reactions by reducing oxygen availability in packaging.

The active agent, especially in higher concentration (4%) also showed to be very effective in improving lipid stability. Aloe vera contains a large amount of bioactive compounds, as polyphenols, which have stronger free radical scavenging abilities. Bioactive compounds of Aloe can act in the second phase (propagation) of lipid oxidation, reacting with radicals present in burger, thus forming stable products and avoiding the formation of undesirable ones (Shahrezaee et al., 2018).

Kanatt and Makwanna (2020) observed a reduction in lipid oxidation in chicken meat stored in active packaging formulated with Aloe vera. They found that lipid oxidation in meat stored in active packaging after 14 days of storage was 2.42 mg MDA kg^{-1} meat, while the value for meat stored in conventional packaging was 2.85 mg MDA kg^{-1} meat in only 3 days of storage. Soltanizadeh and Ghiasi-Esfahani observed that lipid oxidation in beef burger reached values from 0.55 (burger with 5% Aloe vera) and 0.74 mg MDA kg^{-1} meat (control sample – 0% of Aloe vera) after 7 days of storage.

Myoglobin oxidation

MMb content also presented significant changes during storage, as can be seen in Figure 2. At day 0, beef burger presented a MMb content of 12.15%. Control Sample stored in vacuum and burger formulated with 2% Aloe were the treatments that showed the highest initial increasing, presenting in day 3 a content of 23.19% e 24.23% (no statistical difference). Low oxygen partial pressure inside the packaging in the firsts days of experiment may had favored the transformation of Mb in MMb, while high concentrations of Aloe presented an antioxidant effect, reducing MMb formation.

In the following days, due to oxygen permeation through packaging, MMb content in beef stored in vacuum packaging was reduced, especially in burgers formulated with additives. Control sample stored in aerobic packaging (atmospheric pressure) was the treatment which presented the highest MMb content (27.6%) at the end of the storage, followed by control sample stored in vacuum (21.5%). In aerobic packaging, the addition of Aloe vera gel reduced (statistical significance) MMb formation, especially high concentrations of Aloe vera (13.27%). The lowest level found was in burgers formulated with 4% Aloe vera gel and stored in vacuum (12.06%). Microbial growth is the main cause of MMb formation in aerobic packaging.

As can be observed, combining vacuum packaging and Aloe vera gel is an efficient strategy to reduce pigment changes during all the storage period studied, reducing especially changes in vacuum-packaged burger in the firsts days of storage and in aerobic-packaged burger at the end of storage.

Figure 2 – Metmyoglobin changes in beef burger formulated with Aloe vera gel (2 and 4%) and stored in aerobic (atmospheric pressure) and vacuum packaging at 4 °C for 7 days.

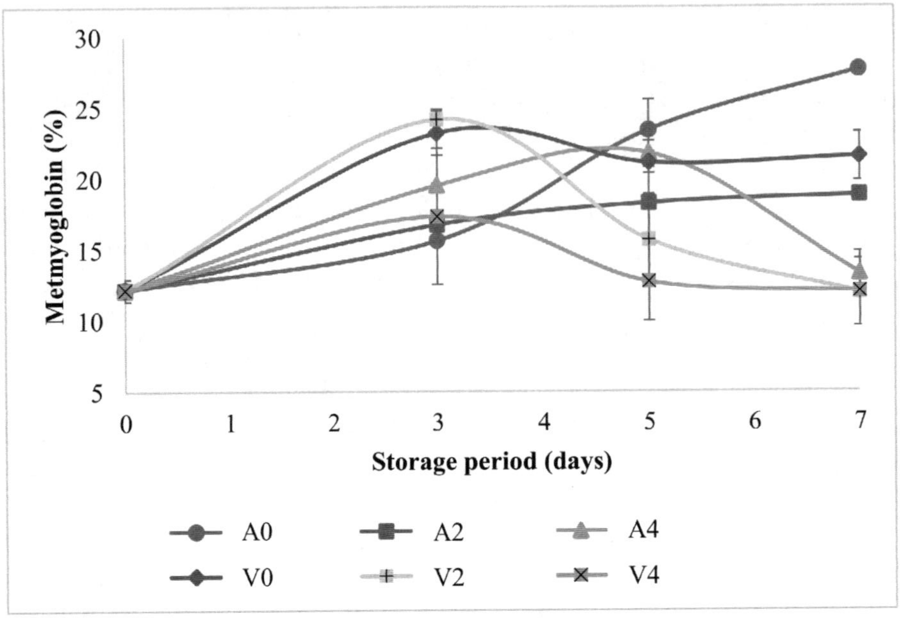

Where A0, A2 and A4 refers to samples stored in aerobic packaging (atmospheric pressure) and formulated with 0 (control), 2 and 4% of Aloe vera gel, respectively; and V0, V2 and V4 refers to samples stored in vacuum packaging and formulated with 0 (control), 2 and 4% of Aloe vera gel, respectively.

Red color changes (a* changes)

Analyzing meat color, particularly a* value, is also very important, since color is a key factor in product purchase decision and read color is seen by consumers as an indicative of quality and freshness. Fresh meat (day 0 of storage) presented an a* value of 30.45, as can be seen in Figure 3.

At the third day of storage, control sample stored in vacuum presented the greatest change in a* value (-12.24 ± 0.07), which coincides with the highest MMb content is this day. At the end of the storage, control sample stored in aerobic (-12.5 ± 2.51) and vacuum (-7.40 ± 1.30) presented the greatest changes in a*, which also corresponds to the greatest changes in MMb. MMb is a brown pigment, thus increasing its content leads to changes in meat color.

The conditions which most improved the maintenance of a* value were high concentration of Aloe and aerobic packaging (-2.15 ± 0.00), and low (-0.21 ± 1.92) and high concentration (-0.21 ± 1.92) of Aloe in burgers stored in vacuum packaging. No statistical differences were found among a* value of these three samples at day 7.

Figure 3 – a* changes in beef burger formulated with Aloe vera gel (2 and 4%) and stored in aerobic (atmospheric pressure) and vacuum packaging at 4 °C for 7 days.

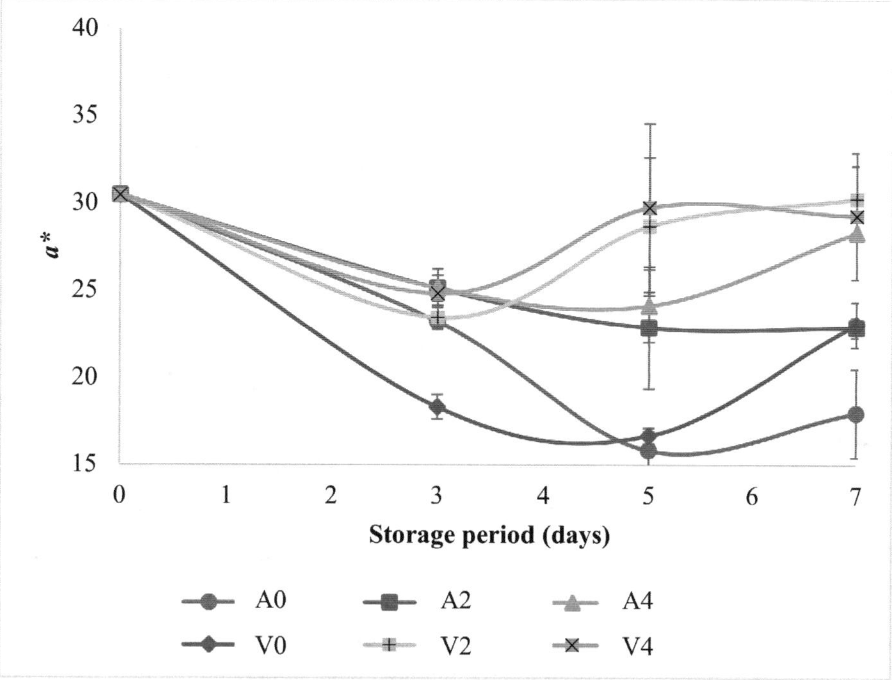

Where A0, A2 and A4 refers to samples stored in aerobic (atmospheric pressure) packaging and formulated with 0 (control), 2 and 4% of Aloe vera gel, respectivelly; and V0, V2 and V4 refers to samples stored in vacuum packaging and formulated with 0 (control), 2 and 4% of Aloe vera gel, respectivelly.

Conclusion

Refrigerated beef burger stored in aerobic (atmospheric pressure) packaging has a short shelf life, especially due to oxidative reactions. Combining vacuum packaging and Aloe vera as natural additive is a great alternative to reduce these deteriorative processes, reducing changes of both lipid and pigment oxidation. As consequence, changes in meat color are also reduced, improving consumer acceptability.

The use of vacuum packaging reduced lipid oxidation in almost 20% at the end of storage, and combining vacuum and 4% Aloe vera reduced changes in lipid oxidation in more than 60%, comparing with aerobic control sample. In pigment oxidation, the use of additives reduced mainly MMb formation in vacuum-packaged burger in the firsts days of storage and in aerobic-packaged burgers at the end of storage, keeping the quality of meat during all the storage period. Moreover, the concentration of additive employed also played an important role, and increasing concentration from 2 to 4% significantly reduced oxidation rate.

The results of this work serve as primary information for the improvement of refrigeration and packaging technologies for beef burger storage, reducing food losses and consequently the environmental impact of the meat chain.

Acknowledgements

The authors would like to thank FAPESC (Fundação de Amparo a Pesquisa dè Santa Catarina) for the financial support.

References

1. AMSA - American Meat Science Association, (2012), "Meat color measurement guidelines", Illinois, AMSA.
2. Decker, E.A., (1998), "Strategies for manipulating the prooxidative/antioxidative balance of foods to maximize oxidative stability", *Trends in Food Science and Technology,* pp.241-248.
3. Falowo, A., Fayemi, P, and Muchenje, V. (2014), "Natural oxidants agains lipid-protein oxidative deterioration in meat and meat products: A review", *Food Research International*, 64, pp. 171-181.
4. Hoffmann, T.G., Amaral, D.P., Angioletti, B.L., Bertoli, S.L., Peres, L.V., Ratto Reiter, M.G., Souza, C.K., (2019), "Potential nanocomposite in food packaging", *Chemical Engineering Transactions*, pp. 253-258.
5. Kanatt, S.R., Makwana, S.H., (2020), "Development of active, water-resistant carboxymethyl cellulose-poly vinyl alcohol-Aloe vera packaging film", *Carbohydrate Polymers*, 227(1).
6. Khalil, A.H., Mansour, E.H., (1998), "Control of lipid Oxidation in Cooked and Uncooked Refrigerated Carp Fillets by antioxidant and Packaging Combinations", *Journal of Agricultural and Food Chemistry*, 46(3), pp. 1158-1162.
7. Lemon, D.M., (1985), "An Improved TBA test for rancidity", *New series,* Circular n° 51, Halfaz, Noza Scotia, Halifax Laboratory.
8. Milani, L.I., (2012), Avaliação da atividade antimicrobiana e antioxidante de extratos de caqui (Diospyros kaky L.) para proteção de produtos cárneos, *Universidade Federal de Santa Maria*, Santa Maria.
9. Repetto, M., Semprine, J., and Boveris, A, (2012), "Lipid peroxidation: Chemical Mechanism, Biological Implications and Analytical Determination", in *Lipid Peroxidation*, Catala, A., ed. London, IntechOpen.
10. Shahrezaee, M., Zoleimanian-Zad, S., Soltanizadeh, N., and Akbari-Alavijeh, S., (2018), "Use of Aloe vera gel powder to enhance the shelf life of chicken nugget during refrigeration storage", *LWT*, 95, pp. 380-386.
11. Soltanizadeh, N. and Ghiasi-Esfahani, H., (2015), "Qualitative improvement of low meat beef burger using Aloe vera", *Meat Science*, 99, pp. 75-80.

609864 Degradable Catechin Particles As an Effective Antioxidant Material

Abstract:

Catechin (CAT) is a polyphenolic compound found in many fruits, vegetables and beverages such as strawberries, black grape, green tea and so on.[1] The natural characteristics such as antioxidant, anti-inflammatory, anticancer, antimicrobial, antifibrotic properties make catechin an intriguing compound for numerous biomedical use.[2-3] Therefore, CAT and its derivatives afford many therapeutic advantages in medical, food and cosmetic industries.[1] Here, we present a single step synthesis of degradable CAT microparticles using trimethylolpropane triglycidyl ether (TMPGDE) as crosslinker at 100, 200, and 300% mole ratios based on CAT molecule in a microemulsion system. The hydrolytic degradation of CAT particles was investigated at physiological conditions e.g., at pHs 1.0, 7.4, and 9.0 at 37.5 °C. A sustainable and long-term CAT release kinetic was obtained for CAT particles prepared at 100% crosslinked TMPGDE with 264 mg/g CAT release within 10 days at pH 7.4 and 37.5 °C. The CAT particles possess potent antioxidant properties with 55.0±0.9 µg/mL gallic acid equivalency in terms of total phenol content and 0.88±0.3 µmol/g trolox equivalent antioxidant capacity. The blood compatibility of CAT particles was investigated via hemocompatibility tests with blood clotting and hemolysis assays and the cytotoxicity of CAT particles was determined on CCD841 normal colon cell line. It was found that CAT microparticles can be used for vascular system up to 10 µg/mL concentration and induce no toxicity to normal colon cell line up to 250 µg/mL concentration for 24 h incubation time. Therefore, the biodegradable CAT particles with natural origin has great potential in different biomedical fields owing to its' sustainable and long-term CAT release capability, antioxidant behavior and lesser toxicity.

References

1. Gadkari PV, Balaraman M, Catechins: Sources, Extractions and Encapsulation: A Review. Food Bioprod Process 2013:93:122-138.
2. Higdon JV, Frei B, Tea Catechins and Polyphenols: Health Effect, Metabolism and Antioxidant Functions. Crit Rev Food Sci Nutr 2010:43: 89-143.
3. Afzal M, Safer AM, Menon M, Green Tea Polyphenols and Their Potential Role in Health and Disease. Inflammopharmacol 2015:23:151-161.

607741 Nutritional Safety of the Oxidized Compounds of Ultra-Processed Foods in the Western Diet

Abstract:

In recent years, a significant increase in the consumption of ultra-processed, calorie-dense, and nutrition-poor food has been noticed in developed countries (Juul et al., 2018; Setyowati et al., 2018). The term "Western diet" has been associated, and eventually, adopted as a term to describe the modern standard American diet. It is characterized by inexpensive ultra-processed foods, high in caloric energy, and lacking essential nutrients needed (Hintze et al., 2012). Even though different processing techniques such as thermal treatments, light exposure, storage, and aging are designed to improve food quality, nutrition, and safety; they may lead to the generation and accumulation of certain chemical compounds able to exert toxic activities against human health (Maldonado-Pereira et al., 2018). This is the case of cholesterol oxidation products (COPs). Exogenous COPs are obtained through the ingestion of animal-derivatives foods, making diet a key factor in the study of these compounds. COPs have been related to different chronic diseases such as Alzheimer's disease, Parkinson's disease, diabetes, obesity, cancer, cardiovascular diseases, among others. Therefore, COPs profile in Western diet's foods is an essential step in the nutritional safety analysis of US diet. The aim of this study was to quantify the oxidative load of most consumed ultra-processed (UP) food in the Western diet by a lipidomic approach. A total of 62 ultra-processed foods 39 Ready to Eat (RTE) items and 23 Fast Foods (FF) meals) were analyzed and results compared with both their nutritional fact labels and USDA database. Secondary oxidation products were measured by the Thiobarbituric acid (TBARs) assay. Quantification of Fatty Acids Methyl Esther (FAME) was performed by Gas Chromatography (GC) means. Profiling of Cholesterol Oxidation Products (COPs) was performed by GC-MS.

RTE fat values were similar to those described in the item's nutritional fact label. Forty six percent of the RTE items (17 out of 39 RTE items) showed a higher value compared to the value shown in the nutritional label. For the FF group, a 74% of the FF meals showed a higher lipid content (for 17 FF meals out of 23 compared to the fat value) compared to what the nutritional information stablished. Overall, SFAs was the group with the highest percentages within the RTE, followed by MUFAs, and lastly, PUFAs. MUFAs was the group with the highest percentages within the FF meals, followed by SFAs, and lastly, PUFAs. For the TBARS assay, RTE items showed a higher MDA content compared to FF meals as it can be observed from the MDA values. However, more FF meals showed "no-presence" of MDA, potentially meaning the conversion of MDA molecules into further oxidation products.

Six phytosterols (Campesterol, Stigmasterol, Brassicasterol, b-Sitosterol, Fucosterol, and b-Tocopherol) were quantified for FF meals and RTE items. b-Sitosterol was the the phytosterol with the highest quantities within the FF meals, and b-Sitosterol was the the phytosterol with the highest quantities within the RTE items. Sixteen FF meals out of 23 showed a higher cholesterol content compared to the value found in either the USDA webpage, the FF's nutritional webpages, or other Food Nutrition webpages used for this analysis (Eat this much (DeMenthon, n.d.) and Fast Food Nutrition (Lenhoff, 2005)). Two items (KFC's mashed potato and McDonald's hotcakes) showed cholesterol values of 90.13 and 268.72 mg/100 g fat, respectively; even when their cholesterol content in each official FF webpage stablished that these items have 0 mg/100 g fat.

Nine RTE items out of 39 showed a higher cholesterol content compared to the value found in the item's nutrition fact label (598.42 ± 126.63 mg/100 g fat – Oscar Mayer's salami, 102.60 ± 18.68 mg/100 g fat – Hellmann's mayonnaise, 276.81 ± 1.83 mg/100 g fat – Praire Farms's butter, 307.81 ± 3.54 mg/100 g fat – Campbell's beans and bacon soup, 300.53 ± 20.51 mg/100 g fat – Meijer's macaroni salad, 2,135.58 ± 168.32 mg/100 g fat – Kroger's chicken noodle soup, 1,247 ± 227.45 mg/100 g fat – Beech Nut's beef and broth baby food, and 313.34 ± 27.54 mg/100 g fat – Gerber's turkey and vegetables. Gerber's strawberry-banana yogurt showed a cholesterol content of 316.80 ± 0.40 mg/100 g fat when the USDA food database said there was 0 mg/100 g fat.

Eight different COPs (7a-OH, 7b-OH, 5,6a-Epoxy, 5,6b-Epoxy, triol, 7-Keto, and 24 methylenecholesterol, and 6-Keto) were determined in FF meals and RTE items by GC-MS operated in SIM mode. In FF meals, total COPs content reached values up to 10.81 ± 1.52 mg/100g fat (roast beef, ham & provolone sandwich – Panda Express). On the contrary, the lowest amount was 0.28 ± 0.059 mg/100g fat (Domino's Pizza – Pepperoni pizza). For RTE items, the highest value for total COPs was Beech Nut's turkey and broth baby food with 14.41 ± 4.37 mg/100 g fat. The lowest COPs contents were: 0 mg/ 100 g fat for both RTE items, Countryside Creamery margarine, and Purple Cow vanilla ice cream.

Cholesterol oxidation by different cooking processes and addition of spices and ingredients was followed by both TBARS and FAME quantifications COPs concentrations were positively associated with cholesterol concentration. However, the association was (correlation = 0.504), indicating that food matrix and food processing play a role in the formation of COPs. Results showed that different processing conditions of UP foods result in the formation of high amount of COPs and other oxidized derivatives such as phytosterols oxidation products (POPs), which were also identified in our samples. UP foods are produced under different and variable conditions. Therefore, an expansion of this data base is vital to know their occurrence in the Western Diet, together with a dietary exposure analysis of these compounds, ensuring food quanlity and human's food safety.

References:

1. Juul, F., et al., *Ultra-processed food consumption and excess weight among US adults.* British Journal of Nutrition, 2018. **120**(1): p. 90-100.
2. Setyowati, D., N. Andarwulan, and P.E. Giriwono, Processed and ultraprocessed food consumption pattern in the Jakarta Individual Food Consumption Survey 2014. Asia Pacific Journal of Clinical Nutrition, 2018. 27(4): p. 840-847.
3. Hintze, K.J., A.D. Benninghoff, and R.E. Ward, Formulation of the Total Western Diet (TWD) as a Basal Diet for Rodent Cancer Studies. Journal of Agricultural and Food Chemistry, 2012. 60(27): p. 6736-6742.
4. Maldonado-Pereira, L., et al., The role of cholesterol oxidation products in food toxicity. Food and Chemical Toxicology, 2018.

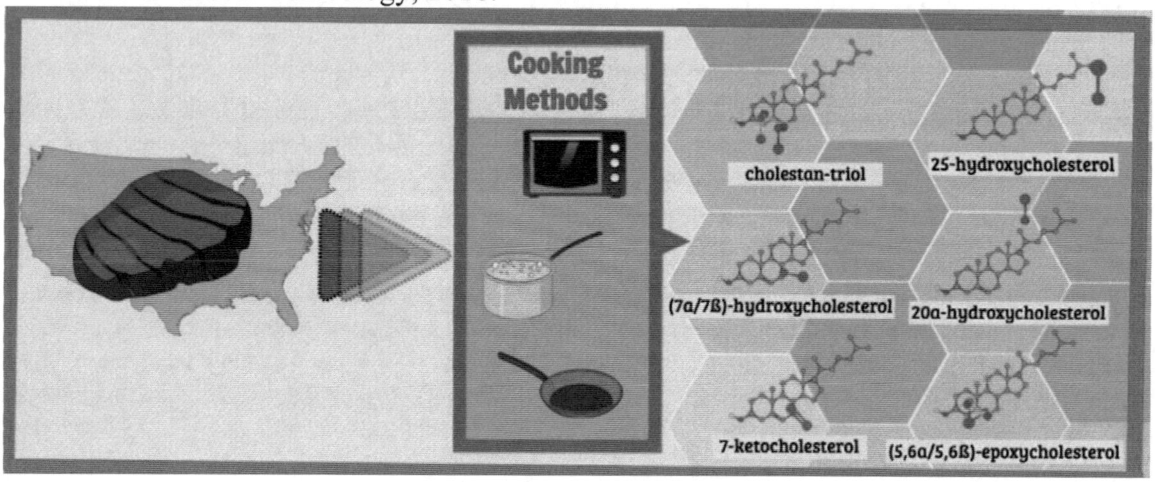

611426 Minimizing the Environmental Impact through Microwave-Assisted Extraction of *Eucalyptus Globulus* Essential Oil

Abstract:

During the last decade, the food industry has shown a growing interest in studying essential oils (EOs) of several plants, herbs, and spices; mainly because of their potential as natural antimicrobials and antioxidants. Although conventional methods as steam distillation (SD) and hydrodistillation (HD) have been widely applied for extraction of EOs; emergent methods, such as microwave-assisted extraction (MAE), have been recently explored to improve the extraction yield and reduce the use of solvents and energy requirements. The aim of this work was to determine the optimum conditions for extraction of *Eucalyptus globulus* essential oil and the environmental impact associated to these process conditions for different extraction processes. Mathematical models for extraction yield and energy requirements were fitted from experimental data. Obtained models were implemented within an optimization problem wherein the objective function was to minimize total cost. Additionally, environmental impact of the different processes was assessed though the Eco-indicator E99 (EI99). *Eucalyptus globulus* EO MAE was performed in a NEOS System equipment evaluating selected conditions of solid/liquid ratio (1:1, 1:3, or 1:5), stirring speed (0, 200, or 400 rpm) and microwave (MW) power output (360, 450, or 540 W). Conventional HD was performed in a three-neck round bottom flask heated by a stirring heating mantle (480 W, 60 Hz) attached to the same condenser and graduated trap utilized during MAE at same conditions of solid/liquid ration and stirring speed. Steam distillation was performed using a pressurized boiler, evaluating different conditions of solid/liquid ratio (1:0, 1:1, or 1:2) and steam generation (6.1, 7.5, or 8.8 g/min). Obtained data were fitted to polynomial models by using Response Surface Methodology to describe yield and energy requirements. EI99 was utilized to compute environmental impact of the extraction processes; this eco-indicator considers the waste of water and total use of energy, as well as which are the variables that most influence the environmental impact. The best experimental yields (1.26 ± 0.01, 1.19 ± 0.02, and 1.06 ± 0.27 %, for MAE, HD, and SD, respectively) were obtained under the following conditions: for MAE a solid/liquid ratio of 1:3, 400 stirring speed and 360 W of MW power; in the case of HD a solid/liquid ratio of 1:5 and 400 stirring speed; for SD a solid/liquid ratio of 1:2 and 6.7 g/min of steam flow. The EI99 for each case was 7.47±0.4, 16.91±0.3, and 50.86±13.2, respectively. The optimal solution was found for MAE at a solid/liquid ratio of 1:2.5, 250 stirring speed, and 360 W of MW power, obtaining a yield of 1.46% and an associated environmental impact of 4.48 Eco-points per gram of EO extracted. In case of HD the optimal solution was found for a solid/liquid ratio of 1:8 without stirring, obtaining a yield of 1.44% and 13.41 Eco-points per gram of EO. For SD the best operation conditions were found for a solid/liquid ratio of 1:10 and 52.37 Eco-points per gram of EO, obtaining a yield of 1.02%. From these results MAE exhibited the greatest EO yield extraction with a low environmental impact, it also showed that the HD can be improved to get a greater EO yield and a lower environmental impact. For the SD solution, although it requires a greater use of water, had a lower environmental impact than the one experimentally obtained. The tested approach allows to optimize different responses, finding an equilibrium amongst the operating process conditions and environmental goals.

604888 Microbial Community Dynamics during the Anaerobic Fermentation of Lignocellulosic Biomass and Wastewater Activated Sludge at Two Temperatures

Abstract:

Lignocellulosic biomass has been considered as an abundant renewable source for the production of biofuels and biochemicals. Robust microbial communities in activated sludges could convert the carbohydrates from lignocellulosic biomass to medium chain carboxylic acids, which can be subsequently utilized as precursors to produce more valuable bioproducts. In this study, wastewater activated sludges were inoculated to pure sugar solutions and sulfite prehydrolysate for anaerobic fermentation, respectively. The fermentation has been conducted under both mesophilic (35 °C) and thermophilic (55 °C) conditions. The result indicated that lactic acid was the major product for pure sugar fermentation at 55 °C, while more medium chain carboxylic acids (such as acetic acid, propionic acid and butyric acid) were generated at 35 °C. Interestingly, it has been found that the addition of peptone could significantly promote the carboxylic acid (especially acetic acid and propionic acid) production rate and affect the acid compositional profiles. For the sulfite prehydrolysate fermentation, similar stimulation effect was observed on acid production at 35 °C, however, the sugars consumption was significantly limited at 55 °C. The reason is that the carbonyl inhibitors in the prehydrolysate showed a strong inhibitory effect at a higher temperature. Additionally, we plan to use the sulfite pretreated aspen as the substrate to investigate anaerobic digestion of woody biomass in our future studies. It is hypothesized that the supplementary of cellulase could accelerate substrate digestion, as well as acid fermentation. To better understand the microbiome-based fermentation process, whole genome sequencing (WGS) analysis will be performed to identify the predominant microbial communities associated with the carboxylic acid productions.

608302 Development of a Plant-Made Therapeutic to Treat Spaceflight Osteopenia

Abstract:

Extended missions in a microgravity environment alter normal physiology in the human body. One critical issue in spaceflight is loss of bone mineral density. In the process of bone homeostasis, specialized cells facilitate bone remodeling with coupled bone resorption and formation. The reduced mechanical loading in microgravity causes resorption to outpace formation, which results in loss of bone mass. Medications such as parathyroid hormone (PTH) (1-34), a peptide fragment of naturally occurring human PTH, stimulates bone formation and may be able to restore bone mass in microgravity.. However, this requires a daily subcutaneous injection for a number of months, and in a limited resource environment, this is impractical. An alternative to transporting an injectable medication is to instead build capacity to produce it during the mission, which we propose to do in transgenic plants. We are producing transgenic lettuce which expresses a fusion protein consisting of PTH (1-34) linked with a fragment crystallizable (Fc) domain of human IgG1 via a flexible linker. The Fc region is intended to eliminate the injection requirement by increasing bioavailability via oral delivery. We constructed a binary vector for expression of this PTH-Fc protein and used it to produce two varieties of transgenic lettuce via *Agrobacterium tumefaciens* mediated transformation of wild type lettuce. From these T0 plants, we have grown T1 plants, and we are in the process of generating T2 plants. We have successfully developed a PCR protocol to detect our gene of interest and are in the process of optimizing it. In parallel, we are developing a protein purification method, along with quantitative and functional assays for lettuce-made PTH-Fc.

609886 Metabolic Features and Proteolytic Pathways of *Lactobacillus Harbinensis* M1 during Soymilk Fermentation

Abstract:

Lactobacillus harbinensis is a novel species that has the potential to improve the quality of fermented product. *L. harbinensis* M1 (M1), previously isolated from naturally fermented tofu whey (NFTW), has been shown to provide probiotic potential and to improve the organoleptic properties of fermented soymilk. However, its lifestyle and contribution to health-related functionality are unclear. This study investigated the metabolic features and proteolytic pathways of M1 involved in soymilk fermentation, to uncover its molecular and functional characteristics. During soymilk fermentation, M1 converted soybean carbohydrates into lactic acid (6.91 mg/mL) and short-chain fatty acids (including acetic, propionic and butyric acids). Proteolytic analysis showed that M1 extensively hydrolyzed the soybean protein into oligopeptides and free amino acids. The soluble peptide fraction (189–451 Da) was increased from 4.6% to 86%. The whole genome of M1 was sequenced for metabolic prediction. Genes encoding enzymes required for carbohydrate/pyruvate metabolism and isoflavone biotransformation are present. The proteolytic system was predicted, including 4 genes encoding extracellular proteases, 64 genes encoding intracellular-peptidases and the Opp/Dpp peptide transport system. Finally, M1 improved the radical scavenging activities and growth inhibition effects on HepG2/MCF-7 cancer cells of soymilk. These findings offered a better understanding of the metabolic mechanisms of *L. harbinensis* M1, which elucidate its role and envisage its application in prebiotic products.

611692 Metabolic Engineering of *E. coli* for the Sustainable Production of Short-Chain Esters

Abstract:

Research Interests: Metabolic engineering, Bioenergy, C tracing and FBA, Cyanobacterial engineering

Teaching interest: Biochemistry, Biotechnology, Genetic engineering, Microbiology

Biological production of industrially valuable chemicals from renewable feedstocks imparts sustainability to the manufacturing sector. With the advent of environmental safety policies and development of sustainable industrialization practices, efforts are being made towards the use of biodegradable organic solvents like lactate esters. In this work, we have constructed engineered *E. coli* to produce ethyl lactate from renewable substrates. As a first step, bioprospecting of ester producing enzymes was done and the following seven were selected: diacylglycerol-transferase, ethanol-o-acyltransferase, acetylxylan-esterase, carbohydrate-esterase; acyl-coenzymeA:ethanol O-acyltransferase, ethyl-ester-synthase-1; and esterase-A. *In vitro* assays were conducted to identify the optimal pH and temperature for these novel enzymes. Recombinant *E. coli* strains were constructed by expressing the esterase and acyl-alcohol transferase in *E. coli* strains engineered for overproducing ethanol and lactic acid. High-cell density fermentations were conducted under microaerobic conditions and the strain expressing esterase-A produced the highest ethyl lactate (3.5±1.5 mg/L) and ethyl acetate (16±8 mg/L) from glucose as a carbon source. Further work is underway, to increase the production rate and titer.

Finally, owing to the faster growth of the microbial host, minimal nutritional requirement of the organism, and extracellular release of ethyl lactate, continuous fermentation and product extraction strategy could be successfully implemented for industrial scale-up using these strains.

612461 Antibacterial Mechanism of Thymol Against *Enterobacter Sakazakii*

Abstract:

Enterobacter sakazakii, an opportunistic foodborne Pathogen can arouse life-threatening infections in neonates, along with the clinical symptoms including necrotizing enterocolitis, bacteremia and meningitis. Due to the antibacterial effects against a variety of pathogenic bacteria, Thymol is the component of essential oils which can be extracted from plants such as *Thymus*, *Ocimum*, *Origanum*, etc. This study aims to uncover the antimicrobial activity of the thymol against *E. sakazakii* and its mechanism of action. The broth microdilution method was adopted to determine the minimum inhibitory concentration (MIC) of thymol.Becides, Membrane potential, intracellular ATP concentrations and intracellular pH (pHin) were measured and the antibacterial effect was observed by confocal laser scanning microscopy (CLSM) and field emission gun scanning electron microscope (FEGSEM), in order to elucidate a possible antibacterial mechanism. The MIC of thymol against *E. sakazakii* BNCC 186088 is 1.25 mg/ml, which inhibits the growth of *E. sakazakii* by breaking membrane integrity. Cell membrane depolarization, a reduction of intracellular ATP concentrations and pHin were observed after treatment, which showed that the cell membrane structures of *E. sakazakii* were apparently impaired by thymol, which means that thymol could destroy the integrity of the cell membrane of *E. sakazakii*, leading to the loss of cell homeostasis. The results indicate that thymol can be used as a food additive to prevent *E. sakazakii* contamination in food.

611026 Modified Feed Spacer Materials for Antimicrobial Dairy Membrane Modules

Abstract:

Downstream contamination of dairy processes by Listeria monocytogenes (LM) is a serious issue because of the associated mortality rate in immune-compromised individuals. Dairy membranes can be modified to include solid-phase antimicrobials to kill or mitigate growth of LM. The feed spacer mesh of spiral wound dairy membranes has been functionalized with Quaternary Amine Compounds (QACs), a family of broad spectrum antimicrobials that are effective against Gram-positive bacteria. The antimicrobial is covalently bound to the polypropylene surface to avoid leaching into the product stream. Growth curve studies with a surrogate have demonstrated the effectiveness of the surface, and well as the necessary robustness to withstand clean-in-place conditions that will be required when implemented in industry. We have also functionalized the surfaces with polymerized phenolic compounds and comparisons were made with QACs. Preliminary results will also be presented in milk and brine systems.

608354 Technical and Economic Feasibility of Phosphorus Recovery As a Coproduct in Soybean Processing Facility

Abstract:

Phosphorus (P) in water bodies is a constant threat to the interconnected food, energy, and water (FEW) systems. Processing of grains produces nutrient rich wastewater that have adverse effects on water quantity and quality downstreams. Overenriched water leads to eutrophication, which can cause hypoxial or anoxial zones, threatening the aquatic life in water bodies, for example, hypoxial zone in Gulf of Mexico, and phosphorus control is a critical factor in abating this issue [1, 2]. Soybean and corn are the two major crops grown in the mid-west, and the processing of these grains produces a significant amount of P as waste that ends up in water bodies. The shift of food processing industries from meat-based to plant-based protein has helped mitigating the nitrogen run-off as well as greenhouse gas emissions. Soybean is a popular choice for the replacement of meat protein due to its high total protein yield and protein quality [3]. This has led to increase in beans processing, which produces wastewater high in phosphorus that enters the wastewater treatment plants. Investments in bean processing are constantly increasing and markets for bean products are estimated to grow in near future in North America to meet the demand of new fast food chains and restaurants offering plant-based meat analogs. These new plants will significantly change the load of nutrient in waste and groundwater. Common methods to remove P from wastewater is aerobic oxidation in the treatment facilities, which is expensive and results in captured P that are not usable for agriculture or other applications that could reduce recovery costs. The objective of this study is to use a chemical method developed previously to recover phosphorus from soybean processing facility [4, 5]. The second part of the study focusses on evaluating the technical and economic feasibility of phosphorus recovery from soybean processing facility on a commercial scale using process simulation. Phosphorus content in every stream in a biodiesel production facility was analyzed using ICP analysis. Soapstock and lecithin streams in the facility were observed to have highest concentration of phosphorus (21395 ppm and 2706 ppm, respectively), and were used to recover phosphorus chemically. The hypothesis of the study is that recovering phosphorus upstream would lessen the load of phosphorus in the wastewater from the plant, and the recovered phosphorus would be applicable as fertilizer on agricultural lands. 1526 MT/day soybean processing facility was simulated in SuperPro Designer, and phosphorus flow in streams was included. In the next part of the study, a separate P recovery unit will be modeled, with experimental data generated. The model simulations would provide the experimentally quantified P that is recovered, the fixed and operating cost of the P recovery unit, energy use, and overall revenues generated due to the addition of a co-product, in comparison to the existing soybean plant. Sensitivity analysis will be performed by varying the P recovery conditions that would identify an optimum cost of recovery based on input requirements.

References

1. Rabotyagov, S.S., et al., *The economics of dead zones: Causes, impacts, policy challenges, and a model of the Gulf of Mexico hypoxic zone.* Review of Environmental Economics and Policy, 2014. **8**(1): p. 58-79.
2. Carpenter, S.R., *Phosphorus control is critical to mitigating eutrophication.* Proceedings of the National Academy of Sciences, 2008. **105**(32): p. 11039-11040.
3. Stice, C., et al., *De-risking Protein Strategies Using a Systems Approach: Novel Analytical Framework.* 2015.
4. Juneja, A., et al., *Techno-economic Feasibility of Phosphorus Recovery as a Coproduct from Corn Wet Milling Plants.* Cereal Chemistry, 2019.
5. Juneja, A., R. Cusick, and V. Singh, *Recovering phosphorus as a coproduct from corn dry grind plants: A techno-economic evaluation.* Cereal Chemistry, 2020. **97**(2): p. 449-458.

594215 Determination of Furanic Compounds from Thermally Processed South African Indigenous Food By GC FID and Their Nutritional Analysis

Abstract:

Processing of food by direct exposure to heat is a good method for killing harmful bacteria but unfortunately the process of heating might also lead to the formation of toxic furan compounds. Reaction pathways including cleavage of sugars and starch present in food have been reported to lead to formation of furan derivatives. The main danger about these types of food (thermally processed foods) is that they are never tested for toxins after preparation, but rather are readily consumed. Therefore, humans are exposed to furanic compounds through the ingestion of foods that have undergone heat treatment as part of the preparation approach in both industrial field and domestic cooking, but the concentrations of such compounds depend on different parameters used by each field. Several studies have been conducted to quantitatively determine the occurrence of furan in different foods using various techniques. However, no information has yet been reported for indigenous South African foods like *Amasonja* (Mopani worms). The current study seeks to develop an analytical method for the determination of furan compounds in Mopani worms, sweet corn and roasted peanuts. The selected foods are all prepared by direct exposure to heat and that might lead to production of toxic furan compounds and possibly posing health risks to consumers.

610966 Algae Turf Filter Photobioreactor Designs, Operation and Testing for Algae Cultivation for Space Explorations

Abstract:

Attached growth cabinet-type reactors were designed for cultivation of *Chlorella vulgaris* on synthetic wastewater as a nutrient source. Two designs are compared in this study for biomass production. Algae cultivation requires biological, chemical and physical factors for algae biomass production. Biological factors in the algae culture inoculum are considered to minimize competition for resources with other predators and microorganisms by decontamination and sterilization. Chemical factors include the nutrients needed for growth at known concentrations determined in a previous study. Physical factors involve illumination for photosynthesis, which is measured as photosynthetic usable radiation (PUR). One reactor design has plates to hold water and algae in modules for water treatment and biomass production. Another design consists of drawers and plates with baffles and a weir to hold water for nutrients for attached algae growth. The modules or drawers are operated in similar to gas absorption column is operated, whereby the air and water flowrates are measured to deliver nutrients and carbon dioxide to the algae. For a laboratory scale set-up, an air pump can deliver air while water in a reservoir at the top was delivered by drop-wise gravity flow through a filter for delivering water. Engineering calculations on the mass transfer coefficient on the gas-water interface and biomass growth kinetic data evaluate the performance of the attached growth photobioreactor, as compared to a suspended growth photobioreactor at the laboratory scale. A system engineering analysis approach for the two designs reveals the essential attributes for operating them in international space station conditions. Biological factors suggested in the literature to enhance the growth of the algae on surfaces are also considered.

598340 Rational Design of Modified-Release Dosage Forms

Abstract:

Oral modified release (MR) dosage forms are developed by altering the kinetics and/or site of drug release/absorption in order to achieve specific clinical objectives, such as improved patient compliance, optimized efficacy and reduced adverse events. This presentation introduces basic design principles and different MR delivery technologies. It also discusses the importance of understanding material, formulation, process and their interplays in rational product and process development along with case studies.

602132 Strategies to Develop IV Formulations of Poor Water-Soluble Compounds and Their in Vitro Evaluation

Abstract:

Although a drug might be developed and formulated for oral delivery, the same compound is commonly administered as a solution via an IV injection/infusion in order to determine absolute bioavailability to support early development. A common challenge to develop IV formulation is the low solubility of drug candidates, risking phlebitis due to precipitation upon injection. Other parameters that need attention include the pH variation (too high or too low) that can cause irritation, and osmolality differences that may cause significant fluid shifts between the intracellular and extracellular spaces. Thus, the development of a stable and safe IV formulation requires an adequate selection of solubilizers and evaluation of the safety of the solution prior to administration.

In this study, a weak acid drug with poor water-solubility required a minimum of 40-fold increase in aqueous solubility for IV delivery. To achieve increased aqueous solubility, multiple strategies were assessed including screening of solvents and co-solvents, in situ salt, and pH control. Phlebitis risk was evaluated by in vitro static dilution using surrogate plasma and observation of flocculation in rat plasma by optical microscopy. In conclusion, a pH-buffered solution containing in situ sodium salt allowed for up to 133-fold increase in aqueous solubility of the compound showing no risk of precipitation while complying with pH and tonicity requirements for IV administration.

602618 Integrated Product & Process Development

Abstract:

During drug product development, science-based design of formulation and manufacturing process is essential for achieving and assuring consistent quality products. This presentation discusses rational product development based on the integrated product and process knowledge. It addresses the importance of fundamental understanding of physicochemical, mechanical and biopharmaceutical characteristics of API, formulation, process and their interactions during product design, development, scale-up and commercial manufacturing with case studies in ensuring consistent quality, robust drug delivery performance and process.

602619 Predictive and Model Based Product & Process Design

Abstract:

Use of predictive technologies and in silico modeling can significantly enhance speed, efficiency and effectiveness of drug product and process design. This presentation will discuss several case studies, in which multidisciplinary sciences and engineering principles, coupled with high power computing, are used to guide formulation development, manufacturability assessment, and drug substance solid form risk assessment.

605621 Rational Design of Amorphous Solid Dispersion

Abstract:

Amorphous solid dispersion (ASD) is a highly effective drug delivery technology for enabling oral absorption of insoluble APIs. This presentation discusses basic principles of formulation design & characterization, and importance of understanding material, formulation, dissolution (LLPS) and stability in rational product development using case studies

602599 Crystal Morphology Modification in a Nucleation Dominant Crystallization: A Tool from Traditional Chemistry

Abstract:

Crystal morphology engineering is an important component of process development in the pharmaceutical industry, since the crystal shape or habit directly influences downstream processing, formulation, and the physical properties of the API. Most commonly the desired habits of crystals are prepared through the modification of crystallization process parameters such as temperature, the crystallization solvent, and/or the addition of a co-solvent or an additive. However, in some cases, these methods do not successfully deliver the desired morphology. Here we present the use of Soxhlet extraction conditions as a valuable tool for morphology modification in a nucleation dominant crystallization process. In this case study, the process to produce a cocrystal between Vertex compound A and coformer X resulted in a fibrous morphology in every attempt under various conditions, leading to mixing and filtration issues and low yields. The use of the Soxhlet extraction method provided very slow mixing of the two compounds coupled with temperature cycling close to the solvent's boiling point to generate a needle morphology, which provided better filtration times, processability and yields, enabling a path forward for further API development.

599684 Process Development to Avoid Meta-Stable Liquid-Liquid Phase Separation during Crystallization of API

Abstract:

Crystallization is a critical unit operation in the production of small molecule active pharmaceutical ingredients (APIs), as it often determines the form and powder properties of the product. Therefore, crystallization process development focuses on identifying conditions that are controlled and scalable. In general, liquid-liquid phase separation or oiling out of product are undesirable as they may lead to uncontrolled nucleation and growth of particles that impacts critical quality attributes, such as particle size and level of impurities. Therefore, understanding liquid-liquid phase equilibria and solubility of the compound is critical in designing crystallization processes.

This presentation will describe how we experimentally investigated the liquid-liquid phase separation that occurred during crystallization of a small molecule API in a ternary solvent system of tetrahydrofuran, acetone and water. Although these solvents are typically miscible, the presence of the API resulted in liquid-liquid phase separation at the seed point under some conditions. In-line process analytical tools such as ParticleViewer and FBRM were incorporated to detect and understand the phenomena. A detailed phase diagram was generated for the crystallization design space as a function of solvent composition, concentration of API in solution and temperature. Along with solubility data, the phase diagram was used to design a process that avoids liquid-liquid phase separation throughout the entire crystallization. In addition, we will describe the impact of wet-milling during seed age that was implemented to ensure that full desupersaturation was achieved, which was critical in avoiding phase separation.

596447 Decoupling Nucleation and Growth for Continuous Crystallization in Microfluidics

Abstract:

1. Introduction

Microfluidics provide several advantages such as high efficiency, easy control and safety over conventional, and mostly batch reaction systems[1]. Therefore it can be seen as a replacement for batch crystallization as one of the most important separation and purification process especially in pharmaceutical industries[2,3].

The aim of this project is the design of a modular and decoupled microfluidic system consisting of separate nucleation and growth sections. The strategy persued in this project is using the two- phase interface in microreactors as a heterogeneous nucleation site. In this regard micro- bubbles are introduced in the system to control the location of the nucleation site and particle aggregates and to limit wall interactions. The velocity of the bubbles determines the residence time and using a gas phase enables a straight forward separation, see Figure 1.

2. Experimental study

2.1. Nucleation section

For designing the nucleation section, in the first step, the generation of micro-bubbles (i.e. nitrogen as an inert gas) in capillaries is studied. The total bubble surface area per unit volume of the reactor is related to the bubble diameter and the number of bubbles present in the reactor volume. Bubble diameter, bubble velocity and bubble generation frequency were extracted from the images captured by the high-speed camera using an inhouse-generated Matlab script. Using this information, the bubble surface area available as a heterogeneous nucleation site is further calculated and the effect of gas and liquid flowrate is investigated, see Figure 2.

This is followed by a systematic study of nucleation of an organic compound (i.e. cooling crystallization of paracetamol) using the developed system employing microbubbles. The particle size distribution is measured at the outlet of the channel subjected to the microbubble residence time and the nucleation and growth kinetics are extracted.

The continuous crystallization experiments provided proof that the micro-bubbles act as a heterogeneous nucleation site. An enhanced nucleation rate compared to the operation without bubbles was observed, and the presence of micro-bubbles results in the formation of more crystals, which indicates that nucleation is faster than without bubbles, i.e. the metastable zone width is reduced, see Figure 3. Furthermore, quantifying the crystal yield showed an increase by more than a factor of 2 in the presence of micro-bubbles.

2.2. Growth section

Further on a growth section is added to the nucleation section which is kept at the constant temperature equal to the outlet of nucleation section to allow formed nuclei to grow. Sample is collected at the outlet of the growth section for further analysis of particle size distribution, see Figure 4.

4. Conclusions

A decoupled crystallization set-up consisting of different nucleation and growth sections is designed using microreactors. Nucleation is enhanced using microbubbles acting as heterogeneous nucleation sites providing lower surface energy interface for the nucleation to start. Further a growth part is added to allow the formed nuclei grow within the constant temperature growth section.

In case of the continuous crystallization of paracetamol the results allow the conclusion that the low-energy heterogeneous surface introduced by the bubbles promotes crystal nucleation. In addition, adding micro-bubbles enables the continuous crystallization for long operating times, as the bubbles act as carriers for the formed crystals and effectively prevent wall deposition and thus clogging.

References

[1] Gunther, K. F. Jensen, Multiphase microfluidics: from flow characteristics to chemical and materials synthesis, Lab Chip , 2006, 6, 1487–1503

[2] S. Flowers, R.L. Hartman, Particle Handling Techniques in Microchemical Processes, Challenges 2012, 3, 194-211

[3] Wu, S. Kuhn, Strategies for solids handling in microreactors. Chim Oggi. 2014; 32(3):62-66.

[4] Fatemi, Z. Dong, T. Van Gerven, S. Kuhn, Microbubbles as Heterogeneous Nucleation Sites for Crystallization in Continuous Microfluidic Devices. Langmuir (2018). doi:10.1021/acs.langmuir.8b03183

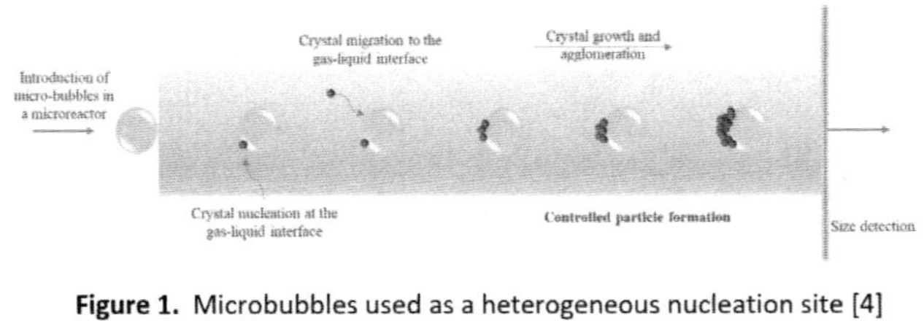

Figure 1. Microbubbles used as a heterogeneous nucleation site [4]

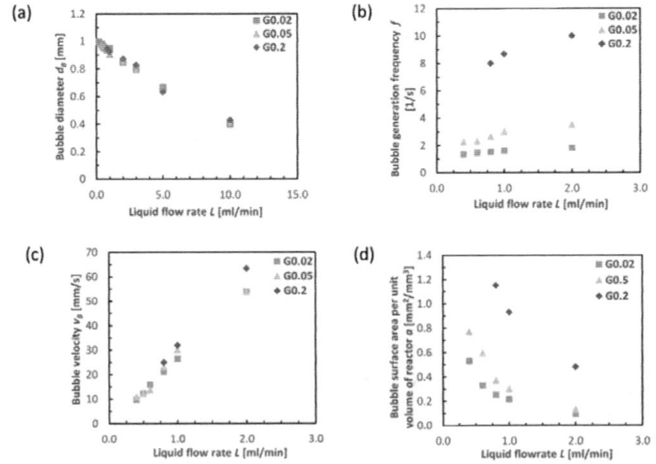

Figure 2. Effect of gas and liquid flow rates on (a) size of the microbubbles; (b) frequency of microbubble generation; (c) microbubble velocity; and (d) microbubble surface area per unit volume of the reactor for [4]

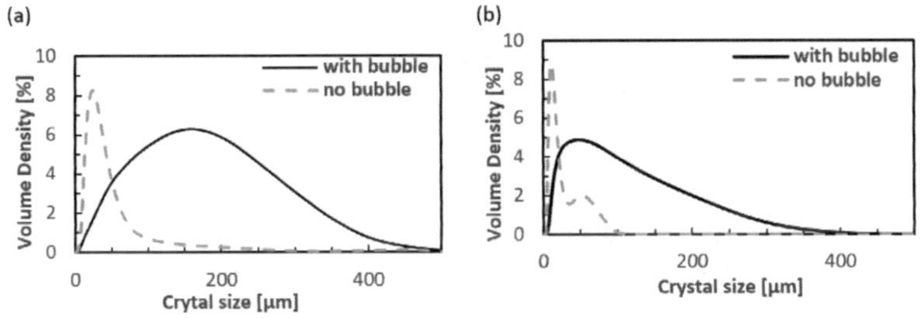

Figure 3- Effect of micro-bubbles on the crystal size distribution for a growth time of 1 minute and a supersaturation of S=1.81 and the gas flow rate G=0.05 ml/min and liquid flow rate of: (a) L=0.5 ml/min, (b) L=1 ml/min

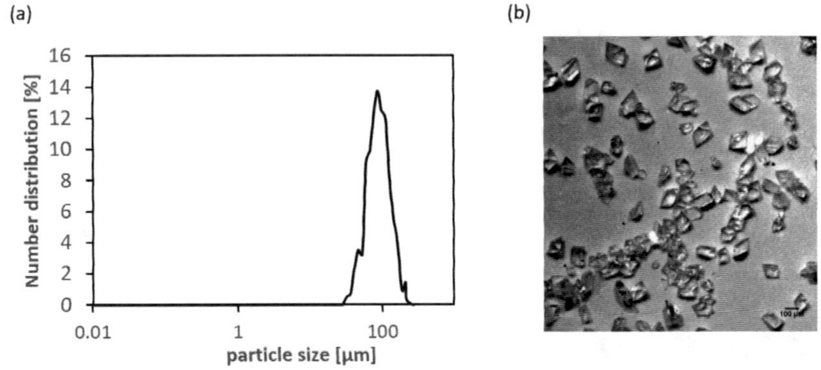

Figure 4- (a) particle size distribution, (b) microscope images of paracetamol crystals for growth section with the length of 1.4m at T=25°C with a supersaturation of S=1.81 and the gas and liquid flowrates 0.05 and 1 ml/min, respectively

597227 Effects of Volume Reduction on Glycine Nucleation

Abstract:

Crystallization in confinement is a technique used both as an approach for nanocrystal development and as way to gain access to unstable crystal polymorphs. To elucidate this, the effect of decreasing volumes in the nucleation of glycine in water was explored. Using a supersaturated aqueous solution of glycine at decreasing volumes from 1000 μL to 25 μL, induction times for primary nucleation were obtained experimentally. Analysis of probability distribution of β-glycine induction times showed that as sample volume decreases, induction time increases. Image analysis revealed that β-glycine formation serves as the primary nucleation event; however, primary nucleation shifts toward γ-glycine nucleation at volume sizes below 100 μL. These results demonstrate that feasibility of accessing metastable and unstable crystal polymorphs with the aid of confinement. However, confinement also increases induction times which relates to very low probabilities of a nucleation event.

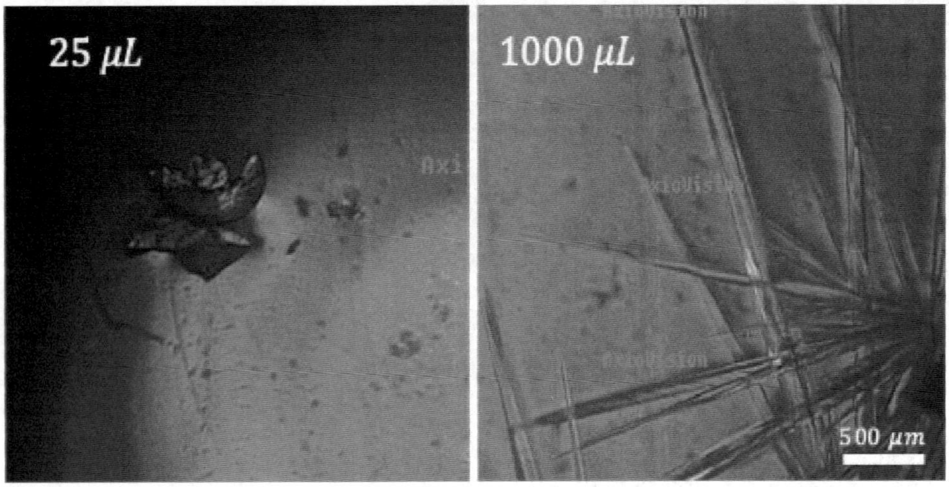

597873 Formation of Solid Solutions As a Key Impurity Retention Mechanism in Solution Crystallization in the Presence of Structurally Similar Impurities

Abstract:

The main purpose of crystallization is to purify the product by removing undesired impurities. In pharmaceutical industry, crystallizations at different steps in the synthesis are relied upon to control the final quality of the active pharmaceutical ingredient (API). During these operations, the main challenge often stems from dealing with structurally similar impurities, which tend to be difficult to remove. Presented herein is an investigation into how this type of challenging impurities are purged mechanistically in the crystallization, using pharmaceutical case studies. Through these examples, it is shown that formation of substitutional solid solutions is the underlying mechanism for impurity retention. Impurities molecularly replace the product molecules in the crystal lattice during the crystallization. This lattice incorporation entrapment occurs unevenly during the crystallization resulting in a gradient in impurity levels throughout the material. A mathematic framework combined with a previously reported SLIP test is used to elucidate this varying impurity entrapment through a so-called *material impurity distribution* (MID). The impurity levels are seen in many cases to change by an order of 20 across the material. The effect of the entrapment can also be observed in the physical properties of the material, e.g. by a significant reduction in crystallinity, increase in solubility and earlier onset melting event.

609881 Design of a Continuous Crystallization for a Drug Substance Using a Dynamically Mixed Reactor

Abstract:

The pharmaceutical industry has predominantly relied on batch processes for manufacturing of drug substance as well as drug products. However, longer campaign times, labor-intensive nature of batch manufacturing and batch to batch variations in the product quality pose significant drawbacks. In view of this, the Pharma industry is turning towards continuous manufacturing, going beyond flow chemistry to continuous crystallization and downstream operations for the synthesis of drug substances. Continuous crystallization offers many advantages such as better control over the solid form and particle size, and the potential for an end-to-end continuous manufacturing process.

The current work describes the design of a continuous crystallization process for a drug substance in order to integrate with upstream and downstream continuous processes. It was not only important to achieve the targeted solid state form (trihydrate, Form I) but also ensure that the filtration rate could be increased to enable integration with the downstream continuous filter and minimize the cleaning frequency. The presented case study discusses in detail the approach, starting from lab-scale batch experimentation to development of a continuous crystallization process at lab followed by scale-up design for a capacity of multi-tons per month.

As a first step, fundamental solubility data was collected followed by identification of critical parameters impacting nucleation and transformation between the various forms. It was observed that the solvent (alcohol) to antisolvent (water) ratio, seed loading and temperature had a significant impact on the solid form. A batch crystallization process was first established by combining antisolvent and cooling crystallization to consistently generate the desired Form I, while ensuring the absence of other solid forms.

The translation of the batch process to continuous was initially attempted in MSMPRs in series but proved unsuccessful as an undesired form started to appear over time despite seeding. It was hypothesized that several more MSMPRs would be required to achieve the desired form but this would occupy a huge footprint and increase hold-up at scale. While plug flow reactors would be suitable, the risk of fouling is high, particularly when flow rates are reduced to achieve sufficient residence time. Scouting for an appropriate technology lead to evaluation of dynamically mixed reactors from AM Technologies®, which were able to handle a slurry and provide a mixing profile close to plug flow decoupled from the residence time. At lab scale, the Coflore® Agitated Cell Reactor (ACR), which is a bench-top mechanical agitated reactor comprising ten small cells, was able to aid in controlled crystallization of the desired Form 1 with well defined crystals and a consistent particle size distribution. Since additional reactors and residence time were required to gradually cool the slurry from 65°C to 25°C, the ACR was combined with three CSTRs in series (forming a mixed suspension mixed process set up). The trials were executed by combining a stream of alcohol containing the dissolved solute (at 65°C) and another stream of hot water containing seeds (at 65°C) in the ACR. The slurry exiting from the reactor was continuously transferred into the CSTRs, where it was subjected to stage-wise cooling in each CSTR. The residence times were designed to achieve maximum crystallization in the ACR and minimize secondary nucleation in the subsequent CSTRs.

After establishing the success and robustness of the continuous crystallization process at lab-scale, the process was scaled up in a 100L, production-scale version of the dynamic mixing reactor - Coflore®

Rotating Tubular Reactor (RTR). Optimization of the process parameters revealed that maintenance of higher temperatures (~65 °C), , enhanced mass transfer at the point of nucleation along with the presence of Form 1 seeds, and higher ratios of anti-solvent were critical for achieving the desired polymorph while low mixing shear aided in crystal growth. The growth of these fine, needle-shaped crystals was essential to achieving a sufficiently high filtration rate for the downstream continuous separation of solids. The RTR, which operates as a ten-stage, actively mixed, self-baffled reactor, provided high wall shear with radial mixing while maintaining a low mixing shear, thus enabling crystal growth and improving the filtration rate while minimizing fouling.

In summary, the work presents key aspects of design and scale-up of a continuous crystallization for a drug substance with regard to both process and technology selection. One of the main advantages of this continuous process was consistency in the solid form, particle size distribution resulting in improved filtration of the slurry, and ability to integrate with upstream and downstream continuous processes.

Internal communication number: IPDO IPM-00637

605997 Improvement of Drug Substance Bulk Powder Properties through Process Crystallization

Abstract:

The existing recrystallization process for a drug substance produced crystalline needle-shaped aggregates with low bulk and tap density and very poor flow properties. The particle properties led to poor performance of the drug substance during formulation. The goal of this work was to develop a process to yield drug substance with improved bulk powder properties for future formulation development and manufacturing work.

The approach taken for this work involved the use of computational and experimental solvent screens to rapidly narrow down the choice of solvent systems and working conditions. The computational work also had the advantage of significantly shortening development timelines.

Ultimately a crystallization process was developed which produced drug substance with enhanced particle morphology and significantly increased bulk density.

611726 Singly-Twinned Growth of Si Crystals upon Chemical Modification

Abstract:

During crystallization a continuum of patterns could emerge due to the interplay of growth kinetics, material or solution chemistry, and crystallographic defects. The coherent twin boundary is widely known to catalyze growth in pristine crystals including polycrystalline Si. Much remains unknown about the impact of changing the chemical environment of the crystallization process through the deliberate addition of trace metallic species – termed chemical modification. Pristine Si has been reported to grow through the classical model of two parallel twin planes acting in concert to enable steady-state propagation of the solid-liquid interfaces. Here, we achieve a new vision on the growth process *via in situ* synchrotron X-ray microtomography and further corroborated by *ex situ* crystallographic investigation. We find that steady-state growth is impossible in chemically-modified alloys that consist of trace (0.1 wt.%) Sr. This is because the Sr modifier poisons the concave re-entrant grooves, thereby de-activating the advantage of the twin-plane re-entrant edge mechanism and leading to a singly-twinned interface. This study may serve as a proxy to chemically-modified crystallization pathways of eutectic Si in Al-Si alloys and, more broadly, as a framework for the crystallization-mediated synthesis of materials.

597305 Elucidating the Influence of Hydroxycitrate on Brushite Crystallization Via Bulk and Interfacial Studies

Abstract:

Kidney stone diseases (nephrolithiasis) affect 12% of the world population and can occur on all ages, sexes and races. Of all the cases, calcium stones are the most prevalent and 20% of these are calcium phosphates (CaP). Brushites, or dicalcium phosphate dihydrate, are one of the types of CaP stones and known to induce calcium oxalate mineralization, and to be converted to less soluble form hydroxyapatite. Kidney stone incidence has increased about three times for the past decades. Efforts in identifying alternative inhibitors of stone growth remain one primary strategy in preventing the increasing occurrence of kidney stones. Once identified and obtained clinical clearance, this may provide patients more choices for therapeutics against pathological calcification. Fundamentally, carboxylates are known to inhibit crystal growth of these minerals and one example is hydroxycitrate (HCA) with have effectively curbed calcium oxalate mineralization. To tackle this goal, we employed a combination of bulk and atomic force microscopy studies to elucidate the impact of HCA on the crystal size, morphology, number density of CaP. Analysis shows a 50 % reduction in the number of crystals formed and an increase in the aspect ratio at a very small inhibitor concentration (Ca/inhibitor=25000). Amplitude and Frequency Modulated AFM studies show that surface roughness increased through time in the presence of the HCA. Changes in the periodic features at near atomic resolution was also observed in the presence of HCA implying possible modifier-crystal surface interactions. Results obtained in this study may serve as a step in understanding the inhibitors' effect on CaP crystals and in designing preventative drugs for kidney stone disease.

602826 Islatravir Penultimate Step Crystallization: Purification of an API in the Presence of High Protein Loading from an Enzymatic Cascade

Abstract:

Islatravir is a novel nucleoside active pharmaceutical ingredient (API) currently in clinical trials for both HIV treatment and prevention. As reported recently in *Science (*Huffman et. al. 2019*)*, the API synthesis under development involves a 9-enzyme 3-step in-vitro cascade without intermediate isolations until the final API molecule is assembled. This process represents an incredibly efficient synthesis of the API molecule, but introduces significant challenges to the downstream purification of the product. Enzymes, host-cell proteins, bacterial endotoxins, residual nucleic acids, and organic impurities/related substances all must be removed to generate a drug substance of adequate purity for human dosing.

The primary challenge in the purification development has been on enzyme/protein removal. The presence of high levels of protein tends to make filtrations very slow, so extractive work-ups are often preferable, but the relatively low solubility of the API in most organic solvents was a formidable obstacle to this approach. Several extractive work-ups were developed but each posed stability concerns.

Additionally, enzyme immobilization was pursued for the enzymes with the highest loadings in the reaction, which improved downstream processing but added a significant cost and supply-chain complexity.

In the final reaction, the API crystallizes from the reaction mixture, and even wth several enzymes removed via immobilization, filtration was still very challenging. Filtration was optimized by adding filter aid (diatomaceous earth), reducing shear in the reaction, adding seed crystals to the reaction, and agitating the batch intermittently during filtration. Optimizing the filtration step to isolate the crude API made it possible to produce quality API without the need to develop an extractive work-up or use immobilized enzymes.

601168 Development of a Crystallization Process for the Final Step of the Biocatalytic Synthesis of Islatravir

Abstract:

Merck recently reported a biocatalytic cascade for the manufacture of Islatravir, a potent HIV reverse transcriptase translocation inhibitor [1]. The API step of the synthesis requires removing protein and other impurities accumulated throughout the synthesis. Also, the low dose of the drug requires API particle size control in the final step to mitigate content uniformity concerns in the tablet. This industrial case study will provide an overview of the development of a pure step that meets purity and particle size targets. The presentation will highlight the development of a novel crystallization process to meet the desired particle size distribution while maintaining a rapid filtration rate of the product crystals. The scale-up crystallization performance and impurity rejection of the process will also be compared to a more 'traditional' seeded crystallization using jet milled seed. The development of the process was motivated by the desire to eliminate the cost and industrial hygiene liabilities associated with dry milling API as seed.

[1] Huffman et al. "Design of an in vitro biocatalytic cascade for the manufacture of islatravir." Science, 2019, 366 (6470), 1255-1259.

598747 Impact of the Process Conditions of an Evaporation to Dryness Operation Performed in a Conical Screw Vacuum Dryer on the Isolated Product Psd: Lab- and Pilot-Scale Studies

Abstract:

Conical Screw Vacuum Driers (CSVD) are frequently used in the pharmaceutical industry for drying heat-sensitive powders to a very low volatile matter content in a low shear mixing environment. This type of dryer can also be used to perform evaporation to dryness operations. Evaporation to dryness operations, routinely executed in laboratory preparations but less frequently at industrial scale, aim solely to recover solids from solutions by complete evaporation of the solvent. A manufacturing process industrialized by UCB-Pharma includes an evaporation to dryness operation executed in a CSVD. It follows a multi-column chromatographic separation, yielding the purity level required and it comes before a re-slurry of the powder in a suitable solvent. After solid-liquid separation and drying, the powder isolated from the re-slurry operation is milled to meet the API granulometry specifications of a mono-modal distribution, characterized by a d(v, 0.5) and a d(v, 0.9) within defined target ranges.

Following a process optimization exercise to increase productivity of the chromatographic unit, the solvents used were changed. This also resulted in a significant change in the typical Particle Size Distribution (PSD) observed after the evaporation to dryness operation. Risks associated to not knowing to what extent the final dry milling operation was able to mitigate those unpredicted changes in the control of PSD, triggered this comprehensive investigations on the CSVD process at different laboratory and manufactory scales. Given the change of performance driven by the solvent changes, an additional aim of this work was also to identify process conditions for the CSVD that might have been able to yield directly material within specifications.

A Design of Experiment (DoE) approach was used to determine the impact of the process conditions and the scale-up of the evaporation to dryness operation investigated here on the PSD of the isolated powder. A first DoE plan was carried out using a lab-scale CSVD (300 g scale) under well controlled process conditions. A second DoE was also reproduced at pilot plant-scale (30 kg scale) using appropriate scale-up criteria for those process parameters identified as critical, like the evaporation rate. The heat transfer characterisation of the equipment at laboratory and pilot plant scale allowed for establishing suitable procedures to control critical scale-up process parameters like jacket temperature and operating pressure.

The complexity of the mass and heat transfer phenomena occurring during the evaporation to dryness were elucidated and rationalised in a multiphase process steps while executing the experimental plan (Figure 1). Each process step was clearly defined by specific rheological behaviours. Evidence of patterns and reproducible mechanistic behaviours were found to be linked to specific operating conditions independently from the scale used for the operation. However, macroscale phenomena linked to the batch size of the operation together with exacerbation of the mixing conditions at scale led to mixing zone segregation. These differences ultimately significantly impacted the reproducibility of the performance of the operation under the same operating conditions at scale.

The complexity of the phenomena and phases involved was rationalised in a model behaviour and optimal process operating conditions and procedures to maximise process robustness. The operation was successfully validated at final production scale (300kg) with material produced according specifications over a number of validation batches.

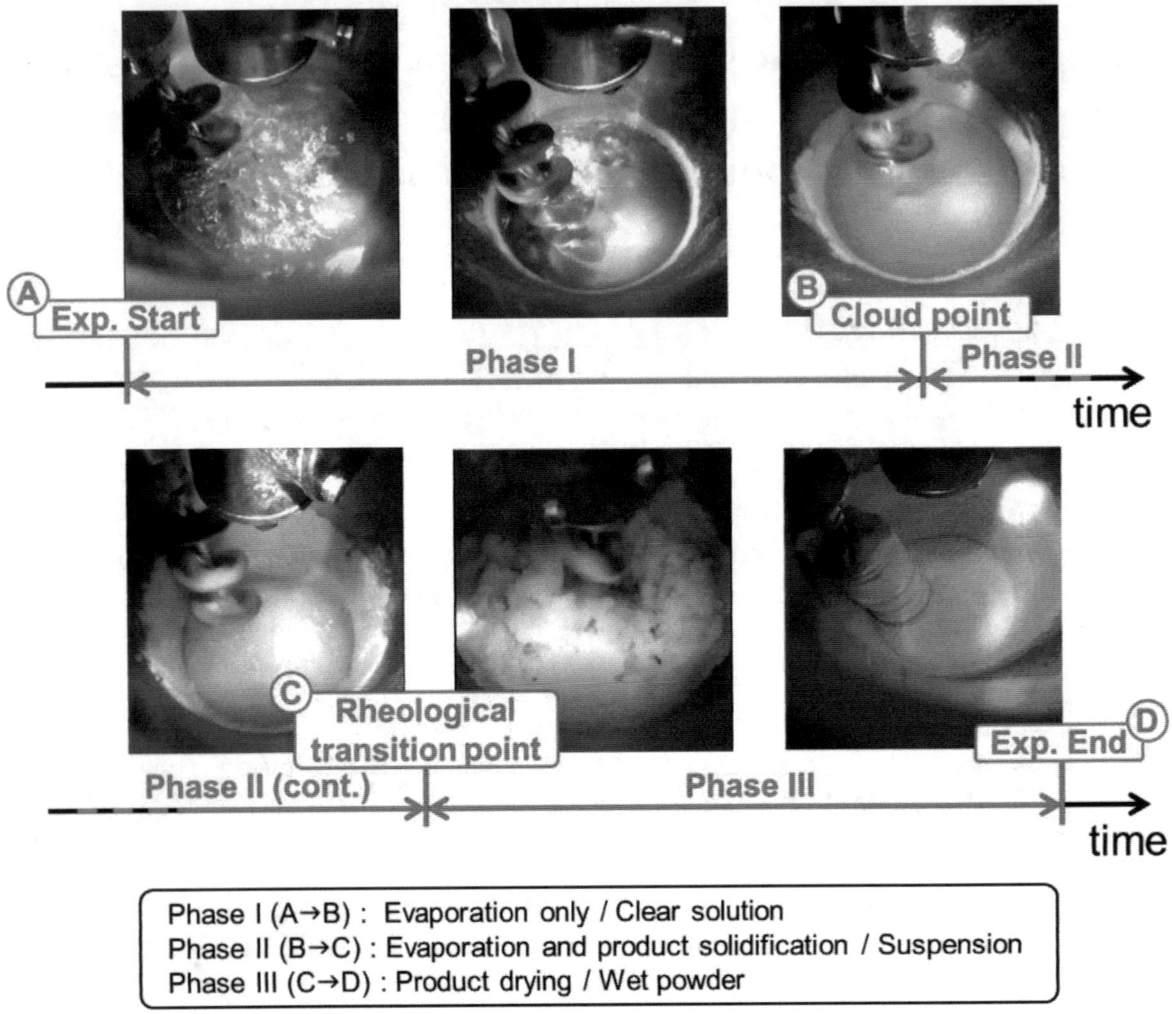

Figure 1 : Observed process phases during the evaporation to dryness experiments

610946 A Generalized Approach for Rapid Aqueous MOF Synthesis By Controlling Solution pH

Abstract:

Metal organic frameworks (MOFs) are a promising class of crystalline materials with potential to significantly advance the fields of catalysis, separations, biotechnology, and sensing, amongst others. Comprised of metal ion nodes and organic linkers, lab-based syntheses have excelled at producing novel MOFs with tuned pore size and chemical activity. However, developing a crystallization technique that meets high throughput scaling requirements and is environmentally benign remains a challenge. Currently, applying classical crystallization techniques to MOF formation are limited due to the multi-component nature. In this work, we approach understanding MOF formation through a reaction-based perspective to better understand the non-classical crystallization pathway. We demonstrate that tuning the reactant speciation, namely labile metal ion nodes and deprotonated organic acid linkers, via pH has a significant effect on MOF yield. Using this approach, we develop a generalizable technique to rapidly produce a range of MOFs, including UiO-66-NH_2, UiO-66, HKUST-1, and ZIF-L, under ambient, aqueous conditions. We further this understanding by tuning the labile metal node and deprotonated linker concentrations to produce space-time yields (STY) exceeding 2250 kg m^{-3} day^{-1}, which is among the highest reported STYs for HKUST-1 and ZIF-L and an order of magnitude higher for aqueous zirconium-based MOF synthesis. Finally, we demonstrate UiO-66-NH_2 completes crystallization in 5 minutes at room temperature with 70% of crystallization of UiO-66-NH_2 occurs within 30 seconds, which is among the fastest reported reaction times in literature. With this work, we demonstrate a generalizable pathway for scaling MOF crystallization to high STY under mild and benign synthesis conditions and lay ground work for understanding the complex multi-step crystallization kinetics of MOFs.

611699 Towards Integrative Mechanistic Models of Mammalian Cell Responses to Anti-Cancer Drug Combinations

Abstract:

A key missing capability in current cancer research is the ability to predict how a particular single cancer cell will respond to a drug cocktail. Yet, it is not even possible to perform this task well for normal healthy cells. This work builds on the hypothesis that first principles, mechanistic models of how cells respond to anti-cancer drugs will improve drug combination response predictions. However, building such single-cell models of complex, large-scale, and incompletely understood systems remains an extremely challenging task. To address this issue, we defined a pipeline to convert structured lists of species, parameters, and reaction types into an SBML (Systems Biology Markup Language) model file and created a model based on one of the largest pan-cancer signaling models in the literature. The previous model (774 species, 141 genes, 8 ligands, 2400 reactions) developed by our lab already incorporated major pathways of receptor tyrosine kinase signaling, proliferation, cell cycle, apoptosis, DNA damage, transcription, and translation. The new model format is available online and generated using open-source tools to accommodate community use and contribution. One key aspect in our effort is our model's compatibility with high performance computing. We can simulate thousands to millions of single cell trajectories, enabling us to study virtual cell population responses to drugs and drug combinations. Another aspect of our model is its ability to accommodate addition of new reactions or pathways. As a proof of concept, we created an even larger model with the addition of Interferon-γ (IFNγ) signaling pathway, resulting in a model with 950 species, 150 genes, 9 ligands, 2500 reactions. All in all, our new model format enables easy modification of large mechanistic models and simulation of thousands of single cell responses to drugs and drug combinations.

610720 Sensitive Image-Based Readout of DNA Barcodes Enables Cell Lineage and History Recording

Abstract:

Multicellular development depends on the molecular history of individual cells and their descendants, but these histories are usually inaccessible to direct observation. Synthetic DNA barcodes allow us to reveal clonal relationship, lineage history, and even molecular trajectory of individual cells over time. Therefore, they can be the key to a wide variety of otherwise intractable questions. However, readout of barcodes presents a tradeoff: Sequencing methods can analyze short barcodes containing a high density of sequence variants, but necessarily lose spatial information. Conversely, imaging methods preserve spatial structure, but can only distinguish lower density sequence features. Here, I present a system for sensitive readout of DNA barcodes that makes high density barcoding and synthetic recording compatible with imaging readout. This system enables reliable in situ detection of barcodes as short as 20bp. By concatenating these barcodes, we made combinatorial lentiviral libraries to investigate clonal relationship of cells in developing nervous system. Further, we showed that probe competition enables precise discrimination of single nucleotide variations in the barcodes. This allowed us to make synthetic memory arrays that record information as precise mutations made by CRISPR base editors and can be readout by microscopy in individual cells. We are now applying these technologies to improve our understanding of cell fate specification in mammalian retina. Decades of study have revealed many aspects of retinal development such as broad developmental potential of retinal progenitors and heterogeneity in their clonal composition. Nevertheless, it remains unclear how the regular structure of retina is built by clones with mostly random composition. Our goal is to make a generalizable system that can comprehensively map the clonal composition and lineage structure in the retina. Combined with multiplexed in situ transcriptional profiling, this system will allow us to link information about lineage, cell fate, and spatial organization of individual retinal cells.

609750 Xenon and Argon Microbubbles for Ultrasound-Guided Therapeutic Gas Delivery

Abstract:

Noble gases like xenon (Xe) and argon (Ar) are known to have cytoprotective effects in treating hypoxic ischemic injuries. There are clinical trials currently underway with Xe which involve inhalation, making the process non-specific, expensive, and prolonged. Our objective was to design a cheaper, more specific method of simultaneous gas delivery and imaging at the injury site. In the last few years, reports have shown local release of Xe gas with ultrasound-activated Xe-containing liposomes; however, liposomes suffer from negligible *in vivo*ultrasound response and low Xe payload due to the gas being trapped within the bilayer. To overcome such problems, we have developed noble gas (Xe, Ar) containing microbubbles (MBs) encapsulated by optimized phospholipid shell compositions capable of high payload and acoustic contrast. MBs contain between 10-100x the amount of gas payload than liposomes, while being extremely echogenic under clinical ultrasound frequencies. Traditionally, MBs contain an insoluble fluorocarbon core gas to prevent dissolution of the bubble in aqueous media. In recent years however, MBs have shown promise as carriers of therapeutic gases, such as oxygen. The first obstacle to the design of these bubbles has been to stabilize the core gas – Xe, Ar, and oxygen are several times more water-soluble than fluorocarbons – to prevent efflux from the bubble into the bloodstream due to the high chemical potential gradient. The other major challenge to this has been the ability to preserve the contrast capability of microbubbles while keeping the therapeutic gas core stable. Phospholipids like distearoylphosphatidylcholine (DSPC, C18:0) have been reported to transiently stabilize other water-soluble gases such as oxygen under closed conditions. We found that the long-chain phospholipid dibehenoylphosphatidylcholine (DBPC, C22:0), with a rigid packing structure and lower permeability, is capable of stabilizing bubbles within a 1-10 µm diameter range when using a high-energy, open-to-atmosphere method like probe-sonication, followed by differential centrifugation for size separation. In addition, due to its lower membrane residence time, shorter-chain DSPC (C18:0) is able to form 1-10 µm bubbles by a lower-energy, closed-system, shaking method. This commercially available shaking method, however, lacks long-term stability and scale up potential. The quantity of gas encapsulated in MBs is determined by gas chromatography-mass spectrometry (GC-MS) and by a novel cryptophane quenching fluorimetry method. These MBs were also found to remain storage-stable for days to weeks in concentrated populations. Xe and Ar microbubbles provided excellent non-linear (bubble-specific) ultrasound contrast, lasting 3-5 min in phantom studies. On intravenous retro-orbital injection into a mouse, a bolus of Xe bubbles provided distinct and high contrast in both cardiac (pre-lung) and renal (post-lung) imaging settings. Ongoing studies are exploring the imaging and therapeutic capability of Xe bubbles in a traumatic brain injury model in pigs. Additionally, preliminary studies indicate the ability of Xe MBs to maintain the viability of a model cell line under cobalt chloride induced hypoxic stress in *in vitro* conditions. In summary, this study presents a matrix of rigorously optimized shell-composition parameters for formulation of pure Xe and Ar microbubbles. We also demonstrate prolonged *in vivo* echogenicity of such bubbles for the first time, making these agents promising carriers for image-guided localized therapeutic gas delivery.

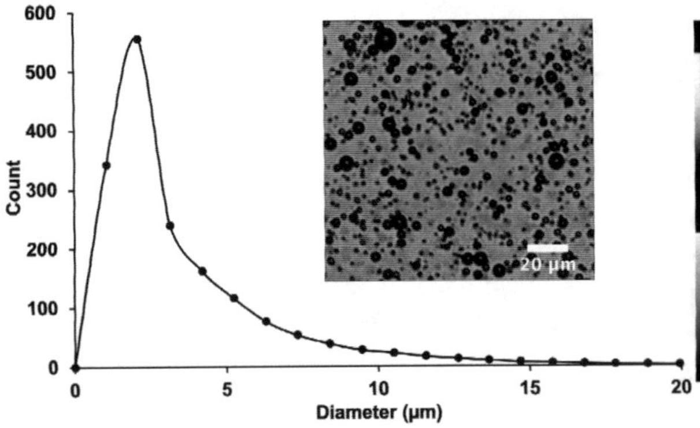

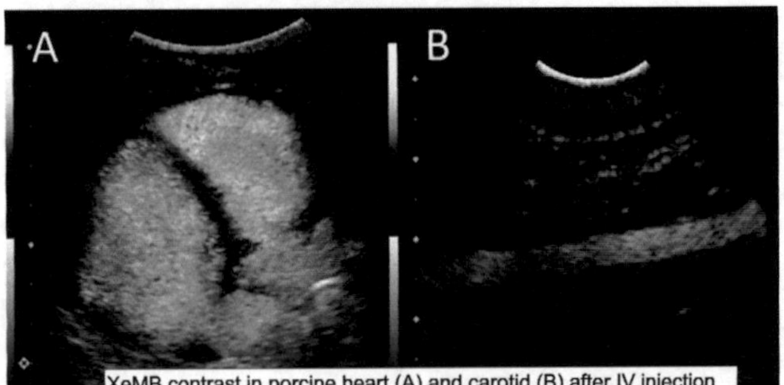

XeMB contrast in porcine heart (A) and carotid (B) after IV injection

608975 Biology-Inspired Trojan Horse Strategies for Drug Delivery and Immunomodulation

Abstract:

Delivering therapeutics to target sites while avoiding off-target accumulation has been the key in designing next generation therapies with desired efficacy and acceptable toxicity. Drug delivery has nevertheless remained a challenge due to the presence of various biological barriers preventing therapeutics from reaching their target sites. To address these challenges, a plethora of drug delivery systems including nanocarrier systems have been widely explored to improve the biological efficacy of respective therapeutics. However, majority of nanocarriers have so far failed to show desired accumulation in target sites, attributed to key hurdles such as poor circulation and the inability to negotiate with other significant biological barriers. Inspired by nature, intrinsic biology has provided many successful solutions to naturally overcome biological barriers, such as circulatory cells, which uniquely tackle hurdles faced by synthetic delivery systems. With this in mind, we have explored biology-inspired Trojan Horse strategies ("cellular hitchhiking") for directing therapeutics to their target sites. In this talk, I will discuss the principles underlying "cellular hitchhiking" and our ongoing efforts in leveraging this concept to address fundamental delivery challenges in cancer chemotherapy, cancer immunotherapy and immunomodulation. This biology-inspired concept represents a unique approach to solve critical problems at the interface of biology, materials, and medicine.

597820 Macrophage Checkpoint Blockade: From Cell Therapy and CRISPR Models to Acquired Immunity

Abstract:

The inhibitory receptor SIRPα on macrophages binds to the 'marker of self' CD47 expressed by all cells to impede phagocytosis. Blocking this macrophage checkpoint to enhance phagocytosis of cancer cells is an emerging cancer therapy, but clinical trials indicate that anti-CD47 monoclonals (i) are ineffective as monotherapies, (ii) have limited success in solid tumors, and (iii) deplete healthy blood cells. Preclinical development of anti-CD47 biologics against liquid tumors in immunocompromised mice do not account for these challenges. Therefore, we used the B16 murine melanoma model, which is syngeneic and weakly immunogenic in immune competent C57 mice, and first deleted CD47 in B16 cells with CRISPR/Cas9. Knockout of CD47 enhanced phagocytosis of B16 cells by primary mouse macrophages more than twofold, but opsonization with a mouse monoclonal against the melanoma antigen Tyrp1 was required for phagocytosis. Knockout tumors were repressed by systemic injections of anti-Tyrp1 at doses that had no effect on control tumors. Approximately 40% of mice survived knockout tumors with treatment, and more than half of these surviving mice rejected tumors when re-challenged. Treated mice generated IgG that recognized B16 surface antigens and functioned as opsonins. To target tumors expressing wild-type levels of CD47, we treated marrow-derived monocytes *ex vivo* with anti-SIRPα and co-injected with anti-Tyrp1, which extended survival while avoiding anemia caused by anti-CD47. Thus, solid tumors can be safely targeted with macrophage checkpoint blockade in combination with pro-phagocytic IgG antibodies – first administered as a combination therapy and later generated endogenously through adaptive immunity.

598055 *N*-Terminal Derivatization-Assisted Identification of Individual Amino Acids Using a Biological Nanopore Sensor

Abstract:

Nanopore technology has been employed as a powerful tool for DNA sequencing and analysis. To extend this method to peptide sequencing, a necessary step is to profile individual amino acids (AAs) through their nanopore stochastic signals, which remains a great challenge due to the low signal-to-noise ratio and unpredictable conformational changes of AAs during their translocation through nanopores. We showed that the combination of an *N*-terminal derivatization strategy of AAs with nanopore technology could lead to effective *in situ* differentiation of AAs. Four different derivatization reactions have been tested with five selected AAs, *i.e.* Ala, Phe, Tyr, His and Asp. Using an α-Hemolysin (α-HL) nanopore, we demonstrated the feasibility of derivatization-assisted identification of AAs regardless of their charge composition and polarity. The method was further applied to discriminate each individual AA in mixture samples using their established nanopore profiles. We envision this proof-of-concept study will not only pave a way for identification of individual AAs but also lead to future applications in protein/peptide sequencing using the nanopore technology.

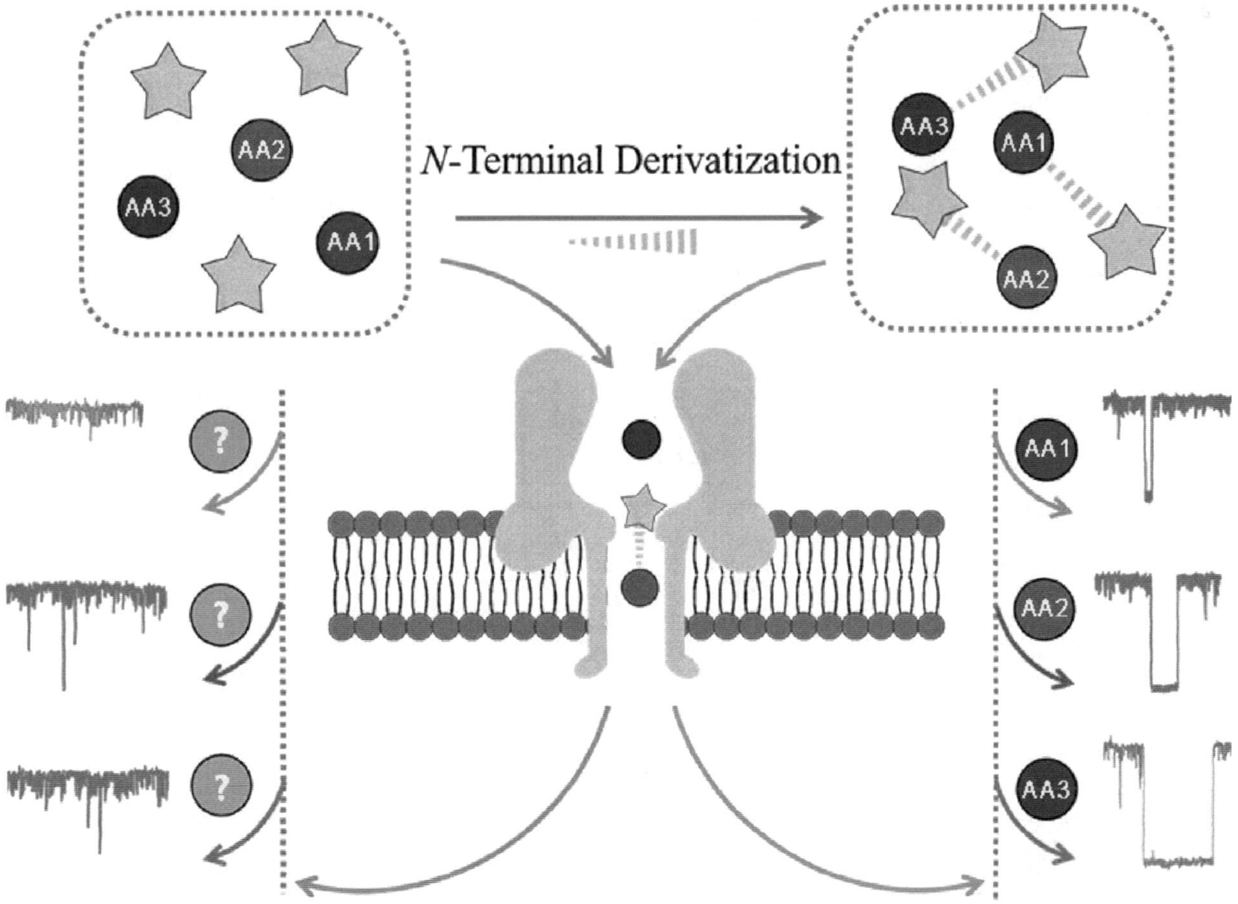

593820 Harnessing Orthogonal tRNA for *De Novo* Generation of Genetic Codes

Abstract:

In living organisms, ribosomal translation transforms genetic information embedded in DNA into actuating components, namely proteins. Though life itself is incredibly diverse at the macroscopic level, at the molecular level, all of life uses the same set of 20 standard amino acid building blocks (with minor exceptions), transfer RNAs (tRNA), and ribosomes to carry out translation. The convergence and association of these interdependent biomolecules is neatly captured in a table known as the 'standard genetic code'. Even after 3.5 billion years of genetic drift, the 'standard genetic code' has been largely refractory to change. In an alternative evolutionary trajectory for life on Earth, would life have settled on using a different set of chemical building blocks and biomacromolecules? In this project, we want to demonstrate that the fundamental language of biological life (namely the 'standard genetic code') is not immutable, and can be rewritten inside living cells. If successful, this technology will be a step towards creation of semi- synthetic life and fundamentally change how we use organisms in biomanufacturing.

Humanity has managed to leverage the extant 'standard genetic code', containing only 20 amino acids, into multi-billion dollar industries. Biomanufactured proteins are used in medicine as therapeutics (e.g. antibodies, insulin), in industrial processes (e.g. production of high fructose corn syrup), and even in the household (e.g. additives to laundry detergents). These examples highlight the extent of functional utility generated from just 20 amino acid building blocks. If we wanted to start mass producing proteins composed of an alternative set of building blocks, we would be faced by the obstacle posed previously: life has evolved and been optimized around using a very specific set of biomolecules and building blocks.

In this talk, **I will be discussing strategies, methods, and results for building a parallel genetic code in living organisms**. This talk will focus on the development of a generalizable strategy capable of producing >20 mutually orthogonal transfer RNA (o-tRNA). I will be covering challenges and solutions for producing stable o-tRNAs in cells, specifically in E. coli. While o-tRNAs can form the basis for a parallel genetic code, we are faced with the challenge of building a new genetic code from the ground-up. I will then shift focus cover strategies for populating the new codon table in the absence of a functional orthogonal translation system. If successful, orthogonal genetic codes (as described) have the capacity to translate up to 32 additional amino acids or non-standard building blocks. (*Fields area: chemical biology and synthetic biology*)

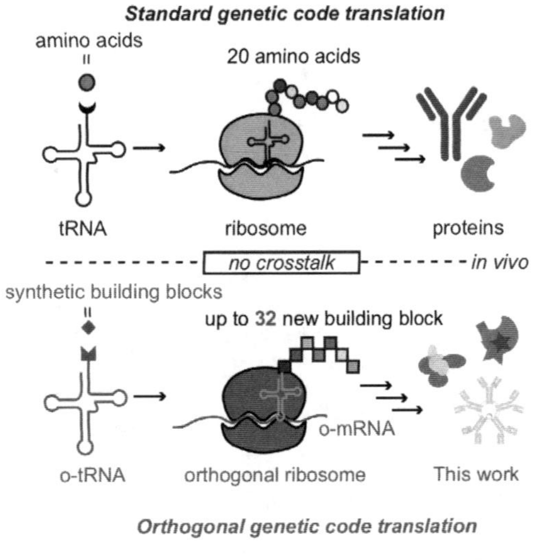

609978 Detangling DNA: Balancing Biophysical Tradeoffs Drives Cellular Reprogramming

Abstract:

Integrating synthetic circuitry into larger transcriptional networks to mediate predictable cellular behaviors remains a challenge within synthetic biology. While significant efforts have been devoted to the design of enhanced synthetic circuitry, less is understood regarding how cellular hardware may impose fundamental performance limitations on integrated circuits. Within the mammalian context, cellular reprogramming continues to generate new cell types, increasingly expanding our perspective of cellular plasticity. Despite improved genetic tools and epigenetic modulations, reprogramming remains a rare cellular event. In this talk, I will describe how I identified molecular roadblocks in reprogramming that arise from tradeoffs between transcription and proliferation rates. My discovery highlights how topological stress impacts the function of gene networks (e.g. native or synthetic circuits) and constrains cellular transitions. I will discuss how this finding opens completely new questions about how the structure of the genome stabilizes cellular identity and suggests strategies to improve the design of synthetic gene circuits.

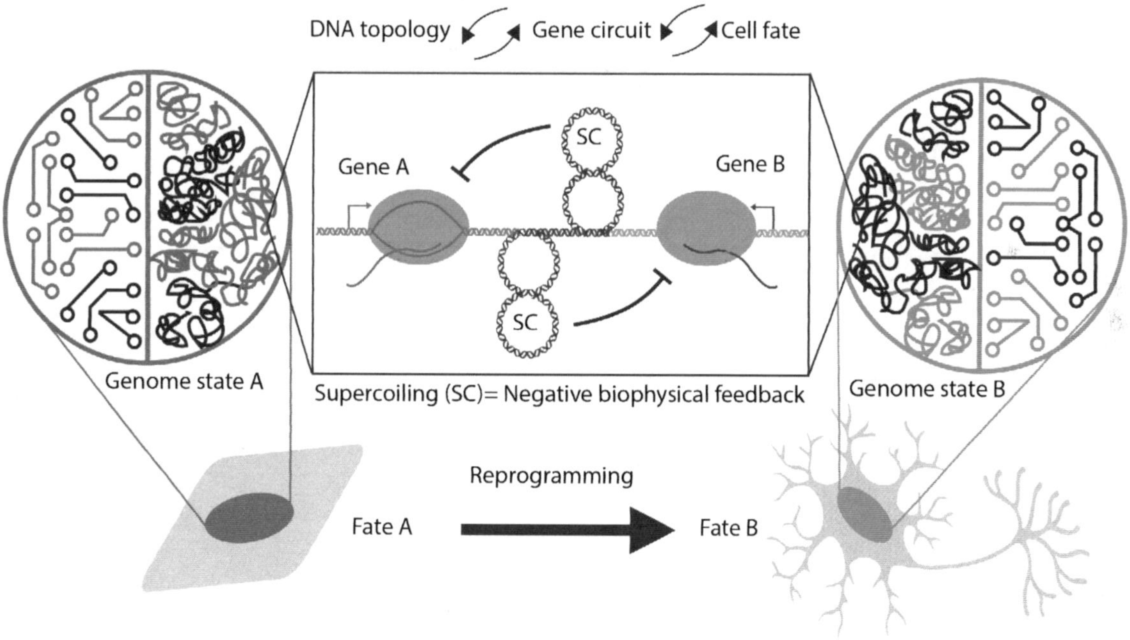

605782 Orthogonal Tuning of Gene Expression Noise Using CRISPR-Cas

Abstract:

The noise in gene expression of cells can cause cellular heterogeneity that leads to the failure of antibiotic- and chemo-therapy, as well as synthetic circuits. To study or control the heterogeneity, previous work has tuned gene expression noise by changing the rate of transcription initiation, mRNA degradation, and mRNA translation. However, these methods are invasive: they require changes to the target genetic components. Here, we propose an orthogonal system based on CRISPR-dCas9 to tune gene expression noise and mean independently in a single cell. Specifically, we modulate the protein expression of a reporter gene in *Escherichia coli* by incorporating CRISPR activation and repression (CRISPRar) simultaneously competing to the same binding site. The CRISPRar uses a single dCas9 that recognizes two different single guide RNAs (sgRNA). We build a library of sgRNA variants with different expression activation and repression strengths. We find that expression noise and mean of a reporter gene can be tuned independently by CRISPRar. Our results suggest that the expression noise is tuned by the competition between two sgRNAs that modulate the binding of the RNA polymerase to a promoter. Our work provides a systematic, quantitative foundation towards designing CRISPR-dCas9 for modulating gene expression noise. The CRISPRar may also change how we tune expression noise at the genomic level. Our work has broad impacts on the study of gene functions, phenotypical heterogeneity, and genetic circuit control.

611576 Development of Modular Tunable Biosensors for Gene Regulation and Autonomous Pathway Optimization

Abstract:

Microbial chemical factories increasingly rely on biosynthetic pathways incorporating toxic intermediates that inhibit cellular growth and product formation. A promising strategy to overcome this challenge is dynamic regulation of production pathways to limit toxic intermediate accumulation, and maintain cells at optimal health for increased production. While dynamic control may be implemented with natural transcription factors as sensors that recognize and respond to a given metabolite, sensors for many intermediates are largely unknown. However, many valuable toxic intermediates and value-added products disrupt cellular pH homeostasis, generating a universal carrier signal for cellular stress. In my lab, we develop modular transcriptional regulators whose sensitivity to this pH and their transcriptional response can be independently tuned for programmable autonomous feedback. Our transcriptional regulators consist of an elastin-like polypeptide (ELP) biosensor fused to a transcription factor to directly respond to general indicators of cellular stress for dynamic control of any biosynthetic pathway. ELPs are engineered proteins that reversibly self-assemble at a critical temperature, pH, and/or ionic strength programmed by their tunable primary sequence. When fused to transcription factors such as orthogonal sigma factors, ELP self-assembly sequesters fused transcription factor, altering gene transcription at targeted promoters in a reversible switch-like fashion. Constructs in *E. coli* demonstrate a 4-fold dynamic range in response to changes in temperature. Similarly, we demonstrate that these sensors respond to small variations in intracellular pH that precede significant cellular damage. The tunable responsive nature of these synthetic regulators make them ideal regulators for pathway optimization, with potentially broad utility across organisms with appropriate transcription factor design

612853 Epigenetic Engineering to Target an Anti-Cancer Gene Module in Breast Cancer (Invited Speaker)

Abstract:

Mounting evidence from genome-wide comparisons of chromosome packaging and gene expression suggests that hyper-repression of groups of genes distinguishes cancerous cells from healthy cells. Certain members of the chromatin system (genomic DNA, RNA, and nuclear proteins) have been implicated as drivers of metastasis and cancer drug resistance. Since the early 1990's small compounds have been used to disrupt hyper-repressed chromatin to simultaneously activate sets of therapeutic genes in cancer cells. However, it is difficult to customize the biological activity of these small compound inhibitors, and they do not directly mediate RNA PolII activity. To address these limitations, our lab has designed transcription-regulating fusion proteins that bind epigenetic marks within chromatin. We have developed the "Polycomb-based transcription factor" (PcTF), a fusion protein that reads histone modifications through a protein-protein interaction between its N-terminal Polycomb chromodomain (PCD) motif and trimethylated lysine 27 of histone H3 (H3K27me3). The C-terminal VP64 domain of PcTF recruits endogenous activators to silenced targets. We identified a set of 104 genes that become consistently activated from 24 - 72 hours after PcTF-overexpression in a triple negative breast cancer cell line (BT-549). PcTF-positive BT-549 cells show poor survival *in vitro*. We propose that the set of 104 genes, which includes tumor suppressor, type I interferon response, and epithelial cell fate genes, represents a 'druggable' anti-cancer gene module. These genes are rarely mutated in basal tumors and therefore may produce a therapeutic effect when they become activated. This talk will review our published results for BT-549, new data from additional PcTF-treated TNBC lines, and transcriptional profiling that suggests PcTF activates a more specific range of genes compared to small compound drugs.

605504 The Effect of Cosolvents on the Loading Efficiency and Release Profile of Anti-CD47 in the Emulsion Solvent Evaporation Technique

Abstract:

Side effects associated with the utilization of antibodies as therapeutics limit their systemic administration at high concentrations often needed for therapeutic impact. Thus, therapeutic antibodies are often considered for targeted delivery through their encapsulation within polymeric nanocarriers. The loading efficiency of antibodies via emulsion-based nanofabrication methods is normally very low. Thus, the fabrication techniques need to be modified to maximize the loading efficiency of antibodies. In this work, we utilized various cosolvents to improve the loading efficiency of anti-CD47, a therapeutic antibody used to block the marker of self signal for initiation of phagocytosis and destruction of atherosclerotic plaques and cancer lesions, via emulsion solvent evaporation technique. Our results demonstrate that a minimum amount of a cosolvent with minimal hydrophilicity can stabilize the antibody in the oil phase; thus, significantly improving the loading efficiency and prolonging the release profile of the antibody. Amongst the cosolvents screened, ethyl acetate was found to be the optimum choice and improved the loading efficiency of anti-CD47 in PLGA nanoparticles to 90% or higher.

610909 Ultra-Deformable Liposomes for Enhanced Drug Delivery

Abstract:

Introduction: Breast cancer is the second leading cause of cancer related deaths of women in the US. The desmoplastic response, in which cancer associated fibroblasts are over-activated, leading to increased extracellular matrix (primarily collagen) production is associated with the most invasive carcinomas [1]. Further, triple negative breast cancer (TNBC) has the highest likelihood of metastasis and poor prognosis when central fibrosis is present [2]. The desmoplastic response leads to increased tumoral solid stress, compression of blood and lymphatic vessels, increased interstitial pressure, decreased perfusion and hypoxia [3]. Accordingly, the desmoplastic response presents a significant challenge to the delivery of anti-cancer therapeutics. Deformable particles, i.e. those with fluid membranes capable of squeezing through small pores, may be uniquely suited to penetrate and deliver drugs to desmoplastic tumors.

Our lab has previously demonstrated that altering nanolipogel elasticity resulted in a Young's modulus-dependent trend in cellular and tumor uptake. These data established that uptake of soft (< 1.6 MPa) were internalized by cells compared to their stiffer (> 13.8 MPa) counterparts [4]. One approach to modulating liposomal deformability is to alter the composition of the lipid bilayer. This can be achieved by incorporating lipids of different acyl chain lengths or degrees of saturation, as well as the use of edge activators (i.e. surfactants). Incorporation of phospholipids with short acyl chain length (less than 14) or multiple double bonds (poly-unsaturated), may disrupt the lipid bilayer and consequently increase fluidity [5]. It is hypothesized that hypo-elastic, ultra-deformable particles are internalized preferentially due to two unique capabilities: (1) internalization via fusion with the cell membrane (a process of low energy dependence) [4] and (2) squeezing through small pores to permeate tissues [6]. Novel formulations developed herein, termed ultra-deformable liposomes (UDLs) have demonstrated increased cancer cell internalization and transendothelial cell permeability.

Materials and Methods: UDLs were formulated with 1,2-dioleoyl-sn-glycero-3-phosphocholine (DOPC), 1,2-dilauroyl-sn-glycero-3-phosphocholine (DLPC; C12) or 1,2-diheptanoyl-sn-glycero-3-phosphocholine (DHPC; C7), and 1% Benzoxazolium, 3-octadecyl-2-[3-(3-octadecyl-2(3H)-benzoxazolylidene)-1-propenyl]-, perchlorate (DiO; a lipophilic carbocyanine dye, which fluoresces when incorporated into a lipid bilayer) for visualization. Liposomes were synthesized using the thin film hydration method. Liposomal size and polydispersity was determined via dynamic light scattering (DLS), using a Brookhaven Zeta-PALS analyzer (Brookhaven Instruments, Holtsville, NY, USA). To study the internalization of UDLs by TNBC cells, MDA-MB-436 or MDA-MB-231 cells were seeded at 75,000 per well in a 24 well plate and incubated overnight at 37C. Cells were treated with 200 nM UDLs for 2-4 hours at 37°C. Cells were rinsed twice with PBS prior to analysis with flow cytometry (Beckman Coulter, CytoFLEX). Quantification of liposome internalization by cells was determined by DiO fluorescence (ex/em: 484 nm), normalizing for background fluorescence. Transendothelial permeability studies were performed using human umbilical vein endothelial cells (HUVECs) seeded on transwell inserts (Corning) and incubated for 2 days prior to use. Then, 200 nM UDLs were added to the upper well of the insert and equal volume of DMEM was added to the bottom of the chamber. Transwells were incubated at 37°C for 4 hours then the bottom chamber media was removed and read on a spectrometer for DiO fluorescence.

Results: Novel UDL formulations using a combination of DOPC and short chain phospholipids (C7 or C10) were prepared and characterized by size, zeta potential and Young's moduli. UDL-C7 and UDL-C10 formulations demonstrated 1.7-fold and 5.4-fold increase in TNBC cell uptake relative to DOPC control liposomes, respectively (**Fig 1a-b**). Further, UDL-C7 demonstrated a 2.3-fold increase in transendothelial membrane permeation, compared to DOPC control (**Fig 1c**). Preliminary data on the inhibition of cell internalization pathways (clathrin or caveolae) show that UDL-C7 liposomes are internalized via fusion, and not receptor mediated endocytosis (**Fig 1d**). Moving forward, we will show that UDLs will increase accumulation within tumors in vivo.

Conclusions: The novel UDL formulations developed herein demonstrated increased internalization by TNBC cells and transendothelial permeability compared to DOPC control liposomes. Similar to our previous findings, these data show that UDL-C7 are internalized predominantly by fusion, not clathrin or caveolae mediated endocytosis.

References: [1] Barsky, S. H., Rao, C. N., Grotendorst, G. R., & Liotta, L. A. (1982). *The American journal of pathology*, *108*(3), 276. [2] Takai, K., Le, A., Weaver, V. M., & Werb, Z. (2016). *Oncotarget*, *7*(50), 82889–8290. [3] Stylianopoulos, T., et al. (2012). *Proceedings of the National Academy of Sciences*, *109*(38), 15101-15108. [4] Guo, P., et al (2018). *Nature communications*, *9*(1), 1-9. [5] Hussain, A., et al. (2017). *International journal of nanomedicine*, *12*, 5087–5108. [6] Cevc, G., Schätzlein, A., & Richardsen, H. (2002). *Biochimica et Biophysica Acta (BBA)-Biomembranes*, *1564*(1), 21-30.

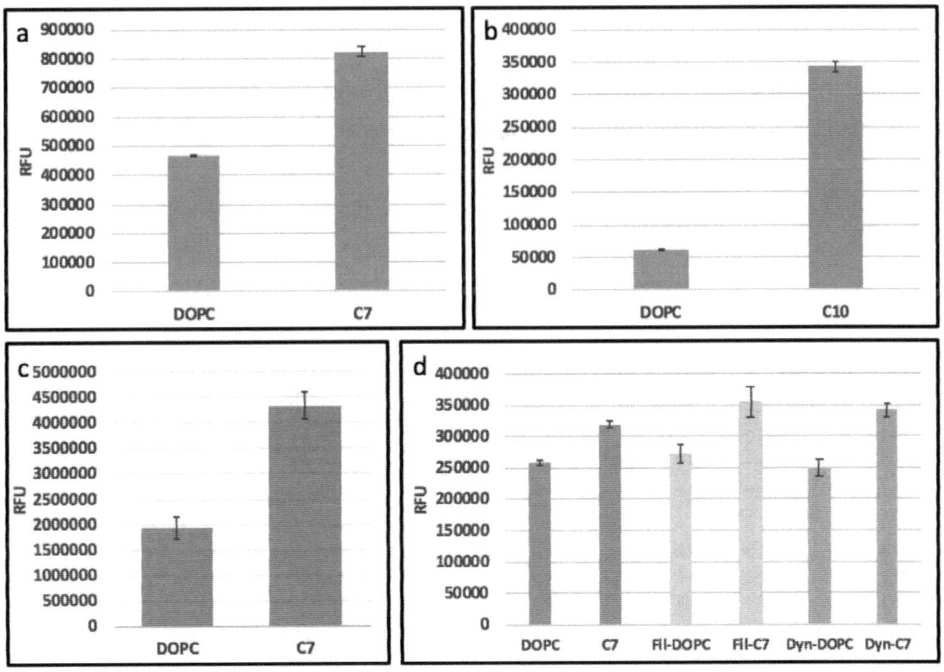

Figure 1. (a-b) Quantification of UDL-C7 and UDL-C10 uptake in MDA-MB-231 and MDA-MB-436 cells, respectively. (c) Transendothelial permeability of UDL-C7. (d) Uptake of DOPC (control) and UDL-C7 after treatment with PBS (blue), Filipin (yellow) or Dynasore (orange).

603817 Cell Membrane-Derived, Core-Shell Structures for the Delivery of Biologics

Abstract:

Vesicles loaded with nanoparticles have found increasing applications for the delivery of biologics. However, issues with stability complicate their synthesis and make it difficult to incorporate proteins in vesicles to guide delivery and biological response. Giant plasma-membrane derived vesicles (GPMVs) are membrane vesiculations that are enriched in membrane proteins. Since GPMVs contain cytoplasmic content, loading cells with nanoparticles prior to GPMV induction can provide a novel approach for the synthesis of nanoparticle-loaded vesicles. Herein, we report the generation of such GPMV-based, core-shell structures for the first time and characterize their composition, physicochemical properties, and cellular interactions.

The cell-derived system used in this study was produced from the A549 alveolar epithelial cells. Cells were first pre-incubated with fluorescent carboxyl-modified silica nanoparticles (50 nm) for six hours, resulting in nanoparticle internalization, and then chemically induced for GPMV generation. Isolated vesicles had a diameter of 5.7 ± 2.3 μm and a zeta potential of -9.9 ± 0.6 mV. Confocal imaging confirmed the presence of nanoparticles inside the vesicles. Vesicles showed the same lipid and protein profile as reported for the cell plasma membrane with only minor changes in their proteome after nanoparticle encapsulation. Stability of GPMVs at 37 °C was examined by monitoring the release of the fluorescent nanoparticles, demonstrating that vesicles remained intact for approximately 48 hours. Nanoparticle-loaded GPMVs were readily endocytosed in two different cell lines (A549 and THP1). In summary, this work reports the generation of the first GPMV-based, core-shell, nanoparticle-vesicle structures. These structures, which contain membrane proteins, are simple to generate, stable, and readily endocytosed by cells, and could find applications in delivery of drugs and imaging agents.

606316 Exploring the Effect of Mucoadhesive Coatings on PLGA Vascular Adhesion and Protein Adsorption for Improved Drug Delivery Systems

Abstract:

Vascular-targeted drug delivery systems require the successful localization and adhesion of drug carriers to diseased sites through the utilization of inflammatory biomarkers. Poly (lactic-co-glycolic acid) (PLGA), a negatively charged copolymer, is the most appealing material because of its tunable biodegradability, biocompatibility, and FDA approval. Unfortunately, the ability of PLGA to adhere to an inflamed endothelium is reduced in the presence of plasma proteins. The surface modification of PLGA with mucoadhesive coatings, such as chitosan (CS) and glycol chitosan (GC), is a potential approach in improving its adhesion by altering the surface hydrophobicity and charge impacting protein adsorption. Here, we aimed to characterize the adhesion and plasma protein coronas of PLGA, CSPLGA, and GCPLGA particles conjugated with a targeting ligand in physiological conditions. Using a parallel plate flow chamber assay, PLGA, CSPLGA, and GCPLGA particles were exposed to an inflamed endothelium in the presence of red blood cells in buffer or plasma. Our results indicate that GCPLGA particles significantly improve adhesion to an inflamed endothelium compared to PLGA suggesting a difference in plasma protein adsorption. The level of adhesion is dependent on shear rate, donor plasma, and ligand density. The biocompatibility of these carriers will need to be examined. Ultimately, this contributes information to drug carrier design, specifically in terms of favorable surface hydrophobicity, charge, and protein coronas obtained.

613343 A Gold-Templating Method for Fabrication of Bile Salt-Based Composite Microparticles with Controlled Geometry for Localized Fat Dissolution (INVITED SPEAKER)

Abstract:

Bile acids and their salts are proposed as therapeutic agents for various diseases, including bile synthesis disorders, liver diseases, and obesity. However, oral or subcutaneous administration of a solubilized version of these drugs can limit their efficacy and impose unwanted side effects. Here, we describe a gold-templating method for fabricating stable, bile salts – cholate or deoxycholate – microparticles. The reduction of gold ions at the oil-water interface in a double emulsion solvent evaporation process enables a gold-bile salt interaction, leading to the formation of bile salt particles. We demonstrated that composite microparticles release cholate/deoxycholate via a surface erosion process, and the size of the particles was linked to the bile salt concentration in the water phase. Importantly, the bile salt particles lyse adipocytes, both *in vitro* and *in vivo*, with minimal side effects, contrary to the FDA approved salt solution that leads to severe inflammation and ulceration. Finally, we show that bile salt-based particles can be loaded with drugs, using rhodamine as a model, allowing for co-administration of the released bile salt with other therapeutic agents. Overall, particle-based cholate/deoxycholate can open opportunities for localized delivery of these salts, which can improve efficacy while minimizing side effects associated with oral and subcutaneous use.

607036 Increasing Isoprenol Production in Engineered *E. coli* through Medium Optimization and Different Fermentation Strategies.

Abstract:

Due to the increasing demand for bio-aviation fuel, different strategies have been developed for its production. One promising bio-aviation fuel is monoterpenes, which have similar characters with current aviation fuels. Monoterpenes can be synthesized from a precursor: isoprenol (isopentenol,3-methyl-3-buten-1-ol). Previous study expressed the mevalonate pathway with a novel IPP-bypass (*AtoB, HMGS, HMGR, MK, PMD*) in *E. coli* DH1 for isoprenol production. This strain can obtain over 60-70% theoretical yield of isoprenol from glucose. In this study, the minimal medium for this engineered *E. coli* strain was further optimized to increase isoprenol titer. Different combinations of carbon sources and nitrogen sources with various concentrations were tested to obtain the optimal C/N ratio in the medium. Different fermentation strategies were also developed to further increase isoprenol titer, such as fed-batch fermentation and in-situ solvent recovery.

601814 Exploiting Inter-Strain Metabolic Cooperation for CO_2 Recycling and Increased Carbon Conversion Efficiency during Biofuel Production

Abstract:

Regardless of the of source of lignocellulosic sugars used, significant carbon loss (~50% weight of the feedstock) in the form of CO_2, a potent greenhouse gas, accompanies biofuel production, limiting the economic viability and environmental sustainability of this approach. This problem is not unique to just conventional ethanol biofuel, since many other advanced biofuel compounds are also derived from precursors produced as a result of pyruvate decarboxylation (e.g. acetaldehyde for ethanol; acetolactate for isobutanol; acetyl-CoA for n-butanol, farnesene and fatty acids) which directly releases CO_2. To overcome this intrinsic limitation we developed a modular, catabolically-orthogonal coculture-coproduction system capable of capturing and fixing CO_2 evolved during biofuel production through inter-strain 'metabolic cooperation'. Orthogonal sugar catabolism was achieved by disabling xylose catabolism (Δ*xylR*; a transcriptional activator required for xylose catabolism) in a 'glucose-to-ethanol' *Escherichia coli* 'specialist' strain (G2E) while also blocking glucose catabolism in a complementary 'xylose-to-succinate' *E. coli* 'specialist' strain (X2S) by inactivating all major glucose uptake systems. In this coculture system, CO_2 emitted by G2E is captured and reutilized by X2S via its fixation by anaplertoic reactions to coproduce succinate. To confirm and characterize the extent of inter-strain metabolic cooperation, ^{13}C-labeling experiments were used to determine both the carbon fixation capacity and overall carbon utilization efficiency, as well as to elucidate other possible exchanges of central carbon metabolism intermediates. Further demonstrating the modularity of this approach, another 'xylose-to-C4-dicarboxylate' *E. coli* 'specialist' strain was also engineered and used to coproduce with ethanol. This flexible and robust CO_2-fixing platform allows for (i) co-utilization of glucose-xylose mixtures without carbon catabolite repression, (ii) reduction in CO_2 emissions during biofuel production (iii) increased overall carbon utilization efficiency, (iv) the potential for facile production of different biofuels and coproducts using modular and interchangeable coculture members. Together, these features provide promising avenue for carbon management and support the prospect of developing carbon neutral bioprocesses.

612316 Metabolic Engineering of *Auroebasidium Pullulans* for Polymalic Acid and Malic Acid Production from Food Processing Wastes

Abstract:

Aureobasidium pullulans is a versatile yeast-like fungus with a large genome and can secrete many hydrolases capable of degrading plant materials. It can also produce many secondary metabolites including poly(L-malic acid) (PMA). The goal of this project is to develop a fermentation process with metabolically engineered strains of *A. pullulans* for PMA production from low-cost food processing wastes such as soy molasses. PMA, a homopolymer of L-malic acid (MA), has unique properties and many applications such as carriers for drugs and flavor compounds. PMA fermentation also provides a novel production route for L-malic acid, a specialty chemical with wide applications in foods and pharmaceuticals and a potential C4 platform chemical with a large market. PMA fermentation followed with acid hydrolysis of PMA can provide a novel process for the production of MA, which currently is produced by chemical synthesis using petroleum-derived feedstocks. We aimed to develop a robust strain capable of producing PMA at high titer, purity, yield and productivity by rational metabolic engineering. In this study, we focused on strategies for strengthening the glyoxylate shunt in *A. pullulans*. First, genes encoding the rate-limiting enzymes in the pathway were overexpressed. A novel transcriptional activator, Cat8, was identified and co-expressed to regulate the glyoxylate shunt. To rewire the carbon flow towards glyoxylate shunt, a CRISPR genome-editing system was used to knock out the node gene in the TCA cycle. Furthermore, ethanol stress, as an operable process parameter, was used to regulate stress-responsive genes in glyoxylate shunt. Compared to the parental strain, the engineered strain developed through these strategies showed a 40% increase in the PMA yield (0.70 g/g) from glucose. PMA production in a fed-batch fermentation carried out in a stirred-tank bioreactor reached the highest titer (>155 g/L MA) ever reported. Techno-economic and life cycle analyses showed that PMA and MA can be produced from food processing wastes or by-products from corn, sugarcane, and soybean refinery industries at an economically competitive cost with significant reduction in emissions and energy consumption compared to current chemical synthesis and corn-based fermentation processes.

609778 Application of Robust Dynamic Flux Balance Analysis Framework to a Wine Fermentation for Understanding and Steering Aroma Formation

- Download/View Extended Abstract

Abstract:

Various commercial *S. cerevisiae* yeast strains are available for use in industrial wine fermentations, each with its own combination of growth characteristics and organoleptic (sensory) profiles. However, the metabolism underlying these differences is complex and not well understood, making manipulation of these strains difficult. One tool for elucidating strain-to-strain differences and identifying key metabolic pathways related to flavor production is the use of genome-scale metabolic models (GSMM). Despite progress in the field, most current models either focus on aerobic systems, contain models that focus on a carbon-limited medium, and/or lack the detailed coverage of Ehrlich and amino acid degradation metabolic pathways that have been shown experimentally to be highly correlated with aroma formation. One way to capture the power of these models is to use dynamic flux balance analysis (dFBA) to predict the flux distribution of all the metabolites within the cell over the course of an entire fermentation. Using this approach, it is possible to test the predictive capability of these models by comparing predictions with experimental fermentation data. Once the models fit dynamic data, they can be used to understand differences between commercial strains and suggest genetic modification strategies towards steering certain aroma formation in strains of interest. Here, in this work, we applied a robust dFBA framework to an expanded GSMM of *S. cerevisiae* over the course of a wine fermentation. Recently, we showed that ester production during alcohol fermentation is regulated by a newly discovered family of alcohol acyl transferases (AATs) in *S. cerevisiae.* However, mysteries still persist regarding the whether other metabolic mechanisms are responsible for ester formation. By applying our expanded GSMM of yeast, which has the most comprehensive representation aroma forming pathways, we can more accurately predict metabolic fluxes for various yeast strains. Furthermore, this model can be calibrated with experimental data to reasonably propose genetic and process engineering strategies to improve the enological performance *S. cerevisiae* strains of interest.

612320 Multi-Plasmid Cotransformation to Combinatorial Engineer Multiple Histidine Kinases in *Clostridium Acetobutylicum*

Abstract:

Clostridium acetobutylicum harbors bulks of functional-like genes which belong to paralogous gene family and are involved in the same metabolic or regulatory pathway. Multigene disruption is key and urgent demand for investigating these genes function to engineer excellent strain, while conventionally performed by successive and time-consuming rounds. Here, we firstly demonstrate a multi-plasmid co-transformation strategy for simultaneously combinatorial disruption of histidine kinases (HKs) in *C. acetobutylicum*, which HKs function in controlling phosphorylation state of transcriptional regulator Spo0A. This strategy reliably produces stable mutants without the limitation of plasmids incompatibility. After optimization, a mutant library (71.0% of theoretical capacity) was generated with an efficiency of 47.5% and 4 weeks, boosting the screening of superior butanol-producing strain. Phenotypic and morphological analysis further elucidated the systematic regulation of HKs on cell growth, solvent metabolism and developmental attributes. Collectively, our strategy will facilitate the genetic dissection of complex regulatory network and the strain development.

612581 "Invited Talk" Metabolic Engineering of *Clostridium Tyrobutyricum* for Butyric Acid Production from Agricultural and Industrial Wastes

Abstract:

Clostridium tyrobutyricum ATCC 25755 is a promising butyric acid-producing strain. However, the relatively high substrate cost, which accounts for ~50% of the overall cost, is limiting the industrial production of bio-based butyric acid. Our research focuses on improving the economic feasibility of butyric acid fermentation via metabolic engineering strategies. For example, sucrose (*scrA, scrB* and *scrK*), xylose (*xylT, xylA and xlyB*) and galactose (*galK, galE, galT,* and *galP*) metabolism-related genes from *C. acetobutylicum* ATCC 824 were successfully overexpressed in *C. tyrobutyricum*, endowing it with the superior ability to co-utilize mixed sugars in cane molasses (sucrose, fructose and glucose), lignocellulosic hydrolysates (glucose and xylose) and coffee ground hydrolysate (galactose, mannose, and glucose), respectively. In addition, the tolerance of *C. tyrobutyricum* to the lignocellulosic hydrolysate-derived inhibitors was improved by overexpressing heat shock proteins and short-chain dehydrogenase/reductase, resulting in an enhanced butyrate production. These studies demonstrated the advantages of these metabolic engineering strategies and the feasibility of economic butyric acid production from agricultural and industrial wastes.

606661 Untargeted Proteomic Study of Disulfide Bond Reduction during Monoclonal Antibody Production in Chinese Hamster Ovary (CHO) Cell (Faculty/Industry Candidate)

Abstract:

Depending on process conditions, monoclonal antibodies (mAbs) produced by Chinese hamster ovary (CHO) cells can undergo partial disulfide bond reduction, which may result in the formation of low molecular weight (LMW) fragments. These partially reduced fragments lack the stability and biological functions of the intact protein. Thus, formation of LMW fragments negatively affects product quality and process productivity. The reduction of disulfide bonds is thought to occur due to host cell proteins (HCPs) that accumulate in the cells. However, mechanistic understanding of this phenomenon remains limited. It is unclear which host cellular pathways contribute to disulfide reduction, and how variations in process conditions affect these pathways.

In this study, we used an untargeted proteomics approach to systematically characterize changes in HCPs under conditions that lead to formation of LMW fragments, and to identify pathways that correlate with markers of disulfide reduction. Cell samples were collected during early stationary phase (days 7, 8, and 9) from fed-batch bioreactors (n=2 per condition) operated under reducing and non-reducing conditions. Antibodies recovered from the non-reducing condition contained no LMW fragments, whereas antibodies from the reducing condition were LMW fragments. These samples were analyzed by liquid chromatography-mass-spectrometry (LC-MS) perform on a quadrupole time-of-flight (TOF) system. Principal component analysis of approximately 3,100 unique monoisotopic peptides showed clear separation between reducing and non-reducing conditions. Annotation of the peptides followed by analysis of covariance (ANCOVA) identified 85 proteins that trend significantly differently between reducing and non-reducing conditions ($p<0.05$). These significant proteins were clustered based on protein-protein interactions using STRING. Significant clusters were associated with carbon metabolism, protein folding, heat shock response, and protein processing in the endoplasmic reticulum (ER). These results are consistent with Fisher's exact test performed on the differentially expressed proteins, which also identified protein folding ($p<10^{-5}$) as the most significantly enriched pathway. Targeted analysis of selected proteins in the oxidative stress response and ER protein processing pathways, including the thioredoxin system, further corroborated that these pathways were elevated under the reducing condition. Taken together, our findings suggest that LMW fragments could form due to disulfide reduction associated with ER oxidative stress. It is possible that the ER stress initially results from accumulation of misfolded proteins, which could further amplify ER stress and additionally cause mitochondrial oxidative stress. Further studies are warranted to investigate, at the molecular level, these possible explanations.

In conclusion, global profiling of HCPs under reducing and non-reducing conditions, followed by multivariate analysis of annotated proteins identified key pathways that are differentially regulated and correlate with disulfide reduction. Prospectively, these proteins could be monitored during process development and scale-up to control product quality.

Wangsanuwat *et al. BMC Genomics* (2020) 21:717
https://doi.org/10.1186/s12864-020-07134-4

BMC Genomics

METHODOLOGY ARTICLE Open Access

Efficient and cost-effective bacterial mRNA sequencing from low input samples through ribosomal RNA depletion

Chatarin Wangsanuwat[1,2†], Kellie A. Heom[1,2†], Estella Liu[1], Michelle A. O'Malley[1,2] and Siddharth S. Dey[1,2,3*]

Abstract

Background: RNA sequencing is a powerful approach to quantify the genome-wide distribution of mRNA molecules in a population to gain deeper understanding of cellular functions and phenotypes. However, unlike eukaryotic cells, mRNA sequencing of bacterial samples is more challenging due to the absence of a poly-A tail that typically enables efficient capture and enrichment of mRNA from the abundant rRNA molecules in a cell. Moreover, bacterial cells frequently contain 100-fold lower quantities of RNA compared to mammalian cells, which further complicates mRNA sequencing from non-cultivable and non-model bacterial species. To overcome these limitations, we report EMBR-seq (Enrichment of mRNA by Blocked rRNA), a method that efficiently depletes 5S, 16S and 23S rRNA using blocking primers to prevent their amplification.

Results: EMBR-seq results in 90% of the sequenced RNA molecules from an *E. coli* culture deriving from mRNA. We demonstrate that this increased efficiency provides a deeper view of the transcriptome without introducing technical amplification-induced biases. Moreover, compared to recent methods that employ a large array of oligonucleotides to deplete rRNA, EMBR-seq uses a single or a few oligonucleotides per rRNA, thereby making this new technology significantly more cost-effective, especially when applied to varied bacterial species. Finally, compared to existing commercial kits for bacterial rRNA depletion, we show that EMBR-seq can be used to successfully quantify the transcriptome from more than 500-fold lower starting total RNA.

Conclusions: EMBR-seq provides an efficient and cost-effective approach to quantify global gene expression profiles from low input bacterial samples.

Keywords: Bacterial mRNA sequencing, mRNA enrichment, rRNA depletion, Low input total RNA

Background

Bacterial species pervade our biosphere and millions of years of evolution have optimized these microbes to perform specific biochemical reactions and functions; processes that could potentially be adapted to develop a variety of products, such as renewable biofuels, antibiotics, and other value-added chemicals [1–5]. Bacterial messenger RNA (mRNA) sequencing provides a snapshot of the genome-wide state of a microbial population, and therefore enables fundamental understanding of these varied microbial functions and phenotypes [6].

However, compared to eukaryotes, mRNA sequencing from bacterial samples has been more challenging for several reasons. First, unlike in eukaryotes, bacterial mRNA does not contain a poly-A tail at the 3′ end that can be used to easily enrich for these molecules during reverse transcription [7, 8]. Further, total RNA isolated from bacterial cells typically contains greater than 95%

* Correspondence: sdey@ucsb.edu
†Chatarin Wangsanuwat and Kellie A. Heom contributed equally to this work.
[1]Department of Chemical Engineering, University of California Santa Barbara, Santa Barbara, CA 93106, USA
[2]Center for Bioengineering, University of California Santa Barbara, Santa Barbara, CA 93106, USA
Full list of author information is available at the end of the article

ribosomal RNA (rRNA), and therefore cost-effective and high coverage sequencing of the transcriptome requires the development of efficient strategies to deplete the abundant 5S, 16S and 23S rRNA molecules [9]. Finally, bacterial cells typically contain approximately 100-fold lower RNA than mammalian cells, and as the starting amount of total RNA when working with rare, non-cultivable, and non-model bacterial species can be limiting, it is a challenge to robustly and accurately capture the transcriptome from small quantities of total RNA with minimal amplification biases [10].

Several commercial kits have been developed to deplete bacterial rRNA from total RNA samples, including the MICROBExpress Bacterial mRNA Enrichment Kit (Thermo Fisher Scientific), the RiboMinus Transcriptome Isolation Kit, bacteria (Thermo Fisher Scientific), and the Ribo-Zero rRNA Depletion Kit (Illumina) [11]. These techniques rely on subtractive hybridization to deplete rRNA and typically work at a scale of hundreds of nanograms to micrograms of starting total RNA. Further, as these commercial kits are only effective on species targeted in the standard probe set, it is challenging to extrapolate these methods to diverse bacterial species [9, 11]. While this limitation of pre-designed kits has been overcome through the development of workflows to generate custom subtractive hybridization probe sets for any species of interest, they still operate at microgram quantities of starting material and either require multiple rounds of hybridization or a series of oligo optimization steps prior to optimal performance [12, 13]. An alternate approach relies on the Terminator™ 5′-phosphate-dependent exonuclease (TEX) (Lucigen) to specifically degrade rRNAs with 5′-monophosphate ends but not mRNAs with 5′-triphosphate ends; however, this method typically has lower efficiencies than other existing rRNA depletion strategies [10, 14, 15]. A more recent method uses complementary single-stranded DNA probes to tile rRNAs that are subsequently degraded by RNase H [16]. The commercial NEBNext Bacteria rRNA depletion kit (NEB) employs a similar strategy and can be applied to as low as 10 ng of starting total RNA. Similarly, another approach uses a pool of tiled single-guide RNAs to direct Cas9 mediated cleavage of rRNA-derived cDNA to deplete rRNA while another approach uses targeted reverse transcription primers designed to avoid capturing rRNAs [17, 18]. However, all these methods require a large array of probes that can be expensive to synthesize and potentially need to be redesigned for distant bacterial species [16–18].

Therefore, in this work we have developed EMBR-seq (Enrichment of mRNA by Blocked rRNA), a new technology that overcomes the limitations of sequencing mRNA from bacterial samples by: (1) Using 5S, 16S and 23S rRNA blocking primers and poly-A tailing to specifically deplete rRNA and enrich mRNA during downstream amplification; (2) Using a single or a few blocking primers for each of the three abundant rRNA molecules, thereby enabling rapid adaptation to different bacterial species and significantly reducing the cost per sample; and (3) Using a linear amplification strategy to amplify mRNA from as low as 20 picograms of total RNA with minimal amplification biases. We applied EMBR-seq to a model *E. coli* system to demonstrate efficient mRNA enrichment and sequencing with increased sensitivity in gene detection. Further, we show that our method accurately captures the genome-wide gene expression profiles with minimal technical biases. Thus, EMBR-seq is an efficient and cost-effective approach to sequence mRNA from low-input bacterial samples.

Results

EMBR-seq uses blocking primers to deplete rRNA

To overcome the limitations described above, we developed EMBR-seq, a new technique to efficiently deplete rRNA from total RNA, thereby enabling cost-effective sequencing of mRNA from bacterial cells. To minimize rRNA-derived molecules in the final sequencing library, we first incubated the total RNA with rRNA blocking primers, designed specifically to bind the 3′ end of 5S, 16S and 23S rRNA, followed by poly-adenylation with *E. coli* poly-A polymerase (Fig. 1 and Methods). To deplete rRNA, EMBR-seq only requires primers at the 3′ end of rRNA, unlike recent methods that tile oligonucleotides along the entire length of rRNA molecules, thereby significantly reducing costs and making our approach more easily translatable to other bacterial species. The blocking primers generate double-stranded RNA-DNA hybrid molecules at the 3′ end of rRNAs, which reduces subsequent poly-adenylation and downstream amplification of rRNA molecules, as the poly-A polymerase preferentially adds adenines to single-stranded RNA [19]. Thereafter, the reaction mixture is reverse transcribed following the addition of a poly-T primer. This primer has an overhang containing a sample-specific barcode to enable rapid multiplexing and reduction in library preparation costs, the 5′ Illumina adapter, and a T7 promoter [20]. After second strand synthesis, cDNA molecules are amplified by in vitro transcription (IVT). However, as only cDNA molecules deriving from a poly-adenylated RNA have a T7 promoter, our technique further amplifies mRNA-derived molecules for sequencing whereas rRNA-derived molecules are excluded from IVT amplification. The amplified RNA from IVT is then used to prepare Illumina sequencing libraries, as described previously (Fig. 1 and Methods) [20–22].

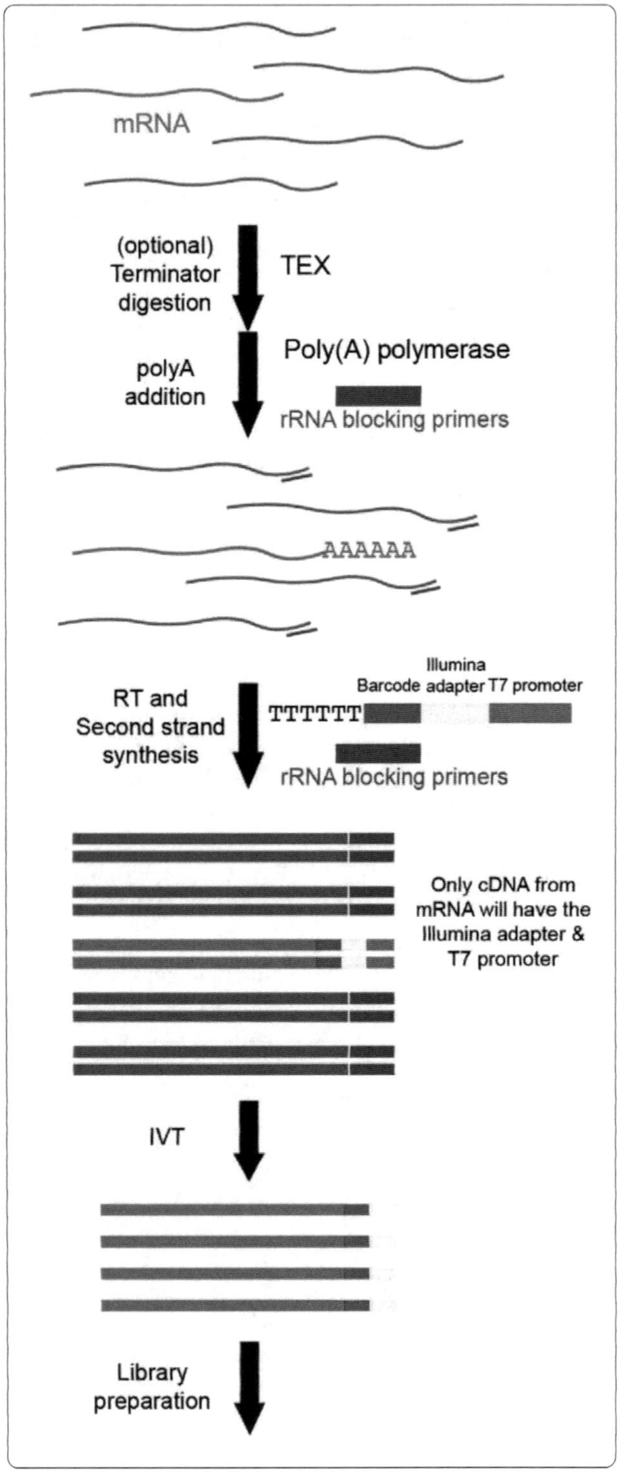

Fig. 1 Schematic of EMBR-Seq. After performing an optional Terminator™ 5′-phosphate-dependent exonuclease digestion, poly-A polymerase and rRNA blocking primers (purple) are added to total bacterial RNA (mRNA in red and rRNA in gray). Blocking primers specifically bind to the 3′ end of 5S, 16S, and 23S rRNAs, resulting in the preferential addition of a poly-A tail to mRNA molecules. Next, reverse transcription is performed using (i) a poly-T primer, which has an overhang containing a sample-specific barcode (blue), 5′ Illumina adapter (yellow), and T7 promoter (green), and (ii) rRNA blocking primers to convert poly-adenylated RNA and rRNA molecules, respectively, to cDNA. The cDNA molecules are then amplified by in vitro transcription, and the amplified RNA is used to prepare Illumina libraries. As the rRNA-derived cDNA does not contain a T7 promoter, these molecules are not amplified during in vitro transcription, resulting in rRNA depletion

EMBR-seq efficiently depletes rRNA to sequence bacterial mRNA

We applied EMBR-seq to total RNA isolated from the exponential growth phase of *E. coli* strain K12 (MG1655). Starting from 100 ng of total RNA, we were able to successfully make Illumina libraries that were sequenced and mapped to the *E. coli* transcriptome. In parallel, we prepared control libraries where total RNA was processed using the EMBR-seq protocol but in the absence of blocking primers. While total RNA from *E. coli* has previously been reported to consist of 95% rRNA [9], our control samples with no blocking primers had approximately 64% rRNA, consistent with previous observations that mRNA molecules are preferentially poly-adenylated compared to rRNA even in the absence of any blocking primers (Fig. 2a) [23, 24]. Importantly, compared to the control samples, we observed a significant increase in rRNA depletion efficiency (from 64 to 16%), with 84% of the mapped reads corresponding to mRNA in samples treated with blocking primers (Fig. 2a). As tRNAs make up another major class of RNA molecules, we analyzed our data to quantify the detection of these molecules [25]. We found tRNA-derived reads to constitute only 0.37 and 1.26% of the mapped reads in the control and EMBR-seq samples, respectively. This expected low detection rate is likely due to the small size of tRNAs, their stable secondary structures and utilization of numerous post-transcriptionally modified nucleotides that are known to interfere with reverse transcription [26, 27]. Overall, these results demonstrate that EMBR-seq achieves a level of mRNA enrichment that is better or comparable to recent bacterial rRNA depletion reports [11–13, 15–18].

In certain applications, such as those where RNA is extracted from non-cultivable bacterial species within natural isolates, the total RNA can be fragmented and of poor quality. To determine if EMBR-seq can still be successfully applied to degraded RNA, we compared the rRNA depletion efficiency of total *E. coli* RNA with two

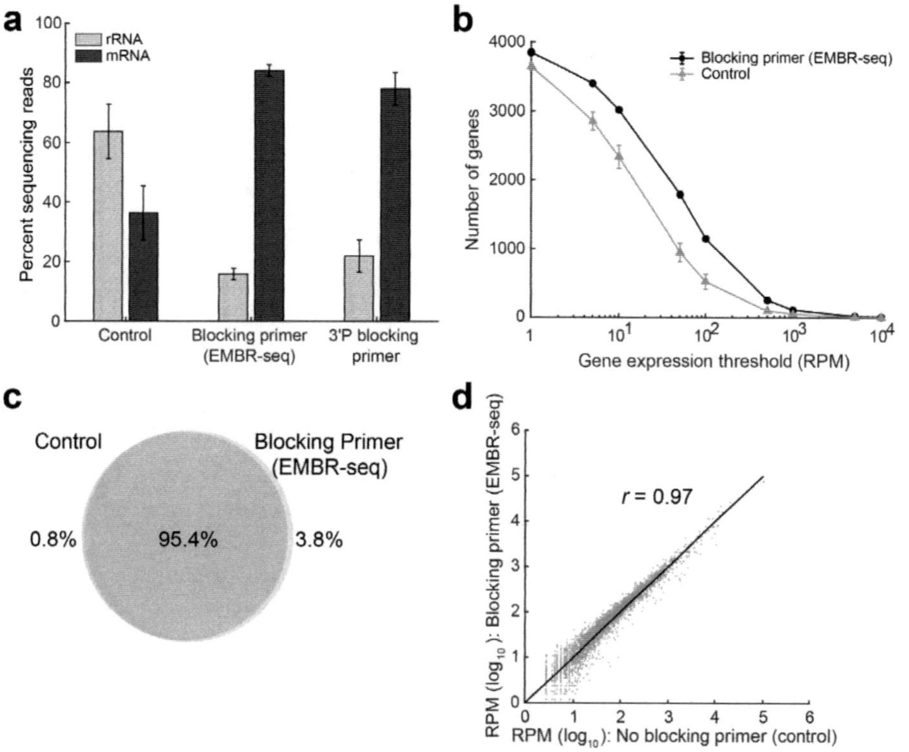

Fig. 2 Blocking primers in EMBR-seq deplete rRNA and provide a deeper view of the transcriptome without introducing technical biases. **a** In the presence of blocking primers, a 4-fold rRNA depletion and more than 2-fold mRNA enrichment is achieved compared to control samples. With the introduction of blocking primers in EMBR-seq, mRNAs account for more than 80% of the mapped reads, which is a greater than 16-fold increase compared to total RNA in *E. coli* cells. The 3′ phosphorylated blocking primers display similar but slightly lesser mRNA enrichment ($n \geq 2$ replicates for all conditions). **b** Comparison between EMBR-seq and control samples in the number of genes detected above different expression thresholds ($n = 3$ for both conditions). For the EMBR-seq group, error bars are of the same scale as the size of the data points. **c** Venn diagram shows that more than 99% of the genes detected in the control samples were also detected when using blocking primers in EMBR-seq. 99.2% of all detected genes were found in the EMBR-seq samples and 96.2% in the control samples. The number of genes detected were calculated by combining data obtained from three control samples and three EMBR-seq processed samples. **d** Gene transcript counts with and without blocking primers are highly correlated (Pearson $r = 0.97$) suggesting that EMBR-seq does not introduce technical artifacts in quantifying gene expression ($n = 3$ for both datasets). These experiments were performed starting with 100 ng total RNA from *E. coli*. Error bars in panels (**a**) and (**b**) represent standard deviations

different RIN (RNA Integrity Number) scores of 7.2 and 2.4. While RNA molecules can be fragmented by several mechanisms, the degraded sample with a RIN score of 2.4 in this experiment was prepared by heating the total RNA at 95 °C for 5 min. Surprisingly, we observed that EMBR-seq depleted rRNA to similar levels of 15 and 17% in the untreated and degraded samples, respectively (Additional file 1: Fig. S1). We hypothesize this occurs because both rRNA and mRNA molecules are fragmented to similar extents, and the increased poly-adenylation of rRNA fragments is matched by a similar increase in poly-adenylation of mRNA fragments, resulting in similar downstream detection of rRNA- and mRNA-derived reads. Thus, these results suggest that EMBR-seq can be effectively applied to sequence the transcriptome of randomly fragmented lower quality total RNA.

We also tested modified blocking primers with a 3′ phosphorylation, designed to prevent Superscript II from reverse transcribing rRNA molecules. As expected, we observed rRNA depletion in these samples as well (from 64 to 22%), with 78% of the mapped reads corresponding to mRNA (Fig. 2a). However, compared to the unmodified blocking primers, these phosphorylated blocking primers were slightly less efficient at rRNA depletion (Fig. 2a). As the 3′ phosphorylated primers prevent polymerase extension, we hypothesize that the reduced rRNA depletion efficiency arises from the small fraction of rRNA molecules that get poly-adenylated, primed by the poly-T primers, and copied through the short 30 bp RNA-DNA hybrid due to the strand-displacement activity of the reverse transcriptase. Therefore, given the reduced efficiency and higher costs of the 3′ phosphorylated blocking primers, all further experiments were performed with unmodified blocking primers.

As an alternate strategy, we also incorporated TEX treatment in EMBR-seq as it has previously been shown to specifically degrade rRNAs with 5′-monophosphate ends but not mRNAs that have 5′-triphosphate ends

[10, 14, 15, 28]. While we again observed rRNA depletion and a corresponding enrichment of mRNA compared to control samples, the effects were less pronounced with a less than 2-fold rRNA depletion, consistent with previous reports (Additional file 1: Fig. S2) [14, 15]. We hypothesize that this reduced efficiency may arise from the additional cleanup step that is necessary prior to treatment with the poly-A polymerase. As a result, we find that blocking primers alone provide the most significant rRNA depletion and mRNA enrichment, and therefore all further experiments were performed without TEX treatment.

EMBR-seq is a cost-effective bacterial mRNA sequencing technology

In designing the steps of EMBR-seq, we wanted to develop a method that is both easily applied and cost-effective. Due to its simplicity, the cost per rRNA depletion reaction in EMBR-seq is ~$0.40, which is at least an order of magnitude lower than other recent rRNA depletion methods and commercial kits [11–13, 15–18] (Additional file 1: Fig. S3a, Table S1, and Additional file 2). The total cost of EMBR-seq, starting from total bacterial RNA to the final Illumina library, was estimated to be ~$36 per sample. However, the total cost per sample decreases as more samples are multiplexed in the same Illumina library. For example, when 96 samples are multiplexed, the cost per sample drops to ~$20, primarily due to the pooling of samples after second-strand synthesis that then requires only a single IVT and Illumina library preparation reaction downstream (Additional file 1: Fig. S3b and Additional file 2). Thus, EMBR-seq is a simple and cost-effective approach to sequence mRNA from total bacterial RNA.

EMBR-seq provides a detailed view of the transcriptome without introducing technical biases

Next, we systematically compared the gene expression profiles obtained from control and rRNA depleted samples to investigate if the use of blocking primers provides a deeper view of the transcriptome without introducing technical artifacts. First, after downsampling sequencing reads to the same depth, we detected 3628 genes in the control samples, while in the mRNA enriched samples we detected 3852 genes, with 99% of the genes in the control samples also detected in the mRNA enriched samples (Fig. 2b, c). Moreover, at different levels of downsampling, we detected more genes using EMBR-seq compared to the control samples (Additional file 1: Fig. S4). This suggests that we can measure the genome-wide gene expression landscape in a more cost-effective way using EMBR-seq. Further, the number of genes detected above different expression thresholds was consistently higher for the mRNA enriched samples compared to the control samples (Fig. 2b). This shows that EMBR-seq is able to detect more genes at different gene expression levels, spanning over three orders of magnitude. Furthermore, we also observed that EMBR-seq derived reads mapped uniformly across the entire length of operons, with modest 3′ and 5′ end bias, suggesting that this method can be used to effectively quantify the expression of genes within operons (Additional file 1: Fig. S5). Finally, we observed that gene expression between the control and mRNA enriched samples were highly correlated (Pearson $r = 0.97$) revealing that the blocking primers do not introduce technical biases in the quantification of gene expression (Fig. 2d). Collectively, these results demonstrate that our new cost-effective method is able to accurately capture the transcriptome of bacterial cells.

EMBR-seq allows mRNA sequencing from low input total RNA

In many practical applications involving non-model and non-cultivable bacterial species, the starting amount of total RNA available for RNA sequencing can be limiting. Therefore, we evaluated if we can successfully deplete rRNA and quantify gene expression from lower amounts of input material. We applied EMBR-seq to 20, 2, 0.2 and 0.02 ng of starting total RNA isolated from the exponential growth phase of *E. coli* strain K12. These starting quantities of total RNA were chosen as they are typically below the sensitivity and detection limit of commercial kits and previously reported methods [11, 17]. As before, we observed a greater than 3-fold depletion of rRNA across the range of input starting material, including at the lowest starting amount of 0.02 ng total RNA, with greater than 77% of the reads in the sequencing library deriving from mRNA molecules (Fig. 3a). Similarly, we observed that the total number of genes detected is higher than that in the control samples and is unaffected by the starting input amount of total RNA, except at the lower starting amounts of 0.2 ng and 0.02 ng total RNA (Fig. 3b). Finally, we also observed that gene expression was highly correlated between different amounts of starting total RNA (Fig. 3c and Additional file 1: Fig. S6). These experiments conclusively demonstrate that we can successfully apply EMBR-seq to quantify gene expression from total RNA starting as low as 20 pg.

rRNA depletion efficiency of EMBR-seq can be further improved through additional blocking primers

In all the EMBR-seq experiments described above, we observed that 13–22% of the mapped reads still derived from rRNA and therefore, we next attempted to further improve the rRNA depletion efficiency of EMBR-seq. Analyzing the mapped coordinates of the rRNA-derived reads showed that while the 3′ blocking primers in

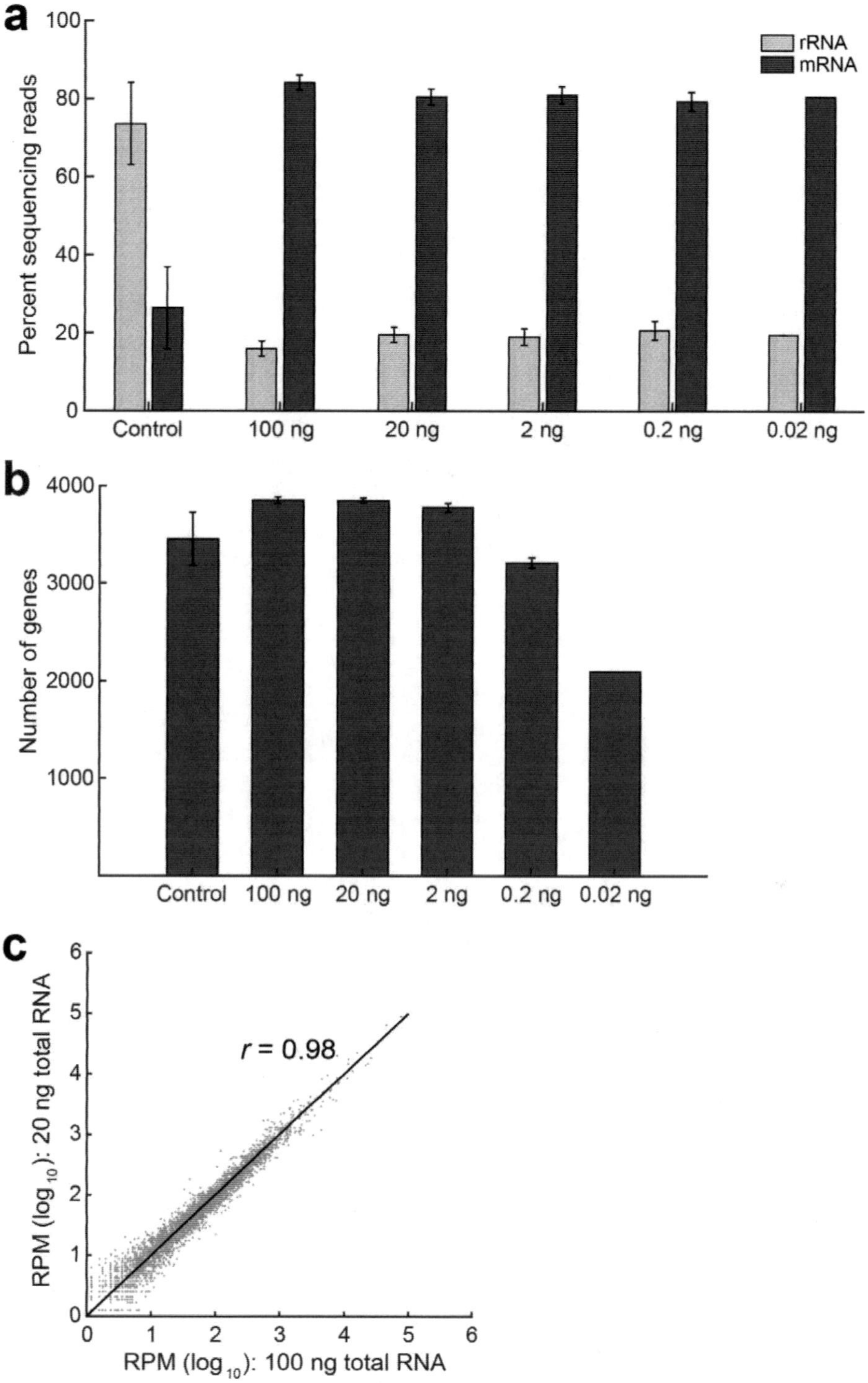

Fig. 3 EMBR-seq can quantify the transcriptome from low input total RNA. **a** Similar levels of rRNA depletion and mRNA enrichment are observed when the starting amount of total RNA is decreased from 100 ng to 0.02 ng ($n \geq 2$, except at 0.02 ng where $n = 1$). The control represents average data of control samples made from different input levels of total RNA. The 100 ng data is reproduced from Fig. 2a. **b** Compared to the control samples, more genes are detected when starting with at least 2 ng input total RNA. Fewer genes are detected when starting total RNA decreases to 0.02 ng. **c** Gene transcript counts are highly correlated (Pearson $r = 0.98$) between 100 ng and 20 ng input total RNA in EMBR-seq. Datasets from lower starting total RNA are also well correlated to the 100 ng samples (Additional file 1: Fig. S6)

EMBR-seq effectively depleted rRNA-derived reads compared to control samples from the 3′ end of rRNA molecules, specific "hotspot" regions along the entire length of the 16S and 23S rRNA were disproportionately abundant in the rRNA capture profile (Fig. 4a, b). We hypothesized that these reads resulted from the combined effects of poly-adenylation of fragmented RNA and biased capture of IVT amplified RNA molecules by random hexamer primers during reverse transcription. This reverse transcription step is part of the final Illumina library preparation protocol where the IVT amplified RNA is first reverse transcribed prior to generation of the Illumina libraries by PCR [20–22]. To minimize reads from these specific rRNA regions, we introduced 3 additional blocking primers per rRNA species that targeted the following hotspot locations: coordinates 107, 682, 1241 on 16S rRNA and coordinates 375, 1421, 1641 on 23S rRNA. We found that these hotspot blocking primers successfully reduced rRNA-derived reads from their target locations in the final sequencing library (Fig. 4a, b). Overall, this resulted in further improvement in the rRNA depletion efficiency of EMBR-seq with only 10% of the mapped reads deriving from rRNA (Fig. 4c). These results demonstrate that EMBR-seq is a versatile technique that can be used to effectively deplete rRNA and can potentially be extended to target and deplete any undesired RNA species.

Discussion

We have developed a new technology, EMBR-seq, to efficiently deplete rRNA from total RNA, thereby enabling a deeper view of the genome-wide distribution of mRNA in bacterial samples. Sequencing bacterial mRNA poses several challenges; for example, the inability to easily enrich mRNA that typically makes up less than 5% of total RNA and the limiting starting amounts of total RNA that may be available when working with non-cultivable bacterial samples [7–9]. Through the use of a single 3′ blocking primer per rRNA species, EMBR-seq efficiently minimizes the downstream amplification of rRNA molecules, thereby enabling a 4-fold depletion of rRNA in the final sequencing library (Figs. 1 and 2a). In the future,

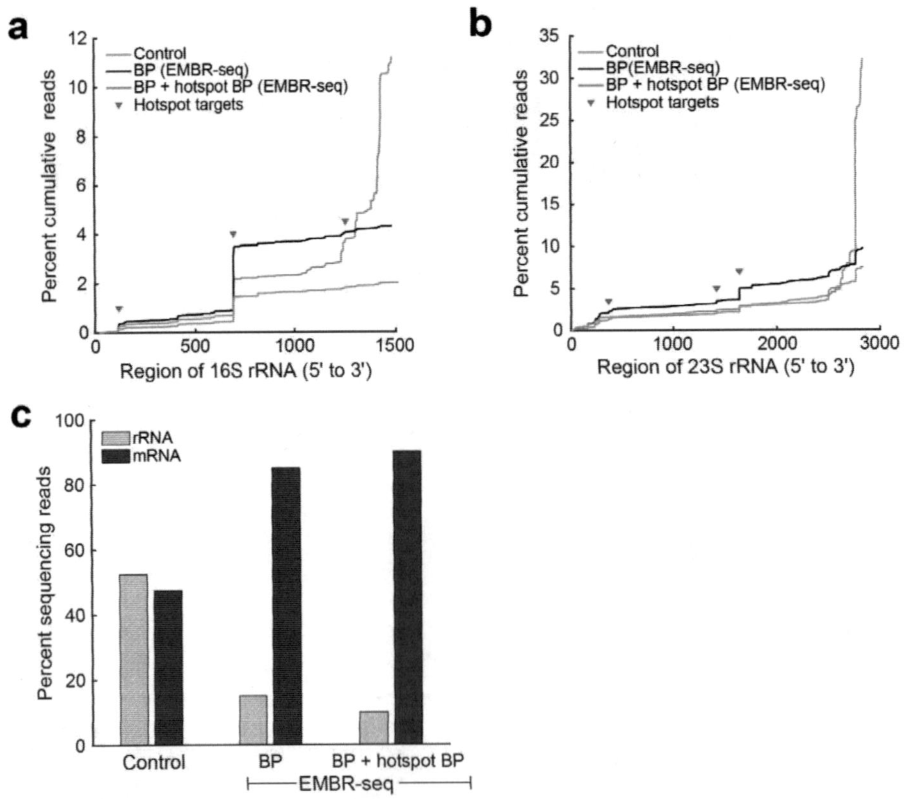

Fig. 4 Additional hotspot blocking primers increase the rRNA depletion efficiency of EMBR-seq. **a**, **b** Cumulative percentage of sequencing reads ordered by mapping location along the (**a**) 16S and (**b**) 23S rRNA subunits, from 5′ to 3′ ends of the transcript. In the control group, the majority of mapped reads are derived from the 3′ end together with a few "hotspot" locations (red triangles) along the gene body (gray lines). In EMBR-seq, 3′ end blocking primers sharply reduce the number of reads derived from the 3′ end (black lines) with the remaining rRNA reads primarily deriving from hotspot locations. Additional blocking primers were designed to minimize poly-A tailing and amplification from the vicinity of these coordinates, resulting in further rRNA depletion (orange lines). **c** rRNA depletion and mRNA enrichment is enhanced upon the addition of hotspot blocking primers. With 3′ end and hotspot blocking primers in EMBR-seq, mRNA molecules account for 90% of the mapped reads

when introducing blocking primers at the 3′ end of an rRNA species, we hypothesize that the rRNA depletion efficiency could be further improved by designing primers with a 5′ overhang or a bulky 5′ modification. As demonstrated in this work, the design of blocking primers at the 3′ end of rRNA molecules efficiently depletes rRNA from high quality total RNA samples; however, certain practical applications can produce degraded and fragmented RNA in which rRNA molecules may be less effectively depleted. While we show that the 3′ blocking primer alone is sufficient to obtain efficient rRNA depletion when starting with degraded *E. coli* total RNA, a more generalized strategy to overcome this challenge in EMBR-seq is to design additional blocking primers per rRNA species, that span the transcript length to minimize amplification of degraded rRNA molecules (Fig. 4 and Additional file 1: Fig. S1).

Starting with total RNA from *E. coli*, we show that efficient depletion of rRNA by EMBR-seq provides higher coverage of the transcriptome at the same sequencing depth (Fig. 2b and Additional file 1: Fig. S4). For example, compared to the control samples, the number of unique genes detected increases from 3628 to 3852 in EMBR-seq (Fig. 2b). In particular, EMBR-seq improves detection of lowly expressed genes below 500 RPM (Fig. 2b). Further, EMBR-seq provides a more in-depth view of the transcriptional landscape without introducing technical artifacts. We find that 99% of the genes detected in the control group are also detected by EMBR-seq, and that gene expression levels between the two groups are highly correlated (Fig. 2c, d).

As EMBR-seq typically uses a single blocking primer per rRNA species, sequence conservation analysis of 16S and 23S rRNA suggests that it is adaptable to other microbial species and complex bacterial communities (Additional file 1: Fig. S7). Recent approaches that employ a large array of probes also achieve a high efficiency of rRNA degradation; however, the need to generate such a large pool of molecules makes it more challenging to extrapolate these methods to evolutionarily distant bacterial species compared to EMBR-seq [16–18]. In addition, the use of just one or a few primers per rRNA species combined with the high level of sample multiplexing reduces cost significantly compared to other methods, enabling cost-effective and high-throughput processing of hundreds of samples simultaneously (Additional file 1: Fig. S3 and Additional file 2). While the cost per rRNA depletion reaction of other reported techniques is an order-of-magnitude higher than EMBR-seq, it is possible that the costs associated with these other methods and commercial kits could be reduced if all reagents and associated steps of the protocol are produced and performed in-house. Finally, beyond depleting rRNA and enriching for mRNA, the approach used in EMBR-seq can potentially also be used to target other high abundance transcripts in total RNA or used to enrich for non-coding RNA in both prokaryotic and eukaryotic systems [28–31].

We also demonstrated that EMBR-seq enables mRNA sequencing of low input RNA samples below the detection limit of commercial kits (Fig. 3a). Bacterial populations frequently contain diverse species, and even isogenic systems have been shown to display substantial cell-to-cell heterogeneity in gene expression that can give rise to dramatic cellular phenotypes [32–37]. Therefore, scaling down bacterial mRNA sequencing techniques to a single-cell level will enable quantification of this variability and provide a better understanding of how transcriptomic heterogeneity regulates cellular function [38, 39]. Over the last few years, a limited number of approaches have been developed to sequence the transcriptome of single bacterial cells. Early proof-of-concept methods were low throughput techniques that sequenced less than 10 single cells and generally suffered from significant technical noise [10, 40, 41]. More recently, Blattman et al. employed combinatorial barcoding to circumvent single cell isolation, enabling high throughput single-cell sequencing of bacterial cell [42]. However, this method did not deplete rRNA, resulting in mRNA detection efficiencies of ~ 2.5–10% (or ~ 200 mRNA per exponential phase *E. coli* cell). As an alternate approach, Imdahl et al. used MATQ-seq to generate sufficient cDNA from individually isolated bacterial cells; however, similar to the previous work, this method also did not deplete rRNA prior to sequencing [43]. In another study, Kuchina et al. combined rRNA depletion with combinatorial barcoding to achieve ~ 5–10% mRNA detection efficiencies in *B. subtilis* [14]. These initial efforts suggest that improved methods could significantly advance single-cell mRNA sequencing in bacteria. EMBR-seq can successfully sequence mRNA from as low as 20 pg of total RNA; therefore, we anticipate that by coupling our rRNA depletion strategy with recent combinatorial barcoding techniques, we will be able to extend EMBR-seq to a single-cell resolution in the future [14, 42].

Conclusion

EMBR-seq efficiently depletes rRNA and provides a detailed view of the gene expression landscape within bacterial samples. As EMBR-seq depletes rRNA using a single or a few blocking primers per rRNA species, this new method is easily adaptable to other microbes as well as an order-of-magnitude cheaper than other reported techniques and commercial kits that frequently use a large array of probes to remove rRNA from total RNA. Finally, EMBR-seq effectively captures the transcriptome from 500-fold lower starting total RNA compared to commercial kits, thereby providing a powerful new approach to investigate gene expression patterns in rare and non-cultivable bacterial species.

Methods

Bacterial strains and culture conditions

Escherichia coli MG1655 (ATCC: 700926) overnight cultures were inoculated into fresh LB medium at 1:50 and grown at 37 °C with shaking (150 rpm). Upon reaching the exponential growth phase, the culture was centrifuged at 3000 g for 10 min. The media was removed and the pellet was resuspended in PBS to a concentration of 10^7 cells per μL. The cells were stored on ice and total RNA extraction was performed immediately.

RNA extraction

Trizol (Thermo Fisher Scientific, Cat. # 15596018) RNA extraction was performed following the manufacturer's protocol. Briefly, 10^8 cells were added to 750 μL Trizol, mixed, and then combined with 150 μL chloroform. After centrifugation, the clear aqueous layer was recovered and precipitated with 375 μL of isopropanol and 0.67 μL of GlycoBlue (Thermo Fisher Scientific, Cat. # AM9515). The pellet was washed twice with 75% ethanol and after the final centrifugation, the resulting pellet was resuspended in RNase-free water.

EMBR-seq

Poly adenylation

100 ng of total RNA in 2 μL was combined with 3 μL poly-A mix, comprised of 1 μL 5x first strand buffer [250 mM Tris-HCl (pH 8.3), 375 mM KCl, 15 mM $MgCl_2$, comes with Superscript II reverse transcriptase, Invitrogen Cat. # 18064–014], 1 μL blocking primer mix (see *Primers*), 0.8 μL nuclease-free water, 0.1 μL 10 mM ATP, and 0.1 μL *E. coli* poly-A polymerase (New England Biolabs, Cat. # M0276S). The mixture was incubated at 37 °C for 10 min. In the control group, no blocking primers were added and 1.8 μL of nuclease-free water was added instead. For EMBR-seq with either unmodified or phosphorylated 3′-end blocking primers, the blocking primer mix was prepared by mixing equal volumes of 50 μM blocking primers specific to 5S, 16S and 23S rRNA. For EMBR-seq with hotspot blocking primers, the blocking primer mix was prepared by mixing equal volumes of 100 μM 3′-end blocking primers with 100 μM hotspot blocking primers, such that the final mixture was 50 μM 3′-end primers (3 primers mixed) and 50 μM hotspot primers (6 primers mixed).

Reverse transcription

The polyadenylation product was mixed with 0.5 μL 10 mM dNTPs (New England Biolabs, Cat. # N0447L), 1 μL reverse transcription primers (25 ng/μL, see *Primers*), and 1.3 μL blocking primer mix, and heated to 65 °C for 5 min, 58 °C for 1 min, and then quenched on ice. In the control samples, the blocking primers were again replaced with nuclease-free water. Next, 3.2 μL RT mix, consisting of 1.2 μL 5x first strand buffer, 1 μL 0.1 M DTT, 0.5 μL RNaseOUT (Thermo Fisher Scientific, Cat. #10777019), and 0.5 μL Superscript II reverse transcriptase was added to the solution, followed by 1 h incubation at 42 °C. The temperature was then raised to 70 °C for 10 min to heat inactivate Superscript II.

Second strand synthesis

49 μL of the second strand mix, containing 33.5 μL water, 12 μL 5x second strand buffer [100 mM Tris-HCl (pH 6.9), 23 mM $MgCl_2$, 450 mM KCl, 0.75 mM β-NAD, 50 mM $(NH_4)_2SO_4$, Invitrogen, Cat. # 10812–014], 1.2 μL 10 mM dNTPs, 0.4 μL *E. coli* ligase (Invitrogen, Cat. # 18052–019), 1.5 μL DNA polymerase I (Invitrogen, Cat. # 18010–025), and 0.4 μL RNase H (Invitrogen, Cat. # 18021–071), was added to the product from the previous step. The mixture was incubated at 16 °C for 2 h. cDNA was purified with 1x AMPure XP DNA beads (Beckman Coulter, Cat. # A63881) and eluted in 24 μL nuclease-free water that was subsequently concentrated to 6.4 μL.

In vitro transcription

The concentrated solution was mixed with 9.6 μL of Ambion in vitro transcription mix (1.6 μL of each ribonucleotide, 1.6 μL 10x T7 reaction buffer, 1.6 μL T7 enzyme mix, MEGAscript T7 Transcription Kit, Thermo Fisher Scientific, Cat. # AMB13345) and incubated at 37 °C for 13 h. Next, the aRNA was treated with 6 μL EXO-SAP (ExoSAP-IT™ PCR Product Cleanup Reagent, Thermo Fisher Scientific, Cat. # 78200.200.UL) at 37 °C for 15 min followed by fragmentation with 5.5 μL fragmentation buffer (200 mM Tris-acetate (pH 8.1), 500 mM KOAc, 150 mM MgOAc) at 94 °C for 3 min. The reaction was then quenched with 2.75 μL stop buffer (0.5 M EDTA) on ice. The fragmented aRNA was size selected with 0.8x AMPure RNA beads (RNAClean XP Kit, Beckman Coulter, Cat. # A63987) and eluted in 15 μL nuclease-free water. Thereafter, Illumina libraries were prepared as described previously [20].

EMBR-seq with TEX digestion

To test the Terminator™ 5′-phosphate-dependent exonuclease (Lucigen, Cat. # TER5120), 100 ng of total RNA in 2 μL was combined with 18 μL TEX mix, comprised of 14.5 μL nuclease free water, 2 μL Terminator 10x buffer A, 0.5 μL RNAseOUT, and 1 μL TEX. The solution was incubated at 30 °C for 1 h and quenched with 1 μL of 100 mM EDTA. The product was purified with 1x AMPure RNA beads and eluted in 10 μL nuclease-free water and concentrated to 2 μL. This TEX digested total RNA was then used as starting RNA in the EMBR-seq protocol described above.

EMBR-seq bioinformatic analysis

Paired-end sequencing of the EMBR-seq libraries was performed on an Illumina NextSeq 500. All sequencing data has been deposited to Gene Expression Omnibus under the accession number GSE149666. In the sequencing libraries, the left mate contains information about the sample barcode (see *Primers*). The right mate is mapped to the bacterial transcriptome. Prior to mapping, only reads containing valid sample barcodes were retained. Subsequently, the reads were mapped to the reference transcriptome (*E. coli* K12 substr. MG1655 cds ASM584v2) using Burrows-Wheeler Aligner (BWA) with default parameters.

Analysis of detection bias in EMBR-seq

E. coli operons were downloaded from RegulonDB [44]. Operons with at least 2 genes were included for this analysis. The data from EMBR-seq libraries with 100 ng starting material was mapped to *E. coli* K12 substr. MG1655 reference genome (ASM584v2). For each read that maps within an operon, the distance of the mapped location from the 3′ end of the operon was calculated, accounting for the read length. Next, the operons were discretized into 50 bins, and all operons with more than 200 unique reads were considered for downstream analysis. The number of reads in each bin was then normalized by the total number of reads in each operon, and the average of the relative reads within each bin was calculated. To compare bacterial data from EMBR-seq to mammalian data from CEL-seq, we downloaded CEL-seq data reported in Grün et al. (GEO Accession: GSM1322290) and performed similar analysis for the mouse genes [45].

Sequence conservation of 16S and 23S rRNA

16S rRNA sequences from 4000 species were obtained from *rrnDB* [46], while 23S rRNA sequences from 119 species were selected from NCBI RefSeq [47]. Next, the last 100 bases from the 3'end of each sequence were aligned using Clustal Omega [48]. Shannon entropy for each aligned base location was then calculated such that the maximal entropy value was 1. Five possibilities were allowed: “A”, “T”, “C”, “G”, and “-”.

Primers

Reverse transcription primers are shown below with the 6-nucleotide sample barcodes underlined [20]:

GCCGGTAATACGACTCACTATAGGGAGTTCTACAGTCCGACGATCNNNNNN(NNNNNN)TTTTTTTTTTTTTTTTTTTTTTTTV

The following five barcodes were used in this study:

AGACTC
AGCTTC
CATGAG
CAGATC
TCACAG

Blocking primers:

5S 5′-ATGCCTGGCAGTTCCCTACTCTCGCATGGG-3′

16S 5′-TAAGGAGGTGATCCAACCGCAGGTTCCCCT-3′

23S 5′-AAGGTTAAGCCTCACGGTTCATTAGTACCG-3′

In the case of the 3′ phosphorylated primers, all blocking primers have a 3′ phosphorylation modification.

Hotspot blocking primers:

16S primer for hotspot at position 107:
5′-GGCACATCCGATGGCAAGAGGCCCGAAGGT-3′.

16S primer for hotspot at position 682:
5′-TCCTGTTTGCTCCCCACGCTTTCGCACCTG-3′.

16S primer for hotspot at position 1241:
5′-CCGTGGCATTCTGATCCACGATTACTAGCGATTCCG-3.

23S primer for hotspot at position 375:
5′-CGCCTTTCCCTCACGGTACTGGTTCACTATCGG-3′.

23S primer for hotspot at position 1421:
5′-TTGCTTCAGCACCGTAGTGCCTCGTCATCA-3′.

23S primer for hotspot at position 1641:
5′-GCAGCCAGCTGGTATCTTCGACTGATTTCAGC-3′.

Each primer is designed to anneal approximately 100 bp downstream of the hotspot. The exact position and length of each primer was adjusted to ensure the T_m was above 65 °C.

Supplementary information

Supplementary information accompanies this paper at https://doi.org/10.1186/s12864-020-07134-4.

Additional file 1 Figure S1. EMBR-seq effectively depletes rRNA from fragmented total RNA. **Figure S2.** Combining TerminatorTM 5′-phosphate-dependent exonuclease (TEX) digestion with EMBR-seq does not improve rRNA depletion. **Figure S3.** Cost associated with performing EMBR-seq. **Figure S4.** Higher number of genes detected in EMBR-seq is not dependent on the sequencing depth. **Figure S5.** Distribution of reads along *E. coli* operons in EMBR-seq. **Figure S6.** Gene transcript count correlation between different input total RNA amounts in EMBR-seq. **Figure S7.** Quantification of 16S and 23S rRNA sequence conservation using Shannon entropy. **Table S1.** Comparison of rRNA depletion methods.

Additional file 2.

Abbreviations

IVT: In vitro transcription; LB: Lysogeny broth; RNase H: Ribonuclease H; TEX: Terminator™ exonuclease

Acknowledgements

We would like to thank members of the Dey and O'Malley groups for helpful discussions. The authors thank Dr. Jennifer Smith for assistance with sequencing Illumina libraries. Sequencing was performed at the Biological Nanostructures Laboratory within the California NanoSystems Institute (CNSI), supported by the University of California, Santa Barbara (UCSB) and the University of California, Office of the President (UCOP). We further acknowledge the Center of Scientific Computing at UCSB for computational facilities that are funded by NSF MRSEC (DMR-1720256) and NSF CNS-1725797.

Authors' contributions
C.W. and S.S.D. conceived the method, C.W., K.A.H. and S.S.D. designed the experiments. K.A.H., C.W. and E.L. participated in data collection. K.A.H. and C.W. analyzed the data. C.W., K.A.H., M.A.O. and S.S.D. wrote the manuscript. S.S.D. guided experimental design and data analysis. All authors have read and approved the final version of the manuscript.

Funding
This work was funded by grants from the National Science Foundation (MCB-1553721) to M.A.O., the UCSB Academic Senate Faculty Research Grant to S.S.D., and the CNSI Challenge Grant Program, supported by UCSB and UCOP, to S.S.D. and M.A.O.

Availability of data and materials
The datasets generated and analyzed in the current study have been deposited to GEO under the accession number GSE149666.

Ethics approval and consent to participate
Not applicable.

Consent for publication
Not applicable.

Competing interests
The authors declare no competing financial interests.

Author details
[1]Department of Chemical Engineering, University of California Santa Barbara, Santa Barbara, CA 93106, USA. [2]Center for Bioengineering, University of California Santa Barbara, Santa Barbara, CA 93106, USA. [3]Neuroscience Research Institute, University of California Santa Barbara, Santa Barbara, CA 93106, USA.

Received: 21 May 2020 Accepted: 9 October 2020
Published online: 16 October 2020

References

1. Otero JM, Nielsen J. Industrial systems biology. Biotechnol Bioeng. 2010;105: 439–60.
2. Dvořák P, Nikel PI, Damborský J, de Lorenzo V. Bioremediation 3.0: engineering pollutant-removing bacteria in the times of systemic biology. Biotechnol Adv. 2017;35:845–66.
3. Peng X, Gilmore SP, O'Malley MA. Microbial communities for bioprocessing: lessons learned from nature. Curr Opion Chem Eng. 2016;14:103–9.
4. Kung Y, Runguphan W, Keasling JD. From fields to fuels: recent advances in the microbial production of biofuels. ACS Synth Biol. 2012;1:498–513.
5. Barajas JF, Blake-Hedges JM, Bailey CB, Curran S, Keasling JD. Engineered polyketides: synergy between protein and host level engineering. Synth Syst Biotechnol. 2017;2:147–66.
6. Creecy JP, Conway T. Quantitative bacterial transcriptomics with RNA-seq. Curr Opin Microbiol. 2015;23:133–40.
7. Mortazavi A, Williams BA, McCue K, Schaeffer L, Wold B. Mapping and quantifying mammalian transcriptomes by RNA-Seq. Nat Methods. 2008;5: 621–8.
8. Proudfoot NJ. Ending the message: poly(A) signals then and now. Genes Dev. 2011;25:1770–82.
9. Giannoukos G, Ciulla DM, Huang K, Haas BJ, Izard J, Levin JZ, et al. Efficient and robust RNA-seq process for cultured bacteria and complex community transcriptomes. Genome Biol. 2012;13:r23.
10. Kang Y, Norris MH, Zarzycki-Siek J, Nierman WC, Donachie SP, Hoang TT. Transcript amplification from single bacterium for transcriptome analysis. Genome Res. 2011;21:925–35.
11. Petrova OE, Garcia-Alcalde F, Zampaloni C, Sauer K. Comparative evaluation of rRNA depletion procedures for the improved analysis of bacterial biofilm and mixed pathogen culture transcriptomes. Sci Rep. 2017;7:41114.
12. Kraus AJ, Brink BG, Siegel TN. Efficient and specific oligo-based depletion of rRNA. Sci Rep. 2019;9:12281.
13. Culviner PH, Guegler CK, Laub MT. A simple, cost-effective, and robust method for rRNA depletion in RNA-sequencing studies. mBio. 2020;11: e00010–20.
14. Kuchina A, Brettner LM, Paleologu L, Roco CM, Rosenberg AB, Carignano A, et al. Microbial single-cell RNA sequencing by split-pool barcoding. Preprint. bioRxiv. 2019. https://doi.org/10.1101/869248.
15. He S, Wurtzel O, Singh K, Froula JL, Yilmaz S, Tringe SG, et al. Validation of two ribosomal RNA removal methods for microbial metatranscriptomics. Nat Methods. 2010;7:807–12.
16. Huang Y, Sheth RU, Kaufman A, Wang HH. Scalable and cost-effective ribonuclease-based rRNA depletion for transcriptomics. Nucleic Acids Res. 2020;48:e20.
17. Prezza G, Heckel T, Dietrich S, Homberger C, Westermann AJ, Vogel J. Improved bacterial RNA-seq by Cas9-based depletion of ribosomal RNA reads. RNA. 2020;26:1069–78.
18. Armour CD, Castle JC, Chen R, Babak T, Loerch P, Jackson S, et al. Digital transcriptome profiling using selective hexamer priming for cDNA synthesis. Nat Methods. 2009;6:647–9.
19. Feng Y, Cohen SN. Unpaired terminal nucleotides and 5′ monophosphorylation govern 3′ polyadenylation by *Escherichia coli* poly(A) polymerase I. Proc Natl Acad Sci U S A. 2000;97:6415–20.
20. Hashimshony T, Senderovich N, Avital G, Klochendler A, de Leeuw Y, Anavy L, et al. CEL-Seq2: sensitive highly-multiplexed single-cell RNA-Seq. Genome Biol. 2016;17:77.
21. Mooijman D, Dey SS, Boisset J-C, Crosetto N, van Oudenaarden A. Single-cell 5hmC sequencing reveals chromosome-wide cell-to-cell variability and enables lineage reconstruction. Nat Biotechnol. 2016;34:852–6.
22. Rooijers K, Markodimitraki CM, Rang FJ, de Vries SS, Chialastri A, de Luca KL, et al. Simultaneous quantification of protein–DNA contacts and transcriptomes in single cells. Nat Biotechnol. 2019;37:766–72.
23. Wendisch VF, Zimmer DP, Khodursky A, Peter B, Cozzarelli N, Kustu S. Isolation of Escherichia coli mRNA and comparison of expression using mRNA and total RNA on DNA microarrays. Anal Biochem. 2001;290:205–13.
24. Westermann AJ, Förstner KU, Amman F, Barquist L, Chao Y, Schulte LN, et al. Dual RNA-seq unveils noncoding RNA functions in host–pathogen interactions. Nature. 2016;529:496–501.
25. Westermann AJ, Gorski SA, Vogel J. Dual RNA-seq of pathogen and host. Nat Rev Microbiol. 2012;10:618–30.
26. Cozen AE, Quartley E, Holmes AD, Hrabeta-Robinson E, Phizicky EM, Lowe TM. ARM-seq: AlkB-facilitated RNA methylation sequencing reveals a complex landscape of modified tRNA fragments. Nat Methods. 2015;12:879–84.
27. Zheng G, Qin Y, Clark WC, Dai Q, Yi C, He C, et al. Efficient and quantitative high-throughput tRNA sequencing. Nat Methods. 2015;12:835–7.
28. Sharma CM, Hoffmann S, Darfeuille F, Reignier J, Findeiss S, Sittka A, et al. The primary transcriptome of the major human pathogen *Helicobacter pylori*. Nature. 2010;464:250–5.
29. Chao Y, Papenfort K, Reinhardt R, Sharma CM, Vogel J. An atlas of Hfq-bound transcripts reveals 3′ UTRs as a genomic reservoir of regulatory small RNAs: Hfq-dependent small RNAs from 3′ UTRs. EMBO J. 2012;31:4005–19.
30. Chao Y, Li L, Girodat D, Förstner KU, Said N, Corcoran C, et al. In vivo cleavage map illuminates the central role of RNase E in coding and non-coding RNA pathways. Mol Cell. 2017;65:39–51.
31. Faridani OR, Abdullayev I, Hagemann-Jensen M, Schell JP, Lanner F, Sandberg R. Single-cell sequencing of the small-RNA transcriptome. Nat Biotechnol. 2016;34:1264–6.
32. Miller-Jensen K, Dey SS, Schaffer DV, Arkin AP. Varying virulence: epigenetic control of expression noise and disease processes. Trends Biotechnol. 2011; 29:517–25.
33. Raj A, van Oudenaarden A. Nature, nurture, or chance: stochastic gene expression and its consequences. Cell. 2008;135:216–26.
34. Balázsi G, van Oudenaarden A, Collins JJ. Cellular decision making and biological noise: from microbes to mammals. Cell. 2011;144:910–25.
35. Gefen O, Balaban NQ. The importance of being persistent: heterogeneity of bacterial populations under antibiotic stress. FEMS Microbiol Rev. 2009;33: 704–17.
36. Russell JR, Cabeen MT, Wiggins PA, Paulsson J, Losick R. Noise in a phosphorelay drives stochastic entry into sporulation in *Bacillus subtilis*. EMBO J. 2017;36:2856–69.
37. Avraham R, Haseley N, Brown D, Penaranda C, Jijon HB, Trombetta JJ, et al. Pathogen cell-to-cell variability drives heterogeneity in host immune responses. Cell. 2015;163:523.
38. Penaranda C, Hung DT. Single-cell RNA sequencing to understand host–pathogen interactions. ACS Infect Dis. 2019;5:336–44.

39. Avital G, Avraham R, Fan A, Hashimshony T, Hung DT, Yanai I. scDual-Seq: mapping the gene regulatory program of *Salmonella* infection by host and pathogen single-cell RNA-sequencing. Genome Biol. 2017;18:200.
40. Wang J, Chen L, Chen Z, Zhang W. RNA-seq based transcriptomic analysis of single bacterial cells. Integr Biol (Camb). 2015;7:1466–76.
41. Liu Y, Jeraldo P, Jang JS, Eckloff B, Jen J, Walther-Antonio M. Bacterial single cell whole transcriptome amplification in microfluidic platform shows putative gene expression heterogeneity. Anal Chem. 2019;91:8036–44.
42. Blattman SB, Jiang W, Oikonomou P, Tavazoie S. Prokaryotic single-cell RNA sequencing by in situ combinatorial indexing. Nat Microbiol. 2020. https://doi.org/10.1038/s41564-020-0729-6.
43. Imdahl F, Vafadarnejad E, Homberger C, Saliba AE, Vogel J. Single-cell RNA-sequencing reports growth-condition-specific global transcriptomes of individual bacteria. Nat Microbiol. 2020. https://doi.org/10.1038/s41564-020-0774-1.
44. Santos-Zavaleta A, Salgado H, Gama-Castro S, Sánchez-Pérez M, Gómez-Romero L, Ledezma-Tejeida D, et al. RegulonDB v 10.5: tackling challenges to unify classic and high throughput knowledge of gene regulation in *E. coli* K-12. Nucleic Acids Res. 2019;47:D212–20.
45. Grün D, Kester L, van Oudenaarden A. Validation of noise models for single-cell transcriptomics. Nat Methods. 2014;11:637–40.
46. Stoddard SF, Smith BJ, Hein R, BRK R, Schmidt TM. rrnDB: improved tools for interpreting rRNA gene abundance in bacteria and archaea and a new foundation for future development. Nucleic Acids Res. 2015;43:D593–8.
47. O'Leary NA, Wright MW, Brister JR, Ciufo S, Haddad D, McVeigh R, et al. Reference sequence (RefSeq) database at NCBI: current status, taxonomic expansion, and functional annotation. Nucleic Acids Res. 2016;44:D733–45.
48. Madeira F, Park YM, Lee J, Buso N, Gur T, Madhusoodanan N, et al. The EMBL-EBI search and sequence analysis tools APIs in 2019. Nucleic Acids Res. 2019;47:W636–41.

Publisher's Note

609928 Vaccine Engineering in the Time of COVID: Rapid and Cost-Effective Development of Stable Clones for the Production of Sars-Cov-2 Vaccine Candidates in HEK293T Cells (Industry Candidate)

Abstract:

The COVID-19 pandemic has crudely demonstrated the value of enabling platforms to develop effective vaccines in a short time. Currently, there are no specific treatments or vaccines approved against COVID-19.

In this work, we describe the development of HEK293T cells engineered for the transient and stable expression of antigens of SARS-CoV-2, the causal agent of COVID-19.

We transfected the HEK293T cells with vectors engineered to produce different versions of antigens inspired in different SARS-CoV-2 proteins (i.e., Spike (S) and Nucleoprotein (NP)) and intracellular Green Fluorescent Protein (GFP) using Lipofectamine 3000. We then selected the transfected cells using puromycin pressure, dilution cloning, and cloning disks. This integrated strategy generated antigen-producing cells in 7 days with a transient expression of ~1 mg/L. Stable pools were produced after 4 weeks, with expression levels of ~1 mg/L. Stable clones with expression levels of ~2.0 mg/L were obtained within 10 weeks. The produced antibodies exhibited the expected functionality; they were selectively bound by anti-SARS-CoV-2 antibodies in ELISA experiments both using commercial reagents and convalescent sera from COVID (+) patients. In addition, we developed perfusion processes for the continuous production of antigens from adherent cultures.

By the combined use of GFP and the set of selection techniques here described, we drastically reduced the time from transfection to stable clone generation without resorting to costly equipment. In outbreaks or emergencies, this platform can remarkably shorten the development of new vaccines and biopharmaceuticals.

611111 Improved Discrimination of Asymmetric and Symmetric Arginine Dimethylation By Optimization of the Normalized Collision Energy in LC-MS Proteomics

Abstract:

Protein arginine methylation regulates diverse biological processes including signaling, metabolism, splicing, and transcription. Despite its important biological roles, arginine dimethylation remains an understudied post-translational modification. Partly, this is because the two forms of arginine dimethylation, asymmetric and symmetric dimethyl arginine, are isobaric and therefore indistinguishable by traditional mass spectrometry techniques. Thus, there exists a need for methods that can differentiate these two modifications. Here, we report that increasing the normalized collision energy (NCE) in a higher-energy collisional dissociation (HCD) cell of a Q-Exactive Plus mass spectrometer increases the generation of characteristic neutral losses (NL) that can distinguish asymmetric and symmetric dimethyl arginine. By analyzing both synthetic and endogenous dimethyl-peptides, we identify an optimal NCE value that maximizes NL generation and also improves methyl-peptide identification. Using two orthogonal methyl peptide enrichment strategies, high pH strong cation exchange (SCX) and immunoaffinity purification (IAP), we demonstrate that the optimal NCE increases improves NL-based ADMA and SDMA annotation and dimethyl peptide identifications by 125% and 17%, respectively, compared to the standard NCE. This simple parameter change will greatly facilitate the identification and annotation of asymmetric and symmetric dimethyl arginine in mass spectrometry-based methyl-proteomics to improve our understanding of how these modifications differentially regulate protein function.

596988 ***Engineering Protein and Biopolymer Complexation for Synthetic Organelle Assembly*** **(Industry Candidate)**

Abstract:

Compartmentalization allows for the spatial organization of intracellular macromolecules and has been co-opted to improve existing biological functions. One strategy for compartmentalization that has recently been acknowledged as a fundamental mechanism for organizing cellular components in eukaryotes is phase separation by protein de-mixing. Phase separated membraneless organelles create distinct environments that are essential to cellular processes ranging from cell signaling to gene expression. These membraneless compartments offer several advantages that make them highly suitable for applications in the production of pharmaceuticals and biocatalysis, including dynamic restructuring of internal constituents and free diffusion of cellular components into/out of compartments. However, the ability to independently engineer the formation and dissolution of membraneless organelles *in vivo* remains a challenge. Many membraneless organelles appear to have the same physical properties as complex coacervates in that they are liquid-liquid phase separated mixtures of oppositely charged polyelectrolytes. Toward this end, we are interested in utilizing the physical phenomenon of complex coacervation and principles underlying the formation of liquid-like biological condensates to create synthetic membraneless organelles. We have investigated the complex coacervation of engineered proteins with biological polyelectrolytes to determine predictive design rules for protein phase separation. We employ these design rules to create synthetic organelles by promoting phase separation of engineered proteins in *E. coli*. We find that coacervation is dependent on protein charge and concentration, and demonstrate the *in vivo* assembly of phase separated compartments that are protein-dense and dynamic.

601805 Directed Evolution of Photoswitchable Enzymes

Abstract:

Protein light switches—or optogenetic actuators—provide a powerful means of placing enzymes under optical control. Most contemporary strategies for using optogenetic actuators, however, suffer from several limitations: (i) They disrupt the structure of the enzyme and can thus interfere with native enzyme-substrate interactions. (ii) They yield dynamic ranges (DR, the ratio of activity in the off- and on-states) that are low and/or difficult to enhance through systematic optimization. (iii) They are modulated by blue or green light, which have heightened phototoxicities, suffer from short biological penetration depths, and, because of their spectral similarity, limit actuation to individual signaling events. (iv) They tend to be limited to specific subsets of enzymes (notably, kinases or recombinases). In this presentation, I will discuss our efforts to address these shortcomings with synthetic biology and directed evolution. In brief, we have developed an approach for using microbial systems to evolve light-sensitive enzymes, and we are using it to extend optogenetic actuation to new enzymes and new colors of light. Our initial efforts focused on protein tyrosine phosphatases, a broadly influential class of signaling enzymes. In early work, we developed a photoswitchable variant of PTP1B with a low—yet strangely influential—dynamic range (2.5-fold). Our approach, which exploits a selection-counterselection screen to find photoswitchable variants, has allowed us to identify beneficial mutations in photoswitchable variants of PTP1B and promising starting points for the assembly of red-light-sensitive enzymes. Broadly, the constructs developed with our platform will support (and, as I will mention, are supporting) detailed studies of intracellular signaling networks in living cells.

613127 Antibodies That Target the Blood-Brain Barrier (Invited Speaker)

Abstract:

The blood-brain barrier presents a major obstacle to brain drug delivery. We have developed several different enabling platforms for the identification of antibodies against blood-brain barrier resident receptors that can be used to ferry drug cargo into the brain. Here we will describe our recent efforts to identify and deploy these antibodies.

608785 Metabolic Modeling and Pulse Amplitude Modulation Fluorometry Elucidate the Interplay between Photosynthesis and Carbon Fixation in a Purple Non-Sulfur Bacterium

Abstract:

Rhodopseudomonas Palustris is a metabolically versatile Purple Non-Sulfur Bacterium (PNSB). Depending on growth conditions, *R. palustris* can operate on either one of the four different forms of metabolism: photoautotrophic, photoheterotrophic, chemoautotrophic, and chemoheterotrophic. *R. palustris* is also a facultative anaerobe, meaning it can operate both aerobically and anaerobically. Furthermore, the organism is capable of fixing nitrogen and subsequently producing hydrogen. This metabolic flexibility and plethora of functionalities have made *R. palustris* a model organism for studying the network of interacting reactions and how it's altered in response to changing conditions. Recently, several works have identified the roles of carbon fixation as a redox balancing mechanism in photoheterotrophically grown PNSB. Furthermore, the involvement of global regulators such as RegA/PrrA was shown to affect a number of metabolic pathways including the electron-sink pathways of CO2 and N2 fixation, hydrogen metabolism, and the energy generating photosynthetic pathway. The aim of this study was to determine the metabolic bottlenecks associated with the use of cyclic photophosphorylation. Mainly, our focus was on how the use of various carbon sources with different oxidation states affected the photosynthetic efficiency. To achieve this aim, a genome-scale metabolic model of *R palustris* was reconstructed, curated, and validated against experimental data. The model was used to generate hypotheses on the effect of growth conditions (i.e. carbon source and light intensity) on the activity of the Electron Transport Chain (ETC), carbon dioxide fixation rate, and Quinone oxidoreductase reaction rates[1]. These hypotheses were then experimentally tested using Pulse Amplitude Modulation (PAM) fluorometry techniques to measure the Electron Transport Rate (ETR) as well as qPCR to determine expression levels of the CO2 fixing components under various growth conditions. This study illustrates (i) the predictive power of genome scale metabolic modeling, especially for the case of metabolically versatile organisms and (ii) the viability of PAM fluorometry for measuring photosynthetic rates in PNSB.

References

1. Alsiyabi, A., Immethun, C.M., Saha, R., 2019. Modeling the Interplay between Photosynthesis, CO2 Fixation, and the Quinone Pool in a Purple Non-Sulfur Bacterium. Sci. Rep. 9, 12638. https://doi.org/10.1038/s41598-019-49079-z

609023 Leveraging Genome-Scale Metabolic Models and Synthetic Biology Tools to Engineer UDP-Sugar Metabolism and Exopolysaccharide Synthesis in *Komagataeibacter Xylinus*

Abstract:

Domestication of alternative host organisms with unique biosynthetic capability is an attractive alternative to extensive rewiring of model organisms. Domestication promises to preserve complex genetic systems associated with tolerance and biosynthesis while enabling genomics-driven metabolic engineering and synthetic biology approaches to optimize and diversify the native products. This is particularly true of natural biomaterials like bacterial nanocellulose (BNC), an attractive starting point for constructing engineered living biomaterials. BNC is found in kombucha tea and nata de coco - and may be used in wound dressings, field electronic transistors, and tissue engineering. Yet, metabolic engineering and synthetic biology tools for tuning BNC biosynthesis are limited. Therefore, we have domesticated a BNC-producing strain, DS12, from active kombucha culture.

Based on 16S genotyping, we identified DS12 as a strain of *Komagataeibacter xylinus*, which is a gram-negative acetic acid bacterium. We then conducted growth and yield experiments, demonstrating that *K. xylinus* DS12 can produce pure, highly-crystalline BNC from a variety of sugars and sugar alcohols. Yet, the current understanding of BNC-producing bacteria does not satisfactorily explain the mechanism behind the varying cellulose yields and properties from different carbon sources. To this end, we have sequenced the genome of *K. xylinus* DS12. Drawing from the genome-scale metabolic model iMR640 based upon *K. xylinus* strain E25, we seek to expand and query a metabolic model with our growth and cellulose production data to elucidate a mechanistic explanation for the substrate preferences.

To facilitate metabolic engineering of BNC biosynthesis, we have successfully transformed *K. xylinus* DS12 and developed a synthetic biology parts toolkit. In order to leverage the power of a modern DNA assembly standard, we have adapted the CIDAR MoClo standard to *K. xylinus* DS12. To do this, we designed a MoClo compatible broad-based BBR1 origin and replaced the ampicillin selectable marker with spectinomycin to avoid observed ampicillin resistance. With these modifications, we were able to characterize MoClo compatible promoters in *K. xylinus* DS12 with flow cytometry. We believe this toolkit has several advantages over existing BioBrick compatible toolkits for BNC producing bacteria. In sum, domestication of *K. xylinus* DS12 is a key step towards rationally designed living cellulosic biomaterials.

610445 Rational Engineering of *Clostridium* for Efficient Production of Renewable Fatty Acid Esters

Abstract:

Production of renewable chemicals through biological routes is considered as an urgent solution for fossil energy crisis. However, endproduct toxicity inhibits microbial host performance and is a key bottleneck for biochemical production. To address this challenge, here we report an example of biosynthesis of high-value and easy-recoverable derivatives to alleviate endproduct toxicity and enhance bioproduction efficiency. By leveraging the natural pathways in solventogenic clostridia for co-producing acyl-CoAs, acids and alcohols as precursors, through rational screening for host strains and enzymes, systematic metabolic engineering, and elimination of putative prophages, we developed strains that can produce 20.3 g/L butyl acetate and 1.6 g/L butyl butyrate respectively, which were both the unprecedented levels in microbial hosts. ***Specifically, the butyl acetate production we obtained is 2400-fold higher than the highest level that has been previously reported in microbial hosts.*** Besides, some of our engineered strains demonstrated great potential for producing other esters as well. Our principles of selecting the most appropriate host for specific bioproduction and engineering microbial hosts to produce high-value and easy-separable endproducts are highly applicable to other bioprocesses, and could lead to breakthroughs in biofuel/biochemical production and general bioeconomy.

611602 Metabolic Engineering of *Clostridium Thermocellum* to Produce Isobutyl Esters from Lignocellulose

Abstract:

Direct fermentation of lignocellulosic biomass for value-added chemical production is promising for an environmentally sustainable and economically competitive biorefinery. *Clostridium thermocellum* is a thermophilic bacterium that has a strong capability of fermenting lignocellulose, but its metabolism has mostly been studied to produce short-chain biofuels such as ethanol. Here, we will present metabolic engineering of *C. thermocellum* to produce medium chain isobutyl esters (i.e., isobutyl acetate, and isobutyl isobutyrate) from lignocellulose. Push-and-pull metabolic engineering strategies involving deletion of the competitive esterase pathway and overexpression of ester production pathway improved the ester production more than 100-folds. This ester biosynthesis platform takes advantages of elevated operation temperature benefiting the *in situ* separation of the volatile ester products and hence enabling an advanced consolidated bioprocessing platform where hydrolytic enzyme production, cellulose hydrolysis, fermentation, and product separation can take place in a single step.

612988 Development of Next Generation Synthetic Biology Toolkits for Non-Model Prokaryotes (Invited Speaker)

Abstract:

Engineering microorganisms for the production of desired chemicals at industrially relevant scales, rates, and titers often requires a combination of adaptive evolution and rational design methods to screen or select for mutants with optimized phenotypes. While useful, these methods are often limited by the ability to introduce non-biased, synthetically designed, and trackable mutations in multiplex and to identify clear gene-to-trait relationships. In model bacterium such as *E. coli*, there is an ever expanding range of cutting edge tools for the introduction of these targeted mutations and rapid genotype-to-phenotype identification. Unfortunately, the adaptation of these methods for use in non-model systems with complex phenotypes that are not easily tractable (e.g. photosynthesis, CO metabolism, biomass degradation, thermophilic growth, etc.) is still lacking due to limited basic genetic tools. I will discuss current progress for the development of expanded genome engineering tools in a variety of non-model, industrially relevant microbes we work with in our group to enable more robust gene editing and accelerated gene-to-trait identification for the production of a variety of fuels and chemicals from a broad range of feedstocks.

607693 Genetically Programmable Microbial Assembly

Abstract:

In nature, most bacteria live in multicellular, often multi-species biofilms or aggregates, which enables emergent properties including resistance to environmental insult, cellular communication, and metabolic flexibility. In this work, we are developing methods to attempt to recapitulate some of these properties in an engineered microbial system. In this presentation, we will demonstrate that by surface display of associative proteins on the surface of *Escherichia coli*, we can program inter-cellular association of bacterial cells, leading to programmable aggregation of the microbes. By choice of the associative protein and control of their expression levels, we are able to control the dynamics, equilibrium state, and stimuli-responsiveness of the underlying aggregates. Finally, we will show that using these aggregates as biocatalysts in multi-step biosynthetic reactions, we are able to increase titers of target chemicals due to channeling of intermediate species within the aggregates. The techniques discussed in this work may have broader applications in metabolic engineering, waste water treatment, and human health.

609146 Amino Acid Cross-Feeding Enables Inter- and Intra-Species Cooperation (Industry Candidate)

Abstract:

Naturally occurring bacterial communities are able to cooperate to complete complex tasks by splitting these tasks between organisms and reducing the metabolic burden placed on each individual strain. The ability to distribute tasks and the metabolic load is of interest for biotechnology applications where dividing labor between populations can address bottlenecks in production and allow one to take advantage of the natural strengths of a specific strain. The potential benefits of creating synthetic communities capable of sharing metabolic load are challenged by the difficulty in co-culturing microbes. One way to combat this challenge is to use pairs of auxotrophic microbes. By growing two microbes with different auxotrophies in a nutrient deficient environment, the two microbes can trade essential nutrients, or cross-feed to enable the mutualistic growth of both organisms. To explore the use of cross-feeding we have assessed the growth of two industrially relevant bacterial hosts in both inter- and intra-species pairs of auxotrophs and examined transcriptomic profiles of successfully cross-feeding pairs.

We have quantified growth of 14 amino acid auxotrophs of *E. coli* and 6 amino acid auxotrophs of *B. megaterium* in both inter- and intra-species pairings. Cross-feeding success was evaluated in three different culture types: semi-solid agar plates, batch, and continuous culture. We observed that cross-feeding occurs both in both inter- and intra-species cross-feeding pairs of bacteria and identified pairs that cross-feed for inter- and intra-species pairs in all three culture types. Using this growth data we have identified two "super donors", strains of both *E. coli* and *B. megaterium* that grow with many other amino acid auxotrophs in both inter- and intra-species pairs.

To characterize how the cells are responding to the co-culture environment and cross-feeding we generated gene expression profiles of three inter-species pairs of auxotrophs that cross-feed successfully in all three of our tested culture types. Supernatant and transcriptomic samples were taken from the cross-feeding pairs in continuous culture, as well as monocultures grown in a minimal media supplemented with their essential amino acid. The data generated via the transcriptomic analysis of the interspecies cross-feeders was used to assess how strains adapt to co-culture with an auxotrophic partner. Preliminary evaluation of differential gene expression between *E. coli* co-cultures and monocultures show that not all strains react the same to co-culture. We also found an upregulation of amino acid transporter and amino acid biosynthesis genes in co-cultured *E. coli* strains as compared to their monoculture counterparts. This work shows that using cross-feeding pairs of bacteria can enable us to build functional synthetic microbial communities by addressing the difficulties in co-culture growth.

608012 Automated Detection of Yeast Genetic Engineering in Whole Genomes and Metagenomes with Prymetime

Abstract:

Yeast genomes can be assembled from sequencing data, but genetic engineering changes often fail to be resolved with accuracy, completeness, and contiguity. Further, searching for engineered sequences in sequence data is currently a manual process. To overcome these challenges, we applied nanopore assembly and short read error correction to create an integrated workflow that achieves accurate whole genome and plasmid sequences of engineered yeasts, automatically annotating synthetic biology parts. We named this workflow Prymetime, "Pipeline for Recombinant Yeast genoMEs That Illuminates Markers of Engineering."

Using Prymetime and a user-specified parts collection, we annotated the engineering within whole genomes of 15 engineered yeasts built from several *Saccharomyces cerevisiae* laboratory strains, *Yarrowia lipolytica* Po1f, and *Komatagaella phaffii* CBS 7435. We show that each pathway and plasmid can be correctly assembled and annotated, even in strains that have part repeats and multiple similar plasmids. Furthermore, the whole genomes are accurate, complete, and contiguous. We found that 40X sequencing depth of both nanopore and short reads was sufficient for accurate assembly, which reduces the resources needed for a complete yeast genome.

Finally, we used our validated workflow to identify engineering in metagenomic data. We combined reads from engineered yeast with a standard metagenomic NGS read set. We were able to correctly assemble and annotate the synthetic biology parts up to a 1:1000 dilution of the engineered yeast reads. Therefore, Prymetime can obtain high-quality *de novo* genome assemblies of engineered yeasts and label the engineering signatures within them, facilitating verification of yeast engineering and identification of engineering within isolates and metagenomes.

Engineered yeast are becoming ever more prevalent as cell factories for fuels, chemicals, and pharmaceuticals. Additionally, evolutionary and combinatorial strain engineering strategies produce more strains than can be reasonably sequenced and manually validated. Prymetime is a step towards overcoming these challenges, as it can be used to verify and detect engineering. This capability can apply to quality control of strain construction, monitoring waste streams for release of engineered organisms, and detection of synthetic biology parts in unknown samples.

608165 Rapid Screening and Identification of Microbial Isolate Collections Using Microwell Recovery Arrays (Faculty/Industry Candidate)

Abstract:

Screening soil and root microbiomes for discovery of microorganisms that can inhibit the survival and growth of pathogenetic bacteria is important for developing improved probiotic and biocontrol agents. With the aid of the microwell recovery array (MRA), a novel high throughput screening tool, libraries containing hundreds to thousands of unique isolates are rapidly screened for pathogen suppression. The MRA is designed to generate random combinations between a fluorescently-labeled bacterial pathogen and a controlled number of microbiome isolates. Isolates that display the highest levels of pathogen antagonism can be extracted, validated with off-chip assays, then sequenced and identified. Here, MRAs were used to investigate *Agrobacterium tumefaciens* interactions with other members of plant microbiomes. *A. tumefaciens* is a key plant biotechnology tool and also the causative agent of Crown Gall disease. While *A. tumefaciens* pathogenesis and intraspecific interactions have been studied extensively, far less is known about *A. tumefaciens* interactions with other members of plant microbiomes. Fluorescently labeled *A. tumefaciens* sp.15955 was combined with non-pathogenic *Agrobacterium* isolates collected from native plant roots at the Konza Prairie Biological Station (Manhattan, KS) to isolate species capable of suppressing *A. tumefaciens* sp.15955. First, an inoculum of *A.tumefaciens* sp. 15955 expressing gfp and root isolates were seeded in microwells at various cellular ratios. After culture, selective recovery of cells from individual wells displaying highest levels of 15955 inhibition was achieved using a high-resolution, light-based extraction system. The interactions are then validated and the isolates are characterized. In our primary data, we have uncovered several candidates for inhibiting *A.tumefaciens* sp. 15955 growth. Discovery of such growth inhibiting isolates will help improve plant productivity by using them as reliable biocontrol agents that prevent Crown Gall disease.

603580 Engineering Colicin Production and Secretion to Eradicate Biofilms (Industry Candidate)

Abstract:

Bacterial biofilm formation is frequently involved in the development of chronic infectious diseases. Inhibiting biofilms is challenging due to their tolerance against conventional antibiotics which are not effective to penetrating biofilm matrix to kill the cells residing in biofilms. Also, metabolically dormant cells (persisters) frequently found in biofilms are difficult to eradicate by antibiotic treatment. Therefore, novel antimicrobial drugs that can effectively kill persisters or inhibit biofilms are needed urgently. Bacteriocins, antimicrobial toxins, are considered as a viable alternative of conventional antibiotics due to their unique cell killing mechanisms that can damage cells by pore-forming on the cell membrane, nuclease activity, and cell wall synthesis inhibition. We previously reported that colicins, *Escherichia coli*-produced bacteriocins, can be synthesized using cell-free protein synthesis and kill persister cells very rapidly. In this study, we hypothesized that colicins production and secretion may be controlled in engineered microbes to eradicate target biofilm cells without affecting other non-target biofilms. We confirmed that cell-free produced colicins kill biofilm cells by more than 100-fold with strain-specific recognition. Next, we constructed genetic circuits encoding lysis, colicin, and immunity proteins to precisely control colicin production and secretion in *E. coli*. The controlled release of the colicin from the engineered microbes significantly repressed target biofilms in multi-species biofilm community without affecting non-target biofilms. This study provides insights that antimicrobial colicins may be used for a promising antibiotic alternative, and engineered microbes via synthetic biology may be developed as living bacterial therapeutics.

608724 Engineering of a Polygalacturonase-Inhibiting Protein As a Pest Control Agent

Abstract:

Fungal pathogens cause extensive plant disease in the agricultural industry, resulting in historical crop failures and diminished food security. Though effective fungicides have been developed, their negative impacts on the environment and animal health have been well documented. A more sustainable and health conscientious method of controlling fungi is desired. Plants produce a group of leucine-rich repeat (LRR) proteins, polygalacturonic inhibiting proteins (PGIPs), to inhibit the pectin-depolymerizing activity of endopolygalacturonases (PGs), one type of enzyme secreted by pathogens that compromise the cell wall and leave the plant susceptible to disease. Here, the inhibitory activity of the best characterized PGIP, PGIP2 from *Phaseolus vulgaris*, against PGs from *Aspergillus niger* (AnPG2) and *Botrytis cinerea* (BcPG1, BcPG2) was reconstituted through a yeast two hybrid (Y2H) system, which was utilized to estimate the activity of various PvPGIP2 truncations and mutants, and establish the sequence-function correlation. We found that truncating PvPGIP2 down to 1/3 of the size still retain the same level of inhibitory activity as the full length PvPGIP2, which may simplify the future engineering efforts for bioproduction and protein evolution. Additionally, the inhibitory activities of PvPGIP2 and mutants on the growth of *A. niger* and *B. cinerea* were examined and confirmed on the pectin agar. This work highlights the potential of using plant-derived PGIPs as an exogenously applied fungal control agent both to plants and postharvest crops while minimally impacting the environment and human health.

612892 Using Fabricated Ecosystems to Develop Approaches for Microbiome Engineering (Invited Speaker)

Abstract:

Microbiomes are important drivers of plant health and beneficial microbial amendments are finding widespread application. However, these often show variable performance and lack sufficient mechanistic understanding to predict the most efficacious conditions. The current lack of control in natural environments makes it challenging to determine causal mechanisms and environmental constraints. To help address this challenge we have recently developed fabricated ecosystems (EcoFABs) that enable mechanistic studies on plant and soil microbiomes. A four-laboratory 'ring-trial' study revealed that these devices result in reproducible plant treatment effects across laboratories. These systems are now being used to investigate plant growth-promoting microbial amendments towards the eventual goal of being able to design microbiomes with predictable field performance.

595432 Ultrasound Enhanced Release of Targeted Liposomes in Cancer Treatment

Abstract:

Introduction: Site-specific drug delivery is an attractive approach in cancer treatment because it can reduce the undesirable side-effects of anticancer therapeutics and increase the accumulation of drugs at tumor sites. To further improve the selectivity of the drug carrier, the surface of these nanoparticles, such as liposomes, are modified with targeting moieties specific to the receptors on the surface of tumor cells. In addition, external triggers such as ultrasound can be used to enhance the release of liposomal contents. Our research focuses on the synthesis of targeted-poly-ethylene glycol (PEG)-liposomes encapsulating the model drug calcein and studying the effects of low- frequency ultrasound (LFUS), applied at different power densities, on calcein release. Three targeting moieties were used in this research, namely, human serum albumin (HSA), arginyl glycyl aspartic acid (RGD), and estrone (ES).

Methods: Unconjugated liposomes were prepared using the thin-film hydration method. 1,2-dipalmitoyl-sn-glycero-3-phosphocholine (DPPC), 1,2-distearoyl-sn-glycero-3-phosphoethanolamine-N-[amino(polyethylene glycol)-2000] (DSPE-PEG(2000)-NH_2), and cholesterol were dissolved in chloroform at a molar ratio of 65:5:30. The thin film was formed by the slow evaporation of the organic solvent under vacuum using a rotary evaporator. The dried lipid film was then hydrated with 2 mL of 50 mM calcein disodium salt solution (pH~7.4). The liposomes were then sonicated and extruded to obtain small, unilamellar liposomes. Any external, unencapsulated calcein was removed using size exclusion chromatography. ES was attached to the lipids prior to the formation of liposomes, while RGD and HSA were conjugated to the liposomes post-formation. ES-modified liposomes were prepared by conjugating ES to DSPE-PEG2000-NH_2. Cyanuric chloride (2,4,6 trichloro-1,3,5 triazine (CC)) was used as a linking molecule to attach ES to DSPE-PEG2000-NH_2. ES was reacted with CC in a 1:1 molar ratio, in the presence of two molar equivalents of triethyl-amine (TEA), a chloride acceptor. The ES-modified lipids were then used to formulate liposomes using the procedure detailed above. RGD and HAS, on the other hand, were conjugated to liposomes via the PEG terminus modified with CC. The liposomes were characterized for size by dynamic light scattering (DLS), estimation of total phospholipid content by the Stewart method, the quantitation of the protein using the bicinchoninic acid (BCA) assay, and finally the release of the model drug by measuring fluorescence using a phosphorescence/fluorescence spectrofluorometer. Low-frequency ultrasound (LFUS) induced release was triggered via 20-kHz pulsed US (20s *on* 10s *off*) at three different power densities (6, 7, and 12 W/cm^2). Finally, the surfactant Triton X-100 was added to the sample to lyse the liposomes (and obtain the maximum release). The recorded fluorescent intensities were used to calculate the percentage of the released calcein using the following equation:

Results: The sizes of 3 batches of control and conjugated liposomes were measured using DLS. For HSA and RGD liposomes, the radius of the nontargeted pegylated liposomes (control) was 83.77 ± 0.91 nm. The average radius of the HSA-PEG liposomes was 84.86 ± 1.81 nm, while the RGD-PEG liposomes showed an average radius of 84.42 ± 1.50 nm. The radius of ES-conjugated, calcein liposomes was found to be 92.9 ± 15.8 nm, while a radius of 96.8 ± 3.8 nm was obtained for their corresponding control liposomes. This confirmed that both types of synthesized liposomes fell within the size category of small unilamellar vesicles (SUV) expected to be efficient for drug delivery as they can make use of the enhanced permeability and retention (EPR) effect. The Stewart assay was used to confirm that both the control and targeted liposomes have similar lipid concentrations before measuring their protein content. The results of the BCA assay showed that HSA and RGD liposomes had a higher protein content compared to control liposomes with no moieties conjugated. HSA-PEG liposomes showed a 3-fold increase in protein content (1.05 ± 0.43 μg/mL) compared to the control liposomes (0.35 ± 0.006 μg/mL); whereas the protein content of RGD-PEG liposomes (2.1 ± 0.114 μg/mL) showed a 5-fold increase compared to the control liposomes (0.41 ± 0.008 μg/mL). As for the online low-frequency US (LFUS) induced release, the normalized-averaged release profiles of control and conjugated liposomes are shown in Figure 1. As can be seen, the curve started with a baseline that measured the fluorescence level in the sample; upon sonication, the level increased due to calcein release. The 20s *on* 10s *off*-cycle mentioned earlier was continued until a plateau was reached, and no more calcein release was observed. Moreover, upon the addition of Triton X-100, the fluorescence level increased slightly above the plateau, indicating that the liposomes have released most of their encapsulated calcein contents. Following the first three pulses, calcein release from HSA-PEG liposomes and ES-PEG liposomes was significantly higher than that of the control liposomes at the three power densities investigated. However, for RGD-PEG liposomes, the calcein release following the sonication at the highest investigated power density (i.e., 12 W/cm^2) showed no significant difference compared to the control liposomes after the first three pulses. This, however, was not the case following sonication with the lower power intensities. When RGD-PEG liposomes were sonicated at 7 W/cm^2, they showed more calcein release compared to the control following the first and the third pulses, but no significant change in calcein release was recorded following the second pulse. Sonication at the lowest power density (6 W/cm^2) showed that RGD-PEG liposomes were more susceptible to US compared to the control liposomes releasing significantly more of the encapsulated calcein, following the first three pulses.

Conclusions: In this study, control and modified liposomes were tested for size, lipid content, conjugation of proteins, and calcein release triggered by LFUS. The DLS findings showed that the modification of pegylated liposomes with HSA, ES, and RGD had no significant effect on the size of the liposomes. Furthermore, HSA-PEG and ES-PEG liposomes were more sonosensitive compared to the control liposomes at the three power densities, showing significantly higher calcein release following the exposure to pulsed LFUS. Using this data, we inch closer to developing a drug delivery system that uses ultrasound as a modality to reduce the side effects of conventional chemotherapy, hence improving the quality of life of patients around the world.

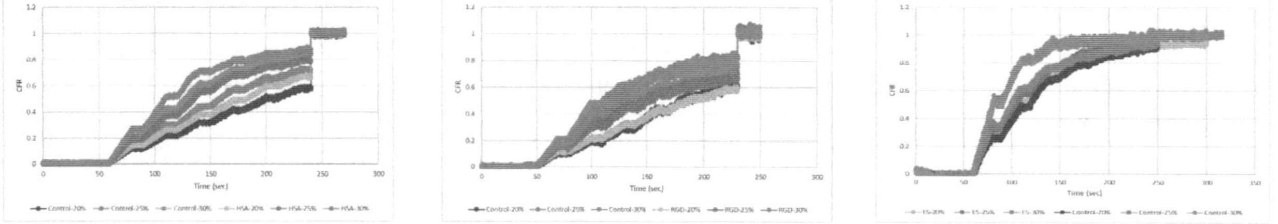

Percent calcein release from (left) HSA-PEG, (center) RGD-PEG and (right) ES-PEG liposomes exposed to pulsed LFUS (20-kHz) at 6, 7, and 12 W/cm^2. Results are the average of three batches (three replicates each).

601464 In Vivo Flow-Regulated Endothelial Glycocalyx Integrity Leveraged for Targeted Intravenous Nanoparticle Delivery

Abstract:

Statement of Purpose: The cell surface-attached extracellular glycocalyx (GCX) layer, a heterogeneous polysaccharide chain, is a major contributor to endothelial cell (EC) function and EC-dependent vascular health, acting as a first line of defense against vascular diseases including atherosclerosis [1]. It largely consists of cell-linked sialic acid and glycosaminoglycans including heparan sulfate, hyaluronic acid, and chondroitin sulfate, acting as a buffer between endothelium surface and blood-flow derived shear forces. Endothelial surface GCX shedding plays a role in endothelial dysfunction, compromising endothelial barrier function and increasing vascular permeability [2], leading to consequences such as increase in lipid influx. A unique surgical procedure was developed by Nam et al. [3] to perform partial left carotid artery (LCA) ligation in mice. This was characterized as a model of disturbed blood flow with characteristics of low and oscillatory wall shear stress. However, this study did not examine GCX expression after acutely inducing disturbed blood flow. The first part of our study sought to elucidate the impact on GCX integrity post-one week of partial LCA ligation in correlation to endothelial dysfunction. The second part of this study examined vessel wall permeability after GCX shedding by examining resultant macrophage infiltration, related to atherosclerosis onset, as well as nanoparticle infiltration, related to targeted drug delivery for early atherosclerosis intervention.

Methods: For the first study, C57Bl/6-background apolipoprotein-E knockout (ApoE-KO) male mice were subjected to a partial LCA ligation procedure [3] to induce disturbed blood flow patterns in the LCA while the right carotid artery (RCA) of each mouse was not surgically intervened, to provide a control in each mouse. Mice were sacrificed post-one-week ligation surgery, the LCA and RCA were dissected, and the GCX was immunohistochemically stained to determine its percent coverage and thickness. Macrophage infiltration was similarly immunohistochemically detected and quantified. For the second study, C57Bl/6 male mice underwent partial LCA ligation surgery. Ultra-small gold nanospheres (GNS) coated with PEG were fabricated as previously described [4,5]. At day 26 post-ligation, anesthetized mice intravenously received 25 mg of biotin-conjugated PEGylated GNS in 150 μL of sterile phosphate buffer solution (PBS). GNS were then allowed to circulate in the mice for 48 hours. The animals were euthanized, fixed, and dissected after 28 days post-ligation surgery, 2 days after GNS treatment. The carotid arteries were then extracted. Fluorescent staining of biotinylated GNS uptake in vessel wall, and GCX expression on the inner surface of the vessel wall, were performed. All results were expressed as mean ± standard error of mean (SEM). Data sets were compared using Graph Pad Prism software and a two-way ANOVA test to analyze statistical significance between groups with an alpha value of 0.05.

Results: GCX coverage of the endothelium was significantly reduced in the LCAs of mice exposed to disturbed flow by partial LCA ligation, compared to control RCA. No differences were found in GCX coverage of RCAs from all cohorts. Regarding inflammation, no difference in macrophage accumulation in carotid arterial walls was observed when comparing the LCAs and RCAs of mice that did not undergo partial LCA ligation surgery. However, macrophage uptake in vessel walls showed a 14-fold increase in the LCAs exposed to disturbed flow following ligation, when compared to control LCAs, while no such statistical difference was observed between the RCAs of the group.

Additionally, GCX degradation in the ligated LCA correlated to approximately 3-fold increase in 10-nm GNS infiltration of the LCA vessel wall in comparison to its control RCA counter-part. This suggests that GCX dysfunction, which coincides with atherosclerosis, can indeed by leveraged and targeted for enhanced nanoparticle-based drug delivery.

Discussion: In summary, it was observed that partial LCA ligation, to acutely disturb blood flow in a mouse vessel, induce blood vessel remodeling, particularly with respect to GCX, and endothelial dysfunction, associated with pro-atherosclerotic recruitment of inflammatory macrophages. Passive targeting of GNS to areas affected by degraded GCX was also observed. The GCX is often neglected even though it acts as a first line of defense against atherosclerosis. Targeting dysfunctional vessels based on GCX shedding offers an innovative approach in early cardiovascular disease therapy and prevention.

Acknowledgements: Funding: National Institutes of Health (KO1 HL125499) and National Science Foundation (DGE-0965843 and CMMI-1846962).

References: **[1]** Mitra, R., et al., *Glycocalyx in Atherosclerosis-Relevant Endothelium Function and as a Therapeutic Target.* Curr Atheroscler Rep, 2017. **19**(12): p. 63. **[2]** Mitra, R., et al., *The comparative effects of high fat diet or disturbed blood flow on glycocalyx integrity and vascular inflammation.* Transl Med Commun, 2018. **3**. **[3]** Nam, D., et al., *Partial carotid ligation is a model of acutely induced disturbed flow, leading to rapid endothelial dysfunction and atherosclerosis.* Am J Physiol Heart Circ Physiol, 2009. **297**(4): p. H1535-43. **[4]** Cheng, M.J., et al., *Endothelial glycocalyx conditions influence nanoparticle uptake for passive targeting.* Int J Nanomedicine, 2016. **11**: p. 3305-15. **[5]** Cheng, M.J., et al., *Ultrasmall gold nanorods: synthesis and glycocalyx-related permeability in human endothelial cells.* Int J Nanomedicine, 2019. **14**: p. 319-333.

608168 Donnan Partitioning Results in Faster Diffusion of Positively Charged Peptide Surfaces through Tumor Extracellular Matrix

Abstract:

One in every six deaths in the world is due to cancer, making it the second leading cause of death globally. To effectively treat cancers, therapies must reach as many cancer cells as possible. However, current therapies against solid cancers fail in part due to their inability to penetrate through the extracellular matrix (ECM) of tumors and reach the cancer cells at therapeutic concentrations. The tumor ECM forms a net negative charged network that hinders the transport of molecules based on their size, shape, and intermolecular interactions. Traditionally, the focus of drug delivery to solid tumors has been to develop delivery systems smaller than the pore size of the tumor ECM and/or with neutral charge coatings to minimize interactions with the ECM for improved transport. Recently, literature suggests that there is an association between the surface charge of carriers with their transport through the tumor ECM—however, the mechanism of how charge-based interactions impact transport and penetration within the tumor microenvironment is not well understood. Here, we used genetically engineered T7 bacteriophage (phage) as a model biological nanoparticle system to display peptides of different charges on its surface and investigated the effect of charge-based transport through tumor ECM.

In contrast to most studies, we found a positively charged peptide "surface" enhanced the penetration, uptake, and retention of T7 phage in tumor-like ECM and tumor tissue when compared to neutral and negatively charged peptides. Transport studies with systematic alanine mutagenesis of T7 show that electrostatic interactions are likely responsible for the higher tumor tissue uptake and retention. Mechanistically, the positively charged particle partitioned into the tumor tissue (with an equilibrium partition coefficient of ~6) due to electrostatic interactions. Further, weak and reversible binding of the phage solute with the tumor bed (with an equilibrium dissociation constant of ~230 nM) allowed for deep penetration within the tumor tissue. Additionally, the positively charged peptide-presenting phage has a high number of intra-tissue binding sites in the tumor microenvironment (~4 μM) that enables almost 100% retention in the tumor tissue up to 24 hours. These results indicate that the net negative tumor ECM can act as a depot for positively charged particles to improve drug penetration and enhance tumor retention. We will also discuss the role of positively charged peptides as anchors to improve intratumor penetration and retention of therapeutics for antitumor efficacy. This work can potentially transform the current paradigm regarding the design of drug delivery systems to ultimately improve therapeutic index and therapeutic outcomes in solid cancers.

609714 Aerosol Therapeutic Delivery in 3D Printed Pediatric Airway Replicas

Abstract:

The current standard method of characterizing aerosol delivery variations between adult and pediatric patients uses idealized throat models in conjunction with a Next Generation Impactor (NGI). The standard pediatric idealized throat groups together children of all ages, neglects variations as the lungs develop, and was modeled based on an adult throat; namely the Alberta Idealized Throat (AIT). There is limited information on aerosol delivery dependency for each age and patient geometry, mostly due to small sample sizes. Advances in 3D printing allow us to develop throat models based on 60 patient computed tomography (CT) scans from healthy pediatric subjects. A Collison jet nebulizer and a PARI Vios Pediatric nebulizer were used in-line with the pharmaceutical standard aerosol characterizing equipment, a Next Generation Impactor (NGI), to quantify deposition profiles variations for patients and the AIT. Results indicate that there are significant differences in particle deposition and mass mean aerodynamic diameter between the IP and the pediatric model for a range of particle sizes tested at 30 L/min. Larger particles deposit in the throat and result in a shift in the particles that are delivered to the lungs for pediatric patients. There is a shift in mass fraction from 0.18 to 0.41, n=3 and p=0.003, of 1 μm particles delivered to the lungs when using a pediatric replica. The underestimation of mass fraction delivered to pediatric patients suggests the possible overestimation of required dosing when developing pharmaceuticals. Additionally, throat deposition of 3 – 6 μm particles indicates that the current fine particle fraction standard is not representative for pediatric patients. Information from trends in delivery based on age and patient geometry will lead to improved fundamental knowledge of pediatric aerosol delivery and guidelines for pharmaceutical development.

Numerical Investigation of Particle Shape and Actuation Flow Rate Effects on Lactose Carrier Delivery Efficiency through a Dry Powder Inhaler (DPI) using CFD-DEM

Jianan Zhao [a], Yu Feng [a,*], Ahmad Haghnegahdar [b], Saurabh Sarkar [b], and Rahul Bharadwaj [b,†]
[a] School of Chemical Engineering, Oklahoma State University, Stillwater, OK 74078, USA
[b] ESSS Inc., Woburn, MA 01801, USA

1. Introduction

Irreversible asthma and chronic obstructive pulmonary disease (COPD) are worldwide pulmonary healthcare concerns, which affected approximately 533 million individuals and resulted in 3.2 million fatalities in 2015 [1]. The administration of medicines *via* inhalation is one of the most popular treatments. Capsule based dry powder inhalers (DPIs) have been widely used to deliver active pharmaceutical ingredients (APIs) attached to micron-sized carrier particles into human respiratory systems. However, such methods are not effective since only a small fraction of the medication, i.e., less than 30%, can reach the designated small airways [2, 3]. Thus, it is necessary to develop a method to increase the delivery efficacy to enhance the curative results and reduce undesired side effects. Previous research has shown that inspiratory flow rate, i.e., the actuation flow rate and aerodynamic particle size distribution (APSD), are key factors that can influence the localized particle deposition patterns in human respiratory systems [4]. Besides, modulating particle shapes is another promising to improve small-airway drug delivery efficacy [5]. In this study, the influence of actuation flow rate and particle shape on lactose carrier dynamics in a representative DPI, i.e., Spiriva™ Handihaler™ [6], has been investigated numerically. A one-way coupled computational fluid dynamics (CFD) and discrete element method (DEM) has been developed, verified, and employed to model the particle-particle and particle-device interactions and the transport and deposition of particles with different shapes in the DPI flow channel.

2. Materials and Methods

The flow channel of Spiriva™ Handihaler™ with a grid inside was reconstructed to simulate the particle interactions, transport, and deposition under various steady-state actuation flow rates (see Fig. 1 (A)). Four actuation flow rates, i.e., 30, 39, 60, and 90 L/min [7] were used to simulate different coordination scenarios between patients and the DPI. The final CFD mesh was generated with a total of 3.7×10^6 unstructured mesh cells consisting predominantly of polyhedral elements. Six prism layers were created to resolve the near-wall gradient accurately. The one-way coupling method was employed for calculating the interactions between particles and the airflow. Using the verified k-ω model [8], the turbulence airflow field in the flow channel was simulated using ANSYS Fluent 2019R3 (ANSYS Inc., Canonsburg, PA). The airflow data was then transferred into Rocky DEM (ESSS, Woburn, MA) to compute the fluid-particle and particle-particle interactions, and the resultant particle trajectories in the DPI flow channel. The surface energy was assumed to be 0.06 J/m^2 between two lactose carrier particles and 10 J/m^2 between the carrier particle and the device surfaces. The high surface energy between particle and device indicates that the particles have a high possibility to deposit on the device once they contact. The Hertz-Mindlin (H-M) model [9] with Johnson-Kendall-Roberts (JKR) cohesion [10] was employed for the calculation of inter-particulate van der Waals force. A total of 15,514 spherical and sphero-cylinder micro-particles with an equivalent diameter of 3.0 μm were injected at time t=0 through the two holes on the capsule (see Figs. 1 (A) and (B)). The total physical time duration simulated was 0.01 s, ensuring the change in DPI delivery efficiency less than 1%.

3. Results and Discussion

As shown in Fig. 1, the simulation results indicate that both particle shape and actuation flow rate can significantly influence the delivery efficiency and distribution of carrier particles emitted at the mouthpiece. Figs. 1 (B) and (C) show that more than 99% of spherical particles deposited in the capsule chamber, including the grid and the cap enclosing the grid. In contrast, the deposition of sphero-cylinder particles is

* Corresponding Author: yu.feng@okstate.edu
† Corresponding Author: rbharadwaj@esss.co

more complex. Specifically, at the flow rates lower than 39 L/min, more than 99% sphero-cylinder particles deposited in the capsule chamber. At the flow rates higher than 60 L/min, a significant amount of sphero-cylinder particles deposit not only in the capsule chamber but also the tube between the grid and the mouthpiece. Such differences are possibly due to the particle resuspension effects of the elongated sphero-cylinder particles. At higher flow rates, the torque acting on the sphero-cylinder is higher than spheres which are able to separate more deposited sphero-cylinders from the device surfaces. The resuspended sphero-cylinders result in a more dispersed deposition pattern (see Fig. 1 (B)). The simulation also predicts opposite trends of delivery efficiency *vs.* flow rate between the sphere and sphero-cylinder particles (see Fig. 1 (C)). Indeed, with the increase in flow rate from 30 to 90 L/min, the delivery efficiency of the spherical particle increases from 27.1% to 43.0%. In contrast, the delivery efficiency of sphero-cylinder particles decreases from 45.3% to 36.3% as more resuspended particles either deposit on the grid or the tube wall connecting the mouthpiece and the grid.

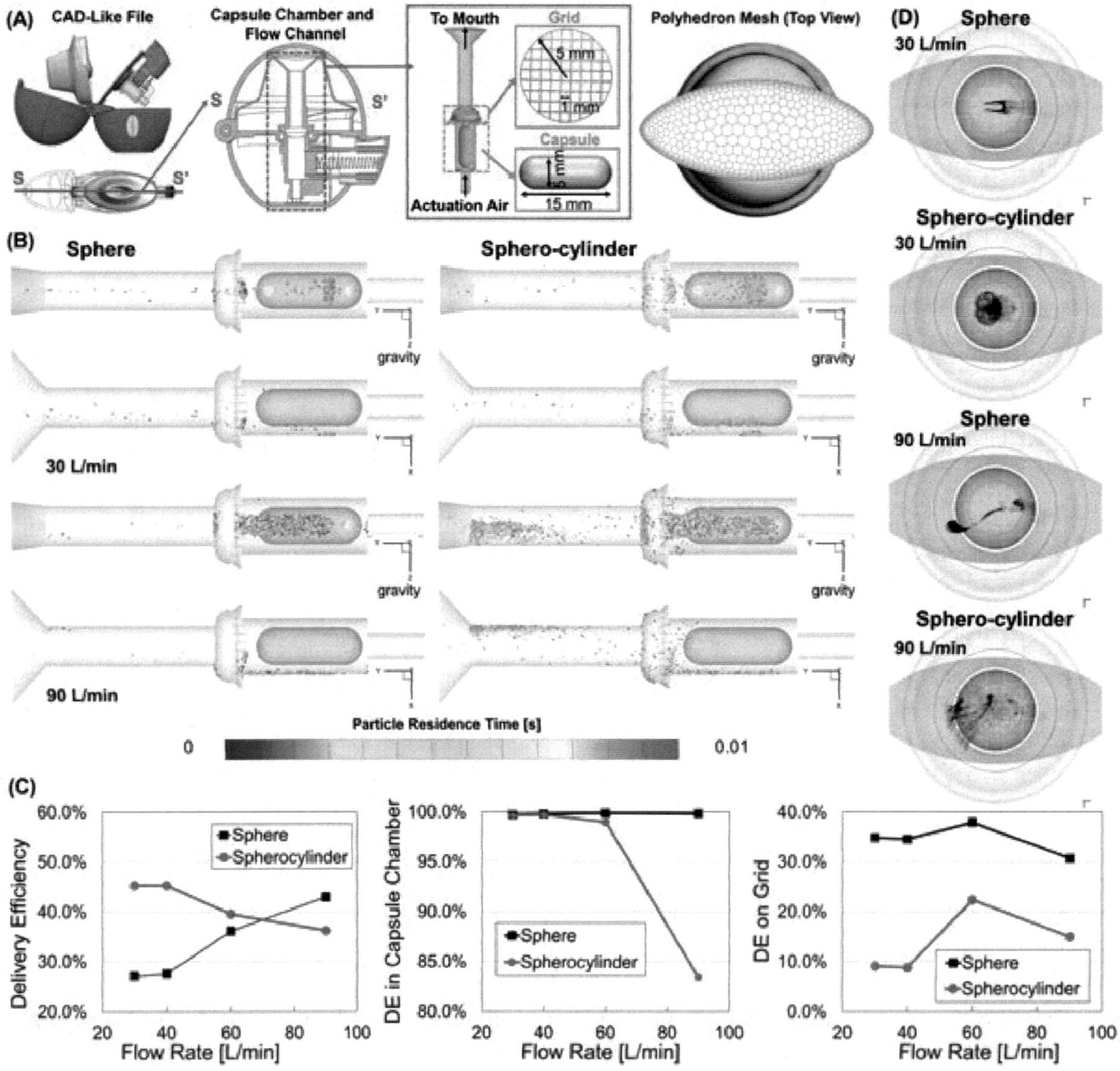

Fig. 1. (A) Reconstructed DPI geometry CFD mesh, including the flow channel, grid, and capsule with pierced holes, (B) Deposition distributions of the spherical and sphero-cylinder particles in the DPI, (C) Comparisons of lactose carrier delivery efficiency and deposition efficiency (DE) between spherical and sphero-cylinder particles through the DPI, and (D) Comparison of emitted particle distributions at the mouthpiece opening.

The emitted particle distribution at the mouthpiece outlet is a crucial factor determining lung deposition distribution. The results in Fig. 1 (D) show that the emitted distribution is affected by both the particle shape and the actuation flow rate. Specifically, the higher flow rate induces more dispersed distribution due to higher particle momentums obtained from the airflow, and the emitted distribution of sphero-cylinder particle is more scattered than spherical particles due to the particle resuspension effect. The findings mentioned above can provide insight into the optimization of DPI performance with selected particle shapes and actuation flow rates.

4. Conclusions

In this study, a validated one-way coupled CFD-DEM approach has been developed and applied to investigate the effects of actuation flow rate and particle shape on the transport and deposition of lactose carrier in a representative DPI. Compared with *in vitro* tests, numerical simulation is less expensive and time-consuming while providing accurate results for reducing the research/development cycle duration of DPI product innovations. From the high-resolution numerical results of particle transport and deposition, it can be concluded that to enhance the delivery efficiency to the mouth front, using spherical lactose carriers is recommended for a high actuation flow rate (90 L/min), while elongated sphero-cylinder carriers are preferred for an actuation flow rate less or equal to 60 L/min. The different resuspension effects induced by the particle shape play a significant role in the particle deposition and distribution inside the DPI flow channel.

5. Future Work

Future work includes: (1) Simulating the interactions between API and carriers to predict APSDs emitted at the mouthpiece, and the resultant delivered dose in human respiratory systems; (2) Employing a two-way coupled CFD-DEM model to simulate the particle dynamics with a rattling capsule; and (3) Investigating the aspect ratio effects of non-spherical particles on the delivery efficiency.

Acknowledgments

The research was made possible by funding through the award for project number HR19-106, from the Oklahoma Center for the Advancement of Science and Technology. The use of Rocky DEM (ESSS, Woburn, MA) as part of the ESSS-CBBL academic partnership is gratefully acknowledged (Dr. Rahul Bharadwaj). The use of ANSYS software (ANSYS Inc., Canonsburg, PA) as part of the ANSYS-CBBL academic partnership is also gratefully acknowledged (Dr. Thierry Marchal).

References

1. GBD., *Global, regional, and national deaths, prevalence, disability-adjusted life years, and years lived with disability for chronic obstructive pulmonary disease and asthma, 1990–2015: a systematic analysis for the Global Burden of Disease Study 2015.* The Lancet Respiratory Medicine, 2017. **5**(9): p. 691-706.
2. Islam, N. and M.J. Cleary, *Developing an efficient and reliable dry powder inhaler for pulmonary drug delivery – A review for multidisciplinary researchers.* Medical Engineering & Physics, 2012. **34**(4): p. 409-427.
3. Kolanjiyil, A.V. and C. Kleinstreuer, *Modeling Airflow and Particle Deposition in a Human Acinar Region.* Computational and mathematical methods in medicine, 2019. **2019**: p. 5952941-5952941.
4. Feng, et al., *An in silico inter-subject variability study of extra-thoracic morphology effects on inhaled particle transport and deposition.* Journal of Aerosol Science, 2018. **123**: p. 185-207.
5. Chow, A.H.L., et al., *Particle Engineering for Pulmonary Drug Delivery.* Pharmaceutical Research, 2007. **24**(3): p. 411-437.
6. Delvadia, R.R., et al., *In Vitro Tests for Aerosol Deposition. IV: Simulating Variations in Human Breath Profiles for Realistic DPI Testing.* Journal of aerosol medicine and pulmonary drug delivery, 2016. **29**(2): p. 196-206.
7. CDER, *Draft Guidance on Tiotropium Bromide.* 2017.

8. Donovan, M.J., et al., *Dry Powder Inhaler Device Influence on Carrier Particle Performance.* Journal of Pharmaceutical Sciences, 2012. **101**(3): p. 1097-1107.
9. Jiang, M., et al., *Experimental and DEM analyses on wheel-soil interaction.* Journal of Terramechanics, 2018. **76**: p. 15-28.
10. Johnson, K.L., et al., *Surface energy and the contact of elastic solids.* Proceedings of the Royal Society of London. A. Mathematical and Physical Sciences, 1971. **324**(1558): p. 301-313.

608246 Oligoteas: A Platform for Multifunctional Antibody Conjugates and Quantification of Intracellular Processes (INVITED SPEAKER)

Abstract:

Antibody drug conjugates (ADCs) are established bioconjugates for tissue-specific delivery of potent therapeutic drugs. As more ADCs get approved or advance to the clinic, more research is needed to understand and improve their potency and safety profile. In this talk, I will discuss new sequence-defined crosslinkers developed by my group, in combination with different site-specific conjugation techniques, to investigate the effect of drug placement on ADC biophysical properties and potency. Additionally, I will also introduce a new synthetic oligothioetheramide (oligoTEA) modality to quantify bond cleavage directly inside of a living cell. This modality involves a reduction-sensitive FRET-based cross-linker that can be readily conjugated to any targeting ligand, particularly antibody-based ligands. This reduction sensitive antibody-probe conjugate was used along with a mass-action kinetic model and live-cell flow cytometry experiments to quantitate the rate constant for intracellular bond cleavage. The analysis described in this work can be performed as a function of any bond type (plug-and-play), endocytosis pathway, and cell type. In this way, intracellular bond degradation will emerge as a new fundamental parameter to understand the intracellular processing of stimuli-responsive drug carriers and enable the rational design of new drug delivery systems.

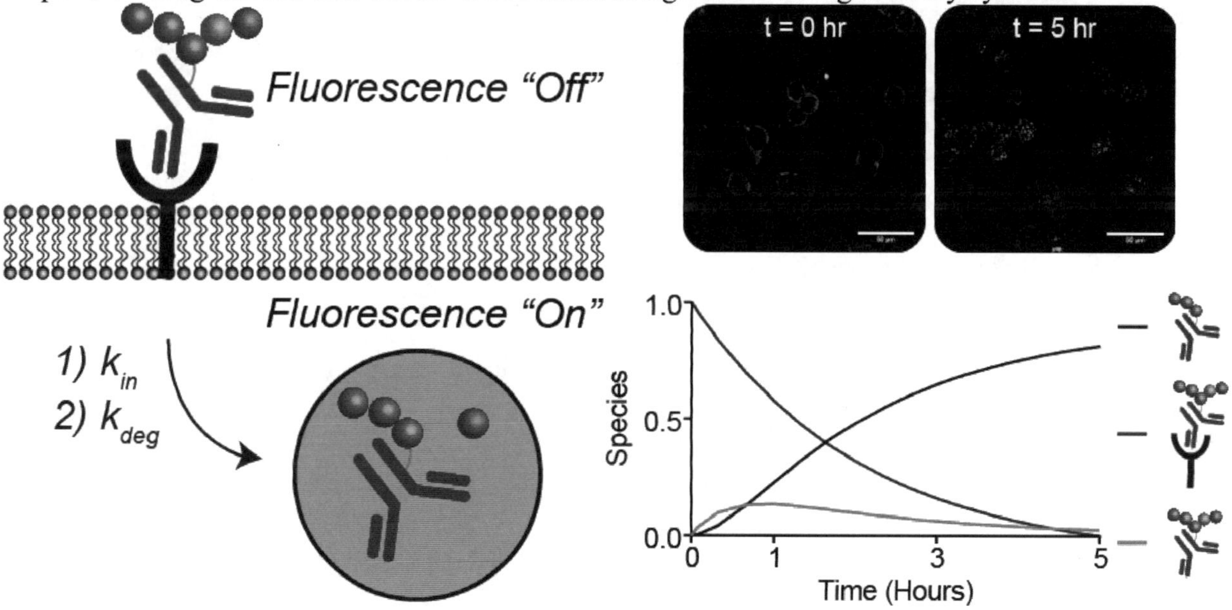

606553 Model-Based Approach for Multivariate Signaling Regulation of Epithelial-Mesenchymal Transition in Pancreas Cancer Cells

Abstract:

Epithelial-mesenchymal transition (EMT) is a normal cell developmental program that occurs aberrantly in numerous carcinomas and promotes chemoresistance. EMT can occur in response to activation of growth factor and adhesion receptors, low-oxygen tension (hypoxia), and exposure to chemotherapeutic drugs. We hypothesize that these cellular perturbations ultimately drive EMT through a conserved set of kinase-regulated signaling pathways and that the strength and duration of signaling through a critical set of those pathways dictate the degree to which EMT will occur. In this work, we model the response of human pancreas cancer cells to a variety of EMT agonists (growth factors) by considering the dynamics of protein phosphorylation along 32 distinct signaling nodes. The most important signaling nodes for EMT were selected through a partial least square regression (PLSR) approach, and those pathways were then used to cast the regulation of EMT in response to cellular perturbations as a control problem.

The overall ligand–signal–phenotype system is decomposed into two subsystems in series: (i) a "signal response" subsystem (f_1), whose inputs are the ligand concentrations and the responses are the relative abundance of the signaling molecules; and (ii) a "phenotype response" subsystem (f_2), for which the signals—the responses from f_1—serve as the stimulus which then stimulates the mesenchymal phenotypes as the response of f_2. This natural decomposition of the system is convenient for modeling the overall system as a convolution of the two subsystems, which also allows us to understand the dynamics of the intermediate signal responses.

Such representation allows us to formulate the determination of the optimal ligand dosage for EMT inhibition as a control problem, where the manipulated variables are concentrations of the ligands—epidermal growth factor (EGF), hepatocyte growth factor (HGF), and transforming growth factor beta (TGF-β); the controlled variables are the indicators of EMT such as the expression and localization of E-cadherin and vimentin, and the circularity of the tumor cell clusters.

In this presentation, we discuss our preliminary results on the identification of the ligand–signal–phenotype system model and on the solution of the control problem. We estimated the model parameters of f_1 (an autoregressive model with exogenous input) and f_2 (a PLSR model) using time-series data of the responses of signaling proteins and of the final phenotype measurements obtained from experiments performed with five different ligand dosages. Subsequently, we determined the optimal ligand dosage trajectory and implemented the prescribed solution via simulation in MATLAB. The results of our study provide a new set of experimentally testable hypotheses for validating the identity of the most important signaling pathways that govern EMT. They also provide a new paradigm for integrating data-driven modeling approaches such as PLSR with dynamic control system models to develop a predictive understanding of how multivariate signaling processes control complex cell phenotypes. The primary benefit of such a paradigm is that it provides a quantitative, model-based framework for using the indicated predictive understanding, in reverse, to determine how best to manipulate the signaling processes to achieve desired phenotypic responses optimally.

603882 Modeling Cellular Signaling and Mesangial Fibrosis during Diabetic Kidney Disease

Abstract:

In the U.S. alone over 250,000 people use dialysis or have received a kidney transplant due to diabetic kidney failure. Although we have come a long way in the treatment of diabetes, kidney failure due to diabetic kidney damage is still prevalent, and the need for increasing our understanding of kidney damage to enable the development of better treatment methods is ever present. The goal of this research is to develop computational models to better understand the kidney damage that occurs due to diabetic kidney disease.

In the kidney glomerulus, the filtration unit of the kidney, lies a network of capillaries that are surrounded by interstitial tissue called the mesangium. In health, the mesangium acts as a support for the capillaries; however, during diabetic kidney disease, the mesangium expands due to excess accumulation of collagen (fibrosis) and causes damage to the cellular environment around it. This mesangial expansion is not only a hallmark of kidneys damaged by diabetes but also many other chronic kidney diseases that lead to kidney failure. There has been a lot of research effort in trying to figure out the cause of the mesangial expansion. Researchers have found high glucose-induced dysfunction in the mesangial cell, a cell native to the mesangium, to be one of the main reasons for mesangial expansion. The mesangial cell dysfunction is mediated by the overstimulation of key signaling and cellular communication molecules such as TGF-beta and Ang II, which play a key role in perturbing the function of downstream collagen metabolism molecules such as MMP and TIMP, leading to the accumulation of excess collagen. The complexity of the interactions necessitates the development of computational models to understand the whole, yet there are few computational models of mesangial expansion and even fewer that study the effect of the mesangial expansion on cellular communication and signaling in the glomerulus.

In this work, we present a multiscale, computational model of mesangial fibrosis to study its effect on cellular signaling and communication. Our computational model considers the impact of mesangial cell mediated mesangial expansion on cellular signaling and communication through multiscale modeling of extracellular matrix remodeling and macromolecular transport. We are incorporating a cellular environment using a cellular Potts type of agent-based stochastic model, modeling mesangial expansion using fundamental biological principles of collagen fiber growth and accumulation, solving macromolecular transport equations defined by continuous reaction-diffusion partial differential equations, and linking them all using Python and CompuCell3D, a multiscale tissue simulation software. Novel and biologically sound, the model captures mesangial fibrosis and provides insight into cellular communication during diabetic kidney damage.

610799 A Discrete Markov Model for the Time Evolution of a Tumor Population with Merging of Adjacent Tumors

Abstract:

Cancer is a disease whose patient outcomes are difficult to predict. In particular, patients occasionally appear to recover, only to suffer a recurrence of the disease after a few years or even decade or more. In melanoma, for example, the disease reappears in patients 10 years or more after treatment at a 7% rate. This recurrence after a long period of dormancy (i.e. medically undetectable) is a feature in many cancers including breast cancer and metastatic melanoma.

We consider a discrete model in which we simulate cancer growth, shrinkage, and metastasis each as a Markov process, with size-dependent Poisson parameters, whose results predict the time evolution of all initial and expected metastasized tumors in patients. This new model complements our parallel existing population balance (partial differential equation) model for large population sizes. Our Markov model's transition probabilities derive from the parameters fit by comparing the continuum model to previous experimental tumor data of melanoma in zebrafish taken in our lab. Our discrete simulation model goes well beyond our continuous model to include the locations of tumors within the body. By doing so, it allows for the natural merging of neighboring tumors via growth and their breakup into smaller tumors via cell-death without the need to introduce separate kinetic processes with their own rate constants that would be very hard to access experimentally.

We shall begin by presenting (1) the probability vs. time of recurrence of tumors of observable sizes after successful operation or treatment and (2) the probability of small metastases growing into tumors of detectable sizes. We shall then present the results of how predictive the model is, by simulating tumor lifetimes and matching the results to a new second set of experimental tumor data. As in the older data, the new data are for the zebrafish melanoma model with immunity and gender as variables, but with far larger initial inoculations than the earlier data. In order to simulate the generation of metastases in the fish, we first derive from extensive data a probability density function where in the fish a newly generated metastasis would appear. We shall then present results concerning three main aspects: (1) how the variance of the fates and lifetimes of a set of tumors varies with the number of tumors followed; (2) how gender differences, as expressed via the experimentally-derived transition probabilities affect tumor fates and the route to tumor disappearance; and (3) how the inclusion of tumor merging changes the observed dynamics of a tumor population and specifically how it affects the probability of the appearance of a tumor of observable size as a function of time.

610810 A 3D Multiscale Model for Prediction of Patient-Specific Platelet Function Under Flow

Abstract:

Atherosclerotic plaque rupture exposes blood to a highly procoagulant surface containing tissue factor, causing platelets to become activated. Activated platelets release soluble agonists such as ADP and thromboxane, causing further activation and platelet recruitment, resulting in thrombus growth. In this work, we have extended our previous 2D multiscale model of thrombosis[1] in a microfluidic channel to develop a fully spatially resolved 3D framework with capabilities to simulate patient-specific phenotypes and blood vessel geometries. The multiscale framework is composed of four modules: Neural Network for platelet signaling, Lattice Kinetic Monte Carlo for platelet motion and binding, Lattice Boltzmann for blood flow field, and Finite Volume Method for tracking soluble agonist concentrations. Thrombus growth dynamics and morphology predicted by the model agree well with experiments. Moreover, the model clearly overcomes artifacts that were present in our previous 2D model[1] by allowing for flow around bound platelets rather than obstructing flow altogether. Furthermore, the model is largely agnostic to the geometry of the channel and can easily be applied to fully resolved simulations of thrombus growth in arbitrary geometries, such as stenoses and bifurcations. To our knowledge, our model is the first of its kind that accounts for patient-specific platelet phenotypes to perform robust 3D simulations of thrombosis given any arbitrary geometry. This framework facilitates the identification of patient-specific thrombotic risks and drug responses.

References:

1. Lu et al., Math. Med. Biol., 2017, Vol. 34(4), pp. 523-546

611849 Modeling the Impact of Heterogeneity in the Early Tumor-Immune Response

Abstract:

Before a tumor can progress to a detectable size, it first must develop ways to evade removal by the immune system. To do this, the tumor creates a favorable tumor microenvironment (TME) that suppresses some immune cells and corrupts others into a tumor-promoting state. Due to local pockets of diffusible factors and phenotypic differences between cells, the TME can exhibit high amounts of heterogeneity, both extracellular and intracellular. As it is difficult to assess this heterogeneity with *in vivo* or *in vitro* experiments, computational models can be used to interrogate the TME system in detail and provide new insights.

Here, we investigate the effects of phenotypic differences among multiple cell types in the TME. We model cancer cells, macrophages, and T cells. Using an agent-based model, we simulate the interactions between these three cell types, along with several diffusible factors present in the TME. Importantly, we model the behavior of individual cells, to examine how differences at the cellular level compound into differences at the population level. In addition, to produce more physiologically relevant scenarios, we utilize mechanistic models of intracellular signaaling for some key pathways related to the immune response. The dynamics of these intracellular signaling responses determine how the cell will behave. Finally, we incorporate heterogeneity in three ways: allowing phenotypic parameters to vary between cells of the same type, varying the xpressions of proteins in the intracellular signaling networks from cell to cell, and varying the levels of extraceullar stimuli. With our model, we are able to examine the ffects of various types of heterogeneity on the initial immune response to cancer growth, with and without various macrophage-based treatment strategies. Altogether, we gain a better understanding of the complexities of the TME.

612072 Integrating Multi-Omics Datasets with Robust Penalized Regression Identifies Context-Dependent Signaling Networks

Abstract:

Cells integrate signals from different external stimuli to decide their fate, such as growth or death. The "context" of a cell – the extracellular and intracellular environments – dictates the structure of signaling networks that determine such cellular responses. Identifying context-specific edges is largely understood as important for predicting and understanding cell decision making logic, however, robust identification of context dependent network structures remains a broadly unsolved problem. We hypothesized that context-dependent network structures could be generated through a combination of robust penalized regression modeling with leave-one-group-out (LOGO) analysis applied to a dataset that captures multiple omic-level dimensions. The LOGO analysis encompasses excluding data from one ligand condition out in the regression model construction and enables detecting ligand dependent associations. The unique dataset we utilized to test this hypothesis was recently generated by NIH-LINCS Consortium, and profiled MCF10A cells (non-transformed, epithelial, breast cell line) with proteomic (proteins), transcriptomic (mRNAs), and epigenetic (chromosomal states) data in response to seven different ligand or ligand combination stimulation conditions (including basal control) that differentially modulate cell proliferation. We then employed the robust lasso-based inference model with LOGO to integrate the three data types and uncover ligand specific sub-networks controlling cell fate decisions. The statistical associations between levels of proteins, phospho-proteins, mRNAs, and transcription factor binding sites on chromosomes are treated as edges of the networks. The incorporation of the LOGO analysis in our computational framework enabled us to find the context dependent sub-networks by comparing the ligand LOGO case to the full data analysis. First, we focused on identifying Interferon-γ (IFNγ), an antiviral and cell death promoting cytokine, dependent networks. Our IFNγ-LOGO analysis revealed two mechanisms in PD-L1 (programmed cell death-ligand 1) expression control through spindle protein FAM83D and histone protein HIST2H2AA3. Another interesting finding of our network-inference model is IFNγ dependent regulation of ACE2, the receptor responsible for SARS-CoV-2 cell entry, through IRF1 (interferon regulatory factor 1) expression. Overall, our novel analysis pipeline integrates multiple data types to delineate context specific signaling sub-networks important in cell growth decision making.

613349 The "Protein Space" Analogy Facilitates an Understanding of Complex Topics in the Evolution of Drug Resistance (INVITED SPEAKER)

Abstract:

Many determinants of drug resistance are proteins that acquire mutations in a stepwise fashion toward resistance to antimicrobial compounds. In this seminar, I utilize the classical analogy of "Protein Space" to discuss research into how adaptive evolution operates in negotiating fitness and resistance, across drug environments, in the evolution of antimicrobial resistance. Using this analogy, I offer a new perspective on concepts in pharmacology (e.g. cross-resistance, druggability) and in evolutionary biology (e.g, evolvability, epistasis). In closing, I emphasize the importance of continued dialogue between modern theory in evolutionary biology and concepts in protein biochemistry, biophysics, and engineering.

612116 Hydrothermal Liquefaction of Unhydrolyzed Solids for Fuels and Value-Added Products

Abstract:

Integrated biorefineries utilizing lignocellulosic biomass feedstock for fuel production will require additional revenue generation from the derived waste streams to demonstrate technoeconomic feasibility. One of the ways of revenue generation will be the efficient valorization of waste streams for the production of value added products. To understand technoeconomic feasibility of integrated bioprocessing, we investigated hydrothermal liquefaction (HTL) of unhydrolyzed solids (UHS) derived from some preprocessing of a lignocellulosic biomass and subsequent valorization of biocrude waste streams at a laboratory scale. Both solid and liquid side products were selectively enriched and their yields were taken into account to estimate revenue generation. A scale-up of the envisioned pilot scale integrated bioprocessing plant, and preliminary process and cost analysis were performed. Biomass processing steps, energy and mass balances, energy recovery, waste stream valorization, and initial assessment of technoeconomic feasibility will be presented.

608027 Understanding of Metabolic Pathways in *Pseudomonas Putida* for Biosynthesis of PHA Using Sugar, Glycerol, Benzoate, and Their Combinations

Abstract:

Efficient utilization of all available carbons from lignocellulosic biomass is critical for economic efficiency of a bioconversion process to produce renewable bioproducts. However, the metabolic phenotypes enabling bacteria such as *Pseudomonas* to utilize mixed carbon sources to generate energy and stored bioproducts remain poorly understood. In this study, *Pseudomonas putida* KT2440 was employed to investigate the polyhydroxyalkanoates (PHA) production from mixed carbon source of glucose/benzoate or glycerol/benzoate. A multi-omics approach involving label-free proteomics and NMR metabolomics was taken to compare the metabolic profiles of different mixed carbon sources. Results showed different patterns of reductant generation and fatty-acid metabolism pathways with different mixed carbon sources, which resulted in enhancing or inhibiting biomass and PHA productions. Overall, our study provides new insights into the metabolic architecture of *P. putida* with different mixture carbon substrates and reveals the potential regulatory metabolic nodes that may guide further system biology design to improve the PHA production.

594776 Lignin-Based Drop-in Jet Fuel Production

Abstract:

Sustainable aviation fuels remain the only near-, mid-, and likely long-term solution for lowering the carbon emissions of commercial and military aviation. Utilizing lignin feedstock for production of high energy density jet fuel along with cellulosic ethanol offers an unparalleled opportunity to enhance the overall operation efficiency, carbon conversion efficiency, economic viability, and sustainability of biofuels and chemicals production. In addition, determination of sustainable aviation fuel's chemical compositions and prediction of their properties is the critical first step for further research and development that eventually lead to the final certification process. Several catalytic processes with bifunctional catalyst Ru/H^+-Y, Ru/Super Lewis acid, and Ru-M/H^+-Y (M= Fe, Ni, Cu, Zn) were recently developed by our lab. Biomass-derived lignin was successfully converted into the lignin-based jet fuel, which consists of mainly paraffinic hydrocarbon species. They can be further classified into several classes, including n-paraffins, iso-paraffins, mono-, di-, and tri-cycloparaffins of which the majority contains carbon numbers in the range of 7 to 20. However, the "impurities" in the lignin wastes such as sugars, furans, and their derivatives were found to suppress metal catalysts and inhibit lignin depolymerization and hydrodeoxygenation reactions. Furthermore, the possibility of meeting drop-in specifications such as the low aromatic concentrations, which illuminates favorable energy content, energy density, possible low emissions, and high-performance characteristics, will be discussed in the presentation.

613064 Toward Feedstock Design for Biomaterial: Novel Biomass Structure Characteristics Determining Properties of Lignin-Based Carbon Fiber

Abstract:

Lignin has been explored extensively as a renewable feedstock for biomaterials, considering its abundance as a major component of plant cell wall and its sustainability as a waste of lignocellulosic biorefinery and paper-making industries. Despite the extensive efforts on defining process-property relationship, it remains largely unknown how lignin biosynthesis and precursor chemistry in planta would impact on the properties of biomaterials. Current knowledge about reducing lignin content to improve saccharification efficiency does not necessarily inform the feedstock for enhanced properties of biomaterials. Such inadequate understanding fundamentally limited the feedstock design and selection to improve carbon material properties toward broader commercial application. Using lignin from a broad range of biomass feedstock for carbon fiber manufacturing, we have fundamentally explored the structure-function relationship between lignin chemistry and carbon fiber properties, for both mechanical and electro-conductive performance. Specifically, lignin extracted from hardwood (sugar maple), softwood (loblolly pine and red cedar), and herbaceous plant (corn stover and switchgrass) were used for carbon fiber manufacturing, considering the very different lignin and biomass structures from these feedstock. Linear regression models were established to define the relationship between carbon fiber mechanical properties (Elastic Modulus, Tensile Strength, and Elongation) with lignin structure characteristics (content, composition, linkage profile and others), along with carbon fiber microstructures (crystallite content and size). The results highlighted that traditional biomass characteristics such as lignin content and composition don't impact the properties of lignin-based carbon fiber. Instead, the content of β-O-4 linkage correlates significantly with tensile strength and elastic modulus of lignin carbon fiber, indicating that the more linear β-O-4 linkage would promote carbon fiber performance. Besides the mechanical property, electro-conductive property is essential for broader energy application of lignin-based carbon fiber, yet the mechanisms controlling the electrocoductive properties of lignin-based carbon fiber is largely unknown. We hereby demonstrated that a higher β-O-4 content also promotes electroconductive performance of carbon fiber. Moreover, we also found that higher β-O-4 linkage profile defines the crystallite content and size of the resultant carbon fiber, which could be due to the better miscibility with the guest polymer. Overall, the study established lignin linkage profile, in particular, β-O-4 content as the new structure determinants for biomaterial property, which could guide the future biomass feedstock development.

596512 Extending the Degradation of Charged Amino Acids for CoA-Free Synthesis of Short-Chain (C3-C5) Diols Using Promiscuous Carboxylic Acid Reductase

Abstract:

Microbial metabolism has been extensively studied and engineered for synthesizing valuable compounds ranging from natural products to bulk commodities. However, many industrially important compounds still cannot be efficiently produced by microbial cell factories. Expanding the pathway repertoire can add new information to the pathway network and compensate the natural microbial metabolism. Here, a novel metabolic platform was established, which is capable of producing industrially important C3-C5 diols, namely 1,3-propanediol (C3), 1,4-butanediol (C4) and 1,5-pentanediol (C5), from charged amino acids through CoA-free steps. Endogenous degradation pathway of charged amino acids were extended to produce corresponding ω-hydroxylic acids, which were then reduced to form C3-C5 diols by a promiscuous carboxylic acid reductase. The generality and potential of this platform was further demonstrated by converting seven different amino acids to C3-C5 diols. To explore the industrial feasibility of the platform, we engineered the *de novo* production of 1,4-butanediol and demonstrated the first *de novo* synthesis of 1,5-pentanediol. Our work showed the great potential of extending charged amino acids degradation for biosynthesis of short chain diols and contributed new knowledge to current pathway networks.

604939 Engineering of Phytosterol-Producing Yeast Platforms for Functional Reconstitution of Downstream Biosynthetic Pathways

Abstract:

As essential structural molecules for plant plasma membranes, phytosterols are key intermediates for the synthesis of many downstream specialized metabolites of pharmaceutical or agricultural significance, such as brassinosteroids and withanolides. *Saccharomyces cerevisiae* has been widely used as an alternative producer for plant secondary metabolites. Establishment of heterologous sterol pathways in yeast, however, has been challenging due to either low efficiency or structural diversity, likely a result of crosstalk between the heterologous phytosterol and the endogenous ergosterol biosynthesis. In this study, we engineered campesterol production in yeast using plant enzymes and enhance the titer of campesterol by upregulating the mevalonate pathway, but no conversion to downstream products was detected upon the introduction of downstream plant enzymes. Further investigations uncovered two interesting observations about sterol engineering in yeast. First, heterologous sterols tend to be efficiently and intensively esterified in yeast, which drastically impedes the function of downstream enzymes. Second, yeast can overcome the growth deficiency caused by altered sterol metabolism through spontaneous adaptive evolution. By employing metabolic engineering, strain evolution, fermentation engineering, and pathway reconstitution, we were able to establish a set of phytosterol-producing yeast strains. In addition, the phytosterol-producing yeast strains were also used to reconstitute the *de novo* synthesis of campesterol-derived specialized phytochemicals. This work resolves the technical bottlenecks in phytosterol-derived pathway reconstitution in the backer's yeast and opens up opportunities for efficient bioproduction and pathway elucidation of this group of phytochemicals.

605200 *De Novo* biosynthesis of a Nitro-Containing Amino Acid

Abstract:

Nitroaromatic compounds serve many purposes, and the development of metabolic pathways for their biosynthesis offers a green chemistry alternative to hazardous industrial synthesis methods. However, investigation of *de novo* nitro-product biosynthesis and the associated enzymes that catalyze nitro-group formation has been limited. Here, we present an integrated heterologous pathway for the production of a stable, non-toxic nitroaromatic product in *Escherichia coli.* Expression of four exogenous enzymes redirects metabolic flux in the shikimate pathway, including a previously uncharacterized amine oxygenase, to a nitro-containing amino acid with titers above 0.5 mM. Integration of our pathway with technology for site-specific incorporation of non-standard amino acids should enable the *in situ* production of nitro-containing proteins.

609518 Shikimate Pathway Refactoring in the Non-Conventional Yeast *Kluyveromyces Marxianus* Enables High Titer Production of 2-Phenylethanol

Abstract:

The non-conventional yeast *Kluyveromyces marxianus* is a promising microbial host for chemical biosynthesis because it is one of the fastest growing eukaryotes, is thermotolerant to temperatures upwards of 50°C, and has the capacity to assimilate a wide range of C5, C6 and C12 sugars. Previously, we have identified a strong phenotype of short and medium chain alcohol production, including the production of 2-phenylethanol (2-PE), an important compound in the fragrance industry with market values upwards of $1,000/kg. Aromatic compounds are almost exclusively biosynthesized from the Shikimate pathway. In this work, we focus on refactoring this pathway to extend the native capacity of *K. marxianus* to produce 2-PE. To do so, we first developed a one-step markerless gene integration system that can simultaneously integrate up to three unique expression cassettes in a single round of strain engineering. The refactoring experiment was enabled by the creation of a set of synthetic promoters derived from three *K. marxianus* genes *TEF3*, *PGK*, and *TDH3* for high, medium and low expression, respectively. The refactored pathway library contains 27 strains with varied expression combinations of $ARO4^{K221L}$, $ARO7^{G141S}$ and *PHA2* integrated at the *ABZ1* locus in the genome. Analysis of the library identified an engineered *K. marxianus* strain with a 4.5-fold increase in 2-PE production, reaching 684.24 ± 72.95 mg/L in shake flask experiment with 25 mL YPD medium containing 2% glucose. Integration of one copy of *ARO10* driven by the synthetic *TEF3* promoter into this high-producing strain further increased titers to about 2 g/L in fed-batch culture. The best strain for 2-PE biosynthesis from the refactored Shikimate pathway library can also be used as a yeast chassis for the biosynthesis of other phenylalanine-derived compounds, and the multigene integration tool advances the ability for rapid engineering of *K. marxianus*.

609758 A Polyketide Synthase Platform to Produce Biofuels and Specialty Chemicals (Faculty Candidate)

Abstract:

Traditionally engineered to produce novel pharmaceuticals, Type I modular polyketide synthases (PKSs) could be engineered as a new biosynthetic platform for the production of *de novo* fuels, commodity chemicals, and specialty chemicals. A significant challenge in PKS design is engineering a partially reductive module to produce a saturated β-carbon through a reductive loop exchange. In this presentation, we first establish that chemoinformatics, a field traditionally used in drug discovery, offers a viable strategy for reductive loop exchanges. We introduced a set of donor reductive loops of diverse genetic origin and chemical substrate structures into the first extension module of the lipomycin PKS (LipPKS1). Product titers of these engineered unimodular PKSs correlated with chemical similarity between the substrate of the donor reductive loops and recipient LipPKS1, reaching a titer of 165 mg/L of short chain fatty acids produced by *Streptomyces albus* J1074 harboring these engineered PKSs. Expanding this method to larger intermediates requiring bimodular communication, we introduced reductive loops of divergent chemosimilarity into LipPKS2 and determined triketide lactone production. We observed a statistically significant correlation between atom pair chemosimilarity and production, establishing a new chemoinformatic method that may aid in the engineering of PKSs to produce desired, unnatural products.

Building upon this work, we expanded to multi-modular systems by engineering the first two modules of lipomycin to generate unnatural polyketides as potential biofuels and specialty chemicals in *Streptomyces albus*. First, we produce 20.6 mg/L of the ethyl ketone, 4,6 dimethylheptanone through a reductive loop exchange in LipPKS1 and a ketoreductase knockouts in LipPKS2. We then show that an AT swap in LipPKS1 and a reductive loop exchange in LipPKS2 can produce the potential fragrance 3-isopropyl-6-methyltetrahydropyranone. Highlighting the challenge of maintaining product fidelity, in both bimodular systems we observed side products from premature hydrolysis in the engineered first module and stalled dehydration in reductive loop exchanges. Collectively, our work expands the biological design space and moves the field closer to the production of "designer" biomolecules.

610550 One-Pot Hydrolysis of Lignocellulosic Biomass Components Via Engineered *Bacillus Subtillis.*

Abstract:

Advent of 21st century has seen surge in demand of energy, the depletion of fossil fuel reserves and detrimental effect of fossil fuels on environment. Therefore, alternative energy sources have been extensively explored to reduce our dependence on non-renewable fuel sources. Agricultural waste is fascinating alternative which is mainly composed of cellulose, hemicellulose and lignin. This untapped resource can be utilised by breaking down complex biopolymers in agricultural waste into fermentable sugars by process called hydrolysis. Both chemicals and enzymes can be used as catalyst in hydrolysis of biomass subunits but chemical hydrolysis produces furfural, hydroxymethylfurfural and other byproducts which can be toxic for microorganisms. Hence, Enzymes are safer and efficient alternative for biomass hydrolysis. Currently, biorefineries utilize purified hydrolases (class of enzyme) for breaking down cellulose and hemicellulose into its sugars, which is not cost effective. Therefore, efficient and cost-effective methods for hydrolysis of the cellulosic and hemicellulosic biomass can play vital role in making biorefineries economically viable. *Bacillus subtillis* strains are enzyme workhorses and are widely used in biotechnology industries for enzyme production. The present work aims at the engineering of *Bacillus subtillis* for heterologous expression and extracellular export of hydrolases (cellulases and xylanases) to reduce the cost incurred by protein purification. Extracellular enzyme export can be facilitated with the help of signal peptide (a short chain of amino acids which help in export of enzymes outside a bacterial cell). Exhaustive study on signal peptides of *B. Subtilis* has been reported by Brockmeier et al. 2006a, 2006b which shows that signal peptides can increase the extracellular enzyme export in *B subtilis.* Different combinations of signal peptides and genes were explored to identify the optimal secretion system. Three signal peptides namely YwmC, SacC and AmyE; two b- glucosidases (Bg) from *Trichoderma resei* (Tr) and *Neosartorya fischeri* (Nf) and two xylanases from *Trichoderma resei* and *Bacillus pumilis* (Bp) were selected. The selection was based upon their activity at the pH and temperature conditions in which the microbials host is still viable. The recombinant and wild type strains of *B. subtillis* WB800N were screened for extracellular hydrolysis of the substrates and comparative analysis was performed on the basis of DNSA assay and p-NPG assay. Recombinant strains with AmyE- Bg (Tr) and YwmC- Xy (Tr) displayed **4.1 fold** increase in cellobiose hydrolysis and **2.2 fold** increase in xylan hydrolysis compared to wild type strain, respectively. Insitu breakdown of xylan was studied for screened strain consisting YwmC- Xy (Tr) which demonstrated production of **7.1 g/L** of xylose from 1% xylan after 72 hours of fermentation. Hence, we, have demonstrated *one pot process* for enzymatic degradation of lignocellulosic biomass components for feedstock generation.Further work is underway to breakdown the biomass components and produce high value products simultaneously.

612980 Engineering Bacteria Phospholipid Pathways for Diverse Fatty Acid Profiles (Invited Speaker)

Abstract:

As the major cell membrane component for almost all organisms, fatty acid (FA) phospholipid biosynthesis represents one of the most active pathways and yields energy-dense molecules that can be used as fuels and chemicals. However, FAs synthesized by common microbial species are straight, long-chain FAs, whose derived fuels and chemicals often have undesirable properties, such as poor cold-flow properties as fuels and low freezing points as lubricants, thus limiting their applications. To diversify the properties and functions of microbially-produced FAs, we engineered the *E. coli* phospholipid pathways to produce FAs with altered structures and drastically different FA profiles. Multiple new FA species were biosynthesized in high abundances and yields by our engineered strains, including terminally-branched FAs, internally-branched FAs, cyclo-propyl FAs, ω-hydroxyl FAs, dicarboxylic acids, etc. These greatly expanded biosynthetic capabilities and FA profiles will facilitate the renewable production of numerous chemicals for practical applications.

605165 Accelerating Antibody Discovery with Cell-Free Systems

Abstract:

The COVID-19 pandemic has highlighted the need for rapid, high-throughput, and robust therapeutic antibody screening platforms. Advances in both synthetic selections and antibody repertoire mining have enabled the rapid identification of hundreds of candidate binders in a single experiment. However, the expression and evaluation of these candidate antibodies still remains a major bottleneck in the discovery pipeline due to labor intensive steps and process throughput mismatches. In this work, we present a newly developed workflow that leverages cell-free protein synthesis (CFPS) and an Echo® 525 acoustic liquid handler to enable the expression and evaluation of hundreds of antibodies in a single experiment. Our workflow consists of a cell-free DNA assembly step to generate antibody expression templates, a modified E. coli CFPS system to express antibodies, and the AlphaLISA protein-protein interaction assay to evaluate binding. Each step of this workflow can be carried out entirely within 384-well plates without ever going into cells, making it highly suitable for automation. The resulting platform is capable of evaluating more than an order of magnitude more antibody candidates in less than half the time when compared to state-of-the-art antibody discovery pipelines. To validate the platform, we expressed 13 different human antibodies, 12 of which neutralize the SARS-CoV-2 virus. We evaluated these antibodies by mapping their target epitopes and evaluating their ability to compete with ACE2, the SARS-CoV-2 target human receptor. Our results are largely consistent with published literature on these antibodies, indicating our workflow is suitable for identifying therapeutic candidates. We expect that these advancements will accelerate antibody discovery and development not only for viral diseases like COVID-19, but for the entire antibody industry.

602378 Characterizing and Enriching Native Membrane Vesicles in *E. coli* Extracts to Improve Cell-Free Glycoprotein Synthesis

Abstract:

Cell-free protein synthesis harnesses endogenous transcription and translation machinery in cell extracts to enable biomanufacturing without living cells. To date, *E. coli* cell-free systems have been used for many applications including decentralized manufacturing of therapeutics and vaccines. Notably, key biopharmaceutical products such as conjugate vaccines rely on protein glycosylation, a post-translational modification that requires membrane-bound machinery. To tailor *E. coli* cell-free reactions to make specific products, soluble heterologous proteins are routinely expressed in the strain prior to lysis. However, membrane-bound functionality has been largely inaccessible in extracts without synthetic membrane mimics. In this work we characterize native membrane vesicles in cell extracts for applications in membrane-dependent protein glycosylation. We first characterize the size and concentration of native vesicles that form during cell lysis. We then determine how extract processing steps impact vesicle properties and create simple methods for modulating concentrations of membrane vesicles and their heterologous cargo. By tuning upstream processing conditions, we increase vesicle concentration in extract by up to 2-fold and use vesicles to enrich a variety of membrane proteins in extract. We show that heterologous membrane-bound glycosylation machinery from the periplasm is associated with membrane vesicles and by applying extract processing improvements, we increase glycoprotein yields by up to 90%. Overall, this work provides valuable characterization of *E. coli* extracts and expands opportunities for engineering membrane-bound functionality in cell-free systems.

600871 A Cell-Free Pipeline for Rapidly Imprinting Complex Methylation Patterns on DNA to Enhance Plasmid Transformation in Bacteria (Industry Candidate)

Abstract:

Genetic tools are critical to study and engineer a diverse assortment of bacteria for next-generation industrial biocatalysts or as cell-based diagnostics and therapeutics. However such tools are utterly lacking in the vast majority of bacteria, with inefficient DNA transformation as one of the prevailing culprits. A principal barrier to DNA transformation is host restriction-modification (R-M) systems that produce a unique and often complex methylation pattern on host DNA and degrade incoming DNA lacking this pattern. Reproducing the host's methylation pattern on a transformable shuttle plasmid has been shown to evade host R-M systems, although all available methods require cellular expression of every methyltransferase thus are extremely laborious and regularly impeded by methyltransferase-dependent cytotoxicity. Here, we describe an alternative approach where every methyltransferase gene is expressed via cell-free transcription-translation (TXTL) to rapidly reproduce and imprint complex methylation patterns upon a plasmid to enhance DNA transformation. We call this method IMPRINT (Imitating Methylation Patterns Rapidly IN TXTL). We first showed that the well-characterized Dam and Dcm methyltransferases from *E. coli* could be efficiently expressed and methylate supplied DNA, thus protecting the DNA from restriction digestion, while the TXTL reaction did not imbue any methylation on its own. We then applied and further optimized IMPRINT to a pathogenic strain of *Salmonella enterica*, where we identified which combination of host methyltransferases are needed to maximize DNA transformation. We are now implementing this approach in a set of probiotic Lactobacilli strains with varying R-M systems to explore strain-dependent barriers to DNA transformation and how DNA transformation can be consistently achieved across strains. Finally, we are applying this method to promising probiotic species that have been previously understudied due to extremely poor or no prior DNA transformation ability. Overall, IMPRINT represents a streamlined pipeline to rapidly boost DNA transformation efficiency across the rich diversity of bacteria by evading prominent yet strain-specific defense systems. This approach thus should help others access a much wider range of bacteria for molecular and synthetic biology experiments, in turn broadening our understanding of bacterial genotype-phenotype relationships and driving more effective bacterial engineering solutions.

611255 Effectiveness of Nuclease Inhibition for Stabilizing Linear DNA in Cell-Free Expression Systems

Abstract:

Cell-free expression (CFE) systems are powerful tools used for applications ranging from environmental sensors to point-of-use diagnostics due to their quick reaction times, affordability, and portability. CFE allows the use of linear DNA templates, which have benefits over plasmids (the typical form of DNA used in CFE) because they shorten the prototyping cycle to a few hours by eliminating the days it takes to clone and replicate plasmids *in vivo*. However, linear DNA added to a CFE system is degraded by endogenous nucleases present in the lysate. This can be detrimental for the implementation of prominent linear DNA-based technologies like aptamers as CFE sensors. Here, we assess the effectiveness of multiple strategies to stabilize two types of linear DNA in CFE systems: PCR-generated gene templates and DNA aptamers. These strategies include use of the competitive inhibitor GamS, non-canonical DNA backbones, longer DNA templates, and oligos encoding Chi sites (short DNA motifs that signal the RecBCD nuclease), some of which have previously been shown to inhibit nucleases from degrading linear DNA templates for gene expression. We found that even with Chi DNA, linear templates with strong synthetic σ^{70} promoters expressed less than 1% of the protein produced from a phage T7 promoter, while the inducible P_{BAD} promoter could not produce detectable protein levels. We also showed that high levels of Chi DNA are unable to stabilize DNA aptamers long enough for them to be used effectively as sensors in CFE systems. These results indicate that the level of protein expressed from linear templates depends heavily upon immediate downstream amplification steps, and the same nuclease inhibition strategies cannot extend aptamer half-life enough for proper functionality. Continuing to explore the limits of nuclease inhibition is critical for enhancing the use of linear DNA in CFE systems and could enable the creation of impactful new sensor platforms.

604189 Rewiring Bacterial Electron Transfer for Living Materials Synthesis

Abstract:

Rewiring Bacterial Electron Transfer for Living Materials Synthesis

Austin J. Graham[1,2], Christopher M. Dundas[1], Gina Partipilo[1], Thomas M. FitzSimons[1,2], Adrianne M. Rosales[1,2], Benjamin K. Keitz[1,2]

[1]McKetta Department of Chemical Engineering, University of Texas at Austin, Austin, TX

[2]Center for the Dynamics and Control of Materials, University of Texas at Austin, Austin, TX

Qualities exhibited by living systems, including self-regulation, self-healing, morphology control, and environmental responsiveness, are highly attractive from a material design perspective. However, traditional biological materials such as biofilms and tissues are generally less robust and more difficult to engineer than synthetic materials. New materials should aim to bridge the characteristics of natural and synthetic systems; however, they will require programming of genetic and transcriptional responses whose outputs can control non-biological material properties. Towards these goals, engineered living materials (ELMs) are a class of materials defined by their incorporation of autonomous cells that use biological functions to control material properties. Our group recently demonstrated that extracellular electron transfer (EET) from the organism *Shewanella oneidensis* can be leveraged to control the redox equilibrium of metal catalysts to synthesize polymers and polymers networks via atom-transfer radical polymerization[1,2]. Using this method, we report the production of hydrogel ELMs in which *S. oneidensis* MR-1 controls the cross-linking of methacrylated hyaluronic acid. Cross-linking by genetic knockouts demonstrated that hydrogel formation was closely coupled to the expression of specific EET proteins, as gelation kinetics and stiffness correlated to the number of removed EET genes. Furthermore, strains complemented with an inducible EET gene showed inducer-dependent cross-linking activity and hydrogel mechanics, including both gel microstructure and bulk properties. Modeling gene expression using an activating Hill Function yielded quantitative prediction of hydrogel storage modulus, demonstrating control over material properties using design principles from synthetic biology. Genetic circuits controlling different EET genes also yielded tunable networks with unique properties depending on the choice of gene. By coupling EET activity to cross-link density, we have engineered a platform that utilizes endogenous bacterial metabolism to control and evolve macroscopic materials in response to environmental cues. Next, we will further demonstrate the general applicability of EET-controlled catalysis for living material synthesis, cross-linking fully synthetic polymer networks using thiol-ene and CuAAC click chemistries. Genetic control over these chemistries will significantly expand the synthetic capability of microorganisms. Finally, these gels will be employed in an extrusion printing setup for injectable applications such as soil stabilization and engineered plant ecology.

1. A.J. Graham, C.M. Dundas, A. Hillsley, D.K. Kasprak, A.M. Rosales, B.K. Keitz. "Genetic Control of Radical Cross-linking in a Semisynthetic Hydrogel." *ACS Biomater. Sci. Eng.* **6**(3): 1375-1386 (2020).

2. G. Fan, A.J. Graham, J. Kolli, N.A. Lynd, B.K. Keitz. "Aerobic Radical Polymerization Mediated by Microbial Metabolism." *Nat. Chem.* (in press, 2020).

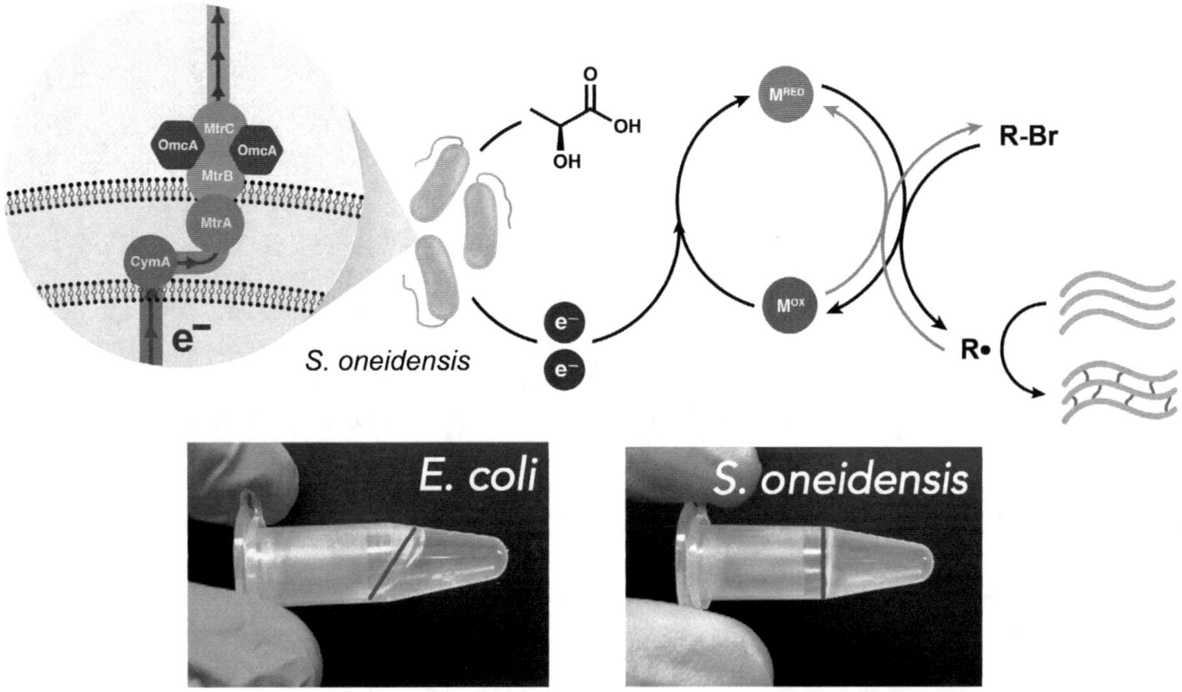

611804 Electrogenetic Design for Bidirectional Communication at a Bioelectronics Interface

Abstract:

Toward bioelectronics technology, there is a compelling opportunity to integrate biological systems with electronic devices, which have become increasingly networked and ubiquitous. However, these two entities are dissimilar in composition and information processing mechanism. Biology is organically composed and often uses molecular recognition for signal response, while electronics are typically inorganic and operate by electron flow. This work presents an approach to interface genetically-programmed bacteria with electronics as a biohybrid system, addressing both challenges of physical coupling by self-assembly and communication compatibility using electrochemically-active molecular cues. First, bacteria express a peptide tailored for inorganics, which is fused to an outer membrane protein for surface display. In this way, the bacteria can selectively assemble onto an electrode surface. Then, the cells feature a simple, modular genetic circuit for electrochemically-activated gene expression. Specifically, in a voltage-dependent manner, a charged electrode biases the oxidation state of redox-active signal molecules. This, in turn, influences gene expression based on redox-specific recognition of the molecular cue. Further, the electrogenetic cells are demonstrated to provide multiple outputs in response to electrochemical stimulation. The cells serve as bioelectronic transducers by 1) relaying the voltage input as small molecule quorum sensing signals to other cell populations and 2) providing redox-active molecular feedback to the electronic infrastructure, which is again detectable through electrochemical analysis. Because the cells are positioned directly on the electrode, this electrochemical signal exchange can occur locally across the interface. This work demonstrates bioelectronic communication where redox-active molecules serve as interconvertible signals that carry electrons between electrodes and simultaneously allow for molecular recognition events within cells. The combination of voltage-mediated redox control and electrogenetic response establishes a bioelectronics dialogue. Bioelectronics connectivity with synthetic biology may yield future applications for programmable biohybrid devices with utility in ecological settings, wearable interfaces, and *in vivo* environments.

613019 Using Synthetic Biology to Engineer Microorganisms That Make and Communicate with Materials (Invited Speaker)

Abstract:

Living organisms have evolved to use minerals in their environment to build protective shells, to generate energy, and as nutrients. To achieve these functions, living systems form interfaces with inorganic materials and use biomolecules to control and monitor the flow of matter across this interface. Inspired by these naturally-occurring systems, my research group seeks to engineer new organisms with tailored abilities to transfer charge and assemble matter across the microbial-material interface. In the first part of my talk, I will describe how we have engineered bi-directional electronic communication between living microbes and non-living systems using synthetic biology. By transplanting extracellular electron transfer pathways into the industrial organism *Escherichia coli,* we can confer upon these cells a molecularly-defined route to both accept and donate electrons to electrodes. Both current production and current consumption shift the metabolism of *E. coli* in well-defined ways, demonstrating that this electronic interface can control intracellular state. In the second part of my talk, I will describe how we have used surface-layer (S-layer) proteins as a programmable material. S-layer proteins form a highly ordered crystalline, yet porous, layer on the outermost cell surface of most species of bacteria and archaea. By engineering S-layers on the surface of different microorganisms, we can create patterned arrays that modulate both the binding and formation of soft and hard materials. The properties of this living hydrogel can be changed dynamically by altering the cell-hydrogel crosslinking. Taken together, this work provides a platform for creating hierarchically-assembled living materials with switchable mechanical properties.

599421 Controlled Release of Antibiotics for the Treatment of Bacterial Infection

Abstract:

Controlled release of antibiotics has the potential to limit exposure of nonpathogenic bacteria to the drug, thus limiting development of secondary infections and slowing development of antibiotic resistance. For example, periodontitis is often treated with systemic antibiotics; while effective in clearing the disease-causing organisms in the mouth, the process exposes the gut microbiota to high concentrations of these drugs and has been associated with increased rates of antibiotic resistance in the relevant organisms. Here, we describe such a technique to deliver antibiotics in a controlled fashion to treat localized aggressive periodontitis (LAP), a disease caused by *Aggregatibacter actinomycetemcomitans* (*Aa*). Not all strains of *Aa* are pathogenic, but those that are most closely related to disease have been demonstrated to produce more of a protein toxin, leukotoxin (LtxA). Our previous work demonstrated that LtxA is able to disrupt model membranes (liposomes), but only when they contain certain lipids, those with a negative spontaneous curvature. We have used this specific membrane-permeabilizing activity to design a "Trojan horse" delivery system for the controlled release of antibiotics to treat LAP. We have demonstrated that LtxA, at physiologically relevant concentrations, efficiently permeabilizes liposomes composed of 1,2-dioleoyl-*sn*-glycero-3-phosphoethanolamine-N-methyl (N-methyl-DOPE), and inclusion of cholesterol enhances this effect. Cell culture medium from an LtxA-producing strain mediates release from the liposomes, but cell culture medium from a non-LtxA-producing strain does not. Finally, we found that antibiotic-containing liposomes inhibit growth of the LtxA-producing strain but not the non-LtxA-producing strain. Control liposomes, composed of POPC do not inhibit bacterial growth. Together, our results demonstrate that N-methyl-DOPE liposomes provide a suitable platform for the specific release of antibiotics only in the presence of the specific disease-causing agent.

603745 Dose-Controllable Light-Activated Drug Release Systems

Abstract:

Drug delivery systems for temporal and spatial drug release using nanocarriers, or so-called on-demand drug delivery systems, are a new therapeutic approach in the medicinal field. Not only can drug release be tightly controlled and negative side effects associated with incorrect dosage minimized, but also the therapeutic efficacy can be prolonged by conserving drug when not needed.[1-2] A photoactive-triggering mechanism of site-specific drug delivery has gained interest within this field of research.[3-4] Although the technique shows potential for repeatable on-demand drug delivery, the lack of understanding of nanostructural changes of the drug-carriers with various parameters of light activation often limit repeated application.

Here we report on novel plasmonic material-coated liposomes (of mixed lipids) for the efficient delivery of methotrexate (MTX), a common drug for cancer and autoimmune diseases. Gold nanorods are used as the plasmonic material on the liposomes, which generate heat during near-infrared laser irradiation. The repetitive drug release from these MTX-encapsulated liposomes was measured using UV-Vis spectroscopy. The reversibility of the liposome structure after repetitive laser irradiation was determined using a single particle electrochemical measurements which measures the frequency of liposome collision events via changes in current that occur when a liposome collides with, and blocks, an electrode surface (2 μm diameter Pt disk).

We find that drug release occurs with each repetitive laser irradiation, whereas the liposome collision frequency (and thus concentration of liposomes) does not change. The collision frequencies of the irradiated liposomes were 0.056 Hz, 0.060 Hz and 0.060 Hz for 0 s, 30 s and 120 s irradiation time, respectively (1.9 W average power) indicating that the liposomes remain intact after irradiation. However, when MTX-encapsulated nanodroplets that underwent phase-transition from liquid to gas bubbles[5] were irradiated, the collision frequencies were 0.13 Hz, 0.107 Hz and 0.023 Hz, for 0 s, 30 s and 120 s, respectively, demonstrating 18% and 82% decrease in collision frequencies for 30 s and 120 s irradiation, respectively. The results indicate the number density after irradiation was maintained for the liposomes but not for the nanodroplets, suggesting reversibility of the liposome structure.

When larger liposomes (mean 3 μm in diameter) were irradiated, the mean current drop per collision increased from 0.2 nA to 0.8 nA, compared to the regular liposomes tested above (230 nm in average diameter). However, the collision frequencies were 0.093 Hz, 0.09 Hz, and 0.103 Hz for 0 s, 30 s and 120 s, respectively, again indicative of the structure reversibility.

The results suggest that the gold nanorod-coated liposomes can be a potential drug carrier for repetitive on-demand drug delivery. In addition, we expect to measure drug dosage released from an individual carrier, leading to precise control of dosage multiple times for long term drug delivery.

604037 Spatial Control of Probiotics Assisted By Magnetic Particles

Abstract:

Engineered probiotics have the potential to diagnose and treat many gastrointestinal (GI) diseases. However, most cellular agents have limited ability to selectively colonize specific targeted regions of the GI tract due to a lack of external control over their location and persistence. Magnetic fields are well suited to provide such control due to their ability to freely penetrate biological tissues, but they are difficult to apply with enough strength to directly manipulate magnetically labeled cells within deep tissue or viscous environments such as in the GI tract. Here, we show that ingestible micron-sized magnetic particles, combined with an externally applied magnetic field, act as *in vivo* magnetic field gradient amplifiers, enabling the trapping and retention of orally administered probiotic *E. coli* within the mouse GI tract. This technology improves the ability of these probiotic agents to accumulate at specific locations and stably colonize without antibiotic treatment. By enhancing the ability of GI-targeted cellular agents to be at the right place at the right time, cellular localization assisted by magnetic particles (CLAMP) adds external physical control to an important emerging class of biotherapeutics.

608008 Internal Liquid Crystal Structures in Nanocarriers Containing Drug Hydrophobic Ion Pairs Dictate Drug Release

Abstract:

Hydrophobic ion pairing is an attractive strategy for encapsulating and controlling release of charged water-soluble species. Hydrophobic complexes formed this way are sensitive to salts and pH, but the underlying mechanisms and dynamics remain unknown. We here demonstrate that a model hydrophobic ion pair formed from the cationic peptide polymyxin B sulfate and the anionic surfactant sodium oleate forms liquid crystal structures inside the cores of nanoparticles formed using Flash NanoPrecipitation. Internal structures were observed by SAXS and TEM following nanoparticle formulation and were found to vary with polymyxin:oleate charge ratio. Upon exposure to phosphate-buffered saline at pH 7.3, release of encapsulated polymyxin from the nanoparticles was induced. The internal structure was either retained or altered during release, depending on the polymyxin:oleate charge ratio. For a formulation containing a four-fold charge excess of oleate relative to polymyxin, the internal structures rearranged into an inverse hexagonal phase. The hexagonal phase formation corresponded with a greatly reduced rate of polymyxin release, suggesting that the polymyxin was incorporated into the center of hexagonally-packed rods. When release tests were repeated using a PBS at pH 2.0 to ensure protonation of the oleic acid and prevention of ion pairing, all internal structures were eliminated and release occurred much faster than at neutral pH, regardless of charge ratio. These findings shed light on the mechanism behind stimulus-responsive drug release from systems containing hydrophobic ion pairs and could lead to rational design of controlled-release formulations by manipulating the formation and dynamics of liquid crystalline phases inside nanocarriers.

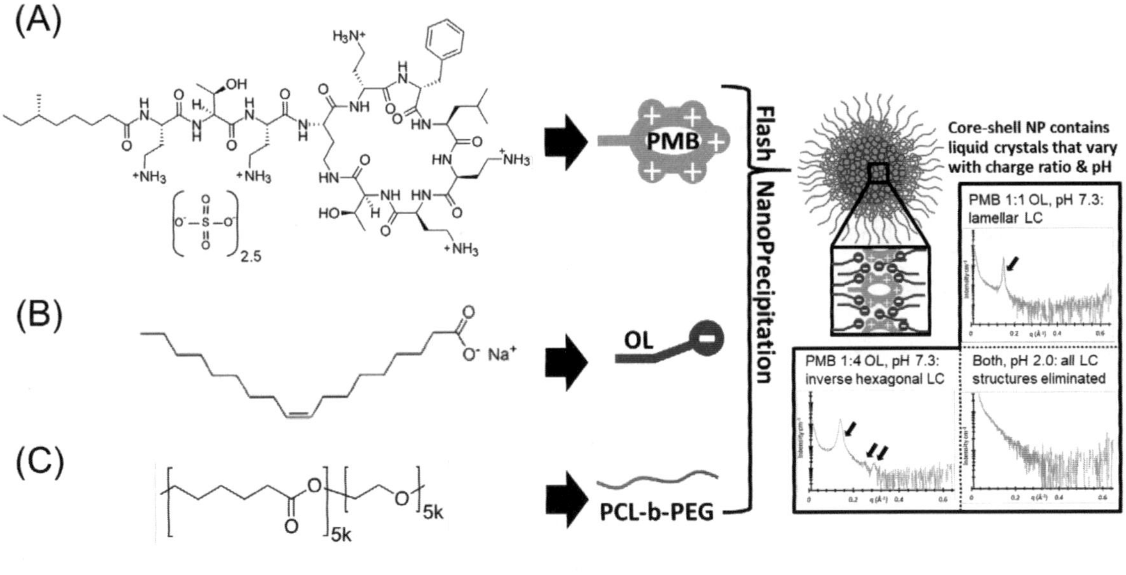

Non-invasive excitation of neural activity via on-demand magneto-thermal drug release

Rohini Guntnur, Gloria Lesly Jimenez Miranda and Gabriela Romero*
The University of Texas at San Antonio

The ability to evoke and inhibit neural activity on-demand is essential to understand and design therapies for neurological disorders such as Epilepsy and Parkinson's disease. Existing technologies for the control of neural circuits offer limited possibilities. Manipulation of neural signaling via direct drug application is restricted by cell-type specificity, which results in poor response to drugs and undesirable side effects, while implantable devices for the direct control of brain circuit dynamics are damaging to biological tissues. Here, we develop a novel pharmacological neural stimulation approach that depends on nanoscale heating effects for the local delivery of neuromodulatory compounds To develop the magneto-thermal platform for neuromodulation, we synthesized magnetic nanoparticles (MNPs) by thermal decomposition of iron oleate precursor. We functionalize the surface of MNPs with temperature-sensitive polymer brushes. Dopamine is loaded within the polymer brushes coating. MNPs loaded with dopamine are targeted to the cell membrane of primary striatal neurons through anti-DARPP32 antibody recognition.

Functionalized MNPs are successfully target the surface of straital neurons labeld with the calcium indicator Fluo-4. Upon exposure to an alternating magnetic field, the heat dissipated by magnetic hysteresis triggers the release of dopamine from MNPs' polymer coating. This liberated dopamine evokes ion influx into the cells detectable by the dynamic enhancement of Fluo-4 fluorescence. By coupling temperature sensitive polymer brushes to the MNPs surface, we are also able to achieve release of neuromodulatory compounds in multiple dosages.

We developed a non-invasive wireless pharmacological neuromodulation approach for the treatment of brain malfunctions. This approach could also allow us to establish a new platform for mapping brain circuits to improve our understanding of complex neural networks. Future clinical translation of magnetothermal pharmacological neural modulation could enable the precise single-cell control of brain circuit dynamics.

This work was supported by the National Institute of General Medical Sciences Grant # SC1GM130542

611094 Silicone Hydrogel Contact Lenses Loaded with Unsaturated Fatty Acids for Extended Release of Cationic Drugs

Abstract:

Contact lenses are potential candidates for the delivery of ophthalmic drugs with high bioavailability and reduced risks of side effects. Studies have shown that the residence time of ocular drugs in the tear film can be increased from 2 minutes for eye drops, to 30 minutes for drug-eluting contact lenses. However, unmodified contact lenses tend to release drugs very quickly; most of the drug loaded is released in less than a few hours. We propose the use of unsaturated fatty acids in silicone hydrogel contact lenses to extend the drug release duration of cationic ophthalmic drugs. Silicone hydrogel commercial contact lenses (ACUVUE TruEye and ACUVUE Oasys) are soaked in fatty acid-ethanol solutions for fatty acid uptake. Lenses are subsequently submerged in a solution of drug in phosphate buffer saline (PBS) for drug uptake. In vitro drug release experiments are performed by immersing each drug-loaded lens in PBS media at 7.4 pH. Drug concentration as a function of time is monitored by using Ultraviolet-Visible Spectrometry at the wavelength of maximum absorbance of each tested drug. The results obtained show that oleic acid-loaded contact lenses significantly extend the release time of cationic drugs from less than 24 hours, to more than 1 month. The drug release of cationic drugs is dependent on the weight percent of fatty acid in the contact lenses, and in the number of double bonds in the alkyl chain of the fatty acid that is loaded in the lens matrix. The addition of oleic acid in contact lenses has a negligible effect on the optical transparency of the lenses, but has a moderate effect on the water content of the lenses depending on the amount loaded. Due to the low ocular toxicity associated with unsaturated fatty acids, contact lenses loaded with these can be promising materials to be used in ocular drug delivery. Future studies on lens characterization and cell viability need to further assess the potential of these biomaterials.

613345 Controlling Controlled Release to Make Medicine That Imitates Life (INVITED SPEAKER)

Abstract:

This talk will outline tools that can be used to rationally design drug delivery formulations with controllable design parameters that produce a given, desired release behavior. This can be used to generate a desired continuous rate of release over time or even controlled bursts of drug with a predefined period of time in between. As just one example, temporal control over the release of specific growth factors can induce robust formation of specific tissues that naturally regenerate via stage-wise processes. Similarly, this concept can also be used to reproduce spatial information that cells (and even tumors) employ to manipulate immunological responses. Collectively, these new tools can effectively mimic the prose and context of local, cell-driven "communication" and have already shown significant promise as next-generation medical treatments in a variety of disease models where current medical treatments have no answer.

610202 Molecular Recognition and *In Vivo* Detection of Temozolomide and 5-Aminoimidazole-4-Carboxamide for Glioblastoma Using Fluorescent Nanosensors

Abstract:

The efficacy of chemotherapeutics such as temozolomide (TMZ) and its metabolic product 5-aminoimidazole-4-carboxamide (AIC) is often affected by the timing, quantity and frequency of dosages. There is strong interest in facilitating the ability to monitor efficacy in individual patients for specific subtypes of cancer. Real-time, dynamic measures of potency may also supplement or in some cases replace reliance on bio-imaging. Towards this end, in this work we develop new, synthetic molecular recognition sites for TMZ, and its decomposition product AIC, grafting them onto near infrared fluorescent nanoparticles capable of forming optode or other biosensor interfaces to monitor drug efficacy in real-time. Infrared fluorescent single-walled carbon nanotubes, wrapped using specific DNA oligonucleotides, are encapsulated within biocompatible, poly(ethylene glycol) diacrylate hydrogels and enable the selective recognition of an anti-cancer drug, TMZ, on U-87 MG human glioblastoma cells. In both solution phase and hydrogel form, the sensors exhibit a fluorescence response to TMZ with a detection limit of 30 M. Furthermore, U-87 cells exhibit no changes in viability for 7 days in contact with the hydrogel. The sensors were used to track the progression of glioblastoma death following TMZ administration. In addition, the sensors were able to monitor *in vivo* AIC and TMZ dynamics in SKH-1E mice. These results enable real-time detection of chemotherapeutic concentration and metabolism at the cellular and subcellular levels, with potential to increase the efficacy of cancer treatments.

606459 An Enzymatic Electrochemical Biosensor for Real-Time Detection of Physiologically Relevant Nicotine Concentrations

Abstract:

The glucose biosensor, built upon the redox enzyme glucose oxidase, is the most commercially successful and studied enzymatic biosensor. However, the lack of available and functionally validated enzymes is prohibiting the development of redox-based sensors for other important analytes. Herein, we present the development and assessment of an electrochemical nicotine biosensor, using genomic screening to identify the gene for a known nicotine catabolizing redox enzyme. The resulting nicotine biosensor demonstrated a specific, sensitive, and stable operational profile with a limit of detection of 27 μM over the range of 0-200 μM. This range is well within the physiological concentrations of nicotine present in smoker urine. Specificity and cross-reactivity were measured against structurally similar compounds to nicotine as well as to known physiological by-products. Our results highlight that this novel enzymatic electrochemical nicotine biosensor possesses operational capabilities for monitoring of nicotine in physiologically relevant conditions. The screening methodology can be generalized for the discovery of enzymes for novel sensor development.

608267 Simulations Guide Optimization of Electroenzymatic Biosensors for Neurotransmitters and Enable Proper Interpretation of Sensor Response *In Vivo*

Abstract:

Electroenzymatic biosensors for neurotransmitters are uniquely compact, sensitive, and selective devices that provide the neuroscience community with a method to monitor fluctuating chemical concentrations in the brain. Many studies primarily consider glutamate, the most common excitatory neurotransmitter, and study the relation between sub-second neurotransmitter dynamics and various stimuli, behaviors, or diseases. Challenges, however, have remained in the optimization of sensor design and in clarifying the significance of assumptions made during *in vivo* use. Accordingly, a number of simulations were carried out to guide the development of glutamate sensors and to test the limitations of the most prominent assumption, that sensor response can be linked to the immediate extracellular neurotransmitter concentration using a predetermined calibration factor. Recommended modifications to sensor design were developed and successfully implemented, improving sensitivity six-fold (up to 320 nA/μM/cm^2) with an accompanying order of magnitude decrease in response time (down to ~80 ms). Since even this improved response time is an order of magnitude slower than the timescale of synaptic neurotransmitter release, it is clearly wrong in many cases to assume that the use of a calibration factor to translate sensor response to neurotransmitter concentration without any signal deconvolution is accurate. To investigate the significance of this assumption and the conditions for which this practice remains reliable, the model was expanded with a new set of partial differential equations to describe the extracellular space of the brain. Considering the effects of variations in the local biological environment bordering a sensor, a range of representative parameters for some biological processes were selected. Comparisons were made between the theoretical responses expected from *in vitro* calibrations, tonic steady-state glutamate presence, and bolus-type glutamate releases that mimic synaptic release. Surprisingly, in steady-state cases, the sensor response could indicate a concentration over twice the actual value, while in cases where sensors were 10 μm or more from the source of glutamate being detected, sensor response was < 50% of the expected value. In some cases, transient concentrations due to a single synaptic-type release of neurotransmitters at a moderate distance from the sensor were not detected by the sensor at all, and when they were, the time course of signal decay did not match the rates of reuptake expected in the brain without significant signal deconvolution. Ultimately, results show that additional characterization of sensor performance *in vivo* may be necessary to enable proper interpretation of sensor data.

611068 Theoretical and Experimental Studies on an Electrochemical Enzyme Immunosorbent Biosensor

Abstract:

Electrochemical immunosensors integrate biological recognition molecules (e.g., antibodies) with redox enzymes (e.g., horseradish peroxidase (HRP) to combine the advantages of immunoassays (extremely high sensitivity and selectivity) with those of electrochemical biosensors (reproducible, quantitative electrical output). Mechanistic mathematical models that describe the multiple mass-transfer and chemical reaction steps that give rise to the electrical output are needed to help design, optimize, and validate electrochemical immunosensors for medical applications. In this study, we developed such a model of an electrochemical immunosensor that used HRP as a reporter enzyme on the secondary and validated it experimentally.

The model consists of a system of non-linear mass-balance equations that describe simultaneous mass transfer and chemical reactions of the reactants (catechol and hydrogen peroxide) that HRP oxidizes, as well as the HRP product, O-quinone, which is reduced at the working electrode to give the amperometric output. A bi-substrate, ping-pong reaction mechanism was used to describe the enzymatic reaction, and the *Butler-Volmer* kinetic model was used to describe the reduction of o-quinone back to catechol at the working electrode's surface. The system of equations was solved numerically using the BVP4C function in MATLAB.

Experimental data for model validation was obtained using electrochemical immunosensors for a surrogate protein antigen (mouse IgG). The immunogens interface was fabricated on the gold working electrode of DropSens screen printed electrode arrays. The steady-state amperometric output of the resulting immunosensors was measured over a range of four independent variables: the concentrations of the two reactants, the voltage of the working electrode, and pH. The experimental results were used to determine optimal operating conditions and to validate the model. The validated model was then used to calculate Damkohler numbers and flux-control coefficients to identify mass-transfer and reaction steps that most strongly affected the magnitude of the immunosensors' output and the output's sensitivity to analyte concentration.

610485 HTL Derived Biochar and Graphene Nanoplatelets for Biosensor Applications

Abstract:

HTL Derived Biochar and Graphene Nanoplatelets for Biosensor Application

Bharath K Maddipudi, Vinod Amar, Hope Dosch, Anuradha Shende and Rajesh V Shende*Department of Chemical and Biological Engineering, South Dakota School of Mines and Technology, Rapid City, South Dakota 57701, United States of America

***Corresponding Author:** Rajesh Shende, Department of Chemical and Biological Engineering, South Dakota School of Mines and Technology, Rapid City, South Dakota 57701, United States of America (E-mail: *Rajesh.Shende@sdsmt.edu*)

Abstract

Biosensors with Au electrode layer immobilized with peptides are currently being developed for the detection of microbial contamination. It is anticipated that the use of a low-*k* dielectric layer modified with graphene will improve the sensor sensitivity. The low-*k* dielectric layer can be modified with the biochar derived from hydrothermal liquefaction of different biomass substrates. In this investigation, HTL of lignocellulosic substrates was performed under non-catalytic and catalytic conditions and by-products were analyzed. The biochar was thermally treated in a tubular reactor under inert environment from 400-1100°C without any pretreatment. In other set of experiments, HTL derived biochar was acid treated and subsequently, thermal activation was performed. After the acid pretreatment and thermal treatment steps, the porosity, the specific surface area (SSA) and the morphology of the char was analyzed using BET and SEM/EDX. The char obtained after thermal treatment was combined with low-*k* dielectric layer and biosensor was fabricated. The sensor output was compared with the biosensor fabricated with low-*k* and graphene nanoplatelets. In addition, the sensor output was correlated with the char materials obtained under different thermal conditions. The results obtained on hydrothermal liquefaction of biomass substrates, characteristics of biochar before and after thermal treatment, biosensor fabrication and sensor output will be presented.

604780 Construction of a Red Emission Bodipy-Based Probe for Tracing Lysosomalviscosity Changes in Culture Cells

Abstract:

A red emission BODIPY-based lysosome-targeted viscosity probe (Lys-VBOD) was synthesized for monitoring lysosomal viscosity in live cells. The increase of viscosity can restrict the twisting of double bond between indole
and BODIPY fluorophore, also restrict the rotating of single bond between quinoline and BODIPY, which lead to the enhancement of fluorescence intensity in high viscosity environment. The changes of fluorescence intensity
exhibit a good linear relationship with viscosity values. Besides, Lys-VBOD is quite stable in various pH solvents with given viscosity, suggesting that Lys-VBOD is suitable for applying in a wide pH range. Moreover, Lys-VBOD
can visualize the changes of lysosomal viscosity in live cells under the stimulus of dexamethasone. BSA titration experiment demonstrated that Lys-VBOD is stable in the presence of macromolecules and proteins. Notably, the
fluorescence lifetime of Lys-VBOD in large viscosity solvents is extremely high. These results indicate that LysVBOD is an excellent tool for detecing viscosity changes in vitro and live cells.

610593 Super Sensitive Cerium Oxide-Based Composite Sensor for the Detection of Hydroxyl Radicals

Abstract:

Free radicals are recognized as essential molecules for upholding the function of normal cells. Immune systems, such as bacteria elimination and virus inactivation, are examples of the crucial role that free radicals play in human cells. Even though free radicals have benefits, an optimum production is key to maintain their advantages, as an excess can cause the development of oxidative stress conditions, such as cancer, Alzheimer's, and Parkinson's diseases. Among all free radical species, hydroxyl radicals (•OH) are the most reactive and dangerous, and they can be used as biomarkers for the detection of those diseases at their initial stages. Thus, a very sensitive analytical sensor is needed to detect and monitor low concentrations of •OH. The integration of an electrochemical technique with a sensing element is regarded as a promising method for •OH detection due to a rapid and direct measurement without the pretreatment of samples. Thus, in this study, a glassy carbon electrode was modified with a sensing element comprised of ultrasmall cerium oxide nanoclusters (<2 nm) grafted onto a superconductive carbon (Figure 1a). XPS results showed 46.3% Ce^{3+} sites, which are those able to react with •OH. Cyclic voltammetry (CV) was implemented to characterize the interaction of the composite sensor with •OH, and they showed that the composite sensor was able to differentiate between solutions with and without •OH by an increased redox response in presence of •OH (Figure 1b). Furthermore, we observed a linear relationship between the redox response and the concentration of •OH in the range of 0.1 mM to 10 mM, with a limit of detection as low as 0.1 μM when using only 0.5 wt% CeO_x loading.[1]

(1) A.C. Alba-Rubio, D.S. Kim, S. Duanghathaipornsuk; Super sensitive sensor for the detection of hydroxyl free radicals with scavenging properties. Application number 62/977,945 (2020).

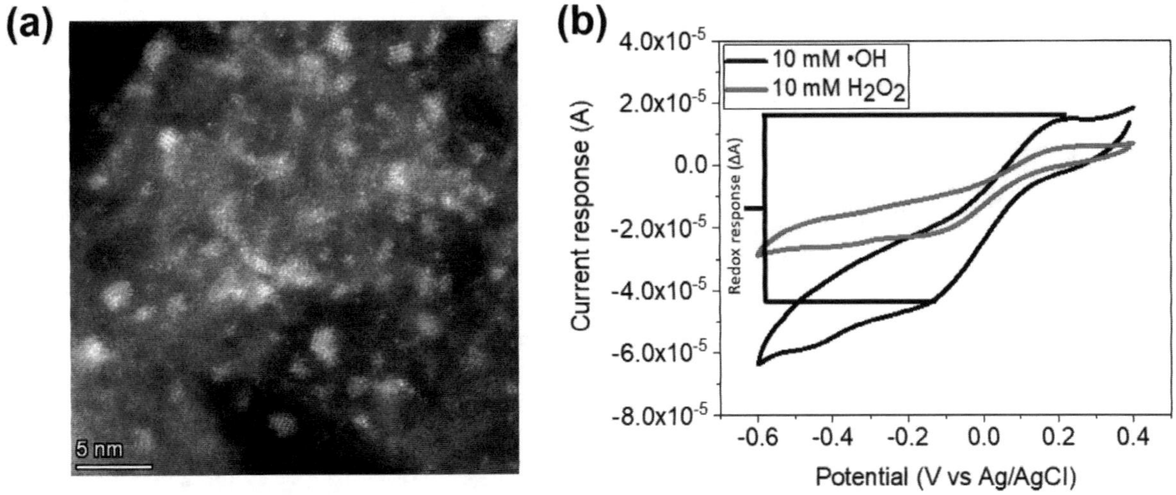

Figure 1. (a) STEM image of 0.5 wt% CeO_x/Carbon and (b) Current response to hydroxyl radicals.

611380 High Throughput Antibiotic Susceptibility Testing with Optical Nanosensors

Abstract:

Infections in a clinical setting can result from biofilm-forming species, which presents two key challenges: increased antimicrobial resistance, and poor diagnostic approaches to categorize antimicrobial susceptibility. Traditional antimicrobial susceptibility assay approaches are built upon planktonic bacteria (bacteria that are growing in solution). Biofilms are bacteria that grow in a community on a surface, and this community structure changes resistance to antimicrobials that is not reflected in these planktonic assays. While current biofilm-based assays exist, they are both time and material intensive, and can still provide antibiotic susceptibility results that may not reflect the actual biofilm response. Here we present our nanosensor based approach to monitor biofilm metabolic response during the administration of antibiotics. We grow biofilms with oxygen-monitoring nanosensors embedded throughout the film, and these nanosensors continuously report out the metabolic state of the biofilm. As we add antibiotics, we can measure what concentration can stop metabolism, and infer the minimum biofilm inhibitory concentration (MBIC) of the specific strain. Our diagnostic approach functions in 96 well plates with minimal materials and time (important for high throughput screening), and works with both lab and clinically relevant strains (demonstrated with Pseudomonas aeruginosa).

610159 Matrix Rigidity Modulates Histone H3 Modifications during Epithelial-Mesenchymal Transition

Abstract:

Epithelial-Mesenchymal Transition (EMT) is a physiological process that is essential during embryogenesis and wound healing and is also a contributor to pathologies including fibrosis and cancer. EMT is promoted by increased extracellular matrix stiffness and is characterized by loss of epithelial markers such as E-cadherin and gain of mesenchymal traits such as increased expression of cytoskeletal proteins and enhanced cell motility. In addition, epigenetic reprogramming, including histone modifications, change chromatin structure and regulate gene transcription during EMT; however, there is limited knowledge available on how mechanical signals control epigenetic remodelling in the context of EMT. Here, we synthesized hydrogels with mechanical properties that span that of normal and diseased mammary tissue and examined the response of mammary epithelial cells to transforming growth factor (TGF)β1-induced EMT. Using immunofluorescence staining and western blotting, we found that matrix stiffness and TGFβ1 together regulate acetylation and methylation patterns of histone H3 that are associated with transcriptional activation and silencing. Furthermore, treatment with contractility inhibitors modulated the levels of histone marks in response to EMT induction cues. Thus, matrix rigidity may regulate EMT process in part through control of epigenetic modifications and gene transcription. These findings may suggest strategies to improve tissue engineering approaches and ways to prevent EMT in pathological processes such as cancer.

597939 An in Vitro Model Utilizing Cell Spheroids and Hyaluronic Acid Hydrogels to Study the Role of Extracellular Matrix Stiffness on Tumor Dormancy in Brain Metastatic Breast Cancer

Abstract:

The ability of breast cancer cells (BCCs) to metastasize to distant organs contributes to approximately 90% of breast cancer related mortalities. These BCCs might metastasize even before the patient is diagnosed with a primary tumor. The common metastatic sites for BCCs are lungs, liver, brain, bone and lymph nodes. Upon arrival to these metastatic sites, these BCCs cells could exhibit either a dormant or proliferative phenotype by sensing the mechanical/biochemical/cellular cues from the microenvironment of that particular site. Among these breast cancer metastases, breast cancer brain metastasis (BCBM) is most aggressive, with a survival rate of approximately 4-9 months. In recent years, tumor cell clusters, specifically, circulating tumor cell clusters, have been shown to possess higher metastatic potential compared to single cells. However, the interactions between BCBM cell clusters and the brain microenvironment are not well understood due to the lack of relevant biomimetic *in vitro* models to study such interactions.

To address this need, herein, we employed hyaluronic acid (HA) hydrogels to mimic the native brain microenvironment, as it is one of the most abundant extracellular matrix component in the brain. Tumor cell spheroids were employed to mimic tumor cell clusters, and in our studies, we utilized MDA-MB-231Br BCBM cells. To investigate the impact of mechanical cues (extracellular matrix stiffness) on cell spheroids, HA hydrogels of varying stiffnesses (i.e., soft - 0.4 kPa, stiff - 4.5 kPa) were prepared. To evaluate the effect of tumor size on dormancy of BCBM clusters, different sizes of spheroids were prepared using 100 – 10,000 tumor cells. In this model, over a period of 7 days, we observed that areas of spheroids composed of 100 and 500 cells remained largely constant and the area of spheroids containing $\geq$ 1000 cells increased on stiff HA hydrogel. On the soft HA hydrogel, spheroid areas remained constant irrespective of spheroid size indicating a dormant phenotype. This was further supported by EdU and Ki-67 staining wherein we observed that the percentage of EdU+ and Ki-67+ cells was significantly higher on the stiff vs. soft HA hydrogel (in the case of spheroids made using 10,000 cells). We also demonstrated the reversibility of observed dormant phenotype on the soft HA hydrogel, wherein spheroids composed of 10,000 cells cultured on soft HA hydrogel for 7 days were transferred to a stiff and soft HA hydrogel respectively and cultured for another 7 days. Ki-67 staining results demonstrated an increase in the percentage of Ki-67+ cells for stiff HA hydrogel compared to spheroids cultured on soft HA hydrogel. Overall, such systems provide useful tools to probe cell-matrix interactions in BCBM and also for testing of drugs.

604665 Biomimicry of the Hypoxic Tumour Microenvironment of Pancreatic Cancer in 3D – on the Design of Novel Studies of Radiotherapy Combined with a Static Magnetic Field

Abstract:

Introduction and Aim:

The interdisciplinary field of biomimetics is emerging to bridge the gap between simplistic 2D cell culture models and complex xenografts, more readily emulating tumour architecture, porosity, cell-cell and cell-matrix interactions, as well as hypoxic regions[1]. More specifically, tissue engineering allows mimicry of bio-physical-chemical and mechanical properties of tumour microenvironments in a finely tuned *in vitro* setting, facilitating an animal free and cost effective platform as a representative tool for treatment screening and resistance profiling for cancer research [2].

Pancreatic ductal adenocarcinoma (PDAC) is a notoriously tragic disease, non-specific symptoms and high metastatic occurrence results in devastatingly low patient prognosis[1-4]. More specifically, PDAC has a 5-year relative survival rate of just 9% [4]. This figure has not significantly improved for over 50 years [6] and predictions that this disease will rise to one of the most lethal cancers in coming decades [7] outline that this is a cancer of unmet needs. The PDAC tumour microenvironment (TME) is extremely complex and hypoxic[8]. Unique disposition of PDAC extracellular matrix results in histological and behavioural hallmarks such as dense desmoplasiaand stromal formation that aid tumour survival and impair treatment delivery[9]. Chaotic blood vessel formation and collapse paired with this desmoplasia contribute to hypoxia, which is directly related to radio-resistance and treatment failure [9]. Thus, this unique and heterogeneous TME requires biomimicry to predict the clinical relevance of evolving treatments.

Radiotherapy for PDAC is traditionally controversial with contradictory clinical trial outcomes, such as the LAP02 2016 trail[10] and the ESPACI study[11] in which no improved overall survival and damage to organs at risk were observed, highlighting a need for further understanding and optimisation. Advances in radiotherapy such as image guided and proton therapy, aim to enhance treatment and reduce damage to healthy tissue in neighbouring organs, facilitating specific tumour targeting and higher dose disposition. Such emerging modalities are associated with varying biological effectiveness compared to conventional therapies, thus there is a need for a reliable and representative system to test delivery strategies. *Moreover, it is evident that realistic 3D models as well as improved and applied therapies are required for PDAC.*

Image guided radiotherapy has inspired a small number of pre-clinical studies investigating the presence of static magnetic field in combination with radiation treatments [12]. However this area is understudied and contradictory. Early research in 2D cell culture suggests an alteration of biological response to radiation in combination with SMF [11-14]. More recently only a small number of studies have been reported [12]. *More specifically, there is an identifiable gap in studying SMF and radiotherapy combinations 3D in vitro systems.*

The BioProChem Group at the University of Surrey have previously fabricated a 3D porous polymeric scaffolding system to support long term PDAC cell growth and *in vivo* properties such as dense cellular masses, initiated collagen-1 growth and the development of hypoxic regions [1-2] for the application chemo-radiotherapy screening [3]. In addition, the research group produced the first scaffold assisted multicellular model for PDAC [15]. *Utilising the 3D porous polymeric scaffolding system, here we aim to replicate the PDAC TME to investigate hypoxia-induced radio-resistance and the cellular response of combining a SMF during exposure.*

Materials and Methods:

Fabrication of polymeric 3D scaffolds employed the Thermally Induced Phase Separation method [1-3]. PANC-1 cells were seeded at $0.5x10^6$ and cultured for 4 weeks before being placed at 5% oxygen in a hypoxic chamber. Radiotherapy exposures were performed using an orthovoltage X-ray unit at the Royal Surrey County Hospital and static magnetic field experiments took place with an MR-Linac at the National Physical Laboratory. Scanning electron microscopy and confocal laser scanning microscopy (CLSM) of multiple scaffold sections, enabled scaffold characterization, allowing analysis of cellular organisation and viability and mapping of (radio-)resistance. More specifically, live/dead analysis as well as Ki67 and acute (HIF-1α) and chronic (PDK1) hypoxic biomarkers quantification/mapping was performed and compared for different SMF-radiotherapy regimes.

Results and Discussion:

This research provides a platform for studying the complex and hypoxic PDAC TME. Growth kinetics and morphological variations in (i) hypoxic and normoxic cultures (ii) different static magnetic field-radiotherapy regimes were identified. These data show for the first time (i) a 3D polymeric scaffold supporting long-term hypoxic PDAC cell culture (ii) a long term post-treatment *in vitro* tissue characterisation for novel SMF-radiotherapy treatment patterns. This system provides a versatile platform to study hypoxia-associated radio-resistance profiling of PDAC in combination with a static magnetic field.

Significance and Impact:

PDAC is a cancer of unmet needs. As advancing technologies evolve, a realistic replica of the unique and extremely hypoxic TME is required for treatment screening and resistance profiling. This novel hypoxic 3D PU scaffold acts as a platform for biomimicry of the complex and hypoxic PDAC TME. *This is the first 3D PU hypoxic scaffold system for PDAC static magnetic field and radiation combination studies.*

Acknowledgements:

This research was supported by the Department of Chemical and Process Engineering of the University of Surrey as well as the National Physical Laboratory, the EPSRC and the Royal Society. E.V. is grateful for a Royal Academy of Engineering Industrial Fellowship.

References:

[1] Totti, S., et al. (2018). RSC Advances, 8(37).

[2] Totti, S., et al. (2017). Drug Discovery Today, 22(4).

[3] Gupta, P., et al. (2019). RSC Adv., 9(71),

[4] American Cancer Society. (2019). Cancer Facts and Figures 2019.

[5] Cascinu, S. et al. (2010). Ann. Oncol, 21, 55–58.

[6] O'Reilly, D., et al. (2018). Pancreatology, 1–9.

[7] Rahib, L., et al. (2014). Cancer Research, 74(11), 2913–2921.

[8] Xie, D., & Xie, K. (2015). Genes & Diseases, 2(2), 133–143.

[9] Pickup, M., et al. (2014). EMBO Reports, 15(12), 1243–1253.

[10] Hammel, P., et al. (2016). JAMA, 315(17), 1844.

[11] Neoptolemos, J. P., et al. (2004). New England Journal of Medicine, 350(12), 1200–1210.

[12] Mohajer, J. K., et al (2018). British Journal of Radiology, 20180484.

[13] Forssberg A. (1940). Acta radiol, 21: 213–20.

[14] Chen, W, F., et al. (2010). Cancer Biotherapy and Radiopharmaceuticals, 25(4), 401–408.

[15] Gupta, P., et al. (2020). Frontiers in Bioengineering and Biotechnology (*in press, accepted manuscript*).

601515 Irradiation Promotes Breast Cancer Cell Proliferation and Invasion in Extracellular Matrix Hydrogels

Abstract:

While radiation therapy (RT) is an efficient treatment for patients with triple negative breast cancer (TNBC), patients continue to experience recurrence after RT. The alteration of extracellular matrix (ECM) of healthy breast tissue induced by radiation and its role on tumor locoregional recurrence are still unknown. We hypothesize that the irradiated ECM may influence pre-metastatic niche formation and lead to tumor cell recruitment and retention. The objectives of the present study were to develop irradiated ECM hydrogels and assess the effects of RT-induced ECM changes on breast cancer cell behavior. This work represents an important step toward elucidating how changes in the ECM after RT contribute to breast cancer recurrence.

Scanning electron microscopy (SEM) was used to visualize irradiated breast tissue changes including the ECM deposition and lipid distribution. In addition, the Raman spectra of irradiated and non-irradiated healthy breast tissue were compared to confirm molecular ECM changes. To investigate the impact of irradiated ECM on tumor cell behavior, murine mammary fat pads (MFPs) were irradiated to a dose of 20 Gy *ex vivo*. Following incubation for 48h post-RT, the resulting control and irradiated MFPs were decellularized and digested to form ECM hydrogels for the encapsulation of murine 4T1 TNBC cells. Mechanical properties of the hydrogels were determined by rheology. The viability and proliferation of tumor cells were examined 48h after seeding by fluorescence microscopy, bioluminescence measurements, and viability measurements. We analyzed how cell-matrix interactions regulate tumor cell invasiveness and intracellular cytoskeletal dynamics by actin, cortactin, focal adhesion, and matrix metalloproteinase (MMP) staining.

SEM images and Raman spectra of irradiated MFPs indicated increased ECM deposition ($p<0.05$), which may provide a platform for cell migration, proliferation and infiltration. Stable hydrogel formation was demonstrated through the increased storage modulus compared to the loss modulus for all conditions. Enhanced tumor cell proliferation in irradiated hydrogels were proved by bioluminescence imaging and fluorescence microscopy ($p<0.01$). Viability staining suggested an increased live to dead cell ratio in RT hydrogels compared to the non-RT control hydrogels. Encapsulated cells were stained with F-Actin, and an increase in alignment and elongation was observed in irradiated ECM hydrogels ($p<0.01$). The invasion of breast cancer cells was confirmed by increased colocalization of actin and cortactin ($p<0.01$).

Our study establishes that the irradiated microenvironment influences tumor cell behavior. The developed ECM hydrogels are important for mimicking how the *in vivo* breast tissue environment responds to radiation damage. Our results suggest that the irradiated ECM promotes tumor cell proliferation and invasion. Future studies will utilize these ECM hydrogels to explore the interactions between tumor and immune cells that may contribute to recurrence following radiotherapy.

613353 Using Physical Models of the Tumor Microenvironment to Identify Invasive Cancer Cells (INVITED SPEAKER)

Abstract:

The Dawson Cell Biophysics Lab uses a single cell biophysical approach to study tumor microenvironment interactions; research is focused on breast and ovarian cancer, which are leading causes of cancer-related death in women. The microenvironment in these tumors includes a multitude of cell types, matrix proteins, and growth factors that profoundly influence cancer cell behavior by providing both physical and chemical stimulation. This microenvironment, which is both dynamic and heterogeneous in nature, plays a critical role in cancer progression from the growth of the primary tumor to the development of metastatic and drug-resistant tumors. Our single cell biophysical approach, can be used to systematically investigate physical and paracrine interactions between cancer and stromal cells, and to identify specific conditions in the tumor that induce tumor-promoting behavior, along with strategies for inhibiting these conditions to treat cancer. This approach has helped us to identify rare populations of invasive and chemoresistant cancer cells, along with changes in the tumor microenvironment that contribute to their formation. In vitro studies of breast and ovarian cancer have primarily focused on adherent cells grown in 2D monolayers; however, 3D culture systems that more accurately mimic the in vivo tissue microenvironment are critical research tools in studying cancer. Our lab has developed 3D organoid culture models to study tumor growth, matrix invasion, and tumor drug response with spatial and temporal control in a single model of the tumor microenvironment. This research will increase our fundamental knowledge of cancer biology and provide critical information about how the 3D microenvironment influences tissue growth and development at all stages. Our long term goal is to use these in vitro systems to understand how heterogeneity develops in the tumor and contributes to the formation of invasive and drug-resistant outlier cell populations. My presentation will focus on lessons we've learned through quantitative biophysical analysis of cells in the tumor microenvironment. I will also discuss current research projects focused on investigating the role of stromal cell aging in cancer progression and mechanisms contributing to chemotherapy and radiation resistance through the formation of polyploidal giant cancer cells.

608967 Dynamic Changes in Epigenome Responding to Environmental Chemical Exposure: A Combination of Experimental and Modeling Approach

Abstract:

Epigenetic changes account for dynamic changes in chromatin that persist over time and are inheritable across generations. Significant advances have been made in past years to understand the epigenome, but limited efforts have been focused on modeling epigenome dynamics. Here, we focus on understanding the effects of a pesticide, ATZ [Atrazine: 2-chloro-4-(ethylamino)-6-(isopropylamino)-s-triazine)] on epigenome and its long-term health implications on central nervous system. Low doses of ATZ exposure has been indicated in Parkinson's Disease (PD) based on population studies, but the underlying molecular mechanism of action remains elusive. Using SH-SY5Y cells as a model system, we demonstrated dynamic and persistent changes in important epigenetic modifications, i.e., DNA methylation, governing gene regulations harnessing the power of immuno-fluorescent staining and live cell tracking. The collective data serves as the foundation to build a mathematical model to investigate the dynamics of epigenome changes in single cell and population. Our model assumptions are (1) methylation occurs in the presence of methyl donor molecules and an enzyme catalyze the methylation behavior; (2) given that a part of the DNA is methylated, the cells senses the need to provide a process of demethylation toward homeostasis; (3) when sufficient demethylation has occurred, the demethylation enzyme synthesis is turned down. We model the dynamics of methylation by incorporating variables of methylated DNA, demethylated DNA, and their corresponding catalytic enzymes. We use the cybernetic approach [*Ramkrishna and Song. Cybernetic Modeling for Bioreaction Engineering. Cambridge University Press, 2018*] to represent the regulatory influence on methylation. The cybernetic goal in this context is maximizing survival which is assumed to be accomplished by maximizing the rate of regulation either by methylation or demethylation based on the prevailing methylated fraction. To this extent, we further consider ATZ effect acting as an inhibitor to DNA methylation and determine its effect on epigenome in both short- (acute) and long- (chronic) term. Our model can faithfully capture the dynamic trend exhibited by various epigenetic changes upon external stimuli and offers a powerful tool to connect experimental data with mechanistic understanding for revealing the underlying molecular forces driving epigenome dynamics and predicting new phenotypic state.

611123 Multi-Omics Reveals That AKT but Not MYC Promotes Reactive Oxygen Species-Mediated Cell Death in Oxidative Culture

Abstract:

Introduction: The altered metabolism of tumors has long been proposed as a therapeutic target. Defining the metabolic vulnerabilities induced by specific oncogenes is crucial for the design and stratification of therapeutics targeting tumor metabolism. Increased glycolytic activity is thought to help satisfy the rapacious demands of highly proliferative cancer cells for biosynthetic precursors including lipids, proteins, and nucleic acids. However, this altered metabolism can leave tumors vulnerable to metabolic disruptions such as starvation of substrates including glucose, asparagine, glutamine, methionine, serine, and others. Therefore, understanding the interplay between oncogenes and metabolism is essential to understand how to design therapeutic strategies targeting tumor metabolism.

Results and Discussion: We tested how AKT and MYC affect the ability of cells to shift between respiration and glycolysis. Using immortalized mammary epithelial cells, we discovered that constitutively active AKT but not MYC induced cell death in galactose culture, where cells rely on oxidative phosphorylation for energy generation. However, the negative effects of AKT were temporary, and AKT-expressing cells recommenced growth after ~15 days in galactose. To identify the mechanisms regulating AKT-mediated cell death, we used metabolomics and found that AKT cells dying in galactose upregulated glutathione metabolism. Proteomic profiling revealed that AKT cells dying in galactose also upregulated nonsense-mediated mRNA decay, a marker of sensitivity to oxidative stress. We therefore measured levels of reactive oxygen species (ROS) and discovered that galactose induced ROS exclusively in cells expressing AKT. Furthermore, ROS were required for galactose-induced death of AKT-expressing cells.

Conclusions: Our findings highlight the importance of oncogene-dependent metabolic vulnerabilities in cancer cells and suggest that therapies targeting tumor metabolism will need to be appropriately paired with tumor genetic profiles.

600039 Single-Cell Transcriptional Dynamics of Retinal Ganglion Cell Diversification

Abstract:

Vision begins in the retina, a neural tissue that lines the back of the eye. The retinal circuitry that processes visual information is composed of six major neuronal classes: rods, cones, amacrine cells, horizontal cells, bipolar cells, and retinal ganglion cells (RGCs), each of which contains numerous types with distinct morphological, physiological and molecular properties. Despite accounting for less than 2% of retinal cells, the RGC class is highly diverse (i.e. contains numerous types), and its axons project the retina's output to the rest of the brain. Recent advances in single-cell RNA sequencing (scRNA-seq) have revealed the existence of 46 distinct RGC types in mice. Here, we interrogate this remarkable diversity during retinal development by decoding the logic of RGC specification and identifying the key molecular players that orchestrate it.

Using scRNA-seq, we profiled mouse RGCs at 5 stages of development: embryonic days 13, 14 (during and slightly after the period of peak genesis), 16 (as RGC axons begin to reach targets) along with postnatal day 0 (as dendrite elaboration begins) and 5 (shortly after they begin to receive synapses). These data were combined with a recently published transcriptomic atlas of adult mouse RGCs [3] to complete a six-time point developmental dataset that enabled the investigation of 75,115 RGCs. As the destructive nature of scRNA-seq precludes direct capture of the lineage dynamics of developing cells, we employ a computational approach termed RNA velocity [1-2] to infer transcriptional dynamics. RNA velocity leverages the relative abundance of exonic and intronic reads within the framework of a dynamical model to infer a cell's future state.

A key assumption of the original RNA velocity framework was that the underlying dynamical process is at steady state, which is violated during early development. To apply RNA velocity to RGC diversification, we use scVelo [2], which applies an expectation-maximization framework to extend RNA velocity estimation to transient processes. We estimate an RNA velocity vector for each immature RGC in our dataset, and use it to resolve developmental asynchrony within each time point and connect RGC states between consecutive time points. We also compare velocity-based estimates of temporal couplings between cells to those based on Optimal Transport [4], an alternative method to infer developmental relationships in time series scRNA-seq data. We then use the velocity estimates to identify genes that contribute most to driving velocity at each time point and evaluate their biological significance through gene ontology enrichment analyses. Assessing how these driver genes change over time will enable understanding of the dynamic processes underlying RGC diversification and maturation. Comparing the temporal variation of these driver genes to that of differentially expressed, transcriptomic cluster-defining genes will reveal the interplay between absolute gene expression and gene expression dynamics. Taken together, these efforts will further our understanding of the transcriptional logic of RGC diversification and identify novel candidate genes for experimental investigation.

[1] La Manno et. al., RNA velocity of single cells, *Nature* (2018)

[2] Bergen et. al., Generalizing RNA velocity to transient cell states through dynamical modeling, *bioRxiv* (2019)

[3] Tran et. al., Single-Cell Profiles of Retinal Ganglion Cells Differing in Resilience to Injury Reveal Neuroprotective Genes, *Neuron* (2019)

[4] Schiebinger et. al., Optimal-Transport Analysis of Single-Cell Gene Expression Identifies Developmental Trajectories in Reprogramming, *Cell* (2019).

610773 Partial Least Squares Regression Modeling Identifies Combination Treatments That Overcome Sprouty2-Mediated Glioblastoma Resistance to Multiple Classes of Therapeutics

Abstract:

Glioblastoma is a devastating form of brain cancer with a median patient survival time of less than 15 months under the current standard of care. Glioblastoma tumors will often recur due to post-translational protein modifications that drive glioblastoma resistance to therapeutic agents. These resistance mechanisms render current treatments inadequate. A quantitative understanding of broad regulators of resistance is essential to the design of therapeutic treatments that improve patient survival. We previously identified the protein Sprouty2 (SPRY2) as a critical regulator of glioblastoma phenotypes. We found that SPRY2 knockdown reduced colony formation by human glioblastoma cells and antagonized the growth of xenografts in mice, suggesting that SPRY2 promotes glioblastoma progression. These findings were surprising because SPRY2 acts as a tumor suppressor in numerous other cancers (e.g., lung cancer). However, therapeutics that target SPRY2 have not yet been developed. We hypothesized that SPRY2 regulates druggable pathways that can be rationally identified using a network level systems biology approach. We explored the role of SPRY2 in modulating glioblastoma cell response to the DNA damaging agents temozolomide and carboplatin, and inhibitors of the EGFR and MET receptor tyrosine kinases. In parallel with measurements of cell death and cell cycle, we assessed dynamic changes in the phosphorylation or abundance of a network of proteins regulating cell survival, apoptosis, and DNA damage response using a multiplexed microbead immunoassay. We constructed a quantitative data-driven model of the relationships between the measured signaling processes and cellular phenotypes using partial least squares regression (PLSR). Our results show that endogenous expression of SPRY2 regulates cell signaling in a manner that drives cancer cell resistance to DNA damaging agents and tyrosine kinase inhibitors. The PLSR model revealed a core subset of SPRY2-regulated kinases that were, surprisingly, common to the response to both DNA damaging agents and tyrosine kinase inhibitors. The model also identified SPRY2-regulated kinases that were unique to the response to DNA damaging agents or tyrosine kinase inhibitors. Model predictions were validated in multiple glioblastoma cell backgrounds. Therapeutic treatment, in combination with an inhibitor of model-identified-kinases of the DNA damage response pathway, augmented cell death in a manner comparable to SPRY2 knockdown. Our results identify novel combination treatments with the potential to overcome glioblastoma cell resistance to a broad range of therapeutics and demonstrate the utility of data-driven modeling approaches for the rational development of combination therapy in oncology.

596606 A Probabilistic Framework for Cellular Lineage Reconstruction Using Single-Cell 5-Hydroxymethylcytosine Sequencing

Abstract:

Lineage reconstruction is central to understanding tissue development and maintenance. While powerful tools to infer cellular relationships have been developed, these methods typically have a clonal resolution that prevent the reconstruction of lineage trees at an individual cell division resolution. Moreover, these methods require a transgene, which poses a significant barrier in the study of human tissues. To overcome these limitations, we report scPECLR, a probabilistic algorithm to endogenously infer lineage trees at a single cell-division resolution using 5-hydroxymethylcytosine (5hmC). When applied to 8-cell preimplantation mouse embryos, scPECLR predicts the full lineage tree with greater than 95% accuracy. Furthermore, scPECLR can accurately extract lineage information for a majority of cells when reconstructing larger trees. In addition, we developed a new single-cell technology to simultaneously quantify both 5hmC and genomic DNA from the same cell. Copy-number variations and single-nucleotide polymorphisms inferred from sequencing genomic DNA enabled us to significantly improve the prediction accuracy of larger trees. We find that for 32-cell trees with more than 10^{26} possible topologies, our integrated single-cell technology improves the prediction accuracy by 31%. Finally, we show that scPECLR can also be used to map chromosome strand segregation patterns during cell division, thereby providing a strategy to test the “immortal strand” hypothesis in stem cell biology. Thus, scPECLR provides a generalized method to endogenously reconstruct lineage trees at an individual cell-division resolution.

613352 Systems Biology of Cell Behavior: Robotics, Machine Learning, Microfluidics, and Multiscale (INVITED SPEAKER)

Abstract:

Cells respond to numerous stimuli in space and time. We developed pairwise agonist scanning (PAS) using 384-well plate robotic assay to create high-dimensional and dynamic response data for neural network (NN) training. Lattice Kinetic Monte Carlo (LKMC) then allows for patient-specific simulation of large cell ensembles (>10,000 cells) in the presence of diffusion, convection, and chemical gradients. We apply this approach to predict human blood clotting and bleeding in response to diverse drug treatments using microfluidic assay conditions. This approach then becomes patient-specific and multiscale with 3D medical imaging of the coronary vasculature and individual NN models of platelet function.

606167 Deposition of Oxide Coatings Using Non-Thermal Atmospheric Pressure Plasmas

Abstract:

Atmospheric pressure plasmas (APPs) have shown great promise in a variety of manufacturing applications. APPs offer unique opportunities for the industrial application of plasmas, due to the versatility of the technique, the ease of scalability, and the opportunity to perform the technique point of manufacturing. This work focuses on depositing zirconia-silica bilayer coatings as a replacement for hexavalent chromium-based coatings. A new non-thermal process to deposit zirconia was developed using APPs enabling processing on temperature sensitive substrates. Electrochemical corrosion performance will be compared to other passivation and barrier coatings. Additionally, we investigate APPs to enable lightweight manufacturing for the automotive industry by depositing a thin film of silica that acts as a water intrusion barrier on bonds joints between dissimilar materials. The adhesion tests performed after water soak meets the industry benchmarks. These processes aim to replace wet chemical processes often used by the automotive and aviation industries. The mechanical properties of the films are tested using lap shear, both before and after a water soak test, which successfully demonstrates the films capabilities as a water barrier. The chemical composition is confirmed using both XPS and RBS, and the films morphology is examined using cross sectional SEM.

608084 Treatment and Extraction of Copper from Electronic Waste Via Induced Morphological Changes Utilizing Supercritical CO_2

Abstract:

The management of electronic waste (e-waste) has become a significant challenge faced by humanity. Globally, the number of discarded computers, phones, and other appliances, often defined as Waste Electrical and Electronic Equipment (WEEE), doubled between 2009 and 2014 to 41.8 million tons per year. Currently, e-waste disposal factories tend to recycle the non-toxic, valuable components of e-waste (metals) and release the toxic components into the surroundings via pyrometallurgy (burning). Chemical treatment, known as hydrometallurgy, is also used and involves large amounts of corrosive acids. As the amount of e-waste grows, the availability of easily mineable elements, particularly metals, decreases, disturbing the overall material cycle. Thus, new, environmentally-friendly approaches to recover valuable components from e-waste are urgently needed. This study focuses on the use of supercritical CO_2 ($scCO_2$) and sulfuric acid, 2 M H_2SO_4, as a novel solvent system for the treatment of e-waste, specifically in the extraction of copper. Printed circuit board (PCB) was used as the e-waste of study and is generally composed of 40% metals, 30% plastics, and 30% refractory material. Through the synthesis of metal and polymer composites, which reduced the complexity of PCB, the effects of $scCO_2$ in the solvent system were investigated. The results showed that $scCO_2$ and acid induced a permanent morphological change in the polymer component, creating voids and causing delamination. This finding was translated to the actual waste PCB system and a novel two-step process was developed. This two-step process involved the pre-treatment of small pieces of waste PCB with $scCO_2$ and 2 M sulfuric acid at 120 °C and 148 atm for 30 minutes followed by leaching of the pre-treated PCB in a solvent containing 2 M H_2SO_4 and 0.2 M H_2O_2 at ambient conditions.Experimental results showed that 82% of the copper contained in the PCB was extracted in under 4 hours. The characterization of the PCB demonstrated that the pre-treatment with $scCO_2$ and acid induced the crystallization of the plastics (polymer component), creating pores and weakening the structure of the PCB, thus enhancing transport of the solvent to the buried metal interfaces. This novel process utilizing $scCO_2$ minimizes intensive physical processing (e.g. grinding of the PCB) and reduces acid usage during the extraction of Cu from e-waste, presenting a greener alternative to the current methods of recycling of metals from e-waste, which tend to be energy-intensive and hazardous.

596648 Annealing Carbon By Pulsed Laser Light

Abstract:

Laser processing of materials is not new, but, using lasers to anneal, bond and otherwise transform carbons is. Fundamental understanding of the dependence upon carbon structure, morphology and chemistry is critical to implementing this technology into manufacturing and processing applications. In this work a Q-switched Nd:YAG laser and a continuous wave CO_2 laser are used to anneal carbon materials. Lasers provide rapid heating and cooling with high temporal control. The extent of transformation is kinetically controlled by time above the threshold temperature for transformation. Enabling the production of carbon structures not possible via traditional furnace annealing. The potential technological importance of laser annealing carbon is demonstrated. Examples of material processing and synthesis not possible via traditional furnace annealing are provided. To resolve the nanostructure changes HRTEM is employed.

Perhaps the most over looked application of laser annealing carbon is the ability to do so continuously and rapidly. Graphitization furnaces operate as a batch processing system. Time from start to completion is on the order of a day due to the slow (~25 °C/min) heating rates, long hold times, and slow cooling of the heavily insulated furnace. Additionally, these furnaces require routine maintenance and replacement of expensive specialty graphite heating elements. CO_2 laser annealing provides equivalent material transformation on the order of seconds and modern CO_2 lasers are designed for years of maintenance free use [1]. Although laser annealing is limited to thinner material due to limited heating depth, materials can be annealed continuously and with potentially high throughput.

The material was heated to 2,600 °C in 1.4 ms Under the action CO_2 laser radiation carbon can be heated to 2,600 °C in 1.4 ms. After 20 s, the structure is equivalent to that obtained from furnace annealing at 2,600 °C for 1 hr. as has been shown elsewhere [2]. In contrast a Q-switched Nd:YAG laser with pulse width ~ 8 ns heats carbon materials to graphitization temperatures with a heating rate on the order of a few 10^{11} °C/s. The peak temperature is controlled by the laser pulse energy; typical temperatures range between 2,500 °C to the C_2 sublimation temperature of 4,184 °C [2,3]. The time at elevated temperature is limited (time above 2,000 °C is 1.5 μs). On this time scale, the long-range material motions are kinetically restricted. Consequently the Nd:YAG laser annealing trajectory deviates from traditional furnace pathways. The limited time at elevated temperature can be used for the purpose of surface modification. Surface modification via kinetically limited oxidation is one potential application. Another application is making connections between carbons without additional material. Laser sintering a mixture of carbon materials could result in a wide range of potential applications.

Heat-Treatment-Temperature (HTT) and resulting carbon structural transformation has been extensively studied. A detailed quantification of HTT and resulting material annealing was provided by Oberlin in 1984 [4]. The four stages outlined in Oberlin's HTT diagram (temperature dependent) are believed to be separated by "very rapid" transitions [8]. So rapid, in fact, that intermediate structures are not observable on furnace heating timescales (several minutes minimum) due to slow ramp rates and have thus been ignored. These intermediate structures remain unexplored and represent an entirely new class of carbon materials. It has been demonstrated that CO_2 laser heating results in equivalent end structure as compared to furnace annealing [1]. The high temporal control of

CO_2 laser annealing allows for exploration of such intermediate structures that can be captured by the rapid heating and cooling timescales. In summary lasers are poised to be instrumental in the advancement of carbon science and technology.

References

[1] J.P. Abrahamson, Pulsed laser annealing of carbon, Ph.D. Thesis, Penn State, 2017.

[2] J.P. Abrahamson, M. Singh, J.P. Mathews, R.L. Vander Wal, Pulsed laser annealing of carbon black, Carbon 124 (2017) 380–390. doi:10.1016/j.carbon.2017.08.080.

[3] H.A. Michelsen, C. Schulz, G.J. Smallwood, S. Will, Laser-induced incandescence: Particulate diagnostics for combustion, atmospheric, and industrial applications, Prog. Energy Combust. Sci. 51 (2015) 2–48. doi:10.1016/j.pecs.2015.07.001.

[4] A. Oberlin, Carbonization and graphitization, Carbon 22 (1984) 521–541. doi:10.1016/0008-6223(84)90086-1.

604429 An All-in-One Approach for Training Deep Learning-Based Control Laws

Abstract:

Model predictive control (MPC) provides a general framework for controlling nonlinear systems. However, the high online computational latency limits the scope of its application for systems with fast dynamics. One way to enable the real-time application of MPC is to approximate the implicit MPC control law with a neural network [1,2,3]. The hypothesis of this approach is that a deep learning neural network can sufficiently approximate the nonlinear behavior of MPC, but at a negligible online computational cost that is significantly better than that of a full-fledged MPC. For linear systems with quadratic cost function, a neural network with rectified linear units as activation functions can exactly represent piecewise affine optimal control laws [4].

An essential aspect of the above-mentioned approaches is the training of large dimensional neural networks. Many researchers follow the "optimize then train" protocol. The idea in these algorithms is to simulate the optimal controller, generate corresponding state and optimal control action data pairs and then use these data to train a neural network via supervised learning to approximate the original optimal controller. Unfortunately, this conventional "optimize then train" protocol is not guaranteed to work on nonlinear systems. For nonlinear systems, it is possible that multiple optimal control actions exit for the same initial state. In this case, the MPC control law is a set-valued function. The non-uniqueness of the state-action data pairs makes it impossible to learn the underlying MPC control law through this two-step approach, as illustrated in Figure 1. Even if the uniqueness of optimal control actions is guaranteed, the optimal control problem might have multiple local optimal control actions, which are typically identified using local solvers. Moreover, for general nonlinear systems, no neural network structure with finite neurons can guarantee an exact representation of the MPC control laws. The training error of the neural network can lead to sub-optimal control actions and state constraint violations. The resulting errors could potentially accumulate over time and eventually making the neural network deviate significantly from the optimal control.

We propose a new "optimize and train" method that combines the steps of data generation and neural network training into one single large-scale stochastic optimization problem. This approach directly optimizes the closed loop performance of the DNN controller over a number of possible initial states. Directly optimizing the control law instead of control actions avoids the difficulty caused by the nonuniqueness of optimal control actions, as shown in Figure 2. Moreover, all constraints are guaranteed to be satisfied for all initial state scenarios considered in the optimization, while this guarantee does not hold for the conventional "optimize then train" method. We also propose methods to ensure constraints satisfaction for all possible initial states. The critical challenge is the solution a large-scale nonlinear stochastic optimization problem. However, tremendous progress has been made recently in the field of stochastic optimization that allows us to design and implement efficient local and global algorithms [5,6]. Besides directly using stochastic optimization solvers, we also proposed a recurrent neural network formulation for nonlinear systems with input constraints and soft state constraints. This approach can take advantage of existing packages (e.g., TensorFlow) in the machine learning community and can potentially tackle extreme-scale problems efficiently with GPUs. The benefits of our all-in-one approach over the conventional ``optimize then train" protocol is illustrated through numerical results.

References:

[1] Parisini, T., & Zoppoli, R. (1995). A receding-horizon regulator for nonlinear systems and a neural approximation. Automatica, 31(10), 1443-1451.

[2] Csekő, L. H., Kvasnica, M., & Lantos, B. (2015). Explicit MPC-based RBF neural network controller design with discrete-time actual Kalman filter for semiactive suspension. IEEE Transactions on Control Systems Technology, 23(5), 1736-1753.

[3] Chen, Y., Shi, Y., & Zhang, B. (2018). Optimal control via neural networks: A convex approach. arXiv preprint arXiv:1805.11835.

[4]Karg, B., & Lucia, S. (2018). Efficient representation and approximation of model predictive control laws via deep learning. arXiv preprint arXiv:1806.10644.

[5] Cao, Y., Zavala, V. M., & D'Amato, F. (2018). Using stochastic programming and statistical extrapolation to mitigate long-term extreme loads in wind turbines. Applied energy, 230, 1230-1241.

[6] Cao, Y., & Zavala, V. M. (2019). A scalable global optimization algorithm for stochastic nonlinear programs. Journal of Global Optimization, 75(2), 393-416.

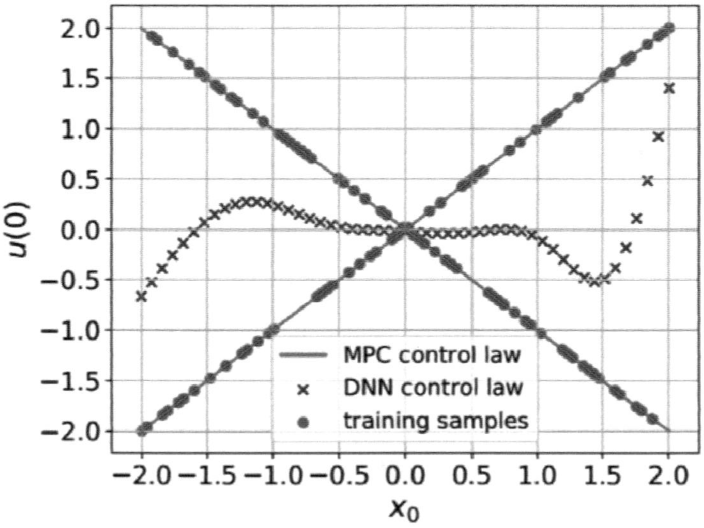

Fig. 1. DNN control law trained with "optimize then train" method

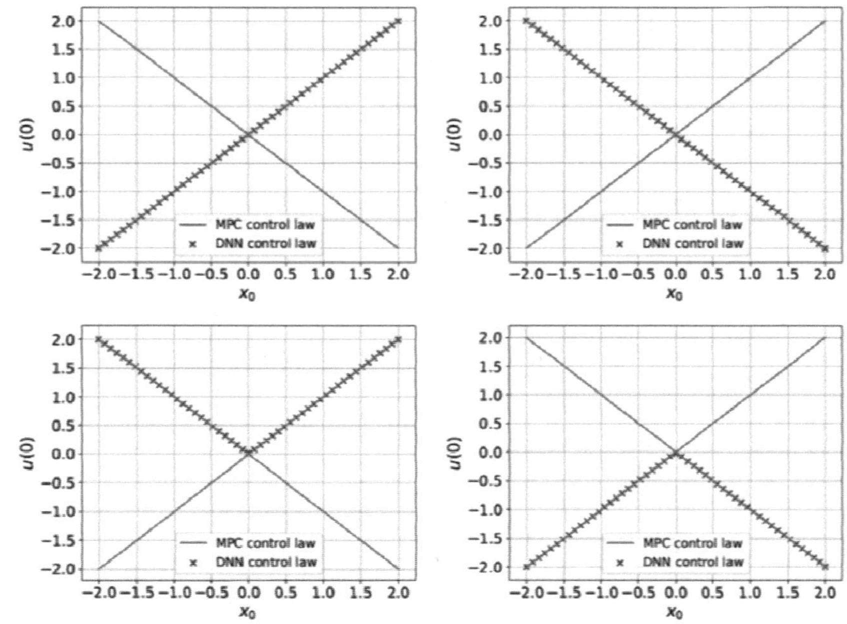

Fig. 2. DNN control law trained with "optimize and train" method

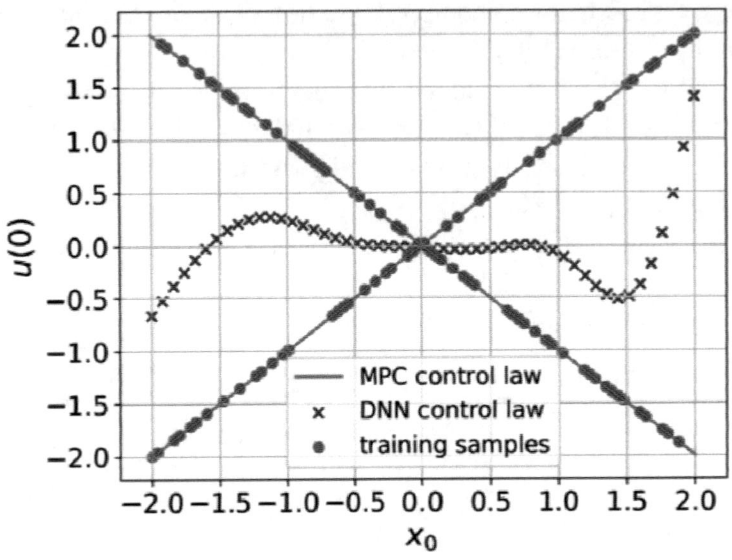

Fig. 1. DNN control law trained with "optimize then train" method

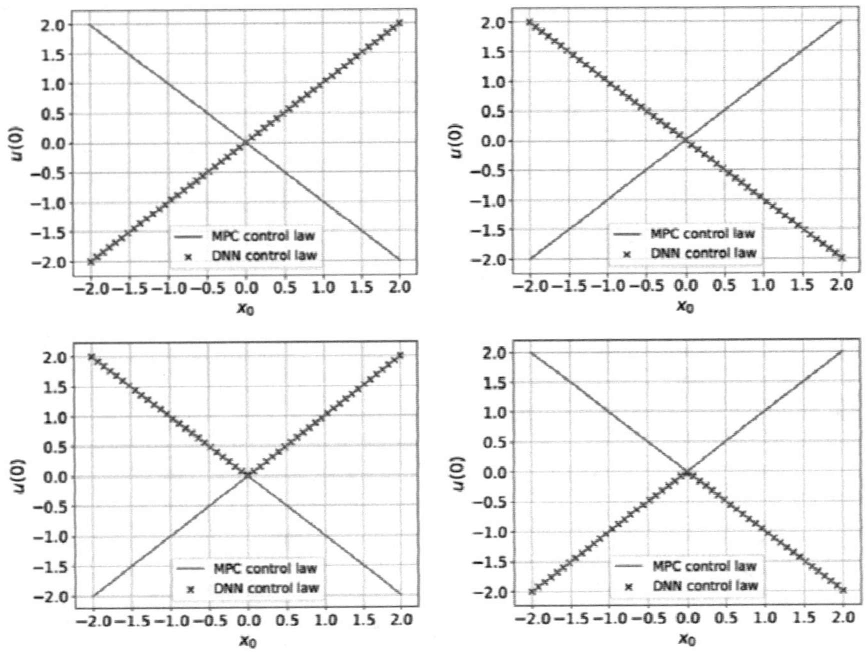

Fig. 2. DNN control law trained with "optimize and train" method

601990 Cloud-Based Control of a Robotic Manufacturing Process for Personalized Medicines

Abstract:

This abstract focuses on an automated production process enabling the manufacturing of personalized drug products. Particularly, we developed a system able to fulfill the needs of small-batch, high-mix, personalized, or even on-demand, production. Recently, the concept of precision medicine has received great attention. Finding ways to manufacture personalized therapies, in a consistent and controlled way is becoming the holy grail in the pharmaceutical field. However, existing manufacturing processes are not suitable for this mass-customization approach. In fact, while traditional systems can repeatably manufacture hundreds of thousands of identical dosage forms, having being designed for one specific product at a time, personalized therapies (*i.e.* extemporaneous formulations) are produced by technicians on a small scale only. Additionally, most of these production methods require labor to monitor, regulate and complete material handling for the equipment in use.

Based on these considerations, the availability of a fully automated architecture (*i.e.* hardware and software) capable of manufacturing personalized pharmaceutical products at scale would be of utmost importance. In this respect, we have developed a modular manufacturing system and a supporting cloud-based software architecture for its control. The modular manufacturing system, referred to as "robotic cluster", consists of *i)* different manufacturing modules, each of them performing specific fabrication steps, and *ii)* a transfer system that moves work-in-progress units (WIPUs) from module to module in a distributed and parallelized way. During the manufacturing process, WIPUs are shuttled into a given module, which performs a specific step of the process. Then, the WIPUs are moved into another module, and this process is repeated for all the following operations, including quality control. The sequence of manufacturing operations for a small batch of WIPUs is performed by moving it through the modules until all needed operations are completed. This sequence can be performed in a fully-planned, fixed-scheduled, a flexible-scheduled, or dynamic manner.

Each manufacturing module is connected to a dedicated control system and to other equipment *via* its control box, enabling the input of commands for the next fabrication steps while transmitting out the data pertaining to the work that has already been performed. All critical process parameters are monitored locally and in the cloud. The strategy pursued here not only ensures the security of the manufacturing system, but also increases its versatility and flexibility. It also makes this architecture suitable for scaling, by deploying many robotic clusters at the same time. Moreover, unlike flexible manufacturing systems described in literature, ours is comprised of a set of swappable units that are readily changed, updated, repaired and maintained. This way, complex operations are divided into simpler steps and for each of them there is a corresponding custom-built module. Therefore, the principal application of this technology is in mass customization, where products can be fully personalized to each order and large quantities of products can be fabricated in parallel to achieve the same scale and throughput of traditional serial manufacturing.

As a proof of concept, we evaluated the suitability of the discussed architecture for the manufacturing of multi-compartment capsules for personalized therapy. The capsules here proposed have a modular structure that consists of multiple inner compartments, which are independent of one another. Each compartment can be filled with different active ingredients and with a range of doses, as well as formulations, for any given drug. Multi-compartment capsular devices are especially advantageous for therapies involving multiple drugs, because they simplify the dosing schedule and have a positive potential impact on overall patient compliance.

Personalized filling of each compartment of the capsule shells was performed thanks to a custom-built robotic cluster, composed of a set of automated filling modules working simultaneously, in parallel. A robotic arm (designed to safely collaborate with human operators) moves the capsular devices form one filling station to the next, and quality control machines check the deposition of the desired amount of active ingredients after each step.

Within this proof of concept, we performed an optimization on the manufacturing steps as a fixed-scheduled sequence. We created a list of variable orders *i.e.* virtual patient prescriptions of varied precise dosages of multiple drugs, and input them into a schedule optimization algorithm. Said algorithm outputs the sequence of robot arm and module actions which maximizes throughput of the cluster. Generally, this algorithm maximizes the amount of work being done in parallel at once (many batches worked on in separate modules), and schedules the order in which the cluster outputs patient-scale batches to minimize the time taken to complete all batches. To ensure quality control we bound the time between manufacturing step and quality control: in-process controls should be measured more-or-less contemporaneously with the action. No WIPU can wait for longer than 10 min - say, sitting in a buffer or idle in a module - before being processed by the corresponding quality control module. For example, we bound the time between a filling module dosing powdered formulations and a weighing module recording the dosed weight. This scheduling optimization algorithm enables the parallelism of the robotic cluster. It is through this novel planning algorithm that we achieve the same scale and throughput of traditional serial manufacturing.

The automated deposition stage allows the rapid and accurate adjustment of the combination and dosage strength of the active ingredients. Cross-contamination is avoided by enclosing each filling station in a dedicated clean room. The capsule filling systems were validated with preliminarily filling trials, which used powders with different particle size distribution and bulk density. We considered 25 batches of 50 capsule each for each powder formulation. In all cases, the amount of active ingredient deposited by the robots had an average coefficient of weight variation (*i.e.* the ratio between the standard deviation and the average value of the weight of the deposited powder) below 3%. These results demonstrate the accuracy and the flexibility of the powder deposition systems. Additionally, we use spectroscopic techniques to check the weight uniformity and identity of the pharmaceutical formulations. Every measurement is collected and saved in the cloud based quality control system, allowing the creation, in real time, of a digital master batch record with all the information relevant to each manufacturing step (*e.g.* time, date, type of machine, weight of drug deposited), which could be accessed only by verified personnel and regulatory entities.

611934 Emerging Cognitive Engineering Approaches to Enhance Control Room Operator Capabilities in Chemical Industries

Abstract:

Control room operator's actions can have significant effect on the throughput, quality and safety of process plants (Baron, 1998). Typically, control room operators monitor the process via a Human Machine Interface (HMI) on which all the necessary information about the process state is presented. The complexity of operating modern chemical plants have increased significantly due to tight mass and energy integrations, deployment of sophisticated control automation strategies, and the need to operate the process in agile fashion in order to capture market and supply chain dynamics. However, there has been no concomitant increase in operations personnel's cognitive ability. Any mismatch between the operators' capability and the process control demands leads to high cognitive workload in human operators, often a precursor for poor performance which in-turn leads to sub-optimal operator actions or at worst accidents. This paper addresses the critical need to understand and enhance control room operator's cognitive capabilities.

Traditional operator performance assessment methods ignore cognitive aspects of performance. Operator's performance is assessed primarily in terms of number of successes and failures, response times, deviation of operator actions from predefined sequence (Lee et al., 2000), etc. However, in addition to overt performance (as done traditionally) it is imperative to consider covert aspects of performance i.e., cognitive workload for a holistic assessment of human performance. Various emerging cognitive engineering based sensors such as eye gaze tracking, Electroencephalography can reveal insights into the operators' cognitive state that determines these covert aspects. Specifically, in this paper we propose an Electroencephalography (EEG) based methodology to assess the operators' cognitive workload, especially during abnormal situation tasks where their active involvement is essential to safe process operations. The potential of EEG to understand cognitive workload for dynamic tasks which evolve with time has received some attention in aviation (Liang et al., 2018) and driving (Morales et al., (2017). However, there are hardly any studies relating to chemical control room operators.

We propose a single dry electrode EEG based methodology to dynamically assess the cognitive workload of operators while they perform various actions to control the process during abnormalities. The proposed methodology is illustrated on a simulated ethanol production plant testbed. A set of 10 different operators were considered in this work. Each operator's performance during the course of experiments involving monitoring and control of the process was studied. In addition to the process data and operator actions, we recorded the EEG data of the operators. Each experiment involved 6 tasks (comprising 5 different scenarios, out of which was repeated) of different abnormal situation. Each operator repeated the same experiment 7-8 times. Overall, 73 experiments were conducted by the 10 operators for a total of 438 tasks. Due to data quality issues, we evaluate results for 8 participants only.

We analyze the EEG signal in frequency domain by applying Fourier transform. Our results reveal that the power spectral density (PSD) of brain waves (4-30 Hz) is a biomarker that can identify the workload of operators. The PSD data of each operator was partitioned into two clustered using k-means clustering. Cluster 1 represents low power spectral density with low magnitude and Cluster 2 showed high power spectra density. We studied the workload variation with the help of cluster fraction over the course of the task as the operators' engagement with the process increased as well as with repetition. To evaluate the cluster variation within each task, a task is divided in three phases. In Phase 0 the operator monitors the process and ensures all the process variables are within the desired limits. During an abnormal condition an alarm is triggered, which leads to Phase 1 where the operator uses his mental model of the process to decide the control action(s) in order to clear the alarm. The portion of the task corresponding to all the control actions taken by the operator is called as Phase2. We observe distinct differences in the cluster fractions across the 3 phases. For a typical operator who successfully completes a task, during the monitoring Phase 0 we found the fraction of Cluster 1 to dominate (about 60%) indicating a correspondence between low workload during monitoring. During abnormality when alarms are triggered, in phase 1 the Cluster 1 fraction decreases (to about 40%) indicating that the operator uses his mental resources to decide the control action, i.e., increased workload. Further in phase 2, as the operator takes various control actions the Cluster 1 fraction further drops significantly (to about 0.20%) until the process becomes normal at which juncture it increases back to the 60% observed in Phase 0. The above patterns of change in the cluster pattern was observed for 81% (243 out of 300) of the tasks conducted by the 8 participants. This brings forth the capacity of the high accuracy of the EEG based cognitive sensor and the proposed machine learning algorithm to accurately estimate the workload of the control operator during normal and abnormal conditions. These distinct patterns were also highly correlated various process metrics (extent of variable deviation beyond alarm limits, time taken to clear alarm, number of standing alarms etc). In this paper, we will describe the proposed experimental and computational methodology and reported detailed results from our human subject experiments.

References:

Baron, RG. (1988). Human Factors in the Process Industries. Human Factors and Decision Making: Their Influence on Safety and Reliability, Symposium for the Safety and Reliability Society. ed. Sayers, BA. (pp.1-9)

Heard, J., Harriott, C. E., & Adams, J. A. (2018). A survey of workload assessment algorithms. IEEE Transactions on Human-Machine Systems, (99), 1-18.

Srinivasan, R., Srinivasan, B., Iqbal, M. U., Nemet, A., & Kravanja, Z. (2019). Recent developments towards enhancing process safety: Inherent safety and cognitive engineering. Computers & Chemical Engineering, 128, 364-383.

Neurosky. (2015). Greek Alphabet Soup – Making Sense of EEG Bands (Last accessed: Nov 10, 2018) http://neurosky.com/2015/05/greek-alphabet-soup-making-sense-of-eeg-bands/

Pfurtscheller, G., Stancak Jr, A., & Neuper, C. (1996). Event-related synchronization (ERS) in the α band—an electrophysiological correlate of cortical idling: a review. International journal of psychophysiology, 24(1-2), 39-46

Liang, Yongqiang, Wei Liang, Jue Qu, and Jie Yang. "Experimental Study on EEG with Different Cognitive Load." In 2018 IEEE International Conference on Systems, Man, and Cybernetics (SMC), pp. 4351-4356. IEEE, 2018.

Morales, J. M., Diaz-Piedra, C., Rieiro, H., Roca-González, J., Romero, S., Catena, A., Fuentes, L.J. & Di Stasi, L. L. (2017). Monitoring driver fatigue using a single-channel electroencephalographic device: A validation study by gaze-based, driving performance, and subjective data. Accident Analysis & Prevention, 109, 62-69.

Sharma, C., Bhavsar, P., Srinivasan, B., & Srinivasan, R. (2016). Eye gaze movement studies of control room operators: A novel approach to improve process safety. Computers & Chemical Engineering, 85, 43-57.

Bhavsar, P., Srinivasan, B., & Srinivasan, R. (2017). Quantifying situation awareness of control room operators using eye-gaze behavior. Computers & Chemical Engineering, 106,191-201.

Iqbal, M. U., Srinivasan, B., & Srinivasan, R. (2019). Electroencephalogram based Biomarkers for Tracking the Cognitive Workload of Operators in Process Industries. In Computer Aided Chemical Engineering (Vol. 46). https://doi.org/10.1016/B978-0-12-818634-3.50233-2

598577 Metabolic Modeling of Ischemic Livers Using Nash Equilibrium

Abstract:

End-stage liver disease claims up to 60,000 lives annually in the US [1]. Many patients become too ill to tolerate liver transplantation, and even if transplantation were indicated, there is a severe shortage of viable donor organs and only 28% of the wait-listed patients receive transplants [2]. These numbers could be reduced dramatically should the donor organ pool be expanded by rendering marginal cases, such as ischemic livers from Donors obtained after Cardiac Death (DCD), transplantable. It is estimated that about 6,000 livers/yr are only marginally damaged by ischemia and could be resuscitated for transplantation [3]. There is evidence from our lab and others that Normothermic Extracorporeal Liver Perfusion (NELP) is a very promising approach for recovering marginal organs that would be otherwise rejected from the donor pool.

In this talk, a mathematical model for liver metabolism is presented in order to study the behavior of cold and warm ischemic livers. The model attributes include (1) the presence of co-factors, (2) electrolyte solution behavior, (3) charge balancing, (4) feedback, allosteric, and other forms of inhibition, (5) explicit enzyme-substrate reactions, (6) up/down regulation of enzymes, (7) effects of mutations and/or re-engineered enzymes, (8) up-scaled genetic information, and in many cases, (9) compartmentalization and transport between cell compartments. The resulting metabolic network is treated as a Nash Equilibrium, in which individual pathways are iteratively solved to quasi-steady-state chemical equilibrium until all feedback and transport variables are converged. Fluxes, concentrations, amounts generated/consumed, pH, energy charge, and other metrics are computed for all metabolites/co-factors, and enzymes in the model. In particular, energy charge and ATP production/consumption are used to measure the viability of the ischemic liver model at various temperatures and experimental data is used to validate the model.

1) Asrani, S.K., et al., *Underestimation of liver-related mortality in the United States.* Gastroenterology, 2013. 145(2): p. 375-82 e1-2.

2) Organ Procurement and Transplantation Network Accessed 15 January 2016; optn.transplant.hrsa.gov]. Available from: optn.transplant.hrsa.gov.

3) Abt, P.L., et al., *Survival following liver transplantation from non-heart-beating donors.* Annals of Surgery, 2004. 239(1): p. 87-92.

599392 Modeling of Copy Number Variability in *Pichia Pastoris*

Abstract:

Development of continuous biopharmaceutical manufacturing processes is an area of active research. This work considers the long-term transgene copy number stability of *Pichia pastoris* in continuous bioreactors. We propose a model of copy number loss that quantifies population heterogeneity. An analytical solution is derived and compared with existing experimental data. The model is then used to provide guidance for stable operating timescales. The model is extended to consider copy number dependent growth such as in the case of Zeocin supplementation. The model is also extended to analyze a continuous seeding strategy. This work is a critical step towards understanding the impact of continuous processing on the stability of *Pichia pastoris* and the resultant products, and, to the authors' knowledge, is the first model of the stability of genomically integrated recombinant products.

603885 Coupling the Mechanisms of Diabetic Kidney Disease By Modeling the Tissues of the Glomerulus

Abstract:

Diabetic kidney disease (DKD) is a complication of diabetes that results in irreversible damage to the kidneys. The primary method of detecting DKD is by detecting albumin in a patient's urine, a phenomenon known as albuminuria. The issue with this method of detection is that albuminuria occurs when damage to the kidney is already done. Improved methods of detection require a better understanding of the underlying mechanisms of DKD. We propose that building a model that captures the known individual mechanisms of DKD and coupling them together may give the insight to build a better method of detecting DKD.

In the literature, multiple physiological changes in the glomerulus are observed in DKD patients. The glomeruli are structures that perform the initial filtration step in producing urine, and the point of failure when proteinuria occurs. In this work, the primary changes of concern are the thickening of the glomerular basement membrane, swelling of the mesangium, and podocyte effacement. Evidence suggest that the renin-angiotensin system (RAS) and TGF-β are connected to the physiological changes associated with DKD. Diabetes leads to systemically elevated levels of glucose in the blood. This in turn leads to dysregulation of the RAS and, via the RAS, upregulation of TGF-β in the podocytes. Evidence further suggests that damage in the mesangium may be linked to upregulated TGF-β in podocytes.

Understanding the crosstalk between cells and the associated tissue damage is instrumental in finding new ways to detect DKD before damage to the kidneys occur. In order to better understand the crosstalk occurring between cells of the glomerulus, we are creating a model that ties together dysregulation within podocytes and mesangial cells by modeling a portion of the cross section of the mesangium. Capturing the effect of tissue structure requires spatial components in the model. To capture how the signaling proteins move through space requires multiple additional phenomena beyond biochemical networks. Several model components were developed to capture spatial effects including the structure of the tissue, fluid flow, and the movement of signaling proteins and glucose through the tissue. The structure of the tissue is modeled with Compucell3D, which uses a cellular Potts model to define how cells interact. The fluid flow is modeled using a CFD simulation in Star-CCM+. The movement of signaling proteins and glucose is modeled using a system of PDEs to solve the diffusion and convection of solutes. Biochemical reaction network models that capture the RAS in both types of cells and the regulation of TGF- β and matrix metalloproteinase-1 in podocytes are included in the simulation. The RAS model predicts the concentrations of angiotensin-I, angiotensin-II, angiotensinogen, and renin in the cells. These models are coupled to the system of PDEs as point sources within the simulation. With these phenomena coupled together, we can make predictions about the importance of cellular crosstalk in the progression of DKD. With a better understanding of the progression of DKD, we can progress towards a better method of detecting DKD.

608169 A Mathematical Model to Study the Effects of Gut Microbes on the Human Host's Energy Balance

Abstract:

The human gut microbiome influences many facets of human health, including weight changes and metabolic disorders such as obesity. While effects of diet on the human gut microbiome are recognized, quantitative relationships among diet, the gut microbiome, and the host metabolism have not been established mechanistically. Many traditional human bioenergetics models link dietary energy intake to the host's energy expenditures, but the microbial metabolism is often neglected.

Here, we developed a mathematical model describing energy flow within the human body in three parts: 1. the upper gastrointestinal (GI) tract tracks energy absorbed via physiological processes; 2. the colon that hosts microbes and absorbs microbe-derived short-chain fatty acids (SCFAs); 3. the host's energy reservoirs that store carbohydrates, proteins, fats, and SCFAs absorbed by the GI tract and oxidize them to satisfy the body's energy-expenditure demands. We evaluated our model and previous models without microbial metabolism by examining the clinical data from Reinhardt et al. (2015) for obese participants on a calorically restrictive diet.

Our model showed increased accuracy when predicting body weight changes while dieting compared with metabolic models that did not include the gut microbiome reactions. Depending on the diet, the SCFAs from microbial production changed the amount of available energy from the ingested food by up to 11%. Diets with higher fiber and protein increased the fraction of available energy from food that the gut microbiome contributed. When in negative energy balance, participants had increased SCFA uptake from microbial processes resulting in an increase in total energy expenditure. Our mathematical modeling framework provides a way to describe how microorganisms mediates the effects of diet on the host's weight changes and energy expenditures. This could be used for the development of individual-specific probiotics/prebiotics, potentially helping with certain health issues associated with unhealthy gut microbiomes.

608630 Spatiotemporal Dopaminergic Modulation of Schaffer Collateral-CA1 Plasticity: A Computational Modeling Approach

Abstract:

Dopamine acts as a critical neuromodulator involved in significantly modulating the hippocampal Schaffer collateral CA1 pyramidal (SC-CA1Py) synaptic plasticity in a dose-dependent manner and thus plays an essential role in shaping hippocampal-dependent learning and memory by regulating the strength of this synapse. Experimentally, it has been shown in a few limited experimental studies that the timing of the activation of the dopamine receptors (D1/D5R) by the pharmacological D1/D5R agonist SKF 38393 relative to high-frequency-stimulation / low-frequency-stimulation (HFS/LFS) of the Schaffer collateral synapse leads to different modulations of the induced long-term-potentiation/long-term-depression (LTP/LTD) of the SC-CA1Py synapse. Although these limited experiments provide insight into few extreme timings of D1/D5R activation relative to the HFS/LFS stimulation, it is not clear how much the LTP/LTD modulation will occur if one activates the D1/D5 receptors at times other than these extreme timings, such as 30 minutes before and after the HFS/LFS stimulation or at different D1/D5 agonist concentrations. Here, we attempt to provide essential insights to fill in the gaps of our current understanding on the D1/D5 receptors mediated modulation of SC-CA1py synaptic long-term potentiation/depression (LTP/LTD) by developing a computational modeling framework, for the first time, that predicts the D1/D5 receptor-mediated spatiotemporal modulation of the HFS/LFS induced LTP/LTD at the electrophysiological level.

We developed a set of phenomenological models to describe the temporal dose-dependent effect of dopamine D1/D5 receptors agonist on the HFS/LFS induced LTP/LTD of the SC-CA1Py synapse using published electrophysiological data from hippocampal CA1 slice experiments. We used an experimentally validated single compartmental biophysiological model in a Hodgkin-Huxley formalism to represent the CA1Py neuron dynamics. The synaptic dynamics, as well as the dynamics of high/low-frequency stimulation (HFS/LFS), induced LTP/LTD of SC-CA1Py synapse were described using published phenomenological models of SC-CA1Py synaptic dynamics and LTP/LTD. We used an approximate Bayesian computation method based on sequential Monte Carlo scheme to estimate the unknown model parameters using the available electrophysiological data from the literature.

We used our model to predict experimentally observed temporal effects of D1/D5R activation by SKF 38393 with different timing relative to HFS/LFS and SKF 38393 concentrations. The predictions from our model show a good agreement, both qualitatively and quantitatively, with the experimental results published in the literature. Based on the results from our model, we conclude that the temporal modulation of the HFS/LFS induced LTP/LTD of the SC-CA1Py synapse by SKF 38393 strongly depends on both the timing of the administration of SKF 38393 relative to the HFS/LFS as well as the concentration SKF 38393. Furthermore, the temporal modulation of the HFS/LFS induced LTP/LTD of the SC-CA1Py synapse by SKF 38393 is independent of the HFS/LFS protocol.

610829 Uncovering Hidden and Observable Features That Contribute to Tumor Aggression and Progression

Abstract:

Tumor aggression is a consequence of subcellular dysregulation, cell decision processes, and the microenvironment. These features collectively drive many hallmarks of cancer, including cell lifespan, enhanced proliferative capacity, increased growth rates, and ability to invade healthy tissue. Additionally, genetic heterogeneity within this dynamic microenvironment enables tumors to build resistance to therapeutic interventions. Treatment mechanisms exploit various microscopic features, such as nutrient uptake rates, cell division rates, and the extent of cell cycle synchronization throughout the tumor. Some of these features can be measured through imaging techniques (e.g., PET scans to understand metabolic demands within a tumor), while others cannot (e.g., heterogeneity between tumor cells and number of subpopulations). We bridge this gap of observables using a computational model--specifically an agent-based model--to understand how cell-level and population-level features contribute to tumor aggression and progression.

Agent-based models (ABMs) are a "bottom-up" computational framework that predict emergent cell population dynamics from individual cell agent rules. We use this framework to simulate tumor microenvironments by integrating heterogeneous tumor cell agents with healthy cell agents in a dynamic spatiotemporal environment that comprises dynamic vasculature. The ABM provides accurate, high-resolution data describing the behavior of individual cells at every timestep that cannot be captured *in vitro* or *in vivo [Yu et al., Frontiers 2020].* We use the established model to generate simulations that vary cell agent characteristics and their decision rule parameters, providing a landscape of data describing emergent tumor cell population dynamics. Through use of various machine learning algorithms, we uncover key intervention strategies--both at the individual cell and population levels--that contribute to tumor progression and aggression. Our work will inform clinically relevant design features that could prove as effective targets for treatments that minimize tumor aggression and progression.

609400 Introducing an Optimization- and Explicit Runge-Kutta- Based Approach to Perform Dynamic Flux Balance Analysis

Abstract:

Genome-scale Models (GSMs) of metabolism have become important tools for the *silico* study and design of metabolism *in silico*. As GSMs are under-defined systems of equations, optimization-based tools are required for their analysis, the most common of which is Flux Balance Analysis (FBA). A perturbation of FBA, dynamic FBA (dFBA), is a tool which allows for the study of a modeled system across time, including biomass and metabolite concentration. Introduced here is a generalized **O**ptimization- and explicit **R**unge-**K**utta-based **A**pproach (**ORKA**) to perform dFBA, which is more accurate and computationally tractable than existing approaches, namely the Static Optimization Approach (SOA) and the Dynamic Optimization Approach (DOA). ORKA is applied to a four-tissue (leaf, root, seed, and stem) model of *Arabidopsis thaliana*, p-ath773, which uniquely captures the core-metabolism of several stages of growth from seedling to senescence while strongly emphasizing plant-scale behavioral agreement between *in silico* results and *in vivo* data. Using ORKA, p-ath773 takes metabolic "snapshots" at hourly intervals throughout the lifecycle of an individual plant. It next shows the transition of metabolism and whole-plant growth, such as the evolution of sulfur metabolism and the diurnal flow of water throughout the plant. Specifically, p-ath773 shows how transpiration drives water flow through the plant and how water produced by leaf tissue metabolism may contribute significantly to transpired water. Investigation of sulfur metabolism shows frequent cross-compartment exchange of a standing pool of amino acids which is used to regulate proton flow. In addition, p-ath773 has shown broad agreement with published plant-scale properties such as mass, maintenance, and senescence. Overall, p-ath773 serves as a scaffold for lifecycle models of other plants to further increase the range of hypotheses which can be investigated *in silico*.

W. L. Schroeder and R. Saha. "Introducing and Optimization- and Explicit Runge-Kutta- based Approach to Perform Dynamic Flux Balance Analysis". *BioRxiv*, Apr. 29, 2020. Available: https://www.biorxiv.org/content/10.1101/761189v2 (doi: https://doi.org/10.1101/761189)

617388 Food, Pharmaceutical and Bioengineering Division Award in Chemical Engineering: Ionic Liquids for Therapeutic Applications

Abstract:

Ionic liquids, the liquid salts comprising organic anions and cations, offer exciting opportunities for several therapeutic applications. Their tunable properties offer an exquisite control over their design features and functions. Starting with biocompatible ions, we synthesized a library of ionic liquids and explored them for various drug delivery applications. Ionic liquids provided unique functionalities including overcoming the biological transport barriers of skin, buccal mucosa and the intestinal epithelium. At the same time, they also stabilized proteins and nucleic acids, and enabled their delivery across these barriers. Ionic liquids also provided unique biological functions including adjuvancy towards vaccines and antimicrobial activity against various pathogens. I will present an overview of the design features of ionic liquids and novel biomedical applications enabled by these unique materials.

617389 Food, Pharmaceutical and Bioengineering Division Early Career Award: Engineering Synthetic Microbial Consortia Inspired by the Herbivore Rumen

Abstract:

Anaerobic microbes work together in complex communities that decompose and recycle carbon biomass throughout the Earth – from our guts to landfills and compost piles. Despite their importance, little information exists to parse the role of each microbial member within their dynamic community. To address these knowledge gaps, we pioneered new techniques to isolate anaerobes from biomass-rich environments (e.g. guts and fecal materials of herbivores), characterize their shared metabolism, and build synthetic microbiomes to drive biomass to renewable chemicals. Herbivore fecal samples were challenged by different types of biomass during cultivation to identify important microbial partnerships; 10 billion metagenomic reads spread across 402 enrichment samples tracked biological diversity as the cultures converged to a set of stable microorganisms. Nearly 200,000 carbohydrate active enzymes were identified from the fecal samples, and 724 genomes were assembled for previously uncultured microbes. Surprisingly, consortia dominated by anaerobic fungi generated more than twice the amount of methane compared to prokaryotic consortia, suggesting that fungi accelerate biomass breakdown and methane release in herbivores. Overall, our analysis points to natural compartmentalization between anaerobes as a means to degrade crude biomass, which can be exploited to harness nature's microbes for sustainable chemical production using synthetic systems.

614032 Reinventing Chemical Manufacturing Using Biotechnology: Manus Bio Approach

Abstract:

Chemical manufacturing is undergoing a period of change, from an era focused on petrochemical-derived chemical synthesis or plant extraction to one which leverages engineered chemical biosynthesis in microbes for biomanufacturing. At Manus Bio, we have developed a robust technology platform from discovery to manufacturing which enables the economical and sustainable production of typically rare complex natural products. Our engineered microbes are capable of converting carbon feedstock to product at high yields and have been adapted to produce a mature pipeline of complex products.

617390 Area 15C Bioengineering Plenary Award: Adventures in the Engineering of Oxidoreductase Enzymes

Abstract:

Oxidoreductase enzymes catalyze a wide range of reactions where electrons are transferred from one molecule to another. These enzymes are involved in a wide range of critical metabolic functions and are also frequently employed in bioengineering applications such as biosensors, biocatalysis, and synthetic biology. In this talk I will highlight contributions from my lab on the engineering of cofactor and substrate specificity, the engineering of new electron transfer mechanisms, and the engineering of oxidoreductases for emerging bioengineering applications.

617391 Area 15D/E Life Sciences Plenary Award: Engineered systems for controlling cellular microenvironments: from designing synthetic extracellular matrices to probing cell responses in disease models

Abstract:

The properties of the microenvironment in which cells reside, from structure to mechanics and biochemical content, increasingly are recognized as important drivers of cell function and fate, including in the onset and progression of disease (e.g., fibrosis and cancer recurrence). Engineering materials to mimic key features of these complex microenvironments offers unique opportunities to probe and direct cellular functions and to study specific diseases in the lab. In this presentation, our recent efforts to design reductionist synthetic mimics of complex collagen-rich microenvironments will be overviewed. Specific applications of these and other engineered systems for the creation of multidimensional controlled cell culture models to study specific diseases will be discussed. Further, the opportunity that 'omics' tools provide for interrogation of cell responses within these engineered systems, from benchmarking versus in vivo and patient data to obtaining unique insights into cellular responses, also will be highlighted. This multipronged approach to understanding cell-microenvironment interactions is providing new tools and insights for addressing currently intractable diseases, including lung fibrosis and late cancer recurrence.

608813 Development of Bio-Based Processes: A Real Challenge

Abstract:

The development of bio-based processes is particularly complex for several reasons:

- *Thermodynamics*. For any process simulation, the thermodynamic model implementation is the keystone of a modeling close to reality. In the case of production of biosourced chemicals, many compounds are not present in the databases available in commercial process simulation software. It is therefore necessary to have, in these process development tools, predictive models that can work with little experimental information (group contribution models, QSPR methods, models based on quantum mechanics, etc.). The studied mixtures are also complex (including molecules with many functional groups, presence of oxygenated compounds...), the liquid phases are very often non-ideal, thus requiring specific models to represent the thermodynamic properties of the mixture. In order to tackle bio-based processes development, it is therefore necessary to have in the modeling tool a very complete thermodynamic calculation server, integrating the most recently developed models and offering the ability to easily implement users' models. This is the case of the *Simulis Thermodynamics* software component, integrated in all ProSim software.
- *Reactions*. As in all processes, reactions are a key step in bio-based processes, which are characterized by very complex reaction patterns, with kinetic models that are often different and more varied than in the industries more traditionally using process simulation. The *Simulis Reaction* software component integrates a very rich library of kinetic models, while allowing the user to simply add new models.
- *Process*. Biosourced processes often integrate unit operations operating in both batch and continuous modes. For batch operations, the technology used is very important because the dynamic of the system is directly linked to this. The user must therefore describe their equipment with great detail. Dedicated simulators are therefore very useful because they allow the user to easily provide, without any code development, the equipment characteristics as well as the recipe. *BatchReactor* and *BatchColumn* are two software packages dedicated respectively to the simulation of reactors and distillation columns operated in batch mode. These models are also particularly useful in the scale-up and scale-down phases of process development. They can be used in the *ProSimPlus* environment, which also offers a complete library of unit operations operating in steady state.

The various concepts presented above will be illustrated during the presentation with examples of practical applications: beer brewing, PHB production, spirits production unit, etc.

608066 A Systems Engineering Approach to Biomanufacturing for Long-Duration Missions on Mars

Abstract:

The Center for the Utilization of Biological Engineering in Space (CUBES) addresses the grand challenges of space systems biology by leveraging partnerships between industry, academia, and NASA to support biomanufacturing for deep space exploration. The scope of the center is to advance the practicality of an integrated, multi-function, multi-organism biomanufacturing system on a Mars mission and showcase a semi-autonomous biomanufacturing of materials, pharmaceuticals, and food under the specifications required for a human exploration mission on Mars [1]. Our pipeline begins with production of biological feedstocks through microbial processing of in situ resources to fix carbon and nitrogen from the atmosphere, decontaminate and enrich regolith, and to transform human/mission wastes to media and feedstocks for utilization by downstream processes. Feedstocks are then used to produce biopolymers and chemicals from media and feedstocks, to recycle products at end-of-life, and to use biopolymers suitable for 3D printing to grow the manufacturing facility infrastructure and provide tools and building materials on demand [2]. Specialized plant and microbial engineering methods are utilized to produce functional food and pharmaceuticals for astronauts along with the recycling of plant wastes. Systems engineering methods are employed to tightly integrate and automate our pipeline and to satisfactorily achieve performance per mission specifications.

In this talk, we will demonstrate the implications of our integrated systems approach in building a process simulation framework as a central feature in the biomanufacturing-driven Reference Mission Architecture. The units are independently studied through experiments and first principles-based models and are connected through input/output relations. More specifically, the first unit consists of a bioreactor for carbon fixation through microbial electrosynthesis (MES) in which acetate is a metabolic product [3]. The latter is fed into a biological nitrogen fixation reactor as well as into a biopolymer (PHB) production reactor, serving as the main carbon source in both cases. Downstream, the fixed nitrogen is utilized for crop cultivation and the produced PHB for 3D printing of necessary tools. The integration aspects as well as the uncertainty regarding system failures and smooth operation are considered within our framework. The production capability of this system is assessed and is driven by the needs of the crew members for survival.

Lastly, since most processes operate in a batch mode, a thorough discussion regarding an optimized scheduling plan under uncertainty is provided. This is a crucial part of the biomanufacturing system design as the objective of maintaining a high quality of life for crew members is of primary importance, given that resources are limited and safety of the operation is critical.

References

[1] A. A. Menezes, M. G. Montague, J. Cumbers, J. A. Hogan, and A. P. Arkin, “Grand challenges in space synthetic biology,” J. R. Soc. Interface, vol. 12, no. 113, p. 20150803, 2015.

[2] A. A. Menezes, J. Cumbers, J. A. Hogan, and A. P. Arkin, “Towards synthetic biological approaches to resource utilization on space missions,” J. R. Soc. Interface, vol. 12, no. 102, p. 20140715, 2015.

[3] M. Kazemi, D. Biria, and H. Rismani-Yazdi, “Modelling bio-electrosynthesis in a reverse microbial fuel cell to produce acetate from CO2 and H2O,” Phys. Chem. Chem. Phys., vol. 17, no. 19, pp. 12561–12574, 2015.

600996 D-Lactic Acid Fermentation and Enzymatic Activities of a Novel Bacterium *Terrilactibacillus Laevilacticus* SK5-6

Abstract:

In this study, *Terrilactibacillus laevilacticus* SK5-6, a novel D-lactate producer, was proven to exhibit a good fermentation performance comparable to the reference D-lactate producer *Sporolactobacillus* sp. A catalase positive *T. laevilacticus*SK5-6 produced D-lactate at higher yield, productivity, and optical purity compared with *Sporolactobacillus laevolacticus*0361^{T}. Investigation on the glucose metabolism for D-lactate along with the activation of the key enzymes was conducted. The results suggested that D-lactate production and its optical purity were controlled by the key enzymes in glycolysis pathway by their allosteric mechanisms. The results suggested that rapid bioconversion of glucose to D-lactate by *T. laevilacticus* SK5-6 was responsible by the expression of these key enzymes during the fermentation stage. Among 5 key enzymes observed, phosphofructokinase (PFK) possessed the lowest specific activity; thus, the conversion of fructose-6-phosphate to fructose-1,6-phosphate (F-1,6-P) was considered the rate limiting step. Pyruvate kinase (PYK) was proven to involve in ATP regeneration; therefore, playing role in driving the glycolysis flux. Lactate dehydrogenase (LDH) was allosterically activated by F-1,6-P. With the highest specific activity of LDH followed by PYK and PFK, respectively, this simultaneously drove glucose flux toward lactate at a high rate. Compensation of yeast extract in the preculture medium by ammonium salts, e.g., NH_4Cl and $(NH_4)_2HPO_4$ was conducted. Complete glucose consumption was observed in the fermentation using NH_4Cl supplemented in preculture medium which yielded a high lactate titer of 102.5 g/L within 48 h with the production yield of 0.84 g/g and the optical purity of 99.74 % enantiomer excess.

605412 D-Lactic Acid Production By Indigenously Isolated Strains of *Terrilactibacillus Laevilacticus* from Soil Samples in Thailand

Abstract:

To date, the limitation of petroleum resources and the concerning consumption of oil-based plastics are being grown. Thus, the development of alternative bio-based raw materials is becoming interested. Lactic acid has long been recognized as a useful reagent that emerged in a wide range of industries. One of the most promising applications is a precursor for manufacturing polylactic acid (PLA) which can be biocompatible and biodegradable material. In order to synthesize PLA, the enantiomerically pure L- or D-lactic acid plays a crucial role to improve the stability, mechanical performance, and hydrolysis resistance of the polymer. Currently, lactic acid production is extensively commercialized via the biotechnological process to achieve high purity of lactic acid isomer. Industrially, L-lactic acid production is higher demand because of its applications in the food and pharmaceutical industries. In contrast, large scale production of D-lactic acid has not yet been commercialized; nevertheless, D-lactic acid production has been gained attention due to the demand for producing thermostable PLA. Therefore, it is interesting to develop a D-lactic acid production platform for economical and industrial aspects. Accordingly, a robust lactic acid-producing strain has been now considered as one of the most important criteria for large scale production of D-lactic acid. Wild-strain isolation strategy has long been employed to serve the industrial requirements, such as high yield, production rate, and optical purity. This study discovered two D-lactate producers from soil samples collected in Thailand by which they consequently identified as *Terrilactibacillus laevilacticus* SBK35-1 and *T. laevilacticus* SBK37-1, gram-staining positive, facultatively anaerobic, and endospore-forming rods with catalase-positive activity. In primary screening of lactic acid production, the results revealed that both strains were homofermentative D-lactic acid bacteria which produced lactic acid concentration of 86.90 g/L and 63.35 g/L, conversion yield of 75% and 60%, productivity of 1.81 g/L·h and 1.32 g/L·h for SBK35-1 and SBK37-1, respectively. Additionally, an optically pure D-isomer of lactic acid (100%) was obtained by these two strains when cultivated in glucose medium (120 g/L) under aerobic condition. Due to little information on this genus, the strain SBK35-1 could be considered as an alternative lactate producer for further study to achieve a suitable condition for effective D-lactic acid production by the genus *Terrilactibacillus* member.

604861 "Invited Talk" the Response of Food-Related Pathogens to Natural Antimicrobials Coupled with Non Thermal Novel Processing Technologies in Structured Food Model Systems

Abstract:

Introduction and Aim:

High quality foods retaining characteristics such as taste, texture, natural colouring and nutritional content, that have been minimally processed and/or are ready-to-eat (e.g. fresh smoothies, fruit juices, fresh fruit and vegetables, cheeses and dairy products), are in increasing demand.[1] Traditional microbial inactivation methods such as pasteurization or heat sterilization cannot meet these demands as the properties of the food are often changed, and some food products cannot be processed with these methods. As such, the food industry is increasingly interested in novel non-thermal processing technologies (e.g. ultrasound, cold pasteurisation, cold atmospheric plasma, natural antimicrobials), which are milder than traditional thermal methods and can maintain the desired fresh-like food characteristics.[2–4] Furthermore, processing efficiency may be increased by synergistically combining non-thermal inactivation methods, as the combined technologies can act as hurdles for microbial growth.[5,6] However, the efficacy and mechanisms of action of combined treatments such as natural antimicrobials and non-thermal processing (e.g. ultrasounds), remains unclear.[7] These technologies could instead represent a mild, sub-lethal stress which could result in stress adaptation, post-treatment survival, and the potential development of antimicrobial resistance.[6,8]

Most studies conducted thus far on the inactivation of food-related pathogens by natural antimicrobials and/or ultrasound are conducted in liquid broths, or in/on specific food products e.g. fruit juice, lettuce leaves [2,9]. However, the chemical composition and rheological/structural properties of foods can vary significantly, potentially impacting the efficacy of such treatments, and studies in real foods are informative only for the specific product studied.[10] Natural antimicrobials may be produced *in situ* by natural microflora (e.g. nisin-producing *Lactococcus lactis*) or added artificially, thus the production rate, quantity and efficacy of natural antimicrobials may also be affected. Furthermore, many factors are known to influence the efficacy of ultrasound treatment i.e. system frequency, viscosity, sonication time, and equipment set-up.[11,12]

A fundamental multi-frequency study into the inactivation effect of ultrasound, alone or combined with nisin, against food-related pathogens in food model systems of controlled rheological and structural complexity is currently lacking: the present work aims to address this gap.

Materials and Methods:

Food model systems were prepared using Tryptic Soy Broth supplemented with 0.6% Yeast Extract (TSBYE), with 0, 0.1, 0.3 or 0.5% Xanthan gum (XG) for a range of viscoelastic immersed systems.[3,4] XG is widely used in the food industry and is stable at a range of temperatures[13] thus was selected for use in the present study. The rheological stability and characteristics of all XG systems was tested by conducting dynamic oscillatory measurements to assess their viscoelastic behaviour.

Ultrasound inactivation of *L. innocua* or *E. coli* stationary phase cells, with/without nisin, was investigated in 0 – 0.5% XG systems (44 kHz, 500 kHz, 1 MHz) with a constant power of 30W and a treatment time of up to 30 min. Cells were grown planktonically in TSBYE and added at 10^6CFU/mL into the viscoelastic systems before inactivation treatments. For nisin-containing systems, nisin was added as a treatment for 30 minutes before or immediately after ultrasound treatment, at a final concentration of 35 IU/mL (a sublethal concentration which allows the observation of combined effects with ultrasound).

Inactivation kinetics were monitored, and morphological changes of cells following inactivation treatments were studied using scanning electron microscopy (SEM). Samples were fixed in a 3% formaldehyde solution and dehydrated in an ethanol/water series before imaging. Flow cytometry methods were also employed to identify the dead and sublethally injured populations using Propidium iodide (PI) and bis-(1,3-dibutylbarbituric acid) trimethaine oxanol (BOX) stains for dead and injured cells respectively.

Results and Discussion:

The inactivation efficacy of ultrasound and nisin, alone or combined, was observed to depend on many factors including themicrobial species, ultrasound frequency, treatment time, treatment order, and system structure.

No inactivation by ultrasound was observed for *Listeria* at any of the conditions tested, although an effect of nisin was observed (as expected, as nisin is known to act on gram positive bacteria). For *E. coli*, inactivation by ultrasound was observed at 500 kHz but not at the other frequencies investigated, with greater reduction in the surviving population as the treatment time was increased. As the concentration of XG was increased i.e. as the system viscosity increased, a reduction in inactivation efficacy was observed i.e. the system structure impacts significantly inactivation by ultrasound. Finally, a combined effect of nisin was observed for *E. coli* inactivation (despite nisin acting only on gram positive cells), but only when nisin was applied before ultrasound treatment.

SEM imaging and flow cytometry analysis of samples following ultrasound and/or nisin treatments show significant differences in cell death, injury and survival on a cellular level for different orders and combinations of treatments, as observed in the inactivation kinetics. Furthermore, ghost cells (i.e. dead cells with an intact cell membrane but no cellular contents) are observed following ultrasound treatment, suggesting that the mechanism of inactivation by ultrasound in this system is physical rather than chemical and involves the partial destruction or disturbance of the cell membrane, causing efflux of the cell contents.

Significance and Impact:

This work presents a systematic fundamental study on the efficacy of non thermal processing and natural antimicrobials as individual or combined treatments on the inactivation of *Listeria* or *E. coli* in/on food systems of different physiochemical and structural complexity. This study sheds light on the combined efficacy and potential mechanisms of novel processing techniques for the inactivation of food-related pathogensin structured food models, and highlights the importance of accounting for structural effects when designing novel inactivation processes for the food industry.

Acknowledgements:

This work was supported by the Department of Chemical and Process Engineering of the University of Surrey, as well as an Impact Acceleration Grant (IAA-KN9149C) of the University of Surrey, an IAA-EPSRC Grant (RN0281J), the National Biofilm Innovation Centre, R-Processing Ltd and the

Royal Society. E.V. is grateful to the Royal Academy of Engineering for an Industrial Fellowship.

References:

1. Troy DJ, Shikha K, Kerry JP, Tiwari BK. Sustainable and consumer-friendly emerging technologies for application within the meat industry : An overview. Meat Sci. 2016;120:2-9. doi:10.1016/j.meatsci.2016.04.002
2. Millan-Sango D, Garroni E, Farrugia C, Van Impe JFM, Valdramidis VP. Determination of the efficacy of ultrasound combined with essential oils on the decontamination of Salmonella inoculated lettuce leaves. LWT - Food Sci Technol. 2016;73:80-87. doi:http://dx.doi.org/10.1016/j.lwt.2016.05.039
3. Costello KM, Gutierrez-Merino J, Bussemaker M, Smet C, Van Impe JF, Velliou EG. A multi-scale analysis of the effect of complex viscoelastic models on Listeria dynamics and adaptation in co-culture systems. AIChE J. 2019;(July):e16761. doi:10.1002/aic.16761
4. Costello KM, Gutierrez-Merino J, Bussemaker M, Ramaioli M, Baka M, Van Impe JF, Velliou EG. Modelling the microbial dynamics and antimicrobial resistance development of Listeria in viscoelastic food model systems of various structural complexities. Int J Food Microbiol. 2018;286:15-30. doi:10.1016/j.ijfoodmicro.2018.07.011
5. Ross AIV, Griffiths MW, Mittal GS, Deeth HC. Combining nonthermal technologies to control foodborne microorganisms. Int J Food Microbiol. 2003;89:125-138. doi:10.1016/S0168-1605(03)00161-2
6. Barba FJ, Koubaa M, do Prado-Silva L, Orlien V, Sant'Ana A de S. Mild processing applied to the inactivation of the main foodborne bacterial pathogens: A review. Trends Food Sci Technol. 2017;66:20-35. doi:10.1016/j.tifs.2017.05.011
7. Zupanc M, Pandur Ž, Stepišnik Perdih T, Stopar D, Petkovšek M, Dular M. Effects of cavitation on different microorganisms: The current understanding of the mechanisms taking place behind the phenomenon. A review and proposals for further research. Ultrason Sonochem. 2019;57:147-165. doi:10.1016/j.ultsonch.2019.05.009
8. Yousef A, Juneja V. Microbial Stress Adaptation and Food Safety. 1st ed. Florida: CRC Press; 2003.
9. Liao H, Jiang L, Cheng Y, Liao X, Zhang R. Application of nisin-assisted thermosonication processing for preservation and quality retention of fresh apple juice. Ultrason Sonochem. 2018;42:244-249. doi:10.1016/j.ultsonch.2017.11.020
10. Smet C, Baka M, Dickenson A, Walsh JL, Valdramidis VP, Van Impe JF. Antimicrobial efficacy of cold atmospheric plasma for different intrinsic and extrinsic parameters. Plasma Process Polym. 2018;15(2):1-12. doi:10.1002/ppap.201700048
11. Gonzalez V, Wood R, Lee J, Taylor S, Bussemaker MJ. Ultrasound-enhanced hair dye application for natural dyeing formulations. Ultrason Sonochem. 2019;52(November 2018):294-304. doi:10.1016/j.ultsonch.2018.11.028
12. Wood RJ, Lee J, Bussemaker MJ. A parametric review of sonochemistry: Control and augmentation of sonochemical activity in aqueous solutions. Ultrason Sonochem. 2017;38:351-370. doi:10.1016/j.ultsonch.2017.03.030

13. Kang KS, Pettitt DJ. Xanthan, gellan, welan and rhamsan. In: Whistley RL, BeMiller JN, eds. Industrial Gums: Polysaccharides and Their Derivatives. 3rd ed. New York: Academic Press; 1993:341-397

594804 Solvent Tolerance of Freeze-Dried Cfps Reactions Enables Materials Applications (Industry Candidate)

Abstract:

Cell-free systems have growing importance as a way to power synthetic biology tools in the field. Lysates may be dried for storage, delivering biochemical activity for sensing or producing molecules on-demand upon rehydration at the point of need. Up to now, cell-free protein synthesis (CFPS) reactions have been studied as aqueous solutions in test tubes or absorbed onto paper or cloth. Embedding biological functionality into broadly-used materials, such as plastic polymers, has long been an attractive goal. Unfortunately, this goal has for the most part remained out of reach, often due to the fragility of biological systems outside of aqueous environments. In this work, we describe and utilize a surprising and useful feature of lyophilized cell-free lysate systems: tolerance to anhydrous organic solvents. To explore this newly discovered feature, a variety of solvents were tested and CFPS reaction components were screened for protein-protective properties. Tolerance to suspension in organic solvents may facilitate the use of polymers to deliver dry cell-free reactions in the form of coatings or fibers, among other processing possibilities. In pursuit of this application, preliminary casting into PLGA polymer is demonstrated as a proof of concept.

607448 Holistic Engineering of Cell-Free Systems through Proteome-Reprogramming Synthetic Circuits

Abstract:

Synthetic biology has focused on engineering genetic modules that operate orthogonally from the host cells. However, the use of orthogonal genetic modules often faces the challenges of varying cellular context, such as growth rate, crosstalk, and noise. A synthetic biological module can instead be designed to reprogram the host proteome, which in turn enhances the function of the synthetic module. Here, we apply this holistic synthetic biology concept by exploiting the crosstalk between metabolic networks in cells, leading to a protein environment more favorable for protein synthesis (Fig. A). Specifically, we show that a local module expressing translation machinery can reprogram the bacterial proteome, changing the expression levels of more than 780 proteins (Fig. B). The integration of the proteins synthesized by the local modules and the reprogrammed proteome generate a cell-free system that can synthesize fluorescent reporters, protein nanocages (Fig. C), and the gene-editing nuclease Cas9 (Fig. D) in different reaction formats, with up to 5-fold higher expression level than classical cell-free systems. Our work demonstrates a holistic approach that integrates synthetic and systems biology concepts. This approach has the potential to achieve outcomes not possible by only local, orthogonal circuits.

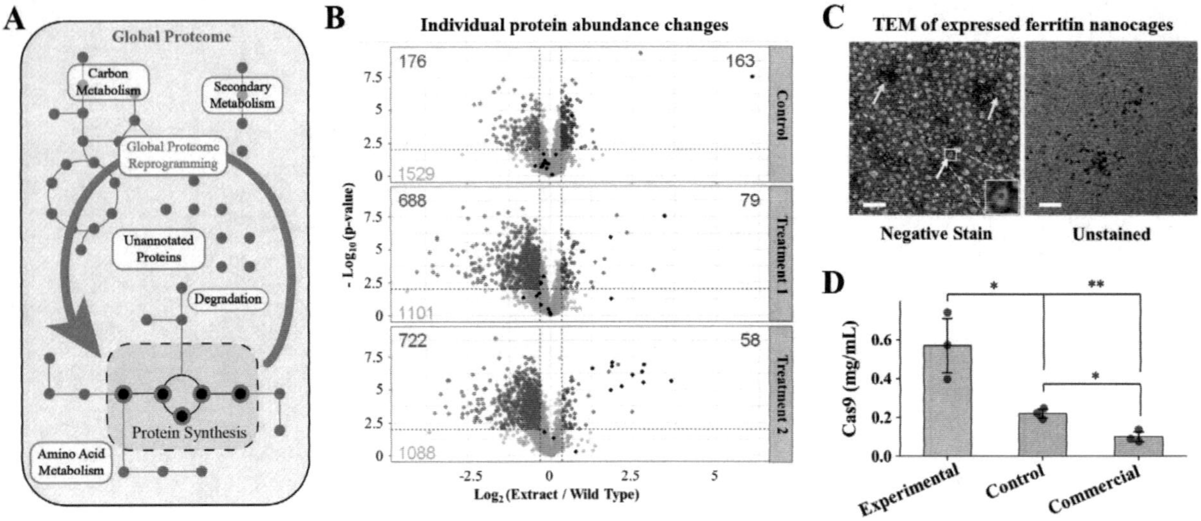

610517 Engineering Yeast for the Manufacturing of Human Proteins (Faculty Candidate)

Abstract:

Microorganisms have tremendous potential for the manufacturing of proteins, boasting fast growth to high cell densities from cheap feedstocks. While microbes are routinely used for the production of industrial enzymes and proteins at lab scale, they are seldom used for the production of therapeutic proteins. Products destined for the clinic have stringent quality requirements and often specific post-translational modifications. Eukaryotic microorganisms like yeast and filamentous fungi are strong candidates as hosts because they can secrete the protein of interest into the culture medium, and are capable of post-translational modifications like glycosylation that are common in higher eukaryotes. Despite lower specific (per cell) productivities, these hosts yield volumetric productivities similar to traditional manufacturing hosts like CHO cells (~10 mg/L/day). Engineering is required, however, to produce high quality therapeutics suitable for the clinic, like the engineering of *Komagataella phaffii* (*Pichia pastoris*) for humanized glycosylation. These engineering efforts may consist of many genetic changes to host cells, with severe impact on growth and cell viability, subsequently reducing volumetric productivity. What is needed are methods to improve the titer and quality of complex proteins without sacrificing cell viability and volumetric productivity. The cost of both manufacturing and development of protein therapeutics could be reduced with microbial hosts by accelerating development timelines and reducing the footprint and length of manufacturing campaigns.

We demonstrate three methods to improve the utility of genetic engineering in manufacturing strains of *K. phaffii*. First, we developed a markerless genome editing tool, eliminating energetically expensive resistance or auxotrophic markers. Second, we applied our engineering tool to humanize the glycosylation in *K. phaffii* without sacrificing cell viability. We utilize native promoters to precisely control the expression of endogenous and exogenous glycosyltransferases and glycosidases, balancing glycan homogeneity with cell viability. Third, we improve the quality and titer of monoclonal antibody production by introducing human chaperones to assist with protein folding and secretion. Chaperones are essential for antibody production in human cells, but themselves are large, complex molecules that require machinery for efficient folding. We demonstrate that despite adding to the recombinant protein load, chaperones work together with glyco-engineering to enable the production of high quality monoclonal antibodies at industrially relevant titers.

601205 Preservation of Quaternary Structure in Thermostable, Lyophilized Filovirus Glycoprotein Vaccines

Abstract:

The filoviruses *Zaire ebolavirus* (EBOV), *Marburg marburgvirus* (MARV), and *Sudan ebolavirus* (SUDV) are some of the most lethal infectious agents known. To date, the *Zaire ebolavirus* vaccine (ERBEVO®) is the only United States Food and Drug Administration approved vaccine available for any species of filovirus. Most licensed vaccines require storage under a continuous cold-chain between 2-8°C to ensure efficacy is maintained and the ERVEBO® vaccine requires cold-chain storage not to exceed -60°C. Such cold-chain requirements, and especially temperatures lower than -60°C, are difficult to maintain in low- and middle-income countries where filovirus outbreaks originate.

To improve the thermostability of filovirus vaccines in order to potentially relax or eliminate these cold-chain requirements, monovalent subunit vaccines consisting of glycoproteins from EBOV, MARV, and SUDV were stabilized within amorphous glasses through lyophilization. Lyophilized formulations and liquid controls were incubated for up to 12 weeks at 50 °C to accelerate degradation. To identify a stability-indicating assay appropriate for monitoring protein degradation and immunogenicity loss during these accelerated stability studies, filovirus glycoprotein secondary, tertiary, and quaternary structures and vaccine immunogenicity were measured. Size-exclusion chromatography was the most sensitive indicator of glycoprotein stability in the various formulations for all three filovirus immunogens. Degradation of the test vaccines during accelerated stability studies was reflected in changes in quaternary structure, which were discernable with size-exclusion chromatography. Filovirus glycoproteins in glassy lyophilized formulations retained secondary, tertiary, and quaternary protein structure over the incubation period, whereas the proteins within liquid controls both aggregated to form higher molecular weight species and dissociated from their native quaternary structure to form a variety of structurally-perturbed lower molecular weight species.

599494 Conformationally Bioactive Bovine Serum Albumin As a Biomolecule in Tissue Engineering (Industry Candidate)

Abstract:

Serum albumin is one of the most abundant plasma proteins and has shown therapeutic potential in bone healing and regeneration. However, the biochemical mechanisms of activity are not well understood. Much of the evidence supporting albumin as beneficial to healing is seen in bone tissue, indicating that biomolecules or mineral ions enriched at healing sites could trigger the bioactivity of albumin. Additionally, we previously detected that bovine serum albumin (BSA) undergoes both an enhanced cell adhesive property and a conformational change when in the presence of calcium ions. The aim of this work is to better understand the biochemical mechanisms of albumin in bone healing and regeneration, as well as strategies for its incorporation into a tissue engineered scaffold.

In this study, we examined BSA bioactivity following adsorption onto TCPS after exposure to varying levels of calcium. The bioactivity was evaluated with respect to calcium exposure concentration, cell seeding density, initial cell attachment, cell viability, and integrin receptor activity. The most bioactive conformations were then covalently bound to a previously studied "blank-slate" polyampholyte hydrogel scaffold to isolate the biochemical impact of albumin while studying cell attachment and proliferation, for incorporation as a future tissue engineering platform.

We previously found that in the presence of calcium, BSA undergoes a conformational change as validated with both molecular modeling and FTIR. To investigate further, we explored osteoblast cell behavior based on varying calcium exposure levels. Cell attachment to BSA adsorbed on TCPS increases with the amount of calcium present, however, when the quantity of calcium gets too high, cell viability decreases rapidly. Cell attachment is partially reduced in the presence of an RGD peptide, but still exhibits high levels of cell attachment, so further integrin blockers were investigated to understand alternative integrin-mediated binding mechanisms. The trends seen on TCPS were validated on the "blank slate" polymer platform, representing a viable path forward for future clinical applications.

The bioactive conformation of BSA in the presence of calcium can facilitate enhanced bone repair through osteoblast cell adhesion and shows promise in aiding regeneration via a tissue engineered scaffold. Additional testing is in progress to isolate the specific bioactive peptide sequence in BSA.

610256 Exploring Sequence-Function Space of a Template-Independent DNA Polymerase for DNA Data Storage and Biorecording Applications

Abstract:

Terminal deoxynucleotidyl transferase (TdT) is a template-independent DNA polymerase that is currently being explored for various applications, including enzymatic DNA synthesis, DNA data storage, and temporal recording of biosignals. TdTs have several unique features that make them a promising tool; they incorporate nucleotide bases probabilistically to single-stranded DNA (ssDNA) substrates, change base incorporation preference in response to their environment, and are catalytically active with several divalent cations. However, TdTs also have properties that can be limiting for various applications. For example, TdTs have relatively slow base incorporation rates compared to other DNA polymerases and cannot effectively incorporate nucleotides into ssDNA secondary structures such as hairpins. While we anticipate that such challenges can be addressed through TdT engineering, there is no adequate knowledge of the relationship between TdT domains and its catalytic properties nor is there a high-throughput screen for TdTs to gain in-depth understanding of sequence-function space. Therefore, we seek to explore the natural diversity of TdTs. To accomplish this, we have selected 10 TdTs that represent the maximum sequence diversity of TdT homologues; our goal is to screen these TdTs for useful and desirable characteristics for recording and data storage applications. These characteristics include favorable kinetics, large base preference biases for signals of interest, and the ability to extend single-stranded DNA primer for long lengths. Our study will correlate differences in TdT characteristics to differences in sequence space that should inform engineering of TdT for a range of applications.

612925 Drug-like Antibodies By Design and Directed Evolution (Invited Speaker)

Abstract:

The biotech industry has seen an explosion in the development of therapeutic antibodies in the last decade, and today most of the best-selling drugs are monoclonal antibodies. The advantages of antibodies as therapeutics – namely their high affinity, specificity, potency, stability, manufacturability and low toxicity – are compelling. Nevertheless, there are many fundamental challenges associated with antibody discovery and development that require key technical advances in order to improve the rational generation of potent antibody therapeutics. I will discuss our progress in addressing some of these challenges, including the design, evolution, selection and characterization of drug-like antibodies with high affinity, specificity, stability and solubility.

599426 Quantitative Analysis of Antisense RNA-Mediated Inhibition and Application in the Fine-Tuning of Biochemical Production (Industry Candidate)

Abstract:

One major challenge for metabolic engineering is achieving the balance between cellular growth and production. Engineering efforts have thus combined gene knockout and knockdown approaches in static or dynamic manners. Genetic knockdown often fulfilled by RNA-based tools (such as asRNA, sRNA or CRISPRi), while most applications are limited to on-off regulations. Experimental and modeling results have suggested the need to fine tune the metabolic flux to further improve strains performance. To advance our knowledge in such direction, we systematically studied the effects of antisense RNA-based knockdown and characterized key factors contributing to different inhibitory levels. We then sought to establish a thermodynamic model to describe and predict such phenomenon based on competitive binding and RNA-RNA interactions. A fine-tuning tool for cellular activity was developed based on this quantifiable inhibition strategy. Finally, we use this tool to fine adjust the biosynthesis pathway flux for 4-hydroxycoumarin by quantifiable knockdown of multiple targets. We believe the develop tool provides an extra layer of control in modifying living systems and is broadly compatible with metabolic engineering strategies.

600093 Metabolite-Responsive Transcription Factors for Feed Activated Pathway Expression in *E. coli*

Abstract:

Metabolic engineering of microbes enables the production of valuable chemicals by introducing heterologous enzymes to the recombinant cell. However, it is well known that the introduced enzymes impose a metabolic burden to the host. In response, common fermentation practice allows cells to dedicate resources to biomass generation before expressing heterologous enzymes from chemically inducible promoters, such as the lac promoter and its IPTG inducer. While chemically inducible promoters have been widely adopted due to their simplicity and pathway-independent nature, they introduce a chemical extrinsic to the process, and the associated cost. We developed a pathway specific expression control strategy that utilizes the galacturonate feed as the inducer, thereby eliminating the need for extrinsic chemicals. Our strategy utilizes a galacturonate-responsive transcription factor to express heterologous enzymes for a glycerate production pathway developed in our lab. We constructed variants of a galacturonate biosensor with a heterologous transcription factor and cognate hybrid promoter, and selected for the best performer through fluorescence characterization. We showed that native *E. coli* regulatory systems have limited interaction with our biosensor and demonstrated simple strain engineering that enhanced biosensor performance. The selected galacturonate biosensor variant was used to express our glycerate production pathway in a feed dependent manner. We confirmed that heterologous pathway enzyme expression is induced in the presence of the galacturonate feed through qRT-PCR. Our control circuit replaced the IPTG induction that was previously used on the glycerate production pathway with no sacrifice to glycerate titer or growth rate. Our work demonstrates that feed activated pathway expression can be used to remove extrinsic chemical inducers from a microbial production system.

609576 Multiomics Analysis of Mitochondrial Versus Cytosolic Compartmentalization of the Isobutanol Pathway in *Saccharomyces Cerevisiae*

Abstract:

Isobutanol is a promising alternative biofuel due to its higher energy density and lower hygroscopicity than ethanol. *Saccharomyces cerevisiae* has been proposed as a suitable host for this process because of its industrial robustness and native capability to produce low levels of isobutanol via valine catabolism; valine biosynthesis and its degradation pathway are separated into two subcellular compartments, the mitochondria and the cytosol, respectively. One engineering strategy to enhance isobutanol production is to localize all the enzymes into a single compartment, however there is debate as to what cellular compartment is optimal for the pathway. In our work, we have equipped yeast with either the cytosolic or mitochondrial localized isobutanol pathway and performed a multiomics analysis to elucidate which compartmentalization strategy favors isobutanol production. We have collected samples for metabolomics, proteomics, and transcriptomics analyses at three different time points during an anaerobic fermentation. Preliminary metabolite data shows a buildup of the first three isobutanol pathway intermediates in the cytosolic localized pathway; these metabolites are upstream of a Fe-S cluster requiring enzyme, Ilv3. Since Fe-S cluster biosynthesis begins with the mitochondrial iron-sulfur cluster (ISC) assembly machinery, we hypothesize that Fe-S cluster loading is limited in the cytosol. We are actively exploring this hypothesis and others to understand why a mitochondria-localized isobutanol pathway has enhanced production in *Saccharomyces cerevisiae.*

609950 Post-Translational Control of Central Carbon Metabolism in Yeast (Industry Candidate)

Abstract:

The baker's yeast *Saccharomyces cerevisiae* is one of the most widely-used microorganisms for industrial chemical production. However, economically-viable production of commodity chemicals with this organism is made challenging by the yeast's overwhelming bias to produce ethanol. Because of the Crabtree effect, fermentation is necessary for the yeast's growth on glucose and other fermentable sugars, making genetic knockouts of the ethanol pathway challenging for growth on many industrially-relevant carbon sources. With dynamic control of the ethanol pathway, production of competing chemicals of interest could be improved without sacrificing biomass accumulation, by splitting the fermentation into a growth stage and a production stage. Our previous efforts to induce the switch from growth to production have focused on transcriptional regulation of the ethanol pathway; while transcriptional circuits can allow for high-fold changes in gene expression, they do not affect proteins already accumulated within cells, creating an inherent lag in circuit response. In this talk, I present a novel genetically-encoded inhibitor of a key enzyme in ethanol production, which allows post-translational control of ethanol production and cell growth. Biochemical and structural characterization of this inhibitor revealed that it inhibits its target with low-nanomolar affinity, with fast association on the order of 10^4 $M^{-1}s^{-1}$, and slow dissociation on the order of 10^{-5} s^{-1}. The inhibitor shows potent, noncompetitive inhibition of its target. Constitutive expression of the inhibitor *in vivo* cuts growth by 1/3 in a wild-type laboratory strain, and virtually eliminates growth in a strain engineered with optogenetic transcriptional control of the ethanol pathway. Finally, inducible expression of the inhibitor shows improved production of chemicals of interest compared to purely transcriptional control of the ethanol pathway, demonstrating the advantage of post-translational regulation of metabolic pathways. This new technology enables faster dynamic control of metabolic enzymes, making it possible to more rapidly shut down essential pathways, leaving a greater pool of substrate to be converted to products of interest.

610673 Genome Editing of *Nicotiana Benthamiana* As an Improved Plant-Based Bioproduction System for Secondary Metabolites (Faculty Candidate)

Abstract:

The wild tobacco relative *Nicotiana benthamiana* is a commonly used plant for manufacturing proteins and reconstituting metabolic pathways which produce complex metabolites such as fragrances or medicines. Importantly, the plant can transiently express heterologous multi-enzyme pathways in just a few days. However, small molecule compounds and their intermediate pathway metabolites produced in *N. benthamiana* are often over-glycosylated, oxidized/reduced, acylated or modified with glutathione by enzymes native to *N. benthamiana*. This unwanted activity has challenged efforts to reconstitute the 11-step pathway to strictosidine (an alkaloid precursor to the anti-cancer drug vinblastine) as well as many other pathways. Therefore, to improve *N. benthamiana* as a bioproduction platform, we are using Cas9-mediated genome engineering to deactivate the endogenous enzymes which are modifying the heterologous metabolites. We have used transcriptomic and phylogenetic analysis of an improved genome assembly to select candidate genes for inactivation from among thousands of possible enzymes and have generated several knockout lines. To further guide selection of knockout candidates, we leveraged cell-free protein synthesis and biofoundry-assisted DNA assembly to express candidate glycosyltransferases and have subsequently measured glycosylation activity towards strictosidine pathway intermediates geraniol and cis-trans-nepetalactol. We anticipate that edited *N. benthamiana* lines will reduce unwanted metabolite derivatization and enable increased production of target molecules.

610502 Development of a Kinetic Model of Lipid Metabolism in *Saccharomyces Cerevisiae*

Abstract:

Studies on metabolic engineering of lipids have transitioned from push-pull-block approaches, where the lipid network can be treated in a black box fashion, to a more targeted approach, where the interactions within lipid metabolism have been probed, and the effect of network perturbations on overproduction phenotypes assessed. Owing to the tightly constrained and interconnected nature of reactions associated with lipid metabolism, assimilation of the results of such studies into a single compendium has proved to be challenging. To address this limitation, here we report the development of a kinetic metabolic model of lipid metabolism in *S. cerevisiae* to assimilate the important modules of lipid metabolism that are either known or suspected to play a role in metabolic engineering efforts of lipid-based biochemicals. The components of the model include fatty acid biosynthesis, glycerophospholipid metabolism, sphingolipid metabolism, storage lipids, lumped sterol synthesis, and the synthesis and transport of relevant target-chemicals, such as fatty acids and fatty alcohols. The model was initially trained on timeseries data of a reference yeast strain, which comprised of the complete lipidomic profile (membrane and storage lipids), as well as secreted free fatty acids (FFAs) or fatty alcohols in the growth media.

In order to assimilate more industrially relevant phenotypes and their associated data into the model, we engineered 3 mutants that have been commonly constructed in literature – i) a FFA overproducer with FFA degradation and activation pathways shut off, ii) an intermediate mutant with overexpression in fatty acid biosynthesis flux, and iii) the above overexpression mutant with a heterologous fatty acyl-CoA reductase (FAR) gene that produced fatty alcohols. Timeseries data from these 3 mutants along with the reference strain enabled the construction of a lipid kinetic model that captured the relevant flux distribution through lipid pathways in overproduction mutants. Finally, to probe the lipid network's response to perturbations, 13 non-lethal single gene deletions in the glycerophospholipid and sphingolipid submodules were constructed in the reference and 3 overproducing mutants, and the lipidomic profiles from these strains were used to train the model thus capturing the lipid system's response to overproduction-relevant mutations.

We performed *a priori* structural identifiability analysis on the model and utilized it to systematically reduce pathways and species within the model. Practical identifiability analysis performed on the model along with the generated lipidomics datasets was used to design further rounds of experiments and also shed light on the blind spots in the model's prediction space. Stability analysis was finally performed on the dynamical system to shed light on the stability of the parameterized mathematical model. Till date, our work represents the richest dataset of lipidomics profiles used to train a kinetic model and has resulted in a model of lipid metabolism that captures engineering-relevant phenotypes. The model is able to successfully emulate the lipid profiles of overproducing mutants reported in earlier studies while providing an explanation for the failure of certain engineering strategies towards overproduction and, to our knowledge, is the first such model to do so. We show how the model can be used to glean insight into the effect of lipid network modifications on overproduction phenotypes and design metabolic engineering strategies for future studies, thus filling a vital gap in the field of lipid metabolic engineering.

613039 Challenges and Opportunities with CRISPR Activation in Bacterial and Cell-Free Pathway Engineering (Invited Speaker)

Abstract:

The development of CRISPR-Cas tools for rapidly programming gene expression promises to accelerate metabolic pathway engineering. The application of CRISPR-Cas tools for transcriptional repression (CRISPRi) in bacterial metabolic engineering is well-established. By comparison, the creation of CRISPR-Cas tools for transcriptional activation (CRISPRa) in bacteria has lagged. Together with our collaborators, we have developed new tools for bacterial CRISPRa by 1) identifying effective transcriptional activators, 2) systematically defining rules for effective bacterial CRISPRa, and 3) creating approaches for encoding complex functions in engineered guide RNAs. We have shown that these CRISPRa tools can drive expression of multi-gene biosynthetic pathways from multiple promoters and in multiple hosts, including *E. coli*, *P. putida* and *E. coli*-derived cell-free TXTL. In this presentation, I will describe our most recent results and outline strategies for integrating CRISPRa and CRISPRi into data-driven workflows to create new capabilities for rapidly optimizing pathways in bacterial and cell-free bioproduction systems.

611167 Optimizing Protein-Polymer Conjugation through Course Grain Simulation and Cell-Free Protein Synthesis: Toward Better 2nd Generation Biologics

Abstract:

Most 2nd generation protein therapeutics are simply PEGylated versions of a prior biologic. Conjugating PEG significantly improves pharmacokinetics and protein stability, however this often occurs at the cost of activity. Indeed there are FDA approved drugs that lose over 90% of their specific activity but are still over 3 times more effective due to the advantages of PEGylation. Here we report our work to integrate course grain simulation results with rapid cell-free protein synthesis screening to determine the optimal locations to attach PEG to proteins that retain activity and further improve protein stability. We also report that traditionally held design heuristics for PEGylation (including solvent accessible surface area and secondary structure) often lead to sub-optimal location choices for PEGylation. However, using the computationally inexpensive approach of course grain simulation is an excellent tool to determine optimal locations for PEGylation and cell-free protein synthesis can be used as a rapid screen to verify computationally determined targets.

597175 Engineering Bispecific Antibodies for Targeted Inhibition of Tumor Metastasis

Abstract:

Introduction: Metastasis is responsible for 90% of deaths from solid tumors. However, current FDA cancer treatments are focused on preventing tumor cell proliferation rather than migration, and the mechanisms driving tumor metastasis remain unclear. Recently, our team reported a synergistic signaling pathway involving the interleukin-6 (IL-6) cytokine and the interleukin-8 (IL-8) chemokine that promotes tumor metastasis. Inhibition of the IL-6 and IL-8 receptors (IL-6R and IL-8R) using a combination of tocilizumab, an anti-IL-6R antibody which is clinically approved for treatment of rheumatoid arthritis, and reparixin, a small molecule drug targeting IL-8R that is currently in phase II clinical trials for treatment of breast cancer, effectively reduces metastasis in mouse models of cancer. However, translation of an antibody/small molecule combination therapy to the clinic is complicated by logistical challenges, such as dosing ratio optimization and increased regulatory hurdles. Moreover, small molecule drugs such as reparixin face significant clinical challenges in terms of specificity. Compared with monoclonal antibodies and small molecule therapies, bispecific antibodies (BsAbs) have higher affinity, potency, and selectivity. In addition, the dual-targeting BsAb approach is also more robust against acquired resistance due to reduced likelihood of mutational escape. In this study, we engineered BsAbs that simultaneously target the IL-6 and IL-8 receptors for targeted inhibition of cancer metastasis. We particularly focused on highly invasive cancer types, such as triple negative breast cancer (TNBC) and pancreatic duct adenocarcinoma (PDAC).

Materials and Methods: We designed two BsAbs (BS1 and BS2) based on the human immunoglobulin G (IgG) platform (**Figure 1A**) using the variable regions of an anti-IL-6R and an anti-IL-8R antibody. BS1 and BS2 were recombinantly expressed in a mammalian cell expression system. We characterized their binding affinities to IL-6R and IL-8R, and determined the degree of inhibition of the IL-6/IL-6R and IL-8/IL-8R interactions on IL-6R- and IL-8R- expressing cell lines via flow cytometry. Blockade of IL-6 signaling was tested on HepG2 cells by flow cytometry, and blockade of IL-8 was evaluated using HTLA cells in the PRESTO-TANGO assay system. We also interrogated the effects of BsAbs on the migration of metastatic cancer cells in a 3D collagen matrix model. Finally, we tested the efficacy of BsAbs in inhibiting cancer metastasis in mouse tumor xenograft models.

Results and Discussion: We successfully produced BS1 and BS2 with high purity (>99%), as demonstrated via SDS-PAGE analysis (**Figure 1B**). Both BS1 and BS2 bind IL-6R and IL-8R, and competitively inhibit the IL-6/IL-6R and IL-8/IL-8R interaction. BS1 has a weaker binding affinity and less potently competes with ligands compared to BS2 due to the monovalent versus tetravalent construction of the respective BsAbs. We also demonstrated that BS1 and BS2 inhibit IL-6 and IL-8 signaling with similar potency to the parent monoclonal antibodies. Both BsAbs greatly reduced motility, persistence, and diffusivity of cancer cell migration in 3D collagen matrices, and they were *significantly more effective* at inhibiting tumor cell migration than combination treatment with the parent monoclonal antibodies or with tocilizumab and reparixin. Finally, BS1 and BS2 achieved >75% inhibition of tumor metastasis in orthotopic mouse xenograft models of TNBC at remarkably low doses, illustrating the significant therapeutic advantage for our bispecific approach.

Conclusions: Our engineered BsAbs simultaneously bind IL-6R and IL-8R and effectively block the IL-6/IL-8 axis to inhibit tumor metastasis. These antibodies present a novel treatment option for highly invasive cancers and promise to improve treatment efficacy, minimize harmful side effects, and synergize with other anti-cancer agents. Moreover, our novel BsAb designs will provide new insights into the development of bispecific constructs for future therapies targeting a broad range of diseases.

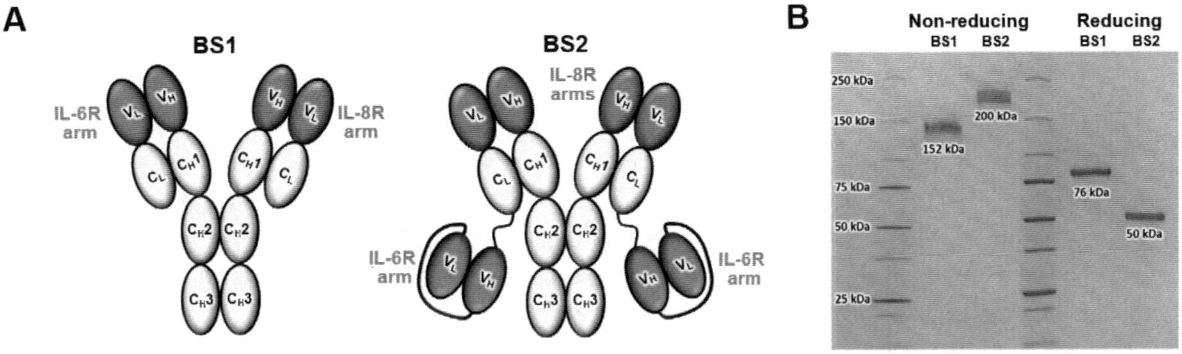

Figure 1 A. Two BsAb formats that simultaneously block IL-6R (magenta) and IL-8R (teal). Variable and constant domains are labeled. B. SDS-PAGE analysis of BsAbs.

597222 Chemical Biology Tools for Identification of Siglec Ligands in Cancer

Abstract:

Cancer immunotherapies, including checkpoint blockade antibodies, represent a breakthrough in cancer treatment with long-term remission observed in select patients. Unfortunately, many tumors remain unresponsive to existing immunotherapies, creating an urgent need for mechanistic studies of additional immune checkpoints that drive cancer progression. Recent evidence suggests that upregulation of select glycoproteins modified with the sialic acid monosaccharide allow tumors to resist treatment by engaging a family of inhibitory receptors called Siglecs on immune cells. However, the molecular identities of Siglec ligands remain poorly understood, preventing therapeutic targeting of these emerging immune checkpoints. I hypothesize that each of the 8 inhibitory human Siglecs bind a discrete, non-overlapping set of glycopeptide ligands, representing a "Siglec code" for immune suppression. My work seeks to decipher this code by defining the molecular identities of Siglec ligands and translate these discoveries into development of a new class of checkpoint blockade antibodies. Specifically, I am using cutting-edge glycoproteomics techniques and a novel synthetic glycopeptide array screening platform to comprehensively elucidate the structural determinants of Siglec binding. Based on promising preliminary data, I expect these experiments to demonstrate that Siglecs bind defined epitopes on cancer cells that can be therapeutically targeted. This work will shed light on a previously understudied mechanism of immune evasion exploited by a significant fraction of tumors, and open the door to a new strategy for immunotherapy applicable to diverse cancers.

609748 Design and Engineering Protein Therapeutics for Targeting Metalloproteinases

Abstract:

The metalloproteinase (MP) family of zinc-dependent proteases, including matrix metalloproteinases (MMPs), a disintegrin and metalloproteases (ADAMs), and a disintegrin and metalloproteinase with thrombospondin motifs (ADAMTSs) plays a crucial role in the extracellular matrix (ECM) remodeling and degradation activities, and thus responsible in several diseases. Specific MPs has shown to play a significant role in progression of several cancers and neurodegenerative disease such as Alzheimer's disease (AD), while other MMPs and ADAMs found to be anti-tumorigenic and neuroprotective. Therefore, it is important to target specific MPs with pathological roles with high selectivity. Protein engineering techniques such as directed evolution can be used to tailor the binding selectivity of protein binders toward pathological MPs. Tissue inhibitor of metalloproetainses (TIMPs) as natural inhibitors of MPs with a wide range of binding affinity offer great scaffolds for developing therapeutics. Engineering and design of TIMP-based scaffolds to improve the selectivity and eliminate the off-target effects requires a comprehensive understanding of the protein structures combined with library design and high throughput screening. We have previously developed a counter-selection strategy to screen TIMP-1mutants that bind selectively to MMP-3 in the presence of MMP-10, an MMP with the greatest sequence and structural similarity to MMP-3. This strategy has improved TIMP-1 mutants with up to 20-fold improvement in binding selectivity, highlighting the significant potential of this approach for development of inhibitors with single-MP selectivity. In ongoing studies, we combine rational design and directed evolution to develop more efficient TIMP-based scaffolds to target specific MPs with high affinity and selectivity. These studies and methodology developed will lay the foundation for developing novel protein engineering strategies for generating therapeutics for MP-related diseases and understanding protein-protein interaction between MPs and their inhibitors.

597791 Directed Evolution of Phenylalanine Ammonia-Lyase (PAL), a Key Enzyme for the Treatment of Phenylketonuria (PKU)

Abstract:

Phenylalanine ammonia lyase (PAL) enzymes are widely found associated with secondary metabolism in plants, bacteria, and fungi. Biocatalytic applications for natural product and fine chemical synthesis has driven the discovery, expression, characterization, and engineering of PAL. More recently, development of PALs for phenylketonuria (PKU) management and cancer therapy has further increase interest in engineering this class of enzymes. Comparative structure and enzyme analysis have informed numerous efforts to improve enzyme stability and alter substrate specificity through rational design. While site-specific mutagenesis has been employed for improving PAL, a more comprehensive mutational landscape has yet to be explored for this class of enzymes. Here, we report development of a directed evolution strategy to engineer PAL enzymes. Central to this approach is a high-throughput enrichment that couples *E. coli* growth to PAL activity. Applying this method to the PAL from *Anabaena variabilis,* used in the FDA approved drug Pegvaliase, for PKU treatment, we identified mutations at residues previously unknown as relevant for function that increase higher turnover frequency almost twofold after only a single round of engineering. This work demonstrates the power our technique for ammonia lyase enzyme engineering and is the first example of improving the specific activity of a PAL toward its native substrate.

Mays, Mohan et al. (2020) ChemComm

603272 Combinatorial Chimeragenesis of Novel Lysins to Treat Antibiotic Resistant Infections

Abstract:

Bacterial pathogens are known to quickly evolve or acquire resistance to new antibiotics, and the global spread of multidrug-resistant bacteria represents a critical threat to public health. Additionally, the decline in approval of new antibacterial chemotherapies is prompting a search for innovative treatment paradigms. The use of bacteriolytic enzymes as biotherapies presents a novel approach to combatting bacterial infections. Antibacterial enzymes, or 'lysins', actively degrade the peptidoglycan of pathogenic bacteria, resulting in rapid cell lysis and death. As a result of this unique mechanism of action, lysins can effectively kill contemporary drug-resistant strains while suppressing, or at least slowing, the development of new resistance phenotypes.

Lysin structures are highly modular, and typically consist of a catalytic domain that cleaves some moiety of the peptidoglycan, and a cell wall binding domain that targets the enzyme to the cell wall. Engineering lysins as an antibacterial therapy typically involves synthetically fusing catalytic domains to known cell wall binding domains to generate recombinant protein chimeras with enhanced activity. Thus far, the approach has been limited to the rational design of a small group of chimeric proteins, resulting in a slow trickle of putative antimicrobial enzymes. A generalizable approach to rapidly screen different combinations of lysin catalytic domains and cell wall binding domains would greatly enhance the throughput of candidate development.

We have developed a pipeline that combines both computational and experimental methodologies to discover, engineer, and screen for novel synthetic lysins with enhanced therapeutic efficacy. While our initial focus is on the critical pathogen *Staphylococcus aureus* (*S. aureus*), our approach is broadly generalizable to the discovery and engineering of bacteriolytic enzymes against any gram-positive bacteria.

We first performed an extensive bioinformatics search using known lysin sequences as seeds to generate a database of approximately 20,000 putative lytic enzymes, comprised of 80,000 unique domains, derived from a diverse collection of over 4,500 bacterial organisms and bacteriophage. We next performed phylogenetic and sequence analysis to identify forty-nine diverse catalytic domains for experimental testing. These catalytic domains represent seven classes of lytic enzymes that target different moieties of the peptidoglycan. We subsequently developed a modular combinatorial chimeragenesis approach to shuffle individual catalytic domain sequences together with cell wall binding domain sequences to generate thousands of unique synthetic lysin sequences. We combined the forty-nine catalytic domains with four linkers and thirteen high-performance cell wall binding domains to generate a 2,548-member library. While our initial library was designed for a simple two-domain fusion, our approach is inherently scalable and capable of generating larger and more diverse multi-domain fusions. We subsequently identified lysin variants with activity against *S. aureus* via a halo-forming screen of our recombinant *Escherichia coli* expression library. Using this approach, we tested over 15,000 individual clones and identified six novel lysins with anti-staphylococcal activity (**Figure**). We are currently characterizing these lysins *in vitro*, and believe they could represent new therapeutic candidates against antimicrobial-resistant infections. Future directions will focus on further enhancing the therapeutic potential of these enzymes through both computational design and directed evolution, and expanding our discovery pipeline to other important bacterial pathogens. We believe our pipeline, and the engineered lysins we have created will provide valuable new tools in the fight against drug-resistant bacterial pathogens.

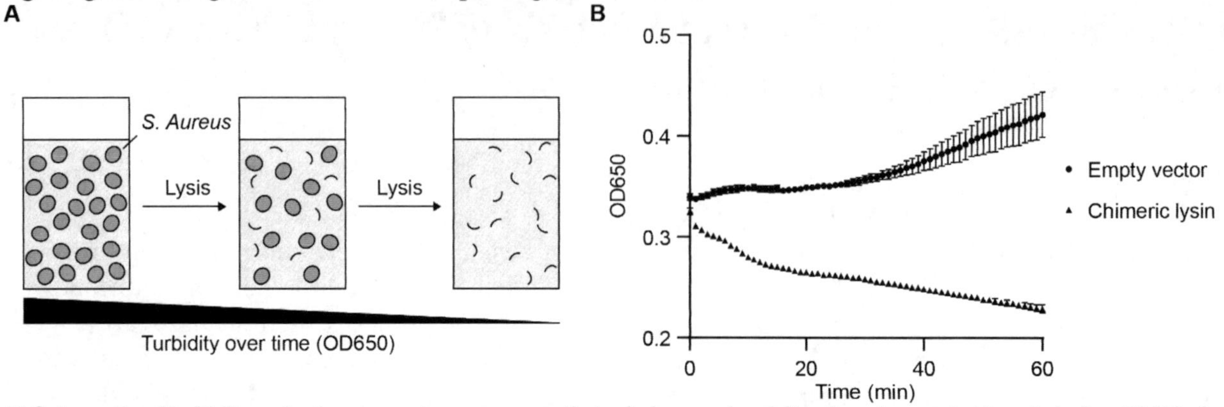

A) Schematic of turbidity reduction assay to measure anti-staphylococcal activity of engineered chimeric lysins. **B**) Whole cell lysates of an engineered chimeric lysin were incubated with *S. aureus* and the turbidity was measured over time via optical density at 650 nm (OD650).

612923 Deciphering and Engineering the Substrate Specificity of Proteases (Invited Speaker)

Abstract:

Proteases are ubiquitous in biology, playing significant roles in initiating, regulating, and terminating cellular processes. The diversity and breadth of protease substrate preference have been harnessed for a variety of applications, including protease therapeutics, protein purification, and mass spectrometry-based proteomics. Furthermore, with advances in protein engineering and synthetic biology, proteases are now being utilized for studying protein-protein interactions, imaging newly synthesized proteins, and performing logic operations inside cells. To successfully repurpose proteases towards these applications, it is often necessary to engineer their catalytic properties using Directed Evolution. In this vein, to effectively isolate protease variants with the desired phenotypes from a large pool, the high-throughput enzyme screening or selection system should ideally exhibit a broad operational range (dose-response or sensitivity over a large variation in input), and a high dynamic range (signal-to-noise ratio). Unfortunately, current protease engineering platforms often require laborious design-build-test cycles to determine, optimize, and operate within these ranges. Failure at any or all three steps prevents one from assessing the functionality of variant libraries in between evolution cycles and lowers the probability of isolating protease variants with the desired phenotypes.

Here we describe YESS 2.0, a highly modular and customizable yeast endoplasmic sequestration screening system suitable for engineering and profiling the specificity of protein-modifying enzymes. By incorporating features to modulate gene transcription, as well as substrate and enzyme spatial sequestration within a versatile and seamless assembly method, YESS 2.0 achieves broad operational and dynamic range. To showcase YESS 2.0, we evolve a TEV protease variant (eTEV) with an 8-fold higher catalytic efficiency to obtain the fastest TEV protease variant to date. Due to the unique features of our system, this phenotype is strictly attributable to an increase in turnover number (k_{cat}). Second, we used YESS 2.0 coupled with NextGen Sequencing to profile the substrate specificity of insulin-degrading enzyme (IDE) and discover IDE substrates with increased activity compared to the prototypical insulin peptide. Lastly, we show for the first time that YESS 2.0 supports calcium-independent sortase-mediated ligations and confirm that residues directly C-terminal of the pentapeptide motif heavily influence the rate of SML. This state-of-the-art platform offers unmatched versatility in profiling and engineering the specificity of protein-modifying enzymes and should enable even more ambitious future undertakings.

610752 Flow Induced Vascularization in a Human Pluripotent Stem Cell (hPSC)-Based Islet-like Organoid

Abstract:

Islets of Langerhans are vital in the regulation of blood glucose levels primarily through the production of glucagon by the alpha cells and insulin by the beta cells. When insulin synthesis is hindered, cells aren't able to take up glucose causing a diabetic state to develop. This can occur from autoimmune beta cell death (type 1 diabetes) or insulin resistance of cells (type 2 diabetes), and both types normally require exogenous insulin replacement. A long-term solution for treatment is islet transplantation, but due to the lack of sufficient donors, this method can't be used on a large scale. An alternative method would utilize human pluripotent stem cells (hPSCs) to create islet like organoids that can be used for implantation. Organoids are in vitro synthesized tissue that replicates the function and structure of in vivo organs. Primary human islets are composed of endocrine cells, stromal cells, and a dense vascular network. The aim of this project is to reproduce these aspects of the islet to more accurately form islet-like organoids that match of the structure and function of islets, with the hypothesis that the inclusion of the vessel network would aid in function of the endocrine cells through the replication of the system that normally exists in vivo.

Forming the organoid utilized a controlled heterotypic aggregation platform that our lab has developed based on a hydrogel platform that efficiently forms spheroids from the hPSC-derived endocrine cells and endothelial cells. This platform utilizes microwells formed from hydrogels that allows bulk aggregation in a standard 12 well tissue culture plate. With the formation of these spheroids, we were able to achieve intra-organoid vascularization with the inclusion of a stromal cells and adipose microvascular fragments under static culture conditions. The resulting gene expression of maturing pancreatic beta cell markers (NKX, PDX1, and Insulin), an islet specific endothelial gene (API), and an endothelial diaphragm fenestration indicator (PLVAP) increased compared to homotypic aggregates of iPSC-derived islets and the initial microvascular fragments. Additionally, glucose responsiveness in the formed triculture organoids were significantly higher compared to the hPSC-islet monoculture aggregates, with the monoculture doubling in insulin secretion to 0.32 micro-insulin units (uIU) per aggregate with a 13mM increase in glucose and the triculture insulin secretion tripling to 0.58uIU per aggregate under the same conditions. Furthermore, we have introduced the organoids after initial aggregation into a microfluidic device that would allow for shear flow through the organoid to drive angiogenesis. We have developed an alginate encapsulation method in the device that sustained the organoid spherical structure while allowing for shear flow across the organoid, which improved network development. The formed vascularized organoids are very capable as a regenerative therapy for diabetes and for disease modeling in a microphysiological system.

607936 Pluripotent Stem Cell-Derived Organoids to Model Fetal Pancreatic Development and Maturation

Abstract:

Type 1 diabetes mellitus is characterized by the autoimmune destruction of insulin-producing β-cells within the pancreatic islets and it affects millions of people around the globe. Human pluripotent stem cells (hPSCs) are an intriguing option to replace the lost β-cells in patients and are currently being used in clinical studies that employ encapsulation to protect the cells from the immune system. Approaches for β-cell replacement therapies follow either hPSC differentiation *in vitro* to mature, functional β-cells, followed by *in vivo* transplantation, or hPSC differentiation *in vitro* towards pancreatic progenitors, which are transplanted for *in vivo* maturation. However, complete maturity of the hPSC-derived β-cells *in vitro* has been difficult to achieve. Therefore, we hypothesize that monolayer induction of pancreatic progenitors from hPSCs, followed by organoid maturation will enhance both maturity and the number of derived β-cells. Organoids provide the three-dimensional environment and signaling that progenitor cells experience during development. Thus, we suggest that this cultivation model could be used to study the signals and pathways involved in β-cell differentiation and maturation. Utilizing culture conditions such as low oxygen (5%), which better mimics physiological oxygen levels available during progenitor development, we have derived a pancreatic progenitor population from hPSCs that expresses high levels of pancreatic and duodenal homeobox 1 (Pdx1) genes, and demonstrate increases in pancreatic progenitor specific markers neurogenin 3 (Ngn3) and NK2 homeobox 2 (Nkx2.2) gene expression over 11 days of monolayer culture, signifying endocrine commitment of pancreatic progenitors. We are further characterizing protein expression in this population with immunostaining, western blot, and functional glucose tolerance tests. We also hypothesize that creating organoids in the form of hollow cysts will provide the best platform to model pancreatic development. This structure most closely recapitulates the structures formed during pancreatic development and has not been very well explored as a pancreatic organoid model. During development, the pancreas forms a hollow branch-like structure where endocrine cells migrate from the epithelial layer and aggregate to form the fetal islets. Therefore, we are devising methods, which include the addition of growth factor-reduced matrigel, soluble factors, and other cell types; to form cyst-like organoids from our D11 cell population with the goal of mimicking this branched structure. Preliminary organoid models, embedded in matrigel, developed both cyst-like and solid structures in culture, which we believe develop from differing cell types within our D11 population; however, the organoids have shown elevated expression of non-pancreatic lineage markers. Yet, these organoids could serve as a promising system to model pancreatic development once conditions and day of culture have been optimized. On the other hand, organoids formed in suspension culture have shown elevated levels of Nkx6.1 and insulin, both expressed by the pancreatic β-cells. Our future work will be to understand how modeling fetal states can enhance organoid maturation and evaluate the function of these organoids with *in vitro* and *in vivo* models of diabetes

611831 An *in Vitro* hpsc 3D MODEL of Embryological LIVER Morphogenesis for Advanced Hepatic Tissue Construction

Abstract:

We aimed to design an *in vitro* model of embryonic liver morphogenesis to mimic the progenitor cell intercommunication regulating liver growth. Liver morphogenesis initiates from the coordinated interactions of the liver progenitor cells that invade the septum transversum mesenchyme (STM) and rapidly proliferate to generate complex tissue. To elucidate the site-specific cellular crosstalk that governs this step we developed a hepatic organoid outgrowth assay to determine how endothelium/hepatic progenitor cells organize to form primitive sinusoidal structures. Initially to model embryonic liver tissue, *in vitro* derived hepatic endodermal (HTE) cells, were replated into ultralow attachment (ULA) round bottom 384 well plates (Corning) to form compact spheroids (n=10). To determine the effect of collective cell migration on nascent sinusoid formation, organoids were embedded into matrigel droplets in an outgrowth assay (n=10). These migratory organoids demonstrated radial, uniform, cord-like projections into the matrix, similar morphologically to primitive hepatic sinusoids that form during liver development between E9 to E10 (mouse). These structures displayed enhanced gene expression consistent with migration, including epithelial-to mesenchymal transition (EMT) markers and early liver differentiation genes more than controls. Immunolocalization of organoids demonstrated liver and biliary markers such as albumin, CK18, Sox9 and HNF6. Endothelium progenitor cells stained positive for VEGFR2 and VWF and co-aligned with hepatic progenitor cells in a radial fashion. The morphogenetic model presented here represents a novel approach to modeling embryonic liver growth, nascent sinusoidal development, and tissue organization, a rate-limiting step during liver development. Further studies will investigate the genomic and biophysical cues that regulate such multicellular events. These studies will improve models to create 3D liver tissue *in vitro* for various liver regenerative medicine applications, including modeling human development, liver disease, and tissue regeneration.

601456 Co-Cultures of Mammary Organoids and Macrophages Provide Insight into Breast Cancer Recurrence

Abstract:

Although breast cancer patients receive chemotherapy, surgery, and radiation treatment, up to a quarter of triple negative breast cancer (TNBC) patients experience locoregional recurrence. Patient data and pre-clinical studies have established that radiotherapy, immune status, and macrophage infiltration are associated with TNBC recurrence. However, there is much about the role of the normal tissue microenvironment in recurrence that remains unknown. Although macrophages have been implicated in facilitating tumor cell recruitment to treated sites, characterization of macrophage polarization following radiation damage has not been undertaken. Determining recurrence associated immune cell infiltration *in vivo* and pairing these studies with biologically relevant *in vitro* models are therefore necessary to elucidate further mechanisms of tumor cell recruitment. In this study, we examined the radiation response of immune cells *in vivo* and applied these observations to an *in vitro* mammary organoid model to analyze the effects of radiation on cell behavior.

To evaluate immune cell recruitment, we irradiated inguinal murine mammary fat pads (MFPs) of 8-12 week old female Nu/Nu and BALB/c mice *in vivo* to a dose of 20 Gy. MFPs were resected 10 days post-irradiation. Tissues were either processed for immunohistochemical characterization or into a single cell suspension for flow cytometry. Cells were stained with F4/80 in addition to CD86+ or CD206+ to determine M1 and M2 macrophage polarization, respectively. To obtain mammary organoids, MFPs were resected, irradiated *ex vivo* to 20 Gy, and digested with collagenase to isolate epithelial cells before self-assembly in low adhesion plates. Control unirradiated and irradiated organoids were then cultured with biologically relevant ratios of M1/M2-polarized RAW 264.7 macrophages. Co-culture experiments were monitored up to 48 hours with live cell imaging. Conditioned media from the co-culture was used for *in vitro* invasion assays with murine TNBC 4T1 cells.

We identified an increase in F4/80+ macrophage infiltration following *in vivo* irradiation, recapitulating observations from previous studies. We also quantified enhanced recruitment of tumor associated M2 macrophages in irradiated sites ($p < 0.05$), potentially implicating the role of wound healing macrophages in tumor cell recruitment. These results were confirmed *in vitro* as tumor cell invasiveness was associated with irradiation and higher M2/M1 macrophage ratios in the co-culture model.

This work describes the interplay between an *in vivo* and *in vitro* model that may elucidate the role of radiation in immune-facilitated recurrence in a biologically relevant environment. This model will provide insight into the mechanisms behind radiation-induced tumor and immune cell recruitment, which will have significant implications for patients suffering from TNBC recurrence.

600439 3D Organotypic Cancer Model: Revealing Vascular Ablation during Pancreatic Cancer Invasion

Abstract:

Although the diagnosis and treatment of cancer in its earliest stages have significantly improved outcomes in many tumors, survival rates in patients with tumors that have spread to distant sites remain dismal. As a result, the vast majority of cancer mortalities stem from metastasis and its complications. Metastasis describes the process where tumor cells escape from the tumor primary sites to enter the blood stream, and arrive at other distant organs to form secondary tumors. Among metastatic cancers, pancreatic ductal adenocarcinoma (PDAC) has remained one of the deadliest metastatic diseases over the past decades owing to the high incidence of metastasis of tumor cells. Notably, patients with PDAC exhibit a high degree of vascular invasion and abundant circulating tumor cells in the bloodstream, which suggest that PDAC cells are highly efficient to gain access to the blood stream. Genetically engineered mouse models for PDAC have made many progresses to understand the tumor biology of PDAC during tumor progression and development. Despite this, the interactions between PDAC cells and blood vessels remain poorly understood, partly due to the lack of *in vitro* models to enable close examination of tumor cells and endothelial cells at the blood vessel interface.

Here, we describe an organotypic PDAC-on-a-chip culture model to examine vascular invasion in PDAC. The model features two hollow cylindrical channels embedded within a three-dimensional matrix [1]. One of the channels was lined with endothelial cells to mimic blood vessel while the other was seeded with PDAC cells to mimic a tumorous pancreatic duct [2]. In this model, PDAC cells from the tumorous pancreatic duct invaded through the intervening matrix and into the blood vessel lumen, induced endothelial cell apoptosis, and ultimately ablated the endothelial cells. A phenomenon we termed vascular ablation. We later confirmed the observation of vascular ablation by tumor cells in our organotypic model by using *in vivo* model of ectopic PDAC growth and genetically engineered mouse model of PDAC. Interestingly, PDAC human patients have also exhibited vascular invasion and tumor-like cells, which populate the lumen side of the blood vessels in previous reports. This confirms the clinically relevant observation of vascular ablation in our 3D organotypic model. Additionally, using small molecule inhibitors and CRISPR technology, we identified the Activin-ALK7 pathway as a critical mediator of vascular ablation by PDAC. This tumor-on-a-chip model thus provides an important *in vitro* platform to reveal previously unappreciated mechanisms by which tumors interact with tumor blood vessels and enable identification of signaling pathway to target tumor invasion and vascular ablation in PDAC.

[1] **Nguyen, D.H.,** Stapleton, S.C., Yang, M.T., Cha, S.S., Choi, C.K., Galie, P.A., Chen, C.S. (2013). *Biomimetic model to reconstitute angiogenic sprouting morphogenesis in vitro*. **Proceedings of the National Academy of Sciences**. 110(17): 6712-6717.

[2] **Nguyen, D.H.,** Esak, L., Alimperti, S., Norgard. R.J., Wong, A., Lee, J.J.K., Eyckmans, J., Stanger, B.Z., Chen, C.S. (2019). *A biomimetic pancreatic cancer on-chip reveals endothelial ablation via ALK7 signaling*. **Science Advances.** 5(8): eaav6789

613357 3D-Printed Models of Organ Development (INVITED SPEAKER)

Abstract:

The development of most organs requires concerted and reciprocal interactions between an epithelial sheet or tube and its surrounding mesenchyme or stroma. Guidance cues in the form of morphogen gradients or aligned fibers of extracellular matrix proteins have been posited to provide directional information that templates the pattern of morphogenesis of the epithelial tissue. For example, the ductal tree of the mammary gland forms stochastically, but with a characteristic bias that has been assumed to result from aligned fibers of collagen within the fatty stroma. To test whether such physical scaffolding can account for the pattern of mammary branching, we carried out quantitative imaging analysis to characterize the 3D architecture of the organ as a function of developmental time. We then used 3D-printing approaches to generate the adipose stroma and collagen-fiber alignment patterns that we found in vivo, and investigated the pattern of epithelial branching that resulted from varying these geometric features. Here, I will describe our imaging and 3D-templating pipelines, discuss how this combination has led to new insights in tissue development, and generalize our findings to future efforts in quantitative developmental biology and tissue engineering.

612711 Portable Paper-based Biosensors and Novel Artificial Olfaction Biosensors Driven by Cell-free Synthetic Biology

Abstract:

Chemical or biological pollution in rapidly urbanizing cities has become a serious problem that disturbs the lives and livelihoods of millions of people, damaging and restricting potable water supply. Therefore, the rapid detection of pollution molecules will play a significant role in health care as well as environmental management. Many attempts, such as the utilization of diverse biological sensing elements, have been designed to respond to polluted chemicals or bacterial biomarkers for the detection. However, the practical applications of the whole-cell- or cell-element-based biosensors has been restricted by several concerns of biosafety, the hard preparation, and tough storage. For these challenges, cell-free protein synthesis technology, which synthesizes proteins using biological machinery *in vitro*, could be employed to broaden the applications of synthetic-biology-based biosensors or functional-protein-based biosensors.

For synthetic-biology-based biosensors, we present a portable *in vitro* biosensing platform for detecting inorganic chemicals or biological organic molecules. The sensing synthetic genetic circuits, coupled with basic cell-free components, were embedded onto papers by freeze-drying, which could be stored and transported at room temperature. These portable testing papers were activated by adding liquid samples for specific inducible reporter expression. To achieve the full availability outside the laboratory, reporter enzymes that could mediate color change by cleaving substrate were selected and demonstrated for *in vitro* paper-based platform. Synthetic genetic circuits respectively responded to two chemicals arsenic ion and AHLs, which were the common industrial pollutant and bacterial communication signal, were demonstrated to effectively sensing in this *in vitro* paper-based platform. Besides, combining the smartphone and image analysis software, the visible output could be easily collected and quantified. Without the intact cells, cell-free systems provide an emerging possibility for synthetic biology-based biosensor, which can be practically applied to environmental management or clinical diagnosis outside the laboratory. In the future, more internet technology such as smartphone and internet-of-thing (IoT) technology could be integrated into the detection platform for data sharing, from which people can learn about the environmental condition from a smartphone with the internet.

Moreover, the controllable protein synthesis conditions of cell-free systems also provide an emerging alternative approach for the effective production of complex proteins, such as the membrane olfactory receptor proteins. The nanomaterials can be applied for the functional folding of proteins in cell-free protein synthesis. Combining with the quartz crystal microbalance platform, diverse natural olfactory receptor proteins can be developed into artificial olfaction, which can be potentially employed to detect toxic odors for environmental management and applications.

611124 Biosensing of Human Hormone Disruptors in Blood and Urine with Protein Folding Using Cell-Free Protein Synthesis

Abstract:

The EPA has listed thousands of chemicals as potential hormone disruptors, however testing these chemicals is currently a lengthy and expensive process. Here we present a colormetric assay to rapidly detect estrogen or thyroid hormone activity of a sample using cell-free protein synthesis. The biosensor is based on protein folding of an allosteric fusion protein that includes a human thyroid or estrogen receptor. When a chemical binds to the hormone receptor during folding, the protein's confirmation changes which can then be detected by the reporter enzyme portion of the fusion protein. Using this approach, known hormone disruptors have been rapidly detected including the weakly binding estrogen disruptor BPA. Using lyophilized cell-free reagents, the sensor can be used in a "just-add-sample" format and has potential as a portable at-home assay. We also report the ability of this sensor to function in human blood and urine and the importance of adding RNAse inhibitors for sensing in these mediums.

611687 Fluorescent Microbial Sensors That Discriminate between Different Radionuclides.

Abstract:

Proper detection and monitoring of nuclear fuel cycle, enrichment, and weapon development activities is critical for supporting warfighter preparation in chemical, biological, radiological, nuclear, and explosives (CBRNE) operations, clandestine activities, and nuclear compliance. Field applications of conventional radiation monitoring and detection systems are restricted by several disadvantages: mandatory proximal placement near a radioactive source for detection, being easily identifiable and avoided, and capacity to only report radioactivity at a specific moment in time. The ability to detect trace amounts of ionizing radiation is a paramount concern for CBRNE operations and observation of clandestine activities in the event of source relocation and contamination. Microbes respond to environmental stress such as low-dose ionizing radiation with morphological and behavioral changes. Responses are known to persist after the stressor is removed. We hypothesize that transcriptional changes can monitor these stress responses and be harnessed to develop biosensors capable of discerning radionuclide type and monitor and report on nuclear fuel cycle, enrichment, and weapon development activities in diverse environmental conditions.

Unique responses of environmental microbes to post exposure to nuclear fuel cycle representative radionuclides were identified through mRNA sequencing analysis. Gene expression analysis showed *Pseudomonas putida and Escherichia coli* responded differently to acute and chronic exposure to plutonium-239, tritium, and iron-55 at a dose rate of 8.7 mGy/d. Highly abundant genes overexpressed by a sole radionuclide were selected for sensor development. Native promoter sequences from candidate genes were inserted into a broad host range plasmid upstream of a fluorescent reporter. Sequence verified *in vivo* sensors were then transformed into *P. putida* and subsequently irradiated by each of the three radionuclides for 24 hours before fluorescence was measured. Preliminary testing revealed, when grown in the presence of the inducing radionuclide, expression of GFP increased. Sensors treated with non-inducing radionuclides showed no significant increase in fluorescence. Future sensor development will include engineering circuits of toehold switch sensors permitting detection of time dependent signals for individual radionuclide sources. In addition to developing passive and autonomous monitoring or clandestine activities, the results of this work will expand upon the limited knowledge of low dose radiation effects in microorganisms.

599613 MRI/S Based Assessment of Donor Efficacy in Stroke Treatment with Human Mesenchymal Stem Cells (Industry Candidate)

Abstract:

Human mesenchymal stem cells (hMSC) have become a potential candidate for cell therapy in stroke treatment due to their strong secretion of trophic factors, immunomodulation and linage-specific differentiation. However, *in vitro* culture expansion leads to cellular senescence and reduced therapeutic functions after transplantation of 2D cultured hMSC, which remains one of the major barriers for biomanufacturing of clinical grade hMSC for stroke patients. To preserve and enhance stem cell properties during hMSC culture, preconditioning strategies such as 3D aggregation and hypoxia culture have been studied widely and show promising results both for *in vitro* and *vivo* hMSC efficacy. However, preclinical studies with preconditioned hMSC for stroke treatment still can provide inconsistent outcomes due to donor age and morbidity. For instance, it is challenging to determine the quality of donors with diseases or from an aged population by efficient *in vitro* cellular assays without long-term expansion, as most clinical procedures mandate short duration culture and limited modifications of cells prior to implant. Magnetic resonance imaging (MRI) provides excellent soft tissue contrast and is commonly applied to stroke diagnosis in both clinical and preclinical studies. However, MR techniques also can provide feedback on the efficacy of stroke therapies applied after the initial diagnosis.

In this study, treatment efficacy between two hMSC donors is investigated in a rat model of transient ischemia created by middle cerebral artery occlusion (MCAO). First, several key *in vitro* characteristics of hMSC with 3D aggregation preconditioning were analyzed. Second, to monitor the *in vivo* therapeutic effect from transplanted hMSC (health donor, compromised donor versus saline-injected immediately after MCAO), biochemical markers were measured longitudinally over 21 d using sodium chemical shift imaging (^{23}Na CSI), proton relaxation-enhanced MR spectroscopy (RE-MRS) and T_2-weighted proton imaging. Applying these techniques at the ultra-high field of 21.1 T provided increased sensitivity, enabling insight into ionic and metabolic homeostasis and displaying the dependence of tissue recovery on donor metrics.

In vitro results showed that hMSC from different donors have similar senescence, proliferation, migration and immunomodulation at passage 5, which is expansion point used for transplantation in MCAO rats. The functional decline in these assays for the compromised donor only can be observed beyond passage 7. *In vivo* results indicate ischemic lesion volume, as determined by ^{23}Na CSI, is significantly reduced by day 3 for the healthy donor hMSC cohort compared to the compromised donor and saline-injected MCAO controls, which both display expansion of the lesion. Percent change in both ^{23}Na and ^{1}H absolute signal within the ischemic lesion also is significantly reduced by day 3 for the healthy donor compared to the compromised donor and saline MCAO controls. Furthermore, widespread systemic effects are measurable as early as 24 h post-transplantation for ^{23}Na signal levels in the contralateral hemisphere, with significant elevation seen for the compromised and saline cohorts compared to the healthy donor, which displays a longitudinally stable and lower sodium signal. RE-MRS results support the ^{23}Na CSI analysis with reduced lactate-to-choline levels by day 3 and increased N-acetyl-aspartate and creatine ratios at day 1 for the healthy donor cohort only. In summary, *in vivo* MRI analysis demonstrates the possibility to evaluate donor efficacy of hMSC in preclinical studies and remedy inaccurate, low passage *in vitro* assessments based on traditional stem cell properties.

This work is supported by NIH (R01 NS102395).

602137 Engineered CRISPR/Cas12a System As a Sensitive Diagnostic Tool for Detecting HCV, HIV, and SARS-CoV-2

Abstract:

CRISPR/Cas (Cluster of Regularly Interspaced Short Palindromic Repeats/CRISPR-associated protein) systems have emerged not only as a precise gene-editing tool but also as a new sensitive diagnostic tool due to additional functionality of some newly discovered Cas proteins. Class II effector proteins, such as Cas12a, are considered as a better alternative to the CRISPR/Cas9 system due to their compact size, specificity, and efficiency. Besides executing successful gene editing, once Cas12a forms a ternary complex with guide RNA (crRNA) and target dsDNA and executes a double-stranded break (referred to as cis-cleavage), it executes nonspecific cleavage of ssDNA. This attribute, known as trans-cleavage, is only activated once bound to a target that has complementary base-pairing to the guide crRNA. When combined with a FRET-based reporter, a fluorophore connected to a quencher via a short oligonucleotide, the presence of the target activator can be validated. Such a phenomenon has been efficiently harnessed by SHERLOCK (Specific High-sensitivity Enzymatic Reporter unLOCKing) and DETECTR (DNA Endonuclease Targeted CRISPR Trans Reporter) to reliably detect nucleic acids.

Though rising as a versatile diagnostic tool that can detect nucleic acids with a high level of precision and accuracy based on complementary base pairing principle, current CRISPR/Cas12a systems are limited to a nanomolar detection limit without an amplification step. We, therefore, sought to overcome this limitation by finding ways to engineer CRISPR/Cas systems to enhance their sensitivity for nucleic acid detection. We extended the 3'- or 5'-ends of the crRNA with different lengths of DNA, RNA, and phosphorothioate DNA. Through selective engineering and design, we achieved amplified Cas12a trans-cleavage activity as high as 3.5-fold with our modified crRNA compared to the wild type crRNA on the target dsDNA eGFP fragment. Employing this phenomenon, we optimized a fluorescence-based detection assay that enabled us to detect nucleic acids without an amplification step with unprecedented sensitivity and with a limit of detection in femtomolar scale. This engineered crRNA-Cas12a was successfully applied towards detecting clinically relevant targets such as HCV, HIV, and SARS-CoV-2.

When combined with an isothermal target pre-amplification step such as RPA/LAMP, we observed an enhance in sensitivity down to a few copies of nucleic acids with our engineered crRNA-Cas12a system. Moreover, we combined this platform with lateral flow assay to create a rapid, portable, and inexpensive point-of-care paper strip tests that can be used by nonprofessionals. As a proof of concept, the paper strip test successfully displayed visibly positive signal when targeting synthetic RNA of SARS-CoV-2 down to 3 copies. Furthermore, such engineered point-of-care diagnostics could be extended to various other infectious diseases.

597876 CRISPR-Cas12a Mediated Universal Electrochemical Biosensing Platform (Faculty/Industry Candidate)

Abstract:

An accurate, rapid, and cost-effective biosensor for the quantification of disease biomarkers is vital for the development of early-diagnostic point-of-care systems. The recent discovery of the trans-cleavage property of CRISPR type V effectors makes CRISPR a potential high-accuracy bio-recognition tool. Herein, a CRISPR-Cas12a (cpf1) based electrochemical biosensor (E-CRISPR) is demonstrated, which is more cost-effective and portable than optical-transduction-based biosensors. Through optimizing the in vitro trans-cleavage activity of Cas12a, E-CRIPSR was used to detect viral nucleic acids, including human papillomavirus 16 (HPV-16) and parvovirus B19 (PB-19), with a picomolar sensitivity. An aptamer-based E-CRISPR cascade was further designed for the detection of transforming growth factor β1 (TGF-β1) protein in clinical samples. As demonstrated, E-CRISPR could enable the development of portable, accurate, and cost-effective point-of-care diagnostic systems.

613020 Building Accessible Hands-on Biology Classroom Activities with Sensors and Freeze-Dried Cell-Free Technology (Invited Speaker)

Abstract:

Hands-on demonstrations greatly enhance the teaching of STEM concepts and foster engagement and exploration in the sciences. While numerous chemistry and physics classroom activities exist, few biology labs are practical and accessible due to the challenges and concerns of growing live cells in classrooms. Here we introduce a platform to develop hands-on molecular and synthetic biology educational activities based on easy-to-use, shelf-stable, freeze-dried, cell-free (FD-CF) reactions, which are simply activated by water. By using DNA which encodes fluorescent proteins to create visual outputs with these FD-CF reactions, we created a variety of engaging modules to teach gene expression, cellular functions, and other basic molecular biology topics that are otherwise difficult to easily teach in a hands-on manner in a classroom. By expanding the platform to incorporate sensor-based components, such as RNA switches, we also developed modules that can teach interdisciplinary biology topics, including biochemistry, biomaterials, and synthetic biology, as well as basic laboratory skills such as pipetting, experimental design, and the scientific method. Our initial kit launched from this platform has already been used in hundreds of classrooms/programs across the country, and initial results and feedback suggest that the activities are accessible, easy to use, educational, and engaging for middle school, high school, and undergraduate students. Overall, the platform introduces low-cost, user-friendly, and hands-on activities that can be used in classrooms to improve the quality of biology education and open the door for student-driven, independent explorations in the life sciences.

609974 Investigating the Effects of Brainstem Neuronal Adaptation on Cardiovascular Homeostasis

Abstract:

Central coordination of cardiovascular function is accomplished, in part, by the baroreceptor reflex, a multi-input multi-output physiological control system that regulates the activity of the parasympathetic and sympathetic nervous systems via interactions among multiple brainstem nuclei. Recent single-cell analyses within the brain revealed that individual neurons within and across brain nuclei exhibit distinct transcriptional states contributing to neuronal function. Such transcriptional heterogeneity, complicates the task of understanding how neurons within and across brain nuclei organize and function to process multiple inputs and coordinate cardiovascular functions within the larger context of the baroreceptor reflex. However, prior analysis of brainstem neurons revealed that single-neuron transcriptional heterogeneity reflects an adaptive response to synaptic inputs and that neurons organize into distinct subtypes with respect to synaptic inputs received. Based on these results, we hypothesize that adaptation of neuronal subtypes support robust biological function through graded cellular responses. We test this hypothesis by examining the functional impact of neuronal adaptation on parasympathetic activity within the context of short-term baroreceptor reflex regulation. In this work, we extend existing quantitative closed-loop models of the baroreceptor reflex by incorporating into the model distinct input-driven neuronal subtypes and neuroanatomical groups that modulate parasympathetic activity. We then use this extended model to investigate, via simulation, the functional role of neuronal adaptation under conditions of health and systolic heart failure. Simulation results suggest that parasympathetic activity can be modulated appropriately by the coordination of distinct neuronal subtypes to maintain normal cardiovascular functions under systolic heart failure conditions. Moreover, differing degrees of adaptation of these neuronal subtypes contribute to cardiovascular behaviors corresponding to distinct clinical phenotypes of heart failure, such as exercise intolerance. Further, our results suggest that an imbalance between sympathetic and parasympathetic activity regulating ventricular contractility contributes to exercise intolerance in systolic heart failure patients, and restoring this balance can improve the short-term cardiovascular performance of these patients.

611043 Single-Cell Immune Landscape of Influenza Infection in the Mouse Lung: Implications of Aging and Genetic Background

Abstract:

Influenza virus infection poses a significant global disease burden, causing an estimated three to five million cases of severe illness and 290,000 – 650,000 deaths each year worldwide. While vaccination has been successful in generating long-lasting immunity to many other viral infections, the current influenza vaccine effectiveness in the U.S. fluctuates from 60% to as low as 10% for a given year, even during flu seasons when the vaccine strain matches the circulating virus. The reason for the underwhelming protection has been partially attributed to problems with vaccine immunogenicity and host-specific factors such as aging, genetic background, and pre-existing immunity. However, due to the complexity of the immune landscape, it has been difficult to comprehensively characterize how these factors affect the immune response. In this work, we demonstrate the utility of mass cytometry (CyTOF), a technology that allows the simultaneous detection of 40 unique biomarkers on single cells, to comprehensively profile the immune response to influenza infection in mice of different ages and genetic background. To support the rapid analysis of such a large high dimensional dataset and reduce human bias, we developed a customizable and automated gating approach based on a probabilistic support vector machine (SVM) classifier to enable efficient, automated classification of over 30 million cells into 24 defined cellular subsets with high accuracy (97%). Analysis of the resulting data revealed significant discrepancies in immune cell infiltration, cellular activation, and viral infectivity due to both aging and genetics. Taken together, we show that such a high-resolution approach, combined with high-dimensional data analysis methods, permits a systems-level exploration of the immune landscape, providing new cellular and molecular insights into the role of host specific-factors during influenza infection.

601098 Modeling Viral Replication and Cytokine Response to Reveal Mechanisms of Inflammation

Abstract:

Influenza remains a global health concern, causing an annual 3 - 5 million infections and 290,000 – 650,000 deaths[1]. The disease presents a significant pediatric burden, causing 374,000 annual hospitalizations of children under 1 years old, globally[2]. Seasonal monovalent vaccines have achieved only 69% efficacy[3], leaving room for improvement through fundamental understanding of the immune response to viral detection. The adaptive immune response, vaccines' primary route of protection, is predicated by the actions of the innate immune response to create an immunocompetent environment during vaccination or infection[4]. Improper innate immune activation has been implicated in a wide range of pathologies from inflammatory bowel[5] and liver[6] diseases to Alzheimer's Disease[7] and beyond.

Immune responses can help or hinder an organism's ability to overcome an infection; excessively inflammatory responses[8] can cause greater tissue damage, higher mortality, and slow recovery[9], while an adequate and well-regulated response to a threat is a prerequisite for survival and the management of the adaptive immune system. Thus, tight regulation of the immune response is critical. The first step to occur during an immune response is the detection of the pathogens, leading to the early, localized, innate immune response[10]. This response to viral infection leads to the production of Type I Interferons (IFNs). Interferons serve to establish an antiviral state by activating and inducing Mx proteins, RNA-activated protein kinase, and the 2-5A system[11]; they also regulate other immune responses by acting on NK cells, T cells, B cells, DCs, and phagocytic cells[12]. Without the initial sensing protein activity, no immune responses would be mounted. Understanding the dynamics of these sensors is vital to quantifying the innate immune response. The presence of influenza virus is primarily sensed by cytoplasmic retinoic acid-inducible gene 1 (RIG-I) and endosomal Toll-like receptors (TLR)[13,14]. RIG-I senses viral RNA in the cytoplasm[15], but is antagonized by influenza A nonstructural protein I (NS1) to varying, strain-specific magnitudes[16,17]; TLR7 is free of this antagonism[18,19], but is only activated after the influenza envelop has been degraded by endosomal proteases[20]. The activation of either sensor leads to the phosphorylation of IRF7 and the production of IFN to act as a signaling cytokine. IFN induces secondary messenger molecules, ultimately leading to the induction of immune activity, inflammation, and antiviral genes[21]

Ordinary Differential Equations (ODEs) have become a common approach in systems biology after their demonstrable success in analyzing the robustness of biological signaling[22–24], the highly dynamic behaviors of NF-kB[25], and Xenopus oocytes' ultrasensitive cell fate binary response[26]. ODE's allow for interpolation of the dynamics between a finite number of time points at which RNA profiles have been measured, based on hypothesis of the mechanisms regulating the system's components.

Here, a novel ODE model was constructed to quantify the innate immune response of human bronchial epithelial cells (HBECs) to influenza A infection. The model incorporates both the classical viral dynamic model[27,28] and the proportionality of sensor protein activity to vRNA levels in the cytoplasm, the first such integration of intracellular immune dynamics and viral replication. The feedback of interferon production on viral replication through the induction of the interferon stimulated gene (ISG) family[28] is included. Sensitivity analysis of the model revealed IRF7 phosphorylation and induction as the primary drivers of IFN production in the model. An NS1 knockout strain was simulated as a validation study, revealing lower cell lethality, lower peak viral titers, and a stronger cytokine response, with only the removal of RIG-I antagonism. This work establishes the first cell-level model of interferon signaling induced by influenza infection that can be used to compare host responses between infections with different infecting agents and antagonism motifs through modification of the sensor proteins' kinetics

1. Influenza (Seasonal). Available at: https://www.who.int/news-room/fact-sheets/detail/influenza-(seasonal). (Accessed: 17th September 2019)
2. Lafond, K. E. *et al.* Global Role and Burden of Influenza in Pediatric Respiratory Hospitalizations, 1982–2012: A Systematic Analysis. *PLOS Med.* **13**, e1001977 (2016).
3. Osterholm, M. T., Kelley, N. S., Sommer, A. & Belongia, E. A. Efficacy and effectiveness of influenza vaccines: a systematic review and meta-analysis. *Lancet Infect. Dis.* **12**, 36–44 (2012).
4. Tritto, E., Mosca, F. & De Gregorio, E. Mechanism of action of licensed vaccine adjuvants. *Vaccine* **27**, 3331–3334 (2009).
5. Knutson, C. G. *et al.* Chemical and cytokine features of innate immunity characterize serum and tissue profiles in inflammatory bowel disease. *Proc. Natl. Acad. Sci. U. S. A.* **110**, E2332–E2341 (2013).
6. Xu, R., Huang, H., Zhang, Z. & Wang, F. S. The role of neutrophils in the development of liver diseases. *Cellular and Molecular Immunology* **11**, 224–231 (2014).
7. Guillot-Sestier, M.-V. & Town, T. Innate Immunity in Alzheimer's Disease: A Complex Affair.
8. Cilloniz, C. *et al.* Lethal Dissemination of H5N1 Influenza Virus Is Associated with Dysregulation of Inflammation and Lipoxin Signaling in a Mouse Model of Infection. *J. Virol.* **84**, 7613–7624 (2010).
9. Peiris, J. S. M., Cheung, C. Y., Leung, C. Y. H. & Nicholls, J. M. Innate immune responses to influenza A H5N1: friend or foe? *Trends Immunol.* **30**, 574–84 (2009).
10. Sun, L., Liu, S. & Chen, Z. J. SnapShot: Pathways of Antiviral Innate Immunity. *Cell* **140**, 436-436.e2 (2010).
11. Trinchieri, G. Type I interferon: friend or foe? *J. Exp. Med.* **207**, 2053–63 (2010).
12. Prchal, M. *et al.* Type I interferons as mediators of immune adjuvants for T- and B cell-dependent acquired immunity. *Vaccine* **27**, G17–G20 (2009).
13. Opitz, B. *et al.* IFN? induction by influenza A virus is mediated by RIG-I which is regulated by the viral NS1 protein. *Cell. Microbiol.* **9**, 930–938 (2007).
14. Wu, W. *et al.* RIG-I and TLR3 are both required for maximum interferon induction by influenza virus in human lung alveolar epithelial cells. *Virology* **482**, 181–188 (2015).
15. Fujita, T., Onoguchi, K., Onomoto, K., Hirai, R. & Yoneyama, M. Triggering antiviral response by RIG-I-related RNA helicases. *Biochimie* **89**, 754–760 (2007).
16. Gack, M. U. *et al.* Influenza A Virus NS1 Targets the Ubiquitin Ligase TRIM25 to Evade Recognition by the Host Viral RNA Sensor RIG-I. *Cell Host Microbe* **5**, 439–449 (2009).
17. Rajsbaum, R. *et al.* Species-Specific Inhibition of RIG-I Ubiquitination and IFN Induction by the Influenza A Virus NS1 Protein. *PLoS Pathog.* **8**, e1003059 (2012).
18. Takeda, K., Kaisho, T. & Akira, S. T OLL -L IKE R ECEPTORS. *Annu. Rev. Immunol.* **21**, 335–376 (2003).
19. Diebold, S. S., Kaisho, T., Hemmi, H., Akira, S. & Reis e Sousa, C. Innate antiviral responses by means of TLR7-mediated recognition of single-stranded RNA. *Science* **303**, 1529–31 (2004).

20. Diebold, S. S., Kaisho, T., Hemmi, H., Akira, S. & Reis e Sousa, C. Innate antiviral responses by means of TLR7-mediated recognition of single-stranded RNA. *Science* **303**, 1529–31 (2004).
21. Le Page, C., Génin, P., Baines, M. G. & Hiscott, J. Interferon activation and innate immunity. *Rev. Immunogenet.* **2**, 374–86 (2000).
22. Kitano, H. Biological robustness. *Nat. Rev. Genet.* **5**, 826–837 (2004).
23. Schoeberl, B., Eichler-Jonsson, C., Gilles, E. D. & Müller, G. Computational modeling of the dynamics of the MAP kinase cascade activated by surface and internalized EGF receptors. *Nat. Biotechnol.* **20**, 370–375 (2002).
24. Shoemaker, J. E. *et al.* Fathead minnow steroidogenesis: in silico analyses reveals tradeoffs between nominal target efficacy and robustness to cross-talk. *BMC Syst. Biol.* **4**, 89 (2010).
25. Hoffmann, A., Levchenko, A., Scott, M. L. & Baltimore, D. The IκB-NF-κB Signaling Module: Temporal Control and Selective Gene Activation. *Science (80-.).* **298**, 1241–1245 (2002).
26. Ferrell, J. E. *et al.* The biochemical basis of an all-or-none cell fate switch in Xenopus oocytes. *Science* **280**, 895–8 (1998).
27. Ho, D. D. *et al.* Rapid turnover of plasma virions and CD4 lymphocytes in HIV-1 infection. *Nature* **373**, 123–126 (1995).
28. Baccam, P., Beauchemin, C., Macken, C. A., Hayden, F. G. & Perelson, A. S. Kinetics of influenza A virus infection in humans. *J. Virol.* **80**, 7590–9 (2006).

610790 Proteomic Profiling Identifies Novel Biomarkers of Aging and Potential Senolytic Therapeutic Targets

Abstract:

Cellular senescence is an irreversible program of cell cycle arrest to restrict proliferation. The irreversible mechanism of senescence constitutes a barrier against cancer cell development. In addition, senescent cells interact with their microenvironment through secretion of numerous inflammatory signals which modulate immune response and wound healing. As senescent cells accumulate in the aging tissue, they contribute to chronic inflammation and age-related diseases. Because of the importance of senescent cells in determining fate of human tissues, organs and lifespan, it is crucial to effectively identify senescent cells with specific biomarkers in order to a) measure the degree of aging and damage in tissue and b) be able to therapeutically target senescent cells for removal (aka senolytic therapies). Here, we used mass spectrometry-based proteomics to identify biomarkers of aging in primary human mammary epithelial cells (HMECs) undergoing senescence. We performed untargeted proteomic profiling on three conditions: 1) primary HMECs undergoing their natural replicative senescence, 2) senescent luciferase-expressing HMECs vs. immortalized telomerase-expressing HMECs, and 3) HMECs induced to senescence by inhibition of nucleotide synthesis. Through statistical analysis, we have identified a group of 38 proteins that are consistently upregulated and downregulated in senescent HMECs. Importantly, we then use this "senescence signature" to predict novel senolytic drug targets through bioinformatic analysis of publicly available gene essentiality screening databases (e.g., the Cancer Dependency Map). In summary, we report a novel senescence signature derived from proteomic analysis that can serve as novel biomarkers of senescence and provide new targets for designing senolytic therapies.

611328 Machine Learning-Driven Vaccine Design Against Highly Mutable Pathogens

Abstract:

Vaccines play a critical role in both preventing and eliminating global pandemics. This is especially true in the case of highly mutable pathogens like HIV and influenza, which present unique challenges to vaccine design. One promising strategy for designing vaccines against highly mutable pathogens is to target regions on the surfaces of the pathogenic proteins that are functionally important (e.g., receptor binding sites for adhering to/entering host cells), and thus cannot be so easily mutated by the pathogen. Antibodies that can bind to these 'conserved' regions on pathogenic proteins are called broadly-neutralizing antibodies, or bnAbs. BnAbs have now been isolated from people naturally infected with many different highly mutable pathogens. However, in most cases it remains unclear how to elicit these same antibodies by vaccination, in part because of the complexity of the key biological processes that occur but also because of the vast design space for rationally designing such vaccines. Design variables include the number of immunizations to administer, the number of Ags (pathogen-like proteins) to administer in each immunization, the concentration of each Ag in each immunization, and the sequences of the Ags. This talk will discuss a novel computational/data science-driven approach to efficiently traverse this vast design space. Our approach features agent-based modeling of cell-cell interactions combined with deep reinforcement learning (DRL) to steer the evolution of antibodies into bnAbs following vaccination. Using a rewards and punishments-based system, results show that DRL is able to rapidly identify optimal vaccination protocols that maximize the formation of bnAbs against highly mutable pathogens.

613348 Towards Novel Angiogenesis Therapies: Explorations of Non-Canonical Angiogenic Signaling (INVITED SPEAKER)

Abstract:

Several diseases are attributed to either deficient or excessive vascularization, so understanding angiogenesis offers great promise for improving vascular therapies. We recently discovered non-canonical, high-affinity binding of platelet derived growth factors (PDGFs) to the primary angiogenic receptor, vascular endothelial growth factor receptor-2 (VEGFR2). Delineating this currently undefined PDGF role in VEGFR2 signaling will provide new tactics for controlling angiogenesis in health and disease. Towards this aim, we computationally model binding and trafficking; experimentally measure signaling and angiogenic function; and molecular interactions. Our mass-action kinetics models predict significant PDGF:VEGFR2 binding when VEGFR2 is overexpressed and during anti-VEGF therapy. Our signaling assays reveal selective activation of VEGFR2 and activation of the common angiogenic signaling molecules: Src kinase (Src), focal adhesion kinase (FAK), phospholipase-Cγ (PLCγ), and phosphoinositide 3-kinase (PI3K). Furthermore, PDGF treatment induces cell proliferation, an angiogenic hallmark, and select PDGFs mediate cell migration. Finally, stable PDGF:VEGFR2 binding conformations are observed via molecular dynamics (MD) simulations. Our modeling, signaling, and functional analyses define the importance of PDGF:VEGFR2 signaling. Further exploration of this non-canonical signaling should drive the development of new treatments for angiogenesis-related diseases.

Acknowledgements: This material is based upon work supported by the National Science Foundation (NSF) and American Heart Association (AHA) under grant numbers: NSF CAREER: 1653925; NSF BPE: 1648454; and AHA:16SDG26940002.

611834 Elucidating Molecular Origins of Drug Selectivity for Rational Design of Selective Drugs

Abstract:

Proteins families that evolve to bind particular ligand also contain diverse members that exhibit a wide range of affinities for that particular ligand. However, this variance in affinity resulting from evolution of the family members is usually at odds with the high sequence and structure similarity shared by its members. In this talk, we discuss three such protein families i.e. human protein kinases, Cannabinoid receptors and Strigolactone receptors in plants. These proteins show several orders of magnitude difference in their ligand affinity but share high sequence and structural similarity. The characterization of the free energy landscapes for ligand binding starting could shed light on the molecular origin of their ligand selectivity mechanisms. In order to achieve this goal, we have develop computational methods for evaluation of their binding mechanism using molecular simulations. These examples served as model system for obtaining generalizable insights about the free energy landscapes of ligand binding that could help design of drugs with desired selectivity profiles.

600211 Computer Aided Generation of Diverse and Synthesizable Molecular Library

Abstract:

Modern drug discovery is extending into chemical spaces with increased complexity. It is desirable to explore these spaces more efficiently. In order to maximize the information gained from testing new molecules, the concept of molecular library has been widely adopted in medicinal chemistry.

Molecular libraries are collections of compounds that have, or expected to have, similar structures or properties. Accessing all molecules in a molecule library can enable the rapid generation of structure-activity relationships during the hit-to-lead or lead optimization phases, which can greatly enhance the efficiency of drug discovery. Small molecule libraries can be constructed in many different ways. A library may be a collection of compounds extracted from the literature with similar functionality;[1,2] or it may be designed by enumerating possible side groups as decorations of a common core scaffold;[3–5] it may also be designed using any one of the increasing number of *in silico* tools for the generation of drug-like compound libraries.[6–8]

Despite the significant progress in molecular library generation, there is a usually a tradeoff between accessibility and diversity – libraries that are constructed by simply substituting functional groups on the same scaffold are usually limited in diversity, and libraries with larger structural diversity tend not to be all accessible through similar reactions. In this work, we leverage informatics tools to maximize the diversity of molecular libraries as well as their synthetic accessibility. Starting from a given molecule, using retrosynthesis analysis and reaction prediction models, we can generate libraries that can be accessed through alternative late-stage functionalization. The diversity of the molecular libraries is assessed by average pairwise similarity, so that the most diverse molecular library can be identified. This work provides insight on expanding the exploration of surrounding chemical space of a molecule with minimal additional effort.

1 U. F. Röhrig, S. R. Majjigapu, P. Vogel, V. Zoete and O. Michielin, *J. Med. Chem.*, 2015, **58**, 9421–9437.

2 B. T. Xin, E. M. Huber, G. De Bruin, W. Heinemeyer, E. Maurits, C. Espinal, Y. Du, M. Janssens, E. S. Weyburne, A. F. Kisselev, B. I. Florea, C. Driessen, G. A. Van Der Marel, M. Groll and H. S. Overkleeft, *J. Med. Chem.*, 2019, **62**, 1626–1642.

3 K. Kannan Sivaraman, A. Paiardini, M. Sieńczyk, C. Ruggeri, C. A. Oellig, J. P. Dalton, P. J. Scammells, M. Drag and S. McGowan, *J. Med. Chem.*, 2013, **56**, 5213–5217.

4 R. Fleeman, T. M. Lavoi, R. G. Santos, A. Morales, A. Nefzi, G. S. Welmaker, J. L. Medina-Franco, M. A. Giulianotti, R. A. Houghten and L. N. Shaw, *J. Med. Chem.*, 2015, **58**, 3340–3355.

5 G. M. Keseru, D. A. Erlanson, G. G. Ferenczy, M. M. Hann, C. W. Murray and S. D. Pickett, *J. Med. Chem.*, 2016, **59**, 8189–8206.

6 R. Gómez-Bombarelli, J. N. Wei, D. Duvenaud, J. M. Hernández-Lobato, B. Sánchez-Lengeling, D. Sheberla, J. Aguilera-Iparraguirre, T. D. Hirzel, R. P. Adams and A. Aspuru-Guzik, *ACS Cent. Sci.*, 2018, **4**, 268–276.

7 M. H. S. Segler, T. Kogej, C. Tyrchan and M. P. Waller, *ACS Cent. Sci.*, 2018, **4**, 120–131.

8 A. Gupta, A. T. Müller, B. J. H. Huisman, J. A. Fuchs, P. Schneider and G. Schneider, *Mol. Inform.*, 2018, **37**, 1700111.

605512 *De Novo* Design of Engineered Cytokine Mimetics As Targeted Immunotherapeutics.

Abstract:

Cytokines are soluble factors that signal through stimulation of their cognate transmembrane receptors on target cells to perform critical biological functions, particularly those related to immune homeostasis. As a result of their essential immune activities, cytokines have great potential as immunotherapeutics, both to activate the immune response to fight diseases such as cancer and infectious diseases, as well as to suppress the immune response to treat autoimmune disorders or for transplantation medicine. However, the pleiotropic activities and unfavorable pharmaceutical properties of natural cytokines have limited their clinical performance.

Recent efforts to re-engineer native cytokines to improve their selectivity and stability have significantly advanced progress in cytokine therapeutic development. However, there are inherent challenges to developing therapeutics from natural proteins. First, most natural cytokines are only marginally stable, hence mutations aimed at increasing efficacy can decrease expression or cause aggregation, making manufacturing and storage difficult. More substantial changes, such as the deletion or fusion of functional or targeting domains, can also dramatically alter pharmacokinetic properties and tissue penetration. In addition, immune responses targeting a variant of a natural protein may cross-react with the endogenous molecule with potentially catastrophic consequences. Finally, the high potency and multifarious functions of cytokines hinder their efficacy and can result in harmful off-target effects and systemic toxicities.

Here, we combine a best-in-class computational protein design software with directed evolution technologies to generate *de novo* cytokine mimetics that circumvent limitations for naturally-derived cytokine drugs. Rather than modifying existing molecules, we adopt a bold new approach to create exceptionally stable proteins with customized receptor interaction properties as robust and efficacious targeted therapeutics. We applied this strategy to develop a mimetic of the interleukin-2 (IL-2) cytokine, which plays a pivotal role in T and natural killer (NK) cell function. We created a hyper-stable *de novo* protein that is biased toward activation of IL-2 signaling pathways on immune effector cells, and showed that this molecule inhibits tumor growth and prolongs survival in mouse models of colorectal cancer and melanoma, while also mitigating the toxic side effects typically associated with IL-2 therapy. We built upon this exciting advance to design *de novo* cytokines that mimic IL-4, which promotes pro-regenerative immune programs following tissue damage. We demonstrated the extreme thermal stability and biased receptor activity of our engineered IL-4 mimetic, and showed that it elicited pro-regenerative immune responses in both cellular and animal models. Collectively, our work pioneers a novel cytokine engineering platform that integrates computational and experimental approaches to establish a general paradigm for advancing the clinical translation of cytokine therapeutics.

606100 Integrated Prediction of Chemical Effects on Important Pathways of Embryo Development Based on in Vitro Embryonic Stem Cell Test and Machine Learning

Abstract:

Currently, there are thousands of chemicals in common use; however, only a small number of them have sufficient toxicity evaluation, and even less information for the specific embryo-related toxicity field: embryotoxicity and developmental toxicity. Embryonic stem cell test (EST) is the only accepted *in vitro*method for assessing embryotoxicity without animal sacrifice. However, EST for regulatory embryotoxicity screening are impeded by its complexity, time-consuming, limited differentiation lineage endpoints, and poor availability for use to develop a reliable prediction model (PM) for embryotoxicity assessments.

A key step in the understanding of embryo development is to construct a systematic and comprehensive model of network with the development-related pathways involved. In our research, we mainly focus on the essential markers in the pathways related to ESCs apoptosis and proliferation, pluripotency, and specific-lineage differentiation. Quantitative structure-activity relationship (QSAR) models which predict the biological activities from chemical structures information is an efficient and less time-consuming strategy compared to experimental-based studies. Therefore, an integrated prediction pipeline for *in vitro*high throughput screening (HTS) based on engineered murine embryonic stem cells (mESCs) expressing a reporter, enhanced green fluorescent protein (EGFP), driven by human promoters for key genes in the pathways associated with apoptosis, pluripotency, and tissue-specific differentiation cultured in novel multi-well plates is developed and used to obtain new embryotoxicity and developmental toxicity data needed for the training of machine learning models such as support vector machines (SVM) and deep neural network (DNN). The application of advanced machine learning algorithms can significantly improve the accuracy and efficiency of the PM for embryotoxicity assessment of compounds with unknown developmental toxicity. This study will provide the technology needed for fast and accurate embryotoxicity assessments. The data obtained from our integrated analysis can provide the information needed for the regulation of industrial and emerging chemicals.

607548 A Computational Approach to Identifying Synergistic Compounds for Treating Antimicrobial-Resistant Pathogens

Abstract:

Antimicrobial resistance (AMR) is increasingly a threat to global public health [1]. AMR-related infections limit the efficacy of lifesaving antibiotic therapy necessary for the treatment of infectious disease and limit the success of advanced surgical procedures like organ transplants [2, 3]. To combat AMR pathogens, either new antibiotics or the cocktails of existing antibiotics with inhibitors of antimicrobial resistance proteins should be explored. Since the development of new antibiotic has been slowed down significantly recently, identifying compounds/antimicrobials with synergistic effects becomes the major approach to addressing the AMR crisis. In this work, we demonstrate a computational approach to identifying plant compounds that have synergistic effect with Fosfomycin to inhibit *Listeria monocytogenes*, a food-borne pathogen that could cause listeriosis disease, especially on immune-compromised people [4, 5]. Experimental results further validate the our finding.

L. *monocytogenes* was found to infect and adversely affect patients' liver and spleen. In addition, *L. monocytogenes* can penetrate blood-brain barrier and blood-placenta barriers to harm the central neural system of pregnant woman and infant [4]. Since *L. monocytogenes* is an intracellular pathogen, antimicrobials used to treat listeriosis should be able to be transported into host cells. Penicillin, ampicillin and amoxicillin were commonly used antibiotics in the treatment of listeriosis [6]. However, antibiotic resistance genes have been continuously found in *Listeria* strains. For example, the strain that has resistance to Penicillin G was isolated from vegetables in 2016 [4, 7]. As a natural product, Fosfomycin was found effective against clinical isolates of *L. monocytogenes* and used as a novel therapeutic antibiotic for listeriosis clinical treatment [8]. In addition, Fosfomycin is able to penetrate the blood-brain barrier and reach clinically relevant concentrations. Thus, it has the potential to eliminate *L. monocytogenes* which would cause neuron damage[9]. However, stronger Fosfomycin resistance was found in the *L. monocytogenes* isolates with Fosfomycin resistance proteins detected [10]. In particular, a resistant gene FosX (LMO1702,402bp) was identified and expressed in *L. monocytogenes* EGDe (strain ATCC BAA-679), a typical well-studied strain [11]. The FosX enzyme catalyzed the hydrolysis of Fosfomycin and resulted in the Fosfomycin resistance in *L. monocytogenes* EGDe [5, 10]. Therefore, there is an urgent need to identify compounds that can inhibit FosX enzyme to revive the efficacy of Fosfomycin to treat *L. monocytogenes.*

Automated molecular docking is the most commonly used computational approach that evaluates the binding of small-molecule ligands like compounds to a target receptor with a known crystal 3D structure [12]. Molecular docking provides an avenue for a high-throughput virtual screening of ligands, and it has been widely implemented in drug discovery research for hit identification[13]. Docking programs have been improved recently to provide more accurate prediction on ligand-target binding by optimizing docking algorithms and scoring functions [14]. Among those existing docking programs, Molsoft ICM was evaluated with 93% accuracy in flexible docking and 90% successful rate in covalent docking. This was significantly better than the performance of Autodock, DOCK, FlexX, Gold, FITTED and MOE [13-15]. Since structures and activities of the FosX protein in *L. monocytogenes* have been well studied, we used an integrated ICM-docking and experimental approach to identify FosX inhibitors that are of synergistic effect with Fosfomycin in treating resistant *L. monocytogenes*. Specifically, automated ligand docking was implemented to perform virtual screening of Indofine natural-product database and FDA-approved drugs to identified potential inhibitors. *In vitro* bacterial growth inhibition test was then utilized to verify the effectiveness of identified compounds combined with Fosfomycin on inhibiting the resistant *L. monocytogenes*. We demonstrated that two phenolic acids, i.e., caffeic acid and chlorogenic acid, were predicted as high-affinity FosX inhibitors from the ligand-docking platform. Experiments with these compounds indicated that: the cocktail of either caffeic acid (1.5mg/mL) or chlorogenic acid (3mg/mL) with Fosfomycin (50mg/L) was able to significantly inhibit the growth of the pathogen [16]. The finding of this work implies that the combination of Fosfomycin with either caffeic acid or chlorogenic acid is of potential to be used in the clinical treatment of *Listeria* infections.

References

1. Velez, R. and E. Sloand, *Combating antibiotic resistance, mitigating future threats and ongoing initiatives.* J Clin Nurs, 2016. **25**(13-14): p. 1886-9.
2. Michael, G.B., et al., *Emerging issues in antimicrobial resistance of bacteria from food-producing animals.* Future Microbiol, 2015. **10**(3): p. 427-43.
3. Davies, J. and D. Davies, *Origins and evolution of antibiotic resistance.* Microbiol Mol Biol Rev, 2010. **74**(3): p. 417-33.
4. Radoshevich, L. and P. Cossart, *Listeria monocytogenes: towards a complete picture of its physiology and pathogenesis.* Nat Rev Microbiol, 2018. **16**(1): p. 32-46.
5. Scortti, M., et al., *Epistatic control of intrinsic resistance by virulence genes in Listeria.* PLoS Genet, 2018. **14**(9): p. e1007525.
6. Temple, M.E. and M.C. Nahata, *Treatment of Listeriosis.* Annals of Pharmacotherapy, 2000. **34**(5): p. 656-661.
7. de Vasconcelos Byrne, V., et al., *Occurrence and antimicrobial resistance patterns of Listeria monocytogenes isolated from vegetables.* Brazilian Journal of Microbiology, 2016. **47**(2): p. 438-443.
8. Lepe, J.A., et al., *In vitro and intracellular activities of fosfomycin against clinical strains of Listeria monocytogenes.* International Journal of Antimicrobial Agents, 2014. **43**(2): p. 135-139.
9. Falagas, M.E., et al., *Fosfomycin.* Clin Microbiol Rev, 2016. **29**(2): p. 321-47.
10. Fillgrove, K.L., et al., *Mechanistic diversity of fosfomycin resistance in pathogenic microorganisms.* J Am Chem Soc, 2003. **125**(51): p. 15730-1.
11. Fillgrove, K.L., et al., *Structure and mechanism of the genomically encoded fosfomycin resistance protein, FosX, from Listeria monocytogenes.* Biochemistry, 2007. **46**(27): p. 8110-20.
12. Meng, X.Y., et al., *Molecular docking: a powerful approach for structure-based drug discovery.* Curr Comput Aided Drug Des, 2011. **7**(2): p. 146-57.
13. Kitchen, D.B., et al., *Docking and scoring in virtual screening for drug discovery: methods and applications.* Nat Rev Drug Discov, 2004. **3**(11): p. 935-49.
14. Lionta, E., et al., *Structure-Based Virtual Screening for Drug Discovery: Principles, Applications and Recent Advances.* Current Topics in Medicinal Chemistry, 2014. **14**(16): p. 1923-1938.
15. Scarpino, A., G.G. Ferenczy, and G.M. Keserű, *Comparative Evaluation of Covalent Docking Tools.* Journal of Chemical Information and Modeling, 2018. **58**(7): p. 1441-1458.
16. Zhang, F., et al., *Synergistic Effect of Chlorogenic Acid and Caffeic Acid with Fosfomycin in Growth Inhibition of a Resistant Listeria monocytogenes Strain.* ACS Omega, 2020. **https://dx.doi.org/10.1021/acsomega.0c00352**.

594532 Liquid Phase Membrane Supported Cyclical Flow Synthesis of Oligonucleotides for Therapeutic Applications

Abstract:

There now exists a consensus that RNA based gene inference techniques such as Anti-Sense Oligonucleotides and siRNA represent a significant therapeutic modality in the pharmaceutical industry. While the applications and understanding of the underlying biology of these systems for use as therapeutic platforms have advanced rapidly, little change in the commercial synthesis of these compounds has occurred. Metric tonnes per annum of oligonucleotides are projected to be required due to the high number of oligonucleotide-containing drugs in late phases of clinical trials and also projected for individual ASO therapies as larger patient populations are targeted in clinical trials following a number of clinical successes in small orphan patient populations.[1][2] The existing state-of-the-art synthesis of oligonucleotides is carried out with the use of a solid phase support and batch sizes currently limited to approximately 5 kg due to the constraints of solid phase synthesis.[3] In contrast, liquid phase supports reduce many of the limitations that are associated with solid phase supports.[4] Liquid phase synthesis of oligonucleotides was established by using linear and branched polyethylene glycol (PEG) as the soluble support[5] Membrane enhanced synthesis of oligonucleotides and peptides using PEG as a soluble support have been shown to be a potential solution for the problems associated with solid phase synthesis.[6][7]

The current work demonstrates proof of concept for a novel liquid phase manufacturing process utilizing cyclical flow synthesis combined with a novel membrane separation (**Figure 1**). This should enable fully scalable cyclical flow synthesis of oligonucleotides with increased productivity within a compact and scalable reactor footprint. This novel process intends to meet the growing demands of the industry and promises reduced capital costs and increased product quality control.

Modifications to soluble supports were utilized to both optimize the nucleotide loading process but also to optimize membrane flux and rejection factor, and hence yield and purity with commercially available membranes. Both liquid batch and phase cyclical flow synthesis of short oligonucleotides were optimized for coupling yield, phosphoramidite consumption, solvent consumption and overall productivity. The ability to use optimum reaction/separation timescales which are inherent to coupling reactions in oligonucleotide synthesis enabled improved system performance and in turn the ability to rapidly supply oligonucleotides on demand.

Acknowledgements:

This publication is supported by Science Foundation Ireland (SFI) through SSPC, The SFI Research Centre for Pharmaceuticals (Grant number 12/RC/2275_P2).

[1] Y. S. Sanghvi, *Curr. Protoc. Nucleic Acid Chem.* **2011**, 1–22.

[2] A. G. Molina, Y. S. Sanghvi, *Curr. Protoc. Nucleic Acid Chem.* **2019**, *77*, 1–17.

[3] A. R. Lajmi, L. Schwartz, Y. S. Sanghvi, *Org. Process Res. Dev.* **2004**, *8*, 651–657.

[4] P. R. J. Gaffney, J. F. Kim, I. B. Valtcheva, G. D. Williams, M. S. Anson, A. M. Buswell, A. G. Livingston, *Chem. - A Eur. J.* **2015**, *21*, 9535–9543.

[5] B. Burcovich, F. M. Veronese, V. Zarytova, G. M. Bonora, in *Nucleosides and Nucleotides*, Marcel Dekker Inc., **1998**, pp. 1567–1570.

[6] M. Mutter, H. Hagenmaier, E. Bayer, *Angew. Chemie* **1971**, *83*, 883–884.

[7] F. Brandstetter, H. Schott, E. Bayer, *Tetrahedron Lett.* **1973**, *14*, 2997–3000.

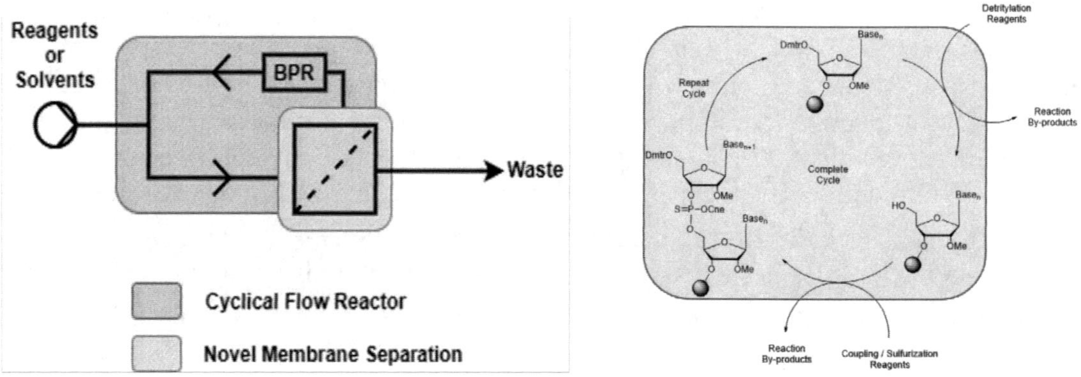

Figure 1: Schematic of system configuration (left) and Overview of cyclical synthetic route (right)

605370 Predict New Cocrystals Via Mechanochemistry

Abstract:

Within the pharmaceutical industry around 60% of recently developed drugs have solubility challenges while 80% of in-production line drugs are classified as being poorly soluble. Solubility is a critical property for drug performance (e.g. bioavailability) [1,2]. Cocrystal formation is one possible solution to overcome solubility problems. Pharmaceutical cocrystals, formed by combining an active pharmaceutical ingredient (API) with a cocrystal former (coformer), are often manufactured in a variety of methods that rely on the utilization of solvents. Pharmaceutical companies use between 25-100 kg solvent per kg of product, causing major water and air pollution. Hence, state-of-the-art solvent-based methods need to be replaced with more sustainable alternatives. Additionally, the formation of a cocrystal requires its components to be soluble in the same solvent. These limitations hinder the processability and development of novel pharmaceutical products.

Mechanochemistry utilizes mechanical forces to induce reactions. The application of mechanochemistry to form cocrystals without solvents is a novel, promising and sustainable method [1,3]. Scientists have exploited the features of mechanochemistry, but only in a random, non-systematic manner for the production of new cocrystals. Indeed, with the enormous number of potential API-coformer combinations, it is currently impossible to predict if two crystalline materials will cocrystallize under mechanochemical forces.

In this work, a high throughput screening approach combined with data analysis (machine learning) has been developed to predict if two crystals will cocrystallize under mechanochemical conditions. Strategically distinct pairs of APIs and crystalline coformers were screened by neat grinding (solventless) in a Retsch MM 400 vibration-mill. Powder X-ray diffraction has been applied to characterize all the samples and to determine the formation of new crystal structures. With the help of these results and molecular descriptors a machine learning algorithm (random forest model) has been successfully developed, effectively predicting the propensity of two molecules to cocrystallize under mechanochemical conditions. Additional cocrystallization events are predicted and experimentally validated. The technical feasibility of further API-coformer combinations can be estimated to create novel cocrystals without any solvent limitations, expanding the scope of mechanochemistry as a technique for drug discovery.

[1] C. B. Aakeröy and A. S. Sinha. Co-crystals. Monographs in Supramolecular Chemistry. The Royal Society of Chemistry, 2018.
[2] M. A. E. Yousef and V. R. Vangala. Pharmaceutical cocrystals: Molecules, crystals,formulations, medicines. Crystal Growth & Design, 19(12):7420–7438, 2019.
[3] S. Lou, Y. Mao, D. Xu, J. He, Q. Chen, and Z. Xu . Fast and Selective Dehydrogenative C–H/C–H Arylation Using Mechanochemistry. ACS Catalysis, 6 (6) , 3890-3894, 2016.

599914 Ratiometric Tuning of Fatty Acid Biosynthesis

Abstract:

Over the past few decades, growing—if sometimes wavering—concern over climate change has fueled research on the microbial biosynthesis of fatty acids, a promising source of precursors for renewable oleochemicals. Unfortunately, despite a multitude of successful efforts to engineer microorganisms to overproduce fatty acids, general strategies for controlling their titer, product distribution (i.e., chain lengths), and unsaturation fraction remain largely absent. With this presentation, I will discuss our efforts to combine a high-resolution transcriptional control system with a detailed kinetic model to control fatty acid biosynthesis in *Escherichia coli*. Our results indicate that independent adjustments to the ratios of FAS components can enable fine control over competing biochemical objectives (e.g., titer and chain length) and suggest that the efficient incorporation of heterologous components often requires compositional re-optimization. Broadly, this work provides an experimental framework for adjusting the FAS of *E. coli* to hit a broad range of production targets (e.g., specific product profiles or fatty acid derivatives).

603966 Characterization of an α-Amino Ester Hydrolase from *Xanthomonas Campestris Pv. Campestris* suitable for Cephalosporin Synthesis but Not Amoxicillin Synthesis

Abstract:

β-lactam antibiotics, especially those derived from penicillins and cephalosporins, are continually the most prescribed antibiotics in the world [1]. Currently, penicillin G acylase (PGA) is the most used enzyme for the biocatalytic synthesis of β-lactam antibiotic and is employed industrially as an efficient alternative to chemical synthesis routes. α-amino ester hydrolases (AEH) have recently gained interest as alternative β-lactam antibiotic acylases to PGA for their enhanced reaction rates and stereospecificity, but AEHs remain largely uncharacterized. To date, only AEHs from *Xanthomonas citri and Acetobacter turbidans* have reported crystal structures and substrate specificities [2,3]. There is interest in expanding the number of characterized AEHs and quantifying their synthetic abilities across a wide range of β-lactam antibiotics. Prior work in our group on AEH from *Xanthomonas campestris pv. campestris* focused primarily on characterizing activity toward commercially less interesting ampicillin [4].

This work characterizes activity, structure, and thermostability of AEH from *X. campestris pv. campestris* as well as two promising quadruple variants [5]. Enzymatic synthesis couples an activated acyl side chain such as phenylglycine methyl ester (PGME) or p-hydroxyphenylglycine methyl ester (HPGME) with a β-lactam nucleus such as 6-aminopenicillanic acid (6-APA) or 7-aminodesacetoxycephalosporanic acid (7-ADCA). Different combinations of these moieties produce unique antibiotics (cephalexin from 7-ADCA and PGME, amoxicillin from HPGME and 6-APA, and cefadroxil from HPGME and 7-ADCA). AEH catalyzes β-lactam antibiotic synthesis as well as acyl donor hydrolysis and β-lactam antibiotic hydrolysis, leading to the formation of unwanted byproducts. Synthesis selectivity is dependent on 6-APA or 7-ADCA concentration and quantified with the parameters α, β_0, and γ used in the reaction scheme developed by Youshko and Svedas [6]. AEH melting temperature was measured with differential scanning fluorimetry (DSF) while molecular weight and oligomeric state of AEH in solution were verified using size exclusion chromatography/multi-angle laser light scattering (SEC-MALLS) and analytical ultracentrifugation (AUC).

AEH was found to be more selective for cephalexin and cefadroxil synthesis than for amoxicillin synthesis. Maximum selectivity, quantified by 1/γ, was found to be between 0.37–0.5 for amoxicillin, 1.41–1.78 for cefadroxil, and 5.6–7.64 for cephalexin. AEH hydrolysis activity had an order of magnitude higher affinity (K_M) and reactivity (k_{cat}) toward PGME over HPGME. These trends indicate AEH's suitability for cephalosporin synthesis over amoxicillin synthesis primarily due to the higher nucleophilic reactivity of 7-ADCA relative to 6-APA.

AEH has found to have two melting points, one at 25°C and another at 40°C. This suggests that the low thermostability of AEH could be due to separate domains unfolding within the AEH monomer. It was also found that 1M NaCl stabilizes AEH in solution and increases the first melting point to >40°C, however, this high salt concentration eliminates AEH activity. SEC-MALLS demonstrates that the molecular weight of AEH in solution is approximately that of the monomer, suggesting the lack of tetrameric AEH that is found in the crystal structure. The presence of AEH as an active monomer in solution is further confirmed with AUC [7].

References:

[1] H. Gelband et al., Wound Healing Southern Africa 8.2 (2015) 30-34.

[2] J.J. Polderman-Tijmes et al., Appl. Environ. Microbiol. 68 (2002) 211–218.

[3] J.J. Polderman-Tijmes et al., J. Biol. Chem. 277 (2002) 28474–28482.

[4] J.K. Blum, A.S. Bommarius, J. Mol. Catal. B: Enzym. 67 (2010) 21-28

[5] J.K. Blum et al., J. Biotechnol. 160 (2012) 214-221

[6] M.I. Youshko, and V.K. Śvedas, *Biochemistry (Moscow)* 65 (2000) 1367-1375.

[7] C. Lagerman, A. Bommarius, Characterization of an α-amino ester hydrolase from *X. campestris pv. campestris* for biocatalytic production of β-Lactam antibiotics, To be submitted.

601320 Biosynthesis of Wax Esters (Industry Candidate)

Abstract:

Microbial synthesis of oleochemicals is a promising path for the cost-effective production of wax esters (WE) from renewable and sustainable feedstocks. In this study, we explored the WE metabolic pathway in both *Escherichia coli* and *Yarrowia lipolytica* for biosynthesis of WE from both sugar and lipid substrates. First, the *ScFAR* from jojoba plant (*Simmondsia chinensis*), the *MmFAR* from house mouse (*Mus musculus*), and the *MhFAR* from *Marinobacter hydrocarbonoclasticus* strain VT8 together with the *AbWS* from *Acinetobactor baylyi* ADP1 were chosen to study the enzyme efficiency and the substrate specificity in the oleaginous yeast *Y. lipolytica* via ARS18-based plasmid expression. All enzyme combinations led to the successful WE production. Co-expression of cytoplasmic MhAFR and AbWS resulted in the high accumulation of intracellular WE. Second, the wild type *Y. lipolytica* ATCC20362 was engineered for WE production by heterologous expression of the MhFAR and AbWS *via* random genome integration. The *E. coli* BL21(DE3) was also used to overexpress two genes via a T7-pET plasmid expression system. To improve WE production, both hydrophilic and hydrophobic carbon sources were tested and optimized in shake flask experiments. Our results indicated that in the presence of exogenous oleic acid or oil-based substrates in the culture medium, the efficiency of WE synthesis in both *Y. lipolytica* and *E. coli* strains rely on external source of free fatty acids rather than endogenous de novo fatty acid biosynthesis. Switching from glucose to a fresh or waste plant oil improved the WE titer by 70-fold for the engineered *Y. lipolytica*, which led to a maximum of 7.6 g/L WE within 120hr of cultivation in flasks. The produced intracellular WE contributed to 57% of the yeast biomass, which is so far the highest level reported on microbial production of WE. The predominant intracellular WE in the presence of hydrophobic substrate as sole carbon source were C_{36}, C_{34} and C_{32}, in order of decreasing abundance, with a large proportion being unsaturated. Co-feeding the glucose with oleic acid, the WE titer of 3.95 g/L and DCW of 19.47 g/L were achieved in engineered *E. coli* after 40hr fed-batch fermentation, which is also the highest titer reported so far for WE produced by a bacterium. Our research results paved the way for further commercial production of WE from a low-cost renewable feedstock.

611577 Repurposing Chloramphenicol Acetyltransferase for a Robust Ester Biosynthesis Platform

Abstract:

Esters are versatile chemicals that have broad applications as flavors, fragrances, solvents, and potential drop-in biofuels. The key enzyme to synthesize esters is alcohol acyltransferase (AAT) that condenses alcohols and acyl-CoAs derived from cellular metabolism under a mild aqueous condition. Particularly, thermostable (≥ 50°C) AAT plays a significant role for developing an industrial bioprocess to produce esters from lignocellulosic biomass but such an enzyme has not yet been reported to date. To enable a robust ester biosynthesis platform, we repurposed chloramphenicol acetyltransferase (CAT) to function like an AAT by harnessing its enzyme promiscuity. Through bioprospecting and rational protein design, we developed a highly thermostable CAT with a broad substrate range, which can function as an AAT for microbial whole cell biocatalyst development. The high thermostability and promiscuity of the engineered CAT make it an ideal platform for microbial biosynthesis of designer bioesters.

612951 Engineering Anaerobic Metabolism of Short-Chain Alkanes in *E. coli* (Invited Speaker)

Abstract:

Natural gas liquids (NGLs) are abundant carbon sources for fuels and chemicals, but their efficient utilization is met with many technical hurdles relating to their low energy density, high cost of transportation and storage, and most notably the catalytic challenges associated with selective and controlled functionalization of these alkanes. While biological routes to alkane functionalization are promising, those requiring oxygen still suffer from significant energy and carbon inefficiencies. In contrast, a variety of anaerobic alkane-degrading bacteria (largely nitrate and sulfate reducers) activate sub-terminal C-H bonds via fumarate addition. Through metabolic engineering, the "alkylsuccinate" products of this reaction can be converted into biofuels or other value-added chemicals, with co-regeneration of fumarate in a cyclic pathway.

Alkylsuccinate synthase enzymes require activation by a partner "activase", which utilizes an iron-sulfur cluster to reductively cleave AdoMet, generating a 5'-dAdo· radical that in turn generates a catalytic glycyl radical on the synthase. Functional, heterologous expression of these finicky enzyme systems has proven difficult. With *E. coli* as the host microorganism, chosen for its ease of genetic manipulation, we have established anaerobic growth and gene expression conditions that support continued biosynthesis of alkylsuccinates using the *mas* gene system from *Azoarcus* sp. str. HxN1. I will describe some of the key experimental considerations and results that led to production of compounds such as 1-methyl-pentylsuccinate (from hexane) and 1-methyl-ethylsuccinate (from propane). I will also discuss other enzyme systems being developed for smaller alkane substrates, steps toward the directed evolution of alkylsuccinate synthases for improved activity and altered substrate specificity, and how these enzymes / intermediates will be interfaced with downstream pathways for biosynthesis of diverse products.

610364 A Low Affinity Humanized CD-19 Targeted Chimeric Antigen Receptor Efficiently Controls B-Cell Malignancy

Abstract:

Chimeric antigen receptor (CAR) engineered T cells have energized the field of cancer immunotherapy with their proven ability to treat hematologic malignancies in the clinic and emerging efficacy in treating other diseases. A CAR is a synthetic receptor that directs T cell activation toward an individual biomarker that a T cell would not ordinarily recognize. The antigen recognition domain that governs this specificity is currently under intense study as molecular parameters including binding affinity, stability, and immunogenicity risk dictate the ability of a CAR T cell therapy to perform. The current gold standard of antigen recognition domains is the CD19-targeted scFv FMC63, which has strong binding affinity, modest stability, and a known immunogenicity risk in humans due to its murine character. This work aims to develop humanized FMC63 variants with decreased immunogenicity risk while maintaining similar efficacy when applied to CAR antigen recognition. Multiple protein engineering strategies were applied to humanize FMC63 *in silico* yielding two lead candidate scFvs. Yeast surface display techniques were employed to characterize expression, binding affinity, and thermal stability of the FMC63 variants, revealing on the order of 1,000-fold weaker binding affinity but 6-7°C improvements in thermal stability of the humanized variants relative to parental. CARs utilizing these scFv variants demonstrated specific activity against CD19+ tumor cells and equivalent degranulation and cytokine production *in vitro*. Tumor rechallenge experiments *in vitro* reveal a binding affinity dependence for CAR T cells to proliferate in the presence of antigen. One CAR using a humanized FMC63 variant performed equivalently to the parental against CD19+ Raji and NALM6 models *in vivo*, demonstrating that a wide range of affinity levels can yield therapeutic impact with a CAR T cell therapy. Ongoing experiments explore the ability of these low affinity CARs to recognize CD19-low Raji variants, further illuminating the affinity-avidity interplay in CAR T cell activation.

610786 Mac-1 Expression Controls If Immune Cells Can Utilize Upstream Migration

Abstract:

Cells of the adaptive immune system such as $CD4^+$ T cells, the innate immune system such as neutrophils, and hematopoietic stem cells (HSPCs), must transit to distal sites such as the lymph nodes, bone marrow, or sites of infection. The recruitment of these immune cells is regulated by a multi-step, leukocyte adhesion cascade of cell rolling, activation, adhesion, and transmigration through the endothelial barrier. After selectin-mediated braking and slow rolling, immune cells migrate along the vascular endothelium through receptor mediated binding to adhesion ligands including Intracellular Adhesion Molecule-1 (ICAM-1) and Vascular Cell Adhesion Molecule-1 (VCAM-1). Previous work by ourselves and others has demonstrated an interesting phenomena where cells such as murine and human T lymphocytes (1, 2), hematopoietic stem and progenitor cells (HSPCs)(3), and marginal zone B-cells (4), can migrate efficiently upstream (against the direction of shear-flow) on ICAM-1 surfaces, or surfaces coated with ICAM-1 and VCAM-1, and endothelial monolayers (HUVECs) which express ICAM-1, using the integrin receptor known as Lymphocyte Functional Antigen-1 (LFA-1). Furthermore, cells undergoing upstream migration transmigrate through the endothelium twice as fast as those going downstream (5). Finally, recent work of ours has furthered the phenomenon of upstream migration by demonstrating that innate immune cells such as neutrophils, which express a second ICAM-1 binding integrin named Macrophage-1 antigen (Mac-1), can migrate upstream when Mac-1 is blocked with antibodies (6).

Here, we describe for the first time how Mac-1 integrin expression can be utilized as checkpoint to direct immune cells either with or against the direction of fluid flow on the endothelial surface using genetic engineering. We demonstrate that myeloid cells of the innate immune system, which cannot normally migrate upstream, can be made to swim upstream by removing Mac-1 function (Figure 1). HL-60 cells and primary neutrophils in which Mac-1 is knocked down are able to migrate upstream on ICAM-1, downstream on VCAM-1, upstream on mixed surfaces of ICAM-1 and VCAM-1 (which mimic the vasculature, and upstream on activated HUVEC endothelial monolayers at shear rates seen in the vasculature. On the other hand, lymphoid cells of the adaptive immune system or stem and progenitor cells, which normally can crawl upstream, can be made to crawl with the direction of shear flow (downstream) by expressing Mac-1. We then demonstrate that KG1a cells, primary HSPCs, and CD4+ T cells genetically engineered to express Mac-1 on their surface cannot migrate upstream on ICAM-1 or any endothelial surface. Finally, by transducing Lifeact into these engineered cells, we demonstrate that the F-actin network is fundamentally different between both upstream and downstream migration and that the presence alone of Mac-1 shifts the actin-network between these modes.

In all, while LFA-1 is the critical integrin mediating upstream motility, Mac-1 has the potential to control migration between the upstream and downstream modes. Therefore, through its expression, which is transient in neutrophils, Mac-1 can not only orient immune cells upstream or downstream but also alter the rate of transmigration. This finding has direct implications in understanding how to manipulate both the direction immune cells trafficking as well as their translocation.

REFERENCES:

1. Dominguez GA, Anderson NR, & Hammer DA (2015) The direction of migration of T-

lymphocytes under flow depends upon which adhesion receptors are engaged. *Integrative biology : quantitative biosciences from nano to macro* 7(3):345-355.
2. Valignat M-P, Theodoly O, Gucciardi A, Hogg N, & Lellouch Annemarie C (2013) T Lymphocytes Orient against the Direction of Fluid Flow during LFA-1-Mediated Migration. *Biophysical Journal* 104(2):322-331.
3. Buffone A, Anderson NR, & Hammer DA (2018) Migration against the direction of flow is LFA-1-dependent in human hematopoietic stem and progenitor cells. *Journal of Cell Science* 131(1).
4. Tedford K, *et al.* (2017) The opposing forces of shear flow and sphingosine-1-phosphate control marginal zone B cell shuttling. *Nature Communications* 8(1):2261.
5. Anderson NR, Buffone A, & Hammer DA (2019) T lymphocytes migrate upstream after completing the leukocyte adhesion cascade. *Cell Adhesion & Migration* 13(1):163-168.
6. Buffone A, Anderson NR, & Hammer DA (2019) Human Neutrophils Will Crawl Upstream on ICAM-1 If Mac-1 Is Blocked. *Biophysical Journal* 117(8):1393-1404.

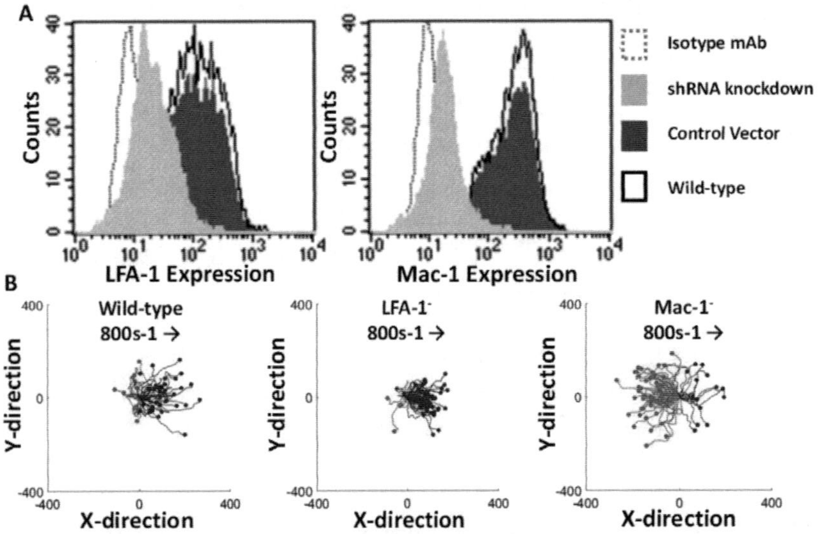

Figure 1: (A) Representative flow cytometry histograms of LFA-1 and Mac-1 HL-60 knockdowns with isotype mAb, shRNA knockdown, control vector, and wild-type expression levels for each knockdown depicted. (B) Cell traces from common origin of HL-60 and knockdowns on ICAM-1 at 800 s^{-1} shear rate. Blue traces in histogram represent cells traveling downstream while red traces represent cells traveling upstream. N=3-4 independent experiments of at least 45 cells analyzed per experiment. *$p < 0.05$ with respect to isotype conditions.

611840 Metabolite-Based Modulation of Dendritic Cells for Developing Effective Immunotherapy

Abstract:

INTRODUCTION – Aging and diabetes associated chronic / slow healing cutaneous wounds have hallmark immune signatures, including *delayed* T-cell and macrophage infiltration, and increased pro-inflammatory immune cell infiltration for *extended periods* (compared to non-aged control). *If there were a method to modulate the local immune responses in a temporal manner then it would be possible to accelerate the wound healing of cutaneous wounds in aged and/or diabetic immune system.* Notably, metabolic pathways control immune cell functions, and modulating these pathways can determine overall immune system response. Therefore, in this work, polymers were generated with metabolite monomers that can deliver metabolites and directly modulate the function of immune cells.

RESULTS – This work for the first time (to the best of our knowledge) demonstrates utilization of metabolites to modulate immune-metabolic responses in wound healing. NMR and FTIR studies were performed to characterize the polymers. Moreover, oil in water emulsions were utilized to generate microparticles of these metabolite-based polymers. These microparticles are capable of encapsulating drugs and release them in a sustained manner (shown via representative drug-molecule rhodamine). The release kinetics of metabolites from these polymers were determined using NMR. Moreover, degradation kinetics of the polymers was determined using weight loss experiments. Importantly, it was demonstrated that these polymers could modulate the function of mouse bone marrow derived dendritic cells and allogenic T-cells in a mixed lymphocyte reaction. Lastly, we demonstrate that these polymers when applied in the presence of tegaderm were able to modulate the wound healing rate in BALB/c mice.

CONCLUSION - In summary, metabolite-based polymers were generated that were capable of controlling the pro- or anti-inflammatory responses of immune cells *in vitro* and *in vivo*. Notably, these polymers were able to lead to healing of cutaneous wounds in mice.

609751 Elucidating the Role of Decellularized Porcine Skeletal Muscle ECM-Silk Composite Systems on Human THP-1-Derived Macrophage Polarization

Abstract:

Traumatic soft tissue injuries often result in permanent loss of skeletal muscle mass known as volumetric muscle loss (VML). This causes a major clinical challenge for military and civilian medicine. Biomaterials can be used to provide the appropriate structural and mechanical properties as well as the release of bioactive and chemotactic signals to address this issue [1, 2]. While it is possible to implant degradable biomaterials at the injury site and see some functional repair, we lack a clear understanding of how the biomaterial can be used to instruct immune cells towards a regenerative phenotype in the skeletal muscle and the role silk-based biomaterials play in this process.

The long-term goal of this work is to engineer decellularized skeletal muscle extracellular matrix (ECM)-silk based natural biomaterials that, upon implantation, modulate *in vivo* tissue behavior and collective cell function for soft tissue repair and rehabilitation. Specifically, macrophages (myeloid cell lineage, with two very plastic phenotypes, M1 pro-inflammatory and M2 anti-inflammatory) dominate the inflammatory response for muscle regeneration and thus recent work has shown that understanding the role biomaterial formulation has on macrophage phenotype will be critical for improved material design to meet clinical needs [3-6]. Our lab utilizes a silk-decellularized ECM sponge platform, which can be tuned to degrade *in vivo* over many months. Previous and on-going work has shown that degradation of silk materials is dependent on the enzyme(s) present and it is known that macrophages and fibroblasts at the injury site are largely responsible for enzymatic activities.

Thus, to obtain a better understanding of macrophage response and enzyme secretion in response to the biomaterial formulation, we investigated the polarization of macrophages in response to two separate silk-ECM composite formats. First, 2D culture of human leukemia monocytes (THP-1s) on silk-decellularized skeletal muscle ECM films were evaluated, with decellularized skeletal muscle ECM coated tissue culture plastic, silk only films, and collagen I coated tissue culture plastic as controls. Using previously described methods [7], THP-1 monocytes are driven towards a macrophage phenotype via treatment with phorbol 12-myristate-13-acetate (PMA) to cause the cells to adhere to the culture surface. Then, adhered cells are either treated with LPS lipopolysaccharide (LPS) and interferon gamma (INF-g) or interleukin-4 (IL-4) and interleukin-13 (IL-13) to polarize them towards either M1 or M2 phenotypes, respectively. Through image analysis and western blots, we evaluated the polarization potential of these cells through pan-macrophage markers (CD68, CD11b), markers for M1 (CD80, iNOS – inducible nitric oxide synthase), and for M2 (CD206 - mannose receptor, CD163) at each point in the polarization process on 2D silk-ECM films and necessary positive and negative control substrates to elucidate the influence of silk and ECM independently and combined on the polarization of the THP-1-derived macrophages.

To determine if the observed trends hold in 3D, adherent THP-1s were cultured in 3D silk-ECM sponges, using silk only sponges and 2D tissue culture plastic culture as the controls. Efficiency of polarization in 3D was evaluated by comparing expression of M1 and M2 markers via western blot to the controls. Ongoing work aims to understand how the biomaterial formulation can be leveraged to instruct and direct macrophage phenotype over time. Additionally, gene expression of key enzymes responsible for biomaterial degradation will be tracked by gene expression analysis and correlated with changes in biomaterial elastic modulus over time. These results will inform future work aimed at designing regenerative constructs for applications in skeletal muscle tissue rehabilitation following traumatic injuries, sports-related injuries, and genetic diseases.

[1] K.H. Nakayama, M. Shayan, N.F. Huang, Engineering Biomimetic Materials for Skeletal Muscle Repair and Regeneration, Adv Healthc Mater 8(5) (2019) e1801168.

[2] S.M. Greising, B.T. Corona, C. McGann, J.K. Frankum, G.L. Warren, Therapeutic Approaches for Volumetric Muscle Loss Injury: A Systematic Review and Meta-Analysis, Tissue Eng Part B Rev 25(6) (2019) 510-525.

[3] J.G. Tidball, Regulation of muscle growth and regeneration by the immune system, Nat Rev Immunol 17(3) (2017) 165-178.

[4] M. Juhas, N. Abutaleb, J.T. Wang, J. Ye, Z. Shaikh, C. Sriworarat, Y. Qian, N. Bursac, Incorporation of macrophages into engineered skeletal muscle enables enhanced muscle regeneration, Nat Biomed Eng 2(12) (2018) 942-954.

[5] A. Urciuolo, L. Urbani, S. Perin, P. Maghsoudlou, F. Scottoni, A. Gjinovci, H. Collins-Hooper, S. Loukogeorgakis, A. Tyraskis, S. Torelli, E. Germinario, M.E.A. Fallas, C. Julia-Vilella, S. Eaton, B. Blaauw, K. Patel, P. De Coppi, Decellularised skeletal muscles allow functional muscle regeneration by promoting host cell migration, Sci Rep 8(1) (2018) 8398.

[6] M. Juhas, N. Abutaleb, J.T. Wang, J. Ye, Z. Shaikh, C. Sriworarat, Y. Qian, N. Bursac, Incorporation of macrophages into engineered skeletal muscle enables enhanced muscle regeneration, Nature Biomedical Engineering 2(12) (2018) 942-954.

[7] A.R.D. Reeves, K.L. Spiller, D.O. Freytes, G. Vunjak-Novakovic, D.L. Kaplan, Controlled release of cytokines using silk-biomaterials for macrophage polarization, Biomaterials 73 (2015) 272-283.

602701 Characterization and Modeling of Regulatory T Cell Enrichment through Immunophenotypic Modulation

Abstract:

Introduction: Regulatory T cells (Treg) are a subset of T cells that suppress aberrant activation of self-reactive effector lymphocytes and are widely regarded as the primary mediators of peripheral tolerance. Cell-based therapy using Tregs effectively treats autoimmune diseases in pre-clinical models and clinical trials are underway to evaluate efficacy in humans. However, sourcing Treg from leukapheresis is inefficient as they circulate at a low frequency in the blood and therefore ex vivo expansion is required to enhance their numbers. Current ex vivo Treg expansion techniques rely on labor intensive purification and diminish Treg function. These requirements for expanding and isolating Tregs ex vivo limit their use for cell therapy. We hypothesized that metabolic modulation of a heterogenous T cell population using β-cylodextrin-rapamycin complexes (CRCs) would efficiently promote the preferential expansion of functional Tregs. Here, we characterize the immunomodulatory properties of CRCs and use our results to develop a kinetic model for manufacturing Treg as an immune cell-based therapy.

Materials and Methods: To test the hypothesis, we first characterized the encapsulation of rapamycin (Rapa) in Mono-(6-amino-6-deoxy)-beta-cyclodextrin (βCD) using spectrophotometry and measured the bioactivity of Rapa encapsulated in CRCs, by phenotypic and metabolic modulation of mouse and human T cells. Metabolic modulation was quantified using (i) cell counts to quantify suppression of T cells stimulated with Dynabeads and (ii) flow cytometry was used to quantify the phenotypic and functional changes of splenic CD4 T cells. To further enrich Treg, we tested combinations of CRC with immunomodulatory growth factor transforming growth factor-beta 1 (TGF-β1) and measured changes in phenotype and cytokine profile using flow cytometry. To validate the findings and model for manufacturing human Treg, we repeated assays with human T cells. To explain the observed effects between TGF-β1 and CRCs, we developed and optimized a kinetic growth model describing cell differentiation and expansion and fit key model parameters using experimental data from mouse and human T cell experiments.

Results and Discussion: CRCs enhanced the solubility of Rapa in water 154 -fold and maintained bioactivity of Rapamycin for at least 30 days in solution. CRCs enriched the fraction by 5-fold of Tregs in vitro from both PBMC-isolated human T cells and mouse splenocyte-isolated T cells. CRCs synergize with TGF-β1 to enhance the enrichment of murine Tregs. The combination also enhanced the fraction of human Tregs over single factor treatments. CRCs reduced interferon-gamma (IFN-γ) expression by T cells and increased the fraction of T cells that expressed tumor necrosis factor-α (TNF-α) without IFN-γ. In human T cells, combining CRC and TGF-β1 further enhanced the enrichment of IFN-γ- TNF-α+ T cells, while in mouse T cells the addition of TGF-β1 increased the fraction of IFN-γ+ T cells, both alone and in combination with CRCs. Computational modeling of T cell expansion and differentiation revealed that CRCs mediate Treg enrichment primarily through suppression of effector T cell expansion, while TGF-β1 does so through promoting Treg differentiation. Strikingly, the model accurately predicts the synergistic effect of CRCs and TGF-β1, and predicts the distribution of T cells by phenotypic subsets, depending on the starting cell population.

Conclusions: The findings from our experiments collectively suggest that the βCD encapsulated Rapa enhances functional Treg in vitro. The increased bioavailability of Rapa with this complex is potentially due to strong interactions with βCD, which was confirmed via spectrophotometry. Compared with DMSO, the CRC could serve to (i) significantly improve the solubility of Rapa without inducing T cell toxicity, (ii) shield Rapa from degradation, thereby increasing half-life and uptake by T cells, (iii) synergize with TGF-β1 to permit preferential expansion of Treg. Additionally, CRCs can act to significantly decrease the production of IFN-γ by effector T cells in culture that may cause naïve T cells to acquire a Th1 phenotype. The enhancement in TNF-α expression may act to prime the Tregs and enhance functionality. Furthermore, the kinetic model predicts the synergism between the CRCs and TGF-β1 in modulating the kinetics of Treg expansion in vitro and provides a basis for applying quality and process control to manufacturing Treg for clinical therapy.

596068 Identification and Characterization of a First Responder Cells: A Dendritic Cell Subtype That Coordinates Adaptive Immune Responses

Abstract:

While it is known that dendritic cells (DCs) are a crucial cell for antigen presentation and initiation of adaptive immunity, DCs are a highly heterogeneous population of cells. It is currently unknown if all cells participate equally in antigen presentation and immune activation. Recent reports have indicated there is a subpopulation of DCs that exhibits highly increased cytokine secretion to toll-like receptor (TLR) signaling but to date these cells have been difficult to isolate and characterize. Using fluorescently labeled polystyrene microparticles (MPs) conjugated with various TLR agonist molecules, we isolated this subtype of cells we call First Responders (FRs). FRs exhibit similar increases in cytokine (TNFα) secretion seen in prior literature and statistically significant increases in MP uptake. FRs also coordinate neighboring innate immune cell activation via paracrine signaling. Furthermore, when adoptively transferred into mice with a model antigen OVA, FRs were both necessary and sufficient to trigger adaptive immune responses, both antibody and CD8 t cells. FRs express CD11c and MHCII, typical markers of DCs, but also express CD11b and SIRPα, typical of cDC2 cells. To understand the phenotype even further, we performed both bulk and single cell mRNA sequencing on FRs and observed increases distinct expression profile with increases in several immune and metabolic genes. After validating the gene expression via flow cytometry, we identified several surface markers unregulated in FRs. The results indicate that FRs have expression of some B cell surface markers, such as CD20 and CD267, but lack CD19 or B220 and also overexpress DC markers such as FTL-3 and CD301b. Due to their importance in innate immune coordination and their crucial role in adaptive immune responses, these cells not only warrant further study but also present an attractive target for vaccine research.

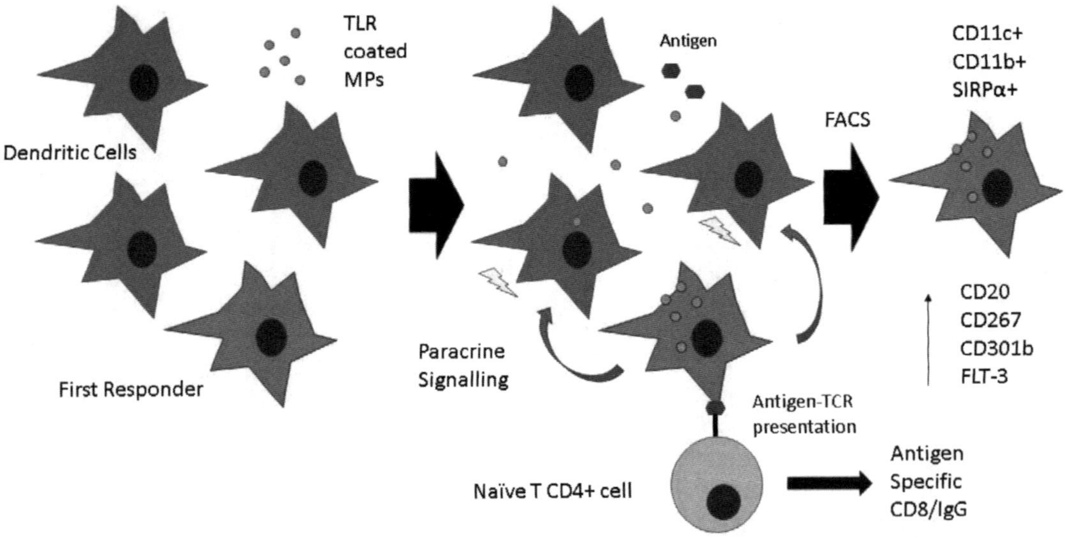

613356 Harnessing Biomaterials to Study and Control Immune Function (INVITED SPEAKER)

Abstract:

Our research combines immunology and biomaterials to understand the interactions between synthetic materials and immune tissues, and to design more selective therapeutic vaccines for cancer and autoimmunity. This presentation will highlight our recent efforts toward these goals combining materials science and bioengineering tools, cell culture, animal models, and samples from human patients. In one example I will discuss new degradable polymer depots that could improve the selectivity of therapies for autoimmune diseases such as multiple sclerosis and diabetes by locally reprogramming the function of lymph nodes – tissues that coordinate immune function. A second area will present the lab's efforts to self-assemble immune signals into modular nanostructures. This rational design approach allows activation of programmable combinations and levels of immune pathways triggered. Modular control over these aspects of immune signaling could help improve the efficacy of vaccines for cancer and infectious disease, and enhance the efficiency of vaccine translation.

608601 Biomedical Potential of Viral and Endogenous Double-Stranded RNA Detection

Abstract:

Detection of viruses, especially the ones related to infectious diseases, has attracted much attention. In recent years, we have seen an increasing number of viral infection outbreaks, such as Middle East respiratory syndrome coronavirus and Zika virus, and they are spreading on a global scale at an increasingly alarming rate. Nucleic acid based methods are the most commonly employed schemes for virus detection. This includes the use of polymerase chain reaction (PCR), which is routinely performed to amplify and detect viral nucleic acids. While nucleic acid-based detection is highly specific, it has the disadvantage that some prior knowledge of the virus is needed in order to prepare the relevant detection methods. Such approach can be limiting when faced with novel or mutated viruses of unknown genomic sequences. Considering that in many situations it is necessary to determine the infection status of a person without having much information regarding the nature of the virus involved, it would be highly desirable to have a primary screening tool capable of detecting many different viruses without knowing their nucleotide sequences. To design these universal virus detectors, it is important to identify biomarkers that are commonly present across many viruses. Recently, the use of long dsRNAs as a universal virus marker has gained popularity.

Long dsRNAs are generated as byproducts of viral replication and are produced abundantly by DNA viruses and positive-strand RNA viruses, and also by a limited number of negative-strand RNA viruses. Moreover, certain viruses such as reovirus, rotavirus, and cypovirus are known to contain dsRNA genome. As these long dsRNAs are a common signature of viruses, they are recognized by mammalian innate immune system and trigger a cascade of immune response reactions. Particularly, dsRNA sensors of the innate immune response system recognize the length and the terminal structure of the dsRNAs rather than their specific sequences, thus these immune response proteins are capable of recognizing a wide range of viruses that generate long dsRNAs.

In this study, we utilize a combination of dsRNA recognizing J2 antibody and reactive poly(pentafluorophenyl acrylate) (PPFPA) to construct a virus detection platform with superior sensitivity. PPFPA attached silicon substrates are first prepared, and the reactive PFP units are utilized to immobilize J2 antibodies that recognize dsRNAs longer than 40 base pairs (bp). We show that these surfaces can successfully distinguish dsRNAs from single-stranded RNAs (ssRNAs), as well as synthetic long dsRNAs (76 bp) from short ones (19 bp), demonstrating their potential in specifically detecting only the long dsRNAs generated by viruses. Furthermore, to enhance dsRNA detection sensitivity, we devise a two-step detection process analogous to sandwich enzyme-linked immunosorbent assay (ELISA) where following dsRNA capture, the bound dsRNAs are then visualized using fluorophore-tagged antibodies that also recognize RNAs' double-stranded secondary structure. By utilizing the developed platform, long dsRNAs can be detected and visualized from total RNA mixture as well as from total cell lysates, which contain a mixture of various abundant contaminants such as DNAs, proteins, and other cellular ss- and dsRNAs. Lastly, we demonstrate the application of the developed platform in detecting long dsRNAs generated by hepatitis C virus (HCV) and hepatitis A virus (HAV) without utilizing information on viral RNA sequences.

Although dsRNAs are strongly associated with immune response to viral infection, increasing evidences suggest that human cells naturally express endogenous dsRNAs that can regulate antiviral machineries in various cellular contexts such as during the cell cycle and response to stressors.

Moreover, accumulation of endogenously encoded dsRNAs is related to the onset of autoimmune and age-related macular degeneration. Lastly, dsRNAs also play a key role during cellular response to chemotherapy where treatment of the DNA-demethylating agent leads to cell death by inducing the transcription of endogenous dsRNA genes, which subsequently activate antiviral machineries. Considering the biomedical potential of dsRNAs, we further develop a small organic molecule to quantify the collective expression level of all types of dsRNAs expressed in human cells.

We investigate the potential of using photochromic spiropyrans as a tool to detect and profile the overall expression of dsRNAs. Our goal is three folds: 1) To establish spiropyran as a molecule that can interact with dsRNAs, 2) To characterize the interactions between dsRNA and spiropyran, and 3) To apply spiropyran to detect changes in levels of endogenous dsRNAs in response to various stressors. We find that the open form of spiropyran, merocyanine (MC), can moderately interact with dsRNAs, which results in protonation of MC to MCH+ and alters the characteristic UV-Vis absorbance spectrum of the compound. By quantifying this change, we can access the amount of dsRNA present in the solution. The effect is greater for the GC pair than that of the AU pair and mixture of the two types of nucleotide pairs also show strong characteristic decrease in the light absorbance. Moreover, using RNases with different substrate specificities, we can enrich and detect long dsRNAs by removing most of the ssRNAs and short hairpin RNAs from the cell extract. Furthermore, we apply our approach to demonstrate the potential usage of MC and elevation of dsRNA expression as a predictive marker for cellular response to the DNA-demethylating agent 5-aza-2'-deoxycytidine (decitabine), which is commonly used to treat colorectal cancer and pre-leukemic disorder myelodysplasia (MDS).

Collectively, our work presents the first example of polymer grafted surfaces as well as small organic molecule to capture and detect long dsRNAs. These findings establish potential of dsRNA detection in biomedical applications ranging from viral infection to cancer

609594 Cell As an Optical Lens: Application to New Class of Chemical Cytometry (Faculty Candidate)

Abstract:

Immune heterogeneities of cells are critically involved in progression and prevention of cardiovascular or neurodegenerative diseases. Thus, a tool that enable the profiling of dynamic antigenic responses of different immune cell populations in terms of expression of chemical efflux and biophysical cell properties could aid in our understanding of past and future unknown microorganisms and diseases. Herein, we developed a new class of chemical cytometry technique that can monitor the multimodal immune cell heterogeneities in real-time. By integrating fluorescent nanosensors with microfluidics, dual functional platform that can directly visualize the target cells and also report their real-time chemical efflux could be efficiently realized. Using the unique nIR lensing effect of cell, both H_2O_2 secretion of the cells and their biophysical information were easily extracted in same time without any labeling and additional optical tools. Instantaneous H_2O_2 efflux of cells could be directly monitored in few 10^{-18} moles resolution by just measuring nIR signals from the platform. For final applications, immune response heterogeneities of different monocytes populations were clearly profiled with high-throughput microfluidic system (~10 cell/min) in non-destructive way. Furthermore, new phenotype correlations between real-time extracellular immune responses and multiple biophysical properties (cell size, eccentricity, RI) of monocyte populations were investigated with exact numerical values and distribution statistics. This is first trial to bridge nanosensors and cytometry for cell monitoring and expected to be used for ideal single cell analysis platform in cell manufacturing and biopharmaceutical industry.

610769 A Novel Platform of Monitoring DNA Hydroxymethylation (5-hmC) Change in Live Cells

Abstract:

DNA methylation is one of the most important epigenetic modifications that play an essential role in development, aging and disease onset. Methylation of cytosine (5mC) is the most abundant form of DNA methylation found in humans and was assumed to be fairly persistent during the life course. Recent studies have shown that the ten-eleven translocation (TET) proteins can catalyze the oxidation of 5mC to form 5-hydroxymethylcytosine (5-hmC) followed by a DNA demethylation process suggesting dynamic nature of DNA methylation. Moreover, 5-hmC is more abundant in central nervous system and has been implicated in many crucial brain activities, including long-term potentiation, synaptic transmission and etc. This exhilarating discovery has motivated engineering efforts to track dynamic changes in 5-hmC in live cells and potentially animals. We have designed and optimized a recombinant protein sensor for probing DNA hydroxymethylation *in situ.* The probe is based on the native 5hmC "reader" domains from SRA and utilized the Bimolecular Fluorescence Complementation (BiFC) technique to improve signal to noise ratio. The probe is shown to have equivalent sensitivity to commercial antibodies but is live-cell compatible. We adopted the probe to track changes in 5hmC during neuron differentiation and recorded 5hmC dynamics at a single cell resolution. 5hmC changes when neurons are exposed to PD-risk factors are also recorded to reveal the potential contributions of 5hmC to the disease development.

611379 Extraction of Interstitial Fluid Using Microneedle Patches: Biophysical and Tissue Properties Limiting Dermal Interstitial Fluid Flow (Industry Candidate)

Abstract:

Interstitial fluid (ISF) is a biomarker-rich fluid found in the spaces between the cells in tissues which has potential for use in point-of-care (POC) diagnostic tests. The ISF in the skin can be accessed using microneedle (MN) patches, but the ISF volumes extracted remain low due to slow fluid transport through the dermis.

In this presentation we will discuss the results from two sets of experiments -one ex-vivo and one in humans- that reveal quantitative information about the biophysical tissue factors limiting ISF flow after treating the skin with MN patches and during application of suction to withdraw the ISF. In the first set of experiments we targeted ex-vivo dermis as a porous fibrous matrix to learn about the types of fibers that limit ISF flow through the dermis. When we treated skin ex-vivo with collagenase and hyaluronidase to degrade skin fibers and increase the amount of ISF extracted, we found that treatment with up to 10 mg/ml of collagenase increased ISF flow rates by up to a factor of 5. We attribute this effect to an increase in hydraulic conductivity produced by the degradation of collagen fibers. We also found that similar concentrations of hyaluronidase – an enzyme previously used in drug delivery to reduce the hydraulic conductivity of the dermis- had a smaller effect than collagenase on the amount of ISF collected.

The second set of experiments, done in human subjects, provides insights about fluid flow through the skin as a living organ with the ability to respond physiologically after application of our MN patches and suction treatment. We found that application of suction of 50 kPa on human skin for periods longer than ~1 hours produces blisters, even after creating micropores on the skin with MNs. A theoretical analysis of flow through the skin during blister formation indicates that the hydraulic conductivity of the epidermis is less than 1/10 that of the dermis. We also estimated fluid fluxes from the dermis into blisters, and found that these fluxes may be larger than 1 $\mu l/min/cm^2$, an encouraging finding that suggests it may be possible to mobilize ISF through the dermis at these rates. Altogether, these results provide relevant mechanistic insights about ISF flow that can guide the design of a MN-based systems for extraction of ISF for POC diagnostics tests.

601503 A Variable Height Microfluidic Device for Traumatic Brain Injury Diagnosis and Prognosis (Industry Candidate)

Abstract:

Traumatic brain injury (TBI) causes significant morbidity and mortality globally partially due to ineffective diagnostic and treatment methods [1]. Assessment of TBI injuries is typically done through neurological examination and neuroimaging techniques. While these methods are capable of identifying direct tissue damage to the brain, they cannot assess the secondary damage stemming from the initial injury. The primary tissue damage sets off a cascade of secondary injuries, such as neuronal cell death, blood brain barrier breakdown, edema, and upregulation of inflammatory markers [1]. Protein biomarkers have been proposed as a way of monitoring the progression of secondary TBI injury and of providing more sensitive diagnostic measures when used in conjunction with imaging and physical examination. FDA approved biomarkers, glial fibrillary acidic protein (GFAP) and ubiquitin c-terminal hydrolase L1 (UCH-L1), can currently only be measured in a hospital laboratory [2].

To make TBI biomarker data actionable in field and pre-hospital settings, we are developing a glass microfluidic device for protein quantification via bead-based immunoassay. The device consists of a single channel with a variable height profile fabricated by slowly lowering a glass wafer into hydrofluoric acid [3]. The device is capable of capturing and separating beads with diameters ranging from 1.5 to 4 μm (Figure 1B). We also developed bead-based QLISAs (quantum dot-linked immunosorbent assay) for GFAP, interleukin-6 (IL-6), and interleukin-8 (IL-8) by conjugating appropriate antibodies to 2.8, 4.5, and 1 μm diameter beads, respectively. The GFAP QLISA is capable of quantifying GFAP in the range of 0.01-50 ng/mL. We used the variable height device to capture the 2.8 μm GFAP assay beads and analyze the fluorescence intensity of buffer samples spiked with varying GFAP concentrations (Figure 1A).

The variable height device is an adaptable microfluidic platform with the potential to quantify multiple biomarkers simultaneously. By using different sized beads for the GFAP, IL-6, and IL-8 immunoassays, the microfluidic platform can analyze a panel of TBI biomarkers [4]. By changing out the bead size and antibodies used for each immunoassay, the variable height device is able to keep pace with the developing field of TBI biomarker discovery and validation. Once a fully point-of-care platform is realized with portable optics, our device has the potential to make TBI biomarker data actionable in both field and hospital settings.

[1] Loane, D. J., & Faden, A. I. (2010). Neuroprotection for traumatic brain injury: translational challenges and emerging therapeutic strategies. *Trends in pharmacological sciences*, 31(12), 596-604.

[2] Papa, L., *et. al.* (2016). Time course and diagnostic accuracy of glial and neuronal blood biomarkers GFAP and UCH-L1 in a large cohort of trauma patients with and without mild traumatic brain injury. *JAMA neurology*, 73(5), 551-560.

[3] de Beer, M. (2020). *Advances in additive manufacturing and microfluidics.* (Doctoral dissertation, University of Michigan, Ann Arbor, MI, USA).

[4] Morganti-Kossman, M. C., *et. al.* (1997). Production of cytokines following brain injury: beneficial and deleterious for the damaged tissue. *Molecular psychiatry*, 2(2), 133-136.

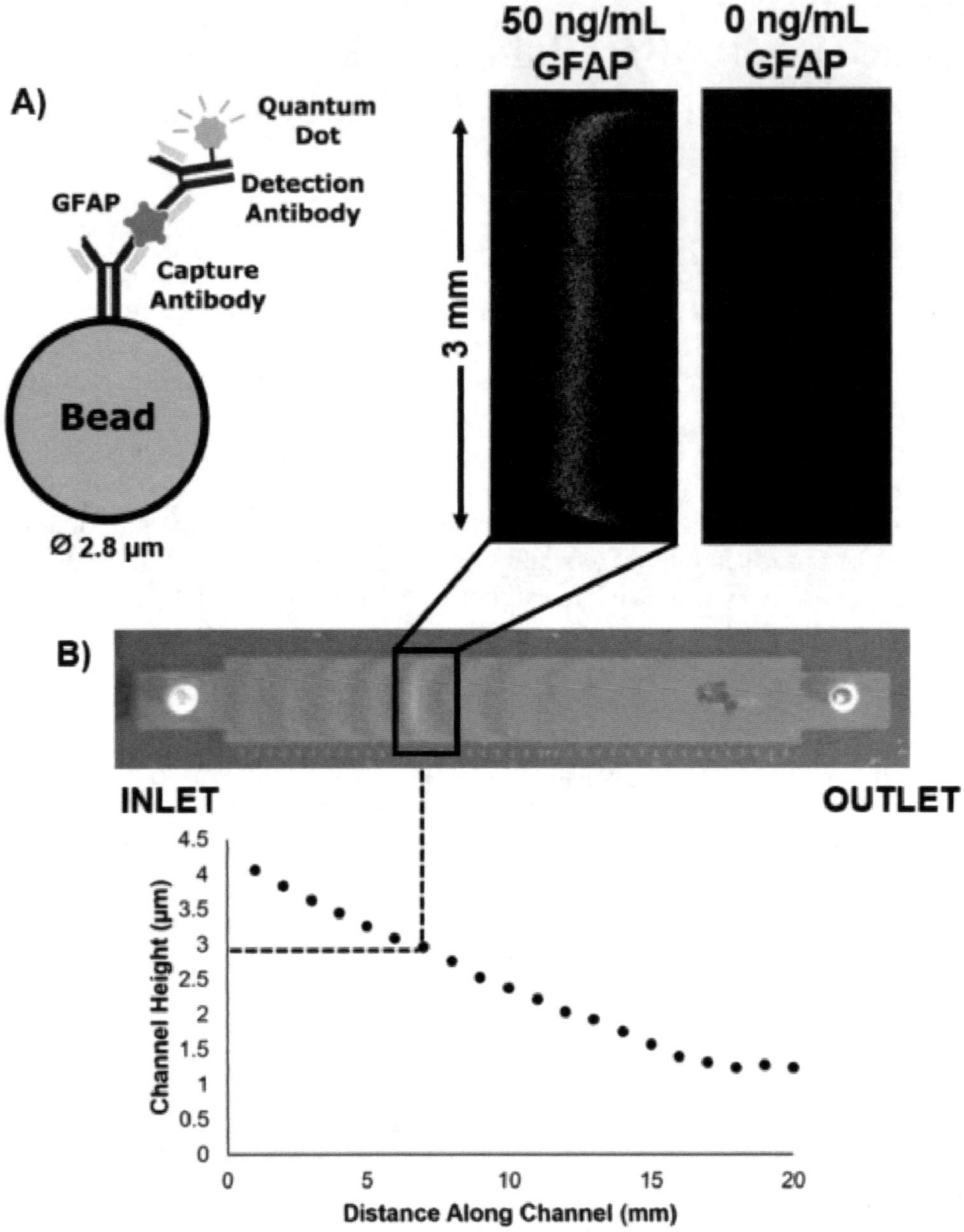

Figure 1: (A) A schematic of the quantum dot-linked immunosorbent assay for quantifying GFAP (left) and representative images of assay beads trapped in the variable height channel after incubation with 50 ng/mL and 0 ng/mL GFAP spiked buffer solutions (right). **(B)** A picture of a variable height channel taken from above, aligned to a graph depicting the channel height between the inlet and outlet.

604774 Multiplexed Detection of Transcription Factors for Liquid Biopsy By Integrating DNA Nanotechnology and Chromatography

Abstract:

The variation patterns of transcription factors (TFs) provide direct information for the states of cell populations, which is of significance for biomedical research and clinical diagnostics. Herein, we show that through multi-channeled isothermal amplification, it is feasible to connect DNA-based signal transduction with chromatography for multiplexed detection of TFs. The described system is referred to as "PAC" which includes three major steps: (i) **P**rotection, which uses DNA-modified magnetic beads to capture TFs and converts the capturing event into triggering signal; (ii) **A**mplification, which receives the triggering signal and generate DNA reporters through multi-channeled extension and nicking of oligonucleotides; and (iii) **C**hromatography, which separates and detects the DNA reporters in liquid chromatography. The quantitative detection of five essential TFs includes p50, p53, AP-1, MITF, and c-Myc is realized in a multiplexed manner, with the lowest detection limit of 0.5 pM. PAC can also provide effective means to measure the above five TFs in real samples, including cultured cells, xenograft tumors, and blood-based liquid biopsy. This study not only established a new solution for multiplexed measurement of TFs for molecular diagnostics, but also paved avenue for bridging the gap between DNA nanotechnology and chromatography.

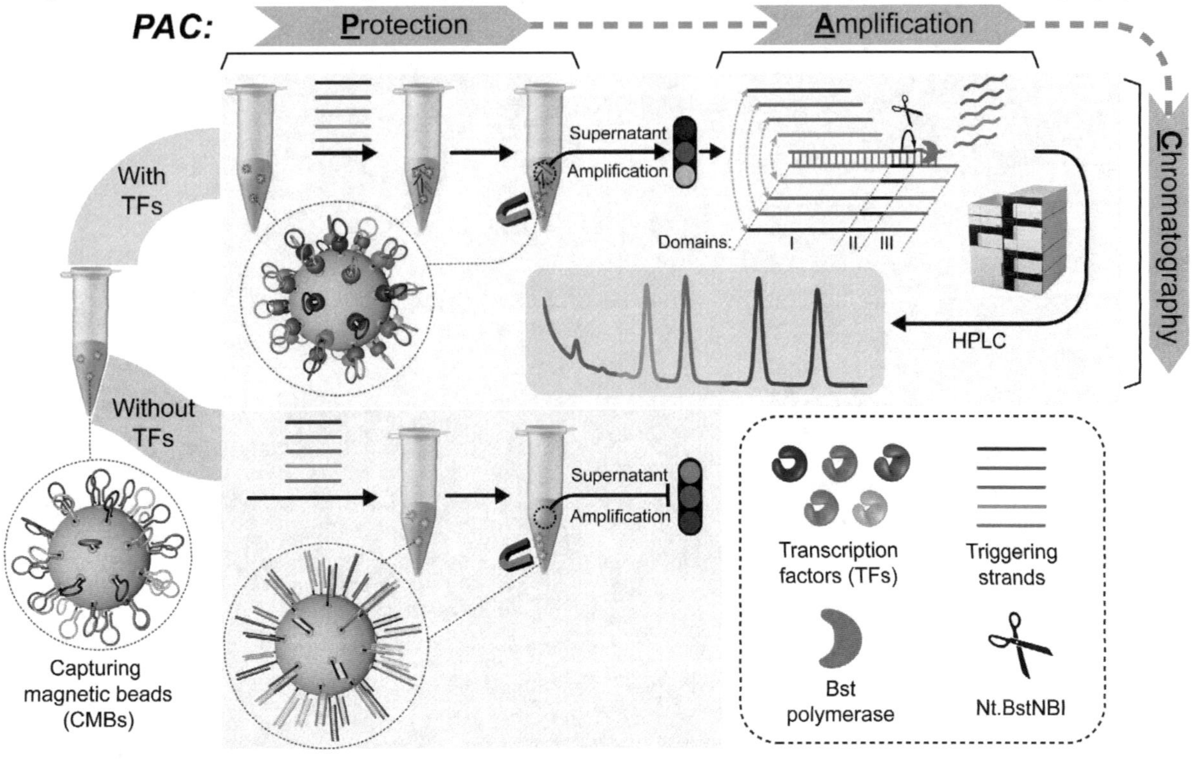

613036 Exploiting Viruses That Kill and Killing Viruses That Exploit (Invited Speaker)

Abstract:

Nature is unparalleled in its structural and functional diversity. In many cases, nature has provided us with a blueprint to overcome gaps in our biomedical arsenal, whether for sensing, actuating or provide therapeutic advances. Nowhere is this more evident than in nature's most abundant "life" forms – viruses, including bacteriophages. We have exploited various molecular machines in bacteriophages to sense and/or kill human pathogens, thereby providing new routes to biosensors, therapeutics and disinfectants. Unfortunately, viruses often exploit us, as we are now all too aware given SARS-CoV-2 and the tragic loss of life and economic destruction caused by COVID-19. We have once again turned to nature's vast arsenal of biomolecules to target viruses, including coronaviruses, among other human pathogens. This includes the use of the anticoagulant heparin, which possesses extraordinary binding to the SARS-CoV-2 spike protein, and heparin analogs and other functional polysaccharides that possess exceptional binding without anticoagulation. In this manner, we have uncovered a unique Achilles heel of many viral pathogens and identify new routes to bring them under control.

610321 ^{13}c-MFA of Human Platelet Metabolism

Abstract:

Female sex hormones, like estrogen, can cause an increased risk for developing blood clots. Taking estrogen in the form of oral contraception increases the risk for venous thromboembolism (VTE) by approximately 3-4 fold. Platelets are blood cells critical for VTE initiation and propagation, yet the relationship between platelet function and sex hormones is unclear. Furthermore, current studies are contradictory, with reports of platelet activation being both inhibited and potentiated by acute exposure to estrogen. Estrogen is an important signaling molecule known to impact metabolism in many human tissues. However, the specific mechanisms by which estrogen might influence platelet metabolism is currently unknown.

Our work aims to characterize platelet metabolism during exposure to estrogen in order to identify the mechanisms by which platelet activation is modulated during response to platelet agonists (thrombin, ADP, collagen, etc.). Specifically, we are combining *-omics* technologies from a systems biology perspective to build metabolic models of platelet metabolism. We have performed intracellular metabolite profiling via mass spectroscopy and utilized carbon isotope labeling paired with metabolic flux analysis to inform our models of platelet central metabolism. We will present the results of these analyses and show flux maps of resting platelet metabolism. Additionally, we will discuss how the presence of estrogen alters carbon flux throughout the metabolism of human platelet cells.

(This project is supported by the National Heart, Lung, and Blood Institute of the National Institutes of Health under award number R61HL141794.)

603883 Computational Modeling of the Gut-Bone Axis and Implications of Butyrate Treatment on Osteoimmunology

Abstract:

The interplay between gut microbiota and the immune system has a pivotal role in the maintenance of bone health. Recently, short-chain fatty acids (SCFAs) produced by gut microbiota have emerged as key regulatory participants in shaping the immune system. Butyrate, the most versatile among SCFAs, has been observed to have local and systemic effects including inducing the differentiation of peripheral regulatory T cells (T_{regs}) in the intestine, blood and bone marrow [1]. T_{regs} are the central actors of the negative feedback component of the immune system. The interaction between T_{regs} and cytotoxic CD8+ T cells suppress the inflammatory status and promote the production of Wnt10b to increase bone anabolism [2]. Studies in the mouse models shows that the ablation of butyrate in the intestine alter the bone marrow density. However, the therapeutic benefit of butyrate in bone anabolism remains poorly understood. Hence, we use computational modeling to describe the butyrate induced differentiation of T_{regs} in the intestine blood and bone marrow that promotes the change of Wnt10b expression in the bone marrow.

We are developing a multi-compartment physiologically based pharmacokinetic model to track and quantify effects of butyrate on T_{regs} in the gut, blood, and bone marrow. The model consists of five species butyrate, naïve CD4+ T cells, T_{regs}, TGF-β, and Wnt10b distributed across three compartments intestine, blood, and bone. We consider an open system with the processes of formation, excretion, differentiation, cell death, and migration to another compartment. We define mass balances on butyrate, TGF-β, and Wnt10b and population balances for the naïve CD4+ T cells, T_{regs}. The microbial fermentation of the dietary fiber modulates the constant production of butyrate in the intestine. The constant supply of butyrate in the intestine depends on the degradation and migration to the blood and bone marrow. The butyrate that absorbs into the blood and bone compartments differentiates the naive CD4+ T cells in the bloodstream and bone marrow. The differentiation rate constant is evaluated based on the dynamic in vitro differentiation of naive CD4+ T cells to T_{regs} in the presence of butyrate. The system is assumed to be at steady state to estimate the unknown parameters for the model based on the in vivo experimental butyrate concentration and percentage T_{regs} of total cells in each compartment. T_{regs} accumulate in the bone and induce upregulation of TGF-β, which suppresses an inflammatory state. T_{regs} also promote Wnt10b secretion that increases net bone formation. The variation of butyrate concentration from homeostasis value changes the percentage of T_{regs}, and the production of TGF-β, and Wnt10b. We estimate the kinetic parameters for the differentiation of T_{regs}, fold changes of TGF-β, and Wnt10b with change of butyrate concentration based on data from published in vivo experimental studies.

Using the model, we analyze experimental data reported in [2] to evaluate the expansion of T_{regs}, TGF-β, and Wnt10b in the bone marrow. The probiotic LGG increases butyrate concentration in intestine and serum blood by 0.18 μM and 0.29 μM respectively. We assume butyrate concentration in bone marrow is same as serum blood. Our simulation result shows 6% increase of T_{regs}, 3.4-fold increase of TGF-β, and 3-fold increase of Wnt10b in the bone marrow consistent with the net change in formation due to stimulus of a probiotic microcrobiota in the gut. We also analyze the pathological condition where butyrate concentration reduces due to dysbiosis. Li et al. [3] reduced the serum butyrate concentration 0.18 μM using antibiotic in mouse model and reported 5% reduction of T_{regs} in bone marrow from homeostasis condition. We set the change of intestine butyrate concentration to -0.11 μM in our model to get 0.18 μM reduction of butyrate concentration in serum blood and observe 5.5% reduction of T_{regs} in bone marrow consistent with the experimental result. The computational approach described here gives insight into the pharmacokinetics of butyrate, biodistribution of T_{regs} in the gut-bone axis, fold changes of Wnt10b in bone marrow, and their contributions to modulating bone formation. This deeper understanding will allow us to explore the potential of modifying butyrate through dietary interventions as a treatment for different pathological conditions related to bone loss such as estrogen deficiency, hyperthyroidism, and vitamin D deficiency.

References

1. Furusawa, Y. et al. *Nature.*7480:446-450, 2013.
2. Tyagi, A. M. et al. *Immunity.*6:1116-1131, 2018.
3. Li, J. Y. et al. *The Journal of Clinical Investigation.* 130(4):1767-1781,

Acknowledgements

Research reported in this abstract was supported by the National Institute of General Medical Sciences of the National Institutes of Health under award number R35GM133763. The content is solely the responsibility of the authors and does not necessarily represent the official views of the National Institutes of Health.

609270 Predicting Enzyme Kinetics and Regulations from Dynamic Flux Balance Analysis through Kinetic Optimization Using Integer Conditions

Abstract:

Stoichiometric models, or Genome-Scale Models (GSMs), are now an indispensable tool for systems biology, yet are limited in their predictive capabilities by their linear natures and general non-consideration of enzyme kinetics and regulation. In contrast, kinetic Models of Metabolism (kMMs) provide not only a more accurate method for designing novel biological systems but also characterization of system regulations; however, the multi-'omics' data required has proven prohibitive to the development of kMMs larger than a few hundred reactions (as opposed to GSMs which may have as many as ten thousand or more reactions). To address the lack of multi-'omics' data required to produce large kMMs, introduce here is a new approach named **K**inetic **OPT**imization using **I**nteger **C**onditions (**KOPTIC**) to make plausible reaction mechanism, kinetics, and regulation predictions. KOPTIC can circumvent the multi-'omics' data requirement and semi-automate kMM construction using an optimization-based approach which relies on *in silico* reaction flux data and metabolite concentration estimates from a dynamic Flux Balance Analysis (dFBA) of a stoichiometric model. As a benchmark for the performance of KOPTIC, a previously published, stochiometric, four-tissue (leaf, root, seed, and stem) metabolic model of *Arabidopsis thaliana* was used, consisting of the core metabolism of Arabidopsis, named p-ath773 (1251 reactions, 1155 metabolites, and 773 genes). The p-ath773 was subject to dFBA using the ORKA approach which simulated plant biomass, reaction rates, and metabolite concentration at each hour throughout a 61-day lifecycle of an *Arabidopsis* plant. This data was then used as the basis for the application KOPTIC. KOPTIC was applied in three distinct iterations to the p-ath773 data where each represents a different set of allowable regulators. The first allowed only within-path regulation, the second within subcellular-compartment, and the third within cell. This resulted in thousands of kinetic predictions, of which some have been verified by biochemical data. In summary, this demonstration of KOPTIC shows that KOPTIC can be useful in the addressing of knowledge gaps by circumventing them using dFBA and optimization.

W. L. Schroeder and Rajib Saha. "KOPTIC: a novel approach for *in silico* prediction of enzyme kinetics and regulation" *BioRxiv*, Oct. 17, 2019. Available: https://www.biorxiv.org/content/10.1101/807628v1 (doi: https://doi.org/10.1101/807628)

610891 A Synthetic Lethal Drug Combination Mimics Glucose Deprivation–Induced Cancer Cell Death in the Presence of Glucose

Abstract:

Introduction: Metabolic reprogramming in cancer cells can increase their dependence on metabolic substrates such as glucose. As such, the vulnerability of cancer cells to glucose deprivation creates an attractive opportunity for therapeutic intervention. Because it is not possible to starve tumors of glucose in vivo, we sought to identify the mechanisms in glucose deprivation-induced cancer cell death and then designed inhibitor combinations to mimic glucose deprivation-induced cell death.

Results and Discussion: Using metabolomic profiling, we found that cells undergoing glucose deprivation-induced cell death exhibited dramatic accumulation of intracellular L-cysteine and its oxidized dimer, L-cystine, and depletion of the antioxidant glutathione. Building on this observation, we show that glucose deprivation–induced cell death is driven not by lack of glucose, but rather by L-cystine import. Following glucose deprivation, the import of L-cystine and its subsequent reduction to L-cysteine depleted both NADPH and glutathione pools, thereby allowing toxic accumulation of reactive oxygen species. Consistent with this model, we found that the glutamate/cystine antiporter xCT is required for increased sensitivity to glucose deprivation. Testing a drug combination that co-targeted GLUT1 (STF-31) and glutathione synthesis (BSO), we found that this combination induces synthetic lethal cell death in high xCT-expressing cell lines susceptible to glucose deprivation.

Conclusions: These results indicate that co-targeting GLUT1 and glutathione synthesis may offer a potential therapeutic approach for targeting tumors dependent on glucose for survival. Future studies will be necessary to evaluate the safety and efficacy of combined STF-31 and BSO treatment for xCT-high tumors *in vivo*. Taken together, our results define a new synthetic lethal drug combination that exploits the glucose addiction phenotype by mimicking glucose deprivation-induced cell death even in the presence of glucose.

References: Joly J. H., Delfarah A, Phung PS, Parrish S, Graham NA. (2020) A synthetic lethal drug combination mimics glucose deprivation–induced cancer cell death in the presence of glucose. J Biol Chem 2020;295:1350–65. https://doi.org/10.1074/jbc.RA119.011471.

612227 Mathematical Modeling of Lipid Metabolism in Bone Marrow-Derived Macrophage Cells Using Cybernetic Control Variables

Abstract:

Metabolism is regulated by a number of factors in the cell. The concerted action of metabolism and regulation gives rise to the cellular phenotype or cellular outcome behavior. The cybernetic approach developed by our group assumes a goal and evolves the dynamics of the system under regulation to determine how each of the variables, e.g., metabolite concentrations, evolve over time, and how the metabolite fluxes are regulated.

The key advantage of cybernetic descriptions of cellular regulation is that they capture the molecular phenomena that control metabolic fluxes in the form of an intuitive regulatory principle. From the cybernetic perspective, regulatory mechanisms at the molecular level are not isolated events. Regulation is a cooperative cascade of molecular incidents that are coordinated to enhance a cell's survival. Regulatory goals, such as maximizing growth [1] or carbon uptake rate [2], provide a causality driven basis for the regulation of individual chemical events. In the absence of high resolution, dynamic data for all cellular events that modulate metabolism, cybernetic assumptions of regulation offer a significant advantage in that they are simple and can robustly predict metabolic phenomena given an appropriate objective function.

While cybernetic models have focused on bacterial systems in the past, we adapt this framework to model the dynamic behavior of prostaglandin (PG) formation in a mammalian cell line, bone marrow-derived macrophage (BMDM) cells. Several kinetic descriptions of PG formation precede this work [3, 4], but none take into account the regulatory phenomena present in PG formation. Our application of cybernetics to macrophages provides a quantitative model of eicosanoid metabolism initiated with the input of arachidonic acid (AA) and using inflammation as the system objective. The cybernetic model provides a robust description of metabolite formation and can be used to predict perturbations to metabolism [5].

Cybernetic models are a robust description of metabolite formation and can be used to predict perturbations to metabolism via various effectors including drugs. Having a more reliable description of PG formation is useful in that it can provide a more predictive description of the action of inhibitory drugs. We show that cybernetic control theory can be broadened to describe objective functions in complex multi-cellular systems and have applications in predicting the response of metabolic networks to drugs.

References

1. Kompala, D. Ramkrishna, N. Jansen, and G. Tsao. Biotechnology and Bioengineering, 28:1044–1055, 1986.
2. Song and D. Ramkrishna. Biotechnology and Bioengineering, 106(2):271–284, 2010.
3. Gupta, M. Maurya, D. Stephens, E. Dennis, and S. Subramaniam. Biophysical journal, 96:4542–4551, 2009.
4. Kihara, S. Gupta, M. Maurya, A. Armando, I. Shah, O. Quehenberger, C. Glass, E. Dennis, and S. Subramaniam. Biophysical Journal, 106(4):966–975, February 2014.
5. Aboulmouna, S. Gupta, M.R. Maurya, F. DeVilbiss, S. Subramaniam, D. Ramkrishna. A Cybernetic Approach to Modeling Lipid Metabolism in Mammalian Cells. Processes. *6*(8):126, August 2018.

597815 Assessing the Importance of Parametric Uncertainty on Flux Balance Analysis

Abstract:

Flux balance analysis (FBA) and associated techniques operating on stoichiometric genome-scale metabolic models have a central role in quantifying metabolic flows and constraining feasible phenotypes. At the heart of these methods lie two important assumptions: (i) the biomass composition neither changes in response to growth conditions nor environmental/genetic perturbations, and (ii) all metabolites incur production and consumption terms that always equal one another. Despite the stringency of these two assumptions, FBA has been shown to be surprisingly robust at predicting cellular phenotypes. In this paper, we formally assess the impact of these two assumptions on FBA results by quantifying how uncertainty in (i) biomass equation coefficients, and (ii) departures from steady-state due to temporal fluctuations propagate to FBA results. The straightforward implementation of Monte-Carlo simulation or chance-constraint programming will yield inaccurate results as in case (i) it is necessary to re-weight the biomass equation so as the molecular weight remains equal to one, whereas in case (ii) metabolite pools must be conserved even under temporally varying conditions. This implies that in both cases conditional sampling of parameter values must be carried out so as the above conditions are satisfied. Our results confirm the importance of standardizing biomass molecular weight to 1 g/mmol. We find that although on average only 4.7% of the uncertainty in the biomass equation coefficients propagates to the biomass yield calculation, not re-weighing the biomass equation magnifies this value to 14.7%. Interestingly, uncertainty in growth-associated ATP maintenance has the greatest effect on biomass yield (28.7%). Furthermore, FBA subject to metabolite pool dynamics was performed while accounting for metabolite concentration coupling relationships and allowing for perturbations in metabolite steady-states. Overall, uncertainty-to-variation of FBA predictions reported here can be interpreted as a cautionary tale and the framework presented in this work can be used for applications where temporal or parameter variability in FBA needs to be assessed.

613347 Metabolomics-Driven Modeling and Characterization of Metabolism (INVITED SPEAKER)

Abstract:

While rapidly advancing technology is making it increasingly easier, cheaper, and more accessible to perform system-scale measurement of metabolic intermediates in biological systems (known as "metabolomics"), our efforts to exploit the resulting data for modeling of metabolic systems have surprisingly lagged behind. One reason for this is that many of the techniques that are viable for systems-scale metabolic modeling assume a steady state in the system, which limits the utility of measurements of metabolite levels. However, we know from time course metabolomics profiles that biological systems are often not at a steady state, and that these metabolic dynamics play a major role in the regulation of metabolism. This regulation is ultimately what defines the metabolic state or output of a system. As a result, it is critical to try to model metabolic dynamics and regulation at the systems scale.

We have recently developed a computational framework, based on the widely-used Flux Balance Analysis framework, that allows for the exploitation of dynamic metabolomics profiles and ultimately the computationally tractable modeling of metabolic dynamics at the genome scale. We will explain our approach and present our most recent developments to enable scale-up and predictive accuracy of metabolic modeling. We will also describe a machine learning-based approach to identification of unknown regulatory interactions in these metabolic models using metabolomics data. This latter approach can be used not only for the implementation of models in our new computational framework, but for the discovery of regulatory interactions for implementation in arbitrary mathematical frameworks for metabolic modeling. It will provide a valuable tool for the development of metabolic models for applications in metabolic engineering, especially with the rising use of non-conventional chassis organisms for industrial applications where regulation is poorly understood (if at all).

610881 Multifunctional Screening of Active Microswimmers Enabled By Dynamic Self-Assembly on an Open-Surface Microgel Array

Abstract:

Image-based screening of *in vivo* small animal models has been critical for system-level investigation to understand many fundamental biological questions, such as gene functions, neurodevelopment, and aging. However, performing whole-animal screening is challenging due to the difficulty to handle and isolate the small, active animals in a high-throughput manner. In addition, it is necessary for small animal handling techniques to provide precise control of microenvironments that stimulate individual samples in order to screen for animals' responses.

In the last decade, various microfluidic techniques have been developed to meet these needs. Taking one of the most-studied multicellular organism, *Caenorhabditis elegans*, as an example, different closed-channel microfluidic systems are designed and coupled with automated microscopy, to isolate large numbers of *C. elegans* in microchannels for high-resolution imaging or in microchambers for animal behavior analysis. Comprehensive forward genetic screening by image-based deep phenotyping and large-scale drug screening are hence made possible. Despite these already impactful applications, however, the adaptation of these microfluidic technologies into general biology laboratories and biotechnology industries are slow. It is mainly because most biology labs are not readily equipped with necessary engineering expertise, including: (i) special cleanroom environment and skills to fabricate microfluidic devices; (ii) sophisticate off-chip controllers and user expertise to operate these microfluidic systems and to achieve automation. Furthermore, a microfluidic device is usually designed only for one screening function, such as mutant sorting based on high-resolution imaging of immobilized animals or freely moving behavior analysis in response to chemical stimuli. Prototyping new microfluidic designs is cumbersome, time consuming and can be expensive. Therefore, a simple screening platform is needed to allow for isolation of active small animals without the need of complicated operating controllers, and to enable multifunctional screening applications without the need of additional device designs.

To address this issue, we present an open-surface platform for rapid isolation and multifunctional screening of active microswimmers, using *C. elegans* as a model. The open-surface device consists of an array of hydrophilic PEG-based microgel pads surrounded by a less hydrophilic plastic surface made of structured Kapton tape. Our device can be easily prototyped using common and readily available materials and fabricated outside a cleanroom. The heterogeneous microgel array enables a novel equipment-free method to rapidly isolate a large number (>100) of active microswimmers on the open surface in 30 seconds by a dynamic self-assembly mechanism. General users with no previous engineering expertise can master the operation within minutes.

The unique open-accessibility of our device allows for individual interrogation of microswimmers and enables various screening functions. We demonstrate its versatile utility by three fundamental image-based screening applications. First, we show that the microgel pad can be used to immobilize active microswimmers by controlled release of anesthetic agents. Consequently, we can achieve high-resolution phenotypical screening of sub-cellular features, such as *in vivo* synapse development with standard fluorescent confocal microscopy. This sets the foundation for most image-based screenings. Second, taking advantage of the open-surface accessibility, we develop an equipment-free on-demand mutant selection workflow. We validate the workflow by performing selective recovery of rare mutant animals within a mixed population based on their fluorescent phenotypes. This workflow can benefit rapid selection of animals in large-scale studies, such as identifying and enriching rare mutagenized populations through forward genetic screening. Third, we demonstrate a multiplexed chemical screening assay through analysis of individual-animal behavioral response. This is performed by individually modifying the chemical microenvironment of each microgel pad with a panel of different concentrations of an anthelmintic drug and screening the heterogeneous paralysis dynamic of isolated *C. elegans*. This function provides a straightforward approach to conduct multiplexed screening assays, which could be instrumental for biomedical applications from elucidating fundamental neuronal mechanism to evaluating drug efficacy. Because of its simplicity and versatility, we envision this platform can be easily adapted in general life science laboratories, and potentially tailored for other model biological systems with simple modifications.

606848 Multi-Contrast Photoacoustic Tomography

Abstract:

Whole-body imaging has played an indispensable role in both preclinical and clinical research by providing high dimensional physiological, pathological, and phenotypic insights with clinical relevance. Yet pure optical imaging suffers from either shallow penetration (up to ~1–2 mm) or a poor depth-to-resolution ratio (~3), and non-optical techniques for whole-body imaging lack either spatiotemporal resolution or functional contrasts. A stand-alone single-impulse photoacoustic computed tomography (PACT) system has been built, which successfully mitigates these limitations by integrating high spatiotemporal resolution, deep penetration, and full-view fidelity, as well as anatomical, dynamical, and functional contrasts. Based on hemoglobin absorption contrast, the whole-body dynamics and large scale brain functions of rodents have been imaged in real time. The absorption difference between cytochrome and lipid has enabled PACT to resolve MRI-like whole brain structures. Taking advantage of the distinct absorption signature of melanin, unlabeled circulating melanoma cells have been tracked in real time *in vivo*.

Assisted by near-infrared dyes, the perfusion processes have been visualized in the brain and internal organs. By localizing the single-dyed droplets, the spatial resolution of PACT has been improved by six-fold *in vivo*. The migration of metallic-based microrobots toward the targeted regions in intestines has been visualized in real time. The integration of the newly developed microrobotic system and PACT realizes deep imaging and precise control of the micromotors *in vivo* and promises practical biomedical applications, such as drug delivery. Genetically encoded photochromic proteins benefit PACT in detection sensitivity and specificity. The unique photoswitching characteristics of different photochromic proteins allow quantitative multi-contrast imaging at depths. A split version of the photochromic protein has permitted PA detection of protein-protein interactions in deep-seated tumors, providing a powerful tool for fundamental tumor study. The photochromic behaviors have also been used to guide photons to form an optical focus inside live tissue, enabling deep tissue photodynamic therapy and deep brain optogenetics.

In addition, a high-throughput, low-cost PA imaging technique—photoacoustic topography through an ergodic relay (PATER)—has been developed to capture a widefield image with a single laser shot using only a single element detector, which is expected to be suitable for portable and wearable applications. As a rapidly evolving imaging technique, PACT promises preclinical applications, clinical translation, as well as consumer electronics.

Multi-contrast Photoacoustic Tomography

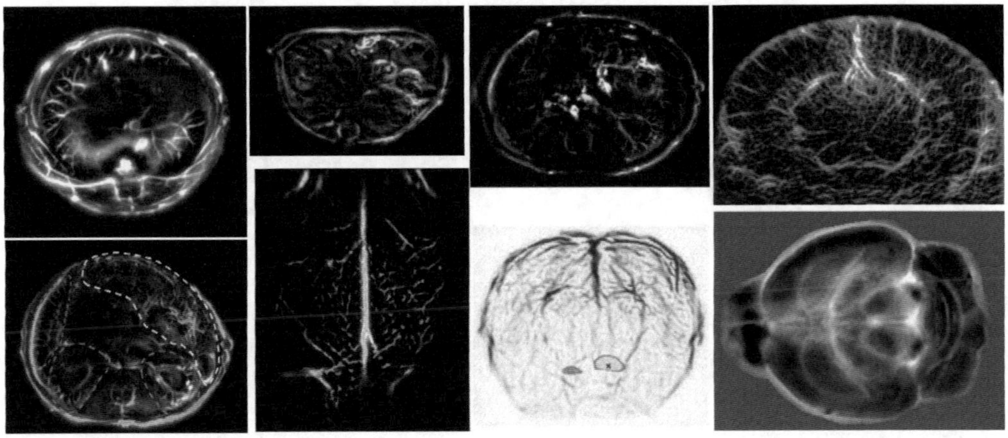

607466 Thermodynamic Analysis of Multivalent Binding of Functionalized Nanoparticles to the Membrane Surface

Abstract:

We present a quantitative model for multivalent binding of ligand-coated flexible polymeric nanoparticles (NPs) to a flexible membrane expressing receptors. The model is developed using a multiscale computational framework by coupling a continuum field model for the cell membrane with a coarse-grained model for the polymeric NPs. The NP is modeled as a self-avoiding bead-spring polymer chain, and the cell membrane is modeled as a triangulated surface using the dynamically triangulated Monte Carlo method. The nanoparticle binding affinity to a cell surface is mainly determined by the delicate balance between the enthalpic gain due to the multivalent ligand–receptor binding and the entropic penalties of various components including receptor translation, membrane undulation, and NP conformation. We have developed new methods to compute the free energy of binding, which includes these enthalpy and entropy terms. We show that the multivalent interactions between the flexible NP and the cell surface are subject to entropy–enthalpy compensation. Three different entropy contributions, namely, those due to receptor–ligand translation, NP flexibility, and membrane undulations, are all significant, although the first of these terms is the most dominant. However, both NP flexibility and membrane undulations dictate the receptor–ligand translational entropy making the entropy compensation context-specific, *i.e.*, dependent on whether the NP is rigid or flexible, and on the state of the membrane given by the value of membrane tension or its excess area.

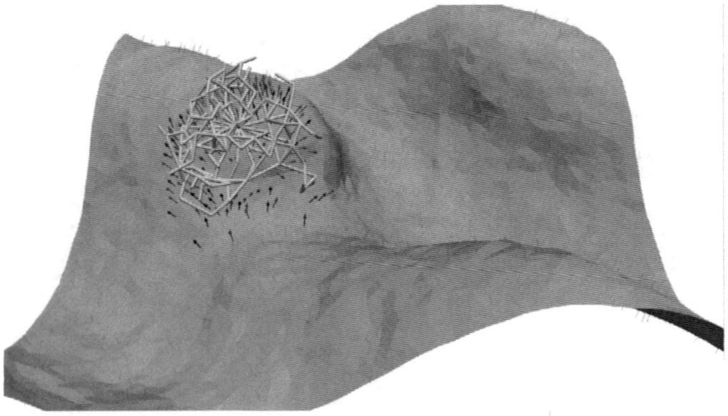

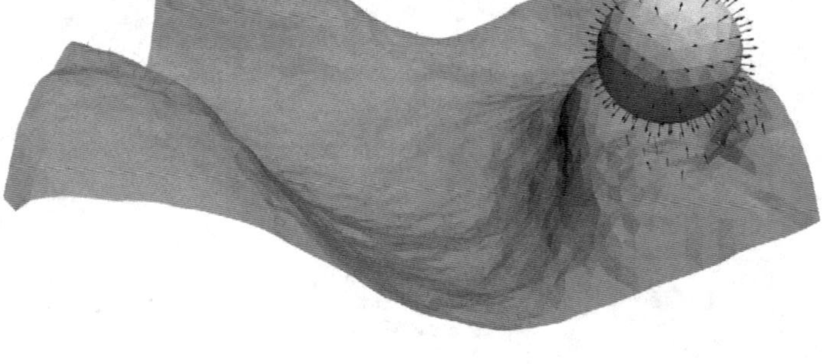

610111 Democratizing Microfluidic Innovation and Commercialization Using Microfluidic Pressure in Paper (μPiP) for Scalable Ultra-Low Cost Manufacturing

Abstract:

Microfluidic engineering has received widespread attention due to the great promise for miniaturizing robust and portable micro total analysis systems (μTAS, or lab-on-a-chip) across a wide range of industries and applications, including point-of-care diagnostics, environmental and public health monitoring, and the detection of biowarfare agents. However, despite significant engineering innovations over the last two decades, market penetration and commercial adoption of microfluidic devices in our daily life remains very low. We believe one reason for this is a lack in a low-cost manufacturing method for producing microfluidic chips. While techniques such as injection molding and laser ablation are capable of scalable production, they can require a six-figure financial investment for both the development and the production phases of commercialization. Such a lack of rapid, scalable fabrication techniques at low-cost is therefore a major bottleneck for enabling innovation and market penetration for researchers who lack access to these expensive methods. Therefore, a low-cost, scalable, robust and reproducible microfluidic device manufacturing method will have significant impact in democratizing innovation, and for allowing stakeholders in both academia and industry to rapidly translate and commercialize microfluidic systems without requiring significant financial investment.

Here, we report a novel low-cost method for scalable manufacturing of microfluidic devices using a combination of precision cut paper and thin PDMS membranes. We call this technique "Microfluidic Pressure in Paper" (μPiP). In μPiP, we utilize a CO_2 laser to rapidly cut hundreds of paper fluidic channels at microscale resolution. We then confine these paper channels between two thin PDMS sheets. Using a combination of corona plasma treatment and high pressure (~5.5 MPa) thermal press, we confine and chemically seal these paper channels within the PDMS membranes. Using this novel workflow, the final μPiP devices are tightly sealed within the PDMS and are void of any air bubbles or structural deformation. We then use a low-cost pressure system to drive fluid through the paper channels in the same way that flows are driven within conventional commercial injection molded chips. Our method, however, can produce hundred-thousands of devices per day for less than a $25,000 USD initial investment and does not require a clean-room. We investigate the fluidic characteristics of these devices using a combination of porous-flow mathematical models for pressure driven fluid flow through polymer-confined paper channels and microfluidic experiments. We demonstrate that, unlike conventional paper-based devices, the hydraulic characteristics of these μPiP devices are not influenced by evaporation or the surrounding relative humidity. We then demonstrate the applicability of μPiP with a range of aqueous buffers and biological samples, including human whole blood. In addition, we demonstrate a wide variety of successful μPiP use cases, including a paper-based microfluidic gradient generator, an H-filter, and a red blood cell deformability assay. We also demonstrate sample preparation operations using μPiP. Finally, we show how to use μPiP to produce conventional PDMS channels using water soluble paper as a sacrificial channel structure. To the best of our knowledge, this is the first time an external flow system has been used to actively drive microfluidic flow through paper-based channels. Our novel fabrication technique democratizes microfluidic innovation by significantly reducing fabrication costs and enables the manufacturing robust microfluidic devices at scale using a workflow that any researcher, regardless of funding, can successfully utilize.

613364 Microdevices for Cardiovascular Disease Investigations (INVITED SPEAKER)

Abstract:

Cardiovascular diseases (CVDs) are the leading cause of death globally. Central to CVDs is inflammation which contributes to tissue damage, fibrosis, and vascular and endothelial dysfunction. Underlying mechanisms remain poorly understood. Current challenges with *in vitro* CVD models include recapitulating the architecture of ventricular and vascular tissue, quantifying contractile function, and tissue dysfunction. Creating *in vitro* systems capable of mimicking both structural and functional properties of the cardiovascular system can be beneficial in mechanistic and drug efficacy studies. We utilize *in vitro* models inspired by *in vivo* vascular and myocardial properties, such as tissue mechanics, cell composition, and tissue architecture. We have utilized angiotensin II (ANG II) to elicit pathological responses in a CVD chip platform. ANG II, an octapeptide of the renin-angiotensin system, is important in cardiovascular homeostasis and pathogenesis, contributing to pathophysiological cardiac remodeling, endothelial dysfunction, and ultimately, heart failure. The goal of this study was to demonstrate that our system could effectively recapitulate features of CVD, specifically cardiac and endothelial dysfunction, by testing the effects of ANG II on the pathological remodeling of engineered cardiac and endothelial tissues. We found that ANG II exposure led to functional decline in cardiac tissues, evident in arrhythmias and decreased contractile stress generation. Further, we found that ANG II contributed to elevated reactive oxygen species levels in our system. Results from this study can be used to design mechanistic studies to better characterize receptor contributions to physiological or pathological states and identify novel pathways for ANG II induced CVDs. We assert that this *in vitro* model is a suitable tool for disease studies and has the potential to serve as a testbed for drug treatment strategies.

605176 Deep Learning Prediction of Interspecies Interactions from Self-Organized Spatiotemporal Patterns of Co-Evolving Organisms

Abstract:

Microorganisms colonizing natural habits such as soils co-evolve to form specific spatial patterns through interspecies interactions. These self-organized patterns are a key ecological phenotype, which provides critical information on their interaction mechanisms. However, conventional network inference techniques that predict interactions based on population size have yet to be extended to account for such spatial heterogeneity. Here we proposed supervised deep learning as a new network inference tool for predicting interaction networks from spatiotemporal patterns of co-evolving organisms. In the absence of biological imaging data that can be used for training deep learning networks, we used *in silico* data generated from high-fidelity agent-based models to determine model structure and parameters. Even though networks were trained under simple configurations where interaction coefficients are assumed to be invariant, the resulting model successfully predicted more complex cases where interspecies interactions vary in space. In the further test against real biological data obtained through imaging experiments of a binary consortium (*Pseudomonas fluorescens* and a mutant of *Escherichia coli*), our model also predicted the dramatic shifts in interactions across different environmental contexts. Through various successful demonstrations in this work, therefore, the combined use of the agent-based modeling and machine learning techniques provides a means to use new type of data - microscopic images - for extracting microbial interactions, therefore presenting itself as a useful tool for the analysis of more complex microbial community interactions.

610932 Microbiota-Derived Metabolites and Bacterial Energetics

Abstract:

Microbial metabolites in the GI tract play a key role in the regulation of bacterial colonization and pathogenesis. Among the important microbiota metabolites is indole, which is well known for regulating bacterial motility, stress responses, biofilm formation, antibiotic resistance and virulence. Our recent work is consistent with the notion that indole segregates cells spatially to repel invaders while recruiting beneficial resident bacteria to growing microbial communities within the GI tract [1]. However, the mechanisms by which indole and other metabolites mediate homeostasis in the microbial community are unclear. I will discuss our latest results that implicate indole in the self-regulation of bacterial energetics. Measurements that show a reduction in the cell membrane potential due to indole will be interpreted in terms of the modifications of enzymatic activity in select protein complexes. I will present our model for how *E. coli* might sense and respond to the dissipation of its potential, based on single molecule measurements of fluorescent proteins. I will conclude with discussions on the potential effects of energetic regulation by metabolites on the colonization potential in bacterial niches.

References

[1] Yang J., Chawla R, Rhee KY., Gupta R., Manson MD, Jayaraman A, and Lele P.P, Proc. Natl. Acad Sci USA, 2020

610957 Exploring the Interspecies Interactions within a Methanotroph-Cyanobacteria Coculture through Genome-Scale Metabolic Modeling

Abstract:

Biogas is comprised primarily of methane (50%~70%) and carbon dioxide (30% ~50%), which can be produced from various waste sources, including landfill material, animal manure; wastewater; and industrial, institutional, and commercial organic wastes. EPA estimates that currently US biogas production potential is 654 billion cubic feet per year, which could displace 7.5 billion gallons of gasoline [1]. It is clear that biogas has immense potential as a renewable feedstock for producing high-density fuels and commodity chemicals. However, the utilization of biogas represents a significant challenge due to its low pressure and presence of contaminants such as H_2S, ammonia, and volatile organic carbon compounds. To tap into this immense potential, effective biotechnologies that can co-utilize both CO_2 and CH_4 are needed.

In our previous work, we have demonstrated that methanotroph-cyanobacteria coculture offer a highly efficient platform to recover the energy and capture carbon from biogas [1, 2]. Through the metabolic coupling of methane oxidation to oxygenic photosynthesis, we have demonstrated significantly improved growth of both the methanotroph and cyanobacteria strain in the coculture. In order to demonstrate this coculture advantage, we have conducted the comparison experiments for three cases using *Arthrosipira platensis - Methylomicrobium buryatense* - as the model coculture system. Case A is the coculture; Case B is the sequential culture of cyanobacterium followed by methanotroph, with the amount of O_2 produced by the cyanobacterium injected into the single culture of methanotroph; Case C simulates the effect of the exchange of *in situ* produced O_2 between the coculture, where the amount of O_2 produced by the cyanobacterium in the coculture was injected into the methanotroph single culture. It clearly showed that both cyanobacterium and methanotroph in coculture (Case A) grew significantly faster than the sequentially operated single cultures (Case B). In addition, the improvement of the methanotroph growth cannot be fully explained by the availability of the extra O_2 produced in coculture (Case C).

In addition, using the model coculture system, we have developed an unstructured model to capture the growth dynamics. Specifically, Monod-like models were developed to capture coculture growth. Two sources of substrate were considered in the model: gas transferred from gas phase and gas (CO_2 and O_2) produced *in situ*. In the unstructured coculture model, we relied on the increased maximum cell growth rates to capture other potential interactions that enhanced the coculture growth.

In an effort to capture the potential interspecies interaction within the coculture, we also developed a structured, dynamic genome-scale metabolic model to capture the dynamic of the cocutlure [3]. By integrating the available knowledge on each strain with data obtained in our own experiments, we used DFBALab to implement the dynamic GEM for the Coculture [4]. For the dynamic GEM, besides the GEMs for each individual strain, the key inputs were the uptake kinetics for different substrates which were provided from the unstructured dynamic model we already developed. The structured coculture GEM was validated using measured product secretion rates and cell growth rates for each strain over time. However, DFBALab doesn't support a common compartment for both microorganisms to freely excrete and update any potential metabolites. Therefore, in the dynamic coculture GEM, we had to adjust growth/non-growth associated maintenance energy to capture the synergic effect within the coculture in order to match the enhanced cell growth rate observed in the experiments.

In order to identify potential "metabolic links" between the methanotroph and cyanobacteria in the coculture through modeling, in this work, we explore a different modeling platform – SteadCom. Although SteadCom only captures steady states of multi-species systems. It enables free exchange of any metabolites that are excreted or consumed by different strains in the system. In this way, we could explore different optimization schemes to identify potential "metabolic links" within the coculture, and then manually add such exchange into the dynamic model in DFBALab to further validate the identified interactions. [5,6].

Reference:

1. Badr, K., Hilliard, M., He, Q.P., & Wang, J., "Understanding the Stability and Robustness of a Methanotroph-Cyanobacterium Coculture through Kinetic Modeling and Experimental Validation." Annual AIChE Meeting. Pittsburgh, PA (2018).
2. Badr K., Hilliard M., Roberts N., He Q.P. and Wang J. (2019), Photoautotroph-Methanotroph Coculture – A Flexible Platform for Efficient Biological CO2-CH4 Co-utilization, IFAC-PapersOnLine, 52 (1), 916-921;
3. Badr, K., Hilliard, M., He, Q.P., & Wang, J., "A Dynamic Genome-Scale Metabolic Network Model for a Novel Methanotroph-Cyanobacteria Coculture" Annual AIChE Meeting. Orlando, FL (2019).
4. Gomez JA, Höffner K, Barton PI. DFBAlab: a fast and reliable MATLAB code for dynamic flux balance analysis. BMC Bioinformatics. (2014),15:409.
5. Chan SHJ, Simons MN, Maranas CD (2017) SteadyCom: Predicting microbial abundances while ensuring community stability. PLoS Comput Biol 13(5): e1005539.
6. Khandelwal RA, Olivier BG, Röling WFM, Teusink B, Bruggeman FJ (2013) Community Flux Balance Analysis for Microbial Consortia at Balanced Growth. PLoS ONE 8(5): e64567.

607982 Continuum Modeling of Chemotactic Migration in Heterogeneous Environments

Abstract:

The chemotactic migration of bacteria—their ability to direct collective motion along chemical gradients—is central to processes ranging from sustaining plant growth, remediating contaminants, sensing and reporting stimuli, and mitigating or causing infection. However, while migration is well-studied in bulk liquid, most bacterial habitats are heterogeneous media, such as soils, sediments, and biological tissues/gels, that are characterized by strong confinement and tortuosity. Unfortunately, how confinement in a heterogeneous space alters the ability of bacteria to migrate remains poorly understood. Here, we extend the classic Keller-Segel model to describe chemotactic migration in heterogeneous media at the continuum scale. Specifically, we incorporate two key effects that are promoted by confinement: (i) suppression of cellular motility due to collisions with the surrounding medium, and (ii) suppression of cellular motility due to collisions with other cells. Our model predicts the formation of localized fronts of cells that collectively propagate over large length and time scales, driven by their self-generated nutrient gradient—remarkably similar to those observed in experiments. Further, the model sheds light on the factors that control front formation, propagation speed, and morphology. For example, we find that these features depend critically on the cell density: for low cell densities, cells initially spread diffusively until sufficient nutrient is depleted to drive front formation, while for high cell densities, a strong enough nutrient gradient forms rapidly, but front formation is hindered by collisions between cells. The propagation dynamics also depend on the relative influence of cellular motility and growth/division: in weak confinement, fronts propagate primarily due to the biased motion of the cells, while with increasing confinement, growth and division increasingly contributes to front propagation. Together, our work provides a framework to predict and control the migration of bacteria, and active matter in general, in heterogeneous environments.

608138 On the Data-Driven Discovery and Calibration of Closures

Abstract:

We explore the data-driven learning of closures in problems involving bacterial chemotactic motion. The data come from agent-based stochastic computations as well as from experiments, and are validated through computational data from established continuum models. Machine learning has long been used to learn unknown PDEs from spatiotemporal data [1] and recently there has been a renewed interest, especially for the case of multiscale simulation data [2]. Starting from neural network architectures that learn a PDE from spatiotemporal data as a "black box", we proceed to learn only parts of this evolution law as a "gray box", and, in particular, the unknown closure terms.

Using agent-based simulation data and experimental data from E. Coli bacterial chemotaxis, we explore the data-driven discovery of the chemotactic terms; find different equivalent parametrizations of them, and also show how to calibrate approximate (qualitatively correct but quantitatively inaccurate) closures to "the truth" in a data-driven manner.

We also show how the procedure can be linked with equation-free multiscale computational techniques (like patch dynamics) to help collect the necessary macroscale data in the most parsimonious way possible. Possibly the most important finding is that several different "equivalent on the data" PDEs as well as several different "equivalent on the data" closures can be identified that are consistent with the observations. We will also discuss the explainability (in terms of humanly understandable terms) of the variables and parametrizations of the discovered closures. The work forms a bridge between analytical/mechanistic understanding, and data-driven "black box" learning of physical process dynamics, allowing for a synergy between the two options.

[1] Gonzalez-Garcia, R., Rico-Martinez, R. and Kevrekidis, I.G., (1998). Identification of distributed parameter systems: A neural net based approach. Computers & chemical engineering, 22, pp.S965-S968

[2] Lee, S., Kooshkbaghi, M., Spiliotis, K., Siettos, C.I. and Kevrekidis, I.G. (2020), Coarse-scale PDEs from fine-scale observations via machine learning, Chaos: An Interdisciplinary Journal of Nonlinear Science, 30(1), p.013141.

598416 (Invited Plenary Talk) Wearable Electrochemical Sensors

Abstract:

Wearable sensors have received a major recent attention owing to their considerable promise for monitoring the wearer's health and wellness. The medical interest for wearable systems arises from the need for monitoring patients over long periods of time. These devices have the potential to continuously collect vital health information from a person's body and provide this information to them or their healthcare provider in a timely fashion. Such sensing platforms provide new avenues to continuously and non invasively monitor individuals and can thus tender crucial real-time information regarding a wearer's health. Our research over the past decade has aimed toward filling the gaps toward providing biochemical information, beyond that given by common wrist-watch mobility trackers. This presentation will discuss recent developments in the field of wearable electrochemical sensors integrated directly on the epidermis or within the mouth for various non-invasive biomedical monitoring applications. Particular attention will be given to non-invasive and minimally-invasive monitoring of metabolites and electrolytes using skin-worn and microneedle amperometric and potentiometric sensors, respectively, along with related materials, energy and integration considerations. The preparation and characterization of such wearable electrochemical sensors will be described, along with their current status and future prospects and challenges.

610185 (Invited Plenary Talk) an Integrated Paradigm for Precision Exposure to Airborne Chemical and Biological Stressors Based on Personal Sensing

Abstract:

Introduction

The current manuscript outlines the fundamentals of a methodological framework for precision exposure assessment and gives concrete examples of its application in different European cities to assess the health risk associated with exposure to fine and ultra fine particles of chemical and biological origin. The work described herein involves a multi-city personal sensor campaign aiming to refine PM exposure using low-cost portable sensor data, exposure and human respiratory tract deposition modelling to estimate the biologically effective internal daily intake on an individual basis and agent-based modelling to estimate exposure and consequent health risk to the community.

Methodology

A prototype monitoring device (IcaruSense) was developed for measuring three size fractions of PM (1, 2.5 and 10 μm), enabling direct assessment of personal exposure. The device is based on an Arduino microcontroller where miniaturized electrochemical sensor-modules are connected. In order to improve sensor performance, correction factors were used to account for diverse environmental conditions and they were calibrated with the help of a well-validated light scattering optical sensor. Study participants wore a physical activity wristband (Garmin Vivosmart 3) that records steps, distance, type of activity, heartbeat and sleeping patterns. Finally, participant positions were recorded using a GPS sensor, integrated into the PM sensor kit. After calibration and validation, the IcaruSense devices were used to capture daily variability of PM exposure. Exposure was further refined by estimating inhalation adjusted intake rate from the continuous time-activity data collected from the physical activity and the GPS sensors. Using the inhalation adjusted intake rate PM deposition across the human respiratory tract (HRT) was readily estimated using the Multiple Path Particle Deposition (MPPD) model.

A city scale agent-based model (ABM) was developed to support this work by expanding the precision exposure estimates from the individual to the community level. The model feeds into population-based exposure assessment without imposing prior bias, basing its estimations onto emergent properties of the behaviour of the computerised autonomous decision makers (agents) that compose the city-system. Population statistics, road and buildings network data were transformed into human, road and building agents, respectively. Informed by literature-extracted relationships between Socio-Economic Status (SES) and human behavioural patterns, the established ABM simulates societal dynamics. Time-use survey outputs were associated with human agent behavioural rules, aiming to model representative to real-world routines. Moreover, real time-geography of exposure data, extracted from the precision sensors campaigns in nine European cities were used to parametrise and enhance the model.

Mobility behaviour and eventually exposure to pollutants change as a result of a human agent-specific decision making. Virtual individuals of different sociodemographic backgrounds use different means of transportation and follow different sequence and types of activities. Behaviours that were not explicitly programmed, arise through the human agents' interactions, enabling the examination of emergent behaviours from the bottom-up. Spatiotemporal trajectories are coupled with spatially resolved Particulate Matter (PM) levels, enabling the assessment of personal and inhalation adjusted exposure, based on type of location and intensity of enrolled activities.

Results and Discussion

The above methodology was applied in the personal sensors campaign of the HORIZON2020 EU Project ICARUS, where exposure and intake to PM of almost 100 individuals in each out of nine European cities were monitored. The integrated hybrid approach outlined above, which couples multi-sensor datasets on an individual basis with advanced respiratory tract modeling, allowed us to significantly differentiate the actual intake of the participants, highlighting larger differences than the ones attributed to the spatial differentiation of air pollution based on fixed stations (difference of ambient PM levels of 50% were translated in intake differences up to 110%). These differences are the result of the differences in PM size fractions that are captured by the sensors and the capability of the HRT model to translate these differences in PM size distribution accounting for variation in respiratory tract physiology among the participants, which is ultimately reflected in intake estimates.

ABM findings indicated that inhalation adjusted exposure differences between 2 individuals living in the same neighbourhood can vary by as much as 87%, due to radically different spatiotemporal behaviours. This major difference was observed for cases where one human agent has a full-time job and often exercises in outdoor courts whereas his/her neighbour is a retired homemaker. For the same reason, ABM runs confirmed that even people who reside in the same dwelling might not necessarily be exposed to similar levels of pollution. Model results indicated that PM inhalation adjusted exposure between flatmates can differ by 61%. Evidence was, also, provided for vulnerable subgroups of population that are disproportionately subjected to high levels of exposure. ABM-retrieved population exposure findings indicated that subgroup median exposure for children, adult males with a full-time job and low income and human agents of a lower educational level was notably higher than the total population median value. The validated model can assess exposure for the entire sociodemographic spectrum of a simulated region, without the constant need for repeated individual monitoring which is often not feasible due to high costs, ethical constraints and complex organisation that is associated with such measurements. By modelling the heterogeneous routines of human agents, the ABM produces detailed information related to the societal system examined and generates data that could be used to fill in the gaps that exist in traditional datasets.

Conclusions

This study presents a new methodological approach to estimate the external exposome, which encompasses the totality of human environmental exposures at an individual and community level. This method helps to integrate substantive considerations of individual status into long-term urban planning and management; it would also facilitate the identification of both physically and socially practical means for reducing life-threatening exposure levels. The methodological paradigm for precision assessment of exposure at the individual and community levels laid out in this work can be used for evaluating the impacts of different public health policies prior to implementation.

595615 Injectable Ventral Spinal Neuroprostheses with Excellent Sensitivity, Fine Motor Control and Chronic Stability

Abstract:

Spinal cord electrical neuromodulation is promising for partial to complete restoration of loss of motor function associated with neuromotor disorders. Current technologies have achieved substantial progress in motor neuron recruitment, but remain limited to dorsal surface or dorsally-inserted devices that either are remote to ventral motor neurons or require physical intervention into the spinal tissue. Here, I will present a spinal neuroprosthesis and minimally-invasive surgical modality for flexible targeting of not only the dorsal, but more importantly the ventral regions of the spinal cord in mice. The spinal devices incorporate a flexible and stretchable design to accommodate strains associated with spinal cord bending in a freely-behaving animal, thereby yielding chronic *in-vivo* stability. The ventrolaterally implanted devices exhibited 1-2 orders of magnitude lower stimulation threshold, finer motor control and higher fatigue resistance than dorsal implants and enabled programmable complex independent left/right hindlimb motion with functional relevance. These results provide a new paradigm for minimally-invasive and chronically-stable spinal cord neuroprosthetics that directly address motor neurons.

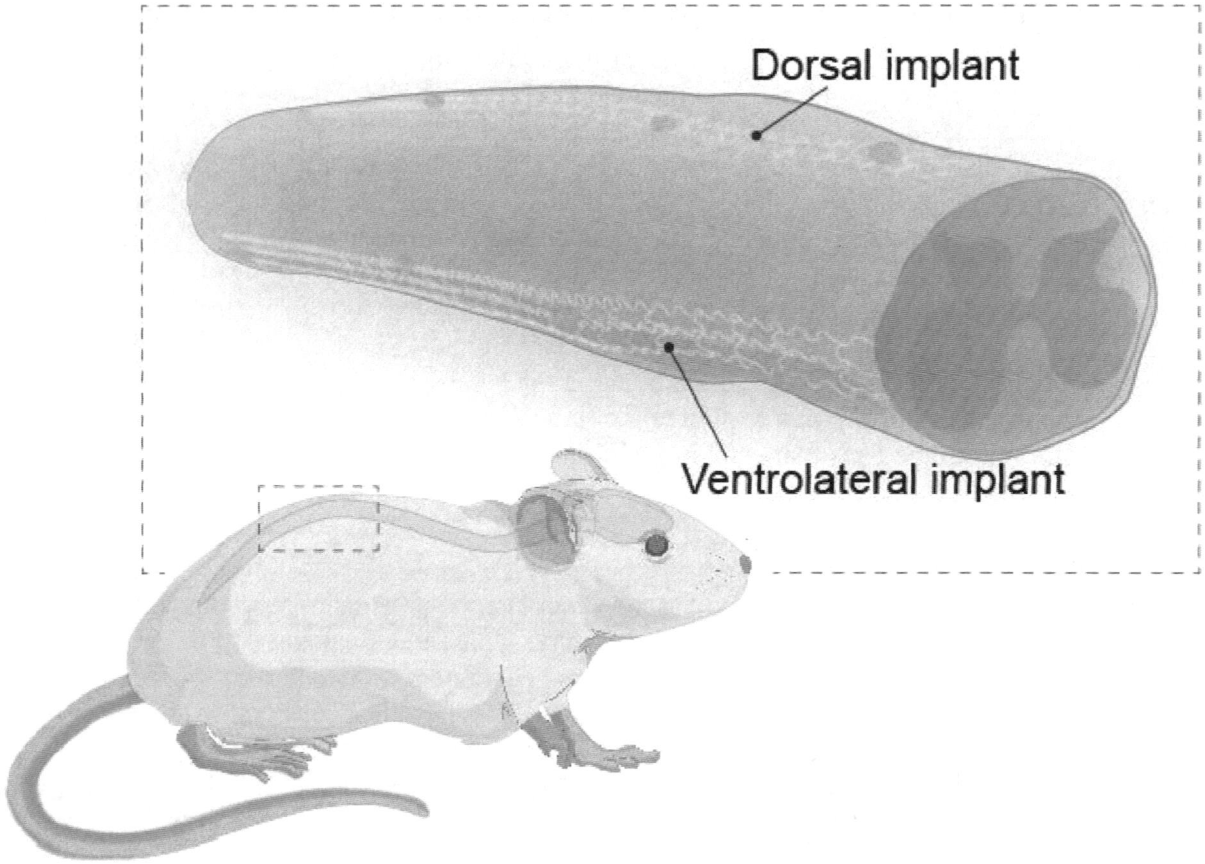

610633 Mac-1 Expression Controls Whether Immune Cells Can Utilize Upstream Migration

Abstract:

Cells of the adaptive immune system such as CD4$^+$ T cells, the innate immune system such as neutrophils, and hematopoietic stem cells (HSPCs), must transit to distal sites such as the lymph nodes, bone marrow, or sites of infection. The recruitment of these immune cells is regulated by a multi-step, leukocyte adhesion cascade of cell rolling, activation, adhesion, and transmigration through the endothelial barrier. After selectin-mediated braking and slow rolling, immune cells migrate along the vascular endothelium through receptor mediated binding to adhesion ligands including Intracellular Adhesion Molecule-1 (ICAM-1) and Vascular Cell Adhesion Molecule-1 (VCAM-1). Previous work by ourselves and others has demonstrated an interesting phenomena where cells such as murine and human T lymphocytes (1, 2), hematopoietic stem and progenitor cells (HSPCs)(3), and marginal zone B-cells (4), can migrate efficiently upstream (against the direction of shear-flow) on ICAM-1 surfaces, or surfaces coated with ICAM-1 and VCAM-1, and endothelial monolayers (HUVECs) which express ICAM-1, using the integrin receptor known as Lymphocyte Functional Antigen-1 (LFA-1). Furthermore, cells undergoing upstream migration transmigrate through the endothelium twice as fast as those going downstream (5). Finally, recent work of ours has furthered the phenomenon of upstream migration by demonstrating that innate immune cells such as neutrophils, which express a second ICAM-1 binding integrin named Macrophage-1 antigen (Mac-1), can migrate upstream when Mac-1 is blocked with antibodies (6).

Here, we describe for the first time how Mac-1 integrin expression can be utilized as checkpoint to direct immune cells either with or against the direction of fluid flow on the endothelial surface using genetic engineering. We demonstrate that myeloid cells of the innate immune system, which cannot normally migrate upstream, can be made to swim upstream by removing Mac-1 function (Figure 1). HL-60 cells and primary neutrophils in which Mac-1 is knocked down are able to migrate upstream on ICAM-1, downstream on VCAM-1, upstream on mixed surfaces of ICAM-1 and VCAM-1 (which mimic the vasculature, and upstream on activated HUVEC endothelial monolayers at shear rates seen in the vasculature. On the other hand, lymphoid cells of the adaptive immune system or stem and progenitor cells, which normally can crawl upstream, can be made to crawl with the direction of shear flow (downstream) by expressing Mac-1. We then demonstrate that KG1a cells, primary HSPCs, and CD4+ T cells genetically engineered to express Mac-1 on their surface cannot migrate upstream on ICAM-1 or any endothelial surface. Finally, by transducing Lifeact into these engineered cells, we demonstrate that the F-actin network is fundamentally different between both upstream and downstream migration and that the presence alone of Mac-1 shifts the actin-network between these modes.

In all, while LFA-1 is the critical integrin mediating upstream motility, Mac-1 has the potential to control migration between the upstream and downstream modes. Therefore, through its expression, which is transient in neutrophils, Mac-1 can not only orient immune cells upstream or downstream but also alter the rate of transmigration. This finding has direct implications in understanding how to manipulate both the direction immune cells trafficking as well as their translocation.

REFERENCES:

1. Dominguez GA, Anderson NR, & Hammer DA (2015) The direction of migration of T-

lymphocytes under flow depends upon which adhesion receptors are engaged. *Integrative biology : quantitative biosciences from nano to macro* 7(3):345-355.
2. Valignat M-P, Theodoly O, Gucciardi A, Hogg N, & Lellouch Annemarie C (2013) T Lymphocytes Orient against the Direction of Fluid Flow during LFA-1-Mediated Migration. *Biophysical Journal* 104(2):322-331.
3. Buffone A, Anderson NR, & Hammer DA (2018) Migration against the direction of flow is LFA-1-dependent in human hematopoietic stem and progenitor cells. *Journal of Cell Science* 131(1).
4. Tedford K, *et al.* (2017) The opposing forces of shear flow and sphingosine-1-phosphate control marginal zone B cell shuttling. *Nature Communications* 8(1):2261.
5. Anderson NR, Buffone A, & Hammer DA (2019) T lymphocytes migrate upstream after completing the leukocyte adhesion cascade. *Cell Adhesion & Migration* 13(1):163-168.
6. Buffone A, Anderson NR, & Hammer DA (2019) Human Neutrophils Will Crawl Upstream on ICAM-1 If Mac-1 Is Blocked. *Biophysical Journal* 117(8):1393-1404.

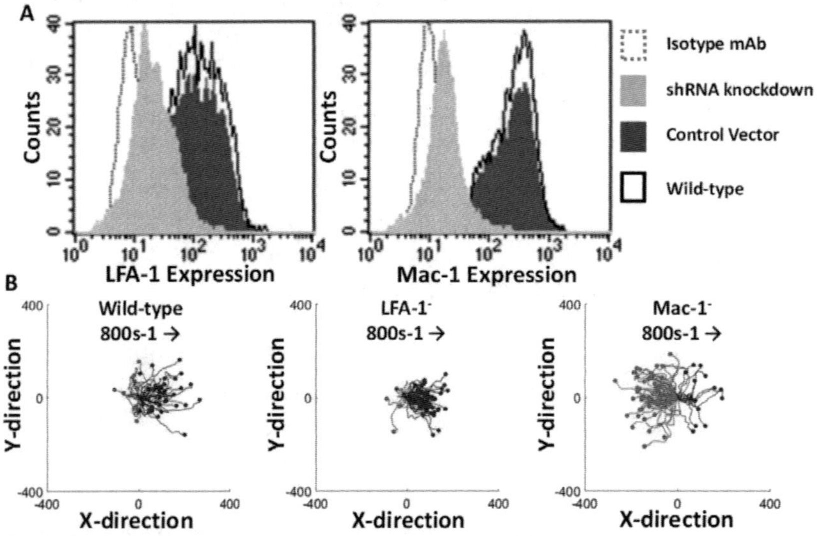

Figure 1: (A) Representative flow cytometry histograms of LFA-1 and Mac-1 HL-60 knockdowns with isotype mAb, shRNA knockdown, control vector, and wild-type expression levels for each knockdown depicted. (B) Cell traces from common origin of HL-60 and knockdowns on ICAM-1 at 800 s^{-1} shear rate. Blue traces in histogram represent cells traveling downstream while red traces represent cells traveling upstream. N=3-4 independent experiments of at least 45 cells analyzed per experiment. *$p < 0.05$ with respect to isotype conditions.

610692 Cell-Free Protein Synthesis As a Tool for Prototyping Metabolic Pathways and Guiding Plant Genome Engineering

Abstract:

Metabolic engineering of organisms to produce sustainable chemicals has emerged as an important part of the global bioeconomy. In parallel, cell-free protein synthesis (CFPS) has matured into a powerful technique for characterizing metabolic enzymes and addressing challenges inherent in engineering life. In one example, CFPS enables *in vitro* prototyping of a pathway containing 9 heterologous steps from glucose to limonene with the goal of shortening design-build-test cycles for metabolic engineering. This effort measured the impact of 54 enzyme homologs, multiple enzyme levels, and cofactor concentrations on pathway performance. Leveraging the open nature of cell-free reactions, we screened over 150 unique sets of enzymes in 580 unique pathway conditions to increase limonene production 25-fold from 23 to 610 mg/L. In a second example, CFPS has proved useful for expressing plant enzymes to guide genome engineering in plants. The wild tobacco relative *Nicotiana benthamiana* is commonly used for manufacturing proteins and reconstituting metabolic pathways which produce complex metabolites such as fragrances or medicines. However, small molecule compounds and their intermediate pathway metabolites produced in *N. benthamiana* are often over-glycosylated, oxidized/reduced, acylated or modified with glutathione by enzymes native to *N. benthamiana*. This unwanted activity has challenged efforts to reconstitute the 11-step pathway to strictosidine (an alkaloid precursor to the anti-cancer drug vinblastine) as well as many other pathways. To guide selection of endogenous candidate genes for knockout by Cas9-mediated mutagenesis, we have built a Golden Gate DNA assembly toolbox adapted for cell-free expression of plant proteins. Using a biofoundry-assisted workflow, we screened expression conditions for dozens of candidate glycosyltransferases and have subsequently measured glycosylation activity towards strictosidine pathway intermediates geraniol and cis-trans-nepetalactol. We anticipate that edited *N. benthamiana lines* will reduce unwanted metabolite derivatization and enable increased production of target molecules.

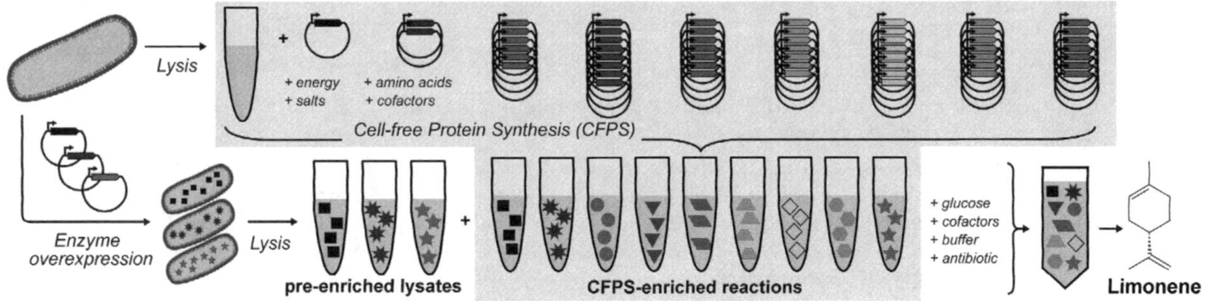

607789 Beta-Carotene Production from Xylose Enriched Syrup of Hydrothermally Pretreated Bioenergy Sorghum Using Engineered *Saccharomyces Cerevisiae* SR8B

Abstract:

β-carotene is bioactive and a precursor for vitamin A synthesis in nature. It can be extracted from plant materials or be produced by chemical synthesis using Wittig reaction. A new bioprocess has been developed to produce β-carotene from the xylose portion of bioenergy sorghum. Bioenergy sorghum was hydrothermally pretreated in a pilot-scale continuous reactor with 50% solids loading at 190°C for 10 min and followed by disc refining. Xylose was extracted from the pretreated energy sorghum by dilute acid hydrolysis. Different temperatures (100°C, 120°C, 140°C, and 180°C), reaction times (5 min and 10 min), and sulfuric acid loadings (0%, 1%, and 2%) were examined to optimize the xylose recovery. The highest xylose yield of 64.87% (17.43 g/L) was obtained from the hydrolysis at 120°C for 5 min with 2% sulfuric acid. The non-concentrated hydrolysate (NCH) was concentrated to 32 g xylose/L (medium-concentrated hydrolysate, MCH) and 66 g xylose/L (high-concentrated hydrolysate, HCH). The engineered *Saccharomyces cerevisiae* strain SR8B was used to ferment the neutralized non-, medium-, and high-concentrated xylose syrup to β-carotene. In HCB, MCB, and NCB cultures, the yeast produced β-carotene titers of 114.50 mg/L, 93.56 mg/L, and 82.50 mg/L, which correspond to specific yeast biomass productions of 7.32 mg/g DCW, 8.10 mg/g DCW, and 8.29 mg/g DCW, respectively. This bioprocess of bioenergy sorghum combining the hydrothermal pretreatment, optimized dilute-acid hydrolysis for the extraction of xylose, and the xylose fermentation with engineered *Saccharomyces cerevisiae* resulted in higher β-carotene yields compared to previous reports in which β-carotene was produced at 2.83 mg/g (Sun et al., 2016), 5.9 mg/g (Verwaal et al., 2007), and 6.3 mg/g (Yan et al., 2012) by engineered *S. cerevisiae* from refined glucose.

References:

Sun, Y., Sun, L., Shang, F., Yang, G., 2016. Enhanced production of β-carotene in recombinant *Saccharomyces cerevisiae* by inverse metabolic engineering with supplementation of unsaturated fatty acids. Process Biochem., 51, 568-577.

Verwaal, R., Wang, J., Meijnen, J., Visser, H., Sandmann, G., van den Berg, J., van Ooyen, A., 2007. High-level production of beta-carotene in *Saccharomyces cerevisiae* by successive transformation with carotenogenic genes from *Xanthophyllomyces dendrorhous*. Appl. Environ. Microbiol., 73, 4342-4350.

Yan, G., Wen, K., Duan, C., 2012. Enhancement of β-carotene production by over-expression if HMG-CoA reductase couples with addition of ergosterol biosynthesis inhibitors in recombinant *Saccharomyces cerevisiae*. Curr. Microbiol., 64, 159-163.

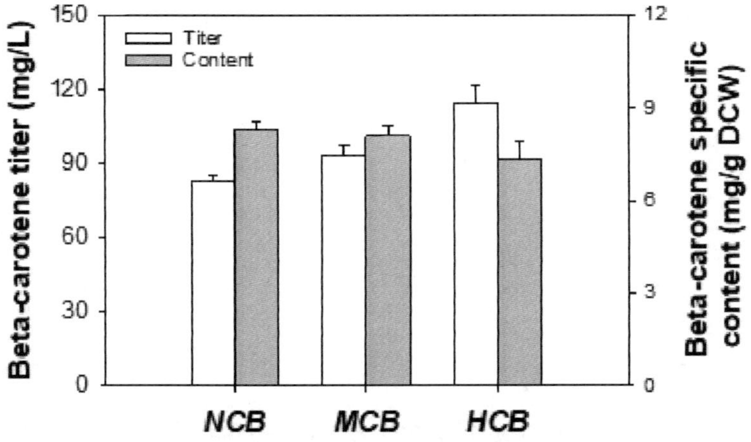

Titer
Content
Beta-carotene titer (mg/L)
Beta-carotene specific content (mg/g DCW)
150
120
90
60
30
0
12
9
6
3
0
NCB
MCB
HCB

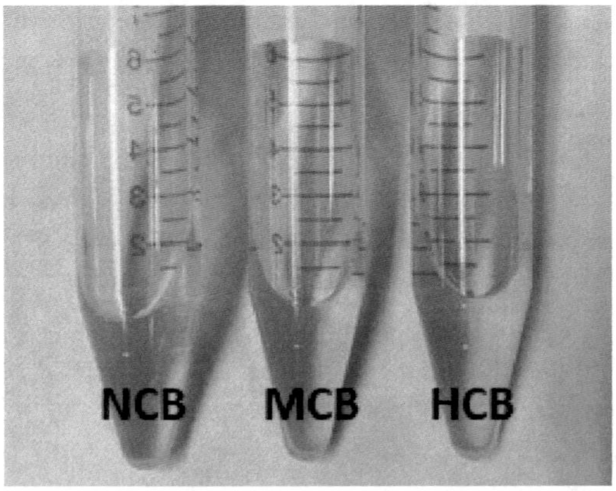

NCB
MCB
HCB

607875 The Biology and Biotechnology of Plant-Microbe Interfaces

Abstract:

Plants have the incredible ability to transform atmospheric CO_2 into a wide array of compounds, from primary and secondary metabolites to complex biopolymers (e.g., cellulose, lignin, starch), all of which support and shape life on our planet. Plant-microbe interactions affect the development, health, and productivity of living plants, while another set of plant-microbe interactions affect the cycling of carbon from dead plant biomass back to the environment. Here, I will discuss two projects that demonstrate the use of systems biology and genetic engineering of non-model bacteria to understand and manipulate these interactions at plant-microbe interfaces. In the first project, the lignocellulosic biomass degradation ability of extremely thermophilic *Caldicellulosiruptor* species was dissected through *in vitro* and *in vivo* characterization of large multi-domain glycoside hydrolase enzymes produced by these species. Proof-of-concept engineered strains with altered biomass degradation ability or metabolic flux to fuel molecules will be presented. In the second project, interactions within the Arabidopsis root microbiome were identified that maintain hormone homeostasis and promote stereotypic root development in diverse bacterial communities. This was traced to a novel auxin hormone degradation locus conserved in *Variovorax* species, ubiquitous members of plant microbiomes. The genes essential to this degradation phenotype, their regulation through a MarR-family transcriptional regulator, and transfer of this pathway to a related bacterial genus will be discussed. These studies highlight a broad array of biological and biomolecular engineering techniques aimed at pinpointing bacterial genes and understanding their specialized biochemical functions at plant-microbe interfaces, with the ultimate goal of engineering these pathways and strains for applications in the bio-agriculture, bio-fuel, and bio-chemical industries.

594120 Chromium Chelated Carrageenan: An Investigation of Physicochemical Properties, Toxicity, and Antimicrobial Potential through Computational and Experimental Approach

Abstract:

Carrageenan (car-) was extracted from *Eucheuma isoforme* and chelated with chromium ions (Cr^{3+}) were computationally and experimentally tested for their physicochemical properties, prediction analysis, toxicity, and antimicrobial potential. The bacterial and leishmanial inhibition studies of car- were conducted through agar-well diffusion method and direct counting-assay whereas Computational techniques such as PatchDock, SwissADME, Molinspiration, and GUSAR were utilized to establish theoretical background to assess its inhibition action on microbial proteins, toxicity and medicinal properties. Cr^{3+} adsorption demonstrated covalent bonding and confirmed through best fit for Freundlich and Dubinin-Radushkevich models. Docking studies proved that car- components have good inhibitory activity for the bacterial and leishmanial proteins. Whereas the prediction analysis and toxicity studies showed that the friendliness of these molecules for the human body is enormous. Thus, it has been proved that car- is a perfect candidate for antibacterial and antileishmanial activity.

610004 Using CFD to Simulate Tank Clean-in-Place Process

Abstract:

In the chemical, food, process and pharmaceutical industries, large tanks are filled, emptied, and refilled with various products many times. Prior to refilling the tank with new product, manufacturers must ensure the previous batch residue is completely removed from the tank's internal surface, as not to affect subsequent product quality. A clean-in-place (or, CIP) system is the typical procedure used to clean a tank after use and before product changeover.

Depending on the product being removed and the equipment size, manual cleaning is usually not attractive economically and a clean-in-place system is desired. A clean-in-place process can be automated and standardized. The process involves installing a dynamic nozzle system, in which water or cleaning solution jets from a nozzle assembly, figures 1 and 2. The nozzles may be forced to move due to the liquid flow (passive) or be motor driven (active). The motion is typically bi-axial, figure 2. This bi-axial nozzle motion and the jetting water/solution form a lattice of track marks on the tank walls which eventually overlap to clean the entire tank's interior. Typical objectives are obtaining a fully cleaned surface while using the optimal amount of water or solution and within the shortest cycle time. Typical input parameters are water pressure, the bi-axial motion of the nozzle assembly, nozzle geometry, and nozzle assembly placement within the tank (e.g. offset from the tank centerline as well as height within the tank). For example, by optimizing the nozzle rotation speeds, flow rate, and nozzle placement it can be ensured that the coverage by the cleaning agent to the mixing tank surface will be maximized. In addition, the impact strength of the impinging jet on the mixing tank surface can be monitored which is directly related to the "mechanical" cleaning action.

Geometrically, this problem spans multiple length scales. The smallest length scale that needs to be captured is that of the jet diameter (mm range). The industrial sized tanks are typically in the multi-meter length scale order of magnitude. Another challenge stems from the variation in time scales that need to be resolved as well. Local jet speeds of multi-hundred meters per second need to resolved stemming in a residence time of around 0.1 seconds form the time of jet departure till impingement. On the other hand, the nozzle assembly is bi-axially rotating at a very low rpm and hence to cover the entire tank with cleaning liquid, multiple "cycles" of rotations are required resulting in physical time in the order of seconds, minutes, or hours.

Due to the large variation in length scales (mm for nozzle vs. several meters for the tank) and time scales (jet flow vs. tank surface coverage) two modeling approaches will be presented. The first approach, tank level, will highlight how the complex bi-axial nozzle motion (combination of spinning and orbiting motion), figure 2, can be controlled to affect the coverage of the mixing tank surface with cleaning agent. A Lagrangian particle tracking approach (Discrete Phase Model or DPM) will be used to model this process combined with a the lagrangian wall film (LWF) model to track and represent the water film spreading on the wall. This model will be very efficient to represent the surface coverage as the nozzles spin. Initial results are shown in figure 3 where the instantaneous wall film thickness is shown on the tank surface after 0.15 seconds.

The second modeling approach, jet level, will focus on the details of the individual jets to accurately predict the impinging force on the mixing tank surface. A detailed Eulerian (Volume of Fluid or

VOF) approach will be used to model the details of the jet. To maintain the fidelity of the narrow jet, coming from an orifice of 0.5 mm, over such a long distance, automatic mesh adaption will be implemented to dynamically refine the water/air interface at least up to impinging with the water surface. The main outcome of this simulation will be the wall shear stress that the jet exerts on the tank walls. Initial results are shown in figure 4 where shear rate on tank wall is displayed.

By combining the two modeling approaches the accurate prediction of the impingement force on the mixing tank can be predicted by the Eulerian (VOF) model while the Langrangian (DPM) model will be used to efficiently predict the entire tank coverage. Finally, the effects of operating parameters like injector rotation speed, flow rate and nozzle placement will be highlighted.

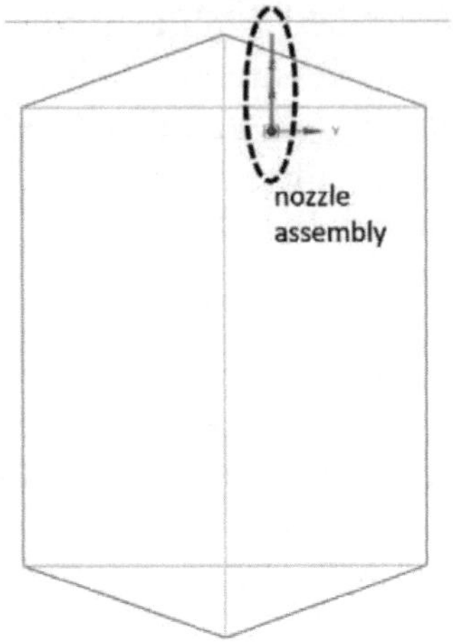

Figure 1, tank with nozzle assembly

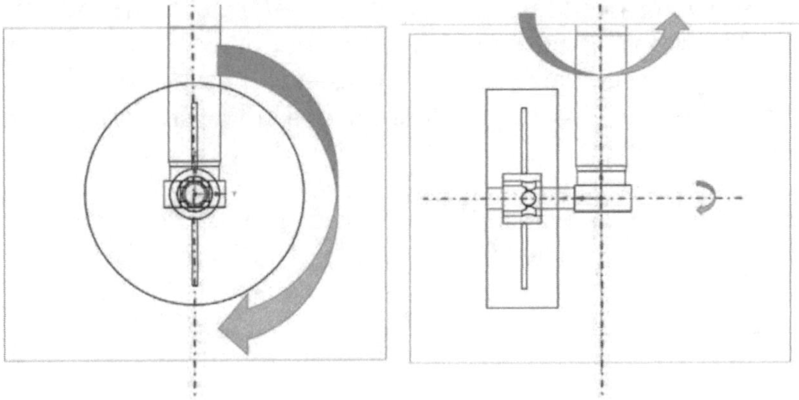

Figure 2, dual axis nozzle assembly

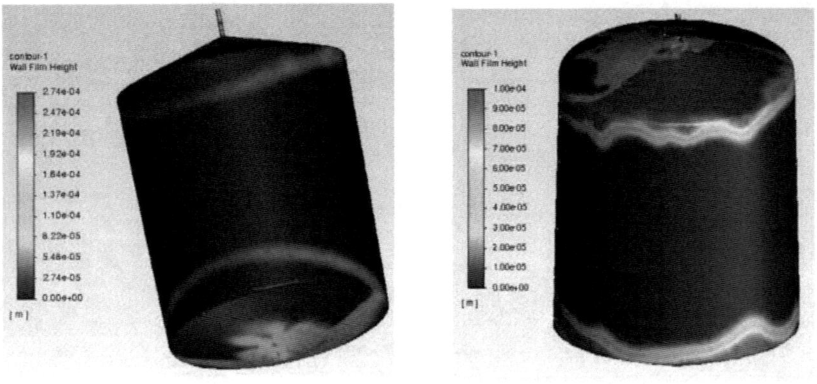

Figure 3, Wall film Height on the tank wall

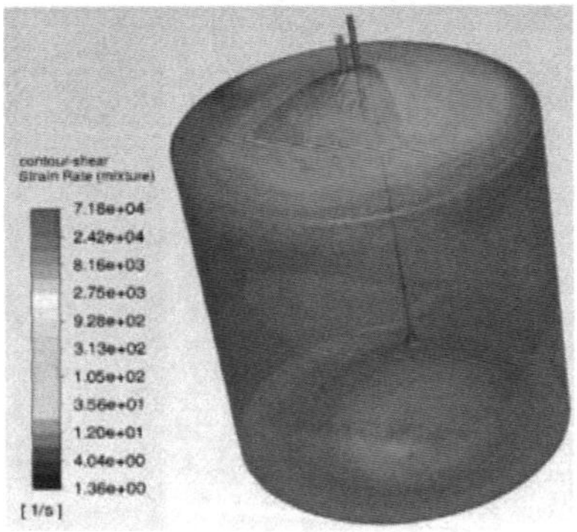

Figure 4, instantaneous wall shear stress on tank external wall

608116 Parameters for *in-Situ* Extraction and Impregnation of Walnut Husk and Beet Root into Polymer Films

Abstract:

Many plants used in food production contain a variety of compounds that exhibit antioxidant properties, however the edible part of the plant frequently contains a fraction of the harvested plant matter by mass. The inedible portion of the plant that is discarded post-processing may contain a wealth of polyphenolic compounds, tannins, and other useful chemicals, that may potentially be recovered. For example, only the kernel of the walnut is typically edible, but the husk is commonly discarded. Likewise in beet root production, while the stem and leaves may be edible, they are typically not preserved, and may be discarded. The antioxidant compounds in both walnut husks and beet roots may be extracted using supercritical carbon dioxide with an ethanol modifier. Supercritical carbon dioxide is a highly effective extraction solvent due to its high diffusivity. Likewise supercritical carbon dioxide may cause polymers to swell or relax, thereby enabling the impregnation of chemicals within the polymer matrix. This paper evaluates the simultaneous extraction and impregnation of compounds present in walnut husk, as well as beet root, into polyethylene films. Process parameters such as proportion of plant material to exaction solvent and temperature were evaluated using supercritical carbon dioxide with 15 wt-% ethanol in an agitated batch extractor. The total phenolic content of the resulting extraction residue was quantified. The polymer films were evaluated by a variety of methods including FTIR, Raman Spectroscopy, UV-VIS spectroscopy, fluorescence spectroscopy, reduction of cytochrome *c.*, visually, and gravimetrically. These results demonstrate that compounds are indeed simultaneously extracted and impregnated into polyethylene films and may indicate effects of operating conditions of the extractor.

607648 Solvent-Polymer Interactions of Ethylcellulose Films for Pharmaceutical Applications

Abstract:

Ethylcellulose (EC) has been widely used as a water-insoluble, rate-controlling polymer for coating oral solid dosage forms, as an excipient for amorphous solid dispersions of the active, for buccal and transdermal drug delivery systems, and for 3D printing applications. The interaction of alcohol/water mixture with EC films will ultimatively impact the drug delivery of the dosage form and thus the therapeutic and prophylactic outcome of the actives. The affinity of the spray solvent for EC can also impact drying rate, secondary drying rate (removal of residual solvent in a secondary batch process), and the glass transition temperature of the spray-dried dispersion. This study is to understand the interaction of alcohol/water mixture with EC films of various molecular weights being cast from various organic solutions. These understandings can also help pharmaceutical scientists design coating formulations for protecting and/or modifying the release of active pharmaceutical ingredients (APIs) with potential alcohol exposure upon ingestion.

Films were cast from organic solution of ethylcellulose polymers with viscosities ranging from 21 to 29 cps. Solvent vapor transmission rates (SVTR) and equilibrium solvent content were determined with a Dynamic Vapor Sorption analyzer (DVS-Resolution). Various EC films were analyzed over a relative partial pressure (P/P_0) range of 0-90%. The vapor program began by exposing the sample to dry air to establish the dry mass. Then, the vapor concentration was increased in 10% P/Po steps to 90% P/Po. The vapor concentration is then decreased in a similar manner to accomplish a full sorption/desorption cycle. Fickian, 1-D diffusion coefficients were determined at each relative partial pressure level.

Solvent vapor transmission rate (SVTR) was measured with two ethanol/water solvent systems (50/50 and 65/35 ethanol/water) at 50% P/P_0 and at temperatures from 20 to 52°C. It was found that the SVTR of ETHOCEL™ film increases at elevated temperatures and higher ethanol concentration of the diffusant. The equilibrium solvent content of ETHOCEL™ films also increases with increasing temperature and ethanol concentration. Additionally, the influence of EC viscosity on SVTR is more pronounced at the higher ethanol level. The diffusion coefficient of ethanol/water vapor is also higher at elevated temperatures and higher ethanol concentration of the diffusant.

In conclusion, ETHOCEL™ film can serve as a good ethanol/moisture barrier but needs to be enhanced when temperature or ethanol concentration increase. Understanding the interaction between organic solvent and EC excipients will help formulation scientists design formulations which can be used to protect APIs, to improve the spray-dry dispersion process and to advance 3D printing processes.

References

1. Dynamic Vapour Sorption for Organic Solvent Sorption by Surface Measurement Systems. https://azom.com/article.aspx?ArticleID=5181

608644 Positively Charged Membranes with Flexible Side Chains to Improve Conductivity

Abstract:

Positively charged membranes have superior selectivity of anions, anti-fouling ability and water flux due to Donnan effect and hydrophilicity, therefore exhibit promising applications as separation membrane for food and bioprocessing. Ion conductivity, though is an important indicator of Donnan effect, usually shows low values in most charged membranes with aromatic polymer materials because of the rigid benzene ring in backbone and short linkage with the quaternized functional groups. Here in this work, two types of flexible side chains with pendent quaternized cations are developed to promote hydrogen bond networks and flexible aggregation of ionic clusters in the charged membranes and thus improve ion conductivity. By means of molecular dynamic simulation, both the N-spirocyclic side chain with -NH-$(CH_2)_x$-NH- spacer and the tri-cations with -O- spacer could induce hydrogen bond networks with water, and the latter side chain could also form hydrogen bond networks within the PBI polymer backbone, which will enhance hydrophilicity and broaden the hydrophilic pathways for selected ion conduction. The trade-off between diffusion coefficient and solubility parameter demonstrates well of the effect of -NH-$(CH_2)_x$-NH- spacer length on the aggregation of ionic clusters, which goes a step further from literatures by merely experimental investigation for the optimal spacer length. TEM and SAXS investigations exhibit good hydrophilic-hydrophobic microphase separated structure in the positively charged membranes with both kinds of functional side chains. High ion conductivity of above 100 mS cm^{-1} at 80 °C is achieved with around 90 % conductivity retained after concentrated KOH treatment for over 600 h. This work provides a universal method to prepare positively charged membranes with high ion conductivity.

605381 Monosodium Urate Monohydrate Crystallization Via Fabricated Uniform Hydrogel Slices

Abstract:

Monosodium urate monohydrate crystallization via fabricated uniform hydrogel slices

Zeqiu Xia, **Xiaobin Jiang***

State Key Laboratory of Fine Chemicals, Research and Development Center of Membrane Science and Technology, School of Chemical Engineering, Dalian University of Technology, Dalian, P. R. China

*Corresponding author: Email: xbjiang@dlut.edu.cn

Abstract

Crystal morphology control is a critical concern for the crystal engineering and relevant applications. For example, urate crystallization is important for the study of the hyperuricemia induced gout. Herein, biomimetic hydrogel material with uniform surface and vertical structure was fabricated, which possessed hybrid pH-temperature responsibility for high-efficiency interfacial nucleation and concentration regulation functions. The developed hydrogel layers were prepared to slice type as a functional micro-platform and then added into the uric acid dihydrate (UAD) crystal solution to inducing monosodium urate monohydrate (MSUM) crystallization. With different dosage of the hydrogel slices (HGS) in the solution, diverse crystal morphologies can be obtained in an efficient manner, such as "flower", "sphere", "sticky spheres", which entirely distinguished from the one prepared via no-HGS. This work shows the advantage of HGS in controlling crystallization procedure and modifying the specific morphology. Considering the feasible fabrication approach of HGS and the tunable amount via HGS area owing to the uniform thickness, the HGS modified crystallization explores a feasible and convenient approach to prepare the biological crystals with different morphologies and diverse hierarchical structure, which is significant for the crystal engineering research from the theoretical and application aspects.

Figure 1 Monosodium urate monohydrate (MSUM) crystal induced by different amount of HGS. (a) Photomicrograph of the crystal when the pH=5.8 with DL-lactic acid. (b) blank control group without HGS. (c) to (e) Added 10 mg, 20 mg, 30 mg HGS, respectively. (bar of the top images: 50 μm; bar of the bottom images: 25 μm)

605378 "Invited Talk" Cefalexin Crystallization Residual Liquor Separation Via Nanofiltration Based Multistage Process

Abstract:

Cefalexin crystallization residual liquor separation via nanofiltration based multistage process

Xiaobin Jiang, Mengyuan Wu, Gaohong He*

State Key Laboratory of Fine Chemicals, Research and Development Center of Membrane Science and Technology, School of Chemical Engineering, Dalian University of Technology, Dalian, P. R. China

*Corresponding author: Email: xbjiang@dlut.edu.cn

Abstract

Pharmaceuticals manufacture is of importance to human life and health. Great amount of pharmaceuticals crystallization residual liquor was produced in the pharmaceuticals manufacture, which lead to adverse influence on the green chemical engineering and environment. Herein, we proposed a multistage separation process based on nanofiltration technology to recovery cefalexin crystallization residual liquor separation. The influence of membranes, pH value, concentration as well as separation stages on the separation performance was explored. The nanofiltration membrane (NF-200D) with strong electronegative charge showed a high rejection of 74.00% for 7-amino-3-desacetoxycephalosporanic acid (7-ADCA) and low rejection of 7.06% for D-phenylglycine in mixed solution. And the increase in the strong electronegative charge of NF membranes could enhance the repulsion between membrane and solute molecules. The separation factor of nanofiltration membrane for 7-ADCA and D-phenylglycine reached a peak value of 3.06 at pH 8.0 and concentration of 10 mmol L^{-1}. Three stage separation, the concentrated 7-ADCA ($C_{7\text{-ADCA}}$=23.12 g L^{-1}) and the permeate D-phenylglycine ($C_{D\text{-phenylglycine}}$ =0.19 g L^{-1}) were obtained. Further improved separation sequence should focus on the preparation of nanofiltration membranes with strong electronegative charge and acid-base tolerance, and the nanofiltration process design with enhanced efficiency of separation process.

Fig. 1. Rejection of NF-150D, NF-200D and NF-500D membranes (a); Schematic diagram of nanofiltration membrane separation mechanism (b).

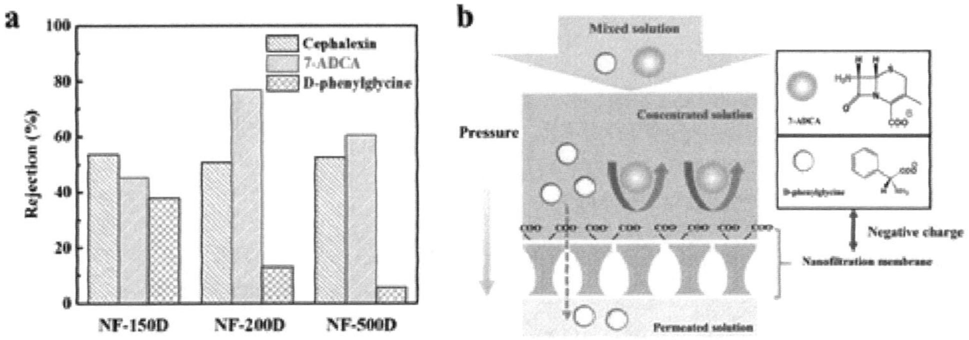

610979 A Modular, Broadly-Applicable Platform for High-Throughput Engineering and Profiling of Protein-Modifying Enzymes

Abstract:

Here we describe Modular YESS (MYESS), a highly versatile version of the yeast endoplasmic sequestration screening (YESS) system suitable for engineering and profiling the specificity of protein-modifying enzymes. By incorporating features to modulate gene transcription, as well as substrate and enzyme spatial sequestration within a versatile and seamless assembly method, MYESS achieves higher sensitivity and dynamic range. To showcase MYESS, we evolve a TEV protease variant (eTEV) with an 8-fold higher catalytic efficiency to obtain the fastest TEV protease variant to date. Due to the unique features of our system, this phenotype is strictly attributable solely to an increase in turnover number (k_{cat}). Second, we used MYESS coupled with NextGen Sequencing to profile the substrate specificity of insulin-degrading enzyme and discover IDE substrates with increased activity compared to the prototypical insulin peptide. Lastly, we show for the first time that MYESS supports calcium-independent sortase-mediated ligations and confirm that residues directly C-terminal of the pentapeptide motif heavily influence the rate of SML. The state-of-the-art MYESS platform offers unmatched versatility in profiling and engineering the specificity of protein-modifying enzymes and should enable even more ambitious future undertakings.

600942 Development of High-Throughput Cell Screening Toolkit for the Directed Evolution of Glycosynthase Enzymes for Bespoke Oligosaccharides Synthesis (Industry Candidate)

Abstract:

Glycans like oligosaccharides play a significant role in the regulation of biological systems which necessitates us to study their structure-function relationship and therefore develop techniques for synthesis of bespoke glycans. Glycosynthases (GSs) are mutant glycosidase enzymes that have been shown to potentially catalyze the synthesis of various oligosaccharides. Till now, a multitude of other GH families have been engineered into glycosynthases by mutating the catalytic nucleophile acid residue in an ad hoc manner to small residues like alanine or glycine. However, the synthesis reactions catalyzed by GSs are highly inefficient and cannot be readily tweaked to alter substrate specificity. Currently available screening strategies are low-throughput which limits the library size of mutant GSs to a few thousand constructs at any given time. Here, we highlight results from a novel high-throughput screening method developed to rapidly screen glycosynthase mutants that utilize sugar azides as activated donor sugar substrates. Lastly, we highlight how our screening strategy can be applied to evolve the substrate specificity of glycosynthase enzymes towards different acceptor or donor groups for synthesis of bespoke glycans like human milk oligosaccharides for commercial applications.

608971 Laying the Foundation for Performing Drug Discovery on the Yeast Surface

Abstract:

Antibodies exhibit a versatile range of molecular recognition properties that make them cornerstones of therapeutics, diagnostics, and basic research tools. However, some antibody properties, including chemical reactivity and covalent antigen binding, are constrained by the narrow range of chemistries encoded in the canonical set of amino acids. In this work, we investigate strategies for leveraging an expanded range of chemical functionality to augment antibody binding properties. We explore this area using the combination of yeast display and noncanonical amino acids. The efforts we describe to augment antibody binding functions with additional chemistries are distinct from efforts to prepare antibody-drug conjugates, where antibody binding and chemical modification occur in very distinct portions of the antibody structure. Enhancing binding function with additional chemistries requires the presentation of ncAAs in or near antibody complementarity determining regions (CDRs). To enable systematic exploration of ncAA presentation sites, we first investigated whether diversification of a single antibody loop would support isolation of binding clones. We prepared a billion-member library containing canonical amino acid diversity and loop length diversity only within the 3rd complementarity determining region of the heavy chain (CDR H3). Antibodies isolated from this library against a series of immunoglobulins from different species exhibit single-species specificity and moderate affinities (double- to triple-digit nanomolar affinities) despite possessing only changes in CDR H3. Because of the narrow location of amino acid diversity in the library, this enabled us to rapidly explore where within antibody structures to place noncanonical amino acids (ncAAs) containing a series of chemically reactive and photocrosslinkable side chains using ncAA-compatible yeast display. Apparent binding affinities on the yeast surface of ncAA-substituted synthetic antibodies vary based on factors including ncAA side chain identity, location of ncAA incorporation in or near antibody CDRs, and ncAA incorporation machinery used. However, numerous substitutions result in clones that retain binding function, confirming that several positions within antibodies support ncAA incorporation despite their proximity to the presumed binding site in CDR H3. We further investigated whether ncAA-containing antibodies displayed on yeast could be chemically modified with bioorthogonal chemistries and what effects these modifications might have on antibody binding. Multiple azide-containing ncAAs supported both copper-catalyzed azide-alkyne cycloaddition (CuAAC) and strain-promoted azide-alkyne cycloaddition without apparent changes in binding function compared to unmodified controls. Alkyne-containing antibodies also appear to support CuAAC without apparent changes in binding function. Finally, we investigated whether antibodies substituted with a photocrosslinkable ncAA would support ultraviolet-mediated crosslinking on the yeast surface. We observed position-dependent photocrosslinking events on the yeast surface consistent with our findings in control solution-based photocrosslinking experiments with the same clones. Taken together, our results highlight the power of integrating the use of yeast display and ncAAs in search of proteins with "chemically augmented" binding function. More specifically, our findings provide one route to productively integrate antibodies with ncAAs by leveraging the use of simple synthetic clones. We believe that these findings are important foundational elements paving the way to performing drug discovery on the yeast surface.

610226 Membrane Protein Engineering for Selective Uptake of MRI Contrast Agents

Abstract:

In this project we use ancestral sequence reconstruction and computational modeling to evolve novel transport function in a human solute carrier (SLC) protein for enhanced uptake of an inert MRI contrast agent. Our newly engineered, non-immunogenic proteins provide exciting opportunities for immune and stem cell therapy – enabling cells to be precisely tracked through space and time with minimal background signal. These membrane transport proteins are moderately conserved throughout mammals (60-80% sequence identity), yet demonstrate diverse and promiscuous substrate activity among closely related species. Substrate preferences are known to be critical for regulating the uptake of hormones, toxins, and drugs throughout the liver, brain, and heart. Unfortunately, the community lacks a mechanistic understanding of this promiscuous transport function. In this study, we explore the individual positions and domains that drive transport activity and specificity by computationally predicting and experimentally testing key mutations. To accomplish this, we functionally characterize transport activity of numerous mammalian SLCs against a panel of imaging agents and, in parallel, compute the phylogenetic relationships between each of these SLCs. Using a Bayesian approach, we then infer the protein sequences for each common ancestor. The experimental substrate transport data are then paired with the phylogenetic analysis resulting in a mutational road map showing a tractable number of mutations necessary for one species to adopt the functional attributes of a distantly related species. Our results showcase a powerful high throughput approach for evolving new function in notoriously challenging membrane proteins.

602859 Engineering a Cell-Free System- Enzyme Fusion for Enhanced Production (Industry Candidate)

Abstract:

Continuous research effort is being put into alternative energy production, such as biofuels, to reduce our reliance on fossil fuels. One of the stumbling blocks in this research is that production *in vivo* is adversely affected by alcohol toxicity and product inhibition. A solution to this problem has been previously proposed by our lab: remove the enzymes from the cells, immobilize them on resin, and run a cell-free reaction. Our previous research has yielded favorable isobutanol production compared previous *in vivo* systems. An issue we encountered was that product inhibition still remained an issue, with aldehyde conversion decreasing as more isobutanol is produced. To solve this issue, we proposed a substrate channeling method of genetically fusing the two ketoisovaleric acid pathway enzymes together. By carrying out this fusion, diffusional length for the intermediate can be reduced and the local concentration of the intermediate in the presence of the isobutanol producing enzyme will increase. To test this theory, we designed four fusion proteins: with either a rigid or flexible linker and two arrangements of our two pathway enzymes. After creating these fusion constructs and ascertaining their activity, stability measurements were conducted to test the fusions' temperature, pH, and isobutanol stability. Using these results, optimized cell-free reactions were conducted, both in solution and immobilized, and the results compared to the previous non-fused system. Future work will examine the cofactor recycling enzyme, formate dehydrogenase, and explore if it can also be fused or co-immobilized to the resin.

612977 A Crowdsourced Understanding of Functionally Potent Antibodies (Invited Speaker)

Abstract:

Antibodies act as a nexus between innate and adaptive immunity: they provide a means to target the potent toxic activities of a spectrum of innate immune effector cells in order to clear viral particles and infected cells. This functional landscape is remarkably complex, with multiple antibody variants directed to multiple epitopes on multiple viral antigens. This diversity of viral recognition characteristics is further complemented by diversity in each antibody's ability to recruit the potent anti-viral effector functions of a suite of innate immune effector cells such as Natural Killer cells and phagocytes. Results from both human and animal model studies have implicated these effector functions in vaccine-mediated protection, but the complexity of these activities has challenged traditional methods of evaluating antibody responses. We demonstrate how high-throughput, high-content platforms for the biophysical and functional interrogation of the innate immune recruiting capacity of diverse virus-specific antibodies are capable of parsing this complex molecular landscape into components that can be used to develop computational models of antibody activity and provide insights into mechanisms of vaccination and new prospects for engineering innovative antibody therapies. This talk will focus on insights resulting from applying artificial intelligence to learn from humoral immune profiles to better understand mechanisms of antibody-mediated protection from infectious disease.

606656 Engineering *Escherichia coli* for Utilization of Ethylene Glycol

Abstract:

Ethylene glycol (EG) is an important chemical that has a wide range of industrial applications, including the manufacturing of polyester plastics such as polyethylene terephthalate (PET). Over 25 million tons of PET are synthesized annually to produce plastic bottles, with EG as a raw material. Due to an increase in global consumption and its persistent negative effect on environmental pollution, the accumulation of PET plastics is a growing concern. If PET wastes can be cost-effectively hydrolyzed, the resulting hydrolyzed PET monomers may be a source of cheap and abundant feedstock to produce useful chemicals.

EG can be a major product when cellulosic biomass is chemically hydrolyzed and hydrogenated. It has the largest market and is expected to reach USD 33.26 billion by 2020. However, the purification of EG from the mixture of other polyols that are generated by the chemical process is not cost-effective. Thus, converting this EG from the cellulose waste and PET waste into microbial biomass or products could contribute to developing a solution for the worldwide concern over the related environmental issues.

The utilization of non-conventional substrates such as EG has not been explored widely in *Escherichia coli.* The wildtype *E. coli* is unable to metabolize EG, although it has the potential genes for this purpose. It can be metabolized aerobically through C2 metabolic pathways via the intermediates glycolaldehyde and glycolate. In this study, we have engineered *E. coli* to enable EG utilization via simultaneous overexpression of L-1,2-propanediol oxidoreductase (encoded by *fucO* gene) and aldehyde dehydrogenase A (encoded by *aldA* gene). Overexpression of *fucO* and *aldA* with a constitutive promoter (PgyrA), improved the cell growth on EG as the major carbon source. EG utilization was further improved with the supplementation of a low concentration of glycerol or amino acids. The engineered strain completely consumed up to 20 g/L of ethylene glycol in 72 hours without any process optimization. Overall, this study provides an *E. coli* strain that can utilize EG efficiently. With future improvements, this strain can potentially produce valuable chemicals from raw EG that are derived from plastic and cellulose wastes.

607977 Rational Approach to Yeast Whole Cell Biocatalyst Design: Compositionally Uniform, High Enzyme Density Display for Improved Cellulosic Biomass Fermentation

Abstract:

As the most abundant biopolymer on Earth, cellulose represents a promising and sustainable feedstock for the production of biofuels and value-added chemicals. Due to its recalcitrant nature, microbes in nature organize cellulases into a multienzyme complex termed cellulosome for efficient cellulose hydrolysis. Inspired by this, researchers have engineered whole cell biocatalysts to display designer cellulosomes using chimeric scaffolds for directed assembly of enzymes. While much engineering effort has been dedicated to improve their performance, most approaches are largely empirical due to a limited mechanistic understanding of the cellulosome assembly process on the cell surface. Our recent study demonstrated that a rational approach can be realized through quantitative characterization of whole cell biocatalysts by revealing structure-performance relationships. The mathematical model predicts enzyme density rather than enzyme proximity as the most important parameter for catalytic performance. Given the heterogeneous nature of biological systems, we hypothesize that harnessing the full potential of cellulosomes will require a concerted effort of increasing enzyme density at the population and single cell level.

In this study, we engineer yeast cells to display anchor scaffolds that bind primary scaffolds, onto which cellulases are docked. The anchor scaffolds are based on the most commonly used yeast surface display platform, Aga1-Aga2, which is known to yield a mixture of non-displaying and displaying cells (i.e. inactive and active biocatalysts). Through careful evaluation of the display system, we speculated low copy number plasmid and plasmid loss as the root cause of limited enzyme density at the single cell and population level. To address this, we replaced the plasmid-based expression with genomic integration, yielding uniformly active whole cell biocatalysts. By adjusting the gene copy number, we were able to further tune the enzyme display level of these biocatalysts. More importantly, the creation of compositionally uniform yeast biocatalysts with high enzyme display levels improved the accuracy of our mathematical model, thereby providing insights into the cellulosome assembly process under the molecular crowding regime. The resulting model enabled rational design of the optimal yeast biocatalyst achieving the highest cellulose to ethanol conversion yield reported to date. This work demonstrates the possibility to rationally design whole cell biocatalysis aided by quantitative tools.

601372 Advanced Cellulosic Biofuel Production Using Cellulosic Emulsions

Abstract:

As global temperatures rise, biofuels are essential sources of renewable energy to reduce greenhouse gas emissions. While corn starch ethanol has been instrumental towards establishing the biofuel industry, its expansion raises concerns of sustainability and competition with food production. Cellulosic biofuels derived from agricultural waste and dedicated energy crops are recognized as more sustainable alternatives. However, cellulosic biofuels have struggled to be economically solvent due to the slow rates and insufficient extent of cellulose hydrolysis achieved with current pretreatment methods, which leads to long process times, waste, and high cellulase enzyme costs. We developed a new pretreatment method for efficient cellulosic biofuel production using cellulosic emulsions that addresses this challenge. We found that preparing cellulose-coated oil-in-water emulsions increase the rate and extent of hydrolysis in simultaneous saccharification and fermentation (SSF) processes. Cellulose pretreated using this new technology hydrolyzes up to 40 times faster than microcrystalline cellulose and generates nearly theoretical yields of ethanol when applied in an emulsified SSF (eSSF) process. Furthermore, we demonstrate the transferability of the eSSF technology to other desired products and more complex systems. Using a strain of *S. cerevisiae* engineered with optogenetic regulators to produce the advanced biofuel isobutanol, we were able to produce cellulosic isobutanol in light-controlled eSSF. This technology addresses the main challenges of cellulosic biofuel production by shortening processing times, lowering enzyme costs, and serving as a flexible platform for the sustainable production of cellulosic biofuels and chemicals.

601041 Bioelectrocatalytic Conversion of N_2: From Chemically Inert Gas to Chiral Chemicals (Faculty Candidate)

Abstract:

Dinitrogen molecule (N_2) is the most abundant natural gas and also the ultimate source of nitrogen for nitrogenated industrial and natural compounds. So far, the fully hydrogenated product, ammonia (NH_3), is the most common product of N_2 fixation.[1] However, NH_3, the regular end-product of N_2 fixation, is a bulk chemical with low added-value. The conversion of NH_3 to nitrogenous compounds with high added-value and complicated structure still requires further tedious chemical synthesis with the use of precious metal catalysts.[2] In our study, an upgraded N_2 fixation system was firstly designed and constructed based on a bioelectrocatalytic N_2 reduction architecture.[3] The NH_3 generated via N_2 reduction catalyzed by nitrogenase was *in situ* utilized by *L*-alanine dehydrogenase to realize the reductive amination of pyruvate and the generation of alanine. The reduced equivalent to support the N_2 reduction and reductive amination of pyruvate was regenerated by the electrochemical method. ω-TA transferred the amino group from alanine to ketone substrates and finally produced desired chiral amine intermediates.[4] On this basis, an H_2 fuel cell module was integrated into the upgraded N_2 fixation system to construct a self-powered H_2/α-keto acid fuel cell. In the new system, the upgraded N_2 fixation was powered by the oxidation of H_2 with high Faradaic efficiency and without the requirement of external electrical energy input.[5] Finally, the end-product of N_2 fixation in our study successfully went beyond NH_3 and reached high-value-added chiral amine intermediates and chiral amino acids. The produced chiral amine intermediates and chiral amino acids have wild application in pharmaceuticals production and biotechnology research.

References

(1) Foster, S. L.; Bakovic, S. I. P.; Duda, R. D.; Maheshwari, S.; Milton, R. D.; Minteer, S. D.; Janik, M. J.; Renner, J. N.; Greenlee, L. F. Catalysts for nitrogen reduction to ammonia. *Nat. Catal.* **2018,** *1*, 490.

(2) Balaraman, E.; Srimani, D.; Diskin-Posner, Y.; Milstein, D. Direct synthesis of secondary amines from alcohols and ammonia catalyzed by a ruthenium pincer complex. *Catal. Lett.* **2015,** *145*, 139-144.

(3) Milton, R. D.; Cai, R.; Abdellaoui, S.; Leech, D.; De Lacey, A. L.; Pita, M.; Minteer, S. D. Bioelectrochemical Haber-Bosch Process: An Ammonia-Producing H_2/N_2 Fuel Cell. *Angew. Chem. Int. Ed.* **2017,** *56*, 2680-2683.

(4) Chen, H.; Cai, R.; Patel, J.; Dong, F.; Chen, H.; Minteer, S. D. Upgraded Bioelectrocatalytic N_2 Fixation: From N_2 to Chiral Amine Intermediates. *J. Am. Chem. Soc.* **2019,** *141*, 4963-4971.

(5) Chen, H.; Prater, M. B.; Cai, R.; Dong, F.; Chen, H.; Minteer, S. D., Bioelectrocatalytic Conversion from N_2 to Chiral Amino Acids in a H_2/α-Keto Acid Enzymatic Fuel Cell. *J. Am. Chem. Soc.* **2020,** *142* (8), 4028-4036.

608651 A Nicotinamide Mononucleotide (NMN^+)-Dependent Redox Cofactor Cycling System Enables Aldehyde Accumulation in *Escherichia coli*

Abstract:

It is challenging to biosynthesize industrially important aldehydes, which are readily consumed by the numerous alcohol dehydrogenases (ADHs) in cells. In this work, we demonstrate that the nicotinamide mononucleotide (NMN^+)-dependent redox cofactor cycling system enables aldehyde accumulation in *Escherichia coli* crude lysates and whole cells. By specifically delivering reducing power to a recombinant enoate reductase, but not to endogenous ADHs, we convert citral to citronellal with minimal byproduct formation (98% and 83% product purity in crude lysate- and whole cell-based biotransformation, respectively). We envision the system's universal application to lower the noise in biomanufacturing by silencing the host's metabolic background.

594447 The Influence of Low Voltage Electric Fields on Lipase Catalyzed Triglyceride Hydrolysis (Industry Candidate)

Abstract:

Lipase catalyzed triglyceride hydrolysis is limited often by slow kinetics due to mass transfer limitations and by challenges of economic enzyme recycling. The principle and sole mechanism for the intensification was assumed to be solely on account of increases in interfacial area and some hydrodynamic turbulence. The possible effect of electrical charge on the lipase activity has hitherto not been considered. In this study, rates of reaction for lipase catalyzed hydrolysis of triglycerides in the presence of microbial lipase derived from *Candida Rugosa* were determined in the presence of steady direct current (DC) electrical field. The reactions were conducted in a fixed rectangular geometry stagnant batch reactor (SBR) fitted with stainless steel planar electrodes. The two immiscible liquid phases, oil and aqueous solution, were located between the electrodes with a quiescent rectangular interface of known area. The rates of percent yield of free fatty acid were determined for applied DC voltages in the range 0-30 Volts imposed between the two electrodes. Comparisons with control conditions in the absence of electrical field, in the absence of enzyme and effect of interface and bulk yield of FFA were also studied. The results showed clear evidence of significant increase in the rates of reaction in the presence of the electrical field. These results also show evidence of an optimum voltage since at the upper end of the voltage range there was significant reduction in the rate enhancement compared with observations at lower voltages. It was concluded that the electric field has a positive effect on reaction rate, independent of interfacial area. This suggested possible changes in enzyme structure due the electrical field, modification of the nature of the enzyme binding at the liquid-liquid interface, or possibly intensification of local transport rates of substrates and reaction products close to the interface. The latter could involve phenomena such as electro-kinetic transport or electro-osmosis.

613022 Engineering an Orthogonal Redox Cofactor System for Cell-Free and Whole-Cell Biocatalysis (Invited Speaker)

Abstract:

Biological production of chemicals often requires the use of cellular cofactors, such as nicotinamide adenine dinucleotide (phosphate) ($NAD(P)^+$). These cofactors are expensive to use *in vitro* and difficult to control *in vivo*. We demonstrate the development of a non-canonical redox cofactor system based on nicotinamide mononucleotide (NMN^+). The key enzyme in the system is a computationally designed glucose dehydrogenase (GDH) with a 10^7-fold cofactor specificity switch towards NMN^+ over $NAD(P)^+$ based on apparent enzymatic activity. We demonstrate that this system can be used (1) to support diverse redox chemistries *in vitro* with high robustness, (2) to eliminate side-reactions in crude lysate-based cell-free biotransformation, (3) to specifically channel reducing power in *Escherichia coli* whole cells from glucose to a pharmaceutical intermediate, levodione, and (4) to sustain the high metabolic flux required for the central carbon metabolism to support growth. In addition, we also established high-throughput selection platforms which allow NMN^+-dependent enzymes to be obtained on demand through directed evolution. Overall, this work demonstrates efficient utilization of a non-canonical cofactor in biocatalysis and metabolic pathway design, and points out a path for rapidly expanding its applications.

606748 Targeted Therapy for Triple Negative Breast Cancer

Abstract:

Triple negative breast cancers (TNBCs) are highly aggressive, metastatic and recurrent. To this date, there is no efficacious targeted therapy. This study aimed to develop an antibody-drug conjugate (ADC) to target epidermal growth factor receptor (EGFR), a surface receptor overexpressed in TNBC, and deliver high-potency drug. Specifically, an ADC was constructed by conjugating anti-EGFR monoclonal antibody with mertansine which inhibits microtubule assembly. The TNBC-targeting specificity of anti-EGFR ADC was evaluated using cell line (MDA-MB-468) and using TNBC subcutaneous xenograft mouse mode. The confocal laser scanning microscopy and In Vivo Imaging System were used to collect the live-cell and live-animal images, respectively. Finally, both *in vitro* toxicity assay and *in vivo* anti-cancer efficacy study showed that the developed ADC significantly inhibited TNBC growth. This study indicated that the anti-EGFR ADC has a great potential to against TNBC.

610596 Developing Islet-on-Chip Model Towards T2D Disease Modeling

Abstract:

In therapy development, inadequate animal models have remained a constant obstacle due to their inaccuracy in representing human physiology and disease, causing many treatments to fail in clinical trials after success in the animal studies. The human-on-a-chip model has become a promising new approach for aiding in treatment testing by mimicking the biochemical and mechanical environment of organs to more efficiently transition from initial testing to clinical trials. This project seeks to use this technique to develop a diabetic islet-on-a-chip model. Diabetes is a widespread disease that develops after the decreased function of pancreatic islets, which produce insulin for blood glucose regulation. The decreased function of islets can stem from an autoimmune elimination of the insulin producing beta cells (Type 1) or a toxic environment damaging the beta cells (Type 2). The current project aims to develop an islet-on-chip model using human induced pluripotent stem cells (hiPSC) that can be used to understand and treat type 2 diabetes (T2D).

The islet-on-chip platform was developed with a modification of a commercially available Micronit microfluidic device, which utilizes three glass slides to form a 2-chamber system partitioned by a customizable middle membrane layer. Both chambers can be individually accessed by a dedicated perfusion channel, which allows the flexibility of multiple flow configurations through the system. The parent design allows for culturing adherent cells on the middle membrane layer, with the flexibility of cell seeding either through perfusion flow or system disassembly and reassembly. Pancreatic islets, on the other hand, lose their function and phenotype upon adherent culture and require retaining 3D conformation. Alternate seeding and culture strategies were investigated to culture primary human islets in the Micronit device, leading to the development of a novel hydrogel-supported islet micropatterning technique on the membrane. The design of the micropatterning was informed by modeling the flow field through the device using COMSOL modeling software. Our islet-on-chip system could retain high viability and glucose stimulated insulin secretion (GSIS) of primary human islets for over 4 weeks of perfusion culture under normal 'fasting' condition. Towards disease modeling of Type 2 Diabetes, islet glucotoxicity was simulated by long term exposure to pathological glucose levels, lipotoxocity by elevated free fatty acids (oleate and palmitate) and glucolipotoxocity by combinations of the two. Extended exposure to cytokines (TNF-alpha, IL-1 beta, IFN-gamma) alone and in combination was used to mimic the complexity of T2D disease condition. Such simulated disease conditions could reproduce many of the known aspects of type 2 diabetes. In parallel, islet organoids are being derived from hiPSCs to replace the primary human islet with a regenerative cell source. Similar to primary islets, the microfluidic culture system could maintain survival, phenotype, and function of the hiPSC-derived islets over extended culture period. Further steps are for inducing the diseased state in the hiPSC-based chip model. These models can be further used to test T2D treatments to determine the reversibility of the toxic states and ability to simulate treating the disease in patients.

609453 Identifying Proteins That Regulate Endocytosis of Nucleic Acid Therapeutics

Abstract:

Nucleic acids can be used to treat diseases of improper protein expression. Messenger RNA (mRNA) and plasmid DNA (pDNA) can be used to increase the expression of a protein, and small interfering RNA (siRNA) can be used to decrease the expression of a target protein. Nucleic acids and other macromolecular therapeutics cannot cross cell membranes due to electrostatic repulsion; therefore, target cells must actively transport nucleic acids into the cell, a process called endocytosis, to have a therapeutic effect. Endocytosis has been classified into six different mechanisms that have been shown to be used by different cell types for different therapeutics. Moreover, the mechanism of endocytosis directly controls the efficacy of the endocytosed material. Clathrin-mediated endocytosis, caveolin-mediated endocytosis, and macropinocytosis have been well studied; however, the regulation of more recently discovered endocytic pathways (Graf1-mediated, Flotillin-mediated, and Arf6-depenedent endocytosis) remains largely unknown. In this work, we used gene expression data and computational methods to determine the regulating proteins among the endocytic pathways. ReliefF was used to find condition-specific differentially expressed genes in NCI-H1299 (human non-small cell lung carcinoma), HEK293 (human embryonic kidney), HeLa (human cervical cancer), and HepG2 (human hepatocytoma) that relate to endocytosis. Fisher's Exact Test for overlap of differentially expressed genes with cellular pathways and combinatorial z-scores were used to identify potential pathways and regulatory molecules of endocytosis, respectively. Computationally predicted regulators of endocytosis were confirmed in cell culture experiments.

603874 A New Methodology for Quantitative and Temporal Measurement of Autophagy.

Abstract:

Autophagy is a multi-step intracellular process by which cells recycle their misfolded proteins and damaged organelles into primary building blocks. The cellular materials are sequestered by double-membrane vesicles called autophagosomes, which fuse with lysosomes to form autolysosomes, which then degrade the sequestered material into fatty acids and amino acids. Autophagy is linked to many diseases, including cancer, neurodegenerative diseases, and infectious diseases. The significance of autophagy in various cellular processes is growing, yet sensitive and temporal quantification of the dynamic autophagic process remains a challenge. Quantifying autophagy temporally allows the fundamental understanding of autophagy and will be useful for biomedical and biotechnology applications. However, many current methods lack the temporal quantification making them inadequate to characterize the dynamic nature of autophagy.

We developed a new single-cell fluorescence microscopy-based methodology to measure all the steps involved in autophagy temporally and quantitatively. The methodology considers the whole autophagic process as a multistep process governed by the rate of formation of autophagosomes (R1), the rate of fusion of autophagosomes with lysosomes (or rate of formation of autolysosomes) (R2), and the rate of degradation of autolysosomes(R3). All three rates can be measured temporally using an instantaneous rate approach, thus characterizing the change in the functional state of the system over time. This methodology would allow systematic characterization and kinetic analysis of autophagy under non-steady state conditions. Therefore, the methodology can be used to study various external perturbations such as small molecules drugs and dynamic perturbations such as viruses. Additionally, the methodology would further assist in the development of new therapies to regulate autophagy accurately.

We validated the new methodology using rapamycin, a well-studied small molecule drug commonly used to regulate autophagy. A tandem green and red fluorescent clonal human cell line was used to monitor autophagosomes and autolysosomes in real-time. We measured the change in the rates of the autophagic steps temporally during rapamycin treatment using the new methodology through direct quantification of autophagosomes and autolysosomes as a function of time. The measured autophagic rates not only provides insights into the effect of rapamycin on autophagy but also the cellular response to the treatment. Furthermore, we are developing a mathematical model that characterizes the autophagic pathway in response to rapamycin treatment based on temporal change in the autophagosomes and autolysosomes data. The predicted autophagic rates from the mathematical model will then be compared to the measured autophagic rates based on methodology indicating the accuracy of the model as well as the methodology. Additionally, the mathematical model characterizes the concentration-dependent effect of rapamycin on the autophagic pathway and can be easily extended to other rapalogs (drugs analogs to rapamycin).

In conclusion, this methodology will lay the foundation for temporal quantification and kinetic analysis of the autophagic system. As a future direction, we will be applying this methodology to study the effect of the various small-molecule drugs as well as viruses such as dengue and Zika virus on autophagy.

603041 Characterization of PMI5011 on the Regulation of Deubiquitinating Enzyme Activity in Multiple Myeloma

Abstract:

Aberrations in the ubiquitin proteasome system (UPS) contribute to the pathological states of several clinical disorders, specifically cancer, thus indicating that targeting proteolytic and regulatory components of the UPS is an efficient strategy for cancer treatment. While inhibition of the proteasome has proven to be effective in the treatment of multiple myeloma (MM), a significant challenge with these proteasome-targeted therapeutics is the heterogeneity associated with cancer cells. This has led to the development of inhibitors for other enzymes associated with the UPS such as the deubiquitinating enzyme (DUBs). While recent studies have suggested that controlled positive regulation of certain DUBs can stabilize autophagy and levels of protein degradation, there are currently no targeted therapeutics that can positively regulate DUB activity. In this work, a botanical extract from *Artemisia dracunculus L* (herein referred to as PMI5011) was characterized for its effects on DUB activity in two model MM cell line (OPM2 and MM.1S) extracts. Prior studies have been informative on the potential of PMI5011 as a potential therapeutic for diabetes; however, they have not investigated the role of the plant extract on DUB activity in cancer cells. The total extract PMI5011 contains several bioactive components of which five have been identified to exhibit bioactivity: DMC-1, DMC-2, davidigenin, sakuranetin and 6-demethoxycapillarisin. An in-depth enzymology analysis was performed using a commercially available fluorescent DUB reporter to demonstrate a significant effect of PMI5011 and three bioactive components (DMC-1, DMC-2, and davidigenin) on DUB activity. Concentration-dependent studies and mathematical modeling revealed that DUB kinetics in the presence of PMI5011, in both OPM2 and MM.1S lysates, followed a heterotropic co-operativity model where PMI5011 bound with the active domains of DUBs to directly or indirectly enhance their activity without affecting the cellular viability. A similar kinetic profile was observed for DMC-1, DMC-2, and davidigenin; however, their influence on DUB activity was found to be cell line dependent at higher concentrations. Interestingly, davidigenin was found to inhibit DUB activity in MM.1S cells but enhance in OPM2 cells. Whereas, a knock-out extract (KOE) missing both DMC-1 and DMC-2 was observed to enhance DUB activity in both OPM2 and MM.1S lines suggesting that these two bioactive components were not the primary compounds to stimulate DUB activity in the PMI5011 plant extract. An mRNA profile revealed that this cell dependent effect of PMI5011 and its components on OPM2 and MM.1S was due to differential expression of specific DUBs such as USP7, USP15, USP14, Rpn11 in OPM2 and MM1S. These findings demonstrating the complexity of the PMI5011 plant extract and its various bioactive components on DUB regulation reveal their therapeutic properties, but also shine light on the allosteric nature and structural specificity of these enzymes.

604882 Developing a Dynamic (PERFUSION BASED) Multicellular Pancreatic Cancer MODEL

Abstract:

INTRODUCTION: Recent advances in tissue engineering have led to the development of scaffold assisted 3D tumour models with better niche mimicking capability in comparison to traditional 2D systems. Additionally, they are easier and faster to use as compared to animal models. Their ability to better recapitulate the tumour niche is attributed to their tunable mechanical properties and structural integrity along with better cell- cell, cell-Extracellular Matrix (ECM) interactions [1]. We have previously reported that poly urethane (PU) based scaffolds can be used as a robust system for the development of robust mono-cellular (cancer cells only) pancreatic cancer model [2]. Furthermore, we have recently reported that our developed PDAC model is appropriate for short and long term chemo-radiotherapy screening [3]. However, in addition to cancer cells and matrix proteins, the tumour niche consists of different cell types, including stellate cells and endothelial cells all contributing to tumour formation, cancer metastasis as well as its response and resistance to treatment. Stellate cells in particular have shown to produce high amounts of ECM, leading to fibrosis-desmoplasia in pancreatic cancer. Thus, recent studies have been focused on the generation of multicellular models of PDAC, primarily spheroid based, which are of limited mechanical integrity [4]. Our group has very recently developed a novel hybrid scaffold assisted multicellular model with cancer cells, stellate cells and endothelial cells [5]. Our scaffold assisted multicellular model captures different levels of desmoplasia, which is a hallmark of pancreatic cancer, along with long term viability of all different cell types [5]. However, *in vivo* cellular niches have additional complexity associated with flow through vascular networks and in the form of interstitial fluid flow which is not recapitulated within a static system. Additionally, mimicking the circulation of immune cells also require a dynamic flow within *in vitro* 3D tumour models. However, very few studies exist till date to assess the effects of dynamic condition of monocellular of multicellular models of PDAC and they are mostly spheroid based and for relatively short time period [6,7].

The aim of this work was to build up on our novel scaffold based multicellular model and develop a long term (> 4 weeks) dynamic perfusion (bioreactor based)pancreatic cancer model, mimicking various physiologically relevant vasculature/interstitial flow levels.

METHODS: PU scaffolds were prepared using the Thermal Induced Phase Separation (TIPS) method. Absorption based surface modification of the scaffolds enabled ECM matrix remodelling with fibronectin and collagen type I [2,3]. Thereafter, hybrid scaffolds with different zonal ECM coatings were fabricated as described in [5], to account for ECM optimisation for different cell compartments. More specifically, a zonal structure with (i) endothelial and stellate cells on the outer side of the scaffold coated with collagen and (ii) pancreatic cancer cells in the inner scaffold coated with fibronectin was designed [5]. The hybrid multicellular scaffolds were then placed in a dynamic (perfusion) bioreactor. Different flow rates in the range of 0.5-5 ml/min were evaluated. Various *in situ* assays for monitoring the cell viability, spatial organisation and ECM production were carried out. More specifically, immunofluorescent assays and visualisation with CLSM, SEM. were carried out at specific time points throughout the cultures for different flow rates.

RESULTS: We have successfully established a zonal multicellular 3D model showing extensive desmoplastic reaction in the presence of stellate cells and cellular migration, mimicking key *in vivo* characteristics of pancreatic cancer. Addition of shear stress and fluid flow within a perfusion bioreactor resulted in discernible changes within the model in comparison to static culture in terms of cellular growth, matrix production and migration.

DISCUSSION & CONCLUSIONS: Our data show, for the first time, the feasibility of PU scaffolds to support a zonal multicellular pancreatic tumour niche growth along with the possibility for a robust ECM mimicry and recapitulation of fibrosis/desmoplasia. We also highlight the importance of fluid flow/ shear stress on *in vitro* cancer model. Different flow rates affect different scaffold compartments/cell types of the tumour microenvironment. Overall, our developed novel advanced pancreatic cancer model is a high throughput tool that can be used for animal free personalized studies and treatment screening of pancreatic cancer.

ACKNOWLEDGMENT: Financial support was received from the Department of Chemical and Process Engineering at the University of Surrey, Impact Acceleration Grant (IAA-KN9149C) from University of Surrey, IAA–EPSRC Grant (RN0281J) and the Royal Society. P.G was supported by Commonwealth Rutherford Post-Doctoral Fellowship. E.V. is thankful to the Royal Academy of Engineering for an Industrial Fellowship.

REFERENCE:

[1] Totti, et al. DDT.2017; 4(22).

[2] Totti, et al. RSC Advances.2018; 8(37).

[3] Gupta, et al. RSC Advances.2019; 9 (71).

[4] Lazzari, et al. Acta Biomaterialia.2018; 78.

[5] Gupta, et al. Frontiers in Bioengineering & Biotechnology.2020; 8:290. doi: 10.3389/fbioe.2020.00290

[6] Tai, et al. Phytomedicine. 2014; 21(4).

[7] Brancato, et al. Acta Biomaterialia. 2017; 49.

613338 Engineered CRISPR/Cas Systems As Diagnostics and Therapeutics (INVITED SPEAKER)

Abstract:

TBD

The Clustered Regularly Interspaced Short Palindromic Repeats/CRISPR-associated (CRISPR/Cas) systems have emerged as potent and versatile genome editing tools with a wide range of applications including understanding disease mechanisms[1-2]. However, clinical translation of CRISPR/Cas systems as diagnostics and/or therapeutics heavily relies on: 1) improving the efficiency and specificity of CRISPR/Cas systems, 2) developing targeted delivery strategies, and 3) engineering the ability to control their activity on demand[3]. Towards these goals, we previously created a light-activatable CRISPR-PLUS technology that allowed spatiotemporal control of gene editing *in vitro*[5]. In parallel, we also developed a peptide-based CRISPR-GPS system for targeted delivery of CRISPR/Cas components into a variety of cell lines[6] and in a xenograft model of human ovarian cancer in mice.

Our lab recently engineered a CRISPR-ENHACE technology by modifying CRISPR RNA (crRNAs) and optimizing conditions that allowed unprecedented enhancement in the sensitivity of CRISPR/Cas12a for nucleic acid detection with improved specificity. While engineering CRISPR-ENHANCE, we discovered a mechanism by which a Cas12a enhances the rate of enzymatic cleavage by 3.5-fold, making it one of the fastest known CRISPR/Cas systems in terms of collateral cleavage activity. We further applied the CRISPR-ENHANCE system with isothermal amplification strategies for detecting prostate cancer in the urine as well as for developing rapid diagnostics for HIV, HCV, and SARS-CoV-2[7]. In this talk, the fundamentals of engineered CRISPR/Cas technologies in Figure 1 will be introduced with a focus on the development of a COVID-19 testing kit using CRISPR-ENHANCE technology.

References:

[1] M. Jinek, et al., Science, 2012, 337, 816-821.

[2] B. Zetsche, et al., Cell, 2015, 163, 759-771.

[3] Komor, et al., Cell, 2017, 169 (3), 559.

[4] S. Q. Tsai, Nat. Biotech., 2015, 33, 187–197.

[5] P.K. Jain, et al., Angew. Chem., 2016, 55, 12440.

[6] P.K. Jain, et al., Nanoscale, 2019, 11, 21317-21323.

[7] L. Nguyen, et al., BioRxiv, 2020, 10.1101/2020.04.13.036079.

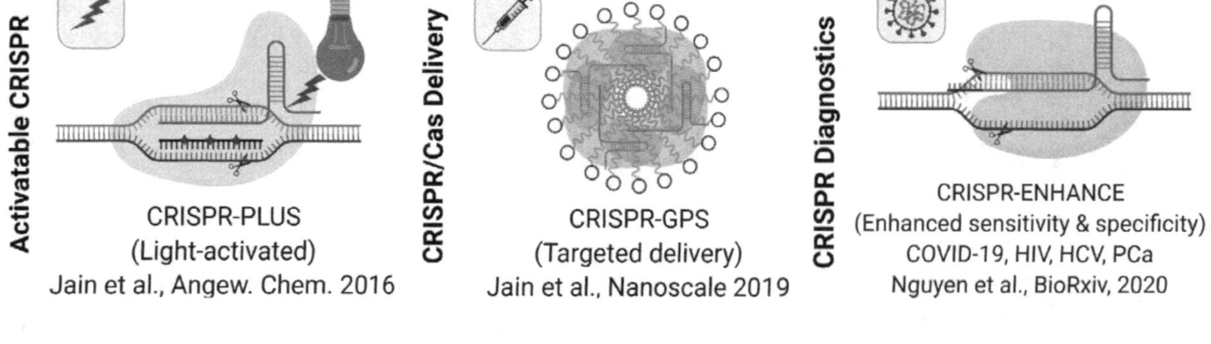

Figure 1: Engineered CRISPR/Cas systems as tools for developing of diagnostics and therapeutics.

610341 Gelpins: A New Strategy for Discrete Yet Cohesive Tissue Assemblies within 'Cut & Assemble' Organ-Chips

Abstract:

Microphysiological systems (MPS) hold the potential to provide benchtop models to investigate fundamental biology and disease while reducing the need for animal models. However, many conventional *in vitro* models fail to fully capture the complex cell-cell interactions, 3D microenvironments, structural organization, or vascularization of multicellular organ systems. A key criterion for replicating physiologically relevant architectures in a dish is the ability to compartmentalize discrete cell populations, extracellular matrix compositions, and/or mechanical properties, without meaningfully restricting auto- and paracrine signaling. Traditionally, compartmentalization within MPS has relied on the use of posts or microtunnels fabricated in silicon-based materials, often requiring expensive lithographic capabilities. Further, these methods are commonly limited to confining discrete tissues in the x-y plane. Towards overcoming these limitations, our previously reported 'cut & assemble' manufacturing technique was expanded to exploit the meniscus pinning effect via GelPins to constrain 3D cell-laden materials within a tailorable MPS[1]. More specifically, GelPins are thin layers of material within our custom microfluidic devices that facilitate the formation of discrete, yet contiguous hydrogels in either the x-y and/or z planes (Figure 1).

Ensuring contiguous, yet heterogenous structure of multicellular structures is paramount for recapitulating complex microenvironments. For this reason, the innervation of organs where effector neuron bodies are isolated, yet synaptically connected, to the distal organ tissue rarely been demonstrated *in vitro*. As a proof of principle, GelPins were utilized to establish an MPS of the cardiac sympathetic nervous system incorporating cell laden hydrogels containing both cardiomyocytes (CMs) and adrenergic neurons (Figure 1). Specifically, cardiac cells and neurons were isolated from neonatal rat pups and cultured *in situ* within a photocrosslinkable gelatin-based hydrogel. Results show neurites extend both above and below the GelPin towards CMs (Figure 1C) after 10 days in culture. Innervation of the cardiac microenvironment was investigated via immunofluorescent imaging at discrete time points throughout the duration of culture and neurite extension quantified using neuron tracing software (Neurolucida®).

The GelPin technology can be additionally leveraged to define boundaries in the z axis by orientating GelPins instead as slats in place of membranes or other means of separating tissue layers (Figure 1B). COMSOL simulations were used *a priori* confirmed that GelPins with a 1:2 and 1:3 width to spacing ratio (resulting in gaps of 400 and 600 μm, respectively), can function as a pressure barrier in the z-direction. The use of GelPins slats allows tailorable compartment heights down to 10s of microns, optimal for developing culture systems with striated tissue layers like the intestinal lumen, dermis, or retina, while ensuring sufficient diffusion across layers. In summary, this is a promising new approach towards assembling the next generation of MPS for benchtop discovery. Coupled with the 'cut & assemble' method, these tools offer robust design flexibility and lower barrier costs to encourage broader use of MPS technology across the scientific community.

1 Hosic, S. *et al.* Rapid prototyping of a multilayer microphysiological system for primary human intestinal epithelial culture. *bioRxiv*, 400721 (2018).

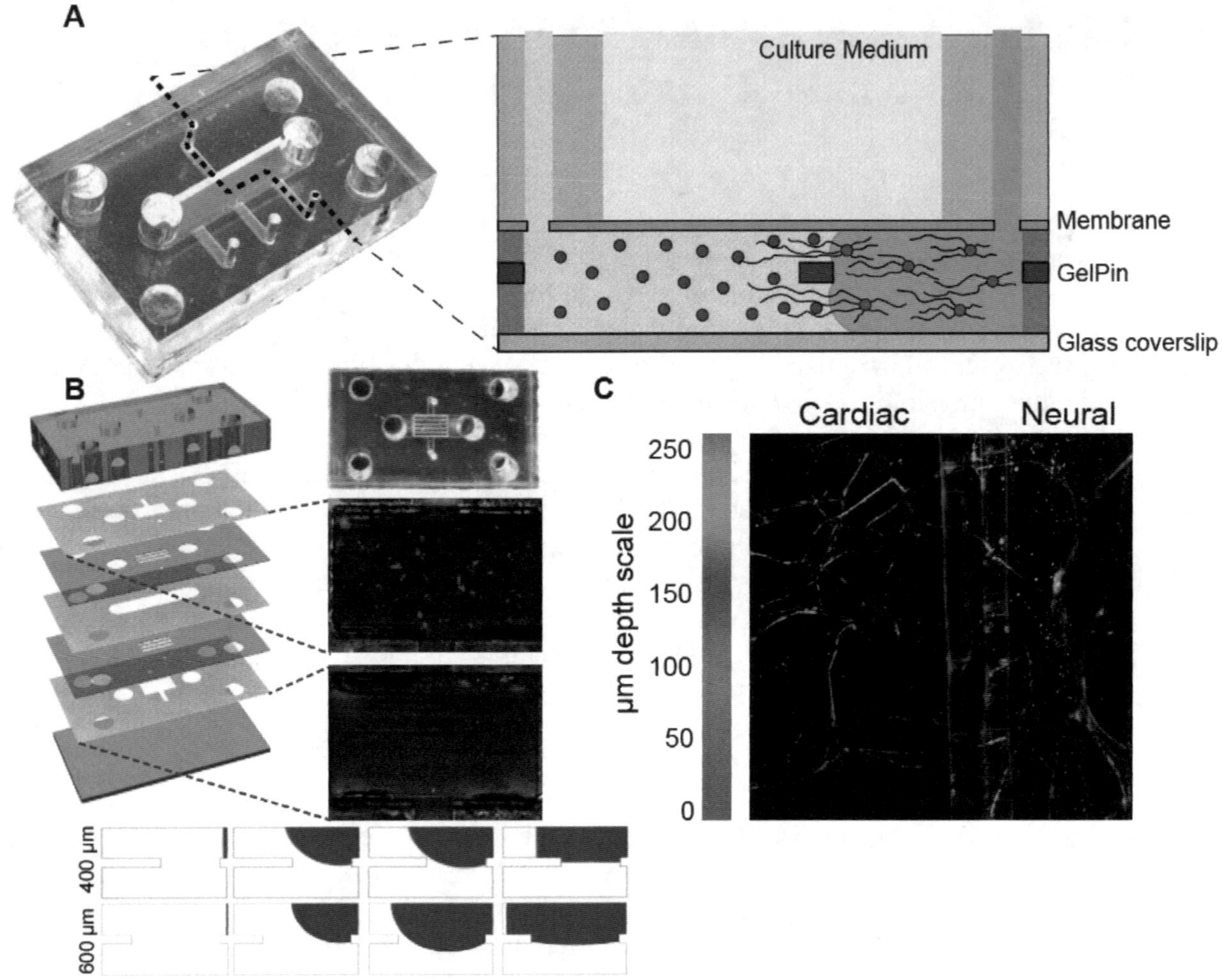

Figure 1. Design of a neuro-cardiac axis MPS. (A) Photograph of organ-chip model and cross-section schematic demonstrating innervation in the basal layer of the device. (B) Layer-by-layer assembly showing how the placement of GelPin slats can create thin tissue layers on-chip and COMSOL modeling showing the meniscus pinning effect with GelPin slats with different spacings to turn the percentage of tissue-tissue or tissue/medium interactions. (B) Color-coded depth map demonstrating the neurons cross both above and below the GelPin into the cardiac microtissue compartment.

608083 High-Throughput Fabrication of Large Breast Cancer Spheroids: Facile Encapsulation of MCF7 Cells in Gelatin Methacryloyl (GelMA) Using a Simple Microfluidic Platform

Abstract:

Cancer is the first or second leading cause of mortality in more than 90 countries worldwide. However, fundamental cancer research and anti-cancer drug development are limited by a lack of accessible and reliable *in vitro* cancer models. A vast body of evidence is now demonstrating that 3D cultures of cancer cell aggregates can recapitulate the response of tumors to anticancer drugs more faithfully than is achieved with conventional 2D monolayer cultures. Cancer spheroids are arguably the simplest and most commonly used 3D model for conducting cancer research and can be fabricated by diverse methods. However, most fabrication methods are based on cumbersome, artisanal protocols that yield heterogeneously shaped and sized spheroid populations. Moreover, these methods result in the formation spheroids that are significantly smaller (50–400 μm diameter) than the actual sizes of clinically relevant avascular tumors.

We have developed a flexible platform for fabricating breast cancer spheroids as cell-laden gelatin methacryloyl (GelMA) structures using a polymethyl methacrylate (PMMA) microfluidic chip connected to a pneumatic pump. The PMMA chip design is a simple T-junction (two inlets and one outlet) that is easily lab-fabricated using a conventional laser cutter. A 5% GelMA solution containing two photo-initiators (0.067% Lap Cellink® and 0.067% Lap Allevi®) and MCF7 cells is fed though one inlet, while mineral oil containing 2% (v/v) Span® 80 is injected through the other. Flow activation then generates spherical drops of cell-laden GelMA in water. Photocrosslinking is then initiated with a high-power blue-violet laser (wavelength of 405 nm) placed in contact with the upper layer of the device at the chip outlet. Upon recovery, the spheroids are exposed again to UV light using an Ominicure® S2000 system (365 nm wavelength).

This simple platform enables the fabrication of micro-spheroids of different sizes. We can produce homogeneous GelMA micro-spheres with diameters between 300 and 1100 μm and low variation coefficients among batch and days (≤15%) simply by varying the diameter of the chip inlet channel and the flow velocity of the GelMA and mineral oil. We have fabricated MCF7-laden GelMA spheres using different initial cell concentrations (from 1 to 3×10^6 cells/mL) and monitored their shape and architecture evolution for 30 days by optical and scanning electron microscopy. We have also evaluated cell viability within the constructs at different timepoints using Live/Dead assays. In general, progressive degradations of the GelMA matrix and cell proliferation were observed within the cell-laden spheres. As expected, denser and more compact cellular aggregates were formed in less time with high-density cell spheres than with spheres initially seeded with the low cell concentrations.

Here, we present an easy, reproducible, fast, and cost-effective method for high-throughput production of homogenous cancer spheroids encapsulated in GelMA. This method may facilitate the production of relevant cancer models for medium- to high-throughput anti-cancer drug development or fundamental cancer research. We envision that the strategies presented here may easily translate to the fabrication of other types of cancer spheroids or to the fabrication of multi-cellular spheroids for cancer modeling or other tissue engineering applications.

609357 Effect of Patterning of ICAM-1 on the Upstream Migration of CD4+ T-Cells Under Shear Flow

Abstract:

Department of Chemical and Biomolecular Engineering, University of Pennsylvania, Philadelphia, PA 19104

Introduction: The leukocyte adhesion cascade is critical in leukocyte recruitment in response to pathogens, infections, and cancer treatments. The cascade consists of 4 stages: the initial capture and rolling of cells through selectin mediated interactions, firm adhesion to the endothelium later through integrins, migration, and transmigration. Recently it has been shown that some leukocyte cell types, including CD4+ T-cells, can migrate upstream under shear flow after arrest. CD4+ T-cells adhere to the adhesion molecules ICAM-1 and VCAM-1 through two different integrin receptors. It has been shown that CD4+ T-cells can migrate upstream on ICAM-1 but downstream on VCAM-1. By combining ICAM-1 and VCAM-1 on a surface, our laboratory previously showed that any amount of ICAM-1 supports upstream migration. The objective of this study was to determine if the patterning of ICAM-1 and VCAM-1 into stripes and patterns had any influence on the direction of T-cell migration.

Materials and Methods: Primary human CD4+ T-cells isolated from blood donations are activated with PHA, followed by Penn-Strep and Interluekin-2 for a week. Patterned PDMS stamps incubated with ICAM-1 and VCAM-1-Fc chimeras are stamped onto a UV activated PDMS coated glass slide. On patterned stamps, a small amount of Alexa-fluor 555 tagged Human IgG is added to visualize stamped patterns. The stamped area is blocked with Pluronic F-127 protein to prevent non-specific adhesion. Experiments are performed over 30 minutes in a flow chamber. Cells are tracked using ImageJ software and trajectories are analyzed in MATLAB. The Migration Index (MI), defined as the cell's migration distance in the axis of flow divided by total migration distance, is calculated from the cell track. A negative value indicates net upstream migration, while a positive value indicates downstream migration. Surface protein concentrations were calculated using two methods: Elisa and fluorescent intensity using a confocal microscope. Fluorescent intensity was determined by imaging FITC tagged anti-ICAM-1 antibodies within the striped surfaces and on uniform surfaces of varying concentrations. Elisa experiments were performed HRP tagged antibody converting slow TMB to measure total chimera concentrations on uniform and patterned surfaces.

Results and Discussion: Primary CD4+ cells were seeded onto four types of patterned surfaces. Two PDMS stamp archetypes were utilized; Uniform stamps provide an even distribution of adhesion proteins on the slide and striped stamps print the adhesion proteins in stripes of 50, 100, or 200 μm width with the equivalent spacing between stripes. Experiments were performed an equal mass mixture of ICAM-1 and VCAM-1 with total 15μg/ml. 10μg/ml stamps without stripes were also included for comparison to previously published results. In addition, three flow conditions were chosen: static, a shear rate of 400 s^{-1}, and a shear rate of 800 s^{-1}. We show that on uniform surfaces, T-cells migrate upstream, consistent with previous results. As a control, no directional migration was seen on these surfaces under static conditions. Upstream migration increases with higher shear rates. On 50μm wide stripes oriented parallel to the flow, CD4+ T-cells migrate downstream under shear flow, rather than upstream as observed on the uniform surfaces. A significant portion of cells interact with or migrate along the edge of the stripe during migration. Widening the stripe to 100μm does not reestablish upstream migration, but rather led to migration that was not statistically different than random migration. While, most cells continue to interact with the edge of the stripe, whether the cell is at the edge or the center of stripe had no overall difference on the direction of migration on 100 μm stripes. On surfaces with 200μm stripes, upstream migration is re-established.

Conclusions: Confining CD4+ T-cells to stripes of 50 microns changes their observed direction of migration under shear flow. Rather migrating upstream as seen on uniform surfaces, T-cells migrate downstream in striped patterns of adhesion proteins. Increasing the width of the stamped stripes to 200μm reestablishes upstream migration, while the intermediate width of 100μm results in an intermediate migration index not statistically different from random migration.

Figure 1. Migration Index of CD4+ T-cells under shear flow on ICAM-1 & VCAM-1 mixed surfaces. Shown (left) each grouping (top to bottom) are different shear rates static, 400 s^{-1}, and 800 s^{-1}. Each grouping contains (top) uniform ICAM-1 and VCAM-1, (middle) 200μm stripes in direction of flow ICAM-1 and VCAM-1, (bottom middle) 100um stripes in direction of flow ICAM-1 and VCAM-1, (bottom) 50um stripes in direction of flow ICAM-1 and VCAM-1 surfaces. Negative value indicates upstream migration. Positive value indicates downstream migration. Mixed surfaces contain equal mass ratio of ICAM-1 to VCAM-1. All striped stamps incubated in total of 15μg/ml chimera proteins. Error bars are of S.E.M. with n = 3 or greater.

Figure 2. Scattergrams of cell tracks from single, sample experiments. Red tracks indicate upstream migration; blue tracks indicate downstream migration. Rows, from top to bottom, are uniform, 200μm horizontal, 100μm horizontal, and 50μm horizontal patterns. Columns, from left to right, are static, 400s^{-1}, and 800s^{-1} flow conditions. All stamps incubated in total of 15μg/ml chimera proteins. Axis range from -400μm to 400μm in both the vertical and horizontal direction.

Figure 3. Examples of cells migrating on stripes. Dot and number represent the end location of each tacked cell. Flow is from left to right. Each image is from 400s^{-1} flow condition with stamps incubated in 15μg/ml chimera proteins. **A)** 50μm stripes. Cells primarily migrate downstream. B) 200μm stripes. Cells primarily migrate upstream.

Figure 1:

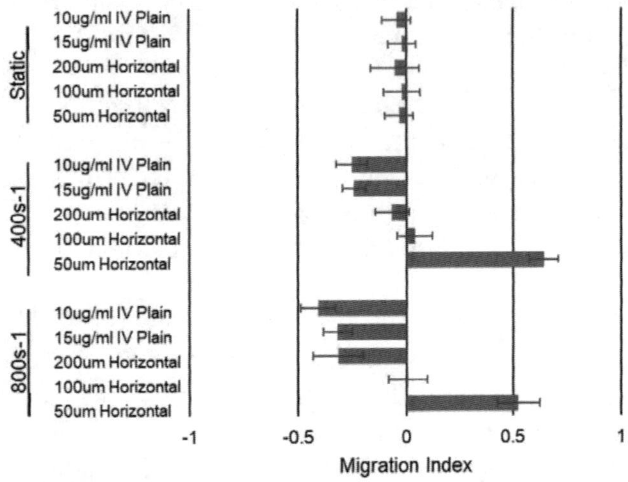

Figure 2:

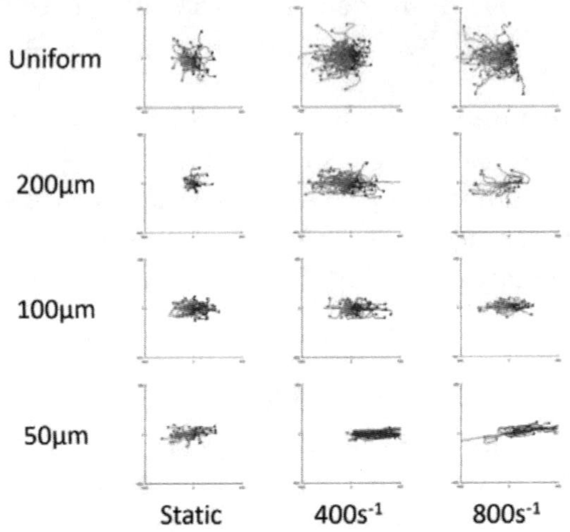

Figure 3:

a) 50μm Stripes Tracked Cells

b) 200μm Stripes Tracked Cells

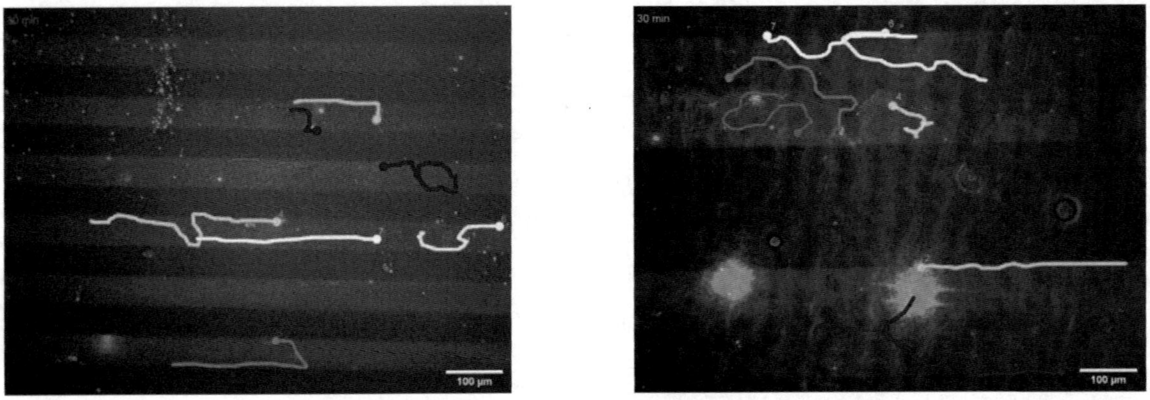

605022 Spatially Controlled Stem Cell Differentiation Via Morphogen Gradients: A Comparison of Static and Dynamic Microfluidic Platforms

Abstract:

The ability to harness the processes by which complex tissues arise during embryonic development would improve our ability to engineer complex tissue-like constructs *in vitro*—a long-standing goal of tissue engineering and regenerative medicine. In embryos, uniform populations of stem cells are exposed to spatial gradients of diffusible extracellular signaling proteins, known as morphogens. Varying levels of these signaling proteins induce stem cells to differentiate into distinct cell types at different positions along the gradient, thus creating spatially patterned tissues.

Here, we describe two straightforward and easy-to-adopt microfluidic strategies to expose human pluripotent stem cells (hPSCs) *in vitro* to spatial gradients of desired differentiation-inducing extracellular signals. Both approaches afford a high degree of control over the distribution of extracellular signals while preserving the viability of the cultured stem cells. The first microfluidic platform is commercially available and entails static culture, whereas the second microfluidic platform requires fabrication and dynamic fluid exchange. In each platform, we first computationally modeled the spatial distribution of differentiation-inducing extracellular signals. Then, we used each platform to expose hPSCs to a gradient of these signals (in this case, inducing a cell type known as the primitive streak), resulting in a regionalized culture with differentiated primitive streak (PS) cells predominately localized on one side and undifferentiated stem cells at the other side of the device. By combining this approach with a fluorescent reporter for differentiated cells and live-cell fluorescence imaging, we characterized the spatial and temporal dynamics of primitive streak differentiation within the induced signaling gradients. We envision that microfluidic approaches to create precisely controlled morphogen gradients will add to the stem cell and developmental biology toolkit, and may eventually pave the way to create increasingly spatially patterned tissue-like constructs *in vitro*.

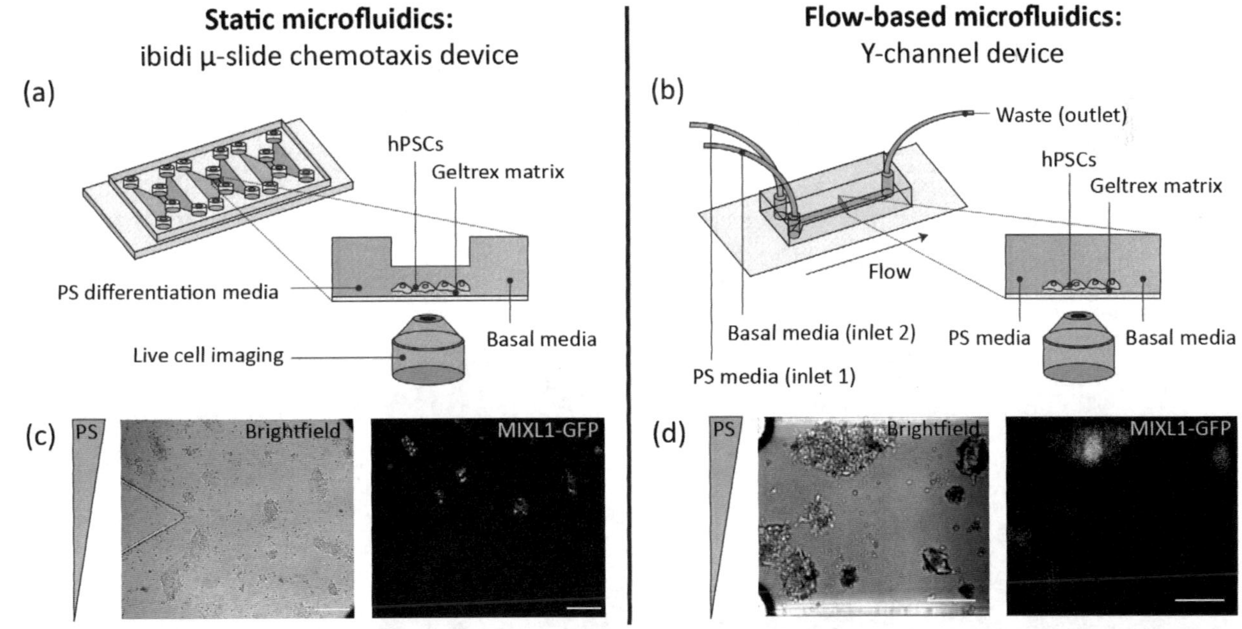

Figure 1. Schematics of (a) static and (b) flow-based microfluidic setups used to generate gradients of soluble signaling factors across MIXL1-GFP hPSCs. Cells undergoing differentiation into anterior primitive streak (PS) express GFP and were observed with live-cell fluorescence microscopy. (b) Brightfield and fluorescence micrographs of MIXL1-GFP hPSCs after 24 h of exposure to a gradient of PS factors in (c) static and (d) flow-based devices.

606245 The Progression and Zonation of Non-Alcoholic Fatty Liver Disease on a Chip

Abstract:

Non-alcoholic fatty liver disease (NAFLD) and its progressive form non-alcoholic steatohepatitis (NASH) affect 25% of the world population. NAFLD is predicted to soon become the main cause of liver morbidity and transplantation. The disease is characterized by a progressive increase of lipid accumulation in hepatocytes, which eventually induce fibrosis and inflammation, and can ultimately cause cirrhosis and hepatic carcinoma.

In this study, we created a patterned model of NAFLD on a chip using a free fatty acid (FFA) gradient of 0 - 2 mM to recapitulate a spectrum of disease conditions in a single continuous liver tissue composed of primary hepatocytes (Fig. 1). We established the NAFLD progression *via* quantification of intracellular lipid accumulation and transcriptional levels of fatty acid transporters and NAFLD pathogenesis markers across the tissue.

While the cell viability did not vary significantly across the five channels of the device, we observed an increasing lipid accumulation across the length of the cell culture chamber. The change in expression levels of several hepatic markers of lipid metabolism, including fatty acid transporter FATP5, showed a significant increase in channel 5 only. The expression levels of other NAFLD markers, such as acyl-CoA synthetase long chain family member 4 (ACSL4), were also significantly upregulated, mostly in channels 4 and 5, where the hepatocytes were treated with > 1 mM FFA.

We then used this platform to investigate the role of oxygen deprivation on lipid accumulation and found that the difference in lipid accumulation among hepatocytes subjected to an physiologically relevant oxygen gradient of 11.2% - 6.9% decreased gradually upon an increase in FFA supplementation, in agreement with the results of recently published computational and in vivo studies.

This progressive high-throughput disease model can be used for hypothesis testing about NAFLD progression and its heterogeneity, including inflammatory cues, as well as for compound screening and efficacy testing of drugs against progressively severe NAFLD conditions. In addition, while we focus on free fatty acids and oxygen as the drivers of NAFLD, the microfluidic platform presented in this work is extensible to simultaneous use of other drivers.

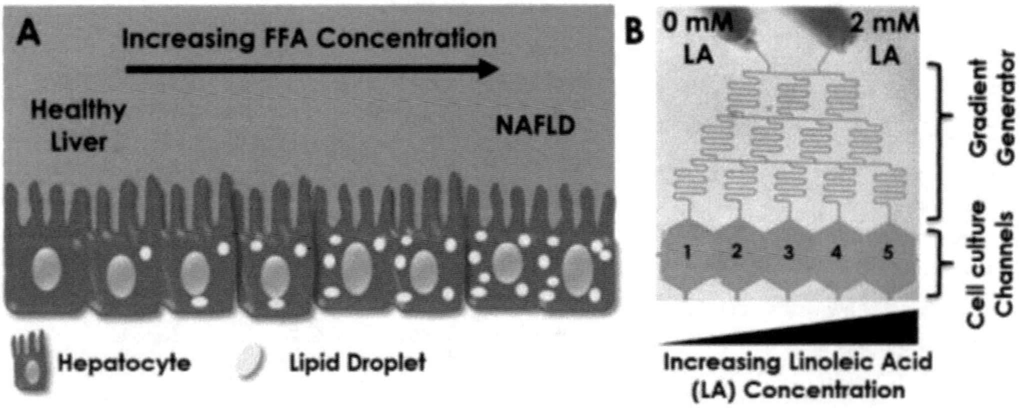

603939 A Microfluidics-Based Approach to Model Drug Transport across 2D and 3D Biological Barriers

Abstract:

Most advanced therapeutics including proteins, peptides, nucleic acids and nanoparticles are administered via subcutaneous or intravenous injections. While these routes offer a logical entry point in the body, injected drugs must overcome numerous biological transport barriers to reach their intended target site. Animal testing is the gold standard of evaluating drug performance in pre-clinical research, but distinguishing delivery efficiency from drug efficacy is not trivial in such complex biological environment *in vivo*. On the other hand, traditional *in vitro* models such as the transwell plate often oversimplify the problem to a basic geometry and static flow conditions. Microfluidic devices can address these issues by capturing essential cellular activities in a controlled manner. In this talk, we discuss how to construct cell monolayers (2D) and cell networks (3D) on a chip with proper flow conditions to study the fate of drugs administered via injections.

Using a microfluidic chip, we created multiple cellular transport barriers that capture diffusion as well as flow-related transport barriers encountered by drugs after injections including diffusion across multi-component cell layers, vascular flow and adhesion to endothelium, among others. To quantify transport, we utilized fluorescent imaging techniques to determine concentration fields for solution-based drugs and number densities for particle-based drugs. Depending on the drug carrier used, we observed various responses of drugs interacting with cellular barriers. Thus, we can not only evaluate drug transport efficiencies, but also investigate their mechanism of action and gain fundamental understandings of the delivery process.

613363 High-Throughput DNA Directed Patterning to Recapitulate Biological Signaling Scenarios (INVITED SPEAKER)

Abstract:

Dissecting the complex action of biological signals that orchestrate single-cell fate decisions as well as tissue function/dysfunction is challenging due to the spatiotemporal parameters modulating their actions. This complexity is further amplified by the fact that these signals— soluble ligands, extracellular matrix-sequestered cues, and cell surface-presented cues—belong to a larger network in which multiple signals are coordinating with each other across time and space. To model and investigate dynamic signaling scenarios *in vitro*, we have developed a DNA- based patterning platform that enables spatial and temporal control over individual and multivariable signaling scenarios. Our method is based on using photolithography to pattern single-stranded DNA oligonucleotides onto a glass substrate. Complementary oligos, to which cells and ligands have been tagged, hybridize to the patterned oligos. This, in turn, enables the recreation of complex biological signaling environments at both the bulk-tissue and single-cell level with micron-scale control. In this talk, I will discuss in depth our method and highlight a few examples of applications—from stem cell to cancer niches—we have pursued.

603880 Biomolecular Ultrasound Imaging of *In Vivo* Phagolysosomal Function

Abstract:

Phagocytosis and lysosomal degradation of pathogens and debris by the innate immune system is critical to organismal homeostasis, and malfunctions of these mechanisms underlie several major diseases. However, the assessment of phagocytic and lysosomal functions *in vivo* is challenging because most imaging agents compatible with non-invasive techniques comprise non-biodegradable synthetic materials that are retained rather than metabolized. Here, we overcome this challenge using an ultrasound contrast agent based on gas vesicles (GVs), a unique class of biosynthetic air-filled protein nanostructures. First, we visualize the temporal dynamics of GVs in circulation and in the liver with ultrafast Doppler ultrasound and non-linear amplitude modulation imaging, respectively. Next, through a combination of pharmacological perturbation and tissue immunohistology, we confirm that systemically administered GVs are primarily taken up by liver macrophages. Then, we demonstrate by live cell optical and acoustic microscopy that internalized GVs are subsequently degraded in the lysosome, resulting in loss of ultrasound contrast. Finally, we parametrize macrophage function by integrating hemodynamic and liver ultrasound time courses into a pharmacokinetic model, and show that rates of phagocytosis and lysosomal degradation are perturbed in two models of liver dysfunction: phagocyte deficiency and non-alcoholic fatty liver disease. Thus, the combination of nanoscale, proteolytically-susceptible contrast agents with quantitative ultrasound imaging enables non-invasive assessment of macrophage phagolysosomal function as a disease-relevant biomarker.

Figure 1. Gas vesicles enable non-invasive visualization of *in vivo* phagolysosomal function. In all experiments, images were acquired at 0.25 Hz and purified GVs (0.35 pmol in 100 μL saline) were injected intravenously 300 s after the start of imaging. **A)** We measured the temporal dynamics of hemodynamic signal enhancement—as a proxy for GV concentrations in the blood—with transcranial ultrafast Doppler ultrasound. Scale bar: 2 mm. **B)** Time course of Doppler signal. Individual trials are shown as light traces. Shaded area represents ±SEM. Thick line represents average of N = 6 biological replicates. **C)** We next applied amplitude modulation (AM) imaging to visualize GV dynamics in the liver. Scale bar: 2 mm. **D)** Time course of AM signal. Individual trials are shown as light traces. Shaded area represents ±SEM. Thick line represents average of N = 5 biological replicates. **E)** We fit our ultrasound data to a two-compartment pharmacokinetic model in order to derive rate constants for the interdependent processes of uptake **(F)** and degradation **(G)**. Rates are normalized to those of age-matched litter-mate control mice. GV concentrations in the blood and liver were represented by our Doppler and AM time courses, respectively. Liver dysfunction was modeled by clodronate-mediated macrophage ablation and by diet-induced non-alcoholic steatohepatitis (MCD). In both cases, uptake and degradation rates were significantly suppressed. Error bars represent ±SD. N=4-6.

Figure 2. GVs are primarily cleared from circulation by liver macrophages and degraded in the lysosome. **A)** Confocal micrographs of liver sections from mice euthanized one hour after injection with Alexa Fluor 647-labeled GVs. Macrophages were labeled with anti-F4/80. **B)** Colocalization of GVs with macrophage areas. Images were segmented in ilastik and quantified in MATLAB. Error bar represents ±SD. N=4 mice, with a total of 60 fields of view analyzed. **C)** Diagram of *in vitro* uptake of GVs. RAW264.7 cells were cultured on fibronectin-coated cover slips and inverted over a dilute suspension of GVs engineered to produce non-linear contrast. After 30 min, the cells were transferred to clean media and allowed to process the GVs for predetermined intervals before loading into an ultrasound phantom. **D)** Cross-propagating amplitude modulation (xAM) images of cell pellets. GV contrast declined in control cells but persisted in cells pre-treated with 100 nM bafilomycin A1 to inhibit lysosomal activity. **E)** Quantification of xAM signal, adjusted for cell density and normalized to the initial timepoint. Error bars represent ±SEM. N=4-8 biological replicates per timepoint.

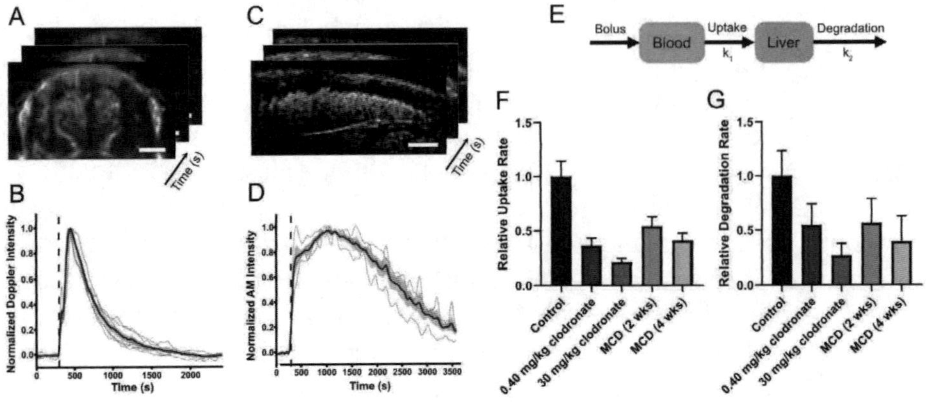

Figure 1. GVs enable non-invasive visualization of *in vivo* phagolysosomal function.

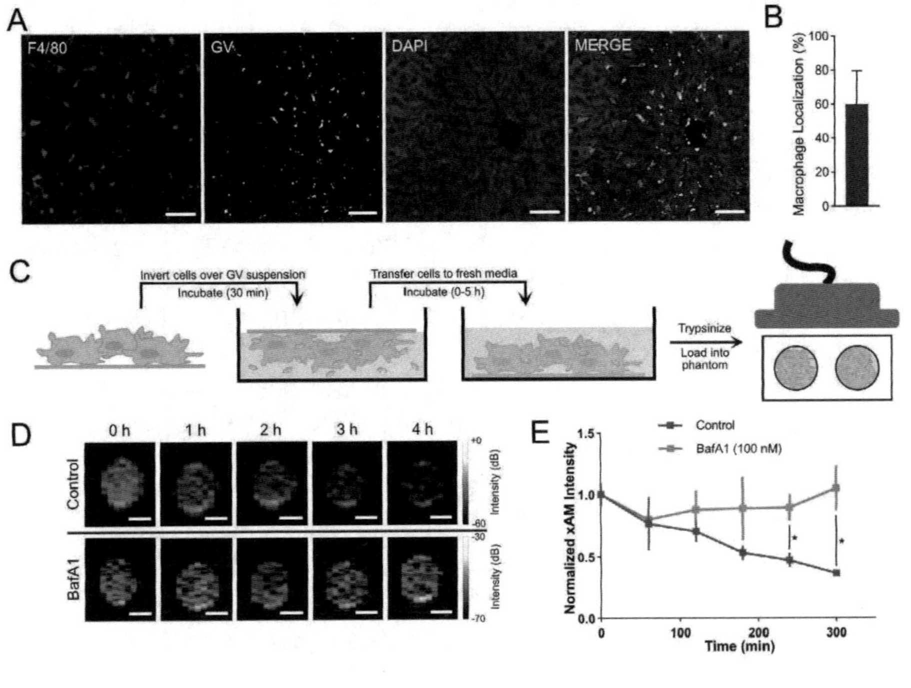

Figure 2. GVs are primarily cleared from circulation by liver macrophages and degraded in the lysosome.

610860 A New Noninvasive Method for Measuring the Size Distribution of Lung Metastases Vs Time in a Mouse Breast Cancer Model As an Input into a Mathematical Model of Tumor Size Evolution

Abstract:

Cancer is a major problem for patients and physicians alike. One such cancer that is particularly harmful for human health is breast cancer. Breast cancer is one of the leading causes of death among women. Triple negative breast cancer (TNBC) is the most aggressive form of breast cancer. TNBC is also one of the few cancers that has several lines that can cause cancer in mice from a mouse disease. In most other cancer studies a human cancer must be given to a mouse to model the disease. This has the advantage of studying cancer in a more realistic animal model and not eliminating the mouse's immune system as would be needed to study human cancer in mice. All these reasons are why TNBC is an attractive cancer model for studies.

TNBC is most lethal once it metastasizes. In TNBC a very common metastasis site is the lungs. Once the cancer enters the lungs, breathing becomes challenging. This metastasis pattern in mice mimicking that in humans allows for analyzing and modeling cancer metastasis in these TNBC mouse models to gain a better understanding in cancer patterns and progression.

Here we used a mouse model, 4T1, of TNBC to study metastasis from the mammary gland to the lungs. Our interest (for our theoretical modeling effort) is to obtain tumor numbers and sizes as functions of time in the mouse lungs after tumor initiation in its mamary fat pad. Existing non-invasive techniques do not come close to providing such results. We describe a new non-invasive computed tomography (CT)-based technique to accomplish this. CT, which is basid on tissue density, cannot normally distinguish tumor from normal tissue, other than by shape analysis by an experienced radiologist. Each mouse was given a CT scan on day 0 before inoculation with cancer cells so that its lungs can serve as their own control. After, cancer cell lines were injected into the mammary fat pad of the mouse and allowed to form primary tumors. CT scans were regularly taken of the lungs to detect remote tumor growth from metastasis. Mice were also placed into two different groups, one group whose immune system did not attack the tumor since the tumor cells were syngeneic, and one group that was treated with anti-PD1 an iummune checkpoint inhibitor, to see if such treatment efffected tumor shrinkage and metastasis.

We hypothesize that as cancerous growths progress the amount of tissue in the lungs increases and the amount of air decreases We analyzed time-gated CT images of the lungs sorteed to reflect the same phase of the breathing cycle. All analysis was performed on images of lungs that were fully empty. For each scan, we separated images of the tissue and air portions of the lungs and calculated the size distribution histograms of the volume segments that comprise it. By comparison these size distribution results of post-inoculation and pre-inoculation scans of the same mouse we develop an algorithm for calculating the size distribution of the additional tissue domains that appear as a result of post-inoculation changes to the lungs. We do this for several sequential scans until mouse death and calculate how these distributions change with time. A small number of mice were given a contrast agent consisting of gold bound to sugar that is known to accumulate more in certain nonlung tumors to see if it alone can distinguish tumor from tissue.

We present the data of mouse lungs as TNBC metastasis size/number histograms as a function of time after tumor cell inoculation, both with and without anti-PD1 treatment. These data are precisely what are needed for our group's mathematical models of how tumor size/number histograms eveolve in time, with the anit-PD1-free mice allowing us to acces parameters for matastasis formation and metastasis growth, and the anti-PD1-treated mice accesing that for the treatment. This is future work. They can be used in future experiments to be predicted by this computational model of TNBC metastasis. Future, experiments will look at mice, e.g., C57 black, for whom these cells are not syngeneic so as to assess their immune response as well as mice undergoing other treatments. Our findings may have implications for future physicians understanding metastasis patterns, the effects of immunotherapy and for incorporating these understandings into treatment plans.

609883 A Low-Cost Muti-Technique Portable Electrochemical Device for Biosensors

Abstract:

Electrochemical devices are commonly used in chemistry and bioengineering laboratories; They are usually expensive and had a significant size. However, researchers such as [1] or [2] had been working on developing portable devices as tools to develop electrochemical measurements with low-cost devices. In [2], the authors use the device as a potentiostat/galvanostat to battery characterization. On [3], the authors introduce a low-cost potentiostat for chemical applications; They developed a potentiostat based on CMOS technology and used a Cyclic Voltammetry (CV) technique to measure chemical redox reactions. Additionally, the biosensors field also makes use of electrochemical measurements; some researchers are working in custom devices to supply their requirements, such as [4], [5]. Hence, we previously developed a potentiostat which was able to perform cyclic voltammetry [6]. His device was our first approach to low-cost portable electrochemical devices.

Later, the device has been continually evolving in the measurements required for the biosensors developed in our research group. Through the use of the developed platform on different projects on the group as a tool to make analysis, other electrochemical measurements were added over time, all using the same platform. At this moment, the device can do staircase voltammetry (SV), linear sweep voltammetry (LSV), cyclic voltammetry (CV), anodic stripping voltammetry (ASV), square-wave anodic stripping voltammetry (SWASV), and chronoamperometry (CA). Besides, we are working to adding more functionalities to the platform. Moreover, we are working on giving the device multiplexing functionalities and an auto range for current measurements on the next version. However, our portable electrochemical device needs a sensing element. So, we developed low-cost electrodes based on screen printing electrodes, such as used by [7]. Our electrodes were made using Printing Circuit Board (PCB) techniques. Briefly, using an FR4 substrate with a copper (Cu) layer of approximately 70μm, we create a three-electrode cell using wet chemical etching, then, we expose the Cu and evaporate a layer of gold over the electrode. Finally, a layer of protective solder mask is used to had a controlled exposition area for the sensor. After that, the electrodes are functionalized according to the experiment and could be used with the platform developed.

Our device is powered by 3.3v, weights about 20g, and its dimensions are 4,45 cms of width, 5,34 cms of length, and around 2 cms of height. It could measure currents from approximately 5 μA up to 10 mA. Also, the techniques of the device could use voltages from the range from -1.6v up to 1.6v. Additionally, the software shows the measured value each 60ms, and it is fully customizable for each procedure that it could do. This device is developed to work with a three-cell electrode designed built by our research group using PCB techniques. Those electrodes had a final coating layer of gold (Au), on which molecules can be functionalized. The use of functionalized electrodes with biological elements such as proteins, antibodies, or enzymes, gives the platform biosensing capabilities. About the device characteristics, Table 1 presents a comparison of our device against other reported on literature and also to commercial tools used in laboratories.

Finally, our device has implemented seven techniques which are common to electrochemical test, even though it could not achieve the range of the commercial tools, the range it had is suitable for biosensors. Due to the small voltage required to the reactions as presented on [9] where they use techniques like EIS and CV to detect different pathogens and shown voltages used for CV in the range of ± 0.700V. Also, authors of [10] present electrochemical biosensors for antioxidant analysis in food. In their review, there are DNA-based biosensors, enzyme-based biosensors, and others. Most of the biosensors reviewed use CV as a technique used to make the measurement.

References

[1] M. D. Steinberg, P. Kassal, I. Kereković, and I. M. Steinberg, "A wireless potentiostat for mobile chemical sensing and biosensing," *Talanta*, vol. 143, pp. 178–183, 2015, doi: 10.1016/j.talanta.2015.05.028.

[2] T. Dobbelaere, "A USB-controlled potentiostat/galvanostat for thin-film battery characterization," *HardwareX*, vol. 2, pp. 1–12, 2017, doi: 10.1016/j.ohx.2017.08.001.

[3] T. Luo, H. Wang, H. Song, and J. B. Christen, "CMOS potentiostat for chemical sensing applications," *Proc. IEEE Sensors*, no. 262005, pp. 0–3, 2013, doi: 10.1109/ICSENS.2013.6688271.

[4] C. Y. Huang, H. T. Huang, and R. T. Yuan, "Design of a portable mini potentiostat for electrochemical biosensors," *Proc. 2017 IEEE 2nd Adv. Inf. Technol. Electron. Autom. Control Conf. IAEAC 2017*, pp. 200–203, 2017, doi: 10.1109/IAEAC.2017.8054006.

[5] R. Pruna *et al.*, "A low-cost and miniaturized potentiostat for sensing of biomolecular species such as TNF-α by electrochemical impedance spectroscopy," *Biosens. Bioelectron.*, vol. 100, no. July 2017, pp. 533–540, 2018, doi: 10.1016/j.bios.2017.09.049.

[6] O. J. F. Segura Crhistian C, "Miniaturization of Cyclic Voltammetry Electronic Systems for Remote Biosensing," *Int. J. Biosens. Bioelectron.*, 2017, doi: 10.15406/ijbsbe.2017.03.00068.

[7] S. Vineeth Raj, J. Stanley, and T. G. Satheeshbabu, "Fabrication of a Configurable Multi-Potentiostat for LOC Applications," *Mater. Today Proc.*, vol. 5, no. 8, pp. 16732–16739, 2018, doi: 10.1016/j.matpr.2018.06.038.

[8] S. Bukkawar, N. Sarwade, and M. Panse, "Polyaniline assisted USB based sensor for determination of benzene biomarker," *Sens. Bio-Sensing Res.*, vol. 22, no. January, p. 100260, 2019, doi: 10.1016/j.sbsr.2019.100260.

[9] E. Cesewski and B. N. Johnson, "Electrochemical biosensors for pathogen detection," *Biosens. Bioelectron.*, vol. 159, no. April, p. 112214, 2020, doi: 10.1016/j.bios.2020.112214.

[10] Y. Ye, J. Ji, Z. Sun, P. Shen, and X. Sun, "Recent advances in electrochemical biosensors for antioxidant analysis in foodstuff," *TrAC - Trends Anal. Chem.*, vol. 122, p. 115718, 2020, doi: 10.1016/j.trac.2019.115718.

	Voltage supply	**Techniques**	**Voltage operation**	**Communication**	**Commercial**	**Size WxH**
Our device	3.3v	SV, LSV, CV, ASV, SWASV, CA	-1.650V to 1.650V	USB	No	4,45cm x 5,34cm
Wireless potentiostat [1]	3.3 V	CA	-0.325V to 0.900V	NFC	No	8,00cm x 5,00cm
USB Potentiostat Galvanostat [2]	5V	CV	-8.00V to +8.00V	USB	No	5,00cm x 5.00cm
USB based sensor[8]	5V	LSV	0.000V to 3.300V	USB	No	19,81cm x 12,19cm
Autolab PGSTAT101	120V	SV, LSV, CV, ASV, SWASV, CA	-10.00V to 10.00V	USB	Yes	9,00cm x 21,00cm
PalmSens4	5V	SV, LSV, CV, ASV, SWASV, CA, EIS	-5.00V to 5.00V	Wireless Bluetooth or USB	Yes	15.70cm x 9.70cm

Table 1 Characteristics comparison between commercial and research devices.

599694 *In silico* Design and Validation of Riboswitches for the High-Throughput Detection of Isoflavone Genistein

Abstract:

Isoflavones are a group of compounds found in soybeans and foods derived from them [1]. These compounds are known to promote health benefits, for example, in the treatment of menopausal symptoms in women, a decrease in the probability of cardiovascular disease and cancer [2]. Of particular interest is the control of symptoms related to hormonal changes associated with menopause due to the absence of definitive treatments [3]. Multiple studies have shown that the administration of isoflavones, mainly genistein, leads to a significant reduction in menopausal symptoms, such as the hot flashes experienced by many women [4]. Also, they appear to induce a sustained cardiovascular response at the uterus, without substantially increasing endometrial thickness [5]. As a result, over the past few years, genistein has attracted much attention from food and pharma companies and consequently, it has been manufactured at industrial scale via chemical synthesis, solid-liquid extraction from soybeans, fermentation processes, and enzymatic synthesis [6][7]. Most of these processes, however, are relatively costly and exhibit low yields, and for this reason, alternatives for higher production rates are of the utmost importance.

Recent efforts have involved the use of microbial factories, which have been enabled by the relative ease for *de novo* gene synthesis and the corresponding heterologous expression [8]. Despite the success of these approaches, the detection of the produced compounds generally involves the use of expensive and tedious liquid chromatography methods [9]. Alternatively, recent works have described novel biosensors for *in situ* determination of the recombinantly expressed compounds including, riboswitches enzyme-based and optical biosensors [10]. Due to their ease of manipulation and specificity, riboswitches have recently emerged as robust alternatives for tracking microbial product production [11]. Riboswitches are segments of RNA responsible for regulating the production of proteins and therefore can be incorporated into processes for highly-sensitive selection and detection of the compounds of interest [12]. Despite the potential of these biosensors, this approach relies on exploring a relatively high number of segments (i.e., aptamers) to search for increased specificity. This process could be lengthy and sometimes might even exceed the transformation limit of the microorganism and the number of attainable experiments. As a result, the associated costs could be prohibitive.

To try to overcome this issue, here, we aimed at accelerating the search for testable aptamers by conducting *in silico* interaction experiments of the candidate RNA sequences with genistein. For this purpose, the secondary structure of the aptamers was predicted via the ViennaRNA package by considering an initial database of 97 nucleotides with 3 regions of folding comprised of 5 variable nucleotides [13]. Also, calculations of the conformational free energy were carried for the active (i,e., ON) and inactive (i.e., OFF) conformations. With the results, it was possible to calculate the probability of occupation of each nucleotide in each variable position of the binding region to the ligand (i.e., genistein) and, therefore, the number of possible useful aptamers were successfully reduced by nine orders of magnitude. Also, terminators (interrupting transcription) and aptamers that are active either in the presence or absence of ligands, were found, which would allow validating the biosensors experimentally in the future. The tertiary structure of the identified aptamers, in conjunction with genistein, was predicted via AutoDock Vina. Additional details of the interactions were explored by Molecular Dynamics (MD) simulations in GROMACS to reduce even further the number of candidates for experimental evaluation [14][15][16].

Bibliography

[1] I. C. Munro *et al.*, "Soy Isoflavones: a Safety Review," *Nutr. Rev.*, vol. 61, no. 1, pp. 1–33, Jan. 2003, doi: 10.1301/nr.2003.janr.1-33.

[2] Barnes, S. (1998). Evolution of the Health Benefits of Soy Isoflavones. *Experimental Biology and Medicine*, *217*(3), 386–396. https://doi.org/10.3181/00379727-217-44249

[3] G. Cheng, B. Wilczek, M. Warner, J. Gustafsson, and B.-M. Landgren, "Isoflavone treatment for acute menopausal symptoms," *Menopause*, vol. 14, no. 3, pp. 468–473, May 2007, doi: 10.1097/GME.0b013e31802cc7d0.

[4] D. C. Knight, J. B. Howes, and J. A. Eden, "The effect of PromensilTM, an isoflavone extract, on menopausal symptoms," *Climacteric*, vol. 2, no. 2, pp. 79–84, Jan. 1999, doi: 10.3109/13697139909025570.

[5] K. Han, "Benefits of soy isoflavone therapeutic regimen on menopausal symptoms," *Obstet. Gynecol.*, vol. 99, no. 3, pp. 389–394, Mar. 2002, doi: 10.1016/S0029-7844(01)01744-6.

[6] W. Mark J., C. Andreas, and H. Paul E., "Process of preparing," US5554519A, 1995.

[7] G. KOCHS and H. GRISEBACH, "Enzymic synthesis of isoflavones," *Eur. J. Biochem.*, vol. 155, no. 2, pp. 311–318, Mar. 1986, doi: 10.1111/j.1432-1033.1986.tb09492.x.

[8] Huang, B., Guo, J., Yi, B., Yu, X., Sun, L., & Chen, W. (2008). Heterologous production of secondary metabolites as pharmaceuticals in Saccharomyces cerevisiae. *Biotechnology Letters*, *30*(7), 1121–1137. https://doi.org/10.1007/s10529-008-9663-z

[9] Q. Wu, M. Wang, and J. E. Simon, "Determination of isoflavones in red clover and related species by high-performance liquid chromatography combined with ultraviolet and mass spectrometric detection," *J. Chromatogr. A*, vol. 1016, no. 2, pp. 195–209, Oct. 2003, doi: 10.1016/j.chroma.2003.08.001.

[10] Mehrotra, P. (2016). Biosensors and their applications – A review. *Journal of Oral Biology and Craniofacial Research*, *6*(2), 153–159. https://doi.org/10.1016/j.jobcr.2015.12.002

[11] Hallberg, Z. F., Su, Y., Kitto, R. Z., & Hammond, M. C. (2017). Engineering and In Vivo Applications of Riboswitches. *Annual Review of Biochemistry*, *86*(1), 515–539. https://doi.org/10.1146/annurev-biochem-060815-014628

[12] K.-M. Song, S. Lee, and C. Ban, "Aptamers and Their Biological Applications," *Sensors*, vol. 12, no. 1, pp. 612–631, Jan. 2012, doi: 10.3390/s120100612.

[13] M. Wachsmuth, S. Findeiss, N. Weissheimer, P. F. Stadler, and M. Morl, "De novo design of a synthetic riboswitch that regulates transcription termination," *Nucleic Acids Res.*, vol. 41, no. 4, pp. 2541–2551, Feb. 2013, doi: 10.1093/nar/gks1330.

[14] S. Mujwar and K. R. Pardasani, "Prediction of Riboswitch as a Potential Drug Target for Infectious Diseases: An Insilico Case Study of Anthrax," *J. Med. Imaging Heal. Informatics*, vol. 5, no. 1, pp. 7–16, Feb. 2015, doi: 10.1166/jmihi.2015.1358.

[15] E. M. Aghdam and M. Esmaeil Hejazi, "Riboswitches as Potential Targets for Aminoglycosides Compared with rRNA Molecules: In Silico Study," *J. Microb. Biochem. Technol.*, vol. 07, no. 05, 2015, doi: 10.4172/1948-5948.S9-002.

[16] R. Penchovsky and C. C. Stoilova, "Riboswitch-based antibacterial drug discovery using high-throughput screening methods," *Expert Opin. Drug Discov.*, vol. 8, no. 1, pp. 65–82, Jan. 2013, doi: 10.1517/17460441.2013.740455.

609798 A Parsimonious and Accurate Method for Determination of Phthalates and Dinch Metabolites in Human Samples, Using High Performance Liquid Chromatography (HPLC–MS/MS)

Abstract:

Abstract Text:

Phthalates and their substitute Hexamoll® DINCH®, are a group of plasticizers with a production volume of millions of tons per year. They are widely used in the manufacture of plastics, to make them soft and flexible, and in personal care products. They can be found in everyday products such as soaps, suntan lotion, soft plastic toys, plastic bottles, raincoats, shoes and food packaging. Due to their endocrine-disrupting properties, some phthalates have been assigned use restrictions since the late 1990s. There is a societal concern due to their toxicity to reproduction and presence in biological matrices of humans with Greenpeace having conducted several studies addressing phthalates in consumer products and the potential health effects emerging from its endocrine-disrupting effects in the early 2000s.

For this study, we analyzed phthalate metabolites urine samples from a pilot study of 20 volunteers. The volunteers cover an age range between 5 and 45 years. Three of the volunteers were children under 10 years, ten volunteers were between 20-30 years old, five between 30-40 years old and 2 older than 40 years old. Eleven were male, and 9 were female. The urine samples were collected in the morning before eating anything. All the volunteers provide their last day description of physical activities and also their last meal description. The morning urine samples, after determination of their full volume, were aliquoted into polypropylene tubes or glassware vials and stored at temperatures −80◦C, prior to sample preparation and analysis.

In the morning urine samples of the pilot study, 14 phthalate metabolites and 2 Hexamoll® DINCH® metabolites were determined by on-line high performance liquid chromatography coupled to tandem mass spectrometry (HPLC–MS/MS) using internal isotope-labelled standards according to the new method that we have develop in our lab. Our analytical method can lead to the quantification of 14 phthalates metabolites (MEP, MBzP, MiBP, MnBP, MCHP, MnPeP, MEHP, 5OH-MEHP, 5oxo-MEHP, 5cx-MEHP, MnOP, OH-MiNP, cx-MiNP, and OH-MiDP) and 2 Hexamoll® DINCH® metabolites (OH-MINCH and cx-MINCH) with very low limits of detection and quantification (0.2 μg/L).

In brief, for sample preparation, aliquots of 900μL of the urine samples, calibration samples and quality control samples were added to 372 μL of ammonium acetate buffer (1 M; pH = 6.0–6.4), 18 μL Glucuronidase K12 from E.coli, arylsulfatase-free (Roche Diagnostics Mannheim, Germany; diluted 1:1 with ammonium acetate buffer) and 67.5μL internal standard solution (of the isotope labelled analogues). Samples were incubated in a water bath for 3 h at 37◦C for enzymatic hydrolyses of the conjugates. After incubation,33.8 μL of acetic acid were added to samples to adjust the pH value. Samples were frozen for at least 3 hrs at −20◦C to precipitate proteins. After that, samples were thawed at room temperature and transferred into a 1.8 ml glass vial for instrumental analysis. For liquid chromatography 500 μL of the above solution were injected into a Thermo Scientific Accela LC Systems. Sample clean-up and enrichment was performed on the Spark Holland On-Line SPE system. Chromatographic separation was performed on a Target-Core Phthalates column C18, 100 x2.1mm, 2.6um. The solvents used were LC-MS grade 100% water with 0.1% acetic acid (solvent A) and 100% acetonitrile with acetic acid with 0.1% acetic acid (solvent B). The 6-port valve was switched to direct the analytical gradient in backflush mode through the clean-up column onto the analytical column. At 18 min the 6port valve was changed back into starting position. The total runtime for one sample was 26 min. Mass spectrometric detection and quantification was performed on a Thermo TSQ Quantum Ultra tandem MS mass spectrometer used in negative ionization mode (ESI-). For each analyte, two ion transitions were recorded. One was used for quantification (quantifier) and the other one for verification (qualifier). For the labelled internal standards, one ion transition was recorded. The mass spectrometer was operated in scheduled multiple reaction monitoring (sMRM) mode. The respective ion transitions are given in Table 2. The precision of the method was determined by analyzing a low and high concentration control material pooled from native urine samples in each analytical batch. Metabolite concentrations and conjugation status thus represented the characteristics of general population urine samples. The accuracy of the method was determined by spiking standard solutions to native urine samples. The limits of quantification (LOQ) of the individual phthalate metabolites were derived based on a signal-to-noise ratio. Accuracy and international comparability of the analytical procedure were ensured and verified by the successful participation in the External Quality Assessment Scheme (EQUAS) of the European HBM4EU project on human-biomonitoring and by obtaining certificates of the RIKILT –Wageningen University & Research Organiser of the Phthalates exercise on behalf of the HBM4EU Quality Assurance Unit offered for the metabolites, MEP, MBzP, MiBP, MnBP, MEHP, 5OH-MEHP, 5oxo-MEHP, 5cx-MEPP, MnOP, OH-MiNP, cx-MiNP, OH-MINCH and cx-MINCH in human urine to the Health and Exposome Research Centre (HERACLES), Center for Transdisciplinary Research and Innovation (KEDEK), Aristotle University of Thessaloniki.

In sum, the presented human biomonitoring method of phthalates and Hexamoll® DINCH® is characterized by higher sensitive in comparison to the existing ones in the literature thus results in reliable results of 16 urinary biomarkers of exposure to plasticizers faster than before (in only one analytical run).

612927 Imaging Striatal Dopamine Release with a Non-Genetically Encoded Near-Infrared Fluorescent Nanosensor (Invited Speaker)

Abstract:

Neurons communicate through chemical neurotransmitter signals that either terminate at the postsynaptic process ("wired transmission") or diffuse beyond the synaptic cleft to modulate the activity of larger neuronal networks ("volume transmission"). Molecules such as dopamine belong to the latter class of neurotransmitters, and have been the pharmacological targets of antidepressants and antipsychotics for decades. Owing to the central role of dopamine over a range of behaviors and psychiatric disorders, real-time imaging of the signal's spatial propagation would constitute a valuable advance in neurochemical imaging. To this end, we present a nanoscale near-infrared fluorescent nanosensor for dopamine and demonstrate its efficacy for imaging dopamine volume transmission in the extracellular space of the brain striatum and prefrontal cortex (Beyene et al. ACS Chem. Neuro. 2017). The sensor is developed from polymers pinned to the surface of single wall carbon nanotubes (SWNT) in which the surface-adsorbed polymer is the recognition moiety and the carbon nanotube the fluorescence transduction element. Excitonic transitions in functionalized SWNT yield up to $\Delta F/F = 4500\%$ near-infrared fluorescence emission in the presence of dopamine (Beyene et al. Nano Letters 2018). We next demonstrate imaging of evoked dopamine release in acute striatal slices, and show altered dopamine reuptake kinetics when brain tissue is exposed to dopamine agonist and antagonist drugs (Beyene et al. Science Advances 2019). Lastly, we discuss how to evolve synthetic molecular recognition for other neuromodulators of interest, such as serotonin (Jeong et al. Science Advances 2019). We characterize our findings in the context of their utility for high spatial and temporal neuromodulator imaging in the brain, describe nanosensor exciton behavior from a molecular dynamics (MD) perspective, and validate nanosensor for use to elucidate dopaminergic signaling variability with disease or pharmacological perturbations at a synaptic scale.

608576 Computational and Experimental Exploration of Sequence-Structure-Function Relationships of a Medium-Chain Acyl-ACP Thioesterase

Abstract:

Enzymes which perform industrially relevant chemistry can provide selectivity otherwise difficult to achieve from chemical synthesis or agriculture. Engineering the acyl-ACP thioesterase in fatty acid biosynthesis has proven an effective approach to tailor the fatty acid distribution toward C_8, C_{10}, and C_{12} products. To date, engineering of acyl-ACP thioesterases with a preference for the corresponding medium-chain substrates has been successful with directed evolution and rational mutagenesis techniques. However, fundamental understanding of the factors driving thioesterase chain-length selectivity remains elusive. Here, we apply a computational method to tune substrate binding and explore the sequence-structure-function relationships of an acyl-ACP thioesterase from *Umbellularia californica*. We used the Iterative Protein Redesign and Optimziation (IPRO) algorithm to computationally design thioesterase variants with improved *in silico* binding to C_8, C_{10}, and C_{12} substrates. The pre-screened thioesterase variants were transformed in an *Escherichia coli* fatty acid production strain to test for changes in fatty acid chain-length distribution from the wild type thioesterase. In this talk, I will describe how we navigated through the design, build, and test cycle to enhance the learnings from each round of mutagenesis.

597328 Codonadjust: A Software for *in silico* design of a Mutagenesis Library with Specific Amino Acid Profiles (Industry Candidate)

Abstract:

In protein engineering, generation of mutagenesis libraries is a key step to study functions of mutants. To generate mutants with a desired composition of amino acids, a codon consisting of a mixture of nucleotides is widely applied. Several computational methods have been proposed to calculate a codon nucleotide composition for generating a given amino acid profile based on mathematical optimization. However, these previous methods need to manually tune weights of amino acids in objective functions, which is time-consuming, and more importantly, lack publicly available software implementations. Here, we develop CodonAdjust, a software to adjust a codon nucleotide composition for mimicking a given amino acid profile. We propose different options of CodonAdjust, which provide various customizations in practical scenarios such as setting a guaranteeing threshold for the frequencies of amino acids without any manual tasks. We demonstrate the capability of CodonAdjust in the experiments on the complementarity-determining regions of antibodies and T cell receptors as well as millions of amino acid profiles from Pfam. These results suggest that CodonAdjust is a productive software for codon design and may accelerate library generation. CodonAdjust is freely available at https://github.com/tiffany-nguyen/CodonAdjust.

601572 Machine Learning Analysis of Protein Aggregates Formed in Container-Closure Systems

Abstract:

Therapeutic protein formulations are exposed to numerous stresses during manufacturing, shipping, storage, and administration that promote aggregation. These different stresses result in aggregates whose morphologies are characteristic of the stress at their root cause. This morphology information can be captured by using flow imaging microscopy (FIM) to image the aggregates in a sample and applying machine learning techniques such as convolutional neural networks (ConvNets) to extract the morphological information embedded in FIM images.

In this study, we describe and demonstrate a novel algorithm that can be used to determine if two protein formulations contain similar particle morphologies and thus were exposed to the same root cause stresses. An algorithm was developed to compare collections of FIM images using ConvNets trained using techniques borrowed from facial recognition and traditional statistical techniques such as nonparametric density estimation and goodness-of-fit hypothesis testing. This algorithm provides a robust statistical method to determine if two samples exhibit significantly different FIM datasets and thus contain particles from different root cause stresses. As a demonstration, this algorithm was used to compare intravenous immunoglobulin (IVIg) formulations that were exposed to freeze-thaw and shaking stresses within several container-closure systems. The algorithm was capable of distinguishing between the populations of IVIg aggregates produced by these stresses in subtly different container-closure systems (e.g. glass vials from different manufacturers). This analysis also revealed that the particle morphologies produced a result of shaking stress were more impacted by the container-closure system than those produced by freeze-thaw cycling.

604209 Docking Refinement and Molecular Dynamics Studies Elucidating the Molecular Mechanism of Attractant Signaling to Dhma By E. coli Tsr (Faculty Candidate)

Abstract:

The study of protein-small molecule ligand interactions is a key subject in understanding protein function in biological systems. Computational modeling and molecular dynamics simulations can be coupled with experimental studies to provide an atomic-level resolution of such interactions and fill crucial gaps by obtaining information that is not accessible from experiments. Docking algorithms are widely used to understand the relationship between the structure and function of interactions within protein-small molecule ligand complexes that govern biological functions. Despite the advances made in computational chemistry, and the variety of docking frameworks to study small molecule : receptor binding, there is still significant room for improvement in obtaining highly accurate computational predictions of the structure of small-molecule : receptor complexes. Additionally, while the use of all-atom molecular dynamics (MD) simulations is increasingly important in the comparison of ligands binding to a receptor, the high resolution of these powerful techniques requires computational resources that scale both with the size of the complex under investigation and the amount of time simulated.

In this study, we show an application of an in-house docking-refinement[1] protocol to reveal a structural mechanism by which the *Escherichia coli* (*E. coli*) Tsr chemoreceptor distinguishes between the (*R*)- and (*S*) enantiomers of DHMA[2], metabolites of norepinephrine, a hormone that is released in high quantities in the gastrointestinal tract. We used the docking-refinement protocol, MD simulations, free energy calculations, and structural analyses to elucidate their binding and to predict which enantiomer induces a stronger signal upon binding to Tsr. The initial docking-refinement studies suggested a difference in Tsr-DHMA interaction and orientation between the two enantiomers. Starting with the binding modes derived from the docking-refinement, we used five 100 ns MD simulations per system to investigate the stability of the putative poses and their effect on Tsr. The MD simulations coupled with principal component analysis (PCA) and dynamic cross correlation (DCC) analysis predicted the piston-like motion by an α-helix of Tsr to be of higher magnitude for (*R*)- than (*S*)-DHMA due to differences in specific interactions of the ligands with residues in the binding pocket. This prediction justified the effort to experimentally separate the commercially available racemic mixture of (*R*)-DHMA and (*S*)-DHMA into the pure enantiomers, which could then be tested for their activity independently. The predicted difference in signaling response between the enantiomers was validated through tests of chemotaxis *in vivo*, showing that the computationally revealed difference between (*R*)- and (*S*)-DHMA binding corresponds to different behavioral responses in *E. coli*. Our study shows that the use of docking-refinement to generate near-optimal initial modes as initial structures for MD simulations can be used to avoid longer simulations[3], with the conclusions of the simulation studies confirmed by experiments[2].

1. Orr AA, Jayaraman A, Tamamis P. Molecular Modeling of Chemoreceptor:Ligand Interactions. Methods Mol Biol. 2018;1729:353-372.
2. Orr AA, Yang J, Sule N, Chawla R, Hull KG, Zhu M, Romo D, Lele PP, Jayaraman A, Manson MD, Tamamis P. Molecular Mechanism for Attractant Signaling to DHMA by E. coli Tsr. Biophys J. 2020;118(2):492-504.
3. Podgorny AR, Ray JCJ. Tasting the Terroir with Tsr. Biophys J. 2020;118(2):279-280.

611903 Investigating the Structure and Dynamics of Human β Defensin Type 3 Interacting with Chemokine Receptor CXCR4 in Lipid Bilayers

Abstract:

Human β defensin type 3 (hBD-3) is a cationic cysteine rich peptide. It belongs to the human innate immune system, mainly secreted from human epithelial tissues and mucosa. It has broad-spectrum of bactericidal activities and also chemotactic activity by directly binding with chemokine receptor CXCR4. It has six cysteine residues that are capable of forming three pairs of intramolecular disulfide bonds which can help restrain its structure (called wildtype). These disulfide bonds are known to break in a particular order in a reducing environment, converting hBD-3 into the linear analog form. The disulfide bonding status of hBD-3 may influence its chemotactic activities comparing the naturally occurring structure (wildtype) with three disulfide bonds connected with the linear-analog structure (analog). hBD-3 may also form a dimer at high concentrations in solution.

CXCR4 is a membrane-bound G-protein coupled receptor (GPCR) involved with chemotaxis in lymphocytes. Because CXCR4 is involved in inflammatory diseases and in the replication of HIV-1, it has become a target of interest for therapeutic control. This study focuses on structural changes that can modify the chemotactic activities of hBD-3. A clear understanding of the function of hBD-3 on CXCR4 is important for development of a novel endogenous treatment. All-atom molecular dynamics simulations have been performed on hBD-3 monomer and dimer in both wildtype and analog forms interacting with the CXCR4 receptor embedded inside lipid bilayers. To consider the lipid membrane effect on the structure and dynamics of the binding, two kinds of membranes were worked on, one is the negatively charged POPG bilayer to represent bacterial lipid membrane and the other is the neutrally charged POPC bilayer to represent normal red-blood cell membrane. The protein-in-membrane system was set up, solvated and ionized using VMD program. The simulations were performed at 310.15 K and 1 atm running NAMD simulations for at least 240 ns. Select simulations were extended for up to 4.8 microseconds on ANTON2 machine. The binding structure and dynamics of hBD-3 with the CXCR4 receptor were studied, including RMSD, RMSF, residue pair distance map, radius of gyration, dynamic network analysis, and dihedral angle correlation matrix. Results up to now support a monomer structure binding with the second extracellular loop of CXCR4. Dimers of hBD-3 binding to CXCR4 have proven to be mostly unstable in microsecond scale simulations, and replacing POPC with POPG lipid membrane can influence the binding stability of hBD-3 with CXCR4.

602860 Computational Design of Binding Proteins for Sars-Cov-2 Using Aubie

Abstract:

Between December 2019 and April 1, 2020, Coronavirus Disease 2019 (COVID-19), caused by the novel coronavirus Severe Acute Respiratory Syndrome Coronavirus 2 (SARS-CoV-2), has infected nearly one million people and caused tens of thousands of deaths globally. In early 2020, vital information has been discovered about the genome, structural proteins and infection mechanism of SARS-CoV-2. Using this knowledge, a mammoth research effort to develop COVID19 therapeutics has begun but has not yet resulted in an approved treatment or vaccine. Hence, there is an urgent global need for the quick development of effective therapeutics to combat SARS-CoV-2.

Antibodies are a major class of therapeutic proteins which can be used against SARS-CoV-2. Antibodies are Y-shaped proteins used by the vertebrate adaptive immune system to bind to any foreign molecule (i.e. antigen) that enters the body. Currently, the global market for antibody applications such as therapeutics and biosensors is valued at over $100 billion. Over the past two decades, significant progress has been made towards the development of computational tools for the engineering and *de novo* design of antibodies to bind a target antigen. The use of such computational antibody engineering tools, in combination with experimental procedures, can expedite the development of COVID-19 therapeutics.

Over the past two years, our lab has developed a computational program called the Algorithm for Ultra-rapid Binding Interactions Engineering (AUBIE) for the high-speed design of binding proteins towards a specific region (i.e. epitope) of a target antigen. AUBIE identifies protein loops from the Protein Data Bank (PDB) that can be positioned into the binding regions of the selected scaffold to have strong interactions, such as hydrogen bonds and pi-pi stacking, with the targeted epitope. While AUBIE can design antibodies, the most commonly used binding protein, it can also engineer binding loops onto other smaller sized scaffolds such as the 10th Type 3 Fibronectin (10Fn3) Domain, nanobodies and anticalins. Computational tools like AUBIE can be used to quickly design binding proteins that can specifically bind a functionally significant virus epitope and hence neutralize the virus.

Like other coronaviruses, SARS-CoV-2 uses the receptor binding domain (RBD) of the spike protein to bind to a host cell surface receptor. The experimental atomic structures of the SARS-CoV-2 RBD, unbound and in complex with its receptor Angiotension Converting Enzyme 2 (ACE2), are now available in the PDB. We have used AUBIE to design neutralizing antibodies, nanobodies and 10Fn3 domains to bind to the ACE2 binding site in the SARS-CoV-2 RBD. Furthermore, we have also used AUBIE's ability to target residue-specific interactions to design broadly neutralizing binding proteins for mutated strains of SARS-CoV-2. This has been done using two approaches which differ in their epitope residue/atom selection for interaction formation. The first approach targets the epitope backbone atoms and the second approach targets the residues that form strong interactions with ACE2. This presentation will describe the predicted binding metrics, such as binding energies and shape complementarity, of the AUBIE designs. Experimental testing of these designs is currently planned and results will be shared as part of the presentation.

612902 Liquid-Liquid Phase Separation of Intrinsically Disordered Peptides (Invited Speaker)

Abstract:

Intrinsically disordered peptides (IDP) are a special class of proteins that do not fold to a unique three-dimensional shape. These proteins play important roles in the cell, from signaling to serving as structural scaffolds. Under pathological conditions, they can self-assemble into structures that are toxic to the cell, and a number of neurodegenerative diseases are associated with the self-assembly of IDPs into fibrillar structures. In addition to forming fibrils, many IDP can also phase separate into a protein rich and a protein depleted phase, a process known as liquid-liquid phase separation (LLPS). The cellular role of LLPS is not fully understood, and it has been suggested that the LLPS process may play a protective role in the cell against pathological fibrillization. I will present molecular dynamics and field theoretic simulations to map out the phase diagram of IDPs (with an emphasis on the Tau protein implicated in Alzheimer's Disease), and compare and contrast the phase separation of biological polymers and synthetic polymers.

605485 Elongated Particles are Preferentially Phagocytosed by Neutrophils- Opportunity for Selective Targeting of Leukocytes in Acute Inflammation

Abstract:

Recent studies have shown that polymeric particles can be used to modulate the behavior of immune cells for the treatment of various inflammatory conditions. However, the impact of physical design parameters on the interaction of particles with different leukocyte subtypes remains unclear. Particle shape is shown to be a crucial factor in the phagocytosis of polymeric particles by macrophages. Elongated particles are reported to experience reduced uptake by macrophages, which leads to an increased circulation time in blood. However, the impact of shape on the interaction of particles with circulating blood leukocytes remains unclear. Neutrophils are the most abundant leukocyte in human blood and the first responders in the case of injury and inflammation. Thus, it is crucial to study the impact of shape on the phagocytosis of particles by neutrophils. In this work, we investigated the phagocytosis of rod-shaped polymeric particles by human neutrophils relative to other leukocytes. Contrary to macrophages and other mononuclear phagocytes, neutrophils were found to exhibit increased internalization of rods in *ex vivo* and *in vivo* experimentation. In whole blood samples, rods were shown to have up to 7-fold increased association with neutrophils compared to monocytes. These results demonstrate that altering the shape of particles can be used to selectively target neutrophils in inflammatory pathologies where they are the key controllers of the disorder.

605998 Relationship between Efflux Pump Expression and Antibiotic Susceptibility in an *in-Vitro* Evolved Drug Resistant *Mycobacterium* to Develop Better Drug Targeting Strategies.

Abstract:

Multidrug resistance poses a significant challenge in the treatment of tuberculosis. To address this epidemic, many approaches are currently being tried, such as repurposing of existing drugs and combination therapy by combining exiting drugs with adjuvants. Through many genomic studies, it is now clear that resistance is a multifaceted phenomenon, and in addition to mutations in the drug target genes, many mechanisms are at play in multidrug-resistant organisms. Understanding the alternate mechanisms of resistance will provide us with novel targets for the development of new bactericidal drugs.

In this work, we have investigated the contribution of efflux mediated resistance in multidrug resistance/ extremely drug resistance (MDR/XDR) *Mycobacterium*. *Mycobacterium smegmatis* mc^2 155 (WT cells), our model organism was subjected to increasing concentrations of a Fluoroquinolone (FQ) – Norfloxacin, and a mutant was isolated, named as Nor^r Mutant. This mutant exhibited increased resistance towards many drugs, including FQ's and other first and second-line drugs from the current tuberculosis regime. Interestingly, no mutations were identified in any of the drug target genes on sequencing such as *gyr*A, *gyr*B, *kat*G, and *rpo*B. However, the mutant exhibited decreased intracellular drug uptake. Alteration in drug transport was also confirmed through drug transport studies using a fluorescent tracer (Ethidium bromide-EtBr). This observation suggested that either the uptake or efflux of the drug was affected in the mutant. Also, diffusivity across the membrane was measured using 1-N-Phenyl-Naphthylamine (NPN) dye, which was found to be similar to that of WT, which further indicated the role of drug efflux in imparting resistance. The expression level of many efflux pumps was measured that showed upregulation of the basal levels of multiple pumps in the mutant.

To investigate efflux-pump mediated resistance, we integrated another copy of selected efflux pump genes in the genome under an inducible promoter. Three efflux-pumps were selected from among those that were differentially expressed in the drug-resistant mutant. Over-expression of each of the pumps led to increased resistance that was proportional to the expression level of the efflux pump. Thus, a quantitative relationship between MIC of various drugs and the expression level of efflux pumps could be established. Next, to identify the regulator(s) leading to this enhanced expression of efflux pumps, the mutant was sequenced on the Illumina platform. While multiple mutations were observed, the mutation in a global regulator was prominent. The gene for the regulator was also expressed under an inducible promoter, resulting in increased susceptibility of the bacteria to multiple drugs, with differential regulation of multiple efflux-pumps. Targeting this global regulator could be an alternate strategy to counter this multidrug-resistant strain.

Thus, combined therapy targeting such transcriptional regulators along with efflux pump inhibitors can be a powerful tool to combat drug resistance in mycobacterium. Overall, this work established how non-target gene-based mechanisms, such as multiple efflux-pumps, contribute to the MDR/XDR phenotype of Mycobacteria. This knowledge will help us to develop alternate strategies to target resistant Mycobacteria.

606984 A Novel, Thrombo-Inflammatory Model to Study Platelet Adhesion and Thrombogenesis *in Vitro*

Abstract:

Introduction: Platelets are responsible for maintaining hemostasis by plugging any interruptions or injuries in the vasculature, preventing blood loss and damage due to vascular injury. However, platelets also play a major role in the formation of pathological vascular thrombosis development, including atherosclerosis, stroke, myocardial infarction, and deep vein thrombosis. While the contribution of platelets in thrombus development has long been established, more recent research has illuminated the impact of the endothelium and the innate immune system in thrombogenesis. More specifically, the endothelium recruits and facilitates binding of platelets, leukocytes, and leukocyte-platelet aggregates under inflammatory conditions. Despite the importance of cells other than platelets, including neutrophils and the endothelium, in thrombosis development, most *in vitro* models studying platelet adhesion and aggregation neglect the role that inflammation and other cell types play in thrombosis development and resolution.

Materials and methods: We developed an *in vitro* flow-based system to more accurately represent the inflammatory environment that facilitates the development of thrombi in blood vessels. Human whole blood is perfused over an activated, damaged endothelial cell monolayer representing an acute vascular injury with underlying inflammation through a parallel-plate flow chamber using a syringe pump. After perfusion, adhesion of both leukocytes and platelets is quantified using fluorescent and transmitted light microscopy and compared to non-inflamed controls.

Results and conclusions: The inclusion of endothelial cell inflammation drastically impacts the level of adhesion of cells to the damaged endothelium in blood flow and in particular, highlights the role of leukocyte-platelet aggregates in thrombosis development. We demonstrate the utility of this model in several ways; first, the *in vitro* model allows us to better understand the mechanism of thrombo-inflammation, specifically by allowing the study of platelet-leukocyte-endothelium interactions in blood flow. Further, this model serves as a platform for evaluating therapeutic efficacy of early, novel therapeutics designed to interfere with platelet or leukocyte adhesion and interactions in thrombus development. Overall, we demonstrated that this model better serves as a sophisticated platform for the study of thrombo-inflammation *in vitro*.

599633 Three-Dimensional Dynamic Culture Improves Extracellular Vesicle Production and Cargo Profile of Human Mesenchymal Stem Cells through Altered Biogenesis

Abstract:

Human mesenchymal stem cells (hMSC) have been shown to exhibit therapeutic potentials in the context of tissue regeneration and immunomodulation. Various clinical trials utilize hMSC as cell therapies in disease models such as stroke, multiple sclerosis, cardiovascular disease and graft-versus-host disease. Studies have shown that the clinical efficacy of hMSC is mainly attributed to the paracrine effects via the hMSC secretome. As part of secretome, extracellular vesicles (EV) have been shown to encapsulate proteins, enzymes and metabolites that contribute to the therapeutic potential of hMSC for various diseases. hMSC-derived EV (hMSC-EV), including both microvesicles and exosomes, exert homing/migratory and paracrine effects to facilitate cell-cell communication and immune response in damaged tissue. Although the mechanisms by which EV mediate tissue regeneration and restore homeostasis are still under investigation (with several clinical trials ongoing), hMSC-EV hold potential as a cell-free therapeutic strategy in regenerative medicine with the high feasibility of biomanufacturing. However, similar to hMSC expansion, scale-up production of hMSC-EV has become a major challenge for clinical research in cell-free therapy. Long-term culture of hMSC results in loss of stem cell characteristics, which may compromise the quality, quantity and payload of EV. Studies have shown low efficiency of EV secretion under conventional monolayer culture of hMSC and altered cargo profiles in EV derived from senescent cells. Inconsistent EV properties with regard to the cargo and functional differences also have been reported. Thus, optimization of large scale production of hMSC-EV with effective therapeutic cargos is critical.

In current study, dynamic three-dimensional (3D) aggregation is applied to hMSC to improve the EV production by biogenesis. hMSC were cultured as 3D aggregates under wave motion to facilitate aggregation and EV secretion. A modified isolation strategy also was utilized to enrich EV efficiently. Compared to monolayer culture, hMSC-EV quantity was improved significantly under 3D dynamic culture. Moreover, the derived hMSC-EV from 3D culture exhibited smaller sizes compared to monolayer culture, indicating higher enrichment of the exosome population among hMSC-EV. Analysis of EV biogenesis markers from 3D hMSC aggregates demonstrated increased activation of the endosomal sorting complexes required for transport (ESCRT) pathway, which may contribute to the altered biogenesis. For functional evaluation, EV from 3D hMSC aggregates enhanced the stress resistance of adult stem cells and rejuvenated aged stem cells expressing cellular senescence after EV application to those cultures. Moreover, 2D and 3D hMSC-EV were shown to modulate the immune response in different patterns determined by T lymphocyte and macrophage phenotype assays. MicroRNA sequencing of EV revealed in detail the cargo profile differences for EV derived from 3D dynamic culture versus monolayer culture, which may explain some of the functional differences. In summary, this study provides a promising strategy for high quality EV production from hMSC with enhanced therapeutic potentials compared to conventional monolayer culture.

This work is in part supported by NIH (R01 NS102395) and NSF (#1743426).

603587 Self-Assembly of Biofunctional Nanohybrids in Complex Morphologies Using Recombinant Proteins

Abstract:

There has been an increasing interest in fabricating hybrid nanomaterials for technological applications in life sciences, because of the advantage that synergy of properties from distinct materials at the nanoscale can be achieved. In particular, protein-based hybrid materials provide biocompatibility and biofunctionality, which further combine useful properties and functionalities from non-protein components such as synthetic polymers or inorganics. Protein cages, which hold hollow structures with a cavity, have received considerable attention as a versatile platform. For example, catalyst or therapeutic cargo materials can be protected by biocompatible thin protein layers with tuned properties through encapsulation into hybrids. Protein cages are engineered from natural proteins (e.g. viral capsid proteins, or artificial protein cages can be engineered to provide specific properties that could be more effective for hybrid material fabrication. In this study, we present a system that assemble protein cages and simultaneously encapsulate nanoparticles, resulting in pomegranate-like protein-nanoparticle hybrid nanomaterials. Amphiphilic recombinant protein building blocks were designed by genetic fusion of elastin-like polypeptides and hydrophilic folded protein domains. Assembly of hollow spheres that mimic vesicles are triggered by a temperature change, while weak non-covalent interactions between polystyrene nanoparticles and the proteins mediate encapsulation nanoparticles in high number densities. Our analysis on the size of hybrid protein cages and the number of encapsulated nanoparticles indicates that formation of the pomegranate-like morphology is achieved by nanoparticle-mixing coacervation followed by maturation of morphology transition. A kinetics study on the hybrid assembly supports the mechanisms of cooperative assembly through weak interactions between nanoparticle and proteins. This platform is applicable as nanoscale tools that combines biofunctionality of proteins with a range of nanoparticle agents or biocolloids such as cells. Encapsulation of fluorescence nanoparticles in functional protein cages was demonstrated in this platform, which will be extended to a wide range of applications for biosensors, bioimaging, and high-throughput screening systems.

613361 The Right Tool for the Right Job: Bioanalytical Methods to Elucidate the Personalized Nature of Cancer (INVITED SPEAKER)

Abstract:

Cancer is a highly heterogeneous disease with significant differences between patients (intertumor heterogeneity) and among cells in the tumor microenvironment (intratumor heterogeneity) motivating the need for a more personalized approach to fundamental cancer research and diagnostics. The analysis of enzymatic activity and cellular behavior in single, intact cells yields more precise information than the bulk analysis of cell lysates and allows researchers to collect both endpoint and dynamic measurements. When examining cell lysates alone, the heterogeneity is averaged across the entire population and valuable information about distinct subpopulations is masked or lost entirely. Emerging technologies, such as microfluidic devices, coupled with novel biosensors have allowed for high-throughput, facile analysis and sorting of individual cells to provide new information on low occurring, rare phenotypes. Microscale systems offer a significant advantage over competing technologies due to reduced cost, ease-of-use, significant reproducibility, biological inertness, and a compatibility with light microscopy. This talk will focus on recent work utilizing a combination of microfluidic devices and peptide-based biosensors to develop novel bioanalytical tools for a more personalized approach to cancer research. It will highlight our efforts to identify drug resistant subpopulations of cancer cells by combining cell permeable, long-lived, enzyme-specific, fluorescent, peptide-based reporters with a microfluidic droplet trapping array and automated image analysis algorithm. It will also explore recent efforts using a microfluidic approach to study the behavior of cancer cells during metastasis and within the tumor microenvironment in particular: how stromal cells enhance cancer cell progression and drug resistance.

604701 *In silico* Discovery of Target Peptide Inhibitors for *C. Diff* Toxin a and B

Abstract:

Clostridium difficile infection is the leading cause of diarrhea and colitis (inflammation of the colon) in North America and Europe. The fraction of the population infected by the disease is increasing as new strains associated with significant morbidity and mortality have appeared. There is a growing concern about the failure of the first line of treatment for *C. diff* infection (metronidazole and vancomycin) and thus, current attention is focused on the need for alternative treatment options like biologic drugs which are safer and more efficacious. A viable and cost-effective strategy is the development of targeted peptide-based therapeutics that prevent and treat *C. diff* infections by deactivating the pathogen while leaving the resident gut microbiota unharmed.

During infection, the *C. diff* pathogen produces two large virulent toxins (Toxin A and B) that share 71% sequence homology. The glucosyltransferase domain (GTD) of these toxins act by binding UDP (uridine diphosphate)-glucose, hydrolyzing it into glucose and UDP, and attaching the glucose monomer to human Rho family of GTPases. Glycosylation of the GTPases by Toxin A and Toxin B GTDs leads to disruption of the cytoskeleton, apoptosis, and cell death.

The objective of this project is to computationally design peptide inhibitors that bind with high affinity and specificity to the glucosyltransferase domain of *C. diff* toxins A and B to inhibit its activity. We have been successful in developing a proof-of-concept disruptive computational approach to design peptides that bind to a biomolecular target with high affinity. Our peptide search algorithm employs Monte Carlo methods, self-consistent mean-field theory, and the concerted rotation (CONROT) technique to search for peptides in the sequence and conformation space that bind to a target protein. The starting input structure for the algorithm requires a reference ligand which is usually an experimentally determined peptide sequence that has a good binding affinity to the target protein.

We have been successful in discovering two peptides inhibitors NPA and NPB with EGWHAHTGGG (K_d 100 nM) as the reference ligand, a peptide identified by the Feig lab using phage display. Simulations of NPA and NPB at the binding site of Toxin A showed that they have a high affinity for the Toxin. We estimated the binding free energy for these peptides and found that they were better binders than the reference ligand. The peptides have shown good neutralization against Toxin A when tested *in vitro* in jejunum cells but showed no effect in colonic cells. Current work is focused on furthering our understanding of the interaction of these peptides with the toxins, redesigning shorter peptide inhibitors for Toxin A and discovering peptides that can bind to Toxin B.

607155 Ironing out the Wrinkles for Botox Therapeutic Applications

Abstract:

Botulinum neurotoxin (BoNT), commonly known as Botox, is widely used to reduce facial wrinkles, but its potent paralytic activity has diverse and important applications that range far beyond its $4 billion global cosmetic market. This exceptionally potent drug has the potential to address a long list of unmet medical needs, such as cervical dystonia, and could additionally revolutionize treatments in critical medical spaces such as acute and chronic pain management. Unfortunately, the development of anti-BoNT antibodies can limit many of these therapeutic uses. To fully capitalize on BoNT's therapeutic potential, therefore, the immunogenicity of this bacterial toxin in human patients must be addressed. Unfortunately, BoNT deimmunization has been an elusive goal for the biotech industry, due in large part to the drug's structural and functional complexity. We have solved this challenge through the combined application of state-of-the-art computational protein design tools and a sophisticated ultra-high throughput functional screen. We designed and constructed a combinatorial library of 300 million deimmunized BoNT light chain candidates, in which the entire library ensemble was computationally optimized for both reduced T cell epitope content and maintenance of stability and activity. From this large, diverse population of molecules, we isolated active variants using a customized high-speed flow cytometric screen for BoNT enzymatic activity. The kinetics of engineered variants were analyzed in vitro, and promising constructs were subsequently tested for immunogenicity in humanized HLA transgenic mice. Ultimately, the best variants were incorporated into full-length toxin and their paralytic activity was evaluated in a murine model of flaccid paralysis. In total, our molecular engineering efforts have created a large panel of deimmunized BoNT candidates with functional, immunological, and pharmacokinetic profiles that are tailored for therapeutic applications.

From this presentation audience members will gain an:

-Introduction to high-impact medical applications of Botox, beyond cosmetics

-Overview of causes and consequences of biotherapeutic immunogenicity

-Discussion of T cell epitope deletion as a strategy to deimmunize biotherapeutics

-Demonstration of state-of-the-art computational design tools for engineering deimmunized biotherapies

-Disclosure of an innovative ultra-high throughput screen for BoNT enzymatic activity

-Presentation of immunological and functional characterization of next-generation BoNT biotherapies

-Example of how humanized HLA transgenic mice can enable pre-clinical immunogenicity analysis of protein drugs

609187 Targeting the Pro-Lymphoma Stromal Microenvironment

Abstract:

The lymphoma microenvironment (LME) is inherently dynamic and complex, consisting of cellular as well as non-cellular compartments. These compartments (i.e. malignant, stromal, immune) facilitate a heterogeneous physico-chemical dialogue that often drives the progression of lymphoma, and subsequent resistance to treatments. Data from animal models suggest that malignant cells are able to hijack and corrupt the normal physiology of stromal cells, thereby rendering them pro-tumorigenic. However, the key effectors of this process remains largely unknown. Using sequencing data from patient-derived xenograft samples and their matched primary tumors, we identified heat shock factor 1 (HSF1) as a critical regulator of this supportive transformation. This discovery of HSF1 within the microenvironment, led us to hypothesize that HSF1 was essential to drive this pro-lymphoma phenotype through stromal reprograming and extracellular matrix remodeling.

To address this, we utilized both in vitro and in vivo model systems to decipher the role of HSF1 in the most common subtype of non-Hodgkin's lymphomas; Diffuse large B-cell lymphomas (DLBCL). To identify the potential mechanism with which HSF1 acted, we utilized an immuno-competent mouse model that had either wild type expression of HSF1 (wt), and a genetically engineered counterpart that was deficient in HSF1 for all cells throughout the animal (ko). Using a mouse HSF1-positive DLBCL cell line (A20), we implanted tumors in the flanks of both the wt and ko animals and tracked the growth kinetics of the tumors for 14 days. For wt animals, we observed the continuous growth of tumors over the 14 days, however, in the ko animals we observed a highly reproducible biphasic growth-shrinkage kinetics, with an inflection point at day 7.

We further investigated the lymphoma microarchitecture with a combination of microscopy and mass spectrometry, and tumor macro architecture using tissue mechanics assessment. Although at day 7, there were no significant difference in tumor sizes between wt and ko, we found significant differences in critical tumor properties, suggesting that the eventual shrinkage of the tumors was a secondary consequence of cascade of factors. Specifically, evaluation of tumor tissue sections revealed that the LME in ko mice exhibited lower cellularity, resulting in part from lower proliferation and increased apoptosis. Furthermore, these changes were accompanied by increases in intra-tumor adipose cells, mesenchymal stromal cells, as well as cytotoxic immune cells, and decreased suppressive immune populations. These findings combined with softer tissue mechanics, functional vascularization within tumors, increased porosity, collagen fiber density and alignment, led us to postulate that HSF1 deficiency within the stromal microenvironment induced a differential regulation of the tumor ECM composition and architecture.

To better understand the extent of this remodeling, we conducted a decellularization process to enrich for ECM proteins followed by mass spectrometry analysis for day 7 tumors derived from wt and ko animals, with spleens from both conditions used as controls. To identify relevant ECM proteins, we first compared our results to the database of ECM proteins within the Matrisome database, then identified the ECM proteins that were significantly enriched and depleted. We found that not only was the abundance of stroma ECM proteins greater within the ko tumors, but we identified differentially regulated ECM modules, some of which involved collagen functionalization by small leucine rich proteoglycans (e.g. Decorin, Biglycan). These results suggested that this differentially remodeled ECM in ko tumors may point to targetable vulnerabilities within the LME.

To evaluate whether we can achieve anti-lymphoma effects based on microenvironmental determinants found in ko LME, we took the list of significantly enriched proteins and identified testing candidates. Of note, we observed that the stroma-derived (mainly fibroblasts) proteoglycan Decorin was highly expressed in ko tumors. To demonstrate potential anti-tumor effects of Decorin, we implanted lymphomas into wt animals and allowed the tumor to establish for five days, after which animals were injected with Decorin for four consecutive days. From these experiments, we observed that injection of Decorin was able to significantly halt the growth of tumors. In addition, the introduction of wt fibroblasts into tumors growing in ko animals, and vice versa, demonstrated varying degrees of phenotypic rescue, suggesting that the stroma-derived ECM do play a role in inducing anti-lymphoma effects.

Together, we show that HSF1 is a critical molecular regulator of the LME, with deficiencies in HSF1 activity inducing an inability of stromal cells to support and sustain the growth of tumors in vivo. Furthermore, targeting the stromal microenvironment may be a viable strategy to induce vulnerabilities and anti-tumor effects within the tumor niche.

596049 Using Drug Loaded Liposomes to Induce Tolerogenic Antigen Presenting Cells

Abstract:

Autoimmune diseases are a growing problem in the western world and are particularly difficult to treat due to the wide range of disease pathogenesis and symptoms. As a result, treatments typically include broad immune suppression that leaves patients vulnerable for other infections. However, all autoimmune diseases originate with self-reactive antigens or self-reactive mimetics being uptaken and presented on antigen presenting cells (APC) and activating downstream adaptive immune cells leading to either a B or T cell or both. Under normal physiological conditions, this antigen presentation occurs in primary lymphoid organs and is accompanied with various inhibitory signals to generate a regulatory T cell (Treg) response that actively depletes any auto-reactive adaptive immune response to the self-antigen in question. This project focuses on delivering a combination of tolerance inducing drugs to APCs to drive a more tolerogenic phenotype to a particular antigen. Namely, three tolerogenic drugs, simvastatin, dexamethasone and SC-514, were loaded into a 200 nm liposome along with antigen protein and one of two toll-like-receptor agonists, GpG (for TLR9) or Flagellin (for TLR5). This strategy allowed for tolerogenic APCs to present either a model antigen (OVA) or MOG peptide (as a multiple sclerosis model antigen) and develop an antigen specific Treg response. We first screened a number of tolerance inducing drugs using bone marrow derived dendric cells (BMDCs) for IL-10/IL-6 and PD-L1/CD80 expression. After identifying three optimal candidates, we developed a liposomal formulation of the inhibitors, OVA and either CpG or FLA, to be injected on successive days to first "prime" APCs with the less activating FLA liposome, then the more active CpG liposome 24 hrs later. The liposome formulation was tested *in vivo* and triggered a two-fold increase in tolerogenic DCs (CD11c+, CD80-, PD-L1+) and a two-fold increase of OVA-specific Tregs (CD4+, CD25+, FoxP3+, OVA-Tetramer+). Furthermore, there was a complete reduction in any OVA specific antibodies or CD8 T cells when compared to non-inhibitor loaded controls. This is strong evidence that our liposome formulation can generate antigen specific T reg responses that can functionally reduce autoimmune responses. We are currently testing in an EAE model to further demonstrate the clinical relevance of this formulation as an autoimmune therapy.

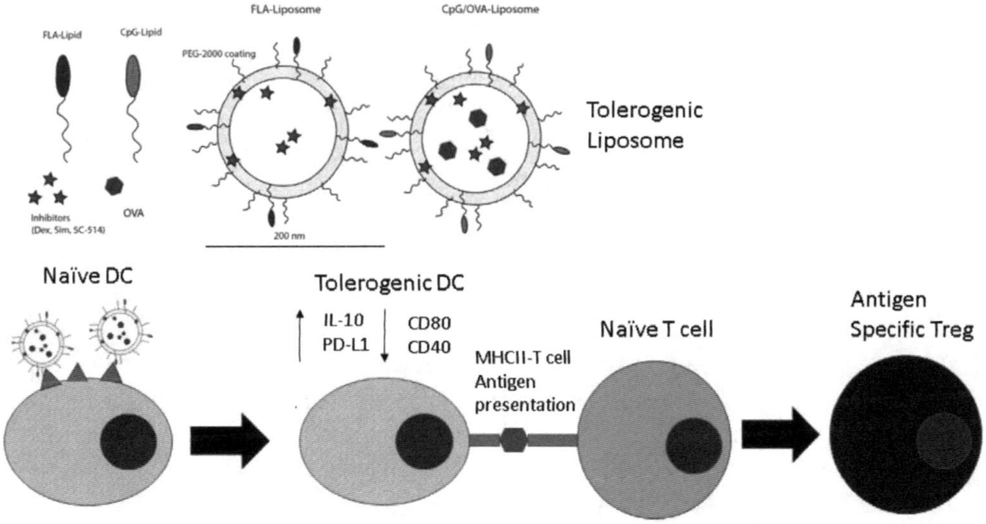

596140 Synergistic Effect of Polyamines Against Lipid Oxidation As *in Vitro* Model Aging System

Abstract:

Introduction

Polyamines are aliphatic compounds with two or more primary amino groups in the molecule, and are often found in fermented foods such as natto and yogurt. In recent years, the polyamines have attracted much attention in preventive effects on aging-related diseases such as atherosclerosis and Alzheimer's disease, as well as life-extending effects.[1] However, the detailed working mechanism of polyamines for this effect remains unclear.

Biological membranes are composed of lipids such as phospholipids and cholesterol, and the oxidation of biological membrane has recognized to lead to the atherosclerosis and aging. Its biological membrane oxidation is thought to proceed by the radical chain reaction mechanism.[2] First, water-soluble radicals (R•) in the blood around the biological membrane react with oxygen into peroxyl radicals (ROO•), which are then transferred to the lipid phase. These peroxyl radicals attack lipids (LH) constituting biological membranes, resulting in lipid radicals (L•). These lipid radicals combine with oxygen to form highly reactive peroxyl radicals (LOO•). This peroxyl radical attacks another lipid continuously, leading to a chain reaction of oxidation. The aging is considered to progress due to damage to biological membranes and accumulation of lipid peroxide (LOOH) during this radical chain oxidation of lipid. In neurodegenerative diseases such as the Alzheimer's disease and Parkinson's disease, which are aging-related diseases, the lipid peroxide as oxidative stress markers have been reported to be increased.[3]

Phenolic antioxidants, such as vitamin E (V_EH), is known to suppress the oxidation of biological membrane lipids.[4] For example, low concentrations of α-tocopherol, V_EH, in plasma have been found in patients with the Alzheimer's disease associated with the aging,[5] suggesting a close link between aging and lipid oxidation. Therefore, the life extension effect of the polyamines seems to be due to their high ability to suppress lipid oxidation. Toro-Funes *et al.* reported the antioxidant effect of polyamines when spermine was added alone to soybean oil.[6] Løvass *et al.* reported that when spermine and spermidine were added to tocopherol-containing fish oil, the antioxidant effect increased with increasing the added amount.[7] However, there are only a few reports on the antioxidant effect of polyamines on lipid oxidation, and the mechanism of this antioxidant effect of polyamines on lipids is still unclear.

In this study, as the first step for verification of lifespan-extending effect by polyamines, we investigated how polyamines act against oxidative stress in vitro homogeneous lipid model system. We performed the oxidation experiments for a co-addition of polyamines and phenolic antioxidants and the combination effects were discussed with synergistic effects in different combinations of antioxidants.

Experimental

Methyl linoleate was used as a model lipid. Spermine (Spe), spermidine (Spd), and putrescine (Put) were used as polyamines. Vitamin E (a-tocopherol) was used as a general phenolic antioxidant within an organism. Oxidation experiments were carried out according to the Rancimat method using a Rancimat apparatus. The reaction tube was filled with 3.0 g of model lipids and the experiments were performed under accelerated conditions of 120°C and aeration of air with a flow rate of 20 L/h. The progression of oxidation was evaluated by collecting the volatile oxidation products formed by the oxidation in pure water and then measuring the electrical conductivity in the collected solution. The amount of antioxidant added to the model lipid was 0, and 7.0 µmol when the oxidant was added alone. For co-addition of V_EH and polyamines, the added amounts of both antioxidants were set at 7.0 µmol.

Results and discussion

The change in electrical conductivity when Spe is used as polyamine is shown in Fig. 1. In the absence of antioxidants, the electrical conductivity increased immediately after the heating started, indicating that the oxidation of lipid proceeded rapidly. In contrast, when polyamine, Spe, was added alone, the rapid rise of electrical conductivity was slightly delayed. The period during which the change in the conductivity is small until the rise, that is, the period during which oxidation proceeds slowly, is called an induction period (IP). In the case of V_EH alone, the IP was longer than that of Spe alone. This indicated that Spe itself has only a small suppression effect of oxidation and VEH has greater effect of oxidation suppression than Spe. The IP was twice as long as that of V_EH alone when V_EH was co-added with Spe. This suggests that Spe can synergistically increase the suppression effect of oxidation by combined with V_EH.

To quantitatively analyze this synergistic effect, the sum of IPs for Spe alone (IP_{Spe}) and V_EH alone (IP_{VEH}), $IP_{sole} = IP_{Spe} + IP_{VEH}$, is calculated. IP_{coadd} in the case of co-addition and IPsole are shown in Table 1. For comparison, similar experiments were performed for different polyamines, Spd and Put, and the obtained IPs are also shown in Table 1. When Put was used as a polyamine, the values of both IPs were comparable, suggesting that there was no synergistic effect. On the other hand, when Spd was used, IP_{coadd} under co-addition conditions was larger, considering that a synergistic effect by Spd addition has appeared. In the case of Spe, the synergistic effect of co-addition was greater than that in case of Spd. The difference between these three polyamines is the number of secondary amine groups in the structure, which indicates that the greater the number of secondary amine groups, the greater the synergistic effect. The suppression mechanism of lipid oxidation involving this secondary amine group is discussed. Since the antioxidant effect of polyamines themselves is small, the reaction of the V_EH with $LO_2•$ radicals into a stable radical ($V_E•$) is considered to occur mainly as antioxidation reaction (Equation (1)). Then, the secondary amine group of polyamines donates hydrogen to $V_E•$ (Equation (2)), which results in the regeneration of V_EH, and the antioxidantion reaction of Eq. (1) again proceeds. We believe that the synergistic effect of oxidative suppression of polyamines was expressed by such an indirect reaction to lipid oxidation.

Conclusion

Polyamine synergistically protects against oxidative stress by supplementing and regenerating other phenolic antioxidants. When compared with other polyamines, the synergistic effect was thought to be closely related to secondary amine groups of polyamines.

----

1. de Cabo R, Carmona-Gutierrez D, Bernier M, Hall MN, Madeo F. The Search for Antiaging

Interventions: From Elixirs to Fasting Regimens. *Cell.* 2014;157(7):1515-1526. doi:10.1016/j.cell.2014.05.031

2. Takahashi A, Shibasaki-Kitakawa N, Noda T, Sukegawa Y, Kimura Y, Yonemoto T. Kinetic Analysis of Co-oxidation of Biomembrane Lipids Induced by Water-Soluble Radicals. *J Am Oil Chem Soc.* 2016;93(6):803-811. doi:10.1007/s11746-016-2821-x
3. Barnham KJ, Masters CL, Bush AI. Neurodegenerative diseases and oxidative stress. *Nature Reviews Drug Discovery.* 2004;3(3):205-214. doi:10.1038/nrd1330
4. Doba T, Burton GW, Ingold KU. Antioxidant and co-antioxidant activity of vitamin C. The effect of vitamin C, either alone or in the presence of vitamin E or a water-soluble vitamin E analogue, upon the peroxidation of aqueous multilamellar phospholipid liposomes. *Biochimica et Biophysica Acta (BBA) - Lipids and Lipid Metabolism.* 1985;835(2):298-303. doi:10.1016/0005-2760(85)90285-1
5. Zaman Z, Roche S, Fielden P, Frost PG, Niriella DC, Cayley ACD. Plasma Concentrations of Vitamins A and E and Carotenoids in Alzheimer's Disease. *Age Ageing.* 1992;21(2):91-94. doi:10.1093/ageing/21.2.91
6. Toro-Funes N, Bosch-Fusté J, Veciana-Nogués MT, Izquierdo-Pulido M, Vidal-Carou MC. In vitro antioxidant activity of dietary polyamines. *Food Research International.* 2013;51(1):141-147. doi:10.1016/j.foodres.2012.11.036
7. Løvaas E. Antioxidative effects of polyamines. *Journal of the American Oil Chemists' Society.* 1991;68(6):353-358. doi:10.1007/BF02663749

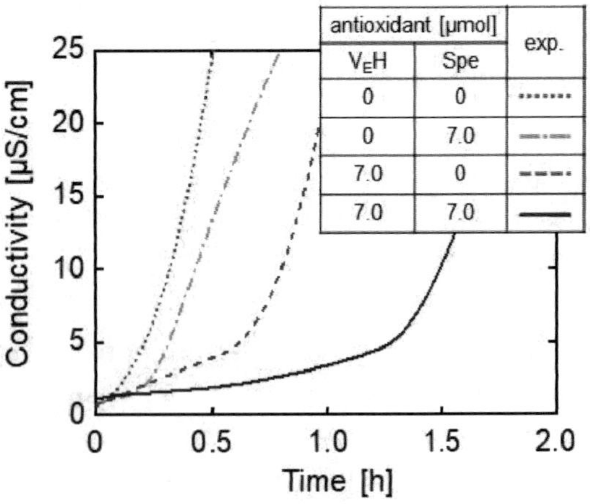

Figure 1 Antioxidant effect of combination of V_EH and Spe.

Table 1 Comparison with IP for V_EH + polyamine

Antioxidant	IP_{coadd} [h]	IP_{sole} [h]
V_EH + Spe	1.40	0.96
V_EH + Spd	1.03	0.88
V_EH + Put	0.82	0.97

$$V_EH + LO_2\cdot \longrightarrow V_E\cdot + LOOH \quad (1)$$

$$V_E\cdot + R^1\text{-}\underset{H}{N}\text{-}R^2 \longrightarrow V_EH + R^1\text{-}\underset{\cdot}{N}\text{-}R^2 \quad (2)$$

613366 Nanoscale Approaches for Targeting Tumor-Associated Macrophages (INVITED SPEAKER)

Abstract:

Tumors are heavily infiltrated with tumor-associated macrophages (TAM), which facilitate tumor progression and, directly and indirectly mount an immune suppressive effect on T cells. TAMs are polarized to M2-like phenotype which are known to promote tumor growth. In contrast, M1 macrophages possess enhanced cytotoxic activity, produce inflammatory cytokines, and have antigen presenting capability. Here, we show that an effective strategy to repolarize the M2 TAMs to M1 phenotype could be devised by tumor-specific inhibition of Colony-stimulating factor-1 (CSF-1) signaling in M2 TAMs. To achieve this, we have developed a completely new platform technology, called supramolecular nanotherapeutics (SNTs) where we designed a CSF-1R inhibiting molecular subunit that can self-assemble with each other and with additional inactive excipients to form nanoscale structures. We have shown that SNTs enabled significant re-polarization of M2 macrophages to M1 phenotype and enhanced phagocytosis of cancer cells. We have also shown that these CSF-1R inhibiting supramolecular nanotherapeutics significantly enhance anti-cancer efficacy without any significant toxicity. This study shows that a successful strategy to overcome immune suppression and improve cancer immunotherapy could be devised by a CSF-1R inhibiting supramolecular nanotherapeutics that can alter the innate immune contexture of tumors by switching the immunosuppressive M2 macrophages to effector M1 macrophages.

603493 A Systematic Study of State-of-the-Art Methods in Crystal Structure Prediction for Organic Hydrates

- Download/View Extended Abstract

Abstract:

Hydrates are co-crystalline materials containing water as one of the molecules in the crystal lattice. The incorporation of water into the crystal lattice produces a unit cell different from that of the anhydrate and, consequently, the physical properties of the hydrate can differ significantly from those of the anhydrate. The existence and stability of hydrates is an important consideration in the development of pharmaceutical products: the prevalence of water during manufacturing and storage can mean that neat forms of an active pharmaceutical ingredient can undergo a phase transition to hydrate form, impacting the effectiveness of the drug. Crystal structure prediction (CSP) methods can in principle be useful in identifying likely hydrates, by undertaking searches for all polymorphs of water and one or more given compounds for a given co-crystal stoichiometry. Minimal information is needed, typically just the chemical connectivity diagram[1], to search for the low lattice energy arrangements of the constituent atoms in space.

Applications of CSP to hydrates have resulted in mixed success so far. In the fifth blind test[2] organised by Cambridge Crystallographic Data Centre, one of the targets was a hydrate but none of the 10 groups that attempted to predict its structure put forward the correct structure within their shortlist. In the sixth blind test[3], only 8 groups submitted predicted structures for the hydrate target, and only one group generated the experimental structure within their shortlist.

In order to gain a better understanding of the challenges that make CSP for hydrates difficult, we present a systematic evaluation of a CSP state-of-the-art method for organic hydrates, in which the lattice energy is partitioned into intramolecular and intermolecular contributions. Intramolecular interactions are modelled via quantum mechanical calculations[4], and intermolecular interactions are divided into electrostatics, modelled using *ab initio* derived distributed multipoles[5,6,7], and repulsion/dispersion interactions modelled using a semi empirical potential. A total number of 107 hydrates extracted from the Crystal Structure Database are minimized locally using the *CrystalOptimizer* algorithm[8] with six models of different levels of sophistication (functional, basis set, use of continuum polarizability). The geometric differences between experimental structures and the corresponding minimization outputs are compared in terms of root mean-squared deviation and a CPU time required. Five of the six models are found to give a good degree of accuracy in more than 95% of cases, but with varying computational costs. A further assessment of the proposed models is undertaken by determining relative energy rankings, which are critical in generating reliable polymorphic energy landscapes.

References:

(1) Pantelides, C. C.; Adjiman, C. S.; Kazantsev, A. V. General Computational Algorithms for Ab Initio Crystal Structure Prediction for Organic Molecules. In *Prediction and Calculation of Crystal Structures Methods and Applications. Topics in Current Chemistry*; Atahan-Evrenk, S.; Aspuru-Guzik, A., Eds.; Springer, **2014**; Vol. 345, pp. 25–58.

(2) Bardwell, David A., Claire S. Adjiman, Yelena A. Arnautova, Ekaterina Bartashevich, Stephan XM Boerrigter, Doris E. Braun, Aurora J. Cruz-Cabeza et al. Towards crystal structure prediction of complex organic compounds–a report on the fifth blind test. *Acta Crystallographica Sect. B: Struct. Sci. Cryst. Eng. Mater.* **2011**, 535-551.

(3) Reilly, A. M.; Cooper, R. I.; Adjiman, C. S.; Bhattacharya, S.; Boese, A. D.; Brandenburg, J. G.; Bygrave, P. J.; Bylsma, R.; Campbell, J. E.; Car, R.; *et al.* Report on the Sixth Blind Test of Organic Crystal Structure Prediction Methods. *Acta Crystallogr. Sect. B Struct. Sci. Cryst. Eng. Mater.* **2016**, *72*, 439–459.

(4) Gaussian 09, Revision A.02, M. J. Frisch, G. W. Trucks, H. B. Schlegel, G. E. Scuseria, M. A. Robb, J. R. Cheeseman, G. Scalmani, V. Barone, G. A. Petersson, H. Nakatsuji, X. Li, M. Caricato, A. Marenich, J. Bloino, B. G. Janesko, R. Gomperts, B. Mennucci, H. P. Hratchian, J. V. Ortiz, A. F. Izmaylov, J. L. Sonnenberg, D. Williams-Young, F. Ding, F. Lipparini, F. Egidi, J. Goings, B. Peng, A. Petrone, T. Henderson, D. Ranasinghe, V. G. Zakrzewski, J. Gao, N. Rega, G. Zheng, W. Liang, M. Hada, M. Ehara, K. Toyota, R. Fukuda, J. Hasegawa, M. Ishida, T. Nakajima, Y. Honda, O. Kitao, H. Nakai, T. Vreven, K. Throssell, J. A. Montgomery, Jr., J. E. Peralta, F. Ogliaro, M. Bearpark, J. J. Heyd, E. Brothers, K. N. Kudin, V. N. Staroverov, T. Keith, R. Kobayashi, J. Normand, K. Raghavachari, A. Rendell, J. C. Burant, S. S. Iyengar, J. Tomasi, M. Cossi, J. M. Millam, M. Klene, C. Adamo, R. Cammi, J. W. Ochterski, R. L. Martin, K. Morokuma, O. Farkas, J. B. Foresman, and D. J. Fox, Gaussian, Inc., Wallingford CT, **2016**.

(5) Nyman, J.; Day, G. M. Static and Lattice Vibrational Energy Differences between Polymorphs. *CrystEngComm* **2015**, *17*, 5154–5165.

(6) Price, S. L.; Leslie, M.; Welch, G. W. A.; Habgood, M.; Price, L. S.; Karamertzanis, P. G.; Day, G. M. Modelling Organic Crystal Structures Using Distributed Multipole and Polarizability-Based Model Intermolecular Potentials. *Phys. Chem. Chem. Phys.* **2010**, *12*, 8466.

(7) Stone, A. J. Distributed Multipole Analysis, or How to Describe a Molecular Charge Distribution. *Chem. Phys. Lett.* **1981**, *83*, 233–239.

(8) Kazantsev, A. V.; P. G. Karamertzanis; C. S. Adjiman; C. C. Pantelides. Efficient handling of molecular flexibility in lattice energy minimization of organic crystals. *J. Chem. Theory Comput.* **2011**, *7*, 1998-2016.

597238 Effects of Inorganic Salts and Volume Reduction on Glycine Polymorphism

Abstract:

Studies have demonstrated that inorganic salts can drive polymorphic selectivity of glycine from the metastable α-polymorph to the stable γ-polymorph. Here, we explore the effect of NaCl and $CaCl_2$ together with volume reduction on glycine nucleation. The salts were added to supersaturated aqueous solutions of glycine at a 2:1 Gly:NaCl/$CaCl_2$ ratio at volumes from 1000 μL to 50 μL. Analysis showed that a primary γ-glycine nucleation event occurs for Gly:NaCl systems, with observed increase in induction times as volume was decreased. However, β-glycine primary nucleation was also detected for Gly:NaCl systems. Interestingly, it was observed that β-glycine nucleation is preferred over the salt-directed γ-glycine nucleation at a certain volume. The increase in induction times combined with a decrease in probability for γ-glycine nucleation may have allowed preferred nucleation of the β-polymorph. On the other hand, no nucleation events were observed for Gly:$CaCl_2$ systems at a time span of 48 hours. These results provide insight on how polymorph modifiers act in confined environments.

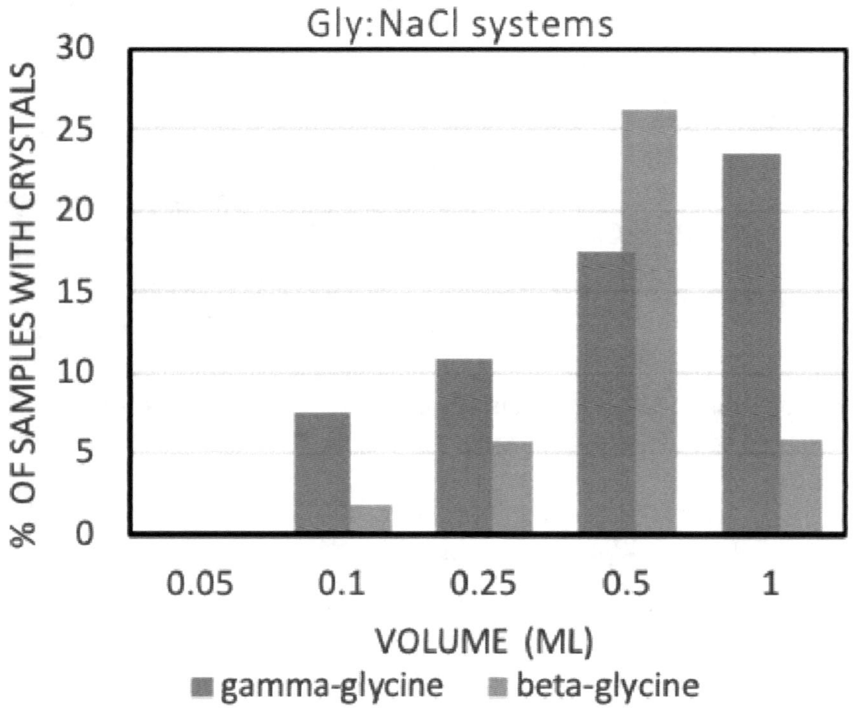

609279 Using Fluid Dynamics and Interface Engineering to Control Polymorphism and Morphology of Organic Molecules.

Abstract:

Crystallization of organic materials as thin films is an area of interest in multiple fields, such as in the pharmaceutical industry, organic electronics and for the multiple applications afforded by metal organic frameworks (MOFs). Over the past decade, significant work has been accomplished in terms of understanding how polymorphism of organic molecules can be enhanced through crystallization as a thin film, through innovative techniques such as confinement, polymer additives, and control of deposition conditions such as the use of electric fields. A deeper understanding of how these polymorphs are formed is critically missing from literature.

In this work, we use a mix of experimental and computational work to study the formation of thin film polymorphs of pharmaceutical compounds and organic semiconductors. The use of interface engineering can influence both the morphology and the polymorphism of acetaminophen, a pharmaceutical compound, and TIPS-pentacene, an organic semiconductor. We also show that changing the flow coating parameters can tune the nucleation and growth rates of organic semiconductors, which then impact the molecule polymorph, crystal size, and crystal orientation. In-situ X-ray diffraction data shows the changes in crystal growth conditions due to changing the fluid dynamics conditions and provides valuable insight in the mechanism of crystal nucleation. Controlling the interface precisely to obtain the desired polymorphs and crystal morphologies will be discussed as well.

609384 Acetaminophen Polymorphism and Texture Control Using Novel Coating Regimes

Abstract:

Spray coating and antisolvent addition have emerged as viable techniques that for use in the continuous processing of pharmaceutical molecules. Through controlling supersaturation during crystallization, these techniques are able to create particle distributions. In this work, meniscus guided flow coating is explored as an alternative to spray coating that creates high surface area crystals for use in pharmaceutical formulations. We have observed the formation of film morphology similar to that obtained using a spray coating technique, and unique from what has been previously described using a flow coating method. Acetaminophen is used as a model compound in this study, where films were fabricated and characterized using polarized optical microscopy and grazing incidence x-ray diffraction (GIXD) to determine particle sizes and crystal orientation on the substrate. Results suggest that the coating regime deviates from traditional evaporative or Landau-Levich coating regimes described in literature. Here we present our current hypothesis for the mechanism of crystallization that produces the observed particulate morphology.

594770 Ternary Phase Diagram and Phase Transformation Behaviors of Various Creatine Phosphate Sodium Hydrates

Abstract:

The thermodynamic stability of creatine phosphate sodium (CPS) in water–methanol system was investigated by the determination of CPS–water–methanol ternary phase diagram. The involved transformation behaviors were explored in solid-state, solutions, and vapor environment by VT-PXRD, Raman, FTIR spectroscopy and DVS. Results show that CPS can exist in three hydrate forms, namely, heptahydrates, tetrahydrates, and dihydrates, in the ternary phase diagram. These three crystal forms become the stable form in turn as the water content of the solvent mixture increases, with no anhydrate form appearing, which was seldom found in current reports. The transition points move towards higher water activity as the temperature increases. Besides, solution-mediated phase transformation could happen between tetrahydrates and dihydrates, which is influenced by solvent composition, temperature, and solid loading. While solid-state phase transformation of them couldn't occur when heating. Heptahydrates and tetrahydrates could transform into each other through adsorption or desorption of the water vapor. The study provides a guide for the control of CPS hydrate forms in theory and in the production to obtain the commercial crystal form with high purity.

601024 A High-Throughput Method for Analyzing Struvite Formation and Morphology: Exploring the Effect of Peptide Additives

Abstract:

Precipitation of struvite ($MgNH_4PO_4*6H_2O$), a slow-release fertilizer, provides a means of recycling phosphate from wastewater streams. Through the implementation of high throughput methods, many different precipitation parameters can be investigated at once. In this work, these reactions occur in small volumes (300 μL or less) in a 96 well plate over the course of 45 minutes. A high throughput platform allows for a minimal use of chemicals and a fast way to compare many reaction conditions. The formation of struvite was investigated by fitting absorbance at 600 nm over time to a first order model with induction time. The impact of peptide addition was assessed through this platform, determining the peptide causes significant changes in the yield parameter and formation constant – allowing for control of precipitation. The impact of struvite seed dosing was also investigated, highlighting the importance of system optimization when peptide is present. Composition of struvite was confirmed through Fourier-transform infrared spectroscopy (FTIR) while morphology and crystal size were analyzed through optical microscopy. Crystals were elongated when precipitated with the peptide present. Finally, the utility of the high-throughput platform was demonstrated with a 2^5 full factorial design to capture the effects and interactions of: magnesium dose, mixing time, seed dose, pH, and temperature.

A Techno-Economic Analysis of Greenhouse Waste and Water Recycling Using Anaerobic Digestion: A Case Study in Food-Energy-Water Nexus

Abstract

The US DOE estimates that about 111 million tons of agricultural crop residues are produced every year(Turhollow, Perlack et al. 2014). The total world production is estimated at nearly 1 billion tons/year(Cherubin, Oliveira et al. 2018). Sustainable agriculture would require us to recover and recycle as much of the materials and energy contained in these wastes. In this paper, we will examine the techno-economics of crop residue recycling using anaerobic digestion for greenhouses in Almeria, Spain which supplies nearly 20% of produce for Spain (About 3 million tonnes/year)(Sayadi-Gmada, Rodriguez-Pleguezuelo et al. 2019). Much of the cultivation is done on a year around basis. Large amount of waste material is produced after every season, which is typically hauled away to landfills where it decays producing unwanted greenhouse gases polluting the atmosphere. The total waste is estimated to contain around 1.35 million tonnes of crop residues and 0.5 million tonnes of rejected produce(Egea, Torrente et al. 2018).

Anaerobic digestion (AD) of the waste represents a potential route to create an integrated and sustainable agricultural model (Kalra and Panwar 1986, Corneli, Dragoni et al. 2016, Sukhesh and Rao 2018). As Shown in attached figure, AD produces mainly 4 product streams all of which has value associated with it. The biogas produced can be used to generate heat and power both of which are needed commodities for greenhouse operation. The waste water produced is rich in nutrients and therefore can supply much needed irrigation water thus reducing the stress on local aquifers and desalination plants. The solid digestate produced can be converted to compost manure after treatment and thus forms a source of organic fertilizer for the next round of crops. The main issue facing the application of this technology is one of economics. Implementation of such a waste recycling system can provide a sustainable solution from the point of view of Food-Water-Energy Nexus as applied to greenhouse agriculture (Haltas, Suckling et al. 2017, Hoolohan, Soutar et al. 2019, Falconer, Haltas et al. 2020).

For the case study, we selected a 100 hectare greenhouse cooperative and carried out an assessment of the economics of constructing and operating an AD system from the crop residues produced. Based on prior studies, the total crop residues was estimated at 4900 tonnes/year (tpy) containing about 3400 tpy of Volatile Solids (VS) and about 500 tpy of water. Additional water needed for dilution in the AD was estimated at 20,000 tpy. The biogas production from the digester was estimated at 1400 tpy containing 660 KW of energy which is roughly 1/3 of the energy content in the waste biomass. The biogas can be used to produce nearly 200 KW of electricity and 460 KW of heat energy both of which are needed in the greenhouse operation and therefore directly recyclable. The wet digestate is sent to a filtration system where much of the water is recovered but is enriched with nutrients. Therefore after some pretreatment, this water can be recycled back for greenhouse irrigation. Likewise the solid digestate is rich in minerals and after treatment, becomes valuable as a compost fertilizer which is also a needed commodity for the greenhouse cultivation.

In this paper, we will present the results of a preliminary economic analysis of the process. We will take a look at the economic feasibility of the process as well as the sensitivity to variables such as waste disposal tipping fee, capital expenditures required and the cost of energy recovered from the AD system.

References

Cherubin, M. R., D. M. D. Oliveira, B. J. Feigl, L. G. Pimentel, I. P. Lisboa, M. R. Gmach, L. L. Varanda, M. C. Morais, L. S. Satiro, G. V. Popin, S. R. de Paiva, A. K. B. dos Santos, A. L. S. de Vasconcelos, P. L. A. de Melo, C. E. P. Cerri and C. C. Cerri (2018). "Crop residue harvest for bioenergy production and its implications on soil functioning and plant growth: A review." Scientia Agricola **75**(3): 255-272.

Corneli, E., F. Dragoni, A. Adessi, R. De Philippis, E. Bonari and G. Ragaglini (2016). "Energy conversion of biomass crops and agroindustrial residues by combined biohydrogen/biomethane system and anaerobic digestion." Bioresource Technology **211**: 509-518.

Egea, F. J., R. G. Torrente and A. Aguilar (2018). "An efficient agro-industrial complex in Almeria (Spain): Towards an integrated and sustainable bioeconomy model." New Biotechnology **40**: 103-112.

Falconer, R. E., I. Haltas, L. Varga, P. J. Forbes, M. Abdel-Aal and N. Panayotov (2020). "Anaerobic Digestion of food waste: Eliciting sustainable water-energy-food nexus practices with Agent Based Modelling and visual analytics." Journal of Cleaner Production **255**.

Haltas, I., J. Suckling, I. Soutar, A. Druckmann and L. Varga (2017). "Anaerobic digestion: a prime solution for water, energy and food nexus challenges." Proceedings of 1st International Conference on Sustainable Energy and Resource Use in Food Chains (Icfes 2017), Including Symposium on Heat Recovery and Efficient Conversion and Utilisation of Waste Heat **123**: 22-29.

Hoolohan, C., I. Soutar, J. Suckling, A. Druckman, A. Larkin and C. McLachlan (2019). "Stepping-up innovations in the water-energy-food nexus: A case study of anaerobic digestion in the UK." Geographical Journal **185**(4): 391-405.

Kalra, M. S. and J. S. Panwar (1986). "Anaerobic-Digestion of Rice Crop Residues." Agricultural Wastes **17**(4): 263-269.

Sayadi-Gmada, S., C. R. Rodriguez-Pleguezuelo, F. Rojas-Serrano, C. Parra-Lopez, S. Parra-Gomez, M. D. Garcia-Garcia, R. Garcia-Collado, M. B. Lorbach-Kelle and T. Manrique-Gordillo (2019). "Inorganic Waste Management in Greenhouse Agriculture in Almeria (SE Spain): Towards a Circular System in Intensive Horticultural Production." Sustainability **11**(14).

Sukhesh, M. J. and P. V. Rao (2018). "Anaerobic digestion of crop residues: Technological developments and environmental impact in the Indian context." Biocatalysis and Agricultural Biotechnology **16**: 513-528.

Turhollow, A., R. Perlack, L. Eaton, M. Langholtz, C. Brandt, M. Downing, L. Wright, K. Skog, C. Hellwinckel, B. Stokes and P. Lebow (2014). "The updated billion-ton resource assessment." Biomass & Bioenergy **70**: 149-164.

607360 Engineering a Purple Non-Sulfur Bacterium to Expand Symbiotic Nitrogen Fixation

Abstract:

The production of nearly 50% of the world's food relies on ammonia fertilizer. While this source of bioavailable nitrogen has led to a marked increase in the number of people that can be fed, the impact on the environment is striking. Ammonia production requires more energy and releases more carbon dioxide than any other industrial reaction, and up to 90% of the product is used to manufacture fertilizers. Furthermore, soil microbes quickly convert ammonia to nitrate, which is easily washed into our water supply. Elevated levels of nitrates can be toxic to humans and animals. Sustainable innovations are required to maintain nutritional security while reducing the environmental damage. When nitrogen-stressed, legumes form an endosymbiotic relationship with bacteria that can fix the critical nutrient. Unraveling the mechanisms responsible for this symbiosis could lead to the expansion of the relationship to cereal crops and a decrease in agriculture's reliance on nitrogen fertilizers. Using the more basic Nod factor-independent relationship between the legume *Aeschynomene evenia* (*A. evenia*) and the photosynthetic bacterium *Bradyrhizobium* sp. ORS278 as a template, the long-term goal is to engineer the closely-related, non-endosymbiotic *Rhodopseudomonas palustris* CGA009 (*R. palustris*) to nodulate the legume while resolving the prerequisites for triggering the symbiotic signal transduction pathway employed by both rhizobacteria and arbuscular mycorrhizal fungi.

Determining *R. palustris*' response to *A. evenia*'s Nodule-specific Cysteine-Rich (NCR) peptides is the first step towards this goal and the focus of this work. NCR peptides are expressed in the nodules of some legumes, such as *A. evenia*, and cause their symbiotic partner to differentiate into nitrogen-producing bacteroids. It has been shown that the NCR peptides from some legumes induce signs of differentiation in their symbiotic partner in vitro, but kill other bacteria. Since *R. palustris* harbors the proteins that have been shown to be essential for the *Bradyrhizobium* sp. ORS278's differentiation in *A. evenia*, the hypothesis is that *R. palustris* and *Bradyrhizobium* sp. ORS278 will respond similarly to the in vitro treatment with the legume's NCR peptides. Changes in both bacteria's ploidy, morphology, membrane permeability, and gene expression, all signs of differentiation, will be presented, in addition to work towards consistently creating knockout strains in the polyploid *R. palustris* using CRISPR.

610606 A Microalgae-Methanotroph Coculture Platform for Fuels and Chemical Production from Wastewater

Abstract:

Wastewater resource recovery facilities are major energy consumers in a community, as well as major contributors for greenhouse gas (GHG) emission. Although anaerobic digestion (AD) is widely employed in wastewater treatment to reduce the amount of solid organic waste and the sludge produced, the use of the produced biogas is mostly limited to heating and electricity generation, while the nutrient rich digestate still requires further treatment. In this work, we propose a waste-to-value (W2V) platform based on a microalgae-methanotroph coculture, which can convert AD-generated biogas into value-added products, while simultaneously removing nutrients from digestate. The coculture platform explores the synergistic interactions within a microalgae-methanotroph coculture to achieve significantly improved productivity of microbial biomass and enhanced nutrient recovery performance.

Using *Chlorella sorokiniana – Methylococcus capsulatus* (Bath) as the model coculture, we demonstrate that the coculture offers a highly promising platform for W2V technologies, which can efficiently recover energy (from CH_4) and carbon (from both CH_4 and CO_2) to produce microbial biomass, while removing nutrients from wastewater to produce treated clean water. Specifically, the coculture demonstrated robust growth on minimally processed AD effluent (via gravitational setting), and showed better growth performance than autoclaved AD effluent. In addition, the nutrient recovery performance of the coculture was compared with microalgae single culture, and the coculture demonstrated significantly improved nutrient removal performance, a direct result of enhanced coculture growth. Next, we showed that the coculture could achieve zero GHG emission without external supply of oxygen, as well as complete removal of inorganic nitrogen and phosphorus from AD effluent. Finally, the potential applications of the wastewater-derived coculture biomass are discussed.

607714 Auxin Degradation By *Variovorax* Maintains Stereotypic Plant Root Development within the Complex Plant Microbiome

Abstract:

Plant-microbe and microbe-microbe interactions within the rhizosphere contribute both positively and negatively to plant physiology, health, and productivity. To test hypotheses about these interactions and define the bacterial genes responsible, we are using a synthetic community of 185 fully genome sequenced bacteria isolated from *Arabidopsis* roots. Through a large screening of *Arabidopsis* plants across various stress gradients (pH, temperature, phosphate, and salt), bacterial colonization patterns robust to these stresses were identified. Then through analysis of subset communities, the genus *Variovorax* was identified as being required to maintain stereotypic root development in these mixed bacterial communities. This ability was traced to *Variovorax* reverting the phenotypic effects of many root growth inhibition (RGI)-inducing bacteria by consuming indole-3-acetic acid (IAA), an auxin plant hormone produced by both plants and bacteria. Using a combination of comparative genomics, RNA-Seq, and screening of a gain-of-function plasmid library in *E. coli*, a single locus conserved in all sequenced *Variovorax* genomes, containing IAA degradation genes, was identified. Using bacterial genetics, this locus was deleted, confirming its role in stereotypic root development. This locus was also expressed in a closely related bacterium from the genus *Acidovorax*, conferring this strain with the ability to degrade IAA and partially revert IAA-induced plant phenotypes. Sub-cloning of genes from this locus have identified the essential genes for IAA degradation. Further, we are investigating the regulation of this locus through a MarR transcriptional regulator. Isothermal titration calorimetry has demonstrated this MarR binds to IAA and related auxins, as well as a promoter element upstream of the degradation genes. Crystal structures of the MarR regulator identified the IAA binding residues and mutation analysis confirms their role in IAA binding. Understanding the *Variovorax* genes required for auxin degradation, and their regulation through the MarR regulator, will enable more carful engineering of this locus in *Variovorax* or related bacteria. The ability to control auxin homeostasis and promote stereotypic root development through this mechanism is of interest for inclusion in biofertilizers and microbial probiotics in agricultural applications.

Reference

Finkel OM†, Salas-González I†, Castrillo G†, Conway JM†, Law TF, Teixeira PJPL, Wilson ED, Fitzpatrick CR, Jones CD, Dangl JL. 2020. A single bacterial genus maintains root development in a complex microbiome. bioRxiv:645655. Doi: 10.1101/645655 († These authors contributed equally to this work)

AUTHOR INDEX